Sovan Lek
Michele Scardi
Piet F. M. Verdonschot
Jean-Pierre Descy
Young-Seuk Park

Modelling Community Structure in Freshwater Ecosystems

Sovan Lek
Michele Scardi
Piet F. M. Verdonschot
Jean-Pierre Descy
Young-Seuk Park
(Editors)

Modelling Community Structure in Freshwater Ecosystems

with 227 Figures and 80 Tables

Springer

Professor Dr. Sovan Lek
UMR 5172, LADYBIO
Laboratoire Dynamique de la Biodiversité
Université Paul Sabatier Toulouse III
Bat. IVR3, 118 route de Narbonne
31062 Toulouse cedex 4
France

Professor Dr. Michele Scardi
Department of Biology
Univ. of Rome "Tor Vergata"
Via della Ricerca Scientifica
00133 Rome
Italy

Dr. ir. Piet F.M. Verdonschot
ALTERRA
Ecology and Environment
P.O. Box 47
6700 AA Wageningen
The Netherlands

Professor Dr. Jean-Pierre Descy
Laboratory of Freshwater Ecology
URBO, University of Namur
Rue de Bruxelles 61
5000 Namur
Belgium

Dr. Young-Seuk Park
UMR 5172, LADYBIO
Laboratoire Dynamique de la Biodiversité
Université Paul Sabatier Toulouse III
Bat. IVR3, 118 route de Narbonne
31062 Toulouse cedex 4
France

Additional material to this book can be downloaded from http://extras.springer.com.

ISBN 3-642-42718-9 Springer Berlin Heidelberg New York
ISBN 978-3-642-42718-3 Springer Berlin Heidelberg New York

Springer is a part of Springer Science+Business Media GmbH
springeronline.com

Softcover re-print of the Hardcover 1st edition 2005

Cover design: E. Kirchner, Heidelberg
Production: Almas Schimmel
Typesetting: Camera ready by authors
Printing: Mercedes-Druck, Berlin
Binding: Stein + Lehmann, Berlin

Printed on acid-free paper 30/3141/as 5 4 3 2 1 0

Foreword

The landmass on which we live is an integral part of our water catchment. Any human activity will inevitably have some consequences on the availability and composition of fresh waters. These consequences are becoming increasingly important and detectable as the human population grows. The problem is to be addressed at the global scale, as frequently, decisions made have inter-regional and international impacts, and must therefore be coordinated. In a number of European Member States, for example, the availability of water resources depends on the activities of other upstream countries. The demand for fresh water in Europe, as well as in the world, is increasing. There is an upward pressure on European water demand for public supplies (drinking water, recreation, etc.), for industry, and for irrigated agriculture. The ecological impacts of different uses are complex, and currently not always predictable. This book should help planners in their decisions on different water management options for human use.

Water, of course, is not only relevant as a resource, exploited for human activities, but it is also relevant to aquatic ecosystems and to their quality. Preservation and restoration of the ecological quality of these ecosystems have a major social impact, as it has been stressed in several European Community actions. For example, those based on the United Nations Convention on Biological Diversity (OJ L309, 13 December 1993) such as the "Communication to the Council and to the Parliament on a European Community Biodiversity Strategy" (COM (1998) 0042) or the Council Directive 92/43/EEC (21 May 1992) on the conservation of natural habitats and of wild fauna and flora.

Water management and environmental policies of different countries in the European Union share a common base, such as the Urban Waste Water Treatment Directive 91/271/EEC, the Integrated Pollution Prevention and the Control Directive 96/61/EEC as well as the Water Framework Directive, EU 2000. Thus policies will be become more and more integrated in the near future. Models for efficient ecosystem managements can and will in the near future aid throughout these processes. The need for general methodologies, based on advanced modelling techniques, for predicting structure and diversity of key aquatic communities under natural and under man-made disturbances will strongly support the implementation of the EU directives. Such tools further will add to the decision and policy making process. It is important to stress that the development of models for the prediction of aquatic ecosystem quality in the context of a European Research and Development (R&D) project can not merely be a scientific exercise, but is a social one as well.

In 1999 the European Commission initiated the R&D project PAEQANN (PAEQANN is the acronym for "Predicting Aquatic Ecosystem Quality using Artificial Neural Networks: Impact of Environmental characteristics on the Structure of Aquatic Communities (Algae, Benthic and Fish Fauna)"; under contract number EVK1-CT1999-00026. The main feature of the PAEQANN project was to provide a unified, common set of tools for 1) checking the river ecology status, and 2) predicting environmental impacts of management action on a European scale.

The project not only provided a significant improvement of our knowledge about the ecological applications of classical statistical techniques, dynamic models, and artificial neural networks and other artificial intelligence techniques, but also a set of predictive tools that is easily applied to real management scenarios. This book disseminates the results of the research activities and makes them available to a wide spectrum of potential end-users. In the first phase of the project, the PAEQANN partners focussed their work in two directions: 1) to produce the databases, and 2) to develop the modelling methodology.

The quality of predictive models is strongly dependent on: i) the quality of the data sets used during the development phase of the model, ii) the strategy used during the model improvement in the calibration phase and iii) the range of ecological conditions represented in the validation phase. For these reasons, it was important that the databases covered a wide range of river types and ecological situations, representing, as far as possible, several eco-regions in Europe. The data in the PAEQANN project did not cover all eco-regions of Europe, but the data collected on community structure and environmental conditions were representative of a wide range of river types. These data were used in different ways to develop and test the models presented here.

In the PAEQANN project, river ecosystem integrity was assessed using the relationship between environmental impacts and organism groups, i.e. community structure as depending upon the environmental variables. The classical statistical, artificial neural network and dynamic models developed were used as predictive tools, and also as tools to explain and understand the complex relationships between variables. These tools can be applied in other river networks throughout Western Europe. They are simple, easy to handle and applicable to stream management and stream policy-making. The PAEQANN partners focussed their work to improve the knowledge on these methods with several algorithms, e.g., multiplayer perceptron with a backpropagation algorithm, self-organising maps, goal function, Bayesian function, and others. Several applications on diatoms, macroinvertebrates and fishes can be found in this book. By including also other scientists this books got a full modelling dimension.

Different end-users can be identified who can and partly will be the recipient of the PAEQANN project results: the scientific community, water professional management (public and private managers and users), nature conservation managers and large people audience. By this book the PAEQANN project results reach firstly the scientific community, which is the most familiar group to the majority of the consortium participants. In the current state of modelling aquatic ecology, this is still the first attempt to a serious validation of the work, a necessary background for large-scale dissemination on the long term. Secondly, the book broadens the scientific scope of the project towards wider developments in prediction. We are confident in this attempt that aquatic community modelling and prediction issues have gained weight among the scientific community. This step also brings tools necessary to tackle future water related problems closer to the international, national and regional water managers and politicians.

Chapter 1 is a review of publications on bioindicators for river quality assessment. Three major aquatic communities (algae, benthic macroinvertebrate and fish) were considered. Chapter 2 aims to review current ecological models that predict community structure in aquatic ecosystems for the selection of the appropriate models, depending on the type of target community. Ecological water management aims to contribute to the value of aquatic ecosystems. Such management requires the understanding of how these ecosystems function, and thus how communities are related to the environment. To learn about the community-environment relationships, data-analytical approaches are explored: *Classical statistical models, Artificial neural networks, Bayesian and Mixture models, Support vector machines, Genetic algorithms, Mutual information and regression maximisation techniques, and Structural dynamic models*. In the following sessions, we summarized these modelling techniques and presented their applications in ecological studies, with their strengths and weaknesses. Chapters 3-5 are designed for papers on different organism groups. Chapter 3 includes 7 papers studied on the modelling of fish communities. They present modelling techniques at several different countries including France (Garonne basin and national scale), Italy, Poland, New Zealand, and Thailand, using self-organizing map (SOM) for patterning communities and multilayer perceptron (MLP) for predicting fish assemblages. Chapter 4 introduces the use of different recently developed techniques to model and predict macroinvertebrates in terms of richness, assemblages or functional groups. In chapter 4 several different techniques were implemented in the models: SOM, MLP, sensitivity analysis, cluster analysis, etc. Chapter 5 presents applications of machine-learning techniques used for classification and prediction of mostly riverine micro-algal assemblages. Two papers are devoted to prediction of planktonic cyanobacteria and other five papers address modeling and prediction of benthic diatom assemblages in rivers, at different scales, with two main objectives: classification of these assemblages as related to environmental gradients, and prediction of community structure. Chapter 6 present 5 papers concerning techniques for exploratory data analysis of aquatic communities: Evaluation of relevant species in communities: Projection pursuit with robust indices A framework for computer-based data analysis and visualisation by pattern recognition, A rule-based vs. a set-covering implementation of the knowledge system LIMPACT and its significance for maintenance and discovery of ecological knowledge, Predicting macro-fauna community types from environmental variables by means of support vector machines. Chapter 7 is designed to introduce the software tool which was developed in PAEQANN project. Models developed in the PAEQANN project were implemented in the PAEQANN tool software. The software is included in the CD-ROM accompanying this book. The main objective of the software is to propose a set of tools for water management and water policies to enable easy assessment of the ecological quality and perturbations of stream ecosystems. These tools will provide information about running water quality as well as community structure, and allow identifying measures which should be taken to restore biological integrity in running waters.

Acknowledgments: Most of the studies in this book were carried out in the framework of the EC-5thFP PAEQANN project, devoted to the use of ANN for predicting aquatic communities in fresh waters. Furthermore, in this book we also invited some selected papers presented in the 3rd Conference for the International Society of Ecological Informatics (ISEI) organized under the direction of the PAEQANN project in Rome, Italy, in August 2002. All papers presented here were peer reviewed at least by two specialists in their research fields. Hopefully, the papers presented here are significant contributions to modelling aquatic communities, and can be considered as a first step toward linking the improvement of water quality through specific management measures (e.g. waste water treatment, habitat restoration, etc.) with the expected improvement in ecological and biological value of running water systems. It will also allow scientists and ecosystem managers to consult the occurrence patterns of organisms in streams based on the database used in the tool, visualise the results of patterning and predicting models with existing data, and provide the possibility to test the new data based on the models developed with existing data. We thank the European Commission for the financial support, and all our colleagues in and outside the consortium for their help in data collection, data provisions, data elaboration, scientific discussion and constructive remarks to finally produce the book. Furthermore, we thank all reviewers who were willing to spend their time on the evaluation of manuscripts and suggested ideas to improve the quality of papers to be published in this book.

August 2004

Piet Verdonschot and co-editors

List of authors

Aguilar Ibarra A, ENSAT (France), aguilar@ensat.fr
Akkermans W, Biometris, Wageningen UR, (The Netherlands) wies.akkermans@wur.nl
Baumeister J, Universty of Wuerzburg (Germany)
Bobbin J, University of Adelaide (Australia)
Campeau S, Université du Québec à Trois-Rivières, (Quebec, Canada), Stephane_Campeau@uqtr.ca
Cataudella S, University of Rome, Rome (Italy)
Cauchie HM, Centre de Recherche Public - Gabriel Lippmann (Luxembourg)
Céréghino R, University Paul Sabatier, Toulouse (France), cereghin@cict.fr
Cha EY, Pusan National University, Busan (Korea)
Cho HD, Pusan National University, Busan (Korea), biology@gsnd.net
Chon TS, Pusan National University, Busan (Korea), tschon@pusan.ac.kr
Ciccotti E, University of Rome, Rome (Italy)
Compin A, University Paul Sabatier, Toulouse (France), compin@cict.fr
Concepcion Villanueva M, ENSAT, Toulouse (France)
Coste M, CEMAGREF Bordeaux, Cestas (France), michel.coste@Bordeaux.Cemagref.fr
de Pauw N, University of Gent, Gent (Belgium), Niels.DePauw@rug.ac.be
Death RG, Massey University, Palmerston North (New Zealand), RG.Death@massey.ac.nz
Dedecker AP, University of Gent, Gent (Belgium), andy.dedecker@UGent.be
Delmas F, CEMAGREF Bordeaux, Cestas (France), francois.delmas@cemagref.fr
Descy JP, Facultés Universitaires N-D de la Paix; Namur (Belgium), jpdescy@fundp.ac.be
Di Dato P, University of Rome, Rome (Italy), pdidato@mclink.it
Ector L, Centre de Recherche Public - Gabriel Lippmann (Luxembourg), ector@crpgl.lu
Gevrey M, Univ. Paul Sabatier, Toulouse (France), gevrey@cict.fr
Giraudel JL, IUT Périgueux, Périgueux (France), giraudel@montesquieu.u-bordeaux.fr
Goedhart P, Biometris, Wageningen UR, (Netherlands), P.W.Goedhart@plant.wag-ur.nl
Goethals PLM, University of Gent, Gent (Belgium), peter.goethals@rug.ac.be
Gosselain V, Facultés Universitaires N-D de la Paix, Namur (Belgium). veronique.gosselain@fundp.ac.be
Hoffmann L, Centre de Recherche Public - Gabriel Lippmann (Luxembourg), hoffmann@crpgl.lu
Horrigan N, University of Adelaide (Australia) nelli.horrigan@adelaide.edu.au
Jeong KS, Pusan National University, Busan (Korea), pow5150@hananet.net
Joo GJ, Pusan National University, Busan (Korea), pow0606@hanafos.com
Joy MK, Massey University, Palmerston North (New Zealand), mikejoy@clear.net.nz
Kim SH, Pusan National University, Busan (Korea)
Kruk A., University of Łódź, Łódź (Poland)
Kwak IS, Pusan National University, Busan (Korea)
Leelaprata W, Department of Fisheries of Thailand Kajutjak (Thailand)
Lek S, Univ. Paul Sabatier, Toulouse (France), lek@cict.fr
Lim P, ENSAT (France), plim@ensat.fr
Liu G, Pusan National University, Busan (Korea)
Maio G, University of Rome, Rome (Italy)
Mancini L, Italian National Institute of Health, Rome (Italy)
Marconato E, Aquaprogram s.r.l., Vicenza (Italy)
Metzeling L, Victorian Environment Protection Authority (Australia)
Moreau J, ENSAT (France), moreau@ensat.fr
Neumann M, University of Jena (Germany) m.neumann@uni-jena.de

Nijboer RC, ATERRA, Wageningen (Netherlands), rebi.nijboer@wur.nl
O'Connor MA, Staffordshire University, Stafford (UK) m.a.oconnor@staffs.ac.uk
Obach M, University Kassel (Germany)
Oberdorff T, IRD (France), oberdorf@mnhn.fr
Park Y-S, Univ. Paul Sabatier, Toulouse (France), park@cict.fr
Penczak T, University of Łódź, Łódź (Poland), penczakt@biol.uni.lodz.pl
Pöppel G, Infineon Technologies AG, Regensburg (Germany)
Recknagel F, University of Adelaide (Australia), friedrich.recknagel@adelaide.edu.au
Rimet F, Centre de Recherche Public - Gabriel Lippmann (Luxembourg), rimet@crpgl.lu
Rohatsch T, University Kassel (Germany)
Salviati S, Aquaprogram s.r.l., Vicenza (Italy)
Scardi M, University of Rome, Rome (Italy), mscardi@mclink.it
Song MY, Pusan National University, Busan (Korea), miysong@pusan.ac.kr
Sricharoendham B, Department of Fisheries of Thailand Kajutjak (Thailand)
Tancioni L, Aquaprogram s.r.l., Vicenza (Italy)
ter Braak C, Biometris, Wageningen UR, (Netherlands), cajo.terbraak@wur.nl
Tison J, CEMAGREF Bordeaux, Cestas (France), juliette.tison@bordeaux.cemagref.fr
Tudesque L, Centre de Recherche Public - Gabriel Lippmann (Luxembourg)
Turin P, University of Rome, Rome (Italy)
Verdonschot PFM, ALTERRA, Wageningen (Netherlands), piet.verdonschot@wur.nl
Wagner R, Limnologische Fluss-Station, Max-Planck-Gesellschaft, Schlitz (Germany)
Walley WJ, Staffordshire University, Stafford (UK), W.J.Walley@staffs.ac.uk
Werner H, University Kassel (Germany) heinrich.werner@uni-kassel.de
Whigham PA, University of Otago (New Zealand) pwhigham@infoscience.otago.ac.nz
Zanetti M, Bioprogramm s.c.r.l., Padova (Italy)

List of reviewers

Akkermans W, Biometris, Wageningen UR, (Netherlands) wies.akkermans@wur.nl
Brosse S, University Paul Sabatier (France), brosse@cict.fr
Céréghino R, University Paul Sabatier, Toulouse (France), cereghin@cict.fr
Chon TS, Pusan National University (Korea), tschon@pusan.ac.kr
Dedecker AP, University of Gent, Gent (Belgium), andy.dedecker@UGent.be
Descy JP, University of Namur (Belgium), jpdescy@fundp.ac.be
Din C, University of Washington (USA), Din@iphc.washington.edu
Elliot A, Centre for Hydrology and Ecology, Lancaster, U.K. alexe@ceh.ac.uk
Eveline Pipp, Universität Innsbruck, Austria, eveline.pipp@uibk.ac.at
Fisher P, University of Konstanz (Germany), Philipp.Fischer@uni-konstanz.de
Gevrey M, Univ. Paul Sabatier, Toulouse (France), gevrey@cict.fr
Giraudel JL, IUT Périgueux, Périgueux (France), giraudel@montesquieu.u-bordeaux.fr
Goedhart P, Biometris, Wageningen UR, (Netherlands), P.W.Goedhart@plant.wag-ur.nl
Goethals PLM, Ghent University (Belgium), peter.goethals@rug.ac.be
Gosselain V, University of Namur (Belgium), vgossela@fundp.ac.be
Grossman G, University Gorgia (USA), grossman@arches.uga.edu
Horrigan N, University of Adelaide (Australia) nelli.horrigan@adelaide.edu.au
Jackson D, University of Toronto (Canada), jackson@zoo.utoronto.ca
Joy MK, Massey University, Palmerston North (New Zealand), mikejoy@clear.net.nz
Köhler J, Institute of Freshwater Ecology and Inland Fisheries, Berlin (Germany), koehler@igb-berlin.de
Moreau J, ENSAT, Toulouse (France), moreau@ensat.fr
Neumann M, Friedrich-Schiller-University of Jena (Germany), m.neumann@uni-jena.de
O'Connor M, Staffordshire University (UK), mo3@staffs.ac.uk
Olden J, University of Colorado (USA), olden@lamar.colostate.edu
Ozesmi U, Erciyes University (Turkey), uozesmi@erciyes.edu.tr
Penczak T, University of Łódź, Łódź (Poland), penczakt@biol.uni.lodz.pl
Recknagel F, University of Adelaide (Australia), friedrich.recknagel@adelaide.edu.au
Rott E, Universität Innsbruck (Austria), eugen.rott@uibk.ac.at
Salski A, University of Kiel (Germany), asa@email.uni-kiel.de
Stevenson J, Michigan State University, U.S.A., rjstev@msu.edu
ter Braak C, Biometris, Wageningen UR, (The Netherlands), cajo.terbraak@wur.nl
Whigham PA, University of Otago (New Zealand) pwhigham@infoscience.otago.ac.nz
Wilson M, Savannah River Ecology Laboratory (USA), Wilson@srel.edu

Contents

	General introduction	1
1.	Using bioindicators to assess rivers in Europe: An overview	7
1.1.	Introduction	7
1.2.	Stream typology	7
1.3.	Diatom ecology and use for river quality assessment	9
1.4.	Typologies, assessment systems and prediction techniques based on macroinvertebrates	12
1.5.	Advantages of using fish as an indicator taxon	16
1.6.	Conclusions	18
2.	Review of modelling techniques	21
2.1.	Introduction	21
2.2.	Conventional statistical models	21
2.3.	Artificial neural networks (ANNs)	26
2.4.	Bayesian and Mixture models	35
2.5.	Support vector machines (SVMs)	37
2.6.	Genetic algorithms (GAs)	38
2.7.	Mutual information and regression maximisation (MIR-max)	39
2.8.	Structural dynamic models	39
3.	Fish community assemblages	41
3.1.	Introduction	41
3.2.	Patterning riverine fish assemblages using an unsupervised neural network	43
3.3.	Predicting fish assemblages in France and evaluating the influence of their environmental variables	54
3.4.	Fish diversity conservation and river restoration in southwest France: a review	64
3.5.	Modelling of freshwater fish and macro-crustacean assemblages for biological assessment in New Zealand	76
3.6.	A Comparison of various fitting techniques for predicting fish yield in Ubolratana reservoir (Thailand) from a time series data	90
3.7.	Patterning spatial variations in fish assemblage structures and diversity in the Pilica River system	100
3.8.	Optimisation of artificial neural networks for predicting fish assemblages in rivers	114
4.	Macroinvertebrate community assemblages	131
4.1.	Introduction	131
4.2.	Sensitivity and robustness of a stream model based on artificial neural networks for the simulation of different management scenarios	133
4.3.	A neural network approach to the prediction of benthic macroinvertebrate fauna composition in rivers	147
4.4.	Predicting Dutch macroinvertebrate species richness and functional feeding groups using five modelling techniques	158
4.5.	Comparison of clustering and ordination methods implemented to the full and partial data of benthic macroinvertebrate communities in streams and channels	167
4.6.	Prediction of macroinvertebrate diversity of freshwater bodies by adaptive learning algorithms	189
4.7.	Hierarchical patterning of benthic macroinvertebrate communities	206

	using unsupervised artificial neural networks	
4.8.	Species spatial distribution and richness of stream insects in south-western France using artificial neural networks with potential use for biosurveillance	221
4.9.	Patterning community changes in benthic macroinvertebrates in a polluted stream by using artificial neural networks	239
4.10.	Patterning, predicting stream macroinvertebrate assemblages in Victoria (Australia) using artificial neural networks and genetic algorithms	252
5.	Diatom and other algal assemblages	261
5.1.	Introduction	261
5.2.	Applying case-based reasoning to explore freshwater phytoplankton dynamics	263
5.3.	Modelling community changes of cyanobacteria in a flow regulated river (the lower Nakdong River, S. Korea) by means of a Self-Organizing Map (SOM)	273
5.4.	Use of artificial intelligence (MIR-max) and chemical index to define type diatom assemblages in Rhône basin and Mediterranean region	288
5.5.	Classification of stream diatom communities using a self-organizing map	304
5.6.	Diatom typology of low-impacted conditions at a multi-regional scale: combined results of multivariate analyses and SOM	317
5.7.	Prediction with artificial neural networks of diatom assemblages in headwater streams of Luxembourg	343
5.8.	Use of neural network models to predict diatom assemblages in the Loire-Bretagne basin (France)	355
6.	Development of community assessment techniques	367
6.1.	Introduction	367
6.2.	Evaluation of relevant species in communities: development of structuring indices for the classification of communities using a self-organizing map	369
6.3.	Projection pursuit with robust indices for the analysis of ecological data	381
6.4.	A framework for computer-based data analysis and visualisation by pattern recognition	390
6.5.	A rule-based vs. a set-covering implementation of the knowledge system LIMPACT and its significance for maintenance and discovery of ecological knowledge	401
7.	User interface tool	435
7.1.	Introduction	435
7.2.	Software aims	436
7.3.	System requirements	436
7.4.	Installing/Uninstalling	436
7.5.	Models implemented in the tool	436
7.6.	How to use the software	439
7.7.	Organisms used in the PAEQANN software	447
8.	General conclusions and perspectives	451
	References	455
	Subject Index	513

General introduction

Lek S*

Historical factors aside, the structure and diversity of aquatic communities in running waters are primarily dependent on a complex of physical, chemical and biotic factors. Physical and chemical variables are themselves heavily dependent on climate and catchment properties (hydrology, geology, topography, etc.) which are in turn influenced by anthropogenic impacts (hydraulic management, land use and agricultural practices, waste water discharge etc.).

During the past several decades, hydrobiological studies have identified the main factors which determine freshwater communities, but very few have been able to establish deterministic links between ecological factors and the structure of key aquatic communities. Ecological theories on determinism of biocenosis structure in streams have put forward the effects of morphodynamic properties of the river channel, and the importance of stream order, related to various channel properties, to watershed surface area and to the contribution of various sources of organic matter. These theories have defined general schemes for explaining longitudinal variations in river systems; however they do not allow prediction of community structure down to the a relevant taxonomic or functional level. A notable exception is the RIVPACS system (Wright 1995), which has been developed in the UK to predict the composition of undisturbed macroinvertebrate assemblages; it clearly demonstrates the possibility to develop approaches for predicting community structure from sets of environmental variables. Similarly, species richness in fish communities may be predicted from watershed surface area, average discharge and net primary productivity (Guégan et al. 1998).

A major difficulty is to distinguish the influence of natural characteristics, including natural disturbances (storms, hydrological variability), from changes due to anthropogenic impacts. Despite these uncertainties, key aquatic communities have been utilised, sometimes for decades, to evaluate the biological quality of streams and rivers. Practical methods for calculating "biotic indices" have been designed, with the requirement that they be sufficiently simple for application in routine surveys. As a result, a variety of standard methods have been suggested and used for assessing water quality in different river orders, often without regard to a reference to the natural state of the community. By contrast, recent approaches based on the concept of ecosystem integrity and biodiversity are more promising, especially for integrated water management.

The aim of this book is to develop general methodologies, based on advanced modelling techniques, for the prediction of the structure and diversity of key aquatic communities under natural and man-made disturbances. This allowed the detection of the significance of various environmental variables that structure these aquatic communities. These have been shown to reveal predictable changes due to natural variability and human disturbances. Natural conditions are described as undisturbed by human activities and man-made disturbances are defined as various pollutants, discharge regulation, etc.

Such an approach to the analysis of aquatic communities made it possible:

- to set up robust and sensitive ecosystem evaluation procedures that will work across a large range of running water ecosystems on a world-wide scale,

---

* Correspondance: lek@cict.fr

- to point out the cause and effect relationships between environmental conditions (physical, chemical, results of management actions) and certain relevant aquatic communities (diatoms, macroinvertebrates and fish)
- to predict biocenosis structure in disturbed ecosystems, taking into account all the relevant ecological variables
- to test ecosystem sensitivity to disturbance
- to explore specific actions to be taken for the restoration of ecosystem integrity

The long-term objective of these investigations was therefore to help to define strategies for conservation and restoration, compatible with local and regional development, and supported by a strong scientific background.

As for scientific, technological and economical objectives, the development of these general methodologies allowed:

- the production of predictive tools that can be easily applied to define the most effective policies and institutional arrangements for resource management;
- the application of the most effective and innovative techniques (mainly Artificial Neural Networks) to identify problems in ecosystem functioning, resulting from ecosystem degradation from human impact, and to model relevant biological resources;
- the full exploitation of existing information, reducing the amount of field work (that is both expensive and time consuming) needed in order to assess the health of freshwater ecosystems;
- the exploration of specific actions to be taken for the restoration of ecosystem integrity;
- the promotion of collaboration among scientists of different countries and research fields, encouraging collaboration and dissemination of results and techniques.

Applied objectives

The principal applied objective of this book was to propose a set of tools for water management and water policies in order to allow the easy assessment of ecological quality and perturbations of stream and river ecosystems. These tools provide information concerning running water quality and community structure (see chapter 7). The assessment tools allow the identification of measures which should be taken to restore biological integrity to running waters. It is hoped that the study can be considered as a first step towards linking the improvement of water quality through specific management measures (e.g. waste water treatment, habitat restoration etc.) with the expected improvements in the ecological and biological value of running water systems.

Scientific objectives

The scientific objectives of this book were:

1. to set up a standardised methodological approach (we have defined a set of technical procedures which will be used in a common framework, firstly to analyse and then to predict the community structure of the ecosystems studied with regard to environmental parameters; each reference site is sampled in a standardised way which will allow direct comparisons of the various sites for regional conservation priorities);
2. to link the environmental characteristics and the community structure at each reference site by using a defined set of parameters and a combination of target groups representing the main functional levels of the ecosystems (rapid assessment procedures were imple-

mented based on the hypotheses that include regulative and functional factors which describe ecosystem functioning in a unifying way);

3. to evaluate, at a functional level, the sensitivity of the ecosystems studied and their response to disturbance through implementation of sensitivity indices and modelling (the main threats on living communities and on local endangered species were identified as predictive models of community structure were built for a set of critical habitats);
4. to investigate the effects of human impacts on the functioning of the ecosystem, i.e. on the composition and change in the groups of structural and functional organisms in comparison with nearby natural reference conditions. Special attention is directed at summarising ecosystem functioning by exploring the chances of community restoration at selected sites submitted to the most common types of disturbance.

In summary, the assessments made and the predictions forecast are hoped to lead to improvements in the physical and chemical characteristics of freshwater ecosystems. More specifically, the tools proposed will be useful in the following:

- implementing different existing water directives in Europe as well as in the rest of the world, such as the Water Framework Directive and the Municipal Waste Water Treatment Directive;
- adapting national or regional legislation or incitement measures, taking into account specific conditions;
- agreeing on e.g. the allowable levels of waste water discharge, according to the capacity of the receiving aquatic ecosystem, providing help to the decision makers and water managers in their actions.

Contribution to the policy

The requirement for data on the status of water resources can be identified on different spatial scales. Our concern is about *an overall assessment of the sustainability of water resources*, at the level of the watershed. River ecology is an important area of research; the applied aspects of this discipline typically address water quality problems through studies of ecosystem function (e.g. water quality models) or through studies of biocenosis structure. The use of ecological indicators has become widespread, and integrated indicators are being developed (e.g. IBI and related approaches). Such indicator systems encompass a number of quality determinants and are also quantity-related. Indeed, the ability of water to support natural life provides a more appropriate and integrated measure of its health than individual chemical, biochemical and physical measurements.

However, *applicable and world-wide methodologies and standards are seriously lacking*. Several approaches based on aquatic fauna and flora have been developed, but further work is needed on their large scale applicability, and also on their value as water management tools in member states. Ecological indicators could be used in combination with the conventional physical, chemical and eco-toxicological indicators, to establish more robust sets of criteria to assess the status of the most common and typical ecosystems found. *Prediction tools* are also extremely helpful in *assessing the success of management and restoration measures*, and the *potential environmental impacts* of major water supply infrastructure development.

This research has helped to develop scientific instruments for the evaluation of the resilience of aquatic systems and impact of different management practices. It has also helped to identify aquatic zones where the anthropogenic influence is minimal and which could serve as reference sites.

The lack of suitable methodologies means that *the potential scope* to improve existing models and to develop new models that have an EU or a world-wide applicability *is clearly extensive.*

From a scientific viewpoint, the prediction of aquatic ecosystem quality is a problem that can be approached at *different spatial scales*, ranging from local to global. The continental scale, as far as Europe is concerned, is the best compromise between homogeneity of environmental conditions and generalisation of the models. The rationale for this choice is that *many forcing functions* that have to be taken into account in ecosystem modelling *can be considered as homogeneous on a European scale* due to the relative homogeneity of land use as well as social and economical conditions. On the other hand, at smaller spatial scales (i.e. regional or national) the spectrum of aquatic ecosystems is not diverse enough to allow for optimal generalisation of the models. Moreover, *the development of ecosystem models on a European scale provides a significant advantage* because modelling is a data-limited activity. In fact, the accuracy and the generality of the results rely on data availability, which is obviously proportional to the number of participants in the development team and on the geographical range of the calibration and validation data base.

The main feature of this book was to provide a unified, common set of tools for checking river ecology status, and predicting environmental impacts of management action on a European scale.

This book provides not only a significant improvement in our knowledge on ecological applications of Artificial Neural Networks and other artificial intelligence techniques, but also a set of predictive tools that are easily applied to real management scenarios. These models have been documented and are distributed in dedicated Web sites (http://aquaeco.ups-tlse.fr), either partially as Java applets (that can be executed on-line) or as stand-alone packages downloaded directly from the web site.

Socio-economic contribution

The land mass on which we live is also an integral part of our water catchment. Many human activities inevitably affect the availability and composition of freshwater. These consequences are increasingly detectable and significant. The issue is addressed at the EU-level, as decisions taken by the players concerned frequently have inter-regional and international impacts, and must therefore be coordinated. In a number of Member States, the availability of water resources depends on the activities of other countries located upstream, as many river basins are transnational. The demand for water in Europe is increasing. There is an upward pressure on European water for public supplies (drinking water, recreation etc.), for industry, and for irrigation. The ecological impacts of the various uses are complex, and currently not always predictable. The results of our programme should help planners in deciding between different water management options for human use, aiming to preserve the quality of the ecosystem, particularly the diversity of aquatic organisms.

Water, of course, is not only relevant as a resource to be exploited for human activities, but is also relevant to aquatic ecosystems and to their quality. Preservation and/or restoration of the ecological quality of these ecosystems has a major social impact, as has been stressed in several Community actions. For instance, those based on the United Nations Convention on Biological Diversity (OJ L309, 13 December 1993) can be cited, e.g. as defined in the "Communication to the Council and to the Parliament on a European Community Biodiversity Strategy" or in the Council Directive 92/43/EEC (21 May 1992) on the conservation of natural habitats and of wild fauna and flora.

This book is organized in eight chapters:

Chapter 1 is a review of bioindicators for river quality assessment summing up current scientific knowledge. A review of publications is classified per group of organisms, i.e. diatoms, macroinvertebrates and fish communities. This chapter is coordinated by L Ector and F Rimet.

Chapter 2 reviews the models for aquatic assessment. Classical and modern modelling methods are presented along with a rich collection of references. This chapter is coordinated by YS Park, P Verdonschot and S Lek.

Chapter 3 concerns the predictive models of the fish community for aquatic ecosystem assessment. It includes 7 original papers relating to fish modelling in Europe, Asia and New-Zealand. This chapter is coordinated by S Lek.

Chapter 4 relates to the use of the macroinvertebrate community for aquatic ecosystem assessment. Nine papers show how the new modelling techniques can contribute to modelling this important group, often used for aquatic assessment. This chapter is coordinated by P Verdonschot.

Chapter 5 focuses on the prediction capacities of diatoms for aquatic assessment. A set of 7 papers shows the models of diatoms on different scales, from small drainage basins to the regional scale as part of Europe. This chapter is coordinated by JP Descy.

Chapter 6 concerns modern modelling techniques in ecological assessment. Six original papers concentrate on recent techniques for patterning and predicting aquatic communities. This chapter is coordinated by YS Park.

Chapter 7 presents a useful tool entirely in graphical user interface (GUI) mode. The installing programme is available on the accompanying CD-ROM, which includes the code source in Visual C++. This tool can be used by managers, as well as scientists. This chapter is coordinated by YS Park and S Lek.

Chapter 8 relates to the discussion and conclusion. It is coordinated by M Scardi.

1 Using bioindicators to assess rivers in Europe: An overview

Editors: Ector L*, Rimet F

1.1 Introduction

Aquatic communities are the first element to be disturbed by modifications of physical or chemical quality of rivers. The study of aquatic organisms is thus very useful to detect and assess human impacts. It is of major interest because they can integrate the variability of ecosystems on different temporal scales, depending on the organisms considered. The use of several aquatic organisms integrating different time scale variations gives a precise idea of the ecosystem's health. That is why for more than a century (Stevenson and Pan 1999) many concepts and tools based on biological aquatic organisms were developed in European countries for river quality assessment and are used by the water managers. Benthic diatoms, macroinvertebrates and fish are the mostly used organisms for these assessment tools.

In Europe, since 2000, the European directive 2000/60/EC has established a framework for a common action in the field of water policy. Precise requirements are given in order to have a homogenous river quality assessment in all the European Community. In particular, the status of a water body must be assessed based on its chemical and ecological status. The ecological status is defined as a deviation measurement between characteristic structural of aquatic flora, macrozoobenthos and age structure of fish fauna and the reference conditions of the same parameter. The reference conditions correspond to a water body with no or minor anthropogenic impacts. It is requested to define ecoregions, and stream types by mean of a determined list of physical parameters (e.g. geology, catchment area, altitude). In each stream type of each ecoregion, reference conditions have to be defined.

The aim of this review is first to inventory the different existing methods used to assess river quality with diatoms, macroinvertebrates and fish in the different European countries.

From this inventory, conclusions will be developed. In particular, the existing methods will be examined to see if they can fulfil the directive requirements. This overview will help the selection of methodologies, this in order to develop new assessment tools for river quality that will match with the directive requirements.

1.2 Stream typology†

Stream typologies form an essential basis for the development of assessment and prediction systems, as required by the EU Water Framework Directive (WFD). A stream type is an ecological entity with a limited internal variation in biotic and abiotic components and which shows a certain biotic and abiotic discontinuity in comparison to neighbouring entities. Such stream types might serve as 'units', within which an assessment system can be

* Correspondence: ector@crpgl.lu

† Verdonschot PFM, piet.verdonschot@wur.nl

applied. The comparison with undisturbed sites of a certain stream type allows the definition and classification of different stages of degradation within that stream type. Assessment and prediction both require sufficiently integrated stream typologies, which should consider both abiotic and biotic criteria. The most prominent abiotic factors are stream morphology, geochemistry, altitude, stream size and hydrology. Such typologies based on several ecological relevant parameters are only available for certain geographic regions in Europe.

Generally, stream typologies can be designed 'top-down' or 'bottom up'. The major difference between a top-down and a bottom-up approach is the reliability of criteria (either environmental parameters or organism groups). In a top-down approach, often abiotic, parameters are chosen on the basis of knowledge and human prejudice. In a bottom-up approach the, often biotic or ecological, parameters are the direct results of ecological analysis. For practical reasons one can start with a top-down approach but a typology should always be verified by a bottom-up ecological analysis.

Hering et al. (2003) presented a review of present available stream typologies in Europe (Table 1). The existing approaches used to define stream types differ greatly between countries and institutions. Some classification systems only use single abiotic parameters (like geochemistry in Greece), others are based on abiotic factors and functional elements (e.g. France), while there are also typologies integrating abiotic factors and the biocoenoses, mainly macro-invertebrates (e.g. Netherlands, Germany). Different European countries are presently dividing their territory into 'sub-ecoregions' or 'aquatic landscape units' (e.g. Austria: Fink et al. 2000). This is a first step to identify and describe stream types in a 'top-down' approach (Hawkins et al. 2000a). Hering et al. (2003) estimated that for the whole of Europe about 100 stream types are present.

Table 1. Stream typology approaches in European countries (Hering et al. 2003).

Country	**General approach of typology**	**Level**	**References**
Austria	abiotic	national	Wimmer et al. (2000), Fink et al. (2000)
Austria	biotic (benthic invertebrates)	national	Moog (2000)
France	abiotic	national	Agences de l'Eau (1998)
Germany	abiotic/biotic	national (regional examples)	Schmedtje et al. (2001), LfU BW (1998), LUA NW (1999a,b)
Greece	climatic, geological and hydrochemical	regional	Skoulikidis (1993)
Iceland	abiotic	national	Gardarsson (1979), Petersen et al. (1995), Friberg and Johnson (1995)
Netherlands	abiotic/biotic	national	STOWA (1992)
Sweden	abiotic/biotic	national	Sandin and Johnson (2000)
United Kingdom	abiotic/biotic	national	NRA (1996), Fox et al. (1996)

1.3 Diatom ecology and use for river quality assessment[‡]

Diatoms are siliceous unicellular algae, with a size contained between a few and more than 500 micrometers. They are worldwide spread, live in many aquatic habitats, and have many life forms. Their short generation time makes them respond rapidly to environmental changes (Stevenson and Pan 1999) and their taxonomic diversity represents a valuable tool to assess water quality as each taxon has precise responses to the environmental factors.

Diatom ecology

A first step to develop monitoring tools for rivers is to define diatom taxa ecology.
The analysis of relationships between diatom communities and pH is one of the major focus in diatom studies. pH preferences were first studied in lakes and rivers by Lowe (1974), and Arzet et al. (1986), Round (1990), Smith (1990), Dixit et al. (1990), Eloranta (1990), Coring (1993), Battarbee et al. (1997), van Dam (1997). Renberg and Hellberg (1982), ter Braak and van Dam (1989), Birks et al. (1990a,b), van Dam et al. (1993) and Håkansson (1993) developed indices and models to reconstruct pH with diatoms.

In similar ways, a halobiont index (Ziemann 1971, 1991) uses salt preferences of diatoms to evaluate water salt concentration in rivers. Others recent studies give salinity classifications of diatoms in lakes and estuaries (Campeau et al. 1999, Cumming and Smol 1993, Gell 1997, Roberts and McMinn 1998, Snoeijs 1994, Underwood et al. 1998, Wilson et al. 1994, 1997).

Lange-Bertalot (1979) determined the ecology of about 100 worldwide abundant freshwater taxa in correlation with defined chemical, physical and saprobiological parameters in the Rhine-Main river system. Biological oxygen demand and oxygen saturation were used to define 4 classes of saprobity.

More recently Denys (1991a,b) defined the autecology of 980 fossil diatoms taxa based on 800 samples taken mainly from cores and also from some outcrops of Holocene deposits along the western Belgian coastal plain. Tolerances and preferences for salinity, pH, trophic state, saprobity, nitrogen uptake, oxygen requirements, intertidal exposure tolerance, current velocity were defined.

van Dam et al. (1994) determined pH, nitrogen, oxygen, salinity, saprobity preferences, trophic state and moisture of 948 diatom taxa of fresh and weakly brackish waters in the Netherlands.

Hofmann (1994) realised a similar work on several lakes, principally in nine alkaline lakes of the Bavarian Alps in Germany. A total of 487 taxa were found and about 200 taxa were described in detail for their trophic and saprobic state, and their preferences for conductivity and pH.

Rott et al. (2003) made a large database of 450 running water sites. About 1000 species from 9 algae classes were listed. Classes of saprobity (Rott et al. 1997) and trophy (Rott et al. 1999) were defined for 650 diatom taxa.

[‡] Rimet F, Ector L, rimet@crpgl.lu

Biocenotic analysis

Diversity indices like the Shannon-Weaver index (Shannon and Weaver 1949) are often used in ecological studies to give a first approximation of the ecosystem quality and the impact of physical or chemical disease. Rank-abundance curves developed by Patrick (1949) and Patrick et al. (1954) can also inform on the ecosystem quality according to the curve shape. But, according to the "Intermediate Disturbance Hypothesis" (Connell 1978, Huston 1979), these techniques can be ineffective to assess ecosystem health as diversity and shape of rank-abundance curves of polluted and undisturbed ecosystem can be similar.

The Differentiating Species System (Lange-Bertalot 1979) takes into account 100 worldwide abundant freshwater taxa. They are filed into 3 classes (resistant, sensitive, ubiquitous). Their relative abundances determine the quality of the site.

The SHE index (Steinberg and Schiefele 1988, Schiefele and Schreiner 1991) is the same method as the Differentiating Species System of Lange-Bertalot (1979), but has been modified to be applicable for the rhithral part of rivers. 386 species were filled in 7 groups of trophic state and pollution resistance.

DAIpo, Diatom Assemblage Index to organic pollution (Watanabe et al. 1988) classifies taxa with pollution tolerance (biological oxygen demand), 226 taxa are integrated in this technique.

Zelinka and Marvan (1961) developed an index to assess water quality with algae (among which diatoms) and macroinvertebrates:

$$ID = \frac{\Sigma_{i=1}{}^{n} A_i.I_i.V_i}{\Sigma_{i=1}{}^{n} A_j.V_j}$$

With: A_j: species abundance, I_j: pollution index of the species, V_j: indicative value or stenoecy degree of the species.

The index of Zelinka and Marvan served as a basis for several indices:

- DES (Descy 1979) 5 classes of sensitivity, 106 species are used.

- IPS, Specific Pollution Index (Coste in Cemagref 1982) 5 classes of sensitivity to pollution, all the species are used (1 to 5).

- SLA (Sládeček 1986), 5 classes of sensitivity (from 4 to 0), 323 species are used.

- ILM, Leclercq and Maquet Index (Leclercq and Maquet 1987a) 5 classes of sensitivity, 210 species are used.

- GDI, Generic Diatom Index (Rumeau and Coste 1988, Coste and Ayphassorho 1991) 5 classes of sensitivity (from 1 to 5) are defined. Determination level is the genus. This index was developed in order to propose an easy usable index for Water agencies. All freshwater species/genus are used.

- CEE index (Descy and Coste 1991) this is a 2 input table, low indicator species horizontally ranked by increasing tolerance, and high indicator species (characteristic of a typological level) vertically ranked by increasing tolerance, 208 species are used.

- TDI Trophic Diatom Index of Schiefele and Kohmann (1993) uses the ecological results of Hofmann's studies (Hofmann 1994).

- EPI-D Eutrophication Pollution Index Diatoms (Dell'Uomo 1996, 2004): the sensitivity of the species is an integrated index from 0 to 4, and the reliability from 1 to 5.

- IDAP Artois Picardie Diatom Index (Prygiel et al. 1996) 5 classes of sensitivity to pollution are defined (1 to 5). This index was developed for the Artois-Picardie (N-W French basin).

- IBD Biological Diatom Index (Lenoir and Coste 1996, Prygiel and Coste 1998, 2000): in order to have a practical index, usable for technicians of French Water agencies near morphological diatom are put together and constitute associated taxa. The ecology of these

taxa can be provided with the software OMNIDIA (Lecointe et al. 1993) and the indices values are calculated automatically. This index is standardized for sampling, preparation, counting the slide and calculation of the index (AFNOR 2000). 209 taxa are taken into account in the index calculation.
- TDI, Trophic Diatom Index (Kelly 1998a, Harding and Kelly 1999): 5 classes of sensitivity to trophic state, and 3 classes of reliability are used. This index is widely used in the United Kingdom, and is part of a suite of techniques used to detect eutrophication in rivers caused by large, predominantly lowland sewage works.

Use of diatoms in the different European countries

These indices were already applied in several European countries. Table 2 summarizes those applications. Some indices like the IBD in France, the IPS in Luxembourg and Spain or the TDI in England are routinely used to assess biological quality of rivers on national networks.

Until now, these currently used techniques do not establish comparisons to the reference conditions, as the European Water Framework Directive requires it. To follow these requirements tools based on the comparison of the existing status to its reference in the same ecoregion and the same stream type should be used. Another approach that could answer the requirements of the directive would be the adaptation of the existing indices to each ecoregion.

Table 2: Most common diatom assessment index used in Europe. With: A: Austria; AND: Andorra; B: Belgium; CH: Switzerland; D: Germany; E: Spain; F: France; FIN: Finland; GB: Great Britain; GR: Greece; HU: Hungary, I: Italy; L: Luxembourg; MK: Macedonia; P: Portugal; PL: Poland.

Assessment system	Country of use	Reference
Zelinka and Marvan index	A: Rott and Pipp (1999)	Zelinka and Marvan 1961
Differentiating species system	CH: Hürlimann et al. (1999) D: Coring (1999) PL: Bogaczewicz-Adamczak et al. (2004)	Lange-Bertalot 1979
Descy index, DES	B, L: Descy and Ector (1999)	Descy 1979
Specific Pollution Index, IPS	F: Coste in Cemagref (1982) PL: Kawecka et al. (1999) L: Descy and Ector (1999), Rimet et al. (2004) FIN: Eloranta (1999) GR: Montesanto et al. (1999), Ziller and Montesanto (2004) HU: Szabo et al. (2004) E: Sabater et al. 1996; Gomà et al. (2004) P: Almeida et al. (1999)	Coste in Cemagref 1982
Sládeček index, SLA	HU: Szabo et al. (2004) P: Almeida et al. (1999)	Sládeček 1986
Leclercq and Maquet Index, ILM	B, L: Descy and Ector (1999) HU: Szabo et al. (2004) P: Almeida et al. (1999)	Leclercq and Maquet 1987a

Diatom assemblage index to organic pollution, DAIpo	MK: Krstic et al. (1999)	Watanabe et al. 1988
Steinberg and Schiefele index, SHE	D: Schiefele and Schreiner (1991)	Schiefele and Schreiner 1991
Generic Diatom Index, GDI	PL: Kawecka et al. (1999) FIN: Eloranta (1999)	Coste and Ayphassorho 1991
CEE index	F: Descy and Coste (1991) B, L: Descy and Ector (1999) AND: Merino et al. (1995) E: Sabater et al. (1996) GR: Ziller and Montesanto (2004) HU: Szabo et al. (2004) P: Almeida et al. (1999)	Descy and Coste 1991
Trophic Diatom Index, TDI	D: Coring (1999) FIN: Eloranta (1999)	Schiefele and Kohmann 1993
Eutrophication Pollution Index Diatoms, EPI-D	HU: Szabo et al. (2004) I: Dell'Uomo (1999), Torrisi (2003), Ciutti (2000, 2001)	Dell'Uomo 1996, 2004
Diatom Index of Artois Picardie, IDAP	F: Prygiel et al. (1996) FIN: Eloranta (1999)	Prygiel et al. 1996
Saprobic Rott Index, ROTT	CH: Hürlimann et al. (1999) A: Rott and Pipp (1999), Rott et al. (2003)	(Rott et al. 1997)
Biological Diatom Index, IBD	F: Lenoir and Coste (1996), Prygiel and Coste (1998), AFNOR (2000) L: Descy and Ector (1999) HU: Szabo et al. (2004) P: Almeida et al. (1999)	Lenoir and Coste 1996, Prygiel and Coste 2000
Trophic Diatom Index, TDI	GB: Kelly (1998a), Harding and Kelly (1999)	Kelly 1998a
Trophic ROTT Index	A: Rott et al. (2003)	Rott et al. 1999
Indice DI-CH	CH: Hürlimann and Niederhauser (2002)	Hürlimann and Niederhauser (2002)

1.4 Typologies, assessment systems and prediction techniques based on macroinvertebrates[§]

The last thirty years, a number of macro-invertebrate assessment methods have been developed in different European countries. Macro-invertebrates are well suited for assessment and quality indication systems since a comparatively large amount of data exists, their identification is relatively simple, and they occur in large numbers in all stream types (Rosenberg and Resh 1993, Davis and Simon 1995).

Until now, most methods indicate the 'quality' of sites and have mainly been used to detect anthropogenic impacts, especially focusing on organic pollution. In addition, systems to indicate eutrophication, acidification and salinization have been developed. Most systems are limited for three reasons:

(1) they are restricted to a single impact factor,
(2) they are only applicable in a restricted geographic range or for a certain stream type,

[§] Verdonschot PFM, piet.verdonschot@wur.nl

(3) they do not permit to take full account of the natural differences to be expected in different streams types.

So, there is a strong demand for assessment systems considering different impact factors in combination, and thus to enable an integrated assessment of streams. This is of special importance because organic pollution, the overriding impact factor on streams in past decades, is declining in most European countries and other impact factors, such as deterioration of stream morphology and eutrophication, are becoming increasingly important. One of the first approaches to assess more than one impact on streams with benthic macroinvertebrates has been the Dutch EKO (Verdonschot 1990), which is now implemented into some Dutch regional water management approaches. Another comparable approach is the British RIVPACS system (Wright et al. 1993b), which attempts to integrate all factors affecting the biocoenosis, based on site comparisons with a database of unimpacted sites, within a habitat classification framework. Table 3 gives a general overview of the assessment methods based on benthic macroinvertebrates most frequently applied in the EU member states (after Hering et al. 2003).

This contribution gives an overview of the use of macro-invertebrates in ecological water management.

Macro-invertebrate assessment

Indices assessment

Assessment techniques applied in Europe have been summarised by Woodiwiss (1964), Nixon et al. (1996) and Knoben et al. (1995). Most contributions dealt with further on, are based on macro-invertebrates. The following review is based on Verdonschot (2000).The first and most traditional biological assessment system was the saprobic system, which focused on species presence in relation to organic pollution (Liebmann 1962). It was quantified by Pantle and Buck (1955) and Zelinka and Marvan (1961) and extended and reviewed by a number of European authors (see amongst others, Knoben et al. 1995). Three techniques dominated in Europe (Sládeček 1973, Newman 1988, Metcalfe 1989), namely:

- Saprobic indices; use the difference in pollution tolerance of aquatic organisms. The tolerance is described in parameters of indicator values (1 to 5), weights (tolerance ranges) and species abundances (Sládeček 1973).
- Diversity indices; use the decrease in species diversity under increasing disturbance/stress. Most widely used is the Shannon-Weaver formula (Shannon and Weaver 1949), which is based on the number of species and their individual abundances. Amongst others, Hellawell (1986) and Boyle et al. (1990) reviewed and evaluateed diversity indices.
- Biotic indices and scores; use both a saprobic index and a diversity measure and thus combine taxa richness and (mostly organic) pollution tolerance (Woodiwiss 1964, Tuffery and Verneaux 1968, BMWP 1979, de Pauw and Vanhoren 1983). They were considerably modified recently by Andersen et al. (1994). An overview is given by De Pauw et al. (1992) and Metcalfe (1989).

Most of these approaches are restricted to the main stressors of organic pollution and to the intrinsic natural value of waters.

Multimetrics and rapid assessment techniques

In recent years rapid assessment techniques and multi-metrics have become popular in the US. The first emphasises a low cost approach through reduced sampling and efficient data analysis. The multimetric approach is more advanced and complex, and uses a number of single metrics to assess environmental degradation (Karr et al. 1986). The Index of Biological Integrity (IBI) was restricted to fish (Karr 1981). Later adaptations included the benthic macro-invertebrate assemblage (e.g. Invertebrate Community Index (ICI); Ohio EPA 1987/1989, Plafkin et al. 1989, Kerans and Karr 1994, Karr 1999), or the macrophytes (Nelson 1990). Barbour et al. (1992, 1996) presented the conceptual base for the multimetrics approach in which the community health is composed of community structure, community balance and functional feeding groups, and in combination with habitat quality, an integrated assessment is obtained. Until now, the metrics remain based on ecological attributes of biological communities. Six major groups of metrics can be distinguished (adapted after Resh and Jackson 1993, Thorne and Williams 1997):

- Richness indices (e.g. no of taxa, no of EPT taxa, no of Chironomidae taxa); often these metrics are considered to be sensitive to organic pollution,
- Enumeration indices (e.g. no individuals, % of the total EPT taxa (sensitive) and chironomids (tolerant), % dominant taxon, no intolerant taxa, % Oligochaeta, sediment tolerant taxa); often these metrics consider an increase in dominance of one or more taxa due to pollution,
- Diversity indices (e.g. Shannon-Wiener Index, sequential comparison index); often these metrics are considered to decrease with increasing disturbance,
- Similarity/loss indices (e.g. no of taxa in common, community loss index, Bray-Curtis index); these metrics use comparisons between sites (reference versus disturbed sites),
- Tolerance/intolerance indices or biotic indices (e.g. Hilsenhoff's family biotic index, BMWP score, ASPT score); these metrics rely on the assignment of (in) tolerance values to taxa and include richness,
- Functional indices (e.g. % of functional feeding groups); these metrics use the alteration in food types under different types of disturbance.

The major assumption is that single metrics increase or decrease along an increase in disturbance. Scores of individual core metrics are aggregated to calculate the multimetric score (e.g. Karr 1981, Barbour et al. 1996). The metrics lack sensitivity to contaminants though these could provide different information (Fore et al. 1995).

Assemblage and community assessment techniques

From the beginning of the eighties, with the upcoming multivariate analysis techniques, ecologists started to explore relationships between taxa lists and accompanying environmental parameters. Wright et al. (1984) used multivariate analysis techniques to classify unpolluted running water sites and to use macro-invertebrate types for assessment and prediction. Verdonschot (1990) conducted a large extensive data collection and multivariate analysis of macrofauna in surface waters in the Netherlands. He described macrofaunal site groups, which are recognised on the basis of environmental variables and the abundance of organisms (so-called cenotypes). The cenotypes are mutually related in terms of key factors, which represent major ecological processes. The cenotypes and their mutual relationships form a web. This web offers an ecological basis for the daily practice of water and nature management (Verdonschot 1991). The web allows the development of water quality objectives, provides a tool to monitor and assess, indicates targets and guides the management and restoration of water bodies.

Assemblage and community approaches focus on almost all components and mutual interactions in the aquatic ecosystem.

Non-taxonomical assessment

Non-taxonomical assessment is defined as assessment based on non-taxonomical characteristics. In fact the taxonomical entities are grouped into non-taxonomical categories. Two examples are the functional group and species trait assessment. Functional group assessment is based upon functional groups, such as the macro-invertebrate functional feeding groups. Cummins and Wilzbach (1985) developed a key to the macro-invertebrate functional feeding groups and mutual score between pairs of groups. Groups can be scored for habitat-organic resource categories. Ratios can be calculated and related to general ranges in three groups of stream orders. Recently, the species traits approach was introduced by Southwood (1977, 1988) and applied in the Upper Rhone (Statzner et al. 1994).

Macro-invertebrate prediction

Wright et al. (1984) used multivariate analysis techniques to classify unpolluted running water sites and to predict community types from environmental data. The results were used in the River Invertebrate Prediction and Classification System (RIVPACS). RIVPACS offers a prediction of the macro-invertebrate fauna to be expected at a given site from a small number of environmental parameters recorded. By comparing the fauna observed (at species or at family level) with the expected or 'target' fauna predicted, a measure of site quality can be obtained (Wright et al. 1989).

The Australian River Assessment Scheme (AUSRIVAS) is based on the RIVPACS model. The difference is that the major habitat are sampled and modelled separately. Furthermore, different models are used for different bio-regions in Australia.

The benthic assessment of sediment (BEAST) (Reynoldson et al. 1995, 1997) is similar to AUSRIVAS/RIVPACS approach, but uses in particular abundances of macro-invertebrates instead of presence/absence.

The sophisticated 'Instream Flow Incremental Methodology' (IFIM; Bovee 1982) and the 'Riverine Community Habitat Assessment and Restoration Concept' (RCHARC; Nestler et al. 1989) use habitat preferences of fish and macro-invertebrates and attempt to predict habitat availability at different flow levels.

Recently, the species traits approach was extended towards a prediction tool. Also Verdonschot and Goedhart (2000) used the macro-invertebrate typology as basis for a prediction tool by using multi-nominal regression analysis.

Table 3. Stream assessment methods most commonly applied in standard monitoring programmes in the EU member states. Country (countries) the method is most frequently used in standard monitoring programmes. A: Austria; B: Belgium; D: Germany; DK, Denmark; E: Spain; F: France; FIN: Finland; GB: Great Britain; GR: Greece; I: Italy; L: Luxembourg; IR: Ireland; NL: Netherlands; P: Portugal; S: Sweden (Hering et al. 2003).

Assessment system	Country	References
Acidification Index	S	Henrikson and Medin (1986), Johnson (1998)
AMOEBA	NL	Ten Brink et al. (1991)
Average Score Per Taxon (BMWP-ASPT)	GB, IR, S	Armitage et al. (1983), Chester (1980), Wright et al. (1984)
Belgian Biotic Index (BBI)	B, P, E, L, GR	De Pauw and Vanhooren (1983), De Pauw et al. (1992)
BMWP Score	UK, S	Armitage et al. (1983), Chesters (1980), Wright et al. (1984)
Chandlers Biotic Score and Average Chandler Biotic Score	(GB)	Chandler (1970), Balloch et al. (1976)
Danish stream Fauna Index (DSFI)	DK, S	Skriver et al. (2000)
EKO	NL	Verdonschot (1990)
EBEOSWA	NL	STOWA (1992), Peeters et al. (1994)
Indice Biologique de la Qualité Générale France (IBG)	L, B	Verneaux et al. (1982)

Assessment system	Country	References
Indice Biologique Global Normalisé France (IBGN)	F, B	AFNOR (1992)
Indice Biotico Esteso (IBE)	I	Ghetti (1997)
K-Index (Quality Index)	NL	Gardeniers and Tolkamp (1976)
Modified BMWP Score (BMWP-ASPT), Spanish version	E	Alba-Tercedor and Sanchez-Ortega (1988)
ÖNORM M 6232	A	Österreichisches Normungsinstitut (1997)
Quality Rating System	IR	de Pauw and Vanhooren (1983), de Pauw et al. (1992)
River Invertebrate Prediction and Classification System (RIVPACS)	GB, IR	Armitage et al. (1983), Wright et al. (1993b)
River Oligochaeta-Chironomidae Index (ROCI Index)	FIN	Paasavirta (1990)
Saprobic Water Quality Assessment Austria	A	Moog (1995), Moog et al. (1999)
Saprobienindex DIN 38 410	D	DEV (1992)

1.5 Advantages of using fish as an indicator taxon**

Fishes live permanently in aquatic environments and they are the longest living freshwater aquatic organisms. They are present in all ecosystems including those where the situation is more or less damaged. Consequently, throughout their lives, fishes integrate the various events, which structure the physical and chemical qualities of their habitats. Furthermore, fishes occupy a wide range of food-web positions and are sensitive to a broad array of human perturbations.

A brief overview of riverine fish community ecology

Patterns of fish species richness in rivers at the global scale (i.e. river basins in different continents) have previously been examined by addressing three of the most widely held hypotheses in community ecology. The species-area hypothesis (Preston 1962, McArthur and Wilson 1963, 1967) explains that species richness increases as a power function of surface area. The species-energy hypothesis (Wright 1983, Wright et al. 1993a) predicts that species richness correlates with energy availability. The historical hypothesis (Whittaker 1977) explains richness gradients by patterns of recolonisation and maturation of ecosystems after glaciation. Factors related to components of river size (surface area and flow regime) and energy availability (net primary productivity) are essential for predicting fish diversity Oberdorff et al. 1995, Guégan et al. 1998), whereas the roles of other factors, even if indubitably acting (e.g. history) are often more marginal (but see Oberdorff et al. 1997).

At the basin scale, longitudinal changes in local assemblage richness and composition have usually been attributed to one of two processes: biotic zonation or continual addition of species downstream. Biotic zonation corresponds to discontinuities in river geomorphology or abiotic conditions promoting distinct assemblages along the longitudinal gradient (Huet 1959, Schlosser 1982, Balon et al. 1986, Rahel and Hubert 1991, Oberdorff et al. 1993, Belliard et al. 1997). For example, species replacement may occur as a result of physiological specialization for temperature. In contrast to the advocates of zonation, addi-

** Oberdorff T, oberdorf@mnhn.fr

tions of species are usually related to environmental gradients having smooth transitions of abiotic factors contributing to nested patterns of assemblage composition along the longitudinal gradient (Sheldon 1968, Rahel and Hubert 1991). Whatever the process (i.e. biotic zonation or species addition) the local species richness usually increases along the upstream-downstream gradient (Grenouillet et al. 2004). This gradual accumulation of species is often attributed to a downstream increase in habitat diversity [e.g. measured as a function of depth, current velocity, substrate composition) (Gorman and Karr 1978, Schlosser 1982, Angermeier and Schlosser 1989) and in environmental stability (Horwitz 1978, Grossman et al. 1985, Schlosser and Ebel 1989, Poff and Allan 1995).

However, the majority of the above studies mainly focused on species richness patterns without addressing explicitly the potential role of environmental factors on the functional aspect of these assemblages. Major exceptions are the studies of Rahel and Hubert (1991), Oberdorff et al. (1993), Belliard et al. (1997), Smogor and Angermeier (1999) which provided support for environmental factors effects on trophic and reproductive attributes of fish assemblages. Therefore, one can reasonably expect that functional attributes of fish assemblages would be related, as for species richness, to natural environmental gradients. However, patterns and processes observed in local fish assemblages are not only determined by local mechanisms acting within assemblages, but also result from processes operating at larger spatial scales (Angermeier and Winston 1998, Oberdorff et al. 1998). The richness and structure of local fish assemblages has been linked to factors ranging from geomorphology and climate (Hughes et al. 1987, Whittier et al. 1988), to richness of regional species pool (Belkessam et al. 1997, Angermeier and Winston 1998, Oberdorff et al. 1998).

This space organisation is in fact more complex because a high majority of the species must obligatorily carry out migrations of more or less great amplitudes between the various habitats of the catchment area for their biological cycle (Lévêque 1995). Thus, at this scale, the quality and the composition of the fish communities depend primarily on the integrity of the river continuum, the heterogeneity of habitats available, and on the accessibility of each element of the hydrosystem.

Effects of human disturbances on fish communities

Human uses can disturb relationships between fish and the environment by direct action on the composition of communities, (e.g., introduction of alien species, chemical and organic pollutions) or indirect action by modification of hydrosystems (e.g. channel and bank modifications, flow regulation and fragmentation) (see Karr and Chu 1999, Sand-Jensen 2001 for reviews).

Use of fish for monitoring in Europe

Fish communities are excellent indicators of aquatic ecosystem health (Karr 1981) and have been used to monitor water quality since the early seventies (Verneaux 1973, 1976a, 1976b). Recently ecologists have developed indicators using a community-based approach (Fausch et al. 1990). One such approach to quantify the impact of human activities on the aquatic ecosystem via fish communities is the Index of Biotic Integrity (IBI), first formulated by Karr (1981, 1991a). Since its introduction, the IBI has been modified for use in other regions and types of ecosystems throughout six continents including Europe (see Hughes and Oberdorff 1999 for a review).

In France, Oberdorff et al. (2001) developed a probabilistic model based on species occurrence to define a fish-based index with nation wide application (Oberdorff et al. 2002a,b; AFNOR 2004). A similar strategy has been adopted by the EC research programme FAME (http://www.fame.boku.ac.at) to develop a fish-based assessment method for the ecological status of European rivers. Models that are able to predict the presence or the abundance of a given species (or set of species) would be useful tools, not only for the ecological insight they could provide, but also because they allow the exploitation of existing databases and help in obtaining estimates of ecosystems quality. Table 4 summarises water assessment methods using fish community.

Table 4. Water quality assessment using fish fauna in Europe: review of methods.

Assessment system	Concept	Variables used	Geographical area	References
Bio-typology	Biocenosis, zonation, species richness	Slope, width	West-Europe	Huet (1959)
Bio-typology	Biocenosis, zonation, species richness, species composition	Distance from the source, width, slope, temperature	West-Europe	Verneaux (1973, 1976a,b)
IBI (Index of Biotic Integrity)	Species richness and composition	12 variables: stream order, abundance classes of species richness, trophic structure, etc.	North American (USA), Western Europe	Karr 1981 Oberdorff and Hugues (1992) Kestemont et al. (2000) Belpaire et al. (2000) Kesminas and Virbickas (2000)
FBI - France	Species richness and composition	7 metrics related to species richness and faunal composition	France	Oberdorff et al. (2002a,b); AFNOR (2004)

1.6 Conclusions[††]

The European Water Framework Directive 2000/60/EC (European Parliament 2000) recommends methods using biological elements for monitoring the ecological status of surface waters. Macroinvertebrates, diatoms and fishes are part of the monitoring methods for the assessment of river quality.

This directive also suggests establishing biological reference conditions for all the types of the different water-bodies in each European ecoregion. These biological references conditions can be established either by sampling carried out in undisturbed sites, or by use of predictive modelling if the reference conditions are impossible to find on the field. A combination of both methods can be envisaged. These reference conditions need to be known in order to measure the deviation between the existing conditions of a site and the reference conditions of the same site if it would not be disturbed by human activities. The Directive

[††] Ector L, Rimet F, ector@crpgl.lu

2000/60/EC recommends this measure (Ecological Quality Ratio) to estimate the quality of water bodies.

Many of the existing biological methods for the assessment of the river quality currently used in Europe are not exactly following the requirements of the directive 2000/60/EC.

- For benthic diatoms, the tools used in Europe by water managers are biotic indices based on species composition, species ecology and relative abundance. Until now no tools make comparisons to the reference conditions. It has been shown that diatom assemblages have regional distribution (Rimet et al. 2004, Soininen 2004). Therefore, the existing tools should be either adapted to each river type in each ecoregion, or replaced by new ones (Ector et al. 2004).

- For benthic macro-invertebrates fauna, the tools used by the authorities in the European countries are more diverse. They are based on the diversity, the presence of sensitive taxa, or the species composition and abundance. Nevertheless the RIVPACS method is assessing the river quality by establishing comparisons to the reference conditions.

- The water managers are beginning to use the fish for a few years to assess river health. The methods are principally based on species composition and abundance, age structure or presence of sensitive species. Fishes constitutes a valuable tool for the assessment of long-term environmental variations and of river continuity.

Most of biotic indices are applicable in many regions because they are based on the ecology of worldwide spread species. They can give a good idea of the river quality but do not take into account the specificity of the precise sampled station; in particular its belonging to a precise ecoregion and typological level. Rott et al. (2003) stated that uncritical use of diatom indication methods without check of regional situation and river-monitoring objectives should be avoided. Indices developed in Western Europe are probably applicable in Eastern Europe but are probably not optimised. In order to improve water quality assessment tools for rivers, biological references conditions for each ecoregion and for each stream type have to be defined and to be taken into account. Finding reference conditions in the field is a critical point. As suggested in the directive, predictive models can be used for reference condition definition. Diatom, macroinvertebrate and fish communities can be estimated with predictive models by mean of physical and chemical parameters (Ector et al. 2004). One of the aims of the PAEQANN project is to develop such predictive models.

2 Review of modelling techniques

Editors: Park YS, Verdonschot PFM, Lek S*

2.1 Introduction

Ecological communities are the expression of complex biological processes (reproduction, nutrition, rest, interspecific relationships, et cetera) and abiotic processes (nutrient cycling, discharge regimes, erosion, et cetera), both being expressed on various scales of time and space. To analyse all these processes (i.e., to include and understand the relationships which exist in the community) and to characterise their relationships using environmental parameters, their degree of importance, and their structuring, require the observation of variables related to the operation of the system. The complexity of the ecological systems often results in complex relations between the biological and abiotic variables, justifying the use of multiple modelling techniques. These models are based on different statistical and simulation techniques, designed to predict community structure from environmental variables.

This chapter aims to review current ecological models that predict community structure in aquatic ecosystems for the selection of the appropriate models, depending on the type of target community. Ecological water management is designed to enhance the value of aquatic ecosystems. Such management requires the understanding of how these ecosystems function, and thus how communities are related to the environment. To learn community-environment relationships, data-analytical approaches are explored: *Conventional statistical models, Artificial neural networks, Bayesian and Mixture models, Support vector machines, Genetic algorithms, Mutual information and regression maximisation techniques, and Structural dynamic models*. In the following sections, we have summarized these modelling techniques and presented their applications in ecological studies, together with their strengths and weaknesses.

2.2 Conventional statistical models

Conventional statistical approaches mainly follow three steps:

1. Samples are clustered into groups on the basis of the biological data.
2. Groups are related to the environmental data, for example by discriminant analysis.
3. The reverse process is used whereby regression techniques use environmental variables to predict the biological communities.

Multinomial logistic regression is an improvement over normal discriminant analysis in carrying out step 2. In the traditional approach, environmental data plays no part in the cluster analysis of step 1. This may be unfortunate. If the environmental data already shows distinct groups as a result of water chemistry processes, it is a shame not to use this information. There is also a statistical reason. If, for example, discriminant analysis is used in step 2 of the analysis, a sample may be misclassified on the basis of the environmental data but on further inspection happens to be a borderline case in the cluster analysis. In this case, it

* Correspondence: lek@cict.fr

would be better to reclassify such a sample and iterate the two steps. A popular rival method for studying community-environment relationships is to use ordination rather than cluster analysis in step 1. By using ordination, the biological data is reduced to continuous gradients rather than to groups. But groups have a simplicity that helps to communicate the results to ecosystem managers. Groups, when accurately described in a typology, can derive meaning and become real, as if they already existed. The cenotypes of Verdonschot (1990) are just one example. For the above reasons the potential benefits of other model-based methods were investigated within the EU funded R&D, Fifth Framework Programme (FP5) PAEQANN project (Predicting Aquatic Ecosystems Quality using Artificial Neural Networks: impact of environmental characteristics on the structure of aquatic communities (algae, benthic and fish fauna)).

Generalized linear models (GLMs) are used to perform regression modelling for non-normal data with a minimum of extra complication compared to normal linear regression. In the statistical analysis of data and observational studies, the identification and adjustment for prognostic factors is an important component. A valid comparison of different treatments requires the appropriate adjustment for relevant prognostic factors. The failure to consider important prognostic variables, particularly in observational studies, can lead to errors in estimating treatment differences. In addition, incorrect modelling of prognostic factors can result in the failure to identify nonlinear trends or threshold effects.

Flexible statistical methods that may be used to identify and characterise the effect of potential prognostic factors on an outcome variable are also described in the following sections. These methods are called "generalised additive models", and extend the traditional linear statistical model. They can be applied in any setting where a linear or generalised linear model is typically used. These settings include standard continuous response regression, categorical or ordered categorical response data, count data, survival data and time series.

Tree-based methods involve dividing the observations into groups that differ with respect to the variable of interest. A tree-based procedure automatically chooses the grouping that results in homogeneous groups that have the largest difference compared to the variable of interest. The tree-based method first divides the observations into two groups. The next step is to subdivide each of the groups based on another characteristic. The process of subdividing is separate for each of the groups. This is an elegant way of handling interactions that can become complicated in traditional linear models. When the process of subdivision is complete, the result is a classification rule that can be viewed as a tree. For each of the subdivisions, the proportion of the variable of interest can be used to predict the effect of that variable. The structure of the tree gives insight into which characteristics are relevant. There are several tree-based methods that differ with respect to the types of variables allowed, the way groups are chosen, and the way groups are split. The most common methods are Classification and Regression Trees (CART) and Chi-Squared Automated Interaction Detection (CHAID).

Partial least squares (PLS) is a method for constructing predictive models when the factors are many and highly collinear. The general idea of PLS is to try to extract from many factors a few underlying or latent factors, accounting for as much of the manifest factor variation as possible while modelling the responses correctly. The overall goal is to use the factor to predict the response in the population.

Multinomial logistic regression (MLGR)

A two-step approach is often used to build a model for the prediction of community composition from environmental data. In the first step, sites are clustered into groups on the basis of the biological data alone. This is generally a so-called hard clustering in which every site is assigned to one organism group only. In the second step, these groups are related to the environmental data. Two important methods to establish this organism group – environment relationship are *discriminant analysis* and *multinomial logistic regression*. Discriminant analysis assumes that the environmental data follows a multivariate normal distribution and that is often not the case, for example for dichotomous or nominal environmental variables. Furthermore, for the MLGR model, the whole range of methods and techniques for GLM is available, such as selection of environmental variables. A third advantage of MLGR is that it directly models the probability of occurrence of each organism group as a function of the environmental data, while in discriminant analysis, this is a by-product of the analysis. MLGR is therefore the preferred method for relating organism groups to environmental data. The results can be used to predict the occurrence of organism groups from environmental variables. MLGR is a well-established technique and is a direct extension of ordinary logistic regression, which itself is a special case of GLM. Further information about multinomial logistic regression can be found in McCullagh and Nelder (1994) and in Hosmer and Lemeshow (1989).

There are in general a lot of environmental variables in ecological studies, and they are frequently correlated. Using all variables in a model then yields unstable estimators of the regression coefficients and thus poor predictions. Some form of selection of environmental variables is therefore necessary. Forward selection, backward elimination or stepwise regression result in only one model, and alternative models, with an equivalent or even better fit, are easily overlooked. A preferable method is to fit all possible regression models and to evaluate these according to some criterion such as deviance. However, the fitting of all possible regression models is very computing-intensive, especially for multinomial logistic models. A practical approach is to perform an iterative model selection. Firstly, the variables are subdivided into a few groups, and model selection is performed within each group. The best predictors from each group are then combined in a new model selection step, which yields a few best variables. With these few variables fixed in the model, the remaining variables are again subdivided into a few groups and the next iteration starts. This eventually results in a number of best candidate models. The predictive power of these candidate models can be assessed by means of leave-one-out, also called cross validation. In the first leave-one-out step, the first observation is temporarily deleted from the data, and the model is fitted to the remaining observations. This model is then used to calculate a leave-one-out prediction for the first observation. In the same way, leave-one-out predictions are obtained for all observations, by subsequently removing them from the data. The mean leave-one-out probability of predicting the correct organism group can then be used as a criterion for choosing among the candidate models.

Another approach, which has been proven beneficial for the prediction of organism groups, is what is called hierarchical modelling. This approach assumes that the effect of certain environmental variables is more or less the same for groups of organism groups. Instead of estimating this effect for every individual group, it is now estimated for groups of organism groups. In this way, a reduction of the number of estimated parameters can be accomplished.

Generalized linear models (GLM)

GLMs are used to carry out regression modelling for non-normal data with a minimum of extra complication compared to normal linear regression. GLMs are flexible enough to include a wide range of common situations, but at the same time allow most of the familiar ideas of normal linear regression to carry over. The essay by Firth (1991) gives a good introduction to GLMs; the comprehensive reference is McCullagh and Nelder (1994). GLM assume a *link linear* relationship with some known monotonic function. Furthermore, responses are not normal, thus generalised linear models assume some known variance function appropriate for the data at hand.

GLMs refer to a wide class of statistical models including log-linear models, analysis of variance, probit analysis, logistic regression and standard multiple regression (Dobson 1983, McCullagh and Nelder 1994). As a result, GLMs have many applications. The GLMs were first given a firm theoretical and computational framework by Nelder and Wedderburn (1972), who assumed distributions in the exponential family. The algorithms for fitting generalised linear models are robust and well established (Nelder and Wedderburn 1972, McCullagh and Nelder 1994).

Generalized additive models (GAMs)

One of the most commonly used statistical models in ecological research is multiple linear regression (MLR) for the quantitative dependent data and the logistic regression model for binary data. The logistic model is used as a specific illustration of GAM. Logistic regression (and many other techniques) model the effects of prognostic factors x_j in terms of a linear predictor of the form $\sum x_j\beta_j$, where the β_j are parameters. The GAM replaces $\sum x_j\beta_j$ with $\sum f_j(x_j)$ where f_j is a unspecified ("non-parametric") function. This function is estimated in a flexible manner using a scatterplot smoother (e.g., local regression; LOESS). The estimated function $\hat{f}_j(x_j)$ can reveal possible nonlinearities in the effect of x_j.

Hastie and Tibshirani (1990) is an excellent reference of GAM, and Chambers and Hastie (1992) cover GAM and LOESS. The building block of the GAM algorithm is the scatterplot smoother. Such procedure is known as "backfitting" and the resulting fit is analogous to multiple regression for linear models. When GAMs are fitted to binary response data (and in many other settings), the appropriate error criterion is a penalised log likelihood or a penalised log partial-likelihood. To maximise this, the backfitting procedure is used in conjunction with a maximum likelihood or maximum partial likelihood algorithm. The usual Newton-Raphson routine for maximising log-likelihoods in these models can be cast in an IRLS (iteratively reweighted least squares) form. This involves a repeated weighted linear regression of a constructed response variable on the covariates: each regression yields a new value of the parameter estimate, which gives a new constructed variable, and the process is iterated. In the GAM, the weighted linear regression is simply replaced by a weighted backfitting algorithm. Algorithm details can be found in Hastie and Tibshirani (1990).

Tree-based regression models

Tree-based methods involve dividing the observations into groups that differ with respect to the variable of interest. A tree-based procedure automatically chooses the grouping that results in homogeneous groups that have the largest difference in proportion to the variable of interest. The tree-based method first divides the observations into two groups. The next step is to subdivide each of the groups based on another characteristic. The process of subdividing is separate for each of the groups. This is an elegant way of handling interactions that can become complicated in traditional linear models. When the process of subdivision is complete, the result is a classification rule that can be viewed as a tree. For each of the subdivisions, the proportion of the variable of interest can be used to predict the effect of that variable. The structure of the tree provides insight into the characteristics that are relevant.

Tree-based methods have several attractive properties when compared to traditional methods. They provide a simple rule for classification or prediction of observations, they handle interactions among variables in a straightforward way, they can easily handle a large number of predictor variables, and they do not require assumptions about data distribution. However, tree-based models do not conform to the usual hypothesis testing framework and there is no assumption of a linear model. This would be a tree with a lot of branches and as many terminal segments (leaves) as there are cases. Normally, some "stopping rule" is applied before this extreme condition is reached. Inevitably this means "impure partitions" occur, but this is necessary to balance accuracy against generality. A tree which produces a perfect classification of training data would probably perform poorly with new data.

There are several tree-based methods that differ with respect to the types of variables allowed, the way groups are chosen, and the way groups are split. The most common methods are CART and CHAID. CART and similar methods allow the response and grouping variables to be either categorical or continuous. CART methods are implemented in SYSTAT version 7, in S-Plus, and in SAS. CHAID and similar methods require the response variable to be categorical. CHAID methods are available in an SPSS add-on module and in the SAS macro %TREEDISC. The best description of regression trees theory can be found in Breiman et al. (1984).

Partial least square regression (PLS)

In such so-called soft science applications, the researcher is faced with many variables and poorly understood relationships, and the object is merely to construct a good predictive model. For example, ecological communities are often used to estimate the biodiversity of the study site and their structures are related to different environmental variables. In this case, the factors are the measurements that comprise the environmental data; they can number dozens or hundreds but are likely to be highly collinear. The responses are community amounts that the researcher wants to predict in future samples.

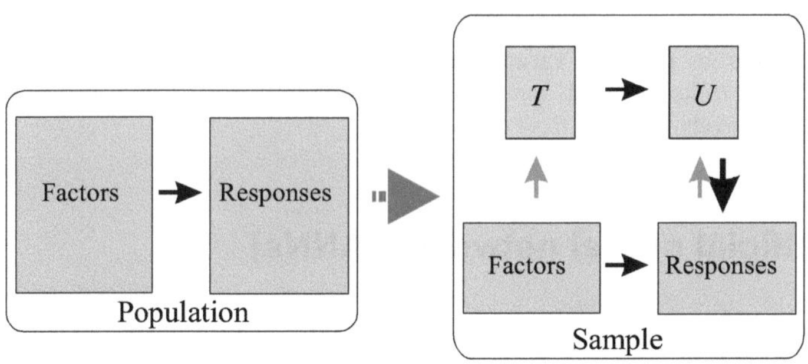

Fig. 2.1. Indirect modelling

PLS is a method for constructing predictive models when the factors are many and highly collinear. Note that the emphasis is on predicting the responses and not necessarily on trying to understand the underlying relationship between the variables. For example, PLS is not usually appropriate for screening out factors that have a negligible effect on the response. However, when prediction is the goal and there is no practical need to limit the number of measured factors, PLS can be a useful tool. The general idea of PLS is to try to extract from many factors, a few underlying or latent factors, accounting for as much of the manifest factor variation as possible while modelling the responses well. Figure 2.1 gives a schematic outline of the method. The overall goal is to use the factor to predict the responses in the population. This is achieved indirectly by extracting latent variables *T* and *U* from sampled factors and responses, respectively. The extract factors *T* (also referred to as *X*-scores) are used to predict the *Y*-scores *U,* and then the predicted *Y*-scores are used to construct predictions for responses. For the details concerning PLS methods, readers can refer to Geladi and Kowalski (1986), Höskuldsson (1988), and Helland (1990).

PLS is a regression technique to solve the linear model in a stepwise fashion, including every predictor variable in the model and bears some resemblance to principal component regression (PCR) in that PCR also creates an orthogonal set of variables. However, in PCR the orthogonal variable extraction is independent of the target variables and a subsequent multinomial regression (MR) step is needed to relate target and explanatory variables, while in PLS, the orthogonal set of variables is constrained to maximise directly the communality of the predictor and response variable blocks. PLS applies this constraint by using NIPALS (Non-linear Iterative Partial Least Squares) rather than digitalisation, to extract factors. PLS calculates an orthogonal set of explanatory variables that are linear combinations of the original variables. Cross validation is used to determine the number of components that yield an optimally predictive model. A cross validated PLS model is usually less subject to errors of over-specification than is a regression model.

Research in ecological sciences sometimes involves explaining controllable and/or easy-to-measure variables (factors), or predicting the behaviour of other variables (response). When the factors are few in number, not significantly redundant (collinear), and have a well-understood relationship to the responses, then MLR can be a good way to turn data into information. However, if any of these three conditions breaks down, PLS can be more efficient and appropriate than MLR.

Malmqvist and Hoffsten (1999) used PLS models to predict community parameters (taxonomic richness, abundance and biomass) of benthic macroinvertebrates at sites exposed to elevated levels of copper, zinc, lead and cadmium resulting from leakage from old mine deposits. They showed that species richness at undisturbed sites was positively related to the size of the catchment, pH, channel width, calcium concentration and the proportion of deciduous trees in the riparian zone. Tegelmark (1998) also developed models for forest ecosystem management with PLS and revealed strong climatic correlation with forest regeneration whereby the main variation could be expressed in terms of latitude and altitude, e.g. positive correlations with temperature sum, length of growing season and humidity, and negative correlation with frost frequency.

2.3 Artificial neural networks (ANNs)

ANNs are powerful computational tools that can be used for classification, pattern recognition, empirical modelling and for many other tasks. Even though most of these tasks can also be performed by conventional statistical or mathematical methods, ANNs often provide a more effective way to deal with problems that are difficult, if not intractable, for traditional computation. In fact, while traditional computation is based on the *a priori* selection of suitable functions or algorithms, ANNs are able to adjust their inner structures to provide optimal solutions, given enough data and a proper initialisation. Thus, if appropriate inputs are applied to an ANN, it can acquire knowledge from the environment, mimicking the functioning of a brain, and users can later recall this knowledge.

ANNs lie in a sort of machine learning middle ground, somewhere between engineering and artificial intelligence (Zurada 1992). They use mathematical techniques, such as mean-square error minimisation, but they also rely on heuristic methods, since very often there is no theoretical background to support decisions about ANNs implementation. Several kinds of ANNs have been developed during the last 10-15 years, but two main categories can be easily recognised, depending on the type of learning process:

- in supervised learning, there is a "teacher" who in the learning phase "tells" the ANN how well it performs or what the correct behaviour would have been;
- in unsupervised learning the ANN autonomously analyses the properties of the data set and learns to reflect these properties in its output.

In the PAEQANN project, both categories of ANNs have been used, with special attention to self-organizing map (SOM) for unsupervised learning, and Multilayer Perception (MLP) with a backpropagation algorithm for supervised learning. Kernel-induced nonlinear models (KINMs) have been developed in recent years and they can be used to perform nonlinear modelling in complex problems found for instance in ecological research. They extract the most informative information from the real-world data sets and establish the nonlinear model by utilizing those partial informative data points in the high dimensional kernel-induced space. The possibility of modelling using unsupervised competitive artificial neural networks (CANNs) and the supervised linear vector quantization (LVQ) network are also described. Applying these two networks in combination may give a powerful tool for fast classification of future observations.

Self-organizing maps (SOM)

SOM was proposed by Kohonen in the early eighties (Kohonen 1982). Since that time, the SOM has been used in a number of different applications in diverse fields and they ware the most well known ANNs with unsupervised learning rules. The algorithm performs a topology-preserving projection of the data space onto a regular low-dimensional space (usually a 2-dimensional space) and can be used to visualise clusters efficiently (Kohonen 2001). This method is recommended for use in an exploratory approach for datasets in which unexpected structures might be found. The SOM approximates the probability density function of the input data, and is a method for clustering, visualisation, and abstraction, the idea of which is to show the data set in another, more usable, representation form. The goal of the SOM is to put the dataset on the map preserving the neighbourhood, so that similar vectors can be mapped close together on the grid.

The SOM consists of two layers: the first (input layer) is connected to each vector of the dataset, the second (output layer) forms a two-dimensional array of nodes (computational units) (Fig. 2.2). In the output layer, the units of the grid (reference vectors) give a representation of the distribution of the data set in an ordered way. Input and output layers are connected by the connection intensities represented in reference vectors. When an input vector $\boldsymbol{x}$ is sent through the network, each neuron k of the network computes the distance between the weight vector $\boldsymbol{w}$ and the input vector $\boldsymbol{x}$. The output layer consists of D output neurons which are usually arranged into a two-dimensional grid in order to improve visualisation. The best arrangement for the output layer is a hexagonal lattice, because this does not favour horizontal and vertical directions as much as the rectangular array (Kohonen 2001). Among all D output neurons, the best matching unit (BMU), which has minimum distance between weight and input vectors, is the winner. For the BMU and its neighbourhood neurons, the weight vectors $\mathbf{w}$ are updated by the SOM learning rule. The training is usually carried out in two phases: at first, a rough training for ordering with a usually carried out in two phases: at first, a rough training for ordering with a large neighbourhood radius, and then a fine tuning with a small radius. This results in training the network to classify the input vectors by the weight vectors they are closest to. The detailed algorithm of the SOM can be found in Kohonen (2001) for theoretical considerations and Chon et al. (1996) and Park et al. (2003a) for ecological applications. After the learning process of the SOM, it is important to know whether it has been properly trained or not, because an optimal map for the given input data should exist. To evaluate the map quality, two criteria, the quantization error for resolution and the topographic error for topology preservation, are commonly measured. The former is the average distance between each data vector and its BMU for measuring map resolution, and the latter represents the proportion of all data vectors for which first and second BMUs are not adjacent for the measurement of topology preservation (Kiviluoto 1996). This error value is, thus, used as an indicator of the accuracy of the mapping in the preserving topology (Kohonen 2001).

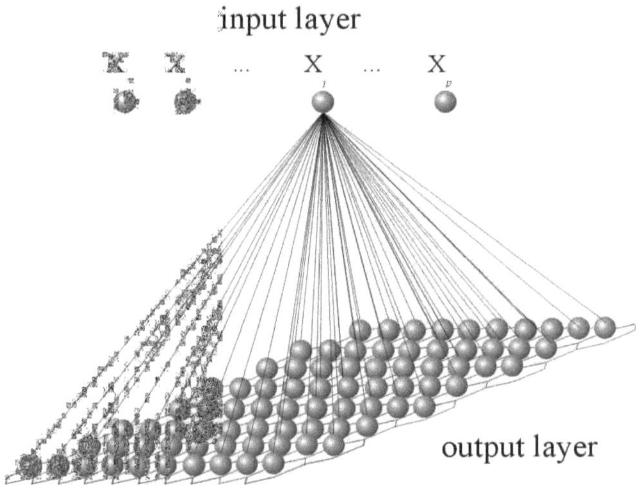

Fig. 2.2. A two-dimensional self-organizing map. Each sphere indicates a neuron in the input layer as well as in the output layer.

On the trained SOM map, it is difficult to distinguish subsets because there are still no boundaries between possible clusters. Therefore, it is necessary to subdivide the map into different groups according to the similarity of the weight vectors of the neurons. To do this, the unified distance matrix algorithm (U-matrix; Ultsch 1993) is commonly used. The U-matrix calculates distances between neighbouring map units, and these distances can be visualised to represent clusters using a grey scale display on the map (Kohonen 2001). In addition, a hierarchical cluster and a k-means cluster methods are frequently used.

During the learning process of the SOM, neurons that are topographically close in the array will activate each other to learn something from the same input vector. This results in a smoothing effect on the weight vectors of neurons (Kohonen 2001). Thus, these weight vectors tend to approximate the probability density function of the input vector. Therefore, the visualisation of elements of these vectors for different input variables is a convenient way to understand the contribution of each input variable with respect to the clusters on the trained SOM. This visualisation method is related to a principal component analysis (PCA), and more directly describes the discriminatory powers of input variables in mapping (Kohonen 2001). Therefore, to analyse the contribution of variables to cluster structures of the trained SOM, each input variable (component) calculated during the training process is visualised in each neuron on the trained SOM map in grey scale.

According to these characteristics of the network, the SOM can be used for clustering without prior knowledge of the number or size of the clusters, and for studying multivariate time series (Ultsch 1999). Cho (1997) showed that the SOM is able to recognise clusters in datasets where other statistical algorithms failed to produce meaningful clusters. The visualisation of clusters is very straightforward and can outperform the results obtained by conventional classification methods. A drawback of the SOM is that the size and the shape of the map have to be fixed in advance. Growing self-organising networks have been proposed in order to deal with this problem (Villmann and Bauer 1998), but this approach remains still to be applied to ecological data.

Multilayer perceptron (MLP)

MLP with a backpropagation algorithm, also called multilayer feed-forward neural networks, is very popular and are used for a wide variety of problems more than other types of neural networks. The MLP is based on the supervised procedure, i.e. the network is built with a dataset where the outputs are known. The MLP is a powerful system, often capable of modelling complex relationships between variables. For a given input, one can predict an output.

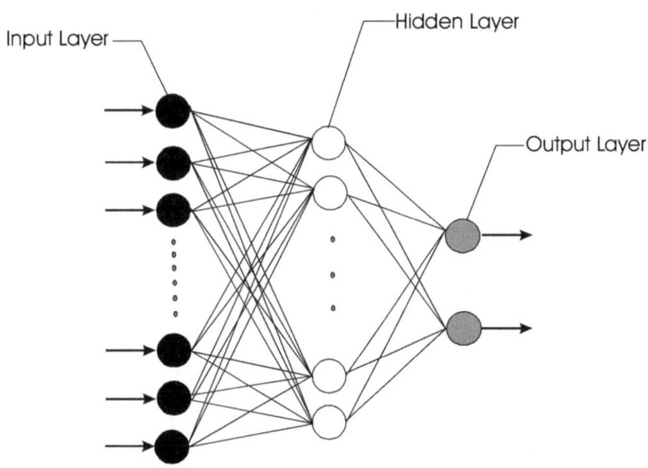

Fig. 2.3. Schematic illustration of a three-layered feed-forward neural network, with one input layer, one hidden layer and one output layer

MLP is a layered feed-forward neural network, in which the non-linear elements (neurons) are arranged in successive layers, and the information flows unidirectionally, from input layer to output layer, through the hidden layer(s) (Fig. 2.3). As can be seen in this figure, neurons from one layer are connected to all neurons in the adjacent layer, but no lateral connection between neurons within one layer, or feedback connection are possible. The number of input and output neurons depends on the number of explanatory and explained variables, respectively. The hidden layer(s) is (are) an important parameter in the network.

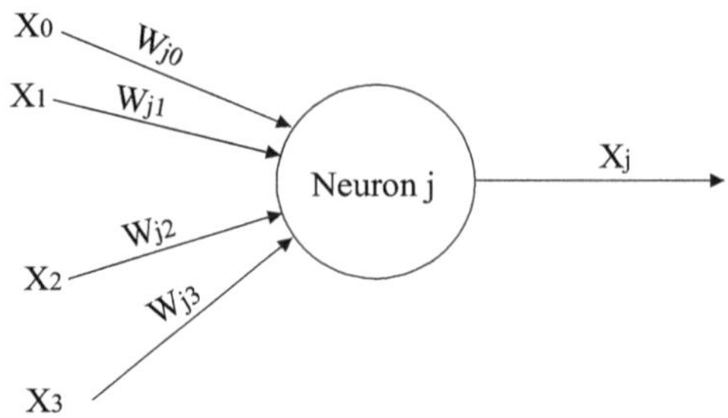

Fig. 2.4. Basic processing neuron in a network. Each input connection value (x_i) is associated with a weight (w_{ji}).

The learning and update procedure of the MLP is based on a relatively simple concept: if the network gives the wrong answer, then the weights are corrected so that the error lessens, thus future responses of the network are more likely to be correct. The conceptual basis of the MLP was presented to a wide readership by Rumelhart et al. (1986a). In a training phase, a set of input/target pattern pairs is used for training, which is presented to the network many times. After training is stopped, the performance of the network is tested. The MLP learning algorithm involves a forward-propagating step followed by a backward-propagating step.

Like a real neuron, the artificial neuron has many inputs, but only a single output, which can stimulate many other neurons in the network. The neurons are numbered, for example the one neuron in Figure 2.4 is called *j*. The *j*th input neuron receives from the *i*th neurons indicated as *x*. Each connection to the *j*th neuron is associated to a quantity called weight. The weight on the connection from the *i*th neuron to the *j*th neuron is denoted w_{ji}. An input connection may be excitatory (positive weight) or inhibitory (negative weight). A net input (called activation) for each neuron is the sum of all its input values multiplied by their corresponding connection weights.

The backward-propagating step begins with the comparison of the network output pattern to the target value, when the difference (or error) is calculated. The backward-propagating step then calculates error values and changes the incoming weights, starting with the output layer and moving backward through the successive hidden layers. The error signal associated with each processing unit indicates the amount of error associated with that unit. This parameter is used during the weight-correction procedure, while learning is taking place. A large value for the error signal indicates that a large correction should be made to the incoming weights; its sign reflects the direction in which the weights should be changed. The adjustment of weight depends on three factors: the error value of the target unit, the output value for the source unit and the learning rate. The learning rate, commonly between 0 and 1, determines the rate of learning of the network.

Before starting the training, the connection weights are set to small random values. Next, the input patterns are applied to the network to obtain the output. The differences between the output calculations and the target expected are used to modify the weights. One complete calculation is called an epoch or iteration. This processed is repeated until a suitable level of error is achieved. Using a parameter called momentum, chosen generally between 0 and 1, enables a local minimum to be avoided.

A testing set of data serves to assess the performance of the network after training is complete. The input patterns are fed into the network and the desired output patterns compared with those given by the neural network. The agreement or disagreement of these two sets gives an indication of the performance of the neural network model. If it is possible, the best solution is to divide the data set with the aim of using two different sets of data, one for the training and the testing stage and the second to validate the model (Mastrorillo et al. 1998). Different partitioning procedures exist according to the size of the available dataset: k-fold cross-validation or hold-out (Utans and Moody 1991, Efron and Tibshirani 1995, Friedman 1997), and leave-one-out (Efron 1983, Kohavi 1995).

The network can be overtrained, that is it loses its capacity to generalise. Three parameters are responsible for this phenomenon: the number of epochs, the number of hidden layers and the numbers of neurons in each hidden layer. It is very important to determine the appropriate numbers of these elements in MLP.

MLPs can be regarded as an extension of the many conventional techniques for understanding complex data, and they have been developed over several decades. Feed-forward ANNs are powerful tools for performing non-linear pattern discrimination. They are especially powerful in pattern recognition and other decision-making, forecasting and signal processing and in the modelling of complex non-linear systems by fitting the network to non-linear data. These are major advantages of using MLP in ecology where the relationships in the data sets are always non-linear and complex.

However, they do not always show high performance abilities. These failures are due to inadequate training, inappropriate architecture for the used dataset, and non-separability of the feature data (inappropriate data). These problems underline the necessity of model calibration for its successful use. One drawback of MLP is that the gradient training encounters multiple local minima. Another comes from the model type. MLP is in fact a black-box type model which means that it is not possible to interpret the phenomena that occur inside the network. The complexity of the model comparing to the classical one as the regression models can also be seen as a disadvantage.

Counterpropagation network (CPN)

CPN proposed by Hecht-Nielsen (1987) is a combined network of the two artificial neural networks: SOM (Kohonen 1982) and the Grossberg outstar (Grossberg 1982). The name "counterpropagation" is derived from the initial presentation of this network as a five-layered network with data flowing inward from both sides (Fig. 2.5). There is literally a counterflow of data through the network. The network is designed to approximate a continuous function defined on a data set and serves as a statistically optimal self-programming look-up table (Hecht-Nielson 1987).

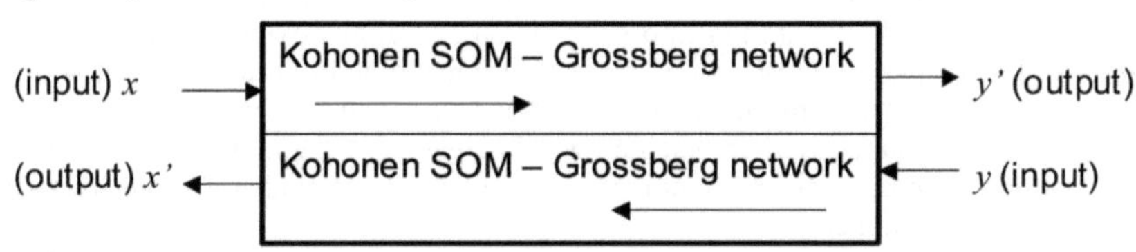

Fig. 2.5. Structure of a full counterpropagation network

The full network is continuous and works best if the inverse function exists (Hecht-Nielsen 1987, Lin and Lee 1996). Although this is an accurate picture of the network, it is complex; thus a simplified forward-only CPN is preferred with no loss of accuracy (Hecht-Nielson 1987).

As with any neural network, there are a few steps that must be performed to test a network's performance. Initially the data for explanatory variables and dependent variables are given to the SOM layer and the Grossberg layer, respectively. For the CPN the learning process occurs in two phases. First, the SOM layer must be trained. The learning process is carried out in an unsupervised mode to follow the general SOM learning rules. After learning the SOM layer, the Grossberg layer should be trained. This is carried out in a supervised mode. The weight vector is updated according to the Grossberg outstar learning rule. Finally, the trained CPN functions exactly as an optimal self-programming look-up table.

Strengths and weaknesses of CPN depend on the geometry and probabilities of the inputs and outputs. If this information is known or estimated, it can be a very good approach. Another advantage is the speed of convergence. Compared to other mapping networks, it typically requires fewer training steps to achieve its best performance. This kind of approach shows how to combine different network architectures, unsupervised network and supervised network, and can produce good results for certain classes of patterns. Finally, the CPN can provide the clustering and predicting values in the same model. However there are also certain limitations, in that a large database is required to train the network, and the predicted values by the network are dependent on the number of groups in the SOM layer.

Competitive artificial neural network (CANN) in combination with linear vector quantisation (LVQ)

Combination network of CANN and a supervised LVQ may give a powerful tool for fast classification of future observations. A CANN consists of two layers, an input layer and an output layer. The input layer consists of a number of input variables; the output layer consists of one neuron for each cluster that will be formed. The researcher has to decide on the number of clusters. The aim of a CANN is to find this number of so-called 'prototype vectors' that can be used to describe these clusters. Starting from the chosen number of initial prototypes (either randomly chosen, or obtained through, for example, K-means clustering), for each input pattern in turn the distance to each prototype is calculated. The prototype it is closest to, is then adjusted somewhat into the direction of this input pattern. This process is iterated until the change in the adjustments becomes smaller than a predetermined threshold (amongst others, Haykin 1999, Hagan et al. 1996).

The LVQ is a network for classification, i.e. it can be trained to learn the relation between a set of input variables and a qualitative output variable. In ordinary statistical language, this means its parameters are iteratively estimated to minimise the prediction errors. Figure 2.6 gives a schematic description of an LVQ network. The circles represent 'nodes' in the network; the squares are indicators for the target variable which the network aims to reproduce. Each square represents one class or group. In addition to the input and output layers, this network has a third, so-called hidden layer. The hidden layer identifies, as it were, a number of 'subclasses' for each class. This feature of the network allows it to have classes formed from non-convex input regions. The number of hidden neurons has to be determined by the researcher.

The first layer in the network is a competitive layer; it is similar to the one described above, except that here the prototypes and input patterns are not scaled to unit length. The second layer merely combines the subclasses into broader classes. The weights of this second layer are not changed during the estimation process. More details again can be found in Haykin (1999) and Hagan et al. (1996).

CANNs can be used to discover clusters of natural sites. For example, think of clustering sites on the basis of environmental characteristics and biotic data. Once a satisfactory clustering has been achieved, an LVQ can then be trained to learn the features of the different clusters from only a subset of the input variables, for example from only variables that are easy to measure. Then future observations could be assigned to one of the clusters using this LVQ, that is to say, using only easy-to-measure variables.

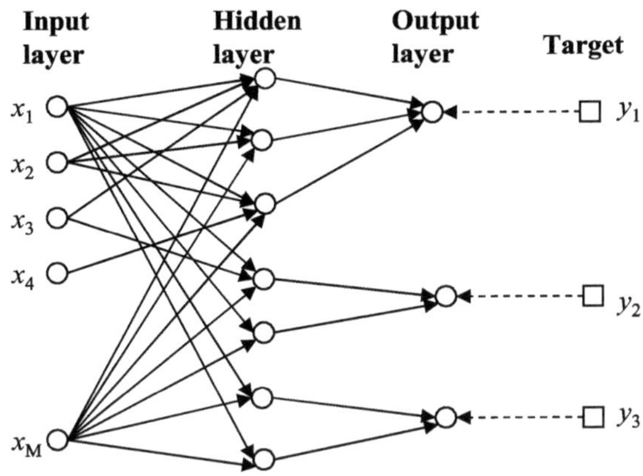

Fig. 2.6. Schematic diagram of an LVQ network (The input layer should be fully connected to the hidden layer)

Kernel-induced nonlinear model (KINM)

KINMs are derived from the principle of statistical learning theory (Vapnik 1995). They have gained more and more attention since the 1990's and have been applied in many areas due to the attractive features and the promising performances. As we know, MLR or its generalised version is widely used for a number of ecological problems. In order to deal with nonlinearity, ANNs are introduced and make great progress towards more accurate models (Lek and Guegan 2000). Nevertheless, there are still some weaknesses in ANN, such as the local minimum of the error surface and the difficulties with generalisation. Therefore, it is very significant to find an alternative or robust method to empirical data modelling in ecological areas.

The main idea behind KINMs is to map the initial data space to the high dimensional feature space by a nonlinear kernel function chosen a priori, also called kernel-induced space. In this space, it is possible to implement a linear regression, which corresponds to a nonlinear regression function in the initial data space.

One of the most important concepts in KINMs is the loss function (***L***), which determines how to penalise the function according to the amounts of error (ε) between the desired and the practical values. Figure 2.7a is the commonly used quadratic loss function and Figure 2.7b is the linear loss function. In KINMs, in order to obtain the sparse distribution of the "support vectors" (explained later), a new type extended from the linear one is developed as shown in Figure 2.7c, the so-called ε-insensitive loss function.

Vapnik (1995) and Gunn (1998) provide more details on both the theoretical and the practical aspects of KINMs. The advantages of KINMs, in comparison to ANNs, are due to the implementation of the structural risk minimization (SRM) principle. This principle has been shown to be superior to the most commonly used empirical risk minimisation (ERM) principles employed in ANN. SRM minimises an upper bound on the VC dimension, i.e. the generalisation error, while ERM minimizes the error on the training data, usually in a quadratic fashion. This difference enhances the power of KINMs when modelling nonlinear complex problems.

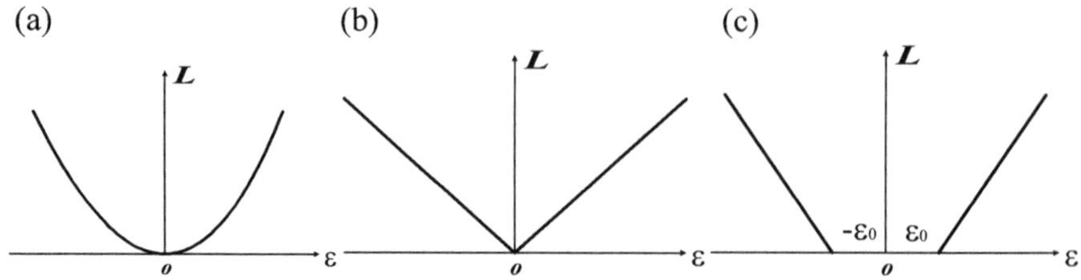

Fig. 2.7. Three types of loss functions. (**a**) Quadratic, (**b**) linear, and (**c**) ε-insensitive

Application of ANNs in ecological studies

Over the last ten years, ANNs have been applied in diverse ways in ecological modelling (Colasanti 1991) and resulted in different applications. Since Chon et al. (1996) applied the SOM for patterning benthic communities, the SOM has become more and more popular for extracting the complexity of ecological datasets in diverse ways: assessment of water quality (Walley et al. 2000, Aguilera et al. 2001), patterning communities (Chon et al. 1996, 2000a, Park et al. 2003a, Foody 1999), evaluation of ecosystems using exergy (Park et al. 2001a), prediction of population and communities (Cereghino et al. 2001, Obach et al. 2001), modelling micro-satellite of fish (Giraudel et al. 2000), and conservation strategies for endemic species (Park et al. 2003b). Giraudel and Lek (2001) compared the ordination capability of the SOM with conventional statistical multivariate analysis showing high performance of the SOM. Recently, Park et al. (2003a) presented a method to relate explanatory variables with the SOM map and contribution of input variables by simply calculating the mean value of each explanatory variable in each output neuron of the trained SOM. This simple technique can provide useful information to understand the nature of the datasets.

The MLP has been implemented in many different research topics of ecological studies in different ways. Typical ecological applications of the MLP include amongst others: *pattern recognition and classification in taxonomy* (Nakano et al. 1991, Simpson et al. 1992, 1993, Boddy et al. 2000), *remote sensing* (Civco 1993, Kimes et al. 1996, Mann and Benwell 1996, Keiner and Yan 1998, Gross et al. 1999, Carpenter et al. 1999), *GIS data analysis* (Silveira et al. 1996), *empirical models of ecological processes* (Brey et al. 1996, Scardi 1996, Aoki and Komatsu 1997, Mastrorillo et al. 1997a, 1998, Brey and Gerdes 1998, Brosse et al. 1999a, Lae et al. 1999, Aoki et al. 1999, Barciela et al. 1999, Scardi and Harding 1999, Recknagel et al. 1997, 2000), *tools for predicting community structure or population characteristics* (Baran et al. 1996, Lek et al. 1996a, Guegan et al. 1998, Giske et al. 1998, Aussem and Hill 1999, Schleiter et al. 1999, Wagner et al. 2000a), *water management* (Kastens and Featherstone 1996), *time series analysis and prediction* (Recknagel 1997, Chon et al. 2000a), *ecosystem dynamics* (Pineda 1987, Chon et al. 2000b, 2001), *flowering and maturity of soybeans* (Elizondo et al. 1994), *changes in the size of animal populations* (Stankovski et al. 1998), *establishment of grasslands* (Tan and Smeins 1996), *energy status of ecosystem* (Park et al. 2001), and *habitat suitability* (Paruelo and Tomasel 1997, Özesmi and Özesmi 1999).

The CPN has been implemented for relational patterning on different hierarchical levels in communities of benthic macroinvertebrates in an urbanised stream (Park et al. 2001b, Chon et al. 2002) and patterning and predicting aquatic macroinvertebrate diversity with quantitative environmental variables (Park et al. 2003c). The latter showed associations between environmental variables and community diversity in the SOM map through the patterning process. These studies showed that CPN as a hybrid model can predict ecological characteristics as well as patterning input variables.

Tang et al. (1998) used the LVQ to identify plankton images and to classify them. Their results showed 95% classification accuracy on six plankton taxa taken from nearly 2000 images and the possibility for a fully automated real time mapping of plankton populations in aquatic ecosystems.

As for hydrological applications, although parametric statistical protocols and deterministic models have been the traditional approaches in forecasting water quality variables in streams, many recent efforts have shown that, when explicit information of hydrological sub-processes is not available, ANNs can then be more effective (Zhu et al. 1994, Maier and Dandy 2000).

A comprehensive overview of ANN applications in ecological informatics has been compiled by Lek and Guegan (2000) and Recknagel (2003). There are also many valuable papers in the three Special Issues of Ecological Modelling: Volume 120 (2-3) in 1999, Volume 146 (1-3) in 2001, and Volume 160 (3) in 2003.

2.4 Bayesian and Mixture models

Bayesian models are based on the principles of Bayes rule or Bayes theorem. It defines the formalism of updating a belief about a hypothesis (or *a priori* probability) in the light of new evidence (e.g. new data). The updated probability is called the posterior probability. The distinctive feature of Bayesian models is the explicit consideration of probability. It is, therefore, a powerful way to increase knowledge about a certain system of the real world by the integrative analysis of probabilities of models and observation data.

The basic applications in Bayesian reasoning were extended for application in time series as well as for cases of interdependent probabilities. Extension of the basic method is the integration of Markov chain theory, Metropolis Hastings algorithm, Monte Carlo methods, information theory, and spatial analysis. Integrated methods are for example the Gibbs sampler, Markov chain Monte Carlo techniques, Bayesian maximum entropy, and Bayesian kriging. This family of methods is used in a wide field of disciplines, e.g. medicine, astrophysics, economy as well as in ecology.

The integration of Bayesian principles into other methods has supported the development of complex Bayesian models in Bayesian networks (BN). In hierarchical BNs, the hierarchical influences of parameters with different probability functions can be modelled. Bayesian belief networks (BBN, also known as belief networks, causal probabilistic networks, causal nets, graphical probability networks, probabilistic cause-effect models, and probabilistic networks) are building the bridge to artificial intelligence by making it possible to integrate expert knowledge into the model. The advantages of BBNs are the ability to represent and manipulate complex models, and the possibility for event prediction based on partial or uncertain data.

A common application of Bayesian models is stock assessment especially in fish ecology and fish management. An application of special practical interest is the determination of stock assessment for the regulation of fish catches and related topics, with a predominance of basic Bayesian models (Adkison and Peterman 1996, Ogle et al. 1996, Kinas

1996, McAllister and Ianelli 1997, McAllister and Pikitch 1997, Punt and Hilborn 1997, Newman 1997, Cow-Rogers 1997, Hammond 1997, Hilborn and Liermann 1998, McAllister and Kirkwood 1998a, 1998b, Pella et al. 1998, Punt and Walker 1998, Robb and Peterman 1998, Smith and Punt 1998, Peterman et al. 1999, Myers et al. 1999, Jon et al. 2000). A lower number of publications in fish management is related to advanced Bayesian models (Lee et al. 1996, Lee and Rieman 1997, Liermann and Hilborn 1997, Vignaux et al. 1998, Kuikka et al. 1999, Meyer and Millar 1999, Patterson 1999, Chen and Fournier 1999, Chen et al. 2000, Helu et al. 2000).

The improvement of Bayesian models by the addition of other methods affected an increase in applications outside fish ecology. Basic Bayesian models are applied for risk and decision analysis (Steinberg et al. 1996, Varis 1997, Qian et al. 2000). A focus of complex Bayesian models is classification and diagnosis of water quality (Trigg et al. 2000, Varis and Kuikka 1997, Walley and Dzeroski 1996, Walley and Fontama 1997 2000). Other applications are related to environmental reconstruction (Vasko et al. 2000) and models about heterogeneous populations (Pinelalloul et al. 1995, Carpenter et al. 1996, Billheimer et al. 1997, Mau et al. 1999, Meyer and Millar 1999, Cottingham and Schindler 2000, Lamon and Clyde 2000).

Mixture models (Titterington et al. 1985) assume that each sample is a member of one of a finite number of classes, and use the data to estimate the parameters of the model. Within each class, a specific group of taxa is linked to a specific environment. The latter implies that knowledge of the environment enables a researcher to predict the group of taxa. Previous research (amongst others, Verdonschot and Goedhart 2000) indicated that the mixture assumption is reasonable and worthwhile pursuing. This approach can be seen as the counterpart of ANNs. ANNs use the data to construct a model with which the number of each taxon can be predicted using environmental variables.

Traditionally mixture models are analysed using maximum likelihood. The latter increases in difficulty with the complexity of the mixture model. It may be impossible to determine if the likelihood has actually been maximised; as confidence intervals and standard errors for the model parameters are unobtainable, it is hard to determine the number classes that have to be used in the mixture, and, the distribution of goodness of fit tests is unknown. Using Bayesian computational statistics, these problems can almost completely be avoided. See Hoijtink and Molenaar (1997), Hoijtink (1998, 2001) for an application of this approach to mixture models where the observed variables are dichotomous.

The use of Bayesian computational procedures for the analysis of data with mixture models does not solve the problem that the number of parameters in the model is rather large compared to the amount of data (the sample size). The consequence is that the predictive validity of the model will probably be rather small if this problem is ignored. Two measures will be taken to control this problem.

Some of the taxa have greater similarities (i.e. a higher probability of living in the same environment) than others. The similarity among the taxa will be quantified into a number of variables that will be used to determine the parameters of a hyper prior for each of the taxa. The result is that similar taxa will receive similar parameter-values. The latter will increase the predictive validity of the model.

The environmental variables are correlated. The latter implies that not all environmental variables are needed to make a distinction among the mixture components. Instead of estimating parameters for each environmental variable, the parameters of a (much smaller) number of discriminant-functions will be estimated. The latter will also increase the predictive validity of the model. The parameters of the informed latent class/discriminant model can be estimated using a procedure based on the Gibbs sampler (Zeger and Karim 1991, Gelman et al. 1995). The selection of the number of classes and the number of discriminant functions can be based on Bayes factors (Kass and Raftery 1995).

The models described above have two features that distinguish them from other approaches. Both could be used in this context: they deal explicitly with the fact that the number of variables exceeds the sample size; in addition, it is a statistical model which implies that it can be used to make inferences with respect to the unknown population from which the sample is obtained. To clarify the latter, both cluster analysis and neural nets are models that describe the structure in the data without reference to a population from which the data are obtained.

2.5 Support vector machines (SVMs)

SVMs are used to predict known class membership from observed variables. In the tradition of machine learning, SVMs are a relatively new and modern tool. An advantage of SVMs over Artificial Neural Networks is that with SVMs the function to be minimised is very well shaped: it is convex and thus has no isolated local minima. SVMs can be used for classification and regression. A full SVM analysis requires three steps, and ideally in each of these three steps, a separate part of the data is used: 1) model selection, 2) fitting, and 3) validation.

The basic SVM distinguishes between two classes with either the value 1 or -1. Classification into more classes is achieved using a combination of several 2-class SVMs. For example, in Figure 2.8 the black dots represent sardines, the open circles are herrings, and the two x-variables are size and weight, respectively. The data in Figure 2.8a are linearly separable, that is, it is possible to draw a straight line, such that all circles are on one side of the line, and all plusses are on the other side. In Figure 2.8b, however, the data cannot be separated by a straight line. In this situation, it would help if a (nonlinear) transformation could be found, such that the transformed data were again linearly separable, or at least 'more' linearly separable (Fig. 2.8c).

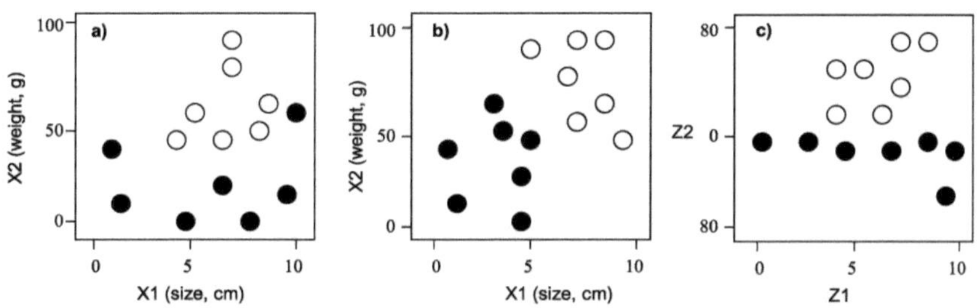

Fig. 2.8. Linear and nonlinear separation of herrings and sardines. (**a**) linear separable herrings and sardines, (**b**) linear separation is not possible, and (**c**) after suitable transformation of $\boldsymbol{x}$, linear separation is again possible.

In choosing a nonlinear transformation and increasing the number of dimensions, one should avoid overfitting. More often it is advisable to allow a certain non-separability and use a punishment or penalty term for the degree of violation of separability. The basic idea of SVMs is to transform the data in such a way that they become 'more', or even completely linearly separable, and to perform a linear separation. Upon returning in the lower dimensional data space X, a nonlinear separation will have been performed. Details of SVMs are given in Chap. 6 in this book by Akkermans and her colleagues.

Morris et al. (2001) compared the SVMs to strongly partitioned traditional radial basis function (RBF) networks for the discrimination of single species of phytoplankton against a background of *N* other species, and showed that SVMs had a greater identification success than the unpartitioned, large single RBF networks. The greatest success was achieved by combining the outputs of the individual networks by means of a `winner takes all' strategy; with RBF, ANNs' identification success dramatically increased, though there was only a modest increase with SVMs. When SVMs trained on one data set were tested with data on cells grown under different light conditions, the overall successful identification rate was low, but when SVMs were trained on a combined dataset identification was high. Akkermans and her colleagues (Akkermans et al. 2004) also applied the SVMs to two datasets of benthic macroinvertebrates, consisting of streams and canals in the Netherlands, for classifying communities. They compared their performances to the results obtained with MLGR and showed that the SVM approach clearly yielded better results. Their results are given in Chap. 6 of this book.

2.6 Genetic algorithms (GAs)

GAs are a part of evolutionary computing, which is a rapidly growing area of artificial intelligence, and they were inspired by Darwin's theory about evolution. The idea of evolutionary computing was introduced in the 1960s by I. Rechenberg in his work "Evolution strategies" and his idea was then developed by other researchers. Holland (1975) first explored GAs, operating on strings of bits called chromosomes. The algorithm is started with a set of solutions (represented by chromosomes) called a population. Solutions from one population are taken and used to form a new population. This is motivated by the hope, that the new population will be better than the old one. Solutions which are selected to form new solutions (offspring) are selected according to their fitness - the more suitable they are the more chances they have to reproduce.

1. [**Start**] Generate random population of *n* chromosomes
2. [**Fitness**] Evaluate the fitness f(x) of each chromosome *x* in the population
3. [**New population**] Create a new population by repeating following steps until the new population is complete
   a. [**Selection**] Select two parent chromosomes from a population according to their fitness (the better the fitness, the greater chance of being selected)
   b. [**Crossover**] With a crossover probability, cross over the parents to form new offspring (children). If no crossover was performed, the offspring is an exact copy of the parents.
   c. [**Mutation**] With a mutation probability, mutate new offspring at each locus (position in chromosome).
   d. [**Accepting**] Place new offspring in a new population
4. [**Replace**] Use new generated population for a further run of algorithm
5. [**Test**] If the end condition is satisfied, stop, and return the best solution into the current population
6. [**Loop**] Go to step 2

Fig. 2.9. Outline of the basic genetic algorithm

GA is based on copying chromosomes and swapping parts of the chromosomes basically by three operations: reproduction, crossover and mutation (Recknagel 2001). Reproduction means that chromosomes are copied according to their objective function where strings with higher evaluations will have a better chance to survive. Crossover means that pairs of chromosomes are recombined by swapping parts of them from a randomly selected point in order to create two new chromosomes. Mutation occurs only occasionally with a very low probability and means that the value in a string position may be changed, for example a 1 is changed into a 0 or vice versa. At the same time, mutation ensures that reproduction and crossover do not loose potentially useful material. The basic GA can be summarised in Figure 2.9.

The GA has been applied successfully in ecological modelling. Recknagel et al. (2000) have applied GA to model the abundance of algae. This optimisation procedure steadily evolves best models for given data that unlike ANN become explicitly available. D'heygere et al. (2003) used the GA to select input variables in decision tree models for the prediction of benthic macroinvertebrates. The specific features of GA make them novel tools for ecological modelling. Inductive models (multiple nonlinear regression functions, ANN designs) and deductive models (rule sets, ecosystem process equations and parameters) can be evolved from databases of individual or classes of ecosystems with high validity (Recknagel 2001).

2.7 Mutual information and regression maximisation (MIR-max)

MIR-max has been developed to cluster datasets based on the information theory. This technique has a visualisation system similar to SOM, but the algorithm is consistently different. The Mir- max technique is based on two separate processes. A major advantage of MIR-max is that it separates the tasks of clustering and ordering into two independent processes (Walley and O'Connor 2001). It first clusters the data into a predefined number of classes using information theory, and then orders the classes in two-dimensional output space using the correlation between corresponding distances in data space and output space. This procedure assumes that the data are interval-valued, so its application to ordinal data involves an approximation. It does, however, provide added benefit to the user in terms of data visualisation. The algorithm can be found in Walley and O'Connor (2001). Walley and O'Connor (2001) applied this technique to clustering and ordering datasets of benthic macroinvertebrates and presented the suitability of the system for use on ordinal or discrete interval-valued data, especially for ordinal data. Rimet and Ector used this technique to explore the complexity of diatom assemblages in the Rhone basin and Mediterranean region. Their results are given in Chap. 5 of this book.

2.8 Structural dynamic models

Models that can account for the change in species composition as well as for the ability of species to change their properties, i.e. to adapt to the prevailing conditions imposed on the species, are called structural dynamic models (Bossel 1992). This type of model was developed by the use of biomass as a goal function in the late seventies and later by the use of the theoretically more correct, exergy, in the mid-eighties (Jørgensen 1997).

Exergy measures biomass and information, thus more developed organisms, will contribute more to the exergy per weight unit than less developed organisms. Exergy is defined as the work the system can perform when it is brought into equilibrium with the environment or another well-defined reference state. If we assume a reference environment for a system at thermodynamic equilibrium, meaning that all the components are: (1) inorganic, (2) at the highest possible oxidation state signifying that all free energy has been utilised to do work, and (3) homogeneously distributed in the system, meaning no gradients, then exergy becomes an expression for the biomass (physical structure) and the information (embodied in the complex biochemical composition of the cells, determined by the genes. In any case, temperature and pressure differences between systems and their reference environments make a small contribution to the overall exergy and for present purposes can be ignored.

The goal function (Table 2.1) describes the development direction of the considered ecosystem. The models describe how organisms will adapt to currently changing conditions and how - if the adaptation process is not sufficient - the present organisms will be replaced by other and better fitted organisms with other properties. This type of model using exergy as goal function has already been applied to aquatic systems (Jørgensen 1997). It should also be possible to use this model type to give the properties of the species present in a river or lake, provided that we know the conditions. From the properties it will in most cases probably be possible to describe the species that have these properties.

Table 2.1. Examples of goal functions.

Goal function	Target system	Reference
Maximum useful power or energy flow	Several systems	Lotka (1956), Odum and Pinkerton (1955)
Minimum entropy	Several systems	Glansdorff and Prigogine (1971)
Maximum ascendancy	Networks	Ulanowicz (1980)
Maximum exergy	Several systems	Mejer and Jørgensen (1979)
Maximum persistent organic matter	Ecological systems	Whittaker and Woodwell (1971), O'Neill et al. (1976)
Maximum biomass	Ecological systems	Margalef (1968), Straskraba (1979)
Maximum profit	Economic systems	Various authors

Changes in the structure can be described by the introduction of a goal function. Exergy is used as a goal function as it describes the distance of the ecosystem (described by the model) from thermodynamic equilibrium = the sum of biomass and information. Structurally dynamic models (Jørgensen 1997) have already been applied in biomanipulation (Jørgensen and de Bernardi 1998), the intermediate disturbance hypothesis (Jørgensen and Padisak 1996), and the succession of phytoplankton species (Jørgensen and Padisak 1996). Exergy has been used most widely as a goal function in ecological models. Exergy has two pronounced advantages as a goal function compared to entropy and maximum power. It is defined far from thermodynamic equilibrium and it is related to the state variables, which are easily determined or measured.

The idea of structural dynamic models is to find continuously a new set of parameters (limited for practical reasons to the most crucial, i.e., sensitive parameters) that are better fitted to the prevailing conditions of the ecosystem. "Fitted" is defined in the Darwinian sense by the ability of species to survive and grow, which may be measured by the use of exergy (Jørgensen 1982, 1986, 1988, 1990, Jørgensen and Mejer 1977, Mejer and Jørgensen 1979). Exergy has previously been tested as a "goal function" for ecosystem development (Jørgensen 1986). However, in all these cases, the model applied did not include the "elasticity" of the system, obtained by using variable parameters, and therefore the models did not reflect real ecosystem properties.

3 Fish community assemblages

Editor: Lek S*

3.1 Introduction

Many groups of organisms have been proposed as indicators of environmental quality. Ideally, a biological monitoring program will integrate multiple assemblages in order to better assess environmental quality (Jackson et al. 2001). Fish are one of the most widely used and useful organisms for measuring water resource quality. They are typically present even in the smallest streams and are easily sampled and identified with the proper equipment and training. The Clean Water Act mandates "fishable" waters and the public widely recognizes fish for their economic and aesthetic value.

A fish community is an assemblage of fish sharing the same area of a stream and interacting with each other. The structure of a fish community is determined by the species present, their relative abundances, their life stages and size distributions, and their distributions in space and time (Meador et al. 1993, Matthews 1998). Natural variability in fish communities can be attributed to differences in land elevation, water temperature, water chemistry, food resources, and physical habitat. The abundance, condition, and species composition of fish communities can be influenced by water and habitat quality that are modified by surrounding land uses (Deacon and Mize 1997). Fish-community data can have a high degree of variability, even when they are collected for the same site several times in one season (Karr 1999).

Fish are a diverse group of organisms and have a wide range of life history requirements. Some fish are sensitive to changes in water temperature, substrate composition, stream flow, or various water chemistry parameters, while others are tolerant to change in their environment. They occupy positions throughout the aquatic food web and characterize a range of trophic levels (planktivores, herbivores, omnivores, invertivores, piscivores). The structural and functional variety of fish communities make them excellent indicators of water quality and provide an integrated view of waterbody condition.

Many Control Agencies have been using fish community data to assess water resource quality for the last decade: IBI's have been developed in USA for streams in the Minnesota, Red, St. Croix and Upper Mississippi River Basins. In Europe, Oberdorff et al. (2001) developed a probabilistic model based on species occurrence to define a fish-based index with national application. The objective of all the methods proposed is to develop biological criteria utilizing fish for all streams in USA and Europe.

Biological assessment is used for several aspects of water resource management, including:

- Long term condition monitoring (status and trends)
- Aquatic life use assessment
- Listing, diagnostics, and effectiveness of implementation
- Problem investigation monitoring
- Effectiveness monitoring

---

* Correspondence: lek@cict.fr

This chapter includes 7 papers:

1. Park et al. present "Visualizing large scale distribution patterns of riverine fish assemblages using unsupervised neural networks". The self-organizing map (SOM), was used to visualize distribution patterns of fish species in rivers, and to evaluate the relative importance of several environmental factors in influencing the organization and structure of fish assemblages. A dataset (40 fish species) of 668 reference sites sampled across all French rivers, i.e. very large scale of data;
2. Gevrey et al. present the capacity for predicting and sensitivity analysis of fish assemblages on the French scale, i.e. by using the same dataset as the first paper;
3. Aguilar Ibarra et al. show for the Garonne basin the diversity patterns of fish assemblages proposing conservation measures by using a self-organizing map;
4. Joy and Death show the capacity of neural network modelling to predict freshwater fish and macro-crustacean assemblages for biological assessment in New Zealand;
5. Moreau et al. compare linear and nonlinear fitting techniques for predicting fish yield in Ubolratana reservoir (Thailand) from a time series data on catch and hydrological features;
6. Penczak et al. use the self-organizing map (SOM) for patterning the spatial variation in fish assemblage structures and diversity in the Pilica River system.
7. Scardi et al. show the predicting fish assemblages in Italian rivers, a neural network case study adapted to the presence/absence data.

3.2 Patterning riverine fish assemblages using an unsupervised neural network*

Park YS†, Oberdorff T, Lek S

Introduction

In temperate river systems, the uni-directional character of water tends to give them a linear structure along a gradient of environmental conditions. In these systems, biological assemblages are organised longitudinally and there is generally an increase in species richness from the source to the river mouth (e.g. measured by the river width, the distance from the source, the stream order, and the size of the watershed). Longitudinal changes in local assemblage richness have usually been attributed to one of two processes: biotic zonation or continual addition of species downstream. Biotic zonation corresponds to discontinuities in river geomorphology or abiotic conditions promoting distinct assemblages along the longitudinal gradient (e.g. Huet 1959, Schlosser 1982, Balon et al. 1986, Rahel and Hubert 1991, Oberdorff et al. 1993, Belliard et al. 1997). For example, species replacement may occur as a result of physiological specialization for temperature. In contrast to the advocates of zonation, additions of species are usually related to environmental gradients having smooth transitions of abiotic factors contributing to nested patterns of assemblage composition along the longitudinal gradient (e.g. Sheldon 1968, Rahel and Hubert 1991). Whatever the process (i.e. biotic zonation or species addition) local species richness usually increases along the upstream-downstream gradient (Huet 1959, Sheldon 1968, Schlosser 1982, Balon et al. 1986, Rahel and Hubert 1991, Belliard et al. 1997, Oberdorff et al. 2001, 2002a). The environmental factors that have been identified to explain this increase in species richness are generally linked (i) to upstream-downstream differences in local habitat characteristics defined by depth, slope, current velocity, temperature and substrate composition (Huet 1959, Gorman and Karr 1978, Schlosser 1982, Angermeier and Schlosser 1989, Rahel and Hubert 1991, Oberdorff et al. 2001, 2002a) or by "dimensionless" hydraulic characteristics such as the Froude number or the Reynolds number (Lamouroux and Souchon 2002, Lamouroux and Capra 2002) and (ii) to an upstream-downstream increase in environmental stability (Horwitz 1978, Schlosser 1982, Schlosser and Ebel 1989).

Resulting from this common feature in riverine fish ecology (i.e., the longitudinal change in fish assemblage structure along the upstream-downstream gradient of a river), some authors have attempted to classify river basins into different biotic zones. The classical studies include the work of Thienemann (1925) who proposed six zones for continental European rivers: spring brook, trout zone, grayling zone, barbel zone, bream zone and brackish-water, each based on the presence of a specific fish species. This elegant concept persisted in the systems devised by Huet (1949, 1954), who proposed longitudinal zonations of rivers based on the occurrence of key species. The Huet zonation consists of four zones, beginning with the headwater and moving to the lowlands (i.e., the trout zone, the grayling zone, the barbel zone, and the bream zone).

---

* This work was supported by the EU project PAEQANN (EVK1-CT1999-00026).
† Corresponding: park@cict.fr

To visualize the organization and structure of fish assemblages, several multivariate techniques have been used depending on the aim of the studies, including multivariate analysis of variance (Bendell and McNicol 1987, Jackson and Harvey 1989), factor analysis (Stevenson et al. 1974, Oberdorff et al. 1993), correspondence analysis (Hughes and Gammon 1987, Strayer 1993, Pusey et al. 1995, Vila-Gispert 2002), cluster analysis (Hughes et al. 1987, Poff and Ward 1989, Johnson and Wichern 1992), principal component analysis (Matthews 1985, Matthews and Robinson 1988, Paller et al. 1994, Vila-Gispert 2002) and canonical correspondence analysis (Taylor et al. 1993, Copp 1992, Koel 1997). These methods are all adversely affected by the non-linear nature of the ecological data, whereas the methods identified subsequently (i.e., adaptive learning algorithms) are not. As alternative methods, adaptive learning algorithms such as artificial neural networks (ANNs) are becoming more and more popular in ecological studies (Lek and Guégan 2000, Rekgnagel 2003). Among the algorithms of the ANNs, the self-organizing map (SOM) shows an ability for classification, abstraction, and visualization, the idea of which is to show the data set in another, more usable, representation (Kohonen 2001), and to efficiently determine patterns of aquatic ecological assemblages (Chon et al. 1996, Brosse et al. 2001, Park et al. 2003a).

In this study, we propose a SOM model as an alternative method to display patterns of fish species distribution in French rivers, and to evaluate the relative importance of several environmental factors in influencing organization and structure of fish assemblages.

Materials and methods

Ecological data

We used data previously analysed by Oberdorff et al. (2001, 2002a). A dataset of 668 reference sites (Fig. 3.2.1) was extracted from the database held by the Conseil Supérieur de la Pêche (Banque Hydrobiologique et Piscicole), covering a period of 13 years of survey (1985-98). The selection of the reference sites was carried out by regional experts (fish biologists) on the basis of water quality map inspection and field reconnaissance. The factors considered in the field inspection included the amount of stream channel modification, channel morphology, substrate character and condition, and general representatives of the sites within the region. The criteria used for selection of reference sites were that the site should belong to the water quality classes 'Excellent' or 'Good' as defined by the Water Quality Index developed by the French Water Agency (Oberdorff et al. 2001). The reference sites were not pristine nor totally undisturbed but were those considered as least impacted within a particular biogeographical region (Hughes 1995).

In the dataset, 40 species were identified (Table 3.2.1). Fish species were assigned to four different trophic guilds (i.e., invertivores, herbivores, piscivores, and omnivores) (Table 3.2.1) to evaluate potential changes in fish trophic structure along the longitudinal gradient. Assignment of fish species to trophic guilds was difficult due to flexibility in feeding habits and changes that occur over an individual's life cycle. In this study, only adult fish were considered to categorize trophic guilds following Froese and Pauly (2003) and Oberdorff et al. (1993, 2002a). Invertivores include generalized insectivores, surface and water column insectivores and benthic insectivores. The more general term of invertivores was used because the fish typically eat crustaceans, oligochaetes, and molluscs, as well as aquatic or terrestrial insects. Piscivores are fish that eat primary fish and a smaller amount of aquatic and terrestrial insects. Herbivores include fish that are planktivores and herbivore-detritivores. Omnivores eat a wide range of plants, detritus, and animal materials with at least 25% plants and 25% animals (Schlosser 1982, Karr et al. 1986, Oberdorff et al. 1993).

Eight abiotic environmental variables were also measured at each site: slope (%), elevation (m), July mean daily maximum air temperature (°C; JulTemp), January mean daily maximum air temperature (°C; JanTemp), stream width (m), mean depth (m), distance from headwater source (km), and catchment area of the basin (km^2). The slope and elevation were derived from topographic maps, and the distance from the source and the catchment area were measured using a digital palimeter on a 1:1 000 000-scale map. A detailed description of all these environmental variables is given in Oberdorff et al. (2001). These variables are known to be the most consistent in structuring fish assemblages under natural conditions.

To find the biogeographical distribution patterns of fish species in French rivers, the assemblage dataset was applied to an adaptive learning algorithm, the self-organizing map (SOM). The densities of species were scaled between 0 and 1 in the range of the minimum and maximum values within a species, after a log-transformation process in order to reduce variations in densities.

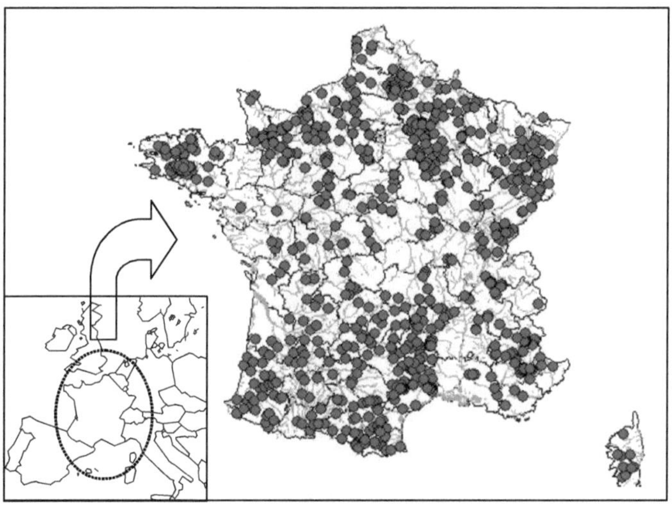

Figure 3.2.1 Map of France showing distribution of all sampling sites.

Statistical analysis

Geographical classification using the self-organizing map

The self-organizing map (SOM) is an adaptive unsupervised learning algorithm and approximation of the probability density function of the input data (Kohonen 2001). The SOM has found wide applications in the fields of data exploration, data mining, data classification, data compression, and biological modelling, due to its properties of neighbourhood preservation and local resolution of the input space proportional to the data distribution. The SOM usually consists of input and output layers connected with computational weights (connection intensities). The array of input neurons (i.e. computational units) operates as a flow-through layer for the input vectors, whereas the output layer consists of a two-dimensional network of neurons arranged on a hexagonal lattice.

Table 3.2.1. Names and trophic guilds of 40 species identified in dataset.

Scientific name	Acronyms	Common name	Trophic guild
Abramis brama	ABB	Bream	Omnivorous
Alburnoides bipunctatus	ALB	Schneider	Invertivorous
Alburnus alburnus	ALA	Bleak	Invertivorous
Anguilla anguilla	ANA	European eel	Piscivorous
Barbus barbus	BAB	Barbel	Invertivorous
Barbus meridionalis	BAM	Mediterranean barbell	Invertivorous
Blennius fluviatilis	BLF	Freshwater blenny	Herbivorous
Blicca bjoerkna	BLB	Silver bream	Omnivorous
Carassius auratus	CAA	Goldfish	Omnivorous
Carassius carassius	CAC	Crucian carp	Omnivorous
Chondrostoma nasus	CHN	Common nase	Omnivorous
Chondrostoma toxostoma	CHT	Soiffe	Omnivorous
Cottus gobio	COG	Bullhead	Invertivorous
Cyprinus carpio	CYC	Common carp	Omnivorous
Esox lucius	ESL	European pike	Piscivorous
Gambusia affinis	GAF	Mosquitofish	Omnivorous
Gasterosteus aculeatus	GAC	Threespined stickleback	Invertivorous
Gobio gobio	GOG	Gudgeon	Invertivorous
Gymnocephalus cernua	GYC	Ruffe	Invertivorous
Ictalurus melas	ICM	Black bullhead	Invertivorous
Lampetra planeri	LAP	Brook lamprey	Herbivorous
Lepomis gibbosus	LEG	Pumpkinseed	Invertivorous
Leucaspius delineatus	LED	Belica	Herbivorous
Leuciscus cephalus	LEC	Chub	Omnivorous
Leuciscus leuciscus	LEL	Dace	Invertivorous
Leuciscus souffia	LES	Varione	Invertivorous
Lota lota	LOL	Burbot	Piscivorous
Micropterus salmoides	MIS	Largemouth bass	Piscivorous
Nemacheilus barbatulus	NEB	Stone loach	Invertivorous
Perca fluviatilis	PEF	Perch	Piscivorous
Phoxinus phoxinus	PHP	Minnow	Omnivorous
Pungitius pungitius	PUP	Ninespined stickleback	Invertivorous
Rhodeus sericeus	RHS	Bitterling	Omnivorous
Rutilus rutilus	RUR	Roach	Omnivorous
Salmo salar	SAS	Atlantic salmon	Piscivorous
Salmo trutta fario	SAT	Brown trout	Invertivorous
Scardinius erythrophthalmus	SCE	Rudd	Omnivorous
Stizostedion lucioperca	STL	Zander	Piscivorous
Thymallus thymallus	THT	Grayling	Invertivorous
Tinca tinca	TIT	Tench	Omnivorous

In the learning process of the SOM, initially the biological data were subjected to the learning network. Then, the weights were trained for a given dataset of the assemblage data matrix. When an input vector $\boldsymbol{x}$ (densities of species) is sent through the network, each neuron k of the output layer computes the summed distance between the weight vector $\boldsymbol{w}$ and the input vector $\boldsymbol{x}$. The output layer consists of N output neurons (i.e., computational units, 35=7×5 in this study) which usually constitute a 2D grid for better visualization.

The form of the output layer is a hexagonal lattice, because it does not favour horizontal or vertical directions as much as the rectangular array (Kohonen 2001). The output neurons are considered as virtual units to represent typical patterns of the input dataset assigned to their units after the learning process. Among all N virtual units, the best matching unit (BMU) which has the minimum distance between weight and input vectors becomes the winner. For the BMU and its neighbourhood units, the new weight vectors are updated by the SOM learning rule. This results in training the network to classify the input vectors by the weight vectors they are closest to. The weight vectors trained in the learning process can be considered as probabilities for each species to occur in each virtual unit. In the same way, the virtual assemblages can be obtained by denormalizing weight vectors. To define clusters between virtual units of the SOM map, hierarchical cluster analysis was used with Ward's linkage method. Based on linkage distances, the map was classified on different scales, which are considered to be differences of assemblages on different scales.

Correspondence between assemblages and environmental variables
During the learning process of the SOM, units that are topographically close in the array will activate each other to learn something from the same input vector. The weight vectors tend to approximate the probability density function of the input vector. Therefore, the visualization of these vectors is convenient to understand the contribution of each input variable to the clusters on the trained SOM (Kohonen 2001, Park et al. 2003a). To analyse the contribution of variables to cluster structures of the trained SOM, the value of each input variable (component) calculated during the training process was visualized in each neuron on the trained SOM map on a grey scale.

To understand relationships between biological and environmental variables, we introduced environmental variables into the SOM trained with biological variables by calculating the mean value of each environmental variable in each virtual unit of the SOM map following Park et al. (2003a). These mean values assigned on the SOM map were visualised with a grey scale, and then compared with maps of sampling sites as well as biological attributes. Differences among patterns defined through the SOM with respect to the environmental variables were analysed using a t-test, analysis of variance and Duncans' multiple range test at different classification scales.

Results

Assemblage patterning

The fish assemblages on the national scale were patterned by training the SOM (Fig. 3.2.2a). After the learning process of the SOM, a hierarchical clustering analysis using the Ward method was applied to find similarities in the units of the SOM map (Fig. 3.2.2b). The numbers in the dendrogram in Fig. 3.2.2b correspond to the number of the units of the SOM map. The weight vector of each unit represents a typical assemblage composition of samples. The different sizes of the black circles stand for the number of samples that fell in each unit of the map ranging from 4 to 64 proportionally. The results show hierarchical classifications of the map units according to the dissimilarities. First, two major clusters (I, II) were considered at a high linkage distance of 2.3, and then at the distance of 1.3 both clusters were divided into two subclusters (IA and IB, IIA and IIB). Finally, six clusters appeared at the distance level of 0.8 (IA-IIBb). These clustering approaches could help to understand the nature of the ecosystem at different scales. The unified distance matrix (U-matrix; Ultsch 1993) was also applied to define the clusters in the units of the map and showed similar results with hierarchical agglomerative clustering. The results of the U-matrix are not presented.

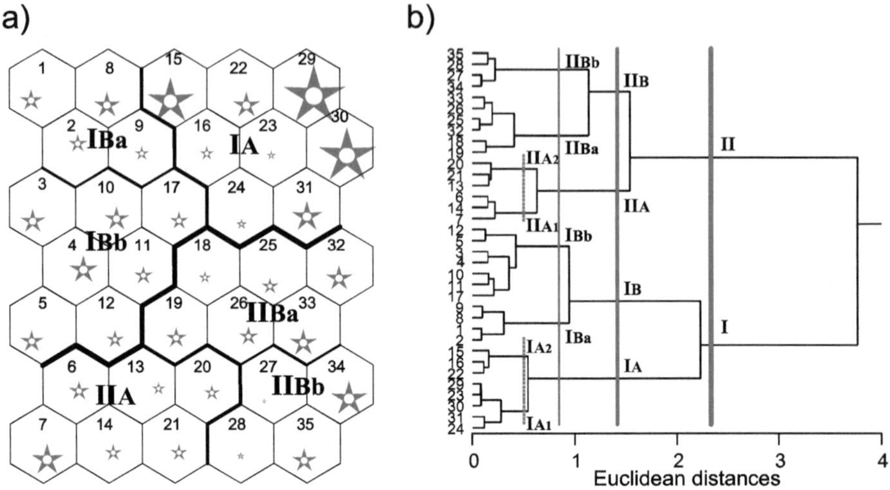

Figure 3.2.2 Classification of fish assemblages on the SOM map (a) and hierarchical classification of SOM units using Ward's algorithm (b). Each unit of the map represents a typical assemblage composition of samples by taking its weight vector. The different sizes of the symboles stand for the number of samples in each unit of the map ranging from 4 to 64 proportionally.

Figure 3.2.3 displays the distribution of each component (fish species) in each unit of the SOM map in a grey scale, by visualizing the weight vectors of the SOM. For convenience of interpretation and to stress the importance of each species in each unit of the map, the weights were rescaled between 0 and 1, representing the probability, assigned to each unit, of each species being observed at sampling sites. Dark represents a high probability, whereas light is low. It also represents the relative importance of each species in each unit of the map. Fig. 3.2.3 shows strong distribution gradients of each species displaying several different distribution patterns. Overall, species abundance and richness were higher in the lower areas of the SOM map, indicating that the classification of the sampling sites on the SOM map is strongly related to species richness. Fig. 3.2.4 shows the differences of species richness in different clusters of the SOM map. Although the highest values for most species werein the lower areas of the SOM map (cluster II in Fig. 3.2.2), different patterns were observed in their distribution. For instance, the species *Salmo trutta fario* (SAT) and *Thymallus thymallus* (THT) were the most abundant in the upper right areas of the map (cluster IA). *Cottus gobio* (COG), *Lampetra planeri* (LAP) and *Salmo salar* (SAS) are in the upper left areas (cluster IBa). *Nemacheilus barbatulus* (NEB), *Phoxinus phoxinus* (PHP) and *Pungitius pungitius* (PUP) were in the left areas (cluster IBb). 10 species including *Alburnoides bipunctatus* (ALB), *Esox lucius* (ESL), and *Leucaspius delineatus* (LED) were in the lower left areas (cluster IIA). *Barbus meridionalis* (BAM), *Blennius fluviatilis* (BLF), *Leuciscus souffia* (LES), and *Lota lota* (LOL) were in the middle right areas (cluster IIBa) and 16 species including *Blicca bjoerkna* (BLB) and *Perca fluviatilis* (PEF) were in the lower right areas (cluster IIBb). Based on these distribution maps, we were able to find the species distribution patterns at different sampling sites. These different distribution patterns were considered to indicate differences of environmental gradients. These characteristics are explained in the following sections in detail.

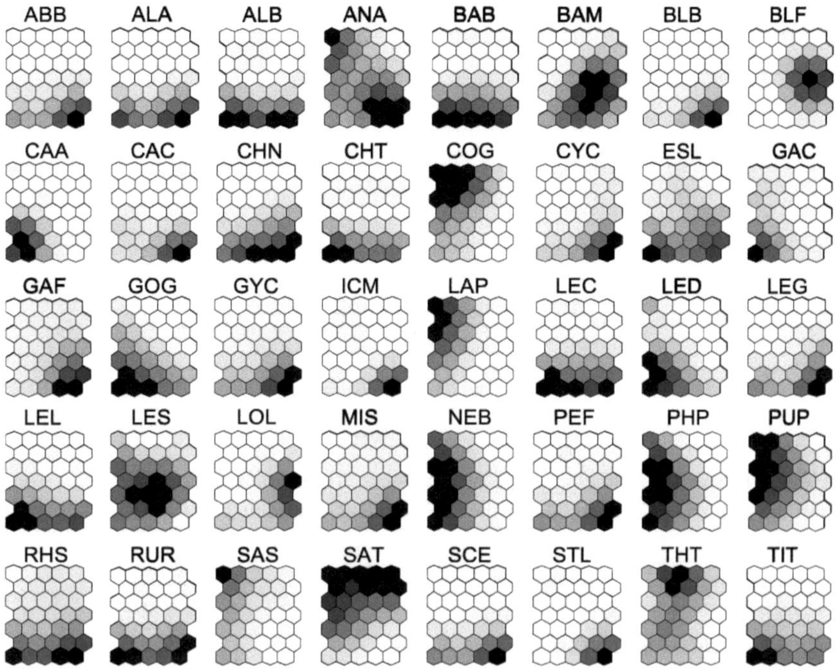

Figure 3.2.3 Visualization of relative abundance of species calculated in the trained SOM in grey scale. The values were calculated during the learning process. Dark represents high values of abundance, whereas light is for low values. The acronyms of the species are presented in Table 3.2.1.

Characteristics of environments

To understand the effects of environmental variables on the fish assemblages and the classification of sampling sites in the SOM, mean values of each environmental variable were calculated and visualized in the SOM map trained with the assemblage dataset (Fig. 3.2.5). Dark represents high values of each variable, and light, low values. Environmental variables showed a clear gradient distribution on the SOM map. The catchment area, the distance from the source, the width and the depth of the sampling areas were the highest values in the lower right areas of the SOM map (cluster IIBb), whereas lower values appeared in the upper areas (cluster IA). In contrast, the slope and the altitude were the highest in upper left area of the SOM map (cluster IA), while lower values occurred in the lower right areas (cluster IIBb). Meanwhile, temperatures in January and July were higher in the middle right areas of the map, although the distribution gradients were not clear.

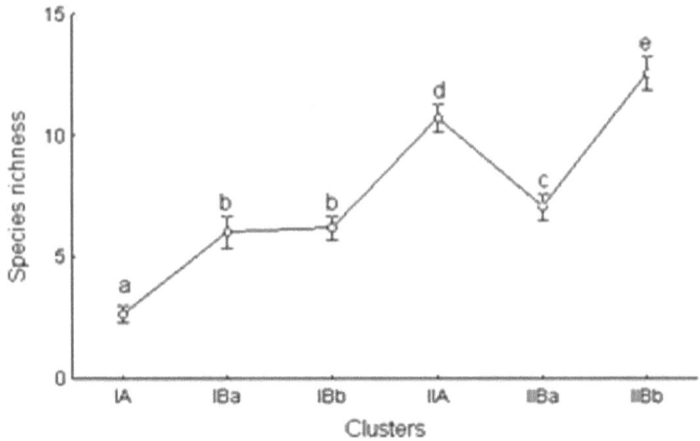

Figure 3.2.4. Species richness at different clusters. The same characters on the error bars indicate no significant difference at the 5% level of confidence using Duncan's multiple comparison test.

Patterns at different scales

Since species are distributed along gradients, their spatial characterization is crucial for understanding ecological functioning at different scales. In this study, fish assemblages were hierarchically patterned at different similarity levels on the SOM map. Therefore, it is worth studying the characteristics of the environmental variables as well as fish assemblage structure at different scales. All environmental variables were significantly different between two large clusters I and II representing the upper watercourses and the lower watercourses, respectively (Table 3.2.2). In the second level of classification with four clusters, clusters IA and IIB were distinctly separated from others in all variables. However, clusters IB and IIA were not clearly distinguished for different variables, indicating intermediate states of gradients of the variables. Finally we considered six number of clusters in the datgaset. For catchment area, cluster IIBb was significantly different from the other groups, indicating that samples in this cluster were mainly from the lower reaches. Furthermore, the variables catchment area, altitude, distance from the source, width, slope and depth showed a gradient from clusters IA to IIBb, representing a gradient from the upper watercourses to the lower ones. However, considering temperatures, cluster IBa showed differences from others, displaying relatively high temperatures for this group. This was due to the characteristics of the sampling sites assigned to this cluster. Most samples came from the southern parts of France, in particular the watershed running to the Mediterranean Sea. Furthermore, these characteristics were reflected in the differences of fish assemblage composition as shown in Fig. 3.2.3 displaying the gradient distribution map of species.

The distributions of four species *Barbus meridionalis*, *Blennius fluviatilis*, *Leuciscus souffia* and *Lota lota*, showing the highest values in cluster IIBa, were limited mainly to the Mediterranean watershed , although they display different distributions on finer scales. Therefore, the patterns identified through the SOM showed the watercourse gradient from the upper right areas (cluster IA) of the SOM map, to the upper left areas (cluster IBa), to the middle left (cluster IBb), to the lower left (cluster IIA), and to the lower right areas (cluster IIBb), excluding cluster IIBa.

Table 3.2.2. Changes in environmental variables at different cluster levels. The numbers in parenthesis are the standard errors of each variable.

Cluster		Environmental variables								N
		Catchm (km^2) [1]	Altitude (m)	Distance (km) [2]	Width (m)	Slope (%)	Depth (m)	Jl.tem. (°C) [3]	Jn tem (°C) [4]	
2	I	89.3	431.1	13.9	5.6	14.4	0.5	17.2	2.2	
		(12.3)	(18.3)	(0.7)	(0.2)	(0.9)	(0.0)	(0.1)	(0.1)	437
	II	2542.1	149.5	86.4	18.4	3.9	0.8	19.3	3.7	
		(474.5)	(9.2)	(7.5)	(1.4)	(0.4)	(0.1)	(0.1)	(0.1)	251
4	IA	56.6	561.5	11.3	5.5	20.7	0.4	17.1	1.7	
		(5.9)	(25.8)	(0.7)	(0.2)	(1.4)	(0.0)	(0.1)	(0.2)	245
	IB	131.1	264.8	17.2	5.8	6.4	0.5	17.4	2.7	
		(26.6)	(20.0)	(1.3)	(0.3)	(0.4)	(0.0)	(0.1)	(0.1)	192
	IIA	479.1	183.1	49.6	10.7	3.2	0.6	19.2	3.4	
		(77.5)	(13.9)	(4.8)	(0.9)	(0.3)	(0.0)	(0.1)	(0.2)	94
	IIB	3777.3	129.5	108.4	23.0	4.3	1.0	19.4	3.9	
		(740.6)	(12.0)	(11.3)	(2.1)	(0.6)	(0.1)	(0.2)	(0.2)	157
6	IA	56.6	561.5	11.3	5.5	20.7	0.4	17.1	1.7	
		(5.9)	(25.8)	(0.7)	(0.2)	(1.4)	(0.0)	(0.1)	(0.2)	245
	IBa	64.1	168.0	12.6	4.5	7.7	0.5	17.2	3.1	
		(10.1)	(18.3)	(1.3)	(0.3)	(0.8)	(0.0)	(0.1)	(0.2)	67
	IBb	166.9	316.7	19.6	6.5	5.7	0.5	17.5	2.5	
		(40.3)	(28.1)	(1.9)	(0.5)	(0.5)	(0.0)	(0.1)	(0.2)	125
	IIA	479.1	183.1	49.6	10.7	3.2	0.6	19.2	3.4	
		(77.5)	(13.9)	(4.8)	(0.9)	(0.3)	(0.0)	(0.1)	(0.2)	94
	IIBa	1194.9	155.4	51.7	13.7	6.2	0.6	19.7	4.3	
		(492.8)	(17.8)	(8.2)	(1.9)	(1.0)	(0.0)	(0.2)	(0.2)	98
	IIBb	8066.6	86.4	202.4	38.4	1.0	1.6	18.8	3.2	
		(1656.2)	(9.9)	(21.9)	(3.9)	(0.1)	(0.1)	(0.2)	(0.2)	59

[1] Catchment area, [2] distance from source, [3] temperature in July, [4] temperature in January

Figure 3.2.6 shows the differences of the densities and proportions of different trophic guilds at different hierarchical levels. The densities of each guild were significantly different between the upper watercourses (cluster I) and the lower ones (cluster II) (t-test, $p < 0.01$), except piscivores (t-test, $p = 0.36$) (Fig. 3.2.6a). Planktivores and omnivores were significantly higher in the lower watercourses than in the upper ones, whereas invertivores were higher in the upper ones (t-test, $p < 0.01$). At the four- and six-cluster levels, invertivores and omnivores showed the highest densities in middle watercourses. However, planktonivores were higher in the lower courses. Meanwhile, considering the proportions of each guild, the differences between the upper and the lower courses were more clearly observed (Fig. 3.2.6b). With four and six clusters, the proportion of the invertivores decreased gradually according to the watercourse gradient, while that of omnivores increased. Piscivores and planktivores also showed high ratios in the lower watercourses.

Discussion and conclusion

In this study, fish assemblages were patterned through an adaptive learning algorithm, the self-organizing map (SOM), according to the distribution similarities of each species. Overall the SOM showed six clusters of fish assemblages, highly related to longitudinal river gradient. These characteristics support the fish zonation theory in European continental rivers (Huet 1954). The results also showed a significant relationship between species assemblages and river size such as catchment area, width, depth, and distance from head water source, supporting the river continuum concept (Vannote et al. 1980).

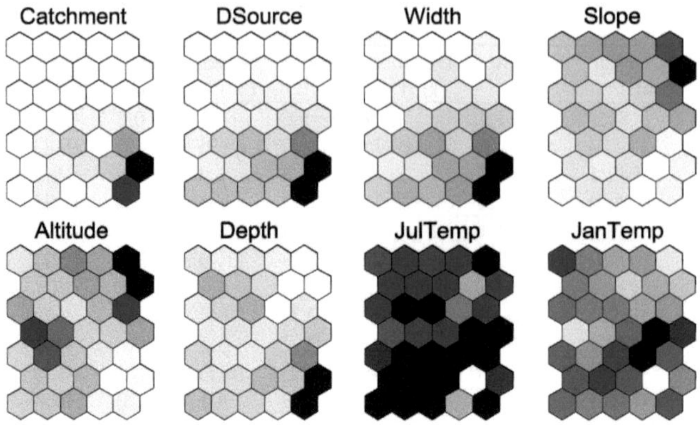

Figure 3.2.5 Visualization of environmental variables on the SOM map trained with fish assemblages. The mean value of each variable was calculated in each output unit of the trained SOM. Dark represents a high value and light a low value. Catchment; catchment area, DSource; distance from source, JulTemp; maximum temperature in July, and JanTemp; maximum temperature in January.

Recently Oberdorff et al. (2001) developed a probabilistic model characterizing fish assemblages of French rivers with environmental variables. They showed that the probability of occurrence is highly dependent on the longitudinal gradient. Our findings on the occurrence patterns for most species agree with their results, although there are some small differences for a few species.

Considering the trophic guilds of freshwater fish in the geographical gradients, planktivores and omnivores were significantly higher in the lower watercourses than in the upper ones, whereas invertivores were higher in the upper ones in French river reference sites. Invertivores and omnivores showed the highest densities in middle watercourses. However, planktonivores were higher in the lower courses. Meanwhile, the proportion of invertivores decreased gradually with the watercourse gradient, while omnivores increased. Piscivores and planktivores also showed high ratios in the lower watercourses. Oberdorff et al. (1993) also presented similar results, indicating that species richness and the proportions of omnivores and piscivores increased with river size, whereas those of invertivores declined downstream.

The River Continuum Concept (Vannote et al. 1980) explicitly predicts changes in fish trophic structure along the longitudinal gradient. This concept suggests a change in community structure and richness from upstream to downstream areas. The main reasons for this include flow regime, temperature, food availability and substrate conditions. This concept attempts to relate the gradient of physical factors that occurs along river systems to changes in assemblage structure and function. According to this hypothesis, available food resources should change along this gradient and should thus be reflected by the trophic composition of the assemblages. Species richness increases with stream size, reaching a maximum in midorder streams, then decreases in large rivers. The lower species richness in headwaters and its decline in large rivers is assumed to be due to reduced environmental variability resulting from interplay between riparian control and water volume (Minshall et al. 1985). These predictions (i.e., a decrease in invertivorous species and an increase in omnivorous species from upstream to downstream) have been confirmed for fish assemblages in French rivers by Oberdorff et al. (1993, 2001).

The major form of environmental variability in stream ecosystems is fluctuations in stream flow (Jackson et al. 2001). Changes in these characteristics alter the physical habitat of streams and rivers, thereby influencing the composition and stability of fish assemblages (Grossman et al. 1998), primarily due to increased mortality and a reduction in recruitment (Jackson et al. 2001). In higher order streams, where the catchment area is larger and thus hydraulic variation is lower, habitat characteristics are more stable and assemblages are able to persist for relatively long periods of time. Maximum diversity is likely to occur in sites where the habitat diversity is enhanced and strong interspecific interactions are mediated by intermediate environmental disturbance (Resh et al. 1988).

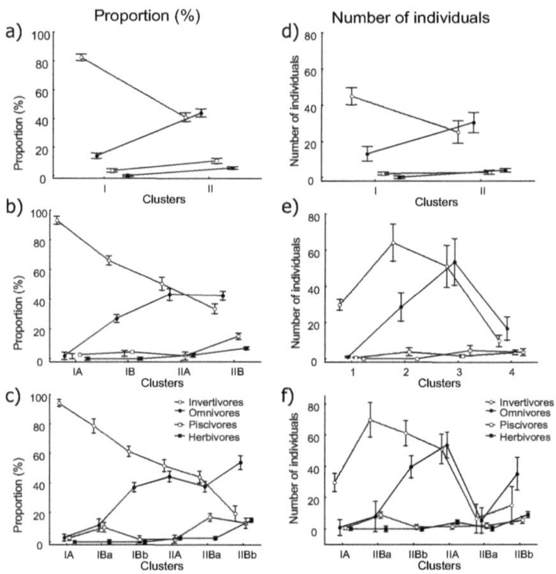

Figure 3.2.6 Proportion (%) (a; 2 clusters, b; 4 clusters, and c; 6 clusters) and abundance (d; 2 clusters, e; 4 clusters, f; 6 clusters) of trophic guilds at different cluster levels.

In conclusion, the characteristics of the distribution patterns of fish assemblages were efficiently visualized on reduced dimensions through the adaptive learning algorithm, SOM. The results confirm major concepts in fish ecology such as the stream zonation and the river continuum concept. Furthermore, the SOM model showed probabilities of occurrences of each species in different environmental conditions. Finally the modelling techniques with the SOM seem to be a powerful analytical tool for identifying habitat and species grouping.

3.3 Predicting fish assemblages in France and evaluating the influence of their environmental variables*

Gevrey M*, Park YS, Oberdorff T, Lek S

Introduction

Fish are valid species for biological monitoring programs, although they have been less widely used than other organisms like diatoms or macroinvertebrates. Fish can be used as indicator organisms for numerous reasons (Karr 1981; Oberdorff et al. 2001): i) they are present in many water bodies, and fish species are relatively easily identifiable; ii) their life-histories are well-known as are their ecological requirements; iii) they represent a variety of trophic levels in various habitat types and iv) their economic aspect plays an important role in their use in biomonitoring programs.

This study was investigated as part of the PAEQANN project (Predicting Aquatic Ecosystem Quality using Artificial Neural Networks, EU project n° EVK1-CT1999-00026, http://aquaeco.ups-tlse.fr/) under the directive of the European Community (European Parliament 2000, directive 2000/60/EC), studying the impact of environmental variables on the structure and the diversity of aquatic communities. Fish assemblages of reference sites were studied over the whole territory of mainland France. As suggested in several studies (Verneaux 1977; Mahon 1984; Oberdorff et al. 1993; Oberdorff et al. 2002a) fish assemblage structures change along an upstream-downstream gradient as proposed by the River Continuum Concept (Vannote 1980). Flow regime, temperature, food availability and substrate conditions vary from upstream to downstream areas. These variations lead to non-linear relationships between the fish assemblage structure and the environmental variables which characterize the river.

Due to their efficiency, artificial neural networks (ANN) with the error backpropagation algorithm are appropriate methods to model non-linear data (Rumelhart 1986). Often compared to multiple linear regression, ANN, which can be used without transformation of the variables, shows higher predictive power (Scardi 1996; Paruelo and Tomasel 1997; Guegan et al. 1998; Kemper and Sommer 2002). Moreover, ANN, which was criticized earlier in its development due to its black-box model type and thus lack of explanatory capacity, has been improved by the introduction of sensitivity analysis methods which are increasingly used to define the most influent variables in ANN models (Lek et al. 1996b; Scardi and Harding 1999; Gevrey et al. 2003).

In this paper we i) examined the capacity of ANN models to predict French fish species richness, trophic guild richness and the occurrence of five relevant species using 8 environmental variables; ii) identified the importance of the predictive environmental variables on the output variables using the sensitivity analysis; and iii) discussed the potential of ANN methods in fish community prediction.

* Funding for this research was provided by the EU project PAEQANN (N° EVK1-CT1999-00026).
* Corresponding. gevrey@cict.fr

Materials and methods

Study area and data collection

The data set was extracted from the database held by the Conseil Supérieur de la Pêche (Banque Hydrobiologique et Piscicole), covering a period of 13 years of survey (1985-98). The data were from 688 least disturbed sites, fairly evenly distributed among French rivers. The fish were sampled by electrofishing during low-flow periods to evaluate fish assemblages throughout France (for details see Oberdorff et al. 2001). The size of each sample site was sufficient to be sure to include the home range of the dominant fish species. Fish were identified to species level, measured and weighed in the field, and then released. Forty species were identified in the dataset and used for the analyses. The total species richness as well as the species richness of 4 trophic guilds (omnivores (OSR), invertivores (ISR), herbivores (HSR) and piscivores (PSR)) were calculated at each sampling site (Table 3.2.1 and see # 3.2 for details).

Local scale environment

Eight abiotic environmental variables were measured at each site (Table 3.3.1): gradient (‰) (derived from topographic maps) (GRA), elevation (m) (derived from topographic maps) (ELE), July mean daily maximum air temperature (TJuly), January mean daily maximum air temperature (TJanuary), stream width (m) (WID), mean depth (m) (DEP), distance from source (km) (measured using a digital planimeter on a 1:1 000 000-scale map) (DIS), and surface area of the drainage basin (km^2) (measured using a digital planimeter on a 1:1 000 000-scale map) (SAD) (see Oberdorff et al. 2001 for details).

Table 3.3.1: Input and output variables from the French fish data used in the MLP model.

Input variables	Code	Output variables	Code
Surface area of the drainage basin (km^2)	SAD	Species richness	SR
Distance from headwater sources (km)	DIS	Omnivores	
Stream Width (m)	WID	Herbivores	
Gradient (‰)	GRA	Piscivores	
Elevation (m)	ELE	Invertivores	
Mean Depth (m)	DEP	Bullhead	
July mean daily maximum air temperature	TJuly	Minnow	
January mean daily maximum air temperature	TJanuary	Barbel	
		Bream	
		Brown trout	

Modelling procedures

From the eight environmental variables described above, three different types of community descriptors were predicted using the ANN method: the total species richness (SR), the species richness of the four trophic guilds (TG), and the abundance of the five most relevant species in river zonation: Brown trout, *Salmo trutta fario* (Linnaeus 1758), Bullhead, *Cotus gobio* (Linnaeus 1758), Minnow, *Phoxinus phoxinus* (Linnaeus 1758), Barbel, *Barbus barbus* (Linnaeus 1758), and Bream, *Abramis* sp. (Linnaeus 1758). Each species is a key member of the groups defined by Park et al. (2003a) (chapter 3.2 of this book) using the Self-Organizing Map method, and each group is strongly related to river zonation.

The predictive models were constructed using the backpropagation algorithm (Rumelhart et al. 1986a). A multilayer perceptron (MLP), also called multilayered feed-forward neural network, trained with a backpropagation algorithm typically comprises three neuron layers linked by connection intensities characterized by a modifiable weight: an input layer, one or several hidden layers and an output layer. The number of neurons in the hidden layers depends on the accuracy of the results required (Smith 1994; Lek et al. 1996b). In the majority of cases, a MLP with one hidden layer is capable of achieving any mapping with a given degree of accuracy (Hornik et al. 1989; Bhat and McAvoy 1992).The input layer contains neurons as independent variables. In our case, it comprises eight input neurons corresponding to the eight environmental variables. The output layer comprises the neurons responsible for producing the results, i.e. the dependent variables to be predicted (SR, TG and the five selected fish species). In this network, signals are propagated from the input layer through the hidden layers to the output layer via the network connections. During the training phase, the network is designed to compare expected and calculated values and to modify connection weights in order to minimize the error of the response, i.e. the difference between expected and calculated values.

To determine the performance of the model, a hold-out crossvalidation procedure was used. A set of the database, composed of half of the data, was used for training the model, a set consisting of a quarter of the data served for model validation and the last quarter was used to test the model. The training set was used to determine the internal parameters of the models, as well as connection weighting, with the best compromise between bias and variance (Kohavi 1995, Geman et al. 1996). The quality of the models was evaluated using the correlation coefficients. Results were also represented by scatter plot of the estimated or predicted values versus the observed values.

To determine the response of the model to each of the input variables separately, a sensitivity analysis of the ANN was performed, using a partial derivative (PaD) algorithm (Dimopoulos et al. 1999; Gevrey et al. 2003). This procedure is based on partial derivatives of the output with respect to each input in order to determine, first, the classification of the relative contribution of each independent variable to the dependent variables and second, the variation of the profiles of the dependent variables for small changes of each independent variable. The results are represented by bar diagrams to illustrate the contribution of each independent variable to model each dependent variable, and by scatter plot of the partial derivatives versus each independent variable to enable direct access to the influence of the independent variables on the dependent variables.

Results

Prediction of species richness

The training results of the model showed high predictability with a correlation coefficient of 0.82 ($p<0.001$). Moreover, the majority of the points in the scatter plot of observed values and predicted values were well-aligned along the diagonal of the best prediction (Fig. 3.3.1). In the validation and testing procedure, predictability was relatively low, but still highly significant ($p<0.001$) with correlation coefficient respectively $r=0.76$ and $r=0.74$ (Fig. 3.3.1).

We can see that the estimated or predicted values are consistently lower than the observed values in the range of higher values: larger than 10 on the x-axis in all cases. This is due to the necessity to stop the learning process to avoid the overfitting of the model, leading to worse prediction of some values and to bad generalization of the model.

Prediction of trophic guilds

After the learning process with 4 trophic guilds (invertivores, omnivores, herbivores, and piscivores), the correlation coefficients obtained for the training set between estimated and observed values were greater than 0.68 ($p<0.001$) (Table 3.3.2) for all TG. Validation of the MLP model on 25% of the dataset revealed a significant correlation ($r>0.64$, $p<0.001$) between predicted and observed values. Finally, in the testing part of the dataset (25%), the correlation coefficients obtained between predicted and observed values were lower, and the minimum and maximum values were 0.56 and 0.74, respectively ($p<0.001$ for both cases). However, for invertivore and omnivore groups, the majority of records were aligned on the 1:1 diagonal (Fig. 3.3.2). Good results were not obtained for piscivores or herbivores even though the correlation coefficients were high (Fig. 3.3.2). The maximum species richness of herbivores was 2. The low representativeness of the species in these two groups led to the worst predictions by the models. Whatever the input, the model has to respond by a 0, 1 or 2 in output. There were not enough data to help the model to learn the output according to the input.

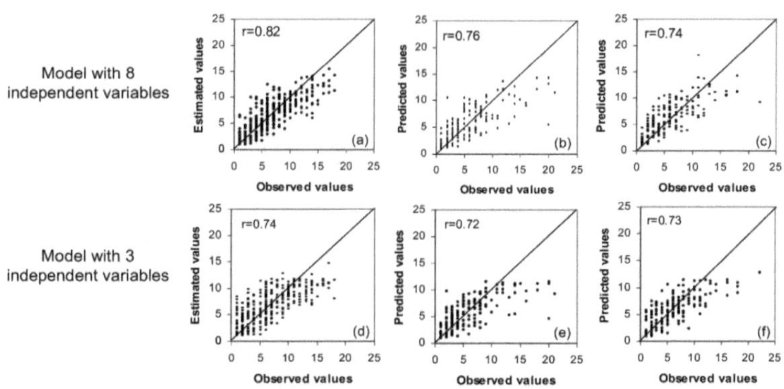

Figure 3.3.1: Recognition performance of the ANN models for species richness, (i) using the eight environmental variables (gradient, elevation, distance from source, surface area, width, depth, July air temperature and January air temperature) with (a) the training dataset, (b) the validation dataset and (c) the test dataset; (ii) (d) using three environmental variables (gradient, distance from source and surface area) with the training dataset, (e) the validation dataset and (f) the test dataset. Scatter plot shows the relationships between observed values and estimated or predicted values: the diagonal solid line indicates the perfect fit line (i.e., y = x)

Prediction of relevant species

The models predicting the abundance of 5 fish species showed significant predictabilities ($p<0.05$), although the correlation coefficients between observed and estimated values were relatively low ranging from 0.33 – 0.7 (Table 3.3.3). The models had difficulty fitting the relationship between fish density and the eight environmental variables. Table 3.3.3 shows that only one model for brown trout gave a correlation coefficient higher than 0.55 in the three datasets (training, validation and test). For the five models, the training and test results were better than the validation results. Nevertheless, the correlation coefficients obtained with the test dataset were higher than 0.5 for brown trout and minnow, slightly lower than 0.5 for bullhead and bream and equal to 0.33 for barbel. Due to the low predictability obtained, the sensitivity analyses were not stable and thus not relevant. Therefore, the results of the sensitivity analysis are not presented in this paper.

Evaluation of influences of environmental variables

The PaD algorithm was applied to the prediction models of SR and each TG. Two kinds of results were then available; the relative contribution of input variables and the response behaviour of the models according to changes in input variables (i.e., the profile of contribution).

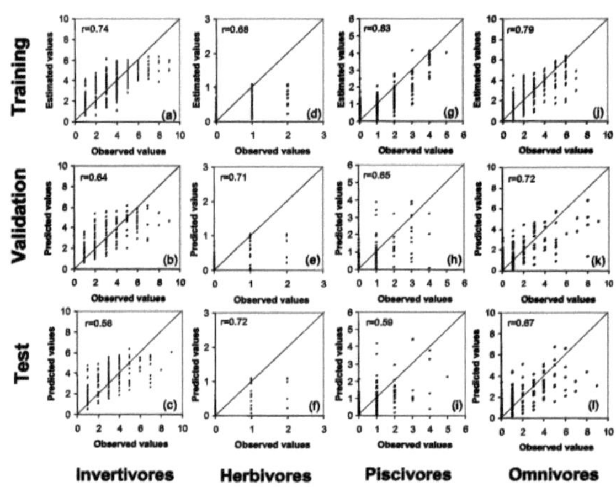

Figure 3.3.2: Recognition performance of the ANN models for the four trophic guilds (omnivores, invertivores, piscivores, and herbivores) using the eight environmental variables (gradient, elevation, distance from source, surface area, width, depth, July air temperature and January air temperature) with the training dataset (respectively a, d, g and j), with the validation dataset (respectively b, e, h and k) and with the test dataset (respectively c, f, i and l). The scatter plots show the relations between observed values and estimated or predicted values: the diagonal solid line indicates the perfect fit line (i.e., $y = x$)

For the prediction of SR (Fig. 3.3.3), the gradient was the variable that made the greatest contribution - over 35%, followed by the distance from the source (17.6%) and the surface area of the drainage basin (12.2%). The other variables made a contribution of less than 10 % (Fig. 3.3.3). Standard errors of contributions of each variable were calculated after ten training procedures (or repetitions) and the variations were very low, testifying the stability of the network models. Based on this sensitivity analysis of input variables, we selected the three variables making the strongest contributions to SR prediction (gradient, distance from the source and surface area of the drainage basin). A new SR prediction model was built using these 3 variables as inputs of the model. The prediction power was less efficient than those of the whole 8 environmental variables. However it was still highly significant and the loss of explanation power in the new model limited. The correlation coefficients, with 3 environmental variables, were 0.74 ($p<0.001$) for the training dataset, 0.72 ($p<0.001$) for the validation dataset and 0.73 ($p<0.001$) for the test dataset (Fig. 3.3.1).

The results of the sensitivity analysis with the PaD algorithm for the trophic guild models showed the relative importance of contribution of each environmental variable to each trophic guild (Fig. 3.3.4). Different trophic guilds were differently influenced by different variables. Invertivores were the most strongly influenced by the distance from the source, omnivores by the gradient, piscivores by the surface area, and herbivores by the distance from the source. Omnivores, herbivores and invertivores were closely linked to the distance from the source, gradient and depth whereas piscivores were strongly influenced by the surface area and elevation. Herbivores were also related to elevation. Standard errors calculated for each variable after ten training procedures were very low, showing the stability of the network models. The contributions of each environmental variable to the model for the four trophic guilds are in agreement with the results of Park et al. (chapter 3.2 in this book) found, from the same dataset, using a self-organizing map algorithm.

Table 3.3.2: Correlation coefficients between observed and estimated values in the training set (n=344 samples), between observed and predicted values in the validation set (n=172 samples) and between observed and predicted values in the test set (n=172 samples) of the MLP modelling procedure for species richness (SR) and the 4 trophic guilds. All correlations are highly significant ($p<0.001$).

Variables	Dataset								
	Training			Validation			Test		
	r	n	p	r	n	p	r	n	P
SR	**0.82**	344	<0.001	**0.76**	172	<0.001	**0.74**	172	<0.001
Omnivores	**0.79**	344	<0.001	**0.72**	172	<0.001	**0.67**	172	<0.001
Invertivores	**0.74**	344	<0.001	**0.64**	172	<0.001	**0.56**	172	<0.001
Herbivores	**0.68**	344	<0.001	**0.71**	172	<0.001	**0.72**	172	<0.001
Piscivores	**0.83**	344	<0.001	**0.65**	172	<0.001	**0.59**	172	<0.001

Table 3.3.3: Correlation coefficients between observed and estimated values in the training set (n=344 samples), between observed and predicted values in the validation set (n=172 samples) and between observed and predicted values in the test set (n=172 samples) of the ANN modelling procedure for the 5 selected species.

Variables	Dataset								
	Training			Validation			Test		
	r	n	p	r	n	p	r	n	P
Brown Trout	**0.6**	344	<0.001	**0.6**	172	<0.001	**0.7**	172	<0.001
Bullhead	**0.5**	344	<0.001	**0.3**	172	0.001	**0.5**	172	<0.001
Minnow	**0.5**	344	<0.001	**0.2**	172	0.006	**0.5**	172	<0.001
Barbel	**0.4**	344	<0.001	**0.3**	172	<0.001	**0.3**	172	<0.001
Bream	**0.6**	344	<0.001	**0.2**	172	0.032	**0.5**	172	<0.001

The prediction models for the various TG showed that only omnivores and invertivores were well predicted by the 8 environmental variables. Therefore, we analyzed the profiles of the contribution of environmental variables on SR as well as on these two trophic groups (Fig. 3.3.5), presenting partial derivatives of each model. The profiles of the contribution are analyzed based on the sign of the partial derivatives. If the partial derivative is negative, the output of the model tends to decrease against an increase of the input variable. Inversely, if the partial derivative is positive, the output variable increases against an increase of the input variable. For example, for the contribution of depth to SR and OSR, the partial derivatives were mainly positive leading to an increase of SR or OSR with an increase of depth. For ISR, the partial derivatives were positive for small depths and became negative for large depths, indicating that ISR increased at shallow depths and decreased for great depths. The partial derivatives of the river gradient for ISR were negative at low gradients and became positive or constant on increasing the gradients, indicating a decrease of the ISR for a decrease of the gradient. The partial derivatives of the distance from the source were positive at low distances and decreased to become constant near zero as the distance from the source increased. Concerning SR and OSR, similar profiles were observed for the gradient but for variable DS, the partial derivatives (even if positive at low distances) increased to reach an asymptote around a DS value of 100km and then decreased slightly. Overall, SR and OSR showed opposite trends to that of ISR along the upstream-downstream gradient.

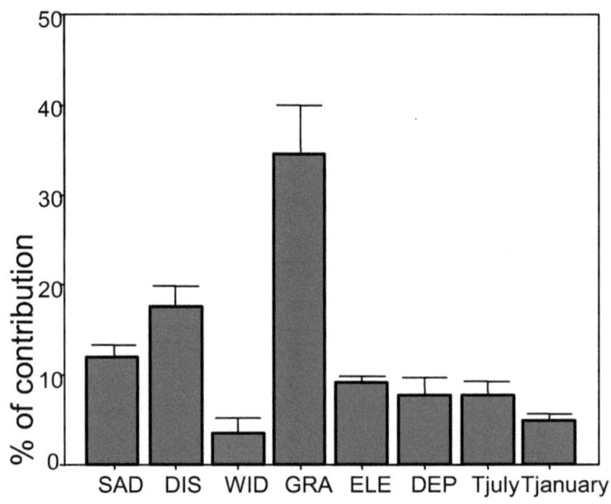

Figure 3.3.3: Percentage contribution of each of the eight independent variables to the prediction of species richness, obtained by PaD algorithm: GRA, Gradient; ELE, elevation; TJuly, July mean daily maximum air temperature; TJanuary, January mean daily maximum air temperature; WID, stream width; DEP, mean depth; DIS, distance from source and SAD, surface area of the drainage basin. Bars indicate the mean of the results from ten models; horizontal lines on the bars represent standard errors of the mean.

Discussion and Conclusion

The complexity of the relationships within an ecosystem has resulted in the development of increasingly sophisticated analytical techniques. In this study the MLP model demonstrated its learning and predicative power as well as its explanatory capacities by presenting a high capability of modelling ecological problems involving non-linear relationships between the data.

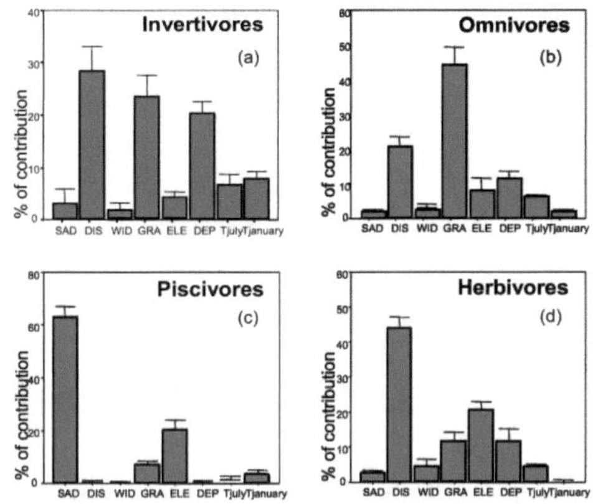

Figure 3.3.4: Percentage contribution of each of the eight independent variables to the prediction of four trophic guilds, obtained by PaD algorithm: a, invertivores, b, omnivores, c, piscivores and d, herbivores. Bars indicate the mean the results of the ten models for each fish trophic guild, horizontal lines represent standard errors of the mean. The names of the environmental variables are explained in Fig. 3.3.3.

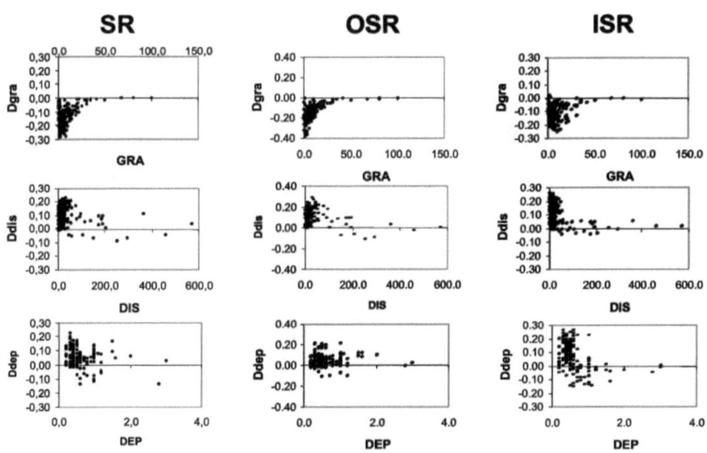

Figure 3.3.5 Partial derivatives of three ANN model responses (SR, omnivores and invertivores) with respect to three independent variables (PaD algorithm, Derivatives Profile); GRA, Gradient; DIS, distance from headwater sources; DEP, mean depth.

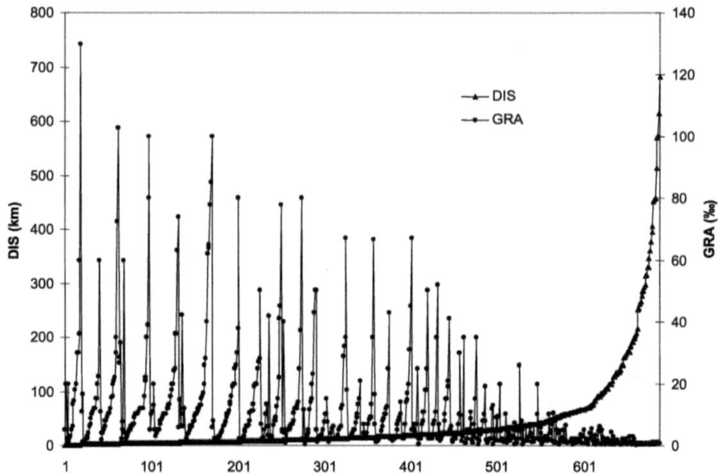

Figure 3.3.6 Relationships between distance from the source (DIS) and river gradient (GRA) in the sampling sites. The sampling sites were ordered on the x-axis based on the gradient from low to high values.

A previous study to explain the fish species richness on a local scale (Garonne River) using three environmental variables (elevation, distance from source and surface area of the drainage basin), explained most of the variability in species richness (Mastrorillo et al. 1998). Here, we showed that local fish species richness of different basins can be relatively well explained by using only the three variables: river gradient, distance from source and surface area of the drainage basin.

Our models were successful in the prediction of the trophic guilds ($r>0.68$ for the training set, $r>0.64$ for the validation set and $r>0.5$ for the test set). Moreover, the MLP models, with the help of sensitivity analysis, also revealed the relative importance of each environmental variable in structuring species richness and trophic guilds.

Looking at the profiles obtained, total species richness and omnivore species richness changed along the upstream-downstream gradient in an opposite way to invertivore species richness. These results are grossly concordant with predictions given by the River Continuum Concept (i.e., an overall decrease in invertivorous species and an increase in omnivorous species from upstream to downstream). Nevertheless, we found an unexpected trend in the relationships between total species richness or omnivore species richness and the river gradient. Total species richness and omnivore species richness are usually assumed to increase with a decrease in river gradient, but we found the opposite trend. The disagreement may be due to the structure of the dataset. Fig. 3.3.6 shows differences of the gradient and the distance from the source at the sampling sites. The sampling sites were sorted based on the gradient from low to high values. A plot with two Y axes (one for distance from source and the other for gradient) and the number of sampling sites on X axis enable the variation of distance from source to be compared to the variation of gradient. Generally gradient values should decrease as the distance from source increases. Observing the curves in Fig. 3.3.8, we can see that the distance from the source increases progressively. However, even if the general trend of the gradient plot is also a progressive decrease, large fluctuations are observed. For a given distance from the source, several gradient values are found, covering a large range (nearly 500 ‰). Although we were looking for a perfect negative regression fit between the distance from the source and the gradient, which would certainly be observed in the study of a single river, it is not surprising to find such results with data coming from several rivers covering various regions. This result highlights the fact that patterns and processes observed in local fish assemblages are not only determined by local mechanisms acting within assemblages, but also result from processes operating at larger spatial scales (i.e., basin and/or ecoregional scale). For example, the richness and structure of local fish assemblages has been linked to factors ranging from geomorphology and climate (Whittier Hughes and Larsen 1988, Nelson et al. 1992, Matthews and Matthews 2000), to richness of regional species pool (Hugueny and Paugy 1995, Angermeier and Winston 1998, Oberdorff et al. 1998). In this study we modeled local fish assemblage structure by using only local scale environmental factors but we believe it is important to also account for the possible sources of inter regional variation in assemblage structure in natural conditions.

It was interesting in this work to use some results from another study which applied another kind of artificial neural network algorithm, the self-organizing map, on the same dataset. The sampling sites were clustered into several groups using the species composition similarities and the groups were representative of the upstream-downstream river gradient (Park et al. chapter 3.2). In this study, we selected one relevant species in each of the groups and tried to predict their abundances from the environmental variables. The results were not a total success even though all predictions were significant. This may be caused by variation of species abundance. Due to the large scale of the dataset, the sample sites differ from each other, showing high variations in species abundances (i.e., absence in many samples and highly abundance in some samples). ANN can predict a zero quite accurately but has more difficulty predicting abundance values. The diversity of the sites can also be a drawback for the ANN which has difficulties in learning dissimilar values.

Our results confirmed the longitudinal variation in the species richness of different trophic guilds, with more invertivores near headwaters and more omnivores and piscivores in midreach locations (Oberdorff et al. 1993). Furthermore, as also previously noticed by Oberdorff et al. (1993) working on eight French rivers we found that total species richness peaks at midreach locations and then declines in the lowermost reaches of the systems, as suggested by the River Continuum Concept. Nevertheless, in their study, Oberdorff et al (1993) could not conclude if this decline in species richness was a natural pattern or a simple effect of man-induced disturbances. As in the present study we noticed the same decline while only "least disturbed sites" were included in the dataset, we are more confident about the ecological validity of the pattern.

In conclusion, the MLP model is an efficient tool for modelling different fish community descriptors using abiotic variables. The predictive power of ANN along with the use of explanatory methods should facilitate the ecologically oriented management of aquatic ecosystems.

3.4 Fish diversity conservation and river restoration in southwest France: a review*

Aguilar Ibarra A*, Lim P, Lek S

Introduction

The evidence of human pressure on freshwater ecosystems has been largely recognised (Dynesius and Nilsson 1994). But it was only recently that water quality guidelines have laid a stronger emphasis on aquatic ecosystem health (Hart et al. 1999). For example, the European Water Framework Directive (EWFD) acknowledges that a good condition of ecosystems is essential for sustainable development (Kallis and Butler 2001). Nevertheless, in order to recognise an aquatic ecosystem in good conditions, we need first to both characterise and identify its biotic communities (Bryce et al. 1999). Consequently, analysing diversity patterns and distributions of aquatic communities has become a critical aspect for water quality management (Boulton 1999, Jenerette et al. 2002).

One important element in aquatic ecosystem management is the spatial characterisation of riverine fish communities. Indeed, fish are considered as indicators of both aquatic quality and aquatic restoration success (Angermeier and Schlosser 1995, Paller et al. 2000, Oberdorff et al. 2001). Furthermore, there is a need to understand how fish assemblages vary within ecoregions, basins or physiographic regions (Naiman et al. 1988, Smogor and Angermeier 2001).

This chapter deals with fish diversity and conservation in the Garonne basin and is heavily based on work by Aguilar Ibarra (2004). We present a review of the research carried out on fish ecology and we discuss fisheries policies and options on fish conservation and river restoration. We organise our chapter as follows: we first discuss why the Garonne basin is an interesting case for studying riverine fish diversity. Second, we review the main results of research on fish ecology in the area of study and the main water fisheries policies. Third, options on fish conservation and river restoration are given, and finally we list perspectives for further management and research.

The Garonne basin as a landscape unit for studying fish assemblages

There are two approaches for considering the Garonne basin as an adequate landscape unit for studying fish assemblages: one refers to fish ecology and another to water policy. From the ecological viewpoint, drainage basins are ecologically relevant regions since stream fishes generally disperse within basins but not among large basins (Angermeier and

* We sincerely thank Tae-Soo Chon whose comments notably improved a former version of this paper. Financial support was provided by the Mexican-French co-operation programme CONACYT-SFERE (No.131742) and the 5th Framework Programme of the European Commission (No. EVK1-CT1999-00026).
* Corresponding: aguilar@ensat.fr

Winston 1998). The potential number of fish species in a basin (i.e. regional species pool) is known as its gamma diversity, which is a function of its alpha and beta diversities. Alpha diversity relates to the number of species in each habitat (i.e. local species richness), and beta diversity is the turnover of species between habitats (Ward and Tockner 2001). Therefore, the understanding of these diversities in a basin and their interactions with humans is of paramount significance to environmental managers (Matthews 1998, Karr 1999). This is thus linked to the water policy viewpoint. Indeed, the Water Framework Directive of the European Union reckons that basins are useful landscape units for water quality management (Kallis and Butler 2001). Furthermore, the Garonne basin is considered an important basin, even on a world scale (Revenga et al. 1998).

The Garonne basin is located in south west France and has a catchment area of 56,536 km^2 . Its hydrological characteristics are related to the basin's altitude, soil types, vegetation, orientation and climate (Gozlan et al. 1998). In this way, the oceanic influence, characterized by warmth and humidity, prevails on the whole basin, diminishing towards the south-east where it faces the Mediterranean influence, with lower precipitation and drier winds (CBAG 1996). Following the classification of Dupias and Rey (1985) the Massif Central and the Pyrenees correspond to ecoregions X and XIII respectively. The other two ecoregions present are the Aquitain (XI) and the Landaise (XII) ecoregions which mainly comprise the flood-plains (Fig. 3.4.1).

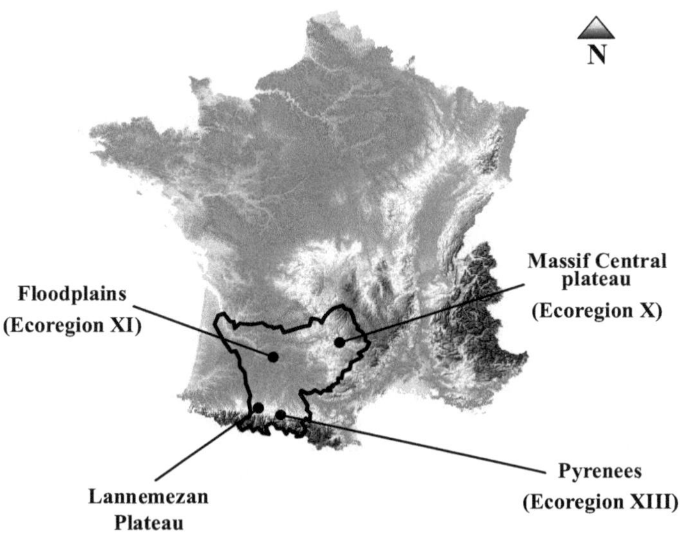

Figure 3.4.1 The Garonne basin in southwest France, showing the ecoregions. Sources: Dupias and Rey (1985) and IGN (2003).

The main channel runs over 525 km from the Pyrenees (in the Maladeta plateau in Spain) to the Gironde estuary in the Atlantic coast, and is the third longest river in France (CBAG 1996). The Garonne bed is composed of a number of facies but gravel and bedrock dominate the channel bed (Sauvage et al. 2003). The flow regime is pluvio-nival since its tributaries come from the Massif Central (pluvial regime) and from the Pyrenees (nival regime), including the Lannemezan plateau. Among the principal tributaries of the Garonne, the Lot river (491 km in length) and the Tarn river (375 km) are the largest, both having their source in the Massif Central plateau. With respect to the Pyrenees, the most important tributary is the Ariege river (150 km). The remaining discharge is given by small rivers draining on the left side of the bank, formed in the Lannemezan plateau, where the Baïse river (180 km) and the Gers river (176 km) are the best known (CBAG 1996). The Garonne is characterised by a very fluctuant flow regime (Etchanchu and Probst 1988). Critical low-water periods last from June through October, while maximum flows are exhibited in February for Winter and in May for Spring. Variations in flow discharge are provoked by upstream dams and variations may reach 30% to 100% of the mean daily discharge on a daily basis (Sauvage et al. 2003). These variations result in a high habitat heterogeneity (i.e. side-arms and oxbows) which is closely related to fish production in the floodplains (Decamps and Naiman 1989, Gozlan et al. 1998).

The biological composition of rivers in the Garonne basin resulted from a rapid degression of glacial activity approximately ten thousand years ago (Persat and Keith 1997). Both the Massif Central and the Pyrenees acted as barriers to fish dispersal, a fact which explains the lower number of fish species (about 45) in the Garonne in comparison with other French basins (Mastrorillo et al. 1998). For example, grayling (*Thymallus thymallus*), although a common fish in Europe, is absent from the Garonne basin (Keith and Allardi 2001). This makes this region an interesting case for studying its fish communities since it does not fit the classical river zonation model of Huet (1959), who described western European rivers as having four zones according to their most important species: brown trout (*Salmo trutta*), grayling (*T. thymallus*), barbel (*Barbus barbus*), and bream (*Abramis brama*).

A review of fish diversity research in the Garonne basin

Extensive field work on riverine fish communities in the Garonne basin has been carried out by the Aquatic Environment Team of the School of Agronomy at Toulouse, France (ENSAT) along with the government fisheries agency (CSP) and local associations of fishermen. They have recorded 41 fish species, belonging to 13 orders and 16 families and whose main features are summarised in Table 3.4.1.

Most of the studies carried out on fish ecology in the Garonne basin have focused on a small scale (i.e. a sector of a river). More research has been conducted on the main Garonne channel than its tributaries. For example, Lim et al. (1985) and Belaud et al. (1989a) tried to understand Huet's zonation in the absence of grayling. Their analysis of fish species composition and abundance between Saint Gaudens and Agen led to the definition of a transitional zone from a Salmoniform-rich area to a Cypriniform-rich area in the Garonne main channel (Cattaneo et al. 1999). Later, Reyjol et al. (2001a) found that this transition was influenced locally by summer water temperature and flow regulation. Gozlan et al. (1998) analysed the relationship between environmental variables and young fish (0+) between Muret and Moissac. They found that partially abandoned channels were important habitats for fish feeding, recruitment and refuge in the floodplains.

The fish community on the Garonne river at Toulouse has been the subject of a few works. On the one hand, Palomares et al. (1993) constructed a preliminary model of trophic interactions (Ecopath II) including not only fish but plankton, macroinvertebrates and benthic producers as well. On the other hand, Hutagalung et al. (1997) noted that the presence of tolerant fish, like roach (*Rutilus rutilus*) and common bream (*A. brama*), induced a higher local species richness in a polluted site than in an unpolluted one. Hutagalung (1998) explains this phenomenon as the result of the diminution of ammonia-related pollutants and of a higher water temperature in the polluted site.

Table 3.4.1 Taxonomic and guild classification of the fish species recorded between 1986-1996 by the Aquatic Envirnoment Team of the School of Agronomics at Toulouse (EAA-ENSAT).

Taxon	Common name	Origin in the basin (i)	Feeding habitat (ii)	Type of food (iii)	Reproductive (iv)
Petromyzontiformes					
Petromyzontidae					
Petromyzon marinus	Sea lamprey	D	B	D	L
Lampetra planeri	Brook lamprey	N	B	D	L
Anguilliformes					
Anguillidae					
Anguilla anguilla	Eel	D	B	Iv*	Pe
Clupeiformes					
Clupeidae					
Alosa alosa	Allis shad	D	W	O	Pe
Alosa fallax	Twaite shad	D	W	O	Pe
Cypriniformes					
Cyprinidae					
Alburnus alburnus	Bleak	N	W	O	PhL
Barbus barbus	Barbel	N?	B	O	L
Carassius carassius	Crucian carp	I	B	O	PhL
Cyprinus carpio	Common carp	I	B	O	Ph
Pachychilon pictum	Albanian roach	I	W	O	PhL
Blicca bjoerkna	White bream	N?	B	O	Ph
Abramis brama	Common bream	N?	B	O	PhL
Chondrostoma toxostoma	French nase	N	B	O	L
Leuciscus cephalus	Chub	N	W	O	L
Leuciscus leuciscus	Dace	N	W	O	L
Phoxinus phoxinus	Minnow	N	W	O	L
Rutilus rutilus	Roach	I?	W	O	PhL
Scardinius erythrophthalmus	Rudd	N?	W	O	Ph
Tinca tinca	Tench	N?	B	O	Ph
Gobio gobio	Gudgeon	N?	B	O	Ps
Pseudorasbora parva	Top mouth gudgeon	I	W	O?	PhL?
Rhodeus sericeus	Bitterling	I?	W	H	Os
Balitoridae					
Barbatula barbatula	Stone loach	N	B	Iv	L
Siluriformes					
Siluridae					
Silurus glanis	Wels catfish	I	B	P	PhL
Ictaluridae					
Ictalurus melas	Black bullhead	I	B	Iv	L
Esociformes					
Esocidae					
Esox lucius	Pike	I?	W	P	Ph
Salmoniformes					
Salmonidae					
Salmo salar	Salmon	D	W	Iv*	L
Salmo trutta fario	Brown trout	N	W	Iv	L
Salmo trutta trutta	Sea trout	D	W	Iv*	L
Oncorhynchus mykiss	Rainbow trout	I	W	Iv	L
Cyprinodontiformes					
Poecilidae					

Gambusia affinis	Mosquito fish	I	W	Iv	V
Gasterosteiformes					
Gasterosteidae					
Gasterosteus aculeatus	3-spined stickel-back	N	W	O	A
Mugiliformes					
Mugilidae					
Mugil cephalus	Lisa	M	W	O	Ps
Perciformes					
Percidae					
Gymnocephalus cernua	Ruffe	I	B	O	PhL
Perca fluviatilis	Perch	I?	W	P	PhL
Stizostedion lucioperca	Pikeperch	I	W	P	Ph
Centrarchidae					
Lepomis gibbosus	Pumpkinseed	I	W	Iv	L
Micropterus salmoides	Black bass	I	W	P	PhL
Blennidae					
Blennius fluviatilis	Freshwater blenny	N	B	Iv	L
Scorpeaniformes					
Cottidae					
Cottus gobio	Bullhead	N	B	Iv	L
Pleuronectiformes					
Pleuronectidae					
Platichthys flesus	European flounder	M	B	Iv	Pe

(i) D= diadromous, I=introduced, N=native, M=marine (Keith 1998, Keith and Allardi 2001).
(ii) B=Benthic, W=Water column (Michel and Oberdorff 1995, Berrebi-dit-Thomas et al. 1998, Oberdorff et al. 2002a, Bruslé and Quignard 2001, Keith and Allardi 2001).
(iii) D=detritivore, H=Herbivore, Iv=Invertivore, O=Omnivore, P=Piscivore (op.cit.).
(iv) A=Ariadnophil, L=Lithophil, Os=Ostracophil, Pe=Pelagophil, Ph=Phytophil, PhL=Phyto-litophil, Ps=Psammophil, V=Viviparous (Balon 1975, Bruslé and Quignard 2001).
? Needs further verification.
* During their freshwater period.

Downstream of Toulouse, Pouilly et al. (1996) applied a model to estimate fish community composition using the most abundant species in this stretch: barbel (*B. barbus*), gudgeon (*Gobio gobio*), chub (*Leuciscus cephalus*), bleak (*Alburnus alburnus*) and roach (*R. rutilus*). Moreover, Belaud et al. (1990) and Bengen et al. (1992) investigated fish ecology of oxbows which are quite numerous and important to fish life cycles. In fact, they demonstrated the importance of the main channel as a migration pathway among oxbows, and that these help to maintain fish diversity in the floodplains.

The implications of a fish elevator in the largest dam on the Garonne river at Golfech have been considered by Belaud et al. (1985), Belaud and Labat (1992) and Bellariva and Belaud (1998). Several diadromous species have been affected by migration barriers along the Garonne river, mainly Atlantic salmon (*Salmo salar*), European eel (*Anguilla anguilla*), lampreys (*Lampetra fluviatilis* and *Petromyzon marinus*) and shads (*Alosa spp.*). However, allis shads (*A. alosa*) have been more deeply studied. For example, Bellariva and Belaud (1998) showed that since 1987, when the fish elevator was fully operational, significant increases have been recorded for allis shad populations.

Trout populations have been extensively studied, mostly in the Pyrenees. For example, many studies have focused on the 'microhabitat method' or 'instream flow incremental methodology' for studying trout populations, (e.g. Belaud et al. 1989b, Delacoste et al. 1993, Baran et al. 1995a,b, 1997). Other methods of analysis have been used, including artificial neural networks (e.g. Baran et al. 1996, Lek et al. 1996b, Reyjol et al. 2001b).

Fish assemblages in headwater streams are almost exclusively composed of brown trout (*Salmo trutta fario*) (Baran et al. 1993a), however, bullhead sculpin (*Cottus gobio*) has been observed in unregulated reaches (Crespin and Usseglio-Polatera 2002). Both species are closely correlated to both high water quality and high habitat quality (Baran et al. 1993a). A higher trout abundance is observed in regulated sectors than under natural flow conditions (Belaud and Baran 1997), being the reduced instream flow the most notable influence (Baran et al. 1995a). Indeed, it has been demonstrated that brown trout has a slower growth and a higher longevity in the Pyrenees than in other French streams, due to harsher environmental conditions (Lagarrigue et al. 2001). In fact, trout abundance depends on a number of factors (Cuinat 1971). For example in the Neste de Louron river, the abundance of trout longer than 180 mm in total length is influenced by mean depth, mean bottom velocity, and the number of refugia (Baran et al. 1993b) and in the Neste d'Aure to the gradient, stream width and the presence of weirs and dams for the whole of the population (Baran et al. 1993a). These obstacles diminish the migrations of trout along the river, leading them to spend the whole of their life cycle in a limited stretch (Baran et al. 1993a). In the Neste d'Oueil, the occasional deteriorations in the population carrying capacity and of the physical habitat may lead to reductions in adult abundance (Gouraud et al. 1999, 2001). The habitat also has an important influence on trout-prey (i.e. benthic invertebrates) interactions (Lauters et al. 1996, Crespin and Usseglio-Polatera 2002), although neither abundance nor diversity of invertebrates are directly correlated with biomass or density of trouts in the Neste d'Aure river (Baran et al. 1993a).

Much less work has been done under a multispecific approach. For example, Dauba et al. (1997) studied the recovery of fish assemblages in the Baïse river. This river was abiotic due to the chemical pollution by ammonia-derived products, but after waste treatment facilities were implemented in the 1970s, both fish assemblages and invertebrates recovered. However, although fish species richness increased from five in 1978 to eight in 1990 the recolonization of the whole river remained incomplete by 1996 (Dauba et al. 1997), as very sensitive species had not yet recovered to their original levels. In another multispecific study, Tourenq and Dauba (1978) showed the changes of the fish fauna composition in the Lot river as a consequence of dams construction. Mastrorillo et al. (1997a) predicted the presence of minnow (*Phoxinus phoxinus*) gudgeon (*Gobio gobio*) and stone loach (*Barbatula barbatula*) on the Ariège river.

On a regional scale (i.e. the whole basin), fish assemblages have been studied by Mastrorillo et al. (1998) and Aguilar Ibarra et al. (2003) who analysed fish species richness by using a back-propagation neural network. Aguilar Ibarra et al. (2003) developed the work of Mastrorillo (1998a) in a deeper way by modelling fish guilds and examining the contribution of environmental descriptors in explaining guild composition. They found that models showed high variability, presumably due to spatial heterogeneity, temporal variability or sampling uncertainty. The area of the catchment basin and annual mean water flow were the most important environmental descriptors of guilds composition; both variables implying human influence (i.e. land-use and flow regulation) on riverine fish.

Other studies were those of Reyjol et al (2003) who compared the longitudinal distribution of fish and invertebrates in the Garonne basin, and Aguilar Ibarra (2004) who applied a Kohonen's self-organising map (i.e. a non-supervised artificial neural network) to presence-absence data of fish, finding three main nested patterns in an aggregated hierarchy: a succession of species along a gradient without defined boundaries, four main zones of fish assemblages, and an upstream-downstream shift of fish communities. Furthermore, the fish assemblages matched the physiography of the Garonne landscape, corresponding to two significant ecological boundaries: one in the upper piedmont and another in the lower piedmont, forming a transitional zone between a mountain assemblage and a floodplain assemblage.

Historical context of water quality and fisheries management in France

The history of management measures with regard to water quality and fisheries management in France is summarised in Table 3.4.2. Domestic effluents were the first form of water pollution in France. Angelier (2001) notes that in the times of Napoleon III in the 19th century, the Seine river was depopulated of fish along 5 km near Paris. This condition later worsened with the contribution of industrial waste during the 20th century (Leynaud and Trocherie 1980).

Table 3.4.2 Historic summary of main events which influenced fish diversity conservation in the Garonne basin.

Period	Main events
19th Century	Pollution from domestic effluent.
Early 20th Century	Organic pollution and industrial waste.
1950 onwards	Intensification of damming and use of fertilizers.
1964	A French water law is enacted in order to improve water quality.
1970s	Fish depopulation in a number of rivers.
1984	The French fishery law is enacted for protecting fish resources.
1988	The application project of the European Water Framework Directive is born, with an emphasis on environmental quality.
1992	A new French water law is enacted in harmonisation with European directives.
1994	Fisheries plans are aimed to encourage natural fish reproduction and habitat improvement.
2002	The French water law is updated with the aim to restore habitats and protect biodiversity.
2005 onwards	Full application of the European Water Framework Directive.

The water law of 16 December 1964 directly addressed the problem of pollution as one major national concern. This law, however, did not imply the protection of the biodiversity nor the integrity of aquatic ecosystems. It was just aimed at reducing the level of pollution entering the riverine systems by setting standards for drinking water. Such an approach proved successful in the fight against point-source pollution, but was poorly suited to integrated management of river ecosystems (Oberdorff et al. 2002a). In spite of the 1964 law, natural ecosystems suffered severe damage, leading to the loss of a number of fish populations in several French rivers during the 1970s (Leynaud and Trocherie 1980). A response to such problems came with the fishery law of 1984, which indicated the importance of safeguarding aquatic environments and fish resources (Levêque 1999).

More recently, French regulations concerning water resources have been harmonised with European directives. This is clearly seen in the new water law of 3 January 1992 as pointed out by Le Roch and Mollard (1996) and Levêque (1999). The 1992 water law stresses the need to preserve aquatic ecosystems with the aim of setting up a policy for the quality of ecosystems, based on three criteria of evaluation: water quality, biological quality, and physical quality of the channel (i.e. river habitats). Piégay et al. (2002) explain that in order to carry out this policy, two management plans have been created: the 'Schéma Directeurs d'Aménagement et de Gestion des Eaux' (SDAGE) and the 'Schémas d'Aménagement et de Gestion des Eaux' (SAGE).

The aim of the SDAGE is to promote balanced management of water resources. It involves the evolution from water management to a management of the aquatic ecosystems in all their forms and all their components (chemical, biological and physical), taking into account their evolution, complexity, and their inter-relationships. The SAGE is simply a plan operating on a local basis (Piégay et al. 2002). The main orientations defined by the SDAGE for the Adour and Garonne basins are: (i) the restoration of water flow by controlling its consumption, (ii) the protection of groundwater aquifers for the supply of drinking water, and (iii) the openning up of watercourses to fish migrations.

This policy has led, since 1994, to a re-organisation of fisheries management plans in order to encourage natural reproduction of fish populations and their conservation, instead of plain re-stocking. According to Armand et al. (2002), this law established a change in direction for French water management, as it shifts the approach from an 'aquatic resources quantity' policy towards an 'aquatic resources quality' policy. Its application thus implies the restoration of habitats and the conservation of the biodiversity.

Management plans are also required for sportfishing. Rivers are managed in France by a system of property rights in which owners of a sector of a river have the exclusive right to fish, and the power to allocate fishing rights in it (Fishery Law of 29 June 1984). However, they must participate in the elaboration of fishing management plans under the supervision of local authorities. In general, permit holders bestow their rights and responsibilities to one of the local associations of fishermen (Associations Agréées de Pêche et de Protection du Milieu Aquatique –AAPPMA), which belong to 93 departmental associations (Fédérations Départamentales), in seven river basin unions (Unions Régionales) These regional associations formed in 1947 the Fishing Union of France (Union Nationale pour la Pêche en France) which to date count nearly two million anglers throughout the country (UNPF 2000). Fisheries management is thus heavily supported by anglers' organisations along with the government fisheries agency (Conseil Supérieur de la Pêche –CSP) which provides technical and scientific advice (Changeux et al. 2001, Armand et al. 2002).

The management plan has to be produced for every region or *Département* and contains two components: one dealing with conservation and restoration of freshwater ecosystems (Plan Departemental pour la Protection du Milieu Aquatique et la Gestion des Ressources Piscicoles –PDPG), and the other dealing with fishing effort (Plan Départamental pour la Promotion du Loisir-Pêche –PDPL). The PDPG establishes fishery management units based on the distribution and stock assessments of indicator species such as salmonids and pike. It encourages measures for protecting natural reproduction of wild stocks and restoring riverine habitats (Changeux et al. 2001). In contrast with former fisheries policies, it tries to prevent 'useless restocking' even though restocking has flourished in France to respond to anglers demands (Armand et al. 2002). Under the PDPL, the number of fishermen is assessed on the basis of their spatial distribution and traveling locations. This spatial distribution allows an estimation of services, accessible areas, training needs and any fishing restriction with respect to available fish stocks in every département (Armand et al. 2002).

A conceptual framework for the management and conservation of fish

In spite of the regulations described above, according to Keith (2000), there are no management plans for conservation of fish diversity, with the sole exception of migratory species. Eight species in four categories on the red list inhabit the Garonne basin (Keith 2000):

- Critically endangered fish: sturgeon (*Acipenser sturgeon*).
- Endangered fish: Atlantic salmon (*Salmo salar*).
- Vulnerable fish: brown trout (*S. trutta*), allis shad (*Alosa alosa*), twaite shad (*Alosa fallax*), eel (*Anguilla anguilla*), pike (*Esox lucius*) and sea lamprey (*Petromyzon marinus*).
- Low-risk fish: soiffe (*Chrondrostoma toxostoma*).

Conservation plans would be preferable when the whole ecosystem is protected or restored, rather than setting up plans for isolated species (Angermeier and Schlosser 1995, Maitland 1995). For example, ichthyoregions can be useful for fish protection and water quality management (Hughes et al. 1987, Oswood et al. 2000). In the case of large basins, Angermeier and Winston (1998) and Smogor and Angermeier (2001) have shown that ichthyoregions should be coincident with physiographical provinces (e.g. mountains, piedmont, plains), because these have distinctive fish fauna. In fact, fish assemblage differences among physiographic provinces in the Garonne river network (Aguilar Ibarra 2004) confirm that heterogeneity, especially in a large basin, should be carefully considered for water management, river restoration and biodiversity conservation (Cowx and Welcomme 1998, Smogor and Angermeier 2001). Thus, we propose a conceptual framework of potential ichthyoregions in the Garonne basin (Fig. 3.4.2).

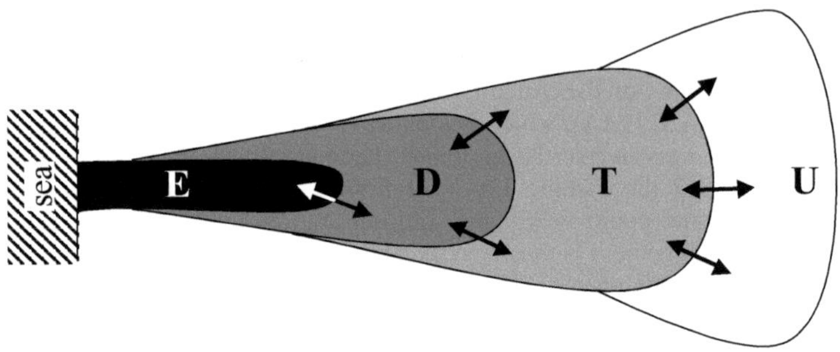

Figure 3.4.2. Conceptual diagram showing fish communities corresponding to the physiography of the landscape. Fish communities located in the upstream of the basin (U) are related to mountain streams, while downstream communities (D) inhabit floodplain rivers. Between both there exists a transition (T) in the piedmont reaches. Further downstream, a hypothetical community is considered to be present in estuarine-influenced waters.

The ecophysiological requirements of downstream fish would not allow them to attain upper reaches (U) or (of the basin, and vice versa, but a transition (T) zone (barbel and grayling zones) could act as a link between the two ends. The upper boundary would correspond to the mountain-piedmont landscape transition, where a change takes place from salmoniform-predominant assemblages to more diverse assemblages where cypriniforms predominate (Rahel and Hubert 1991, Reyjol et al. 2001a). The lower boundary would relate to the lower piedmont-floodplain boundary (Rahel and Hubert 1991, Smogor and Angermeier 2001), where an even more diverse assemblage dwells, including cyprinids, ictalurids and percids. Further downstream, we predict the existence of another transition zone between fresh and brackish waters coming from the estuary (E). The latter, in fact, has been identified elsewhere as the flounder zone by Keith and Allardi (2001), but remains to be verified in the Garonne basin. Diadromous species would pass throughout the whole gradient in order to complete their life cycles.

One practical use of physiographic ichthyoregions for water quality management would be the detection of deviations from the normal composition of a community due to climatic changes (Tonn 1990) or anthropogenic disturbances on the aquatic-terrestrial interface (Oswood et al. 2000, Smogor and Angermeier 2001). For example, Aguilar Ibarra (2004) showed patterns of fish diversity highlighting the presence of discontinuities in the longitudinal profile of rivers. Even when the causes of such discontinuities are difficult to determine, fish guilds may show the relationship between environment and community structure. Indeed, as land use implies non-point source pollution, economic activities may deteriorate the primary productivity and the concentration of nutrients would have severe effects on fish assemblages (Berkman and Rabeni 1987, Harding et al. 1998). Under such circumstances, generalists will be favoured with the increase of organic matter, thus, omnivorous fish and some planctivores would be better adapted than specialists, such as invertivores. However, the increment of suspended matter may also cause sedimentation, which may be negative even for the best adapted species. The influences on a particular trophic group will also have consequences on other groups of the trophic chain, such as carnivores (Schlosser 1990, Oberdorff et al. 1993).

Table 3.4.3 Introduced fish in the Garonne basin (alphabetical order). Interrogation point means it is not sure whether it was introduced. Sources : Keith (1998), Bruslé and Quignard (2001), Keith and Allardi (2001).

Scientific name	Common name	Origin	Date of introduction	Observations
Carassius auratus	Goldfish	Asia	18th century	Very invasive
Carassius carassius	Crucian carp	Eastern Europe	?	Rapidly expanding
Cyprinus carpio	Common carp	Central Europe	Roman Era	Very common
Esox lucius	Pike	Western Europe	?	Popular sportfish
Gambusia affinis	Mosquito fish	North America	1920s	Rapidly expanding
Gymnocephalus cerna	Ruffe	East and North Europe	20th century	Very rare
Hypophthalmichtys molitrix	Silver carp	Asia	1970s	No natural reproduction
Ictalurus melas	Black bullhead	North America	1870s	Intensive competitor
Lepomis gibbosus	Pumpkinseed	North America	1877-1885	Very resistant species
Leuscaspius delineatus	Rain bleak	Central and Eastern Europe	?	Very rare
Micropterus salmoides	Black bass	North America	1890s	Very rare
Oncorhynchus mykiss	Rainbow trout	North America	1880s	Popular sportfish
Pachychilon pictum	Albanian roach	Eastern Europe	1980s	Accidentally introduced
Perca fluviatilis	Perch	Western Europe	19th century	Popular sportfish
Pseudorasbora parva	Stone moroko	Asia	1970s	Accidentally introduced
Pungitius pungitius	Nine-spined stickelback	Western Europe	?	Very rare
Rhodeus sericeus	Bitterling	Central and Eastern Europe	?	Very rare
Rutilus rutilus	Roach	Central and Western Europe	?	Very common
Salvelinus fontinalis	Charr	Switzerland	19th century	Only in mountain lakes
Salvelinus namaycush	Lake trout	North America	1880s	Only in mountain lakes
Silurus glanis	Wels catfish	Central Europe	1850s	Only in large rivers
Stizostedion lucioperca	Pikeperch	Central Europe	20th century	Popular sportfish
Thymallus thymallus	Grayling	Western Europe	?	Very rare

Other causes of discontinuities in the longitudinal distribution of fish in rivers are intensive agricultural practices (Rahel and Hubert 1991, Harding et al. 1998), urbanisation (Hutagalung et al. 1997, Wang et al. 2000) and the presence of dams (Tourenq and Dauba 1978), which create lotic-lentic environments along a river (Ward and Stanford 1983a). Dams and weirs have been built since the 14th century in the Garonne basin, but in the last few decades both damming and fertilizer use have increased (Steiger et al. 1998, Semhi et al. 2000), leading not only to concerns on fish conservation but on river restoration as well (Decamps and Naiman 1989, SMEAG 2003).

A word should be devoted to introduced species. Exotic fish are especially successful in human-disturbed habitats (Moyle and Light 1996, Angermeier and Winston 1998) so an increase in extension or frequency of these conditions may boost introduced populations to the detriment of native species. Although introduced fish are rarely the direct cause of local species extinctions (Moyle and Light 1996) they may lead to a reduction in beta-diversity, that is to say, a regional-scale homogenization of fish assemblages (Rahel 2000), and thus decreasing global fish diversity (Angermeier 1994). Table 3.4.3 lists the exotic species recorded in the Garonne basin.

River restoration options

Restoration of regulated flows involves reconnecting their lateral and longitudinal components (Cowx and Welcomme 1998, Ward et al. 1999). Lateral connectivity is particularly important in the Garonne basin since a number of oxbows exist in its floodplains, serving as refuges for a diverse fish fauna (Belaud et al. 1990, Bengen et al. 1992). In fact, dredging activities downstream from Toulouse have lowered the river level, substituing typical riffle stretches with molassic paving (Pouilly et al. 1996), diminishing the surface of riparian forests (Steiger et al. 1998), and reducing the number of refuges for fish (Gozlan et al. 1998). Therefore, actions should involve the regeneration of vegetation along the riparian zone of rivers in order to improve the environmental conditions for a number of fish species (Wichert and Rapport 1998, Paller et al. 2000). Indeed, the riparian forest offers shadow refuges for many fish and provides a diversity of food sources (Wang et al. 2000). Furthermore, vegetation on river banks could reduce the impact of non-point-source pollution, acting as a biofilter against water pollutants (Hendry et al. 2003).

Longitudinal restoration includes three actions, mainly the implementation of minimum flow, the openning up of waterways for fish migration, and the rehabilitation of spawning grounds. In the first case, a minimum flow is defined as the minimum water quantity necessary for supporting fish life. Minimum flow has been applied since 1998 in the Pyrenees by hydropower facilities (Baran et al. 1997). However, even though minimum flow has shown favourable results, this method has been criticised (e.g. Barinaga 1996, Poff et al. 1997) because it aims primarily to favour selected species, and because it diminishes the natural transport of sediment by a river. In this respect, Ward et al. (1999) and Tockner et al. (2000) propose a higher diversity of flow levels set up by the authorities, representing a more "natural" flow regime. Allowing the migration of fish is another measure of restoration in longitudinal connectivity of rivers (Cowx and Welcomme 1998). Fish ladders and by-passes have been built since the 1980s on several dams of the Garonne channel. These devices, along with implementation of restocking programmes have allowed Atlantic salmon to come back after decades of absence in the region (Keith and Allardi 2001). In fact, the passage of diadromous species in the Garonne river is a prioritary goal of recent management plans (Piégay et al. 2002). However, fish ladders may favour salmonids but they make no difference for other diadromous species without jumping behaviour (Larinier 2001). Thus, a fish elevator is a more efficient solution, like the one set up on the Garonne river at Golfech, allowing the passage of diadromous fish such as *Alosa alosa*, *A. fallax* and *Anguilla anguilla* (Bellariva and Belaud 1998). Even when fish passages are effective, spawning grounds for a number of species also should be in good condition to achieve a successful restoration of diadromous fish populations (Cowx and Welcomme 1998, Crisp 2000).

Perspectives

The example of the Garonne basin can help us to provide some suggestions for other large basins in Europe or the world with respect to fish conservation and river restauration. Perspectives for further study and management of large basins should be directed:

- To gain a better comprehension of processes functioning on scales extending from local to regional.
- To set up monitoring programmes for ecological quality according to physiographic ichthyoregions in order to detect changes in fish distributions and abundances.
- To understand more complex ecological processes, like trophic relationships, population dynamics, and recruitment and recolonisation processes in ecosystems under human pressure.
- To study the ecological and economic consequences of introduced species.
- To produce an integrative ecological index of quality for water management, including several biotic components, such as fish, invertebrates and plants.
- To promote reforestation efforts of over the whole basin, but especially of the riparian zones.
- To coordinate the various users of a basin under an integrated management approach.
- To apply multidisciplinary research, including aspects of ecology and economics related to water quality and biodiversity use.

3.5 Modelling of freshwater fish and macro-crustacean assemblages for biological assessment in New Zealand*

Joy MK †, Death RG

Introduction

Biological assessment of flowing water has undergone a conceptual change from the use of biological indicators of water quality to the assessment of 'biological quality' (Wright et al. 2000) or 'biological integrity' (Karr 1991). This change from biological indicators to the assessment of biological integrity marks a shift towards ecosystem level evaluation and these approaches are often referred to as measures of 'ecosystem health' (Chessman et al. 1999). These new concepts come from a recognition that water quality is the result of many factors including biological interactions, flow regimes and habitat structure and that these are all dependant on modification by human activities (Karr 1991, 1995). It follows from this movement to an ecosystem approach that impacts on the biological condition of rivers exposed to human impacts can be judged by comparing the river biota with that from relatively unimpacted reference ecosystems. The prerequisite is that the sites occur in similar geomorphological and climatic settings referred to as reference conditions (Chessman et al. 1999). Thus, we can assess the condition of human disturbed sites by measuring their structural attributes and then comparing them with relevant reference conditions that are pristine or, at worst, relatively undisturbed (Hughes et al. 1986). This reference site method of river assessment has been in use for some time although different approaches have been used in different parts of the world. In the USA, a reference site approach has been applied with 'indices of biotic integrity' (e.g. Karr 1981). Whereas in the United Kingdom and more recently Australia, multivariate reference site predictive models have been used (Simpson and Norris 2000).

The predictive reference condition approach (RIVPACS: River InVertebrate Prediction And Classification System) and its derivatives developed originally by Wright et al. (1984) and later advanced by Reynoldson et al. (1995), Simpson and Norris (2000) and Hawkins et al. (2000b) have been successfully applied to streams worldwide. The output from the models is a list of taxa expected to be at a site in the absence of human impacts predicted from a suite of environmental variables. The predictions come from a database of least impacted sites selected to cover all stream types within a region. In the bioassessment of a site, the final output is a measure of the relationship between the fauna collected at a site and that predicted. That is the observed number of taxa (O) is compared to the number expected (E)

* This work was funded by a grant from the sustainable management fund (New Zealand Ministry for the Environment grant No. 5099 "Nga Ika Waiora": stream health evaluation tool) and the Auckland Regional Council. Thanks to Allison Hewitt for assistance with all fieldwork and to Auckland Regional Council staff Brent Evans, Chris Hatton, David Glover, and John Maxted. For assistance with site access, information, and accommodation, we thank Conservation Department staff Chris Roberts, Thelma Wilson, Jo Ritchie, and Mike McGlynn.

† Corresponding: mikejoy@clear.net.nz

in the absence of stress as a measure of departure from expected conditions, and is thus, a measure of biological impairment. Although the predictive models described above have been developed using macroinvertebrates they have the potential for use with other biotic groups (Reynoldson and Wright 2000) and have been developed for use with fish (Joy and Death 2002), diatoms (Chessman et al. 1999) and stream habitat features (Davies et al. 2000).

New Zealand has a limited native fish and macro-crustacean fauna of 39 recognised species, dominated by Galaxiidae and Eleotridae as well as a non-migratory crayfish and a diadromous shrimp. The fauna is characterised by a high proportion of diadromous species such that there are marked longitudinal trajectories of fish distribution with species richness reducing with elevation (Joy et al. 2000). These distributional patterns negate the application of bioassessment methods using fish employed elsewhere in the world such as the index of biotic integrity (IBI). These index approaches would be problematical in New Zealand because they rely on relationships between community metrics and habitat quality and would not account for the overriding longitudinal distributional patterns caused by diadromy (McDowall and Taylor 2000). Migratory fish and crustacea can however be used in bioassessment if the method used takes into account their longitudinal distribution patterns. A predictive modelling approach using reference sites allows for the migration driven distributional patterns to be incorporated in the bioassessment of fish and macro-crustaceans (Joy and Death 2002).

We took a reference site predictive modelling approach to the use of fish and macro-crustaceans in the assessment of biological quality in the rivers of Auckland, New Zealand. To achieve this we modeled the presence or absence of 10 fish and 2 macro-crustacean species using individual artificial neural network models, one for each taxon based on environmental variables. To make the predictions from the model independent of human impacts we used only environmental variables unlikely to be influenced by human impacts and used only data from minimally disturbed reference sites. The predictions from these models were combined to predict the fish and macro-crustacean assemblage to be expected at sites.

Methods

Reference sites

The sites designated as reference sites were those that represented the best available natural condition within the region, that is, sites with open access to the sea and the least evidence of human disturbance. These reference sites are used as a standard against which to assess the health of other sites with potential impacts. The reference site selection process involved two phases. The first was essentially 'desk based' and was performed in consultation with staff from the local regulatory authority using topographic maps, global information systems (GIS) and local knowledge. From this initial phase, 165 potential reference and 35 impacted sites were selected and these were sampled over the Austral summer of 2000-2001 (Fig. 3.4.1). The 35 test sites had impacts includind migratory barriers in the form of weirs and dams, eutrophication, and high densities of introduced piscivorous fish.

The second phase took place post sampling and included updated information from site visits. To further refine the reference site dataset the sites were ranked based on the following criteria: 1) unimpeded access to the sea, 2) indigenous forested catchment, 3) mature exotic forest catchment (Hughes et al. 1986). The sites below the 25th percentile of the rankings were discarded leaving 118 reference sites for further analysis.

The aim of the model was to enable the determination of the biological condition of sites with respect to reference conditions; thus, only predictor variables that are uninfluenced or least influenced by human activity were included. Sixteen predictor variables were selected from the suite of variables available for each site from GIS databases or measures made during sampling (Table 3.5.1).

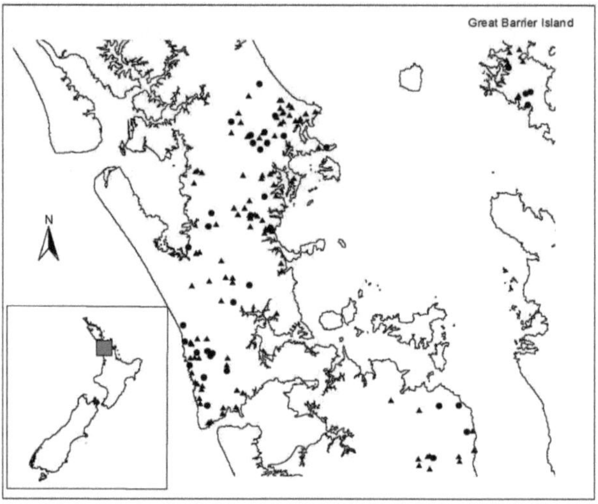

Figure 3.5.1 Site map showing the reference sites (triangles) and test sites (circles) surveyed in the Auckland region between January and May 2001. Inset shows location in New Zealand.

Fish and macro-crustacean data

Fish and macro-crustacean communities were sampled using overnight trapping. This is the most efficient sampling method for New Zealand fish found in the small, low gradient, low water clarity, streams common in this area (McDowall 1990). Although other sampling methods may have been more suitable at a few of the sites, the use of a consistent sampling method for all reference sites is imperative for data used in predictive modelling. At each survey site, two types of fish trap were used. The traps consisted of five pot type 'Gee minnow traps' and three large fyke nets. The pot traps were metal with 5 mm mesh (220 mm diameter) and two of the three nylon fyke nets (660 mm diameter hoops, 3 m long, 3 m gate) at each site had 12 mm mesh and the other had 2 mm mesh (see McDowall 1990 for details). Traps were not baited and were positioned where possible to cover all microhabitats over approximately 50m of the stream reach. Fyke nets were positioned with the entrance facing downstream and the gate angled across the stream. After 24 hrs in situ the traps were retrieved and fish and crustaceans were removed. Juvenile eels (< 300-mm length) were treated as separate operational taxonomic units (elvers) in the analysis because they are known to shift habitat during ontogeny (Hayes et al. 1989, Glova 1998). This process was applied only to eels, because there is no evidence of an ontogenetic shift of within-species habitat requirements for the other species. All fish and crayfish data were presence/absence only not abundance.

Table 3.5.1 Physical and chemical variables measured at the 118 reference and 35 test sites in the Auckland region between January 18 and May 2001 and selected for use as predictor variables in ANN models.

		Reference sites			Test sites		
Variable	Foot note	Min	Mean	Max	Min	Mean	Max
Geographic Variables							
Altitude (m)	1	1.0	47.9	220.0	2.0	62.1	200
Longitude (grid)	1	26397	26641	27282	26379	26662	27293
Latitude (grid)	1	64419	65074	65611	64522	65082	65560
Distance to coast (km)	2	0.1	15.3	79.5	0.3	17.5	79
Channel variables							
Mean width (m)	3	0.8	3.3	31.8	0.8	3.8	17.3
Mean depth (m)	4	0.15	.42	1.96	0.02	0.47	1.42
Median substrate	5	0.0	9.1	39.1	0.0	6.2	40
Leaf litter	6	1.0	2.1	4.0	1.0	2.0	4
% Pool	7	0	75.3	100.0	0	59.4	100
% Riffle	7	0	12.7	70.0	0	24.3	100
% Run	7	0	10.8	90.0	0	15.7	100
Cobble packing	8	0	1.8	3.0	0	1.9	3
Water variables							
Temperature (° C)	9	12.9	16.5	24.0	13.5	16.8	20.9
Conductivity (μS/cm^{-1})	9	100.00	216.07	974.00	99.4	222.2	897.0
Velocity	10	0	0.2	1.3	0	0.2	1.0
pH	11	7.1	8.1	8.9	6.3	8.0	8.6

1. Obtained from 1:50,000 NZMS topographic maps.
2. Geographic information systems (ARCVIEW) using 1:50,000 vector data
3. Mean of 5 measures over length of reach fished.
4. Mean of the maximum depth measured at the 5 points above.
5. Median substrate size index from 50 – 100 stones collected at random over the reach surveyed and measured in 11 classes (Wolman 1954).
6. Leaf litter visually assessed (0 = absent, 1 = rare, 2 = sparse, 3 = common, 4 = abundant)
7. Visually estimated at site
8. Subjectively assessed at site after moving substrate (1 = loosely packed; 4 = tightly packed)
9. Measured at time of fishing with YSI model 85 meter.
10. Calculated from time taken for a slug of dye to travel 20m over length fished
11. Orion Quickcheck model 106 pocket meter

Predicting fish and macro-crustacean assemblages from habitat data using ANN models

Artificial neural networks are derived from a simple model of the structure and function of the brain, and are characterised by their ability to 'learn'. This is achieved by comparing actual and desired outputs during the model-training phase. In the training phase an algorithm modifies the internal parameters (weights) until the performance of the network, in this case prediction success, is maximized. For this case the presence or absence of each taxon was predicted using the back-propagation algorithm (Rumelhart et al. 1986a). The architecture of the layering has been described by other authors (Lek et al. 1995; Manel et al. 1999). The first layer, called the input layer, comprises 16 cells representing each of the environmental variables. The second or hidden layer, is composed of a further set of neurones the number of which depends on the reliability required and the structure that best optimises bias and variance (Lek et al. 2000). In this application we used a network with a single hidden layer of three neurones (more layers and more neurones did not improve performance) trained through 100 iterations (SAS 1999). The third layer or output layer consisted of a single neurone responsible for the prediction of presence or absence of the taxon from the environmental variables. We constructed one model for each of the species and the models were later combined to predict assemblages.

Model evaluation

For each site output values in the range of 0.5 - 1 were interpreted as presence and values 0 - 0.5 as absence. We used several methods to assess the performance of the individual taxa models. First, all taxa models were assessed based on prediction success, which is the overall percentage of sites at which the presence or absence of each taxon was correctly predicted. This comparison involved using all 118-reference sites as training data and provided a lower boundary for the error probabilities (Fielding and Bell 1997). As an independent test, we employed *k*-fold-partitioning. In this partitioning method we randomly divided the reference sites into a training set of 80% (98 sites) of the sites and an independent validation set of 20% (24) of the sites (Manel et al. 1999). This process was repeated five times and the results pooled giving 120 sites for assessment of models. To deconstruct overall prediction success into separate elements, matrices of confusion were derived, after (Fielding and Bell 1997), in which true presence, false presence, true absence, and false absence were identified. From these values we calculated a range of performance measures: 1) sensitivity (percentage of true presence correctly identified), 2) specificity (percentage of true absence correctly identified), 3) false positive (percentage of actual absences wrongly predicted as being present), 4) false negative (percentage of actual presence wrongly predicted as being absent), 5) positive predictive power (percentage of true positives that were real) and 6) negative predictive power (percentage of predicted absences that were real) (Fielding and Bell 1997; Manel et al. 1999).

Relationships between taxa and environmental variables

The habitat variables associated with the predictions from the ANN models were assessed by obtaining pairwise Spearman rank correlations using SAS (1999) between predictions of presence for each of the 12 taxa and the 16 predictor variables. To visualise these relationships between the predictions from the models and the environmental variables the probability of capture for each taxon was plotted against environmental gradients for each of the variables. For this, the environmental variables were arranged in ascending order then split into five groups of 24 sites. The mean probability of occurrence was calculated for each of the groups and this was repeated for all taxa.

Calculation of the expected number of taxa and O/E values

To identify the assemblages expected to occur at sites, the individual models (one for each taxon) were combined using the following protocol of Wright et al. (1984). The probabilities of the predicted taxa are summed to give the expected number of taxa (*E*). The number of species actually captured at a site, providing they were predicted to occur is the observed number of taxa (*O*). The ratio of the observed to the expected number of taxa (*O/E*) and taxonomic composition is the output from the model used for assessing the biological quality of the test sites (see Moss et al. 1987). To calibrate reference site *O/E* variability the reference sites were run through the ANN models and *O/E* ratios were calculated using the process described above.

Test sites

After satisfying the validation criteria the 12 trained ANN models were applied to assess the status of 35 sites with potential impacts. Observed over expected ratios were calculated for the test sites using the process described above for reference sites. Analysis of variance was used to compare *O/E* ratios for test and reference sites using SAS (1999). To evaluate how the actual occurrence of individual taxa related to the predicted patterns of occurrence, the number of sites each taxon was predicted to occur at was plotted against the number of sites where it was observed.

Results

A total of 11 967 fish and macro-crustaceans from 23 species were caught at 200 sites during this study. Of these, 16 fish and the 2 crustacean species (a diadromous shrimp and a non-migratory crayfish) were native (see Table 3.5.2 for names). Eleven fish species, including all non-native species, were found at less than 2% (4) of the sites and were not used in any of the analyses. These rare species were not included because most were introduced and their rarity made association of habitat variables problematical. At the 118 reference sites, ten fish taxa (including elvers) and two crustacean species were used in analyses after the removal of rare species. The most common was the longfin eel occurring at 88% of the sites, and the rarest was the torrentfish at 4% of the sites (Table 3.5.2.).

Table 3.5.2 The twelve fish and crustacean species found at more than 2% of the 200 potential reference sites and the number of the 118 reference sites at which they were present.

Scientific name	Common name	Code	present	absent
Anguilla australis	shortfin eel	ANGAUS	25	93
Anguilla dieffenbachii	longfin eel	ANGDIE	105	13
Anguilla spp.	Elvers	ELVERS	7	111
Cheimarrichthys fosteri	Torrentfish	CHEFOS	5	113
Galaxias fasciatus	banded kokopu	GALFAC	73	45
Galaxias maculatus	Inanga	GALMAC	51	67
Gobiomorphus basalis	Crans bully	GOBBAS	51	67
Gobiomorphus cotidianus	common bully	GOBCOT	23	95
Gobiomorphus huttoni	redfin bully	GOBHUT	42	76
Paranephrops planifrons	koura	PARANE	72	46
Paratya curvirostris	Shrimp	PARATA	69	49
Retropinna retropinna	common smelt	RETRET	7	111

Fitting and validating models

Training data

The overall prediction success rate was high for the training dataset with 93% of the predictions correct (ranging from 78% to 100%) (Table 3.5.3). The mean values for sensitivity, specificity, and were also high showing that prediction success was not linked to species prevalence (Table 3.5.3, Fig. 3.5.2). False positive and negative rates were also low but suggested some linkage to prevalence with the most prevalent taxon longfin eel having the highest false negative rate (43%).

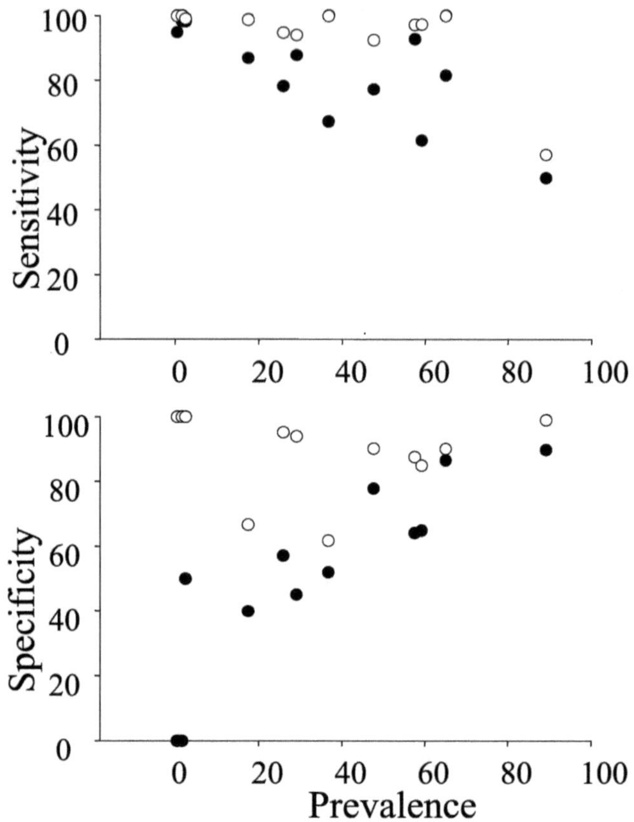

Figure 3.5.2 The relationship between species prevalence and A) and specificity (percentage of true presence's correctly predicted) and B) sensitivity (percentage of true absences correctly predicted) for predictions of the presence of 12 fish and macro-crustacea taxa using ANN models. Open circles are 118 reference sites used in training and filled circles are 120 validation sites pooled from 5 fold partitioning.

Validation data

The average prediction success results for the independent data set pooled from the 5 fold partitioning also revealed a high overall success rate of 80% ranging from 64% to 98% (Table 3.5.3). However, inspection of the alternative assessment measures revealed a linkage between prevalence and prediction success. Common smelt and elvers occurred at only one or two sites in this dataset and their presence was not correctly predicted (i.e. sensitivity, false positive and negative predictive power = zero) (Table 3.5.3). The percentage of true absence correctly predicted (specificity) was considerably higher (mean 81%) than prediction of true absence (mean 52%) revealing that the model was better at predicting presence than absence (Fig. 3.5.2). This bias however, may relate to the imbalance in prevalence, as there are more rare than common species (mean prevalence = 36%, Table 3.5.3).

The values for false positive, false negative, positive predictive power and negative predictive power revealed similar patterns with prediction success linked to prevalence.

Relationships between taxa and environmental variables

There is a complex mixture of associations between predictor variables and taxa (Table 3.5.4). Two variables latitude and pH, however, appeared to have little influence on any predictions. The highest correlation coefficients were between the three bully species and elevation and distance inland; the two migratory bully species were negatively correlated while the non-migratory Cran's bully was positively correlated with distance inland and elevation. Torrentfish show a negative relationship with the percentage of pool and a positive association with the percentage of run and riffle, while in contrast redfin bullies show opposite associations with percentages of pool and run. Banded kokopu show an affinity for small streams evidenced by the negative coefficients for width and depth.

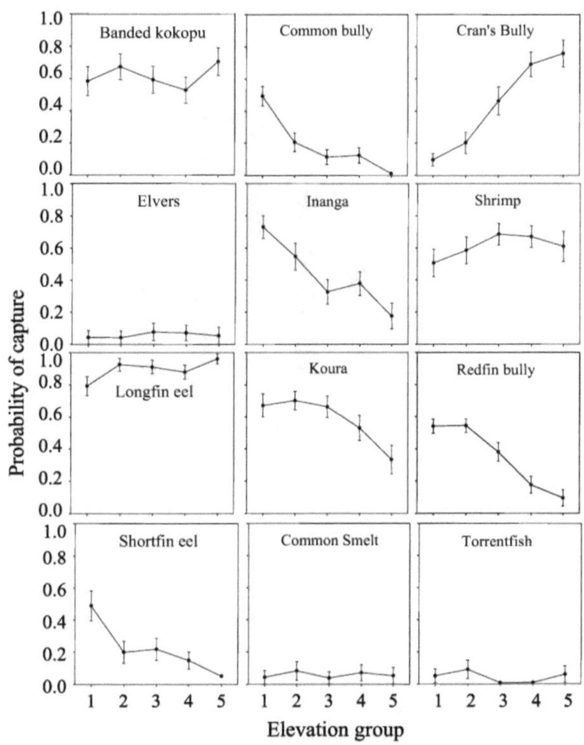

Figure 3.5.3 Mean probability (± SE) of capture of 12 taxa in five groups of 24 sites over an elevational gradient from the sea (left) to 260 m a.s.l. (right). The mean elevation in each of the five groups is 6, 20, 39, 61 and 105 m a.s.l. respectively

To visualise the relationships an example, Fig. 3.5.3 shows the probabilities of capture for each of the taxa plotted against the elevational gradient. The three bully species (common, Cran's and redfin), shortfin eel and the crayfish (koura) showed a strong relationship with the elevational gradient. Five taxa showed reducing probability of capture with elevation, they were common and redfin bullies, inanga, koura, and shortfin eel. The non-migratory Cran's bully and longfin eel showed increasing probability of capture with elevation. There was no discernable relationship with elevation for the rare taxa: smelt, torrentfish and elvers or the more abundant banded kokopu.

Calculation of observed over expected ratios

All reference sites were run through the ANN models and predictions were used to calculate *O/E* ratios. The distribution of these *O/E* ratio values for the reference sites approximated a normal distribution (Fig. 3.5.4). The *O/E* ratios were centered on unity with a mean of 0.99 (standard error 0.02) and ranged between 0.6 and 1.41. The plot of observed versus expected species number (Fig. 3.5.5) shows a strong relationship between the two as revealed by the slope and intercept very close to 1 (observed taxa richness = - 0.11 + 1.02 predicted taxa richness; $r^2 = 0.64$).

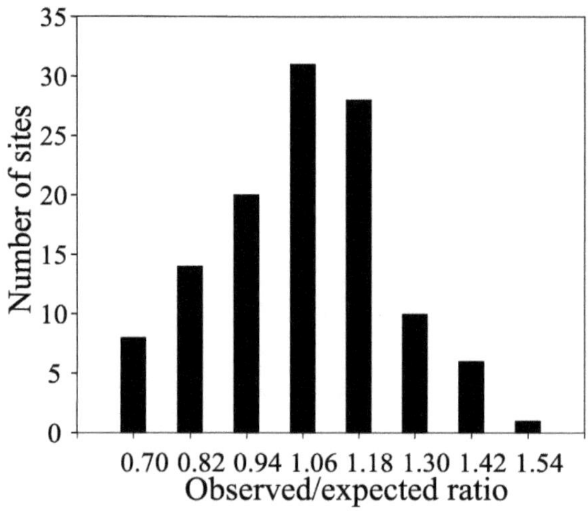

Figure 3.5.4 The distribution of *O/E* ratios from the 118 reference sites using predictions from the ANN models for the twelve taxa.

Assessment of test sites

The 35 test sites with potential impacts were run through the ANN models and predictions were used to calculate *O/E* ratios. The mean *O/E* ratio for the test sites (0.69, SE = 0.04) was significantly lower than the reference site *O/E* mean ($F_{1,152} = 62.9$, $P < 0.0001$). The plot of observed versus expected number of sites revealed the individual taxa that occurred at fewer sites than expected and thus, contributed to the lower overall *O/E* ratios (Fig. 3.5.6). The two galaxiid species (inanga and banded kokopu) as well as the two macro crustaceans (koura and shrimp) and the two migratory bullies (redfin and common) occurred at considerably fewer sites than expected. Three taxa however, occurred at slightly more sites than expected they were elvers, the non-migratory Cran's bully and shortfin eels.

Table 3.5.3. Prediction assessment results for the ANN models. For each assessment measure the left column contains the results for the 120 pooled validation data set and the right column contains the results for the 118-reference sites used as training data for ANN model. See text for details on assessment measures.

	Prevalence %		% correct classification		Specificity		Sensitivity		False positive		False negative		Positive predictive power		Negative predictive power	
GALFAC	65	62	85	93	82	100	87	90	12	90	18	0	74	82	92	100
GOBCOT	18	19	79	90	87	99	40	67	38	67	13	1	88	88	61	97
GOBBAS	29	43	66	94	88	94	45	94	35	94	12	6	60	96	90	93
ELVERS	5	6	95	100	95	100	0	100	0	100	5	0	100	100	0	100
GALMAC	48	43	78	92	77	93	78	90	18	90	23	7	81	93	78	91
PARANE	58	61	68	91	93	97	64	88	26	88	7	3	25	78	99	99
ANGDIE	89	89	89	92	50	57	90	99	9	99	50	43	8	92	99	92
PARATA	59	58	64	89	62	97	65	85	26	85	38	3	33	76	90	99
GOBHUT	37	36	64	78	67	100	52	62	32	62	33	0	84	66	45	100
ANGAUS	26	21	76	95	78	95	57	95	30	95	22	5	93	99	38	81
RETRET	2	6	98	100	98	100	0	100	0	100	2	0	100	100	0	100
CHEFOS	3	4	98	99	98	99	50	100	33	100	2	1	99	100	50	80
Min	1	4	64	78	50	57	0	62	0	62	2	0	8	66	0	80
Max	89	89	98	100	98	100	90	100	38	100	50	43	100	100	99	100
Mean	36	37	80	93	81	94	52	89	22	89	19	6	70	89	62	94

Table 3.5.4. Spearman rank correlations > 0.30 between predictor variables and probability of capture for each taxon. Probabilities are from the neural network model using all reference sites (variables ranked by number of correlations > 0.30). (* indicates non-migratory taxa).

Predictor variable	GALFAC	GOBCOT	GOBBAS*	ELVERS	GALMAC	PARANE*	ANGDIE	PARATA	GOBHUT	ANGAUS	RETRET	CHEFOS
Distance inland		-0.75	0.75			0.45	0.35	-0.41	-0.71			
Altitude		-0.57	0.59		-0.48				-0.72	-0.4		
Width	-0.6	0.36			0.41			0.56				
Conductivity		0.4	-0.36				-0.42			0.43		
Temperature					0.44						0.37	
Velocity		-0.39					0.35		-0.36			
Longitude						0.46						-0.4
Depth	-0.3										0.5	
% Pool									0.45			-0.4
% Riffle				0.39								0.35
% Run									-0.42			0.35
Substrate size										-0.37		
Leaf litter												0.36
Cobble packing							0.54					
Latitude												
pH												

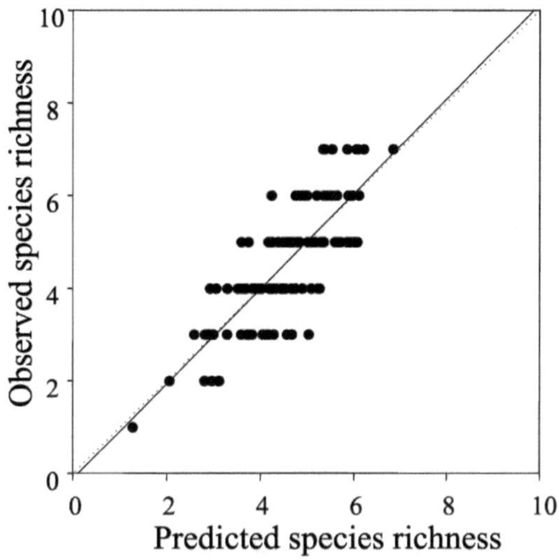

Figure 3.5.5 Observed species richness plotted against expected species richness from the 118 reference sites using predictions from the ANN models for the twelve taxa. The dotted line has a slope of 1.

Discussion

This project was conceived to make available a practical tool for the assessment of a significant part (fish and macro-crustaceans) of the biological integrity of stream sites in a region of New Zealand. It allows for an objective, statistically robust, site-specific prediction of the fish assemblage to be made without the requirement for extensive analysis. This is because the sophisticated statistical knowledge used to create the model is not required for its implementation. The model can be made available for use by others requiring only the input of environmental variables of a site to give a list of expected taxa. A RIVPACS type predictive model of biological quality using fish has been developed in New Zealand based on biotic site groups (Joy and Death 2002) and individual species discriminant function analysis (DFA) (Joy and Death, submitted). The ANN models reported here were applied to the same data used in the DFA models above (Joy and Death, submitted) and similar results were obtained. Discriminant function analysis was slightly less accurate than ANN using all training data but DFA was marginally more accurate than ANN when considering holdout validation data (Joy and Death, submitted). This similarity of results suggests that the relationships being modelled are linear.

The species-specific models we have developed appear to be methodologically, and ecologically robust for comparing reference and impacted sites. The species that occurred at fewer test sites than expected (Fig. 3.5.6): inanga, banded kokopu koura, shrimp, redfin and common bullies are the same species that have been observed to have reduced densities and/or sensitivity to impacts in other studies (Dean and Richardson 1997; Richardson 1997; Richardson et al. 1998, 1994, 2001; Rowe et al. 1999a,b 2000). The species occurring at more sites than predicted (Cran's bullies and shortfin eels) have similarly been shown to be tolerant to factors associated with human impacts (Richardson et al. 1994; Rowe et al. 1999b, 2000). The correlations between the taxa and environmental variables from the reference data also revealed ecologically realistic associations. Examples include the reducing probability of capture with elevation and distance from the sea for the diadromous species, and the association of banded kokopu with small streams (Joy et al. 2000; McDowall 1990).

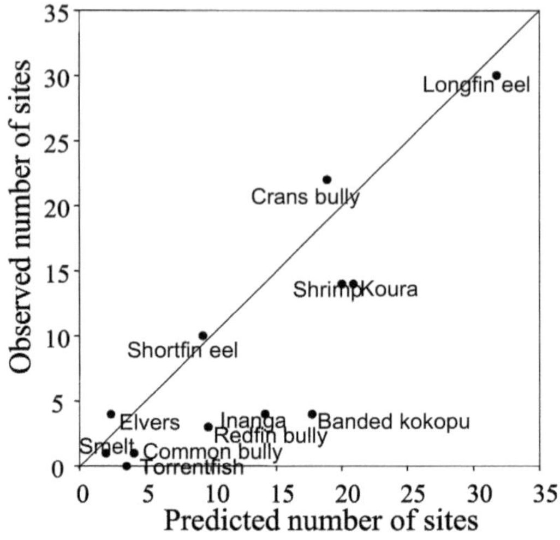

Figure 3.5.6 Predicted number of sites plotted against observed number of sites for the 35 test sites, the diagonal line represents the line of perfect agreement.

The effectiveness of predictive models has in the past generally been judged based on prediction success alone (Mastrorillo et al. 1997a,b; Oberdorff et al. 2001). Recent work by Fielding and Bell (1997) and Manel et al. (1999) have revealed the importance of assessing the different elements of prediction success separately and our results support these recommendations. In this study, the models were much better at predicting absence (mean specificity 81%) than presence (mean sensitivity rate 52%) when using validation data (Table 3.5.3; Fig 3.5.2). This is despite the apparently high percentage of correct predictions. However, the effect of prevalence on predictions was much more noticeable with the 20% *k* fold validation data than the training data. There are two potential explanations for this pattern. First that the reduction in number of training sites reduces the model precision, especially for the rarer taxa through a lack of experience of the conditions at these sites. The second explanation is that there was some overtraining of the total data set. Overtraining occurs when the network attempts to model noise in the data rather than real patterns (Walley and Fontama 1998, Lek et al. 1996b).

The models generally performed well when applied to independent validation data suggesting that the predictions are accurate. The validity of the models was further supported by the ecologically meaningful associations that were found between taxa predictions and environmental variables (Table 3.5.4). We are thus confident that the *O/E* ratios produced from the models provide an accurate assessment of the expected biological condition at a site in the absence of human impacts. The predominantly diadromous fauna of New Zealand has negated the application of other bioassessment tools such as the index of biotic integrity (McDowall and Taylor 2000). This ANN model for site assessment takes into account not only conditions at the site but also from the site to the coast as the diadromous species are dependent on access to and from the sea as well as proximal habitat and catchment quality. Thus, an *O/E* ratio indicates biological quality of the waterway at many scales over the whole catchment from the source to sea.

Conclusions

In this study, we developed an empirical model capable of predicting the stream fish and macro-crustacean assemblages expected to occur in the absence of human impacts using a suite of environmental variables. Expected assemblages based on data from reference sites were compared to those observed at a number of potentially impacted sites, and the deviation between the two measures provided a measure of the magnitude of degradation of biological quality. This process also identifies individual taxa responsible for the deviations between observed and expected assemblages, which then potentially allows for diagnosis of the relationship between individual taxa and impacts. This diagnosis process can be achieved by correlating individual components (taxa) of the biological structure with known impacts.

The model developed here provides a rapid and powerful technique for assessing the biological condition of the fish and macro-crustacean fauna of a stream and can potentially identify targets for management or rehabilitation. Therefore, it has potential for assisting with stream management in New Zealand because the stream is assessed from headwaters to the coast and migration induced trajectories of occurrence are taken into account. Furthermore, the method applied here also has potential for application in other countries with a high proportion of diadromous species.

3.6 A Comparison of various fitting techniques for predicting fish yield in Ubolratana reservoir (Thailand) from a time series data*

Moreau J†, Lek S, Leelaprata W, Sricharoendham B, Concepcion Villanueva M

Introduction

Understanding and predicting the biological productivity and the resulting fishery production is a key goal of fisheries scientists and managers. Numerous studies have attempted to predict the possible catches using characteristics of the water bodies: area of the drainage basin of the rivers, floodplain areas, morpho-edaphic index, depth, coastal lines, primary production (e.g. Henderson and Welcomme 1974, Marshall 1984, Crül 1992).

Diverse multivariate techniques have been used for this purpose including several methods of ordination and canonical analysis, univariate and multivariate regressions (Ryder 1982; Welcomme 1986, De Silva et al. 1991, Payne et al. 1993, Bernascek 1997). With these conventional techniques, one problem is that relationships between ecological variables are, most often, not linear whereas most methods are based on linear principles. Therefore, non-linear methods were also used and have proved helpful in ecological sciences as demonstrated by Lek et al. (1995) and Laë et al. (1999).

In inland waters, the influence of the hydrological conditions at year t-1 on the catch at year t was documented by Welcomme (1985, 1986) based on time series data for the Kafue floodplain and the Central Delta of Niger in Mali (Western Africa). The time lagged relationship was attributed to the life span of one year most of the African fishes inhabiting these ecosystems, which is also the age of capture. Before that, Welcomme and Hagborg (1977) had identified that variations of the flooded area between low and high water level as a key factor regulating potential fish catch in tropical floodplains and shallow water bodies.

A multispecific fishery is established in Ubolratana reservoir (North-East Thailand). The Royal Department of Fisheries of Thailand has collected actual catch data since the impoundment of the dam, in 1965, whereas the Royal Department of Irrigation has recorded monthly maximum and minimum area, water level and shoreline variables of the reservoir. This provides a good opportunity to develop predictive models of actual catch for a single reservoir based on time series data. The aim of this paper is therefore, to compare various linear and non-linear methods of prediction of the dependent variable (Catch of Clupeid and other fish species, referred to here as littoral fisheries) from the independent ones (the hydrological features).

---

* This work has been carried out under the EU INCO/DC Project number ERB 3514 PL 96 16 95, "FISHSTRAT". Thanks are due to Dr Pinit Sihapikutgiat, the coordinator of the FISHSTRAT project in Thailand for his helpful support and encouragement to produce this contribution. A preliminary version of this paper was part of the BSc thesis of two students from I.N.P. Toulouse: Bérangère Dudognon and Myriam Aissa. We are also grateful to the anonymous referees for helpful comments and advice on this contribution.

† Correspondence: moreau@ensat.fr

Material and methods

Study site and data

Damming the Pong River, 500 km North-Eastern of Bangkok, Thailand, in January 1965 has created Lake Ubolratana (Fig. 3.6.1). It has a maximum area of 41 000 ha and a simultaneous average depth of 16 m at maximum water level (182 m above mean sea level, MSL). The shoreline development (the ratio between the shore length and the length of a circle of the same area) is 6.16. During the dry season, the water level commonly drops to about 175 m above MSL and the surface area decreases to about 16 500 ha. Floods events during the rainy season cause nutrients to be leached from the soils and facilitate the decomposition of organic matter. These events increase the primary production and the resulting fish production and provide critical spawning habitats.

A littoral fishery exploits endemic fish riverine populations, which have colonised the littoral shallow zones of the lake since its impoundment and some introduced species, mainly *Oreochromis niloticus* (Pawaputanon 1987; Benchaken et al. 1989, De Iongh and Van Zon 1993). In addition, a fishery oriented towards *Clupeichthys aesarnensis* using lift nets and lights started in 1972 and continued to flourish until 1990. This clupeid is a pelagic fish with a short life span (about 1 year).

The actual catch increased from 1965 to the end of the seventies (Fig. 3.6.1) i.e. immediately after the dam closure (Van Densen and Morris 1999) but since then, it has been stabilised with important year to year variations. In the 1990s, the clupeid catch dropped drastically and the authorities banned the utilisation of lights in 1993. This fishery is now of marginal importance.

Modelling methods

Actual catch (tonnes yr^{-1}) of clupeids and of other fishes (littoral catch) at year t have been considered as dependent variables. Following the ideas of Welcomme and Hagborg (1977), and Welcomme (1986) the independent variables are "simple" hydrological parameters pertaining to year t-1: maximum area; difference between maximum and minimum area (surface area range); shoreline development. The data set is available on request from any of the authors. In order to examine the serial dependencies in fish catches, the partial autocorrelation function (PACF) was used (Box and Jenkins 1970). The seasonal patterns of time series can be examined via correlograms. The correlogram (autocorrelogram) displays graphically and numerically the autocorrelation function, that is, serial correlation coefficients (and their standard errors) for consecutive lags in a specified range of lags (e.g., 1 through 15). Ranges of two standard errors for each lag are usually marked in correlograms but typically the size of autocorrelation is of more interest than its reliability because we are usually interested only in very strong (and thus highly significant) autocorrelations.

According to the results shown in Fig. 3.6.2 (the strong possible influence of catch the year before on the catch at year t) the lag of 1 was used in all subsequent models

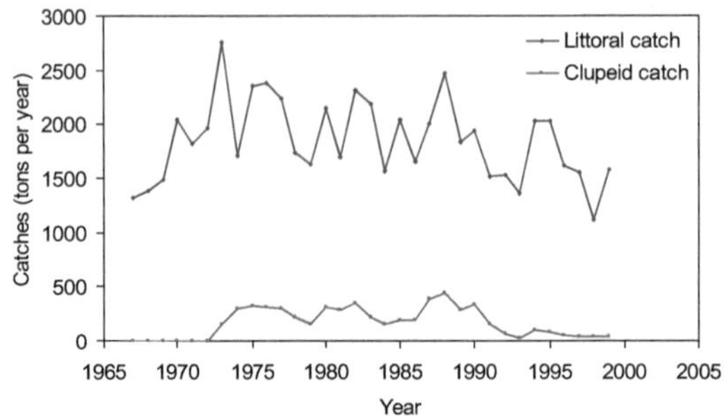

Figure 3.6.1 Variations of the actual catch in Ubolratana reservoir from the closing of the dam to 1999 for littoral fish and clupeid catches.

The present contribution will therefore deal separately with the Clupeid fishery (until 1992 only) on the one hand and the littoral fishery on the other. The latter uses various types of fishing gear by both permanent and seasonal fishermen, the exact number has not been regularly monitored and seems to be permanently increasing mostly during the last twenty years (W. Leelaprata, pers. data).

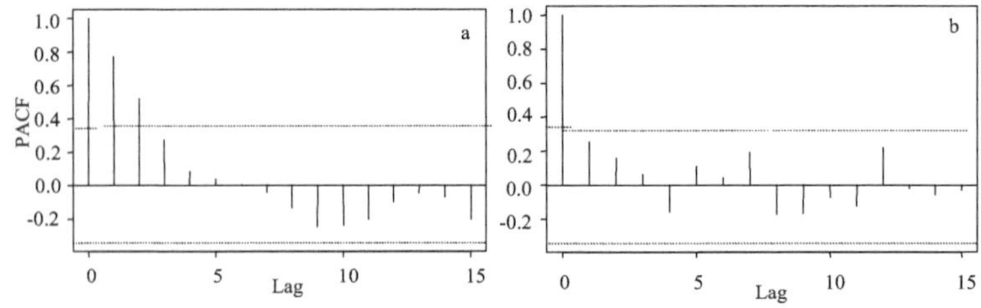

Figure 3.6.2 Partial auto-correlation functions analysis for clupeid fisheries (a) and littoral fisheries (b): the dotted line (ordinate ± 0.35 shows the limits of 95% of confidence interval)

The shoreline development and the area of the reservoir are highly correlated (r= 0.98). It is believed however that the shoreline development has to be incorporated as an independent variable to the possible models has it expresses the availability of breeding area for strictly littoral species which do not benefit from surface availability for breeding but rather look for shore availability (Balon and Coche 1975, Welcomme 1985, Payne 1986, Kolding 1994). A close relationship also exists between the maximum area and the surface area range (r = 0.88). However, some variability does exist as the slope can be variable with the water level as recorded with the hypsographic curve (The Royal Department of Irrigation of Thailand, available from the first author upon request).

The relationship between hydrological characteristics and the fishing yield were studied with multiple regression analysis (James and McCulloch 1990). The diagnosis of the Student residuals (normality and independence) was used to test the validity of the determination coefficient obtained (Tomassone et al. 1983). Aiming to improve the model's performance, we also used generalised additive models (GAM) (Hastie and Tibshirani 1990). The GAMs are a generalisation of multiple linear regression and generalised linear models. They are non-parametric regression methods, which model the dependent variable as an additive sum of unspecified functions of covariates. Least squares and maximum likelihood methods used in multiple linear regression and generalised linear models are replaced by quasi-likelihood methods which rely on local scatter plot smoothing methods. Here, we used the locally weighted smoother of Cleveland (1979), currently called "loess" in the Splus statistical computing language. The loess smoother first computes a defined percentage of the nearest-neighbours to the target point. A tricube kernel, centred at the target point, becomes zero at the furthest neighbour. The smoother at the target point is the fitted value from the locally weighted linear fit, with weights supplied by the kernel. One of the major advantages of this method is that it automatically shows the dependence of the response on each of the predictors.

The regression tree analysis (also referred to as the CART : Classification And Regression Tree) was also performed. This method was advanced by Breiman et al (1984) in the statistical literature. CART has proven to be a useful tool for identifying non-linear patterns of variability (Magnuson et al. 1998, Emmons et al. 1999, Rejwan et al. 1999, De'ath and Fabricius 2000, De'ath 2002). A tree regression analysis does not make assumptions regarding the linearity or homoscedasticity of the variances and it automatically accounts for interactions among variables. Although decision trees are intuitively attractive, there are several difficulties that can limit their use and performance in several instances such as the possible instability of the tree structure itself. Small changes in the data set used to grow the tree model can cause significant effects on the shape and on the predictive capabilities of a tree (De'ath 2002). Tree regressions analyses results in functions encompassing the complexity between the dependent and independent variables while handling a lot of data. The strength of the tree regression analysis is also for prediction of covariate importance under broad environmental variations. In the present exercise, implementing the following procedures enhances the performance of the tree. First, a tree model is grown on a learning example and the selection of the best-sized tree is performed by cross validation (Breiman et al. 1984). A transfer procedure is then implemented based on the comparison of distances between the empirical distribution of the predictor variables at different states of the response variable. The efficiency of this procedure applied on a tree model is tested using a test sample to estimate the error rate in classifying future observations.

Finally, a three-layer feed-forward artificial neural network (ANN) was used such as in Laë et al (1999). A neural network is created by designing the so called layers. The first one connects with the input variables and is called the input layer. Here it comprises 4 neurones (the independent variables). The last layer connects with the output variable and is called the output layer of only one neurone (the dependant variable). The layer between the two previous layers is called the hidden layer. Each of the neurons of a particular layer is connected to the neurons of the neighbouring layers and all connections are fed-forward; that is: they allow information transfer only from one layer to the next consecutive one. No feedback connections are permitted in these "feed-forward" networks. The back-propagation algorithm which has been used here in order to train the network to provide the proper outgoing signal of the output layer (the predicted value of the fish yield in this study) is documented in Laë et al. (1999) following Rumelhart et al. (1986a).

In order to compare the predictive performance of the different statsitical methodologies an application was made on the whole database (32 units). Then, a cross-validation was operated to justify the predictive quality of the various methods by implementing a leave-one-out procedure as used for ANN models (see Jain et al. 1987 for detail). This validation procedure is useful in cases where the quantity of observations is limited.

The S Plus software was used for analysis using GLM, GAM and CART; and the ANN was carried out using the software package Matlab for Windows. The output originating directly from the use of the software mentioned here is available on request from one of the authors (S.L.).

Results

Multiple regression analysis (MLR)

A MLR was performed in order to check if a significant correlation could be obtained with this classical method (Figs. 3.6.3 and 3.6.4). With the three hydrological variables and the littoral catch the year before, we obtained a determination coefficient (r^2) of only 0.33 for the littoral catch. For the clupeid fishery, the determination coefficient was only 0.62.

The two equations to predict the littoral and the clupeid catch are respectively:

Littoral catch (t) = 1021 - 87.7 Maximum surface area (t-1) + 105.8 Shoreline (t-1) – 6.7 Smax-Smin (t-1) + 0.22 Catch (t-1)

Clupeid catch (t) = -31.16 - 1.48 Maximum surface area (t-1) + 3.65 Shoreline (t-1) + 1.9 Smax-Smin (t-1) + 0.78 catch (t-1)

In the case of the littoral catch, the model as a whole is significant ($p < 0.05$) whereas individual variables are not significant. On the other hand, for clupeid catch the model is highly significant ($p< 0.01$) and this also holds true for the catch of the year before. ($p < 0.001$)

In order to completely fulfil the requirements of the MLR method (i.e. normal distribution of variables considered) the fish yield and the independent variables were transformed into their log10 but the MLR showed determination coefficients very close to those observed before log transformation for both fisheries.

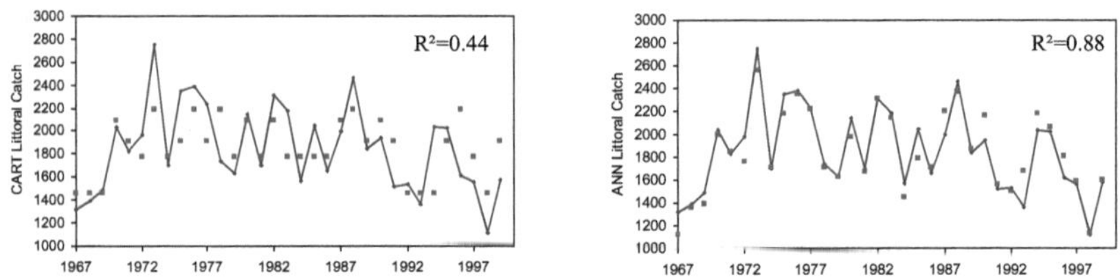

Figure 3.6.3 Comparison of the observed and fitted trends of variations of the littoral catch using multilinear regression, generalized additive model, tree regression analysis, ANN.

The non linear fitting techniques and models

When using the nonlinear techniques (GAM, CART) the determination coefficients between the predicted and observed values of the catch were computed and compared to the same values resulting from the MLR analysis. The computed determination coefficients were all much higher than when implementing the multi-linear regression analysis (Figs. 3.6.3 and 3.6.4). When using the GAM, the determination coefficients were 0.67 for the littoral fisheries and 0.82 for pelagic fisheries; When using the regression tree analysis the determination coefficient was only 0.44 for the littoral fishery and 0.80 for the Clupeid fishery. In the tree regression analysis (Fig. 3.6.5), the dominant parameters to explain the variations of the littoral catch are the maximum surface area and the surface area variations the year before. For the Clupeid fishery, the catch the year before and the surface variations are the main variables to consider.

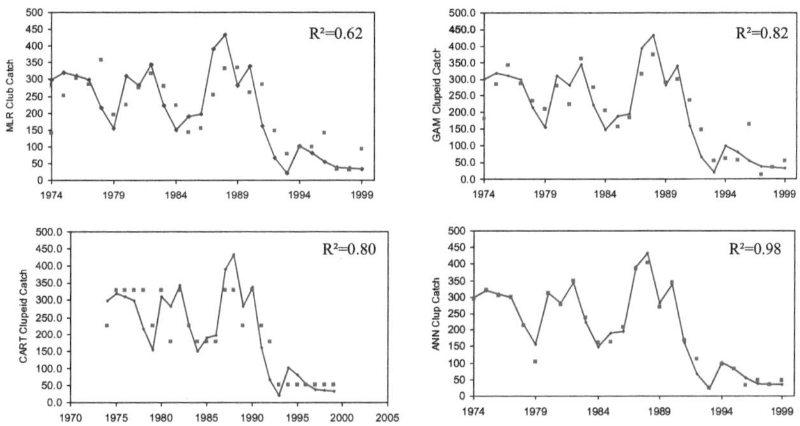

Figure 3.6.4 Comparison of the observed and fitted trends of variations of the actual catch of clupeids using multilinear regression, generalized additive model, tree regression analysis, ANN .

Comparing the percentage of explained variances (i.e. the determination coefficients) observed when using tree and MLR quantified the importance of non-linear relationships and interactions between catch and the concerned parameters. These non-linear relationships are more important for the Clupeid catch than for the littoral one.

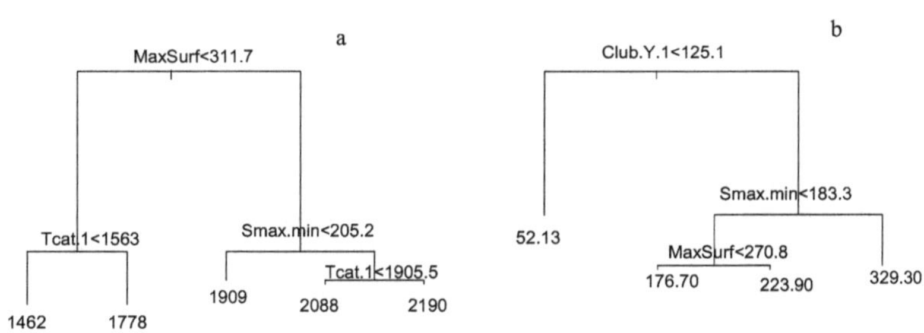

Figure 3.6.5 Tree regression models for littoral (a) and clupeid (b) catches. The number at the extremity of the tree is the quantity predicted (in tons/year).

For the ANN analysis, in order to avoid possible over-fitting, several tests were carried out with different configurations of the neural network (change in the number of neurones in the hidden layer). The best configuration that had a minimal dimension and which gave satisfactory results was retained (in the present work, three neurones in the hidden layer).

Again, in order to avoid overfitting, the number of iterations was limited to 500 and the leave-one-out procedure was used to determine the predictive quality of the ANN model. Owing to the small size of the dataset, the leave-one-out validation procedure was used to test the performance of the model. The resulting determination coefficient between observed and predicted values was 0.88 and 0.98, respectively for the littoral and clupeid catches.

When using the ANN, the partial derivative algorithms (Dimopoulos et al. 1995, 1999) quantify the relative contribution of the hydrological variables in the model (Fig. 3.6.6). For littoral catch, the highest contribution is about 70% (S Max) and 50% (Shoreline and Smax-Smin) whereas the catch the year before contributed only about 10 %. For the clupeid fisheries, the main contributions are 35 % for the catch at year (t-1) and 20 % for the difference between maximum and minimum surface area.

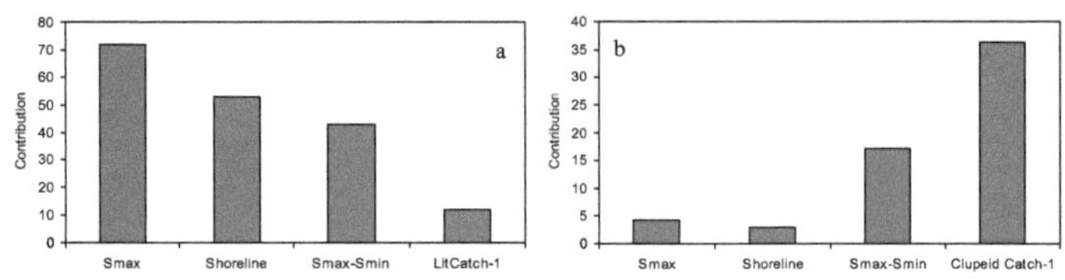

Figure 3.6.6 Relative contributions of input variables to explain the littoral (a) and clupeid (b) catches in artificial neural network models (partial derivative analysis).

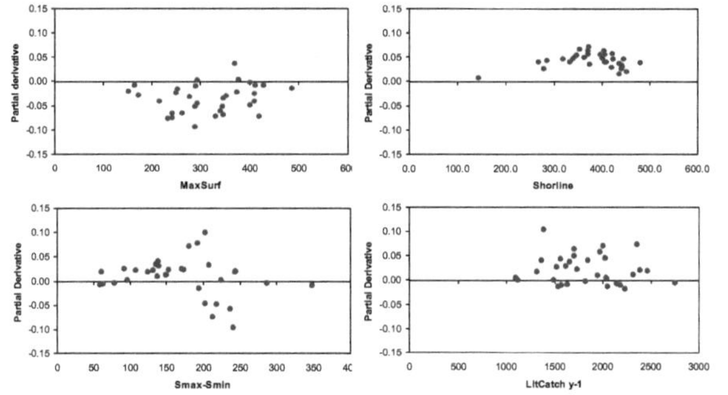

Figure 3.6.7 Contribution profiles (sensitivity analysis) of each of the four dependent variables for the prediction of the littoral catch. The values cover the whole range of variations of each independent variable under consideration. The observed value, i.e. the partial derivative of the annual catch, was plotted vs. each of the independent variables.

The sensitivity analysis, as summarized in Figs. 3.6.7 and 3.6.8 suggests that the influences of all the variables are complicated and nonlinear. The negative values of the partial derivative (y axis) for maximum surface area for both littoral and clupeid catches confirms the negative impact of this variable. When low values are recorded, maximum and minimum surface area has a positive effect on both catches whereas negative impacts can be noticed for the highest values of this variable. The partial derivatives of littoral catch related to shoreline are positive for all values due to the already identified positive effect of this variable (see Fig. 3.6.6). However, for the clupeid catch, the values of partial derivatives are more or less equally distributed around the zero line, which confirms that shoreline development has no observable influence on clupeid catches (see also Fig. 3.6.6). The positive values of partial derivative of littoral catch (y axis) for the majority of the values of littoral catch in year minus 1 (x axis) shows that the increase of the catch the year before contributes to the increase of current littoral catches. For clupeid catches, the influence of the catch the year before is more complicated to identify. It seems that there are positive effects for the low values of the catch the year before, but negative effects for high values.

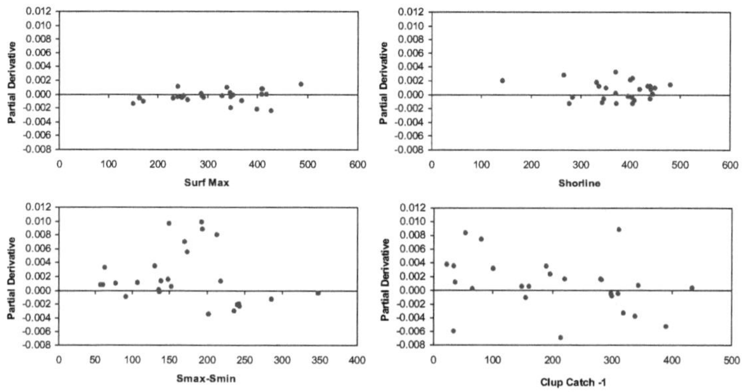

Figure 3.6.8 Contribution profiles (sensitivity analysis) of each of the four dependent variables for the prediction of the catch of Clupeids. The values cover the whole range of variations of each independent variable under consideration. The observed values of annual catch were plotted vs. each of the independent variables.

Discussion

The fisheries statistics of the Royal Department of Fisheries of Thailand are collected on a very regular basis and have been shown in several cases to be suitable for the kind of exercise carried out here (De Silva et al. 1991, Moreau and De Silva 1991). They were therefore considered to be reliable.

The methods used here have so far only been applied to databases involving several water bodies and, at least to our knowledge, time series data involving a multispecific fishery in a single water body have not yet been considered for such exercises except in a marine environment by Jarre-Teijmann et al. (1995) on Peruvian anchoveta (*Engraulis ringens*), and by Cisneros et al. (2000) to forecast one year in advance the annual spawning biomass of Pacific sardine (*Sardinops caeruleus*) of the California Current. Moreover, Watters and Deriso (2000) carried out a regression tree analysis in order to estimate time series of abundance indices of *Thunnus obesus* in the Central and Eastern Pacific Ocean from catch per unit effort data of Japanese long--line fisheries.

The fishing effort has not been considered here because reliable data are unavailable. The main reason is the permanently unknown and variable number of "migrant fishermen" who come to the lake from other reservoirs for a short and variable periods of time in a completely unpredictable pattern in order to try to get, even very temporarily, higher financial incomes. This trend seems to have increased during the recent years because of the economic crisis which led workers from Bangkok and other large cities to return to the countryside (W. Leelaprata, personal observations). In addition, the aim of this contribution was to provide a useful tool for fisheries managers and /or decision makers who need to predict the actual catch without long and costly surveys of fishing effort and fish population dynamics but by using, instead, easy to measure ecological parameters which can, at least partly, explain the population dynamics in the reservoir.

The two fisheries considered here rely on different hydrological patterns: The pelagic catch is strongly related to the surface area variation and to the catch the year before, whereas the littoral catch relies to a large extent on the total area and to the difference between the maximum and minimum water area. This last result is in agreement with previous findings of Welcomme and Hagborg (1977) and Welcomme (1986) on floodplain fisheries. They discussed the importance of the area to be flooded during the rainy season for feeding and reproduction of shallow water fish. Tropical fish usually have a short life span, very often about 1 year and it helps to understand the influence of the flooded area at year t-1 on the actual catch at year t. Another finding here is the relative importance of the shoreline development. This comes from the fact that several littoral fish populations seek very shallow littoral water to breed. The variations of availability of such areas is taken into account more accurately by considering the shoreline than any other parameter related to the surface area as can be seen from the hypsographical curve, as already mentioned .

The main limit of the present exercise is the relatively low number of observations (32). It comes from the fact that they are time series data which are always difficult to collect for such a long period of time. Several datasets exist for the marine environment (one of the oldest and most documented is from for plaice from the North Sea, *Pleuronectes platessa in* Daget and Le Guen 1975) whereas those concerning tropical inland water fisheries are very few: for instance Lake Victoria (Pitcher and Hart 1995), Lake Tanganyika (Petit 1996) or Lake Kariba (Kolding 1994; Moreau 1997).

As already observed by Laë et al (1999), the advantage of the ANN over the MLR and even over the generalised additive model and the CART is the ability of the ANN to directly take into account any non-linear relationships between the dependent variable and each independent variable. The back propagation procedure of the ANN provides greater predictive power than MLR, especially for the training calculations. The result is that for any coming year, the yield would be computed by introducing four independent variables pertaining to the year before, including the actual catch. All this holds true for other non-linear fitting techniques: the GAM and TREE regression method.

Neural networks tend to be more successful with "large" data sets for fitting complex non-linear interactions, which was not really the case here, whereas the GAM, and to some extend the CART model can get by with few observations (Rejwan et al. 1999). In the case of GAM, it comes from the fact that they explicitly focus on lower interactions (Garrison 1991). In addition, the fitting of neural network models requires some experience and effort from the investigator.

Conclusion

For a proper prediction of the fisheries in the Ubolratana reservoir, and most likely in other similar Thai reservoirs in which a clupeid population is exploited, it is essential to distinguish the littoral fisheries and the pelagic ones. This comes basically from the fact that those two fisheries exploit target fish whose demography is not regulated by the same ecological factors, at least at the current level of investigation.

The relative importance of environmental variables assessed when using both CART and ANN methods is in accordance with ecological factors reported in previous studies (Hagborg and Welcomme 1977, Welcomme 1986). Both models are able to reproduce the catch on the basis of the ecological variable introduced. Moreover, the predictive power of GAM, CART and ANN overpasses the capabilities of more common techniques. Consequently, these methods can be used either as predictive tools or as explanatory tools.

3.7 Patterning spatial variations in fish assemblage structures and diversity in the Pilica River system*

Penczak T†, Kruk A, Park YS, Lek S

Introduction

The study of fish assemblages in the Pilica River has not yet been undertaken despite data being available for various reaches. At the beginning we tried to analyse the data by detrended correspondence analysis (DCA), but obtained ordinations were not clear. On the two-dimensional scatterplot, sites from the main channel and sites from lower courses of the largest tributaries formed a long bowed "black cloud" on which single sites and their symbols were not visible.

Besides, there are well-known limitations for indirect gradient analysis, such as strong distortions with non-linear species abundance relations (horseshoe effect, arch effect, missing data, noise, redundancy, outliers, disjointed data matrix, etc.) (Gauch 1982; Kenkel and Orloci 1986; ter Braak 1987; Jongman et al. 1995; Guegan et al. 1998; Palmer 2000). These limitations can be avoided by applying the artificial neural network (ANN) which has already been successfully used in ecology (Chon et al. 1996; Guegan et al. 1998, 2001; Lek and Guegan 1999, 2000; Brosse et al. 2001; Park et al. 2001a). In this study we used the self-organizing map (SOM) with an unsupervised learning algorithm, which was used in a few studies to reveal relationships within ecological communities (Chon et al. 1996, 2000a; Giraudel et al. 2000; Giraudel and Lek 2001; Brosse et al. 2001; Park et al. 2001a). The conclusions drawn from these works indicate that the SOM algorithm is fully usable in ecology as a technique for analyzing data and for community ordination (Chon et al. 1996).

The fish fauna of the Pilica River was selected for the study because it is one of the best known in Poland (Penczak 1988, 1989; Penczak et al. 1995, 1996). Also some comparative research on its stability and variability has already been done (Penczak and Kruk 1999, 2000) and this fact creates a chance for determining which method used for data ordering gives a more clarified and closer-to-reality picture of the assemblage structure in the river system.

In this fish fauna inventory study (1992-95) fish sampled at each site were not only counted but also weighed. It is well known that the potential energy of an ecosystem is not distributed proportionally between species (Odum 1980). Out of tens or hundreds of species forming a community, relatively few, named dominants, exert a major controlling influence, because they are ecologically successful in a given environment. Expressing the importance of populations in an ecosystem is difficult in energy units, but it is much easier in biomass. Density can be effectively used only for comparing populations of species of similar body size (Acarina, Chironomids, etc.).

The aim of the study is to show how two different methods, SOM and DCA, can be advantageous to analyse complex, non-linear fish population datasets. Here we try to test if

* Thanks are addressed to Łukasz Głowacki for improvement of the English and preparation of the DCA scatterplots. Collection of fish samples was possible thanks to the Polish Anglers Association.
† Correspondene: penczakt@biol.uni.lodz.pl

clusters of neurons distinguished by the SOM on the basis of fish biomass only, also differ in terms of species diversity, community dominance and assemblage stability.

Methods

Study area

The Pilica River is the longest (336 km) tributary of the Vistula (Fig. 3.7.1) in Poland. The dam of the man-made reservoir filled in 1975 is located 198. km from the source. In the main channel fish were sampled on: 14.06-9.09.1994 and 11.07-22.09.1995, while in tributaries: 03.07-5.09.1992, 20.09-22.09.1993 and 13.06-20.10.1994. The Pilica River tributaries in which fish samples were collected are marked with their names, or – if very small – with numbers (Fig. 3.7.1). In order to provide detailed information on the distribution of 118 sites in the tributaries we include Table 3.7.1, in which every site has a successive number, a symbol for marking it on a self-organizing map (SOM) and on a DCA scatterplot, as well as a full name of the tributary and name of its receiving river. In the main channel, 63 sites were distributed proportionally, at a distance of about 5 km from each other, and on SOM and DCA scatterplots they are marked with an asterisk and successive numbers; the numeration starts from the source and is the same as in the former paper (Penczak et al. 1995). Sites from tributaries have numbers as in Penczak et al. (1996). In these papers, site characteristics are included.

Because data on discharge are not available for individual sites we use the product of depth and width (DW) to show the difference in channel dimensions and thus approximately in amount of water carried as discharge is correlated with both depth and width (Allan 1995).

Table 3.7.1 Sites located on the Pilica River tributaries. Site abbreviations used on SOM and DCA scatterplot are in the column "Symbol". * no fish at a given site

No.	Symbol	Tributary	Receiving river	No.	Symbol	Tributary	Receiving river
1	Kr1	Krztynia	Pilica	61	M61	Mogielanka	Pilica
2	Kr2	Krztynia	Pilica	62	Cz62	Czarna Woda	Pilica
3	Kr3	Krztynia	Pilica	63	Cz63	Czarna Woda	Pilica
4	Kr4	Krztynia	Pilica	64	U64	Uniejówka	Pilica
5	Kr5	Krztynia	Pilica	65	Z65	Zwlecza	Pilica
6	Kr6	Krztynia	Pilica	66	Z66	Zwlecza	Pilica
7	KrB7	Białka	Krztynia	67	Z67	Zwlecza	Pilica
8	KrB8	Białka	Krztynia	68	ZJ68	Jeżówka	Zwlecza
*9	KrZ9	Żebrówka	Krztynia	69	Ku69	Kurzelówka	Pilica
10	KrZ10	Żebrówka	Krztynia	70	Cw70	Czarna Włoszczowska	Pilica
*11	KrZ11	Żebrówka	Krztynia	71	Cw71	Cz. Włoszczowska	Pilica
12	KrZ12	Żebrówka	Krztynia	72	Cw72	Cz. Włoszczowska	Pilica
13	KrZ13	Żebrówka	Krztynia	73	Cw73	Cz. Włoszczowska	Pilica
14	B14	Białka	Pilica	74	CwC74	Czarna	Cz. Włoszczowska
15	B15	Białka	Pilica	75	CwC75	Czarna	Cz. Włoszczowska
16	Bt16	tributary 1	Białka	76	CwCs76	Czarna Struga	Cz. Włoszczowska
*17	Lu17	Luciąża	Pilica	77	CwCs77	Czarna Struga	Cz. Włoszczowska
18	Lu18	Luciąża	Pilica	78	Cwt78	tributary No 3	Cz. Włoszczowska
19	Lu19	Luciąża	Pilica	79	*Ck79	Czarna Konecka	Pilica

20	Lu20	Luciąża	Pilica	80	Ck80	Czarna Konecka	Pilica
21	Lu21	Luciąża	Pilica	81	Ck81	Czarna Konecka	Pilica
22	Lu22	Luciąża	Pilica	82	Ck82	Czarna Konecka	Pilica
23	Lu23	Luciąża	Pilica	83	Ck83	Czarna Konecka	Pilica
24	Lu24	Luciąża	Pilica	84	Ck84	Czarna Konecka	Pilica
25	Lu25	Luciąża	Pilica	85	Ck85	Czarna Konecka	Pilica
26	Lut26	tributary 2	Luciąża	86	Ck86	Czarna Konecka	Pilica
27	Lut27	tributary 2	Luciąża	87	Ck87	Czarna Konecka	Pilica
28	LuP28	Pródka	Luciąża	88	Ck88	Czarna Konecka	Pilica
29	LuS29	Strawa	Luciąża	89	Ck89	Czarna Konecka	Pilica
30	W30	Wolbórka	Pilica	90	Ck90	Czarna Konecka	Pilica
31	W31	Wolbórka	Pilica	91	Ck91	Czarna Konecka	Pilica
32	W32	Wolbórka	Pilica	92	CkK92	Krasna	Czarna Konecka
33	W33	Wolbórka	Pilica	93	CkK93	Krasna	Czarna Konecka
34	W34	Wolbórka	Pilica	94	CkK94	Krasna	Czarna Konecka
35	W35	Wolbórka	Pilica	95	CkK95	Krasna	Czarna Konecka
36	W36	Wolbórka	Pilica	96	CkCt96	Czarna Taraska	Czarna Konecka
37	WMi37	Miazga	Wolbórka	97	CkCt97	Czarna Taraska	Czarna Konecka
38	WMi38	Miazga	Wolbórka	98	CkCt98	Czarna Taraska	Czarna Konecka
39	WMi39	Miazga	Wolbórka	99	CkCt99	Czarna Taraska	Czarna Konecka
40	WMi40	Miazga	Wolbórka	100	CkCtt100	tributary No 4	Czarna Taraska
41	WPB41	Bielina	Piasecznica	101	CkP101	Plebanka	Czarna Konecka
42	WPB42	Bielina	Piasecznica	102	D102	Drzewiczka	Pilica
*43	WP43	Piasecznica	Wolbórka	103	D103	Drzewiczka	Pilica
44	WP44	Piasecznica	Wolbórka	104	D104	Drzewiczka	Pilica
45	WP45	Piasecznica	Wolbórka	105	D105	Drzewiczka	Pilica
46	WP46	Piasecznica	Wolbórka	106	D106	Drzewiczka	Pilica
*47	WMo47	Moszczanka	Wolbórka	107	D107	Drzewiczka	Pilica
48	WMo48	Moszczanka	Wolbórka	108	D108	Drzewiczka	Pilica
49	WMo49	Moszczanka	Wolbórka	109	D109	Drzewiczka	Pilica
50	WMo50	Moszczanka	Wolbórka	110	D110	Drzewiczka	Pilica
51	G51	Gać	Pilica	111	Dt111	tributary No 5	Drzewiczka
52	G52	Gać	Pilica	112	DM112	Młynkowska	Drzewiczka
53	Lb53	Luboczanka	Pilica	113	DMt113	tributary No 6	Młynkowska
54	Lb54	Luboczanka	Pilica	114	DW114	Węglanka	Drzewiczka
55	R55	Rokitnica	Pilica	115	DW115	Węglanka	Drzewiczka
56	R56	Rokitnica	Pilica	116	DW116	Węglanka	Drzewiczka
57	M57	Mogielanka	Pilica	117	DW117	Węglanka	Drzewiczka
58	M58	Mogielanka	Pilica	118	*Dt118	tributary No 7	Drzewiczka
59	M59	Mogielanka	Pilica				
60	M60	Mogielanka	Pilica				

Fish sampling

Fish were caught from a boat or while wading, by two people, each operating an anode dip-net. Full-wave rectified, pulsed 230 V and 3-10 A DC current was taken from a 3 kW generator. Single electrofishing at each site in accordance with Becklemishev's rule was done (Penczak 1967; Backiel and Penczak 1989). Species relative abundance and biomass, calculated on the basis of a catch per unit effort (CPUE), were assessed from a 100 m long reach when wading in shallow streams, and from a 500 m long one when drifting in a boat along a bank. In the main channel we collected 16,504 specimens representing two lamprey and 31 fish species, and in tributaries 4,134 specimens belonging to one lamprey and 29 fish species. In the Appendix fish and lampreys are ordered according to reproductive guilds (Balon 1990).

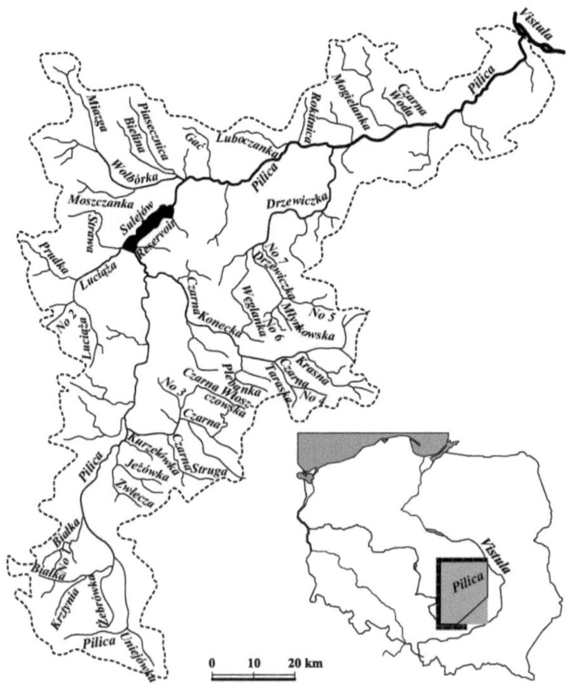

Figure 3.7.1 The Pilica River system. Investigated tributaries are marked with names or numbers.

From the original data set, referring to lamprey and fish biomass, species occurring on less than 3 (out of the total 181) occasions: spirlin *Alburnoides bipunctatus*, rudd *Scardinius erythrophthalmus* and bitterling *Rhodeus sericeus* were excluded as well as 7 fishless sites (Table 3.7.1). The input data matrix was created for the remaining 34 species (columns) and 174 sites, i.e. sample units (rows). The data were log-transformed and then normalized on a (0, 1) scale.

Data analysis

Self-organising map (SOM)

Artificial neural networks (ANN) are realistically models of the functional properties of the human brain. The SOM is their most popular type used for unsupervised learning, and to reduce a high-dimensional space to fewer dimensions, visualize and interpret the data structure with nonlinear relations (Kohonen 1982, 1995, 2001).

The SOM consists of two layers: input and output layers, connected with weight vectors. The input layer receives input values from the data matrix, whereas the output layer consists of output neurons which are usually arranged into a two-dimensional grid for better data visualization. In this study, the output layer of the SOM consists of 24 neurons (virtual units) arranged into a 6 x 4 hexagonal lattice. During the learning process, the SOM weights are modified to minimize the distance between weight and input vectors. The learning process is usually done in two phases: first rough training for ordering with a large neighborhood radius, and then finetuning with a small radius. This results in training the network to classify the input vectors by the weight vectors they are closest to. The detailed algorithm of the SOM can be found in Chon et al. (1996), Giraudel et al. (2000), and Park et al. (2001a, 2002) for ecological applications.

The map obtained after learning represents all the observations assigned to neurons so that similar ones are located close to each other and far from those that are dissimilar. For each neuron, information relating to species composition is available.

Detrended correspondence analysis

Detrended correspondence analysis (DCA) was used in a modified version of DECORANA (Hill 1979). It is free of lax criteria for stability and a bug in the rescaling algorithm. The bug caused sensitivity of ordination results to sample order, mainly on the third axis and higher. These problems have been corrected in the PC-ORD Multivariate Analysis that we used (MacCune and Mefford 1992).

Eigenvalues in DCA cannot be interpreted as proportions of the variance explained (Palmer 2000) but the minimum value recommended for data interpretation is $\lambda_1 = 0.20$ (Matthews 1998), and for axes 1 and 2 they appeared to be remarkably crossed. Our choice for scaling axes was the raw scores, which is arbitrary and dependent on the units of the original data.

Community indices

The additional advantage for SOM would be the situation where the most distant clusters of neurons, differing from each other in fish biomass, would also differ in other community parameters which were not investigated directly with SOM. In order to do this, after the learning process, the SOM lattice was partitioned into clusters by the use of the unified distance matrix algorithm (U-matrix) (Ultsch 1993). The additional parameters studied were: number of species, diversity index and community dominance index.

Biodiversity was assessed with the Shannon index (H'):

$H' = - \Sigma p_i \ln p_i$, where p_i is the proportion of individuals, being members of the ith species, in a community, and S is the number of species present at a site.

The CDI was calculated as: $CDI = 100 * (n_1 + n_2)/N$, where n_1 = number of the most abundant species, n_2 = number of the second most abundant species, and N = total number of all species (Krebs 1994). Thus the CDI determines the percentage of abundance contributed to a community by the two most abundant species.

The significance of differences in parameters studied between neuron clusters was assessed with the Kruskal-Wallis test and the Tukey test adopted for non-parametric post-hoc comparisons.

Results

Community patterns

The number of neurons in the chosen map is seven times smaller than the number of sites. The trained SOM showed three main clusters according to the U-matrix distances: 1) neurons AB, 2) CD, and 3) EF (Fig. 3.7.2). The area of U-matrix corresponding with the middle cluster CD is dark which means that differences between its neurons are big (Fig. 3.7.2). The two most distant clusters AB and EF are homogeneous within themselves and should differ from each other if the classification done by SOM is efective. Indeed, they differ essentially from each other in the character of sites assigned to them (Fig. 3.7.3). Cluster AB is dominated by sites from small streams, while EF – by sites from the middle and lower courses of the main channel and lower courses of the biggest tributaries (Figs 3.7.1, 3.7.3, Table 3.7.1). In cluster AB, neurons A1-2 and B1-2 contain sites from the headwater area of the Pilica River as well as mountain sections and tributaries of the Czarna Konecka (Figs 3.7.1, 3.7.3), whereas A3-4 and B3-4 (right side of SOM) contain lowland stream sites.

Neurons EF contain 47 of the Pilica River sites and 11 located in the lower courses of its biggest tributaries (Drzewiczka, Czarna Włoszczowska, Czarna Konecka and Luciąża) (Figs 3.7.1, 3.7.3).

Cluster CD contains sites located in the middle course of the biggest tributaries and 9 sites of the main channel (Figs 3.7.1, 3.7.3).

Differences in the size of assigned sites are clearly presented in Fig. 3.7.4A prepared without use of the artificial neural network. The product of stream depth and width shows a clear vertical gradient, which results in differences in species composition. For example neurons E1-2 and F1-2 located close to each other on the SOM contain exclusively Pilica River sites (Fig. 3.7.3) and have the same dominant species: the pike *Esox lucius* and roach *Rutilus rutilus* (Table 3.7.2). Their mean biomass is expressed in thousands of grams. These sites occupy four different neurons because of differences in some remaining species: in E1 bream *Abramis brama* is a co-dominant; in F1 – chub *Leuciscus cephalus* and ide *Leuciscus idus*, while in F2 there is also barbel *Barbus barbus*.

Neurons B1-4 are characterised by the occurrence of loach *Barbatula barbatula*, pike, stickleback *Gasterosteus aculeatus* and perch *Perca fluviatilis*, and in B2-4 gudgeon *Gobio gobio* and roach, but in B1-2 the highest mean species biomass is expressed in tens of grams, while in B3-4 in hundreds (Table 3.7.2). Additionally in B4 crucian carp *Carassius carassius* attained high biomass.

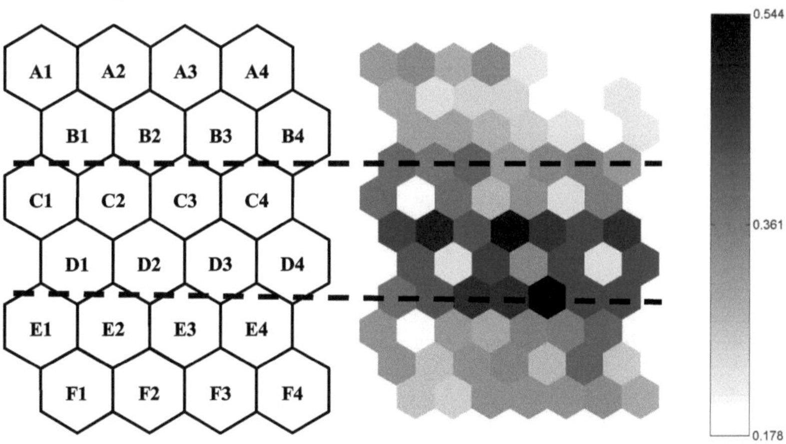

Figure 3.7.2 A self-organizing map formed by 24 hexagons representing neurons. Clusters of neurons were distinguished on the basis of the U-matrix: 1) neurons *AB*, 2) *CD*, and 3) *EF*. The shading intensity indicates the level of activation.

Table 3.7.2 Mean biomass of 34 fish species calculated for samples in each neuron: ◦ <10 g, ○ tens, • hundreds, •• thousands of grams, ••• > 10 kg in CPUE

/	AALBU	AANGU	AASPI	ABRAM	BBARBA	BBARBU	BBJOE	CAURA	CCARA	CCARP	CGOBI	CNASU	CTAEN	ELUCI	EMARI	GACUL	GCERN	GGOBI	LCEPH	LDELI	LIDUS	LLEUC	LLOTA	LPLAN	MFOSS	PFLUV	PPHOX	PPUNG	RRUTI	SAURA	SFARI	SGLAN	SLUCI	TTINC
1																													○					
2																																		
3																																		
4																													◦					
1																																		
2																													○					
3																													◦					
4																													•					
1																													••					
2																													○					
3																													○					
4																													○					
1														•															••					
2														•								•							••					
3																													○					
4																													•					
1				•										•															••					
2														•															••					
4														•												•			••					
1														•					•		•								••					
2						•								•					•		•								••					
3														•							•					•			•••					
4														•					•										••					

After the learning process, the importance of 34 species is shown along the SOM (Fig. 3.7.6). This is particularly useful information illustrating how the biomass of a given species is distributed between neurons, how its importance changes in neighbouring neurons, which species co-occur and thus have the same environmental demands. An example of such co-occurring species are pike, chub, perch and roach because their highest biomass was recorded in neurons EF (Fig. 3.7.6). Again, the differences between neighbouring neurons EF result from the changes in occurrence, increase or decrease in biomass of additional species, e.g. bleak *Alburnus alburnus*, bream, barbel, ide, dace *Leuciscus leuciscus*, burbot *Lota lota* and zander *Stizostedion lucioperca* (Fig. 3.7.6).

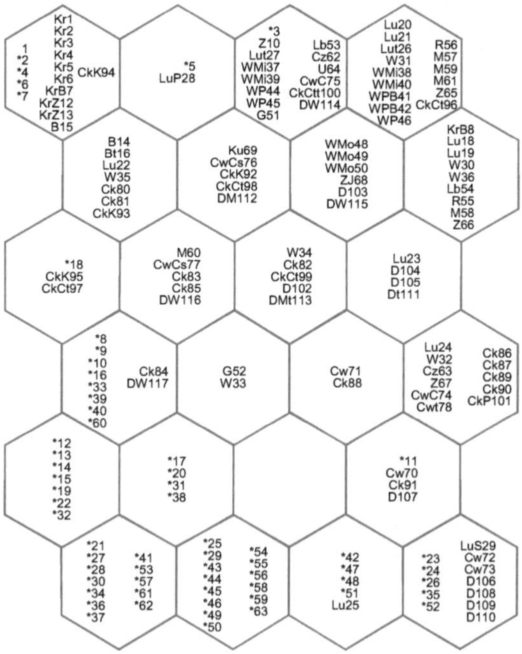

Figure 3.7.3 The 174 Pilica River and its tributaries sites mapped on the self-organizing map (Fig. 3.7.2). The main channel sites are marked with asterisks, and abbreviations for the sites in tributaries are explained in Table 3.7.1.

Community indices

Significant differences between the most distant clusters were also found in selected community parameters, of which some were based on fish number and were not revealed by SOM.

A clear vertical gradient was noted in the total fish biomass. The highest mean values were noted in neurons F2 and F3, the latter of which contains four Pilica River sites located downstream of the dam and one located in the estuary of its large tributary, the Luciąża River (Figs. 3.7.3, 3.7.4B).

The average participation of lithophils (obligatory riverine, reophilic species) in total fish biomass was highest in cluster AB on SOM because brown trout *Salmo trutta* was the main dominant at 16 sites of neuron A1 (Table 3.7.2). Cluster EF contains even more lithophils (Fig. 3.7.6) but their representatives were rare in comparison to other species and thus they contributed little to the total biomass (Fig. 3.7.4C). Instead, the biomass of phytophils in clusters CD and EF significantly exceeded cluster AB (Fig. 3.7.4D). Their biomass increased successively and significantly from the top to bottom of the SOM. The clear gradient in dominance of these two groups of species demonstrates the high quality of the classification done by the SOM as the neural network did not know which reproductive guild a given species belongs to.

The number of species increased with channel size, and the difference between clusters AB and EF was highly significant (Figs. 3.7.4A, 3.7.5A). Similarly, the Shannon biodiversity index displayed significantly lower values in cluster AB (Fig. 3.7.5B), contrary to the Community Dominance Index (CDI), which is based on the abundance of the first two dominants and was significantly higher in cluster AB. The highest mean of CDI was calculated for A3 where stone loach, stickleback and minnow *Phoxinus phoxinus* were exclusive dominants (Fig. 3.7.5C, Table 3.7.2). Again, highly significant differences in the above indices were recorded between sites assigned to the most distant clusters though the neural network did not have information on fish number.

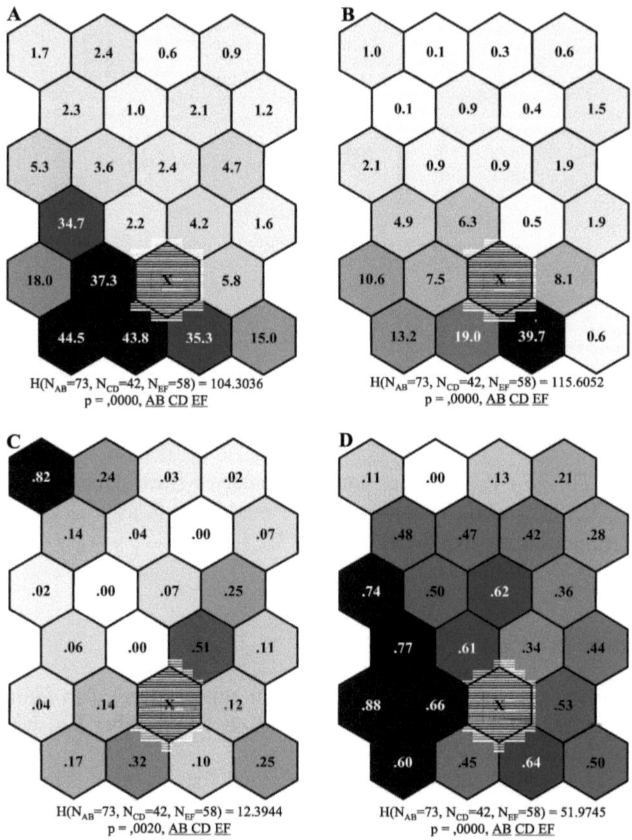

Figure 3.7.4 Selected parameters not analysed by the SOM, compared between neuron clusters in order to assess the effectiveness of classification done by the SOM. The presented values are means calculated for sites assigned to a given neuron. Dark (normalised for a given species) represents high values. Explanation for statistics: between clusters of neurons underlined with the same line no significant difference was recorded. A. Product of depth and width. B. Total fish biomass. C. Participation of lithophils in total fish biomass. D. Participation of phytophils in total fish biomass.

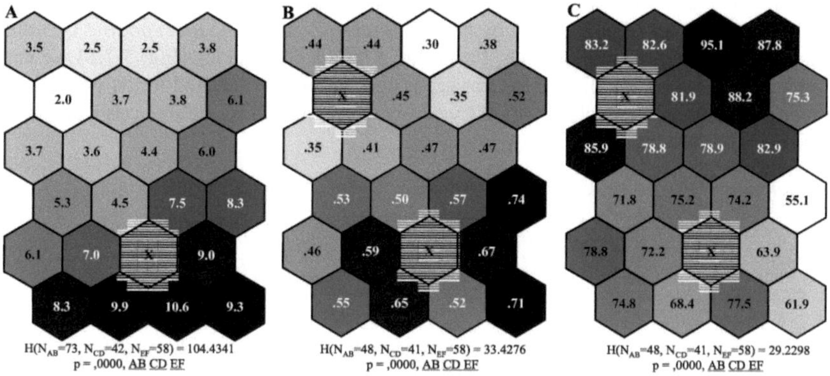

Figure 3.7.5 Selected community parameters not analysed by SOM, compared between neuron clusters in order to assess the effectiveness of classification done by SOM. The values presented are means calculated for sites assigned to a given neuron. Explanation for statistics: between groups of neurons underlined with the same line no significant difference was recorded. A. Number of species in a sample. B. Shannon biodiversity index. C. Community Dominance Index (CDI). In B and C sites with less than 3 species were excluded.

DCA analysis

The scatterplot prepared for 174 sites is interpretable by axes 1 and 2 because $\lambda_1 = 0.55$, and $\lambda_2 = 0.31$, and the importance of groups of site scores are weighed by both axes (Fig. 3.7.7). Eigenvalues calculated for axis 3, $\lambda_2 = 0.19$, is on the edge of the minimum threshold established for the value (Matthews 1998).

53 out of 63 Pilica River sites are ranked higher by the second axis score and form the clearly distinguished group A in multivariate space (Fig. 3.7.7). The ten remaining main channel sites, among them those located in the source part (*1–*7), are spread in multivariate space determined by positive parts of axes 1 and 2, with three exceptions (*2, *3 and *5); site *2, located farther from others in the multivariate space, is represented only by brown trout in the scatterplot. Group A also comprises 19 large tributary sites as well as three from small streams (LuS29, Cz63, DW117).

Site group B, located on the crossing of the main axes, corresponds to neurons C2, C3, C4, D2 and D4, though the latter contains also a few other sites belonging to the DCA groups C and D (Figs. 3.7.3, 3.7.7). Group C, containing 13 sites, has sites situated in neurons B2, B3 and B4, and additionally one in C2 and one in C4. Group D has one Pilica site (*2) and 27 tributary sites that occur exclusively in cluster AB in the SOM: all 15 sites from neuron A4, 9 located in A3, two in B4, and one in B2 (Figs. 3.7.3, 3.7.7). The sites spread in the central and marginal space appointed by positive lengths of axes 1 and 2 are in neurons A1, with the exception of *2. Other single sites, with five exceptions, are spread over cluster AB on SOM (Figs. 3.7.3, 3.7.7).

The species score obtained from DCA analysis indicates which species are ranked with axes. Axis 1, which can be named "a small river" or "cluster AB", is best characterized by a succession of minnow, brown trout, ten-spined stickleback *Pungitius pungitius*, brook lamprey *Lampetra planeri*, stickleback, stone loach, gudgeon, sunbleak *Leucaspius delineatus*, mud loach *Misgurnus fossilis*. Axis 2, which can be named "the Pilica River" is ranked highest in succession of brook lamprey, eel *Anguilla anguilla*, brown trout, gudgeon, roach, pike, bream, burbot, silver bream *Blicca bjoerkna*, ruffe *Gymnocephalus cernuus*, barbel, dace, perch, ide, bleak and chub. In the DCA graph for species (Fig. 3.7.8) we can see that species typical for small streams are separated, and those situated in the positive part of the scatterplot occur in reaches settled by brown trout, and those below axis 1 in small streams with lower velocity.

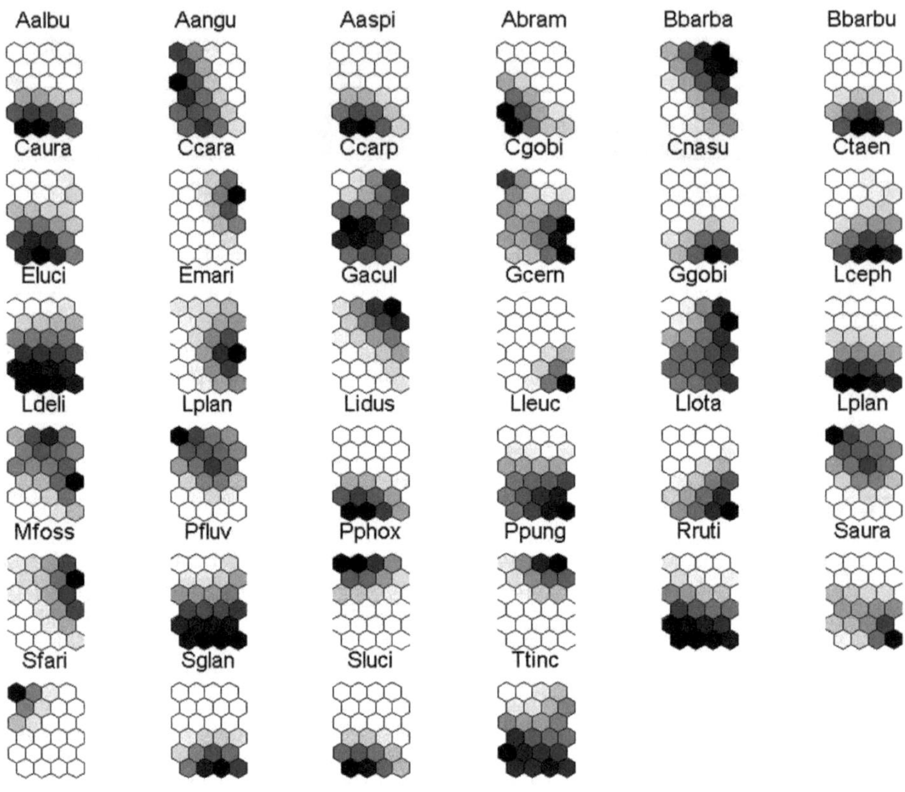

Figure 3.7.6 Importance of 34 fish species in the self-organizing Kohonen map (SOM).

Discussion

After comparing SOM to the DCA scatterplot we can infer that the former method effectively solves difficult high-dimensional and nonlinear problems (Kohonen 1995, 2001), which was underlined with emphasis by scientists applying it to ecological studies (Chon et al. 1996, 2000a; Giraudel et al. 2000; Lek and Guegan 2000; Giraudel and Lek 2001; Brosse et al. 2001; Park et al. 2001a). ANN is a very convincing method because using biomass for expressing abundance of fish populations gave an excellent grouping into neurons of a high number of sites on the topological map, which shows how useful this method is for classification of very large data sets (Kohonen 2001).

On the DCA scatterplot containing 174 samples, most sites lying on one another have to be separated, and if a given group forms a big "data cloud", a reader does not know which symbol belongs to a given mark (Fig. 3.7.7).

Another limitation of the DCA when compared to SOM in the study is that 53 Pilica River sites are together (group A, Fig. 3.7.7), but from the SOM analysis we know that these sites differ in fish community structure as adjoining neurons, containing the main channel sites, have some dominant species in common, but differ in additional species that enrich and differentiate these assemblages. Such moderate changes induced assignation of these samples to different neighbouring neurons though were not revealed by DCA (Fig. 3.7.7), including the direct gradient analysis reported in a previous paper (Penczak et al. 2002).

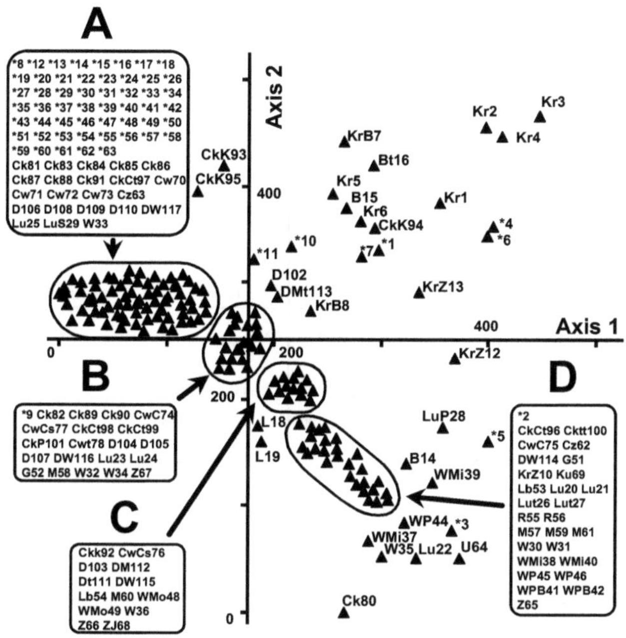

Fig. 3.7.7 Sites ordered on the basis of fish population biomass data in the Pilica River system (DCA). The Pilica River sites are marked with asterisks, sites in tributaries are marked with symbols explained in Table 3.7.1.

In general, there are similarities in results obtained with both methods, however the SOM constitutes a more reliable data representation than other clusters and gradient analyses if different ecological characteristics are applied (Chon et al. 1996; Brosse et al. 2001; Giraudel and Lek 2001; Park et al. 2001a). The method provides "a realistic image of the spatial assemblage of populations without using *a priori* knowledge about their organisation" (Brosse et al. 2001). Such precision in ordering data is very useful for understanding fishery problems important for management, protection and rehabilitation (Hickley and Tompkins 1998).

It is also very promising for the SOM method that the sites were grouped (Fig. 3.7.3) on the basis of one factor only, i.e. the biomass of fish populations, but the diversity indices analysed as well as the number of species, CDI, DW are also significantly different between clusters (Figs. 3.7.4, 3.7.5) though most of these variables were not known by the neural network. The fact that the SOM can clearly order fish samples on the topological map without employing environmental parameters is also important in view of Rose's research (2000), which revealed that qualitative relationships between environment quality and fish populations can be very elusive.

Coming to species distribution in multivariate spaces determined by SOM and DCA methods (Figs. 3.7.6, 3.7.8) we can conclude that DCA was less precise in separating a few species only, such as brown trout, brook lamprey, bullhead *Cottus gobio*, sunbleak and eel. However, the main dominants, typical for the Pilica River, distinguished by high biomass, were located in cluster EF (SOM) and group A of sites in the DCA scatterplot (Figs. 3.7.6, 3.7.7, 3.7.8).

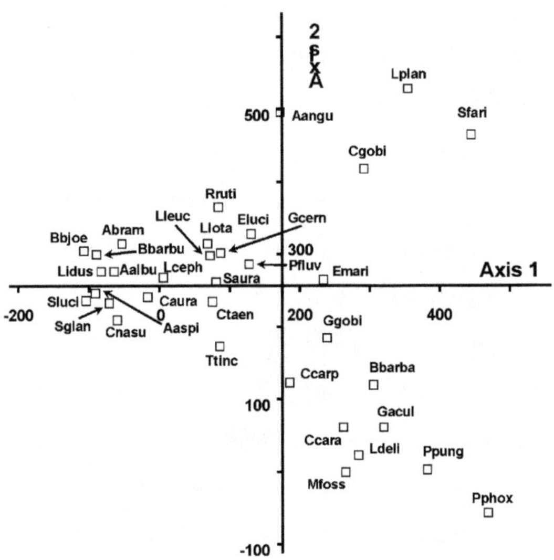

Figure 3.7.8 Species sampled in the Pilica river system on multivariate space of DCA.

Conclusion

The SOM is a meaningful tool for analysing associations among fish communities, excellent in reducing many dimensions of the input data. Very subtle differences in sites and species grouping are hardly possible to obtain with DCA and presumably other gradient analyses, hence authors recommend the self-organizing Kohonen map for studying fish assemblages in large river systems with a huge number of samples.

Very useful information is provided by simultaneous analysis of sites and fish species biomass mapped on the self-organizing map. Distribution of fish species on the SOM facilitates the study of which species are present in samples assigned to a given neuron and thus have similar environmental demands.

Appendix

List of fish species recorded in the Pilica river system; reproductive guilds according to Balon (1990)

Nonguarding and open substratum egg scattering (A.1)

	Scientific name	Common names
Pelagophil (A.1.1.)	*Anguilla anguilla* (L.)	Aangu / eel
Lithopelagophil (A.1.2)	*Lota lota* (L.)	Llota / burbot
Lithophils (A.1.3)	*Leuciscus cephalus* (L.)	Lceph / chub
	Phoxinus phoxinus (L.)	Pphox / minnow
	Alburnoides bipunctatus (Bloch)	Abipu / spirlin
	Aspius aspius (L.)	Aaspi / asp
	Chondrostoma nasus (L.)	Cnasu / nase
	Barbus barbus (L.)	Bbarbu / barbel
Phytolithophils (A.1.4)	*Leuciscus leuciscus* (L.)	Lleuc / dace
	Leuciscus idus (L.)	Lidus / ide
	Blicca bjoerkna (L.)	Bbjoe / silver bream
	Perca fluviatilis L.	Pfluv / perch
	Gymnocephalus cernuus (L.)	Gcern / ruffe
	Rutilus rutilus (L.)	Rruti / roach
	Alburnus alburnus (L.)	Aalbu / bleak
	Abramis brama (L.)	Abram / bream
Phytophils (A.1.5)	*Esox lucius* L.	Eluci / pike
	Scardinius erythrophthalmus (L.)	Seryt / rudd
	Tinca tinca (L.)	Ttinc / tench
	Carassius carassius (L.)	Ccara / crucian carp
	Carassius aureatus gibelio (Bloch)	Caura / giebel
	Cyprinus carpio L.	Ccarp / carp
	Misgurnus fossilis (L.)	Mfoss / mud loach
	Cobitis taenia L.	Ctaen / spined loach
	Sabanejewia aurata (Filippi)	Saura / goldside loach
Psammophils (A.1.6)	*Gobio gobio* (L.)	Ggobi / gudgeon
	Barbatula barbatula (L.)	Bbarba / loach

Nonguarding and brood hiding (A.2)

	Scientific name	Common names
Lithophils (A.2.3)	*Salmo trutta* L.	Sfari / brown trout
	Lampetra planeri (Bloch)	Lplan / brook lamprey
	Eudontomyzon mariae Berg	Emari / Ukrainian lamprey
Ostracophil (A.2.4)	*Rhodeus sericeus* (Bloch)	Rseri / bitterling

Guarding and clutch tending (B.1)

	Scientific name	Common names
Phytophils (B.1.4)	*Leucaspius delineatus* (Heckel)	Ldeli / sunbleak
	Silurus glanis L.	Sglan / wels

Guarding and nesting (B.2)

	Scientific name	Common names
Ariadnophils (B.2.4)	*Gastreosteus aculeatus* L.	Gacul / stickleback
	Pungitius pungitius (L.)	Ppung / ten-spined ststickleback
Phytophil (B.2.5)	*Stizostedion lucioperca* (L.)	Sluci / zander
Spelophil (B.2.7)	*Cottus gobio* L.	Cgobi / bullhead

3.8 Optimisation of artificial neural networks for predicting fish assemblages in rivers*

Scardi M†, Cataudella S, Ciccotti E, Di Dato P, Maio G, Marconato E, Salviati S, Tancioni L, Turin P, Zanetti M

Introduction

Fish assemblages are among the most sensitive and reliable indicators of the ecological status of stream and rivers (Fausch et al. 1990). Fish assemblages are able to integrate over both time and space the biological response to ecological processes more effectively than other biotic components (Harris 1995). Sampling fish fauna, of course, is not as simple as sampling other organisms, but in spite of this problem indices of biotic integrity based on fish have been developed and are now widely accepted (Karr 1981; Karr et al. 1986). Targeting fish fauna in environmental monitoring activities is effective not only from the ecological point of view, but also in the light of the need for straightforward communication with decision-makers as well as with other stakeholders. In fact, fish are probably the most direct and intuitive expression of aquatic ecosystem quality (McCormick et al. 2000).

Therefore, it is not surprising that composition, abundance and age structure of fish fauna are considered as some of the main biological quality elements for the classification of the ecological status of surface water in the EU Water Framework Directive (i.e. Directive 2000/60/EC of the European Parliament and of the Council of 23 October 2000 establishing a framework for Community action in the field of water policy).

This Directive also states that biological reference conditions have to be established for each type of water body. These reference conditions are based on community structure and take into account all the biological quality elements, thus including fish fauna as well as benthic macroinvertebrates and aquatic flora. Hence, modeling fish assemblage composition on the basis of biotic and abiotic environmental descriptors will play a major role in the implementation of the Water Framework Directive and, more generally, in the management of aquatic ecosystems.

Predicting fish fauna as well as other biotic assemblages is not only relevant to the definition of reference conditions that are aimed at the evaluation of environmental quality. In fact, it is also an important achievement in scientific research, e.g. as a framework for studies on species interactions, and it can be very useful for a number of other applied tasks. In particular, species composition models may support environmental management by simulating different environmental scenarios and pointing out the most critical factors that need changes or regulation. Sensitivity analyses of the species composition models play a relevant role in this kind of studies.

* This chapter has been supported by the EU 5th Framework Programme PAEQANN project ["Predicting Aquatic Ecosystem Quality using Artificial Neural Networks: impact of environmental characteristics on the structure of aquatic communities (algae, benthic and fish fauna)", URL: http://aquaeco.ups-tlse.fr/], under contract EVK1-CT1999-00026.

† Correspondence: mscardi@mclink.it

Even though the idea of modeling fish fauna composition on the basis of environmental variables is not new (e.g. Faush et al. 1988), only recently Artificial Neural Networks (ANNs) have been applied to this problem. ANNs have been used to predict fish species richness (e.g. Guegan et al. 1998) as well as density and biomass of single fish populations (Baran et al. 1996; Lek et al. 1996a,b; Mastrorillo et al. 1997b) and ecological characteristics of fish assemblages (Aguilar Ibarra et al. 2003). As far as fish assemblage composition at the river basin scale is considered, only a few models have been developed so far, either using conventional statistical methods (e.g. Oberdorff et al. 2001) or ANNs (Boët and Fhus 2000; Joy and Death, # 3.5; Olden and Jackson 2001). A very useful introduction to the ecological applications of ANNs can be found in Lek and Guégan (1999).

ANNs and other modelling techniques that have been developed and formerly applied in other disciplines have often been introduced into ecological applications with no modification. In most cases this was not a problem and very useful results were obtained anyway. However, in ecological modelling adaptations of the modelling techniques are sometimes required in order to fit particular needs or to properly exploit the available information. This is certainly the case of species composition models, as the data that are involved in this kind of application cannot be regarded as mere numbers, because each species has a different ecological "meaning", which in turn depends on its coenotic context.

This chapter will present a case study about fish assemblages from some river basins in north-eastern Italy, showing how the above-mentioned problem can be tackled by developing ecologically enhanced ANNs.

Data set

The ANN models presented in this study are based on a data set that included sampling sites from several river basins in the Veneto region (north-eastern Italy), as shown in Fig. 3.8.1. The data set consisted of 264 records and it comprised two groups of variables. The first group included the variables to be predicted by the models, i.e. 34 fish species, whereas the second group embraced 20 predictive environmental variables, as shown in Tables 3.8.1 and 3.8.2 respectively.

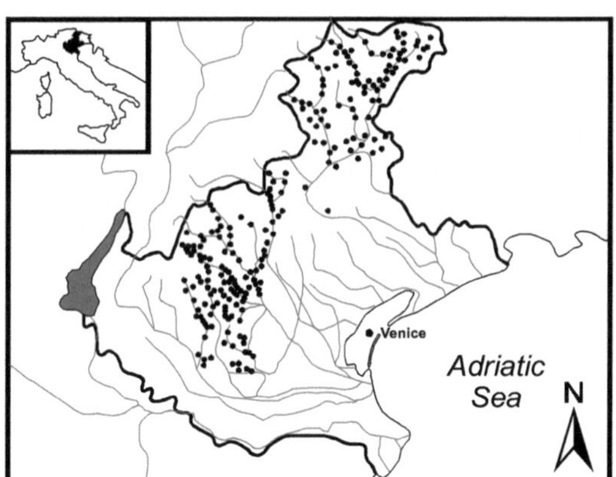

Figure 3.8.1 The sampling sites (black dots) were located in several river basins in the Veneto region (NE Italy).

Fish were collected by means of electrofishing gear. Either direct current or pulsed direct current electrofishing devices were used in streams and small rivers, while these tools were supported by nets when only part of larger rivers was sampled. Basically, in the latter case the electrofishing area was closed by means of nets that also acted as a sampling device.

Table 3.8.1 List of the fish species in the Veneto data set. Modeled species are on white background, while species that were excluded (see text) are on grey background. Italian names are shown in parentheses for those species that do not have an English name.

N	Scientific name	English name
1	*Salmo (trutta) trutta (Linnaeus 1758)*	Sea Trout
2	*Leuciscus cephalus (Linnaeus 1758)*	Chub
3	*Padogobius martensii (Günther 1861)*	(Ghiozzo di fiume)
4	*Scardinius erythrophthalmus (Linnaeus 1758)*	Rudd
5	*Esox lucius (Linnaeus 1758)*	European Pike
6	*Rutilus erythrophthalmus (Zerunian 1982)*	(Triotto)
7	*Alburnus alburnus alborella (De Filippi 1844)*	Bleak
8	*Cottus gobio (Linnaeus 1756)*	Bullhead
9	*Tinca tinca (Linnaeus 1758)*	Tench
10	*Cobitis taenia (Linnaeus 1758)*	Spined loach
11	*Phoxinus phoxinus (Linnaeus 1758)*	Minnow
12	*Anguilla anguilla (Linnaeus 1758)*	European Eel
13	*Knipowitschia punctatissima (Canestrini 1864)*	(Panzarolo)
14	*Salmo (trutta) marmoratus (Cuvier 1817)*	Marble Trout
15	*Sabanejewia larvata (DeFilippi 1859)*	Italian Loach
16	*Ictalurus melas (Rafinesque 1820)*	Black Bullhead
17	*Lepomis gibbosus (Linnaeus 1758)*	Pumpkinseed
18	*Barbus plebejus (Bonaparte 1839)*	Italian Barbel
19	*Chondrostoma genei (Bonaparte 1839)*	South Europe Nase
20	*Gasterosteus aculeatus (Linnaeus 1758)*	Three-spined Stickleback
21	*Carassius auratus (Linnaeus 1758)*	Crucian Carp
22	*Gobio gobio (Linnaeus 1758)*	Gudgeon
23	*Leuciscus souffia (Risso 1826)*	Blageon
24	*Thymallus thymallus (Linnaeus 1758)*	Grayling
25	*Lampetra zanandreai (Vladykov 1955)*	Po Brook Lamprey
26	*Gambusia holbrooki (Girard 1859)*	Eastern mosquitofish
27	*Barbus meridionalis*	Meriditerranean Barbel
28	*Micropterus salmoides (Lacepede 1802)*	Large-Mouthed Bass
29	*Perca fluviatilis (Linnaeus 1758)*	Perch
30	*Abramis brama (Linnaeus 1758)*	Common Bream
31	*Cyprinus carpio (Linnaeus 1758)*	Common Carp
32	*Salvelinus fontinalis M.*	Brook Char
33	*Oncorhynchus mykiss (Walbaum 1792)*	Rainbow Trout
34	*Salmo (trutta) hybr. trutta/marmoratus*	Sea Trout-Marble Trout hybrid

Two fish taxa, namely *Oncorhynchus mykiss*, i.e. the rainbow trout, and *Salmo (trutta)* hybr. *trutta/marmoratus*, i.e. a sea trout - marble trout hybrid (on grey background in Table 3.8.1), were excluded from the models, as their distribution only partly depends on environmental variables. In fact, the distribution of the first taxon is linked to the artificial release of reared juveniles, while that of the second taxon is clearly not independent of the distribution of the two parent species and is probably associated to problems in species identification too.

Some of the available records refer to sampling activities that were carried out at the same site at two different times, thus representing the local interannual variability of both the fish fauna and the environmental variables.

The fish fauna composition was described using binary variables, i.e. presence or absence of each taxon. Quantitative data, although available in most cases, were not considered for model development as they were not sufficiently accurate because of the combined effects of varying efficiency of the electrofishing gear and morphodynamic heterogeneity of the sampling sites. The environmental variables were coded in different ways, either as quantitative or semi-quantitative data, and all the non-binary variables were normalized by rescaling them in the [0,1] interval.

Table 3.8.2 Environmental descriptors used as input (i.e. predictive) variables in the models.

1	elevation (m)
2	mean depth (m)
3	runs (area, %)
4	pools (area, %)
5	riffles (area, %)
6	mean width (m)
7	boulders (area, %)
8	rocks and pebbles (area, %)
9	gravel (area, %)
10	sand (area, %)
11	silt and clay (area, %)
12	stream velocity (score, 0-5)
13	vegetation covering (area, %)
14	shade (%)
15	anthropogenic disturbance (score, 0-4)
16	pH
17	conductivity (μS cm^{-1})
18	gradient (%)
19	catchment area (km^2)
20	distance from source (km)

The whole data set was divided into three subsets for training, validating and testing the ANN models. The training data set included 50% of the records (n=132), whereas the validation and the test data sets included 25% each (n=66). Every record was assigned to a different subset after sorting all the records according to the elevation of the sampling sites. Starting from the highest elevation, the records were divided into the above-mentioned subsets by assigning uneven records to the training subset and by assigning each couple of successive even records to the validation and test subset, respectively. This way, the records in each group of four were assigned to the ($1x$) training, ($2x$) validation, ($3x$) training and ($4x$) test data subset, with x ranging from 1 to 66. This break up strategy allowed a homogeneous allocation of records for different elevations classes among the three subsets, thus stratifying the procedure on the basis of the most relevant environmental variable.

Neural network training

The most common type of ANN, i.e. the multilayer perceptron, was used for modeling the fish fauna composition. The error back-propagation algorithm (Rumelhart et al. 1986a) was used for training the ANNs, both in its original formulation and in a modified version that will be described later in this chapter. Other training algorithms were not tested because the theoretical advantages they might provide (e.g. quicker training) are not really relevant for ecological applications.

ANNs with 20 input nodes, 32 output nodes and 17 nodes in the hidden layer were selected after a set of empirical tests involving ANNs with different numbers of nodes in the hidden layer (from 10 to 40 nodes). The architecture selected was the one that provided the minimum overall error with respect to an independent test set. However, the selection of the number of nodes in the hidden layer was not a critical issue, as the differences among the models were negligible. Sigmoid activation functions [i.e. $f(x)=1/(1-e^{-x})$] were used both in the hidden and in the output nodes of all the ANNs that have been trained and used in this study.

In order to prevent overtraining, i.e. to avoid that the ANN "learned by heart" the fish fauna composition at each known site while losing its generalization ability, different strategies were adopted. The first strategy involved stopping the training procedure early. In other words, the training procedure was terminated as soon as the error, computed on the basis of the validation set only, ceased to decrease monotonically (obviously, the validation set records were never used as training patterns). The second strategy was based on the random selection of a subset of training patterns at each epoch during the training procedure. This way it was not possible for the ANN to be influenced by the order in which the training patterns were submitted (thus possibly memorizing them). Finally, white noise in the [-0.01,0.01] range was added to each input, i.e. predictive variable. Such a small random perturbation of the input values, also known as *jittering*, favored the generalization of an ANN model because the latter learned how to associate each output pattern with a set of input intervals rather than with a single input pattern (Györgyi 1990).

The accuracy of the ANN predictions was expressed by the percentage of Correctly Classified Instances (CCI), while the significance of the deviation of the ANN predictions from a random model was tested by means of the K statistics (Cohen 1960; Fielding and Bell 1997). Details about the computation of CCI percentage and K statistics are provided in the Appendix.

Model selection

A few different basic options are available for developing models of species distribution using ANNs. The first option is to train a different model for each species, another is to train a single model that is able to simultaneously predict the distribution of all the species. A further option is to split the species list into two or more subsets on the basis, e.g. of trophic characteristics, and to train a model for each subset. In the latter case, however, the number of possible models is very high and selecting the best combination is not a straightforward task.

If only the first two options are considered, the selection of the best approach may be based on empirical tests, but there are also some theoretical considerations that should be taken into account.

In fact, when modeling the distribution of a complex set of species, such as a fish assemblage, an ANN model that predicts more than a single species is able to learn not only the distribution of each species, but also some information about interactions among species. Of course, ecologists know that this kind of information is relevant, but in many cases their theoretical knowledge about species interactions is not adequate, as it is often based on hypotheses, personal observations, etc. Therefore, it is not easy to exploit such knowledge in modeling applications using conventional statistical methods (e.g. logistic regression). Since ANNs are able to learn from data, they are also able to learn by themselves what is relevant in species interactions and this may enhance their predictive ability.

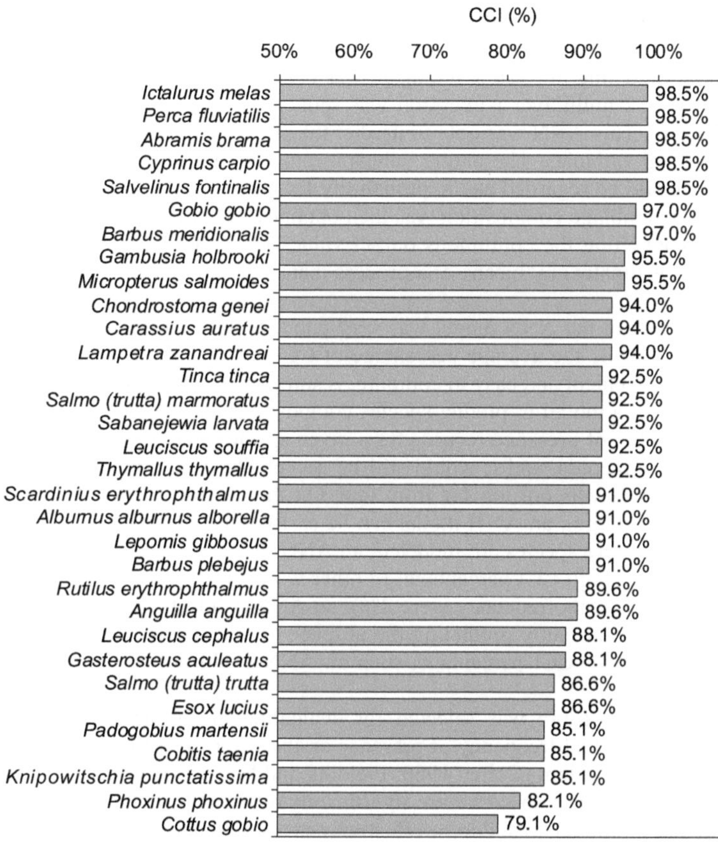

Figure 3.8.2 Percentages of Correctly Classified Instances (CCI) for the 32 modeled species. Species are sorted in descending CCI order.

Given a species assemblage containing *s* species, 2^s different combinations of species presence and absence data exist. In the case of our data set, 2^{32}=4 294 967 296 different patterns are theoretically possible, but only 131 different patterns were actually found in 264 observations. This is clear evidence for the non-independence of different species responses to environmental factors and for the role that biotic interactions play.

Even though simultaneously modeling all the species in a community or in an assemblage is theoretically more efficient, there are practical constraints that may hinder this approach. In fact, the complexity of the ANN structure grows very rapidly with the number of species to be modeled, and the need for training data grows proportionally. Moreover, the set of predictive environmental variables used by the model might be more relevant to some species than to others, and this would impair the model response. In the case of fish assemblages, however, the overall number of species is usually not too large and the species response to environmental variables is rather homogeneous. Therefore, a single model approach was selected in our study.

A conventional training procedure

The first attempt at modeling the fish assemblage was based on a very conventional ANN approach, as a 20-17-32 multilayer perceptron was trained using an ordinary error back-propagation algorithm. This ANN was able to predict the presence of all the species on the basis of environmental variables. The output values it returned ranged in the [0,1] interval and therefore they could be regarded as the probability for each species of being observed. The predicted fish assemblage composition was then obtained by setting a 0.5 threshold for each output, thus converting the continuous output values into binary values (i.e. species presence or absence estimates) by means of a process that is closely related to defuzzyfication.

The overall accuracy of the ANN model was very good, as the CCI ranged from 98.5% to 79.1% (Fig. 3.8.2), while the average percentage of CCI was 91.6%. The percentage of CCI, although very convenient and easy to compute, is sometimes a misleading criterion for evaluating the ability of a model to predict species composition. In fact, it would be really appropriate if the number of presence records for a given species were exactly the same as the number of absence records, and it would still be acceptable if the ratio between presence and absence records was not too far from one. On the contrary, when the ratio becomes too small (or too large), an ANN model can be easily affected by a significant bias. For instance, when very rare species are modeled, an ANN that always returns null outputs can easily provide a very high CCI percentage. In other words, if a species were present in 2 out of 100 records (i.e. if its frequency were 2%), an ANN would be very easily able to provide 98% of CCI by constantly predicting the absence of that species. Needless to say, notwithstanding a very high CCI percentage, such an ANN could not be considered as a true model.

Therefore, another procedure was selected for evaluating the accuracy of the ANN model in the light of the actual frequency of presence or absence record for each species. In particular, the K statistics system (Cohen 1960; Fielding and Bell 1997) was applied to test whether the predictions for each species were significantly different from those of a random model or not. The ANN model was able to effectively predict 20 species out of 32, i.e. in 20 cases the K statistics was significantly different from zero (p=0.95), whereas it failed in the remaining cases (table 3.8.3).

It was evident, however, that the ability of the ANN to predict species presence and absence was strictly related to species frequency. In fact, the maximum frequency among the 12 species with non-significant K statistics was 8.71%, and 10 of them had frequencies lower than 5%. Thus, the model failed to predict several rare species, while it was quite accurate in predicting more frequent species (Fig. 3.8.3).

This result, of course, was not surprising. An ANN learns from examples, and it is obvious that it cannot learn how to correctly predict the presence of a species if the latter is only present in a few records. In these cases no ANN, or any other model, can associate the species response to patterns in the variation of predictive variables. Obviously, exactly the same problem would occur if a model were trying to predict an almost ubiquitous species.

The lack of information about the distribution of rare species is usually related to the way data are collected. In many cases the sampling effort is evenly distributed over the studied region (e.g. a river basin), because the main purpose of the sampling is the characterization of the fish assemblage composition. Therefore, stenotopic species are only found in a limited number of samples and not enough data are available about their relationships with environmental variables. A similar problem would also arise for really ubiquitous species, although in practice it is not common that a species is present in almost all the records in a data set. Moreover, density and population structure data usually provide useful hints about the environmental gradients that play a role in defining the distribution of ubiquitous species. As far as assemblage composition modeling is concerned, however, the practical effects of the lack of information about the relationships between environmental variables and species *absence* are exactly the same as those of the lack of information about the relationships between environmental variables and species *presence*.

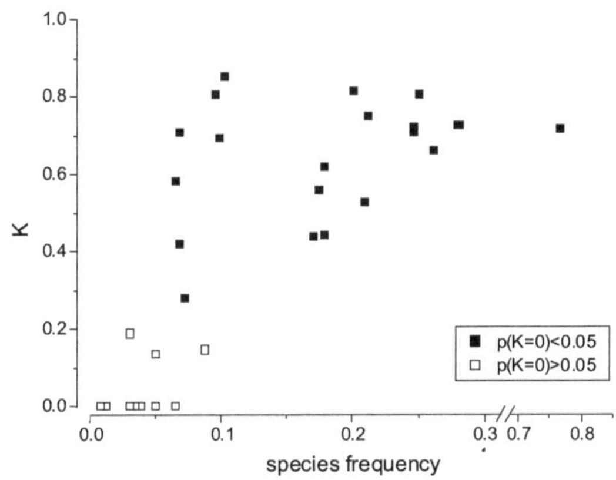

Figure 3.8.3 Conventional ANN model: K statistics vs. species frequency. The model is not reliable as far as rare species are concerned, whereas it works much better with more frequent species.

Problems in error computation

Even though no modeling technique can actually fill the gaps in the available information, it is certainly possible to improve a model by exploiting that information in a more effective way.

A conventional ANN training procedure is driven by the minimization of the Mean Square Error (MSE). As soon as the MSE becomes smaller than a previously defined value, the training procedure is stopped, assuming that the agreement between ANN output values and target (i.e. known) values is good enough. The early stopping procedure that was used in this study involves a similar role of the MSE, although the latter is minimized with respect to a validation data set that is independent of the training data set. In particular, the MSE is computed by comparing the continuous ANN outputs with the binary target values.

This approach makes perfect sense when continuous quantitative variables are involved (e.g. biomass, concentration, etc.), but it is not adequate when species composition is taken into account. There are at least three reasons for this inadequacy and they are probably not as obvious at they should be.

Table 3.8.3 Conventional ANN model: observed and predicted frequency by species (sorted in descending order of observed frequency) and K statistics (significant values are marked with asterisks).

	observed frequency	predicted frequency	K	
Salmo (trutta) trutta	76.5%	83.3%	0.719	*
Leuciscus cephalus	28.0%	31.1%	0.727	*
Padogobius martensii	26.1%	36.4%	0.660	*
Scardinius erythrophthalmus	25.0%	28.0%	0.806	*
Esox lucius	24.6%	31.1%	0.709	*
Rutilus erythrophthalmus	24.6%	26.9%	0.723	*
Alburnus alburnus alborella	21.2%	25.8%	0.748	*
Cottus gobio	20.8%	19.3%	0.528	*
Tinca tinca	20.1%	25.0%	0.816	*
Cobitis taenia	17.8%	15.5%	0.619	*
Phoxinus phoxinus	17.8%	11.4%	0.442	*
Anguilla anguilla	17.4%	12.9%	0.560	*
Knipowitschia punctatissima	17.0%	12.1%	0.440	*
Salmo (trutta) marmoratus	10.2%	9.8%	0.853	*
Sabanejewia larvata	9.8%	11.0%	0.696	*
Ictalurus melas	9.5%	12.5%	0.807	*
Lepomis gibbosus	8.7%	0.8%	0.148	n.s.
Barbus plebejus	7.2%	2.7%	0.280	*
Chondrostoma genei	6.8%	5.7%	0.709	*
Gasterosteus aculeatus	6.8%	6.4%	0.419	*
Carassius auratus	6.4%	0.0%	0.000	n.s.
Gobio gobio	6.4%	7.2%	0.583	*
Leuciscus souffia	4.9%	0.0%	0.000	n.s.
Thymallus thymallus	4.9%	0.4%	0.137	n.s.
Lampetra zanandreai	3.8%	0.0%	0.000	n.s.
Gambusia holbrooki	3.4%	0.0%	0.000	n.s.
Barbus meridionalis	3.0%	0.8%	0.190	n.s.
Micropterus salmoides	3.0%	0.0%	0.000	n.s.
Perca fluviatilis	1.1%	0.0%	0.000	n.s.
Abramis brama	0.8%	0.0%	0.000	n.s.
Cyprinus carpio	0.8%	0.0%	0.000	n.s.
Salvelinus fontinalis	0.8%	0.0%	0.000	n.s.

Firstly, when a threshold function is applied for discretizing the ANN outputs, the real contribution of each single error to the MSE strongly depends on the output value. For instance, if the target value for a given species is 0 (i.e. absence), a 0.495 output value would contribute $(0.495\text{-}0)^2$=0.245025 to the overall MSE, although it would result in a perfect agreement when the output value is transformed into a binary value by passing it to the threshold function (0.495<0.5 would be transformed into 0, i.,e. absence). A very similar output value, like, for instance, 0.505, would provide an almost identical contribution to the overall MSE $(0.505\text{-}0)^2$=0.255025, but it would be in disagreement with the target value after applying the threshold function (0.505>0.5 would be transformed into 1, i.,e. presence).

Secondly, the potential contribution of each modeled species to the MSE is identical and it varies between 0 and 1. Although this makes perfect sense from a computational point of view, it fails to capture the real effect of different errors in different contexts, because it does not weight each error according to its impact on the characterization of the species assemblage structure. In fact, a wrong prediction about a single species might have a limited effect on the overall composition of the predicted assemblage if the latter included many other species, while it might completely change the assemblage structure if the latter included only a few species. In other words, each species has an ecological "meaning" that depends not only on its ecological characteristics, but also on the way the species combines with other species, i.e. on the assemblage structure.

Finally, the efficiency of sampling is usually not homogenous, even within a single study. For instance, it is much more likely that a species, although present at a given site, escapes from sampling devices in a large river than in a small stream. Therefore, the contributions of different species to the error computation should not be simply added to each other, as in the case of MSE.

In conclusion, species presence and absence data are not to be used as mere numbers (i.e. as 0s and 1s) in the error computations that are needed for optimizing species composition models. As a consequence, the MSE is not an appropriate measure of the error in such models.

An enhanced training procedure

Several options exist for implementing an ecologically sound procedure for error computation, although not all the problems that were mentioned in the previous section can be solved. Since it is clear that the role of each species depends on other species, i.e. on species assemblage structure, a binary similarity coefficient may provide a simple yet effective way to measure the difference between the model outputs (predicted assemblage) and the target values (observed assemblage). This solution leads to a different problem, i.e. the selection of the most appropriate similarity coefficient. However, this is a common problem in ecological multivariate data analysis and most ecologists are acquainted with it and are certainly able to select a suitable coefficient. In our case study, we were able to assume that the fish assemblage composition was recorded very accurately at every sampling site. This implied that species absence in samples might be regarded as reliable information. Therefore, a symmetrical similarity coefficient that slightly emphasized differences in species composition was selected as a measure for model errors. In particular, the Rogers and Tanimoto (1960) similarity coefficient (S_{jk}) was chosen and transformed into a dissimilarity coefficient (D_{jk}), which was monotonically related to the error in the species composition prediction:

$$S_{jk} = \frac{a+d}{a+2b+2c+d} \qquad D_{jk} = 1 - S_{jk}$$

In the above formula a and d are the number of species whose presence (a) or absence (d) are correctly predicted, whereas b and c are the number of species present that are not predicted by the model and vice-versa.

The conventional ANN training procedure was then modified in order to use the mean dissimilarity between model outputs and validation patterns (i.e. samples) as the criterion for controlling ANN learning. In particular, the training procedure was halted as soon as the mean dissimilarity began to increase. This allowed an optimal generalization of ANN learning, which only takes place during the first part of the training procedure, i.e. while the error (the dissimilarity, in this case) decreases monotonically (Fig. 3.8.4).

The results of this enhanced training procedure were almost identical to those of the conventional procedure in terms of CCI percentages, but they showed a substantial improvement when other criteria were taken into account. In fact, while the average value for the CCI was 91.8%, i.e. only 0.2% higher than the one obtained by conventional training, the differences between predicted and observed species frequencies, as computed on the basis of the whole test set, were substantially smaller than in the case of conventional training (2.2% and 3.5% in absolute values, respectively).

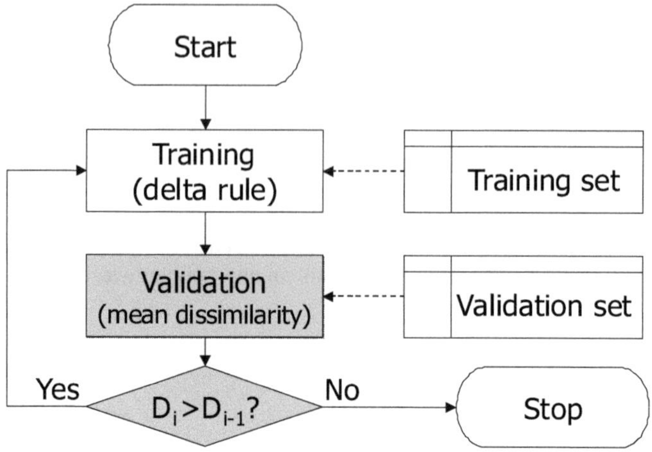

Figure 3.8.4 The training procedure for the enhanced ANN model. The modified steps are shown on grey background.

However, the most important advantage of the modified training procedure over the conventional one was in its ability to obtain better predictions for those species whose frequency was smaller than 10% (Table 3.8.4, but see also Table 3.8.3).

Moreover, the only species whose presence was never predicted by the model were the two rarest species, namely *Cyprinus carpio* and *Salvelinus fintinalis*, while the conventionally trained model was not able to predict the presence of 9 species out of 32.

Finally, the K statistics results were on average much higher than in the conventionally trained model (0.59 and 0.42, respectively), and only 5 out of the 7 less frequent species were associated to K values that were not significantly different from zero. This implied that the enhanced model was unable to predict only 5 species, while the conventionally trained model failed with 12 species.

In order to summarize the differences between the conventional (MSE-based) ANN model and the enhanced (dissimilarity-based) one, it is useful to compare the K statistics species by species, as shown in Fig. 3.8.5. The small boxes show the K values for the conventional model (solid boxes) and for the enhanced one (white boxes), while the whisker on the left of each box indicates the lower end of the confidence interval of the K statistics (the upper one is not relevant in this case, so it was omitted). Obviously, the K statistics is not significantly different from zero (at a probability level p=0.95) if the left whisker intersects the vertical axis at K=0. The boxes on the vertical axis with no whisker on the left show those cases in which the K statistics was not computed because the model always predicted the absence of the corresponding species. The species have been sorted according to their frequency, shown in parentheses on the right of each species name.

It is very easy to notice that there were no cases in which the conventional training provided higher K values than the enhanced model, but the most striking difference between the two models can be observed for the less frequent species. In fact, the enhanced model led to dramatic improvements in the predictive ability and in several cases the K statistics for the enhanced model was significant, while it was not significant or not even computable for the conventional model.

In the case of the enhanced model only five species were associated with values of the K statistics that were not significant, while twelve species were in that situation when the conventional model was used. It is interesting to noe that the largest changes in K values were observed for species whose frequency ranged from 3% to 9%. These species, that cannot be considered as truly rare species, are certainly associated with particular physical, chemical and biotical conditions and play a relevant role in defining the ecological characteristics of the fish assemblage.

Table 3.8.4 Enhanced ANN model: observed and predicted frequency by species (sorted in descending order of observed frequency) and K statistics (significant values are marked with an asterisk).

	observed frequency	**predicted frequency**	**K**	
Salmo (trutta) trutta	76.5%	74.6%	0.726	*
Leuciscus cephalus	28.0%	24.6%	0.805	*
Padogobius martensii	26.1%	22.0%	0.767	*
Scardinius erythrophthalmus	25.0%	23.5%	0.836	*
Esox lucius	24.6%	21.2%	0.754	*
Rutilus erythrophthalmus	24.6%	21.6%	0.765	*
Alburnus alburnus alborella	21.2%	19.7%	0.790	*
Cottus gobio	20.8%	12.5%	0.640	*
Tinca tinca	20.1%	17.4%	0.824	*
Cobitis taenia	17.8%	15.2%	0.675	*
Phoxinus phoxinus	17.8%	14.0%	0.615	*
Anguilla anguilla	17.4%	13.3%	0.721	*
Knipowitschia punctatissima	17.0%	13.6%	0.665	*
Salmo (trutta) marmoratus	10.2%	9.1%	0.876	*
Sabanejewia larvata	9.8%	8.3%	0.794	*
Ictalurus melas	9.5%	8.3%	0.829	*
Lepomis gibbosus	8.7%	2.3%	0.375	*
Barbus plebejus	7.2%	4.5%	0.603	*
Chondrostoma genei	6.8%	4.5%	0.709	*
Gasterosteus aculeatus	6.8%	3.8%	0.601	*
Carassius auratus	6.4%	1.9%	0.415	*
Gobio gobio	6.4%	4.5%	0.603	*
Leuciscus souffia	4.9%	2.3%	0.476	*
Thymallus thymallus	4.9%	1.5%	0.458	*
Lampetra zanandreai	3.8%	1.5%	0.485	*
Gambusia holbrooki	3.4%	0.4%	0.195	n.s.
Barbus meridionalis	3.0%	1.5%	0.560	*
Micropterus salmoides	3.0%	1.1%	0.490	*
Perca fluviatilis	1.1%	0.4%	0.497	n.s.
Abramis brama	0.8%	0.4%	0.394	n.s.
Cyprinus carpio	0.8%	0.0%	0.000	n.s.
Salvelinus fontinalis	0.8%	0.0%	0.000	n.s.

Conclusions

Predicting the species composition of fish assemblages on the basis of environmental descriptors is a feasible task that can be carried out either by means of conventional probabilistic models (e.g. Oberdorff et al. 2001) or by means of ANNs (e.g. Aguilar Ibarra et al. 2003; Joy and Death # 3.5; Olden and Jackson 2001). ANNs have been successfully used in these applications, as they allow exploitation of heterogeneous sources of information in a very effective way (Scardi and Harding 1999). Moreover, ANNs may be easily enhanced and adapted to specific modeling tasks (Scardi 2001), as they are entirely empirical tools.

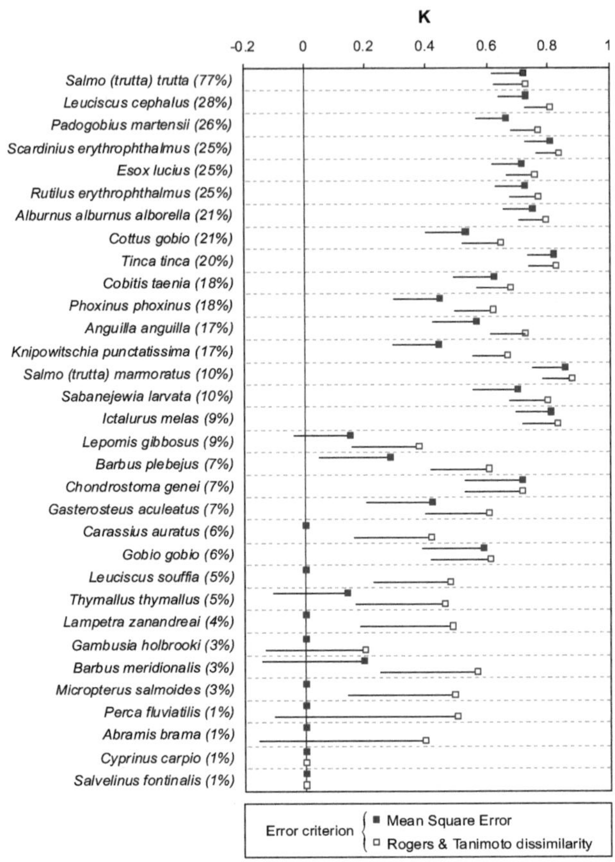

Figure 3.8.5 A comparison of K statistics values for the conventional model, using Mean Square Error as the error criterion (black squares), and the enhanced model, using Rogers and Tanimoto (1960) dissimilarity instead (white squares). The line on the left of each square shows the lower limit of the confidence interval of the K statistics. Therefore, when the line (or the symbol) intersects the vertical axis at K=0 the K statistics is not significantly different from zero (p=0.95).

Even though ANNs are the most effective tools for modeling species composition (Olden and Jackson 2002), they cannot solve problems that arise from a lack of relevant information. In fact, in many cases the only predictive variables that are readily available for the modeler are those that can be obtained from cartographic records or direct observation. Other sources of information that involve sampling and laboratory analyses are usually less abundant and therefore play a secondary role. Moreover, species distribution data are also scarce, and distributed in space according to the local resources for monitoring activities rather than on the basis of a suitable and consistent sampling design. Therefore, predicting the species assemblage composition is not feasible without compromises. For instance, accurate ANN models can be trained on a regional scale, or focus on species assemblages simpler than communities. Our application, dealing with fish assemblages in northeastern Italian streams and rivers, belongs to this category and is certainly an example of successful modeling that can be used in practical applications. For instance, our model can be considered as a generator of expected fish assemblages, i.e. of biotic reference conditions in the light of the EU Water Framework Directive.

In particular, our model predicts the assemblage structure on the basis of environmental descriptors that are mainly (but not exclusively) focused on the geo-morphological characteristics and is based on data from real assemblages, as observed in a number of real sites. Therefore, the predicted assemblage is not just the one that is considered present at a theoretical pristine site, but a compromise that represents the more likely biotic response given a number of existing constraints, mainly related to the long term anthropogenic impacts on pristine ecosystems (e.g. changes in land usage, introduction of exotic species, modification of river banks, etc.). In regions where pristine conditions have not existed for several centuries, this is probably the only meaningful way to define reference conditions.

The ANN models presented here are not only an achievement in applied ecological research, as they also point out more general problems in species distribution modeling and provide solutions for them.

The most general scientific issue that emerged from our work is that very rare and very frequent species cannot be effectively modeled unless enough information is available. This obviously does not happen in many real studies, in which the only acceptable solution should be based on several species-specific sampling designs, i.e. on multiple sampling designs tailored to fit the distribution of each studied species.

Another relevant scientific issue that was highlighted by our work was the need for adequate error measurements in ecological applications. In fact, conventional criteria like MSE may fail when applied to data that are not strictly quantitative, like species presence and absence data. These data are binary from a formal point of view, but they cannot be treated just as sequences of 1s and 0s. Each species contributes to the assemblage structure in a way that depends simultaneously on its ecological characteristics and on the composition of the assemblage. Therefore, some errors in predicting species composition might be more relevant than others. For instance, in many upstream sites the only fish species is *Salmo trutta trutta*, which is also very frequent as a member of much more complex assemblages in other sites downstream. It is obvious that not predicting its presence in an upstream site would be a much more severe error than not predicting its presence elsewhere.

Using a binary dissimilarity coefficient instead of MSE as the criterion for measuring prediction errors provided a significant enhancement of a conventional ANN model. Even though the functioning of the error back-propagation algorithm was not changed, the modified training procedure relied on the minimization of the mean dissimilarity as a criterion for stopping the learning phase, thus allowing optimal generalization of the model. In other words, the enhanced training procedure did not change the way the ANN model learned, but it changed the conditions for stopping its optimization.

In our application the Rogers and Tanimoto (1960) dissimilarity was used, because we were confident about the reliability of our absence data and because we wanted to stress differences rather than resemblances between assemblages. In different situations, however, other coefficients would prove more adequate. For instance, if absence data are not completely reliable (e.g. because of avoidance of the catching net) an asymmetric dissimilarity that only takes into account presence data, like that based on Jaccard's coefficient (Jaccard 1908), could be more appropriate.

The enhanced training procedure not only improved the overall accuracy of the species composition predictions, but it also significantly increased the ability of the model to correctly predict the occurrence of rare species, thus mitigating the effects of the unbalanced availability of information about rare species that was previously mentioned.

In order to obtain further improvements of species composition models, however, changes in the modeling strategies should be coupled with the optimization of the sampling strategies. In fact, modeling rare or ubiquitous species is only feasible if adequate information is available, such as the ratio between the number of absence and presence records in training and validation data set which should be as close to one as possible, while the variability of the environmental descriptors within each subset, i.e. within the presence or absence subsets, should be maximum. Therefore, *ad hoc* sampling designs that significantly deviate from the usual monitoring approaches are needed. This shortcoming is not specific to ANNs, as it obviously affects any modelling technique.

The enhanced ANN model presented in this chapter was incorporated into the software tool that was published as one of the deliverables of the PAEQANN project and that can be found in the CD attached to this book. Therefore, the readers will be able to experiment the model on their own, to check its results and compare the predictions it provides with those of other models.

Appendix

Both the percentage of Correctly Classified Instances (CCI) and the *K* statistics (Cohen 1960; Fielding and Bell 1997) are based on the confusion matrix, i.e. on a 2 x 2 contingency table in which the predicted presence and absence of a taxon are compared with their observed counterpart. In particular, if each case is expressed as a proportion p_{ij}, then the confusion matrix will be:

		Predicted	
		1	0
Observed	1	p_{11}	p_{12}
	0	p_{21}	p_{22}

and the sum of its elements will be 1. The CCI percentage will then be computed as:

$$CCI\% = 100 \cdot \sum_{i=1}^{2} p_{ii}$$

The *K* statistic can be easily computed from the same confusion matrix. The observed (P_o) and expected (P_e) proportion of agreement between observed and predicted data are the basis for the *K* statistics computation:

$$K = \frac{P_o - P_e}{1 - P_e}$$

In particular, P_o is closely related to CCI%, whereas P_e depends on the number of cases in all the elements of the confusion matrix:

$$P_o = \sum_{i=1}^{2} p_{ii} \qquad P_e = \sum_{i=1}^{2}\left(\sum_{j=1}^{2} p_{ij} \cdot \sum_{j=1}^{2} p_{ji} \right)$$

In order to test the significance of the deviation from zero of the K statistics, the standard error s_{K0} has to be computed, because the ratio between K and s_{K0} is distributed as the standardized normal variate Z. The standard error s_{K0} can be obtained as:

$$s_{K0} = \frac{\sqrt{P_e + P_e^2 - C}}{(1 - P_e) \cdot \sqrt{n}} \qquad Z = \frac{K}{s_{K0}}$$

where n is the number of cases considered in the confusion matrix and C can be obtained as

$$C = \sum_{i=1}^{2} \left[\sum_{j=1}^{2} p_{ij} \cdot \sum_{j=1}^{2} p_{ji} \cdot \left(\sum_{j=1}^{2} p_{ij} + \sum_{j=1}^{2} p_{ji} \right) \right]$$

It is very important, however, to remember that the standard error s_{K0} is not exactly the same as that needed, for instance, to compute the two-sided confidence interval for K.

4 Macroinvertebrate community assemblages

Editor: Verdonschot PFM*

4.1 Introduction

Ecological communities are the expression of complex biological processes (reproduction, nutrition, behaviour, interspecific relationships, etc.) and abiotic processes (such as nutrient cycling, discharge regimes, erosion-deposition), both being expressed on various scales of time and space. Analysing all these processes, i.e. including and understanding the relationship which exists in the community, and characterising their relationships with environmental parameters, their degree of importance, and their structuring, requires the observation of variables related to the functioning of the ecosystem. The complexity of ecological systems often implies complex relations between biological and environmental variables. This justifies the use of sophisticated modelling techniques. Such models are often based on different statistical and simulation techniques, designed to predict community structure from environmental variables.

Macroinvertebrates are well suited indicators for the state of ecosystems and the processes ocurring in them, because (i) there is a large amount of data available, (ii) the identification of the composing taxa is relatively easy, and (iii) macroinvertebrates occur in high numbers in all types of surface waters (among others, Rosenberg and Resh 1993, Davis and Simon 1995). The use of macroinvertebrates in modelling has developed strongly over the last 20 years (Wright et al. 1984, Reynoldson et al. 1995, Nestler et al. 1989). With the introduction of multivariate analysis techniques this use even accelerated (among others, Wright et al. 1989). Prediction techniques were a recent and logical follow-up, using environmental variables as input and either macroinvertebrate communities or species as output (Verdonschot and Goedhart 2000, Nestler et al. 1989).

The use of macroinvertebrate - environment relationships during the last centuries can be summarised into four major steps:

- multivariate/pattern/association analysis: macroinvertebrate groups or taxa are grouped based on, for example, pattern analysis, clustering or similarity calculation;
- (multiple/logistic) regression: the macroinvertebrate groups or taxa are related to the environmental data by using either regression or ordination techniques;
- reverse regression: the environmental data are related to the macroinvertebrate groups or taxa whereby again regression techniques are used to predict the occurrence of the groups or taxa based on the environmental data; and
- validation: the model developed is validated by an independent dataset.

The strength of future models will take the first three steps together in a single approach such that several disadvantages of clustering, regression and ordination can be overcome. The present chapter introduces the use of different recently developed techniques to model and predict macroinvertebrates in terms of richness, assemblages or functional groups.

---

* Corresponding: piet.verdonschot@wur.nl

Dedecker and co-authors (section 4.2) use backpropagation algorithms to induce predictive models on a macroinvertebrate dataset of the Zwalm river basin in Flanders. They show that these models are in general quite robust with a rather high predictive reliability.

Di Dato and co-authors also use a neural network approach to predict the benthic macroinvertebrate fauna composition in rivers (section 4. 3). The authors show that the best results in reproducing the similarities among sites were obtained by selecting the taxonomic units that are routinely used in the computation of biotic index variables to be predicted.

Gevrey and co-authors compare (section 4.4) the ability of a multiple linear regression, a general additive model, a partial least square model, a regression tree, and an artificial neural network to predict Dutch macroinvertebrate species richness and functional feeding groups using environmental variables. The variability in predictive ability of five methods tested here indicates that there is no single best predictive method. Furthermore, macroinvertebrate species richness is always more accurately predicted than functional feeding groups.

In section 4.5 Nijboer and co-authors compare benthic macroinvertebrate data from streams and channels with three techniques: a Self-Organizing Map (SOM), non-hierarchical clustering and canonical correspondence analysis. Non-hierarchical classification shows less overlap in taxon composition compared to the SOM classification. By plotting the environmental variables on the SOM, similar gradients in the environment are obtained like those resulting from the canonical correspondence analysis. The two techniques are complementary.

Park and co-authors (section 4.6) apply a counterpropagation neural network for patterning benthic macroinvertebrate data. The prediction performance of the counterpropagation neural network is compared with that of multilayer perceptron with a backpropagation algorithm. Both methods show high predictability of species richness and diversity index, although the backpropagation algorithm shows relatively higher values than the counterpropagation neural network.

In section 4.7 Park and co-authors use a combination of two unsupervised artificial neural networks (the SOM and the adaptive resonance theory) to construct a hierarchical patterning of benthic macroinvertebrate communities. The resulting hierarchical grouping in macroinvertebrate communities reflects the environmental impacts and appears to be useful for recognizing pattern changes in community development caused by environmental disturbances.

Compin and co-authors (section 4.8) describe the relationships between biological and environmental variables using SOM, and then use a backpropagation algorithm as a nonlinear predictor, to predict richness of ephemeropterans, plecopterans, trichopterans and coleopterans on the basis of a set of four environmental variables. They show that this prediction can be a valuable tool to assess disturbances.

Kwak and co-authors (section 4.9) use artificial neural networks to pattern community changes in benthic macroinvertebrates in an urbanized polluted stream. The authors show that the model developed could be used as an alternative tool for identifying community changes.

Finally, in section 4.10 Horrigan and co-authors use three techniques for the classification and prediction of stream macroinvertebrate assemblages in Victoria, Australia. The SOM are applied to localize ecological regions in the stream system identified by distinctive clusters of macroinvertebrate communities. The multilayer perceptron neural networks are applied to predict the occurrence of clusters of macroinvertebrates and genetic algorithms are applied to describe the ecological regions by predictive rule sets. The resulting rule sets reveal specific environmental conditions determining the spatial occurrence of macroinvertebrate assemblages in landscapes.

4.2 Sensitivity and robustness of a stream model based on artificial neural networks for the simulation of different management scenarios*

Dedecker AP†, Goethals PLM, de Pauw N

Introduction

In recent years, ANNs have been increasingly used for predicting data in ecological and aquatic sciences (Lek and Guégan 2000). Examples can be found on water quality modelling (e.g., Schleiter et al. 1999) and on relating community characteristics with environmental variables (e.g., Lek et al. 1996b, Recknagel et al. 1997, Maier et al. 1998, Wagner et al. 2000 a,b, Schleiter et al. 2001). ANNs are thus suited for the modelling of ecosystems which are known to be very complex and often non-linear (Lek et al. 1996b, Rumelhart et al. 1986b). In this study back-propagation ANN algorithms were used to induce predictive models on a dataset collected in the Zwalm river basin in Flanders, Belgium. Often, model validation is based on a dataset that consists of measurements similar to the model training dataset. In this way, the model performance can only be optimized and assessed for predictions of river conditions that are in the same range as those in the collected dataset. Data-driven models are thus less useful for predictions in river restoration management. Most relevant predictions can be classified as 'extreme' simulations for the induced ANN models and are therefore less useful for most decision support purposes.

The aim of this paper was to test the sensitivity and robustness of the ANN models for these 'extreme' values. Since these 'extreme' values are not present in the collected dataset, the use of ecological expert knowledge is recommended. To introduce this expert knowledge in the ANN models, a virtual dataset, containing 'extreme' values, was created. The ANNs were also assessed for their applicability in simulations of different river restoration scenarios. In this way, the sensitivity and robustness of the models were assessed from a theoretical and practical point of view.

Materials and methods

Study area

In general, Flanders, Belgium, is a rather flat region. However, the Zwalm river basin is characterized by a number of differences in altitude, making it quite a unique ecosystem within the Flemish region (Soresma 2000). The Zwalm river basin is part of the hydro-

* The first author is a recipient of a grant of the Institute for the Promotion of Innovation by Science and Technology in Flanders (IWT)

† Corresponding: andy.dedecker@UGent.be

graphical basin of the Upper-Scheldt (Carchon and de Pauw 1997). It drains an area of about 11,650 ha and its total length is 22 km (Fig. 4.2.1). Since 1999, the water quality in the Zwalm river basin has considerably improved due to investments in sewerage and wastewater treatment plants during the preceding years (VMM 2000). Nevertheless, some parts of the river are still polluted by untreated urban wastewater and by diffuse pollution originating from agricultural activities (Goethals and de Pauw 2001). Besides, still numerous structural and morphological disturbances exist (e.g., weirs for water quantity control, artificial embankments, etc.) (Carchon and de Pauw 1997).

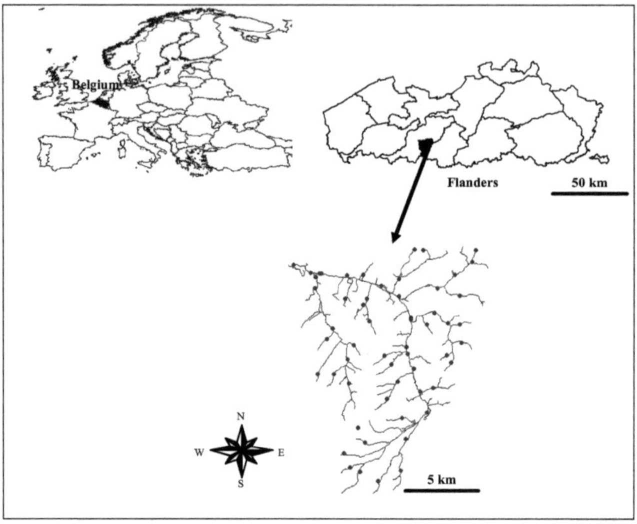

Figure 4.2.1 Location of the Zwalm river basin in Flanders, Belgium.

Data collection

To build and to assess the ANN models, data were used from 60 sampling sites fairly evenly distributed across the Zwalm river basin. Each site was examined twice over a two-year period (2000-2001). In this way, 120 sets of observations were available. At each site 15 environmental variables were recorded (Table 4.2.1). Besides physical-chemical measurements, observations were also made about the structural characteristics. Certain structural characteristics (meandering, hollow river beds, deep/shallow variation and artificial embankment structures) were monitored visually (Dedecker et al. 2002). Flow velocity was determined by timing the transport of a float over a distance of 10 m. Field measurements were made for temperature and dissolved oxygen (OXI 330/SET), pH (Jenway 071) and conductivity (WTW LF 90). Suspended solids were measured spectrophotometrically in the laboratory (Dedecker et al. 2002). Macroinvertebrates were collected by means of a standard handnet during five minute kick sampling within a river stretch of 10 m (NBN 1984) and by *in situ* exposure of artificial substrates (de Pauw et al. 1994). The objective is to collect a representative sample of the macroinvertebrates at the examined site (de Pauw and Vanhooren 1983).

Modelling technique

Modelling was carried out after rescaling the data. This rescaling was applied because the input variables are of a different order of magnitude. The variables are rescaled to be included within the interval [-1, 1] by using the following equation:

$$V_n = 2 \times \frac{(V_0 - V_{\min})}{(V_{\max} - V_{\min})} - 1 \qquad (4.2.1)$$

in which V_0 and V_n are respectively the original and rescaled value of the variable for a sampling point, V_{min} and V_{max} are the minimum and maximum values of that variable in the original dataset. Also the targets are rescaled over the interval [-1, 1] to adapt to the transfer function used (tangential sigmoid) in the output layer. In this way, the network will be trained to produce outputs in the range [-1, 1]. Afterwards, these outputs were converted back into the same units which were used for the original targets.

Table 4.2.1 Abiotic input variables, units and range of the variables used in the ANN model.

Variables	Units	Range
Temperature	°C	10.6 – 15.5
pH		6.96 – 8.10
Conductivity	µS/cm	393 – 1078
Suspended solids	mg/l	0 – 949
Dissolved oxygen	mg/l	3.25 – 10.8
Depth	Cm	3 – 170
Fraction of pebbles	% of river bed	0 – 100
Shade	%	0 – 100
Water plants	Absent / Present	0 – 1
Width	Cm	25 – 950
Flow velocity	m/s	0.03 – 1.92
Meandering	6 classes (1 = well developed to 6 = absent)	1 – 6
Hollow river beds	6 classes (1 = well developed to 6 = absent)	1 – 6
Deep/shallow variation	6 classes (1 = well developed to 6 = absent)	1 – 6
Artificial embankment	3 classes (0 = absent; 1 = moderate; 2 = intensive)	1 – 6

For ANN modelling, a multilayer feed-forward neural network was used. The processing elements in the network, the neurons, are arranged in a layered structure. The network typically comprises three types of neuron layers: an input layer, one or more hidden layers and an output layer. The input layer connects with the input variables. In our case, it comprises fifteen input neurons corresponding to the fifteen environmental variables, respectively. The last layer, called the output layer, comprises a single neuron which corresponds to the dependent variable to be predicted (the presence/absence of the macroinvertebrate taxa). The layer between the input and the output layer is called the hidden layer. The network configuration is approached empirically by testing various possibilities and selecting the best solution for the prediction of the presence/absence of the macroinvertebrate taxa (Dedecker et al. 2004). Training was carried out by using a training data set to adjust the connection weights in order to minimise the error between observed and predicted values. This training was performed with the back-propagation algorithm described by Rumelhart et al. (1986b). The model validation was based on tenfold cross-validation (Witten and Frank 2000). For assessment of the model predictions, the percentage of Correctly Classified Instances (CCI), in other words the prediction success or matching coefficient (Buckland and Elston 1993), as well as Cohen's kappa (Cohen 1960) were calculated. Manel et al. (2001) recommended Cohen's kappa for the assessment of presence/absence models in ecology. Cohen's kappa is a measure of the proportion of all possible cases of presence or absence that are predicted correctly after accounting for chance effects. It is a very useful assessment method when predictions of very rare or very common taxa have to be evaluated. The ANN models were implemented with the neural network extension of the software package MATLAB 5.3 for MS Windows™.

Results

Testing the sensitivity and robustness of the ANN models for 'extreme' values

To test the sensitivity and robustness of the ANN models for 'extreme' values, two examples for Asellidae (Crustacea) were elaborated. Asellidae were chosen as a representative taxon because of their highly variable presence in the headwaters of the Zwalm river basin and their use as bio-indicators in river quality assessment (MacNeil et al. 2002). Predictions were made for 'extreme' values of depth and flow velocity, which are important variables influencing the presence/absence of Asellidae. The distributions of flow velocity and depth for the collected sites are shown in Figs. 4.2.2 and 4.2.3 respectively. The ranges of flow velocity and depth are given in Table 4.2.1. 82.5% of the monitored sites have a flow velocity lower than 0.7 m/s. Only two sites situated in the southern part of the Zwalm river basin have a flow velocity of more than 1.4 m/s. 74% of the sites have a depth greater than 30 cm. The sites with a depth of over 100 cm are mainly situated upstream of the weirs and just before the Zwalm river discharges in the Scheldt.

Since these 'extreme' values of depth and flow velocity were not present in the collected dataset, the use of ecological expert knowledge was recommended. To introduce this expert knowledge in the ANN models, two virtual datasets, containing 'extreme' values of depth and flow velocity respectively, were created. Based on information found in the literature, variables were set to an optimal level. Only depth and flow velocity were set to an extremely low and an extremely high level, respectively. In both cases, Asellidae are expected to be absent (Gledhill et al. 1993). To test the sensitivity of the ANN model, the virtual dataset was randomly implemented in the dataset for training. Validation was carried out on a virtual dataset consisting of three sites and on the original dataset, merely consisting of measurement data (a validation set without 'extreme' sites). Based on the optimisation study of the ANN model design (Dedecker et al. 2004) training was performed by means of the gradient descent back-propagation algorithm (Hagan et al. 1996). Table 4.2.2 indicates for the variable flow velocity, that three 'extreme' training sites were needed to predict Asellidae absence in the three 'extreme' validation sites. However, the percentage of Correctly Classified Instances (CCI) and Cohen's kappa (CK) decreased with the original validation set from 80.0 to 74.9 and 0.59 to 0.47 when the number of 'extreme' sites at the end of the training set increased.

A similar procedure was followed for studying the effect of the variable depth. Two randomly implemented 'extreme' training sites were sufficient to predict the absence of Asellidae correctly in the three 'extreme' validation sites (Table 4.2.3). The percentage of CCI and CK decreased with the original validation set from respectively 80.0 to 74.9 and 0.59 to 0.48 when the number of 'extreme' sites in the training set increased from zero to three (Table 4.2.3).

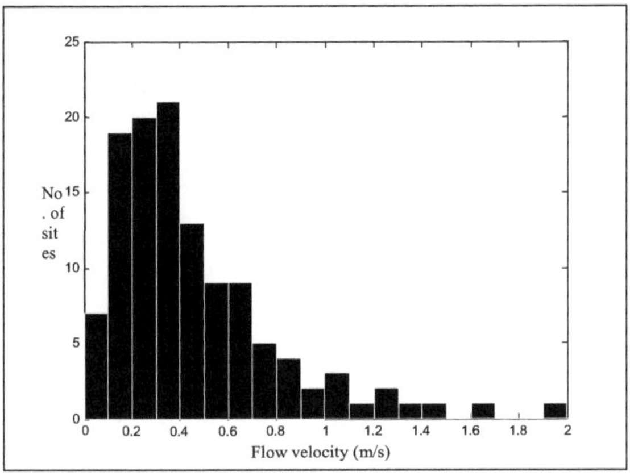

Figure 4.2.2 Distribution of flow velocity (m/s) for the collected sites in the Zwalm river basin.

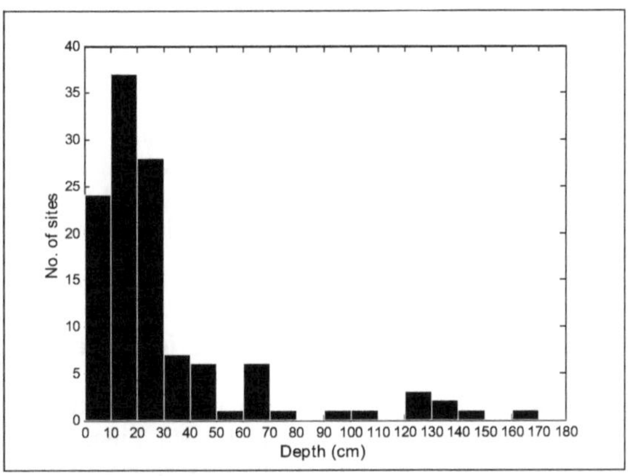

Figure 4.2.3 Distribution of depth (cm) for the collected sites in the Zwalm river basin.

Table 4.2.2 Predictions of Asellidae with 'extreme' values of flow velocity. Results of the ANN models for the 'extreme' validation set (3 sites) and the original one (12 sites) (validation set without 'extreme' sites). The 'extreme' sites were located randomly in the training set. CCI: Correctly Classified Instances; CK: Cohen's kappa (see text for detail).

No. of 'extreme' sites in the training set	**'Extreme' validation set**	**Original validation set**	
	% predicted absent	CCI	CK
0	0.0	80.0	0.59
1	0.0	73.9	0.47
2	0.0	73.7	0.47
3	100.0	74.9	0.48
4	100.0	74.9	0.47

Table 4.2.3 Predictions of Asellidae with 'extreme' values of depth. Results of the ANN models for the 'extreme' validation set (3 sites) and the original one (12 sites) (validation set without 'extreme' sites). The 'extreme' sites were randomly located in the training set.

No. of 'extreme' sites in the training set	'Extreme' validation set	Original validation set	
	% predicted absent	CCI	CK
0	0.0	80.0	0.59
1	0.0	75.5	0.50
2	100.0	74.9	0.49
3	100.0	74.9	0.48

Sensitivity analysis was performed to determine the impact of both input variables on the predicted probability of presence of Asellidae (Figs. 4.2.4, 4.2.5). To perform sensitivity analysis, an experimental approach was used (Lek et al. 1996a). A range of variation of a single independent variable to the model was applied while the others were held constant. The sensitivity analysis was done with and without one to five 'extreme' sites in the dataset. These 'extreme' sites included very low values of depth and very high values of flow velocity. In both cases, Asellidae were expected to be absent. If more 'extreme' sites were used in the dataset, the curve for flow velocity moved to the left (Fig. 4.2.4). In this way, the threshold of 0.5 was obtained by a lower level of flow velocity. For the variable depth, the inverse trend was obtained. If more 'extreme' sites were used, the threshold of 0.5 moved to a higher level of depth (Fig. 4.2.5). The combined effect of the two input variables on the predicted probability of presence of Asellidae is shown in Fig. 4.2.6. If only the collected dataset was used to perform the sensitivity analysis, the network always gave a predicted probability of one (Fig. 4.2.6a). That means, Asellidae could be expected at the whole range of both variables. If 'extreme' sites were added, the cumulative effect of Fig. 4.2.4 and Fig. 4.2.5 was obtained. The predicted probability decreased for the combination of high values of flow velocity and low values of depth.

Model simulations of practical river restoration scenarios

The ANN models tested were assessed for their practical applications to simulate different restoration management scenarios in the Zwalm river basin. First, the effect on Asellidae of the installation of six weirs (1-6) in the Zwalm river basin was simulated (Fig. 4.2.7). Because the six weirs are already present in the Zwalm river basin, a simple rule was used for the prediction of the presence/absence of Asellidae without weirs. The simple assumption used was that the site upstream of the actual weir would have the same content of dissolved oxygen, depth, width and flow velocity as downstream of the actual weir, while the situation downstream of the weir was not altered. When comparing Figs. 4.2.7a and 4.2.7b, representing the observed and estimated values for Asellidae before installation of the six weirs, one can notice that good predictions were made (CCI = 76.5%; CK = 0.54).

For the prediction of the presence/absence of Asellidae after weir installation the actual data could be used. When the network was trained without weirs in the training dataset, Asellidae was predicted present upstream and downstream of all weirs (Fig. 4.2.7d). The same predictions were obtained when training was performed with weirs in the training dataset, except for the site downstream of weir 1 where Asellidae was predicted absent (Fig. 4.2.7e). The predictive performances for both networks were CCI = 75.7% and CK = 0.50. Finally, 'extreme' data was added to the training dataset in which no weirs were included (Fig. 4.2.7f). The same predictions were obtained. Asellidae was predicted present upstream of each weir (CCI = 75.7% and CK = 0.50).

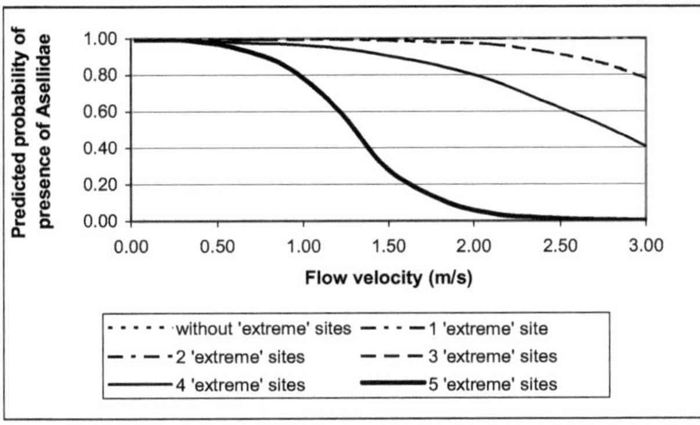

Figure 4.2.4 The impact of flow velocity (m/s) on the probability of the presence of Asellidae. The sensitivity analysis is performed for different numbers of 'extreme' sites in the dataset.

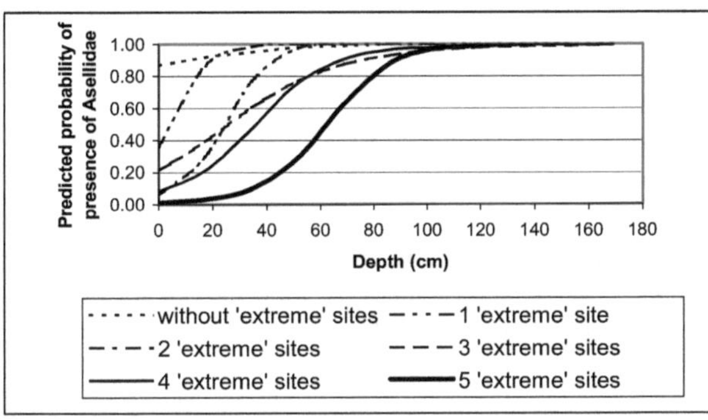

Figure 4.2.5 The impact of depth (cm) on the probability of presence of Asellidae. The sensitivity analysis is performed for different numbers of 'extreme' sites in the dataset.

In a second case study, the effect on Gammaridae (Crustacea) of the installation of six weirs in the Zwalm river basin was simulated (Fig. 4.2.8). Gammaridae were chosen as a representative taxon because of their variable presence upstream and downstream of a weir and their use as bio-indicators in river quality assessment (MacNeil et al. 2002). A similar procedure was followed as for the Asellidae. The same simple rule was used for the prediction of the presence/absence of Gammaridae without weirs. Before the weirs were installed, the prediction performances were respectively 74.9% and 0.36 for the CCI and the CK. For the prediction of the presence/absence of Gammaridae after weir installation the actual data could be used. When the network was trained without weirs in the training dataset, Gammaridae was predicted absent upstream of all weirs (Fig. 4.2.8d). Only upstream of weir 6 Gammaridae was predicted present (CCI = 70%; CK = 0.31). When training was performed with weirs in the training dataset, Gammaridae was predicted absent upstream of all weirs (Fig. 4.2.8e). The predictive performances for the network were CCI = 74.9% and CK = 0.33. Finally, 'extreme' data was added to the training dataset in which no weirs were included (Fig. 4.2.8f). Gammaridae was predicted present upstream of weirs 1 and 6 and absent upstream of weirs 2, 3, 4 and 5 (CCI = 71.7%; CK = 0.34).

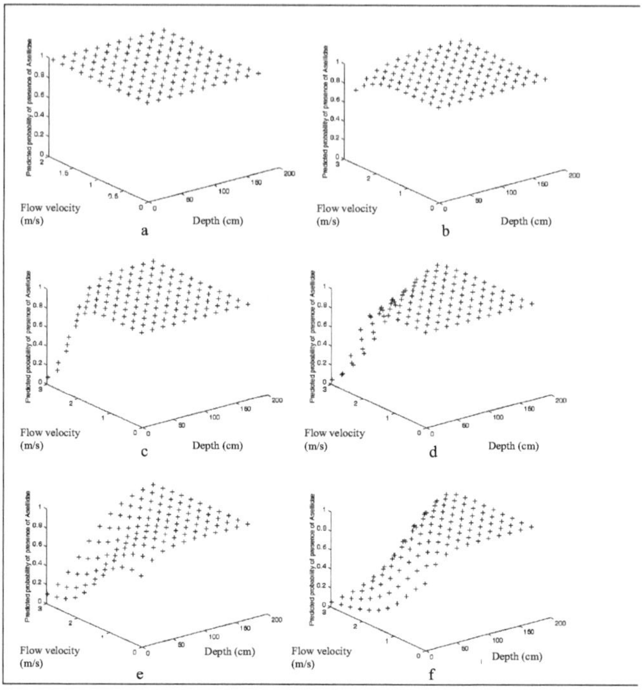

Figure 4.2.6 The combined effect of flow velocity and depth on the predicted probability of presence of Asellidae. The sensitivity analysis is performed for different numbers of 'extreme' sites in the dataset (a = without 'extreme' site; b = 1 'extreme' site; c = 2 'extreme' sites; d = 3 'extreme' sites; e = 4 'extreme' sites; f= 5 'extreme' sites).

In a third case study, the effect of an improvement of the water quality and structural characteristics on the Limnephilidae (Trichoptera) was simulated (Fig. 4.2.9). Limnephilidae were selected because of their sensitivity to pollution. These vulnerable macroinvertebrates were only found in the upper reaches of the watercourses in the Zwalm river basin. Since Limnephilidae is a rare taxon in this basin (the frequency of occurrence was only 10%), training was performed by means of the Levenberg-Marquardt back-propagation algorithm (Hagan et al. 1996) and based on an optimisation study of the ANN model design (Dedecker et al. 2004). The Levenberg-Marquardt algorithm is similar to the quasi-Newton method in which a simplified form of the Hessian matrix is used.

Comparing Fig. 4.2.9a and 4.2.9b, which represent the observed and estimated values for Limnephilidae, one can see that the predictions were good (CCI = 86.7%; CK = 0.43). Because of this, three sites near the city of Zottegem in the north of the Zwalm river basin were selected. Using a virtual dataset, the variables were set to an optimal level, representing a good water and structural river quality. A good water quality could be obtained in this area if diffuse pollution originating from agricultural activities and could be minimized and if more households were connected to the sewerage system in the future. Where no connection to the sewerage system would be possible, individual small-scale wastewater treatment plants could be proposed. No improvement of the structural quality (meandering, hollow river beds and development of pools and riffles) was needed because quite good values were already obtained. The presence of Limnephilidae was predicted well for the three sites and the overall prediction performance of the network was still good (CCI = 86.7%; CK = 0.56).

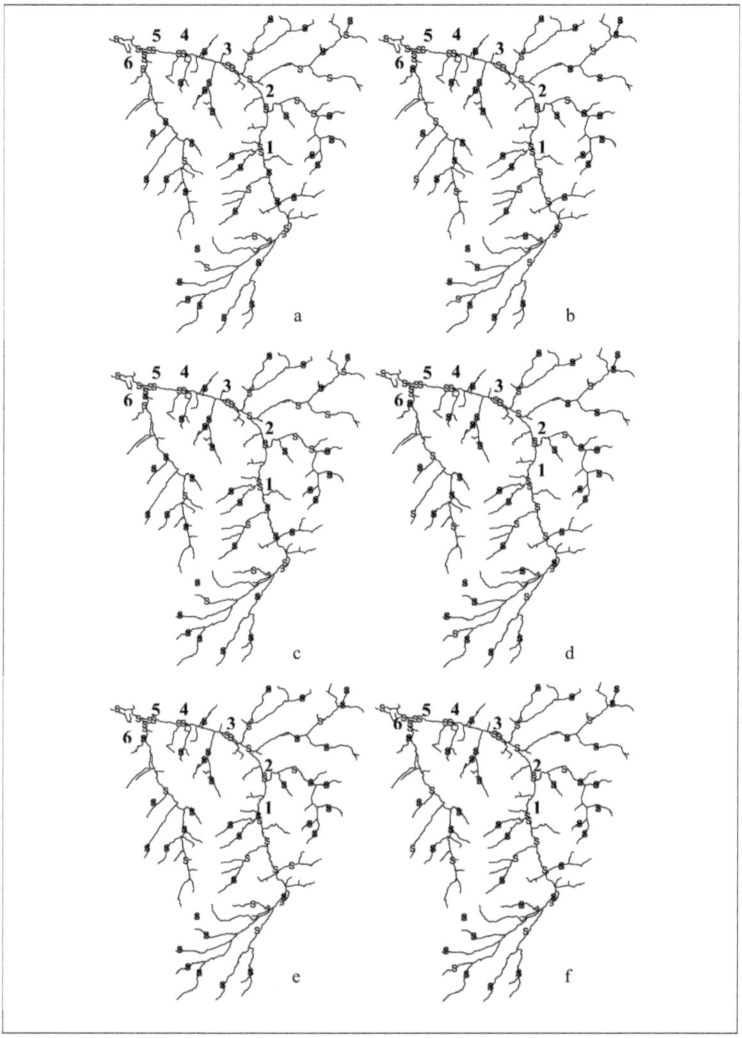

Figure 4.2.7 Distribution plots of Asellidae (Crustacea) in the Zwalm river basin. Black marks indicate the absence of Asellidae, white marks their presence. The six maps indicate what effect the installation of six weirs (1-6) in the Zwalm river basin can have on the Asellidae populations (a = simple assumption without weirs; b = ANN simulations without weirs; c = measurements August/September 2000 (with weirs); d = ANN simulations, training without weirs; e = ANN simulations, training with weirs; f = ANN simulations, training without weirs with addition of 'extreme' data).

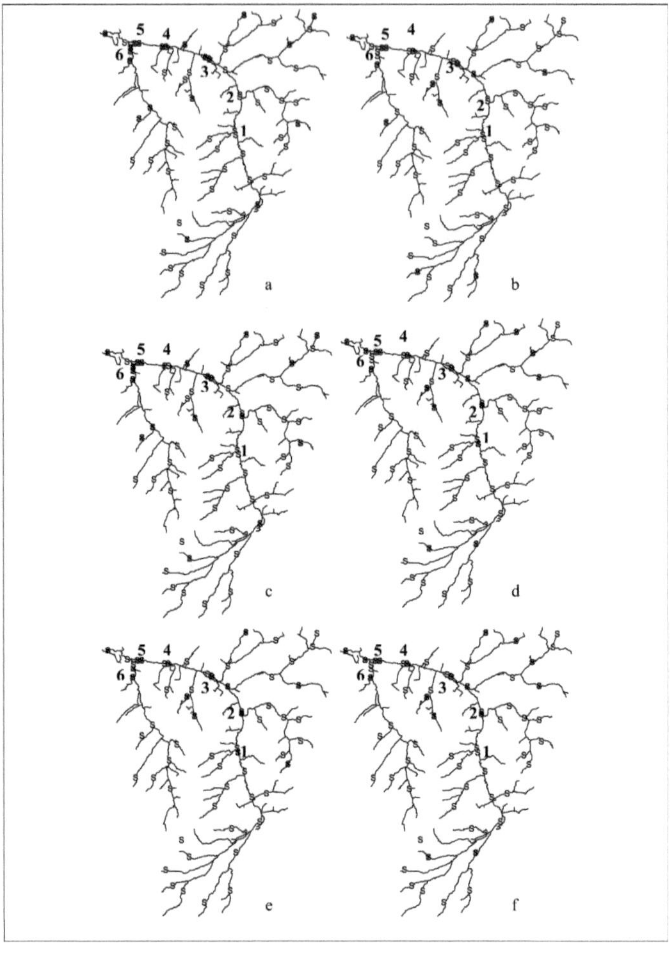

Figure 4.2.8 Distribution plots of Gammaridae (Crustacea) in the Zwalm river basin. Black marks indicate the absence of Gammaridae, white marks their presence. The six maps indicate what effect the installation of six weirs (1-6) in the Zwalm river basin can have on the Gammaridae populations (a = simple assumption without weirs; b = ANN simulations without weirs; c = measurements August/September 2000 (with weirs); d = ANN simulations, training without weirs; e = ANN simulations, training with weirs; f = ANN simulations, training without weirs with addition of 'extreme' data).

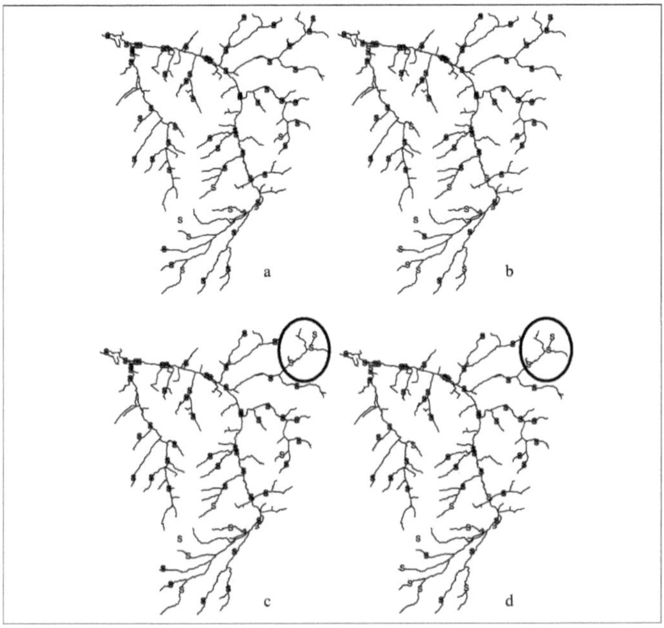

Figure 4.2.9 Distribution plots of Limnephilidae (Trichoptera) in the Zwalm river basin. Black marks indicate the absence of Limnephilidae, white marks their presence. The four maps indicate what effect an improvement of the water quality and the structural quality (3 encircled sites) in the Zwalm river basin could have on the Limnephilidae (a = measurements August/September 2000; b = ANN simulations for August/ September 2000; c = simple prediction of the impact of the improvement of water and structural quality; d = ANN simulations of improvement of water and structural quality).

Discussion

As mentioned by Lek and Guégan (1999), the ANN models are built up solely from the examples presented to the model. These examples are together assumed to completely contain the information necessary to establish the relation. In this way, simulations of the current situation in the Zwalm river basin, represented by the data measured, are reliable. The performance of the predictions of the selected macroinvertebrate taxa based on the abiotic characteristics of their aquatic environment were rather good (CCI = 86.6% and CK = 0.50). However, if predictions for practical river restoration scenarios have to be made, data describing these scenarios are not always included in the collected data. In this way, the collected data and induced models are less useful as such for predictions in river restoration management. However, the implementation of a virtual 'extreme' dataset in the training set containing relevant ecological expert knowledge can lead to better model validation. Consequently, model performance cannot only be optimized and assessed for predictions of river conditions that are similar to those in the data collected but also for the predictions of river restoration scenarios. Nevertheless, two problems can occur when using an 'extreme' dataset. To build this virtual dataset, first information has to be available. Expert knowledge and information found in the literature can be used. Second, the amount of 'extreme' sites in the virtual dataset should not be too high since the overall predictive power of the ANN models can decrease. As could be expected, the presence/absence of Asellidae in the 'extreme' validation set is predicted well when the number of 'extreme' sites in the training set is increased. Two and three 'extreme' sites for depth and flow velocity respectively are sufficient to predict absence of Asellidae in the three 'extreme' validation sites. In this way, the implementation of a virtual dataset can be very useful for predictions of scenarios out of the data range. As mentioned before, a disadvantage is that if '0' (absence) is introduced in the output, the network will try to predict '0', but it is bound to make some mistakes and predict '0' where '1' is expected. In this way, the overall predictive power of the ANN models decreases when a relatively large virtual training dataset is applied. Good predictions are thus possible for the current conditions in the Zwalm river basin using the collected dataset and for 'extreme' situations using a virtual one. However, a problem still exists where both current and 'extreme' conditions have to be reliably predicted. In the ANN models developed, the threshold to predict macroinvertebrates as present was set at 0.5 (output < 0.5, Asellidae = absent; output ≥ 0.5, Asellidae = present). If only the variables flow velocity or depth are taken into account, based on the sensitivity analysis, Asellidae should always be predicted as present if no 'extreme' sites are used. If 5 sites containing extremely high values of flow velocity are added, it should be possible for the network to predict Asellidae as absent. However, the range of flow velocity in the collected dataset is 0.03 to 1.92 m/s. Only two sites have a flow velocity of more than 1.40 m/s. In this way, even if 5 'extreme' sites are added to the dataset, Asellidae will be predicted as present for most of the sites in the collected dataset based on the variable flow velocity. The range of depth in the collected dataset is 2.5 to 170.0 cm. If sites containing extremely shallow depths are added to the collected dataset, the curve and the threshold of 0.5 move to the right (Fig. 4.2.5). Based on the variable depth, the network should predict Asellidae as absent if depth is very low.

A first practical case for the predictive models developed was the simulation of the effect of weir installation on the taxa Asellidae and Gammaridae. Goethals et al. (2001) used classification trees to perform the simulations of the effect of weir-removal on Asellidae. They predicted first the situation where weirs were already present, after that they predicted the presence/absence of Asellidae after weir-removal. Predictions after weir installation and before weir-removal (and vice-versa) are very similar with the two models.

The CCI before and after weir-removal for classification trees were 76.7% and 78.3% respectively. With the ANN models they were 75.7% and 76.5% after and before weir installation respectively. CK was 0.50 and 0.54 with both models for respectively after and before weir installation and before and after weir-removal. With both models 'width' is an important variable to predict the presence/absence of Asellidae. The only rule generated by the classification trees was: 'if width is more than 3.5 meters then Asellidae are present, while absent in the narrower streams' (Goethals et al. 2001). If the width is more than 3 meters, the ANN model predicts Asellidae to be present in 97% of cases. Although the predictive power was rather good, the ANN model was not able to predict the effect of the installation of the six weirs.

Based on the ANN models, the Asellidae continue to colonize the sites downstream of weirs 1 to 6. This could be explained by the sensitivity analysis (Figs. 4.2.4-6). The values of flow velocity and depth, which are important variables upstream and downstream of a weir, are not so extreme downstream of the six weirs. In this way, the ANN model predicted Asellidae as present downstream of the six weirs. Based on ecological expert knowledge however, the Asellidae would not colonize river stretches downstream of the weirs because rather fast current conditions are not tolerated by Asellidae (Gledhill et al. 1993).

However, Asellidae were found downstream of the six weirs. This can most probably be explained by drift from the sites upstream of the weirs where the flow conditions are better suited to Asellidae.

Another explanation can be that the habitat is not so extreme downstream of the weirs. It is thus still suitable for Asellidae. Unlike Asellidae, fast current conditions suit Gammaridae (Gledhill et al. 1993). For this reason, the ANN model predicts Gammaridae almost always absent upstream of the weirs and present downstream of the weirs. Based on the sensitivity analysis, Gammaridae are expected to be absent if the watercourse is too wide. Before weir installation the Gammaridae were predicted to be present upstream and downstream of weirs 1, 2 and 6 while absent upstream and downstream of weir 5 because the flow velocity was rather low and the variable 'width' exceeded 6.5 meters. But, no reason could be found why Gammaridae were predicted absent upstream of weirs 3 and 4 in spite of the river conditions suitable for Gammaridae. For both practical applications, the positive influence of the addition of 'extreme' data on the training dataset is negligible (Figs. 4.2.7f, 4.2.8f). Data-driven models are sufficient here to obtain good model predictions. However, the addition of 'extreme' data has no negative influence on the predictions. On the other hand, data-driven models are not appropriate for the prediction of the 'good ecological status' in the Zwalm river basin as mentioned in the European Water Framework Directive. This is illustrated by the sensitivity analysis of the variables flow velocity and depth (Figs. 4.2.4-6). If the sensitivity analysis was performed only with the collected data, Asellidae were always expected to be present. No difference was made for the 'extreme' situations. However, if 'extreme' sites, including expert knowledge, were added to the dataset, the 'extreme' situations could be better predicted.

Based on an optimisation study of the ANN model design (Dedecker et al. 2004), in a third application, training was performed by means of the Levenberg-Marquardt back-propagation algorithm since Limnephilidae is rare taxon in the Zwalm river basin. The percentage CCI was rather high (86.7%) which is quite normal for the prediction of very rare taxa since the models tend to learn that very rare taxa are always absent. Manel et al. (2001) mention that the prediction success (the CCI) may be affected by the frequency of the organism being modelled. Unlike the CCI, the effects of the frequency of occurrence on Cohen's kappa appear to be negligible. Cohen's kappa was considered 'moderate' (0.43). According to Manel et al. (2001) Cohen's kappa values of 0.40-0.60 for presence/absence models are considered to indicate 'moderate' model performance. If training is performed by means of the gradient descent back-propagation algorithm a percentage of CCI and a Cohen's kappa of 90% and 0 respectively are found. The apparently high percentage of CCI contrasted with the very low Cohen's kappa. A high overall prediction performance was obtained since Limnephilidae were predicted absent for all sites. It can be concluded that the percentage of CCI does not always give a reliable evaluation of the suitability of appling models for management purposes. However, the CCI added to Cohen's kappa is a good way to analyse presence/absence data.

ANN models are in general quite robust with a rather high predictive reliability. The reliability of the models has to be assessed via simulations made by ecological experts who can deliver knowledge that is often not included in the database used for the model induction. For example, Limnephilidae are predicted present in three sites in the northern part of the Zwalm river basin when environmental parameters are modified in order to reach a good water quality. Although the river water quality is suited to this sensitive taxon, they cannot reach this part of the river basin without migration. To increase the model feasibility with regard to simulations for river restoration management, spatial-temporal expert-rules will also have to be included. Migration barriers along the river and the migration kinetics of the organisms in water, on land and in the air can provide important additional information. Annual measurement campaigns will improve the database with regard to the information content of the data. Further optimisation of the ANN models could also be tion content of the data. Further optimisation of the ANN models could also be obtained by the selection of more appropriate input variables (d'Heygere et al. 2002).

Conclusion

To predict different river restoration scenarios, 'extreme' datasets were added to the original one. Therefore, ecological expert knowledge was used. The presence/absence of Asellidae in the 'extreme' validation set was predicted well when the number of 'extreme' sites in the training set increased. However, the overall predictive power of the ANN models decreased when a relatively large virtual training dataset was applied. Three case studies have shown that ANN models are in general quite robust with a rather high prediction reliability. For very extreme situations, addition of 'extreme' data to the training dataset can be very useful. However, for practical applications, the positive influence of the addition of 'extreme' data to the training dataset is negligible. Although the addition of 'extreme' data has no negative influence on the predictions, data-driven models are often sufficient to obtain good model predictions for practical applications. As a conclusion, adding extreme data improves the reliability of ANN models significantly for predictions under similar conditions. The addition of only one 'extreme' case is however insufficient to obtain a significant improvement of the predictive performance in similar cases, while the addition of too many 'extreme' cases in the training set decreases the general predictive performance of the models.

4.3 A neural network approach to the prediction of benthic macroinvertebrate fauna composition in rivers*

Di Dato P†, Mancini L, Tancioni L, Scardi M

Introduction

Predicting the composition of benthic macroinvertebrate fauna in rivers is not a trivial task, both because of the number of species to be modelled and because of the complexity of biotic and abiotic relationships that determine their distribution. However, the composition of the benthic macroinvertebrate fauna usually provides very useful insights into the ecological quality of lotic systems, as these organisms are very sensitive to disturbance. Benthic macroinvertebrates are relatively sedentary and long-lived, with life cycle durations ranging from a few months to 2-3 years, and they show a wide range of adaptations to local environmental conditions. They represent a continuous monitoring system of the water body where they are living, but they are also very easy to collect and to identify, at least at an intermediate taxonomic level. Therefore, benthic macroinvertebrates are widely used as biological indicators (Hellawell 1986) and, in particular, they have been used for many years as a source of information for computing several biotic indices that are now used worldwide to assess biological water quality (e.g., Metcalfe 1989, Resh et al. 1996, Lammert and Allan 1999). In this study, the Italian IBE index (Ghetti 1997), derived from the Extended Biotic Index proposed by Woodiwiss (1981) was used as a reference for selecting ecologically homogeneous taxa.

Several different biotic indices have been developed, as they had to be suited to ecoregional characteristics in order to provide correct diagnoses of the riverine ecosystem quality. Most indices, however, share the same rationale that is based on the identification of sensitive taxa and on the recognition of the ecological role of other taxa. The main advantage of this approach with respect to more thorough community structure analyses lies obviously in its simplicity. In fact, even people with a limited taxonomic background can be easily trained to carry out rapid surveys aimed at the computation of biotic indices. A more complex approach to the assessment of the ecological status of streams and rivers is based on the prediction of the whole community structure. In the case of benthic macroinvertebrate fauna, different modelling techniques based on ecological knowledge and monitoring data are now available. In the United Kingdom, the work by Wright et al. (1984) led to the prediction of community types on the basis of environmental data by means of a multivariate analysis procedure. This appraoch was then extended and used in the River Invertebrate Prediction and Classification System (RIVPACS) (Wright et al. 1993b), which provides estimates of the ecological quality at a given site by comparing the observed macroinvertebrate fauna composition with the expected one.

* Funding for this research was provided by the EU project PAEQANN (N° EVK1-CT1999-00026).

† Corresponding: pdidato@mclink.it

The RIVPACS approach has also been adapted to other ecoregions. For instance, the Australian River Assessment Scheme (AUSRIVAS) (Simpson and Norris 2000) is based on the RIVPACS approach, although it has been expanded and adapted to each Australian ecoregion. Another method that is closely related to RIVPACS and AUSRIVAS is the benthic assessment of sediment (BEAST) (Reynoldson et al. 1995), that is based on quantitative data about macroinvertebtare fauna instead of presence/absence data only. Even though the RIVPACS approach proved to be very effective, it has limits related to the non-linearity, complexity and dynamic nature of biotic responses to environmental characteristics. Moreover, the development of an assessment system based on the RIVPACS rationale requires a considerable amount of work and thorough statistical analyses.

A new generation of empirical techniques for analysing and modelling complex ecological data in a more simple and straightforward way is now emerging. Among these new modelling methods Artificial Neural Networks (ANNs) play a relevant role and represent a useful tool when relationships among data are unknown and/or non-linear. ANNs learn from examples and do not require *a priori* theoretical models, nevertheless they are able to model complex temporal and spatial patterns and to reproduce the behaviour of very complex systems (Recknagel and Wilson 2000). During the last 10 years, ANNs have been applied to various ecological fields (see, for instance, Lek and Guegan 2000), including studies relating community characteristics with environmental variables (e.g., Chon et al. 1996, Recknagel 1997, Recknagel et al. 1997, 1998, Guégan et al. 1998) and modelling habitat suitability (e.g., Paruelo and Tomasel 1997, Ozesmi and Ozesmi 1999). As for the particular case of macroinvertebrate fauna, Pudmenzky et al. (1998) and Walley and Fontama (2000) recently developed ANN approaches that are aimed at the same goals and ecoregions as AUSRIVAS and RIVPACS respectively.

Our study was focused on a benthic macroinvertebrate data set provided by the Latium Regional Environmental Protection Agency and it is aimed at testing different strategies for modelling the presence or absence of macroinvertebrate benthic taxa on the basis of environmental variables, using ANN models.

Materials and methods

Our data set is based on 153 sampling sites, distributed over 76 rivers in the Latium region (Central Italy), where macroinvertebrate fauna was sampled between 1998 and 2000. The hydrographic characteristics of the study area are highly variable, as a consequence of the very diverse origin and evolution of the river basin. The main river in the area is the Tiber, which is the second longest river in Italy, flowing from the north-eastern Appenine mountains through Central Italy and Rome to the Tyrrhenian Sea. All the rivers and streams in the area studied are located in the Tiber basin, with the exception of those in the Liri-Garigliano basin, which is located in the southern part of the Latium region.

The macroinvertebrate benthic fauna was collected at each sampling site by means of a small dredge. The dredge consisted of a handle, a rectangular frame (25 x 40 cm) and a cone-shaped net. The net was made of nylon and mesh size was 0.5 mm. The net had a cup-shaped detachable jar at its closed end that facilitates the collection of the organisms sampled. The sampling sites were dredged from bank to bank to cover all the microhabitats using a technique called "kick sampling". According to this technique the dredge, placed on the bottom of the river with the mouth against the water flow, was dragged along a fixed transect. At the same time the operator scrambled the substrate with his feet in order to direct the benthic organisms towards the net. The fauna collected was preliminarily sorted *in situ*, but an in-depth study by stereomicroscope was then carried out in the laboratory on material fixed in alcohol (70%). The taxonomic analyses led to the identification of 174 taxa. At each sampling site 11 environmental variables were also recorded (Table 4.3.1) some of them were derived from maps or from Geographical Information Systems (elevation, distance from source, gradient), while others were measured in the field (watershed drainage area, water flow, structure of sediment in terms of granulometric classes). The whole data set included 153 records for 11 predictive environmental variables and 174 taxa.

Four modelling strategies, based on different model structures and different complexity levels in the model outputs were selected:

- Strategy A: a single model for all the taxa that were present in more than 5% of the samples (65 taxa out of 174);
- Strategy B: a separate model for each taxon that was present in more than 5% of the samples (65 taxa out of 174);
- Strategy C: a single model for all the taxa present in more than 20% of the samples (19 taxa out of 174). Before adopting this strategy, we checked if the smaller subset of taxa preserved the information contained in the data set based on 65 taxa. A Principal Coordinate Analysis (PCO) (Gower 1966) using Jaccard's dissimilarity (Jaccard 1908) matrices and a Mantel test (1967) were carried out to compare the results obtained with 19 and 65 taxa.
- Strategy D: a single model for only 8 major taxa, which were selected on the basis of their ecological properties. In particular, we selected the taxonomic groups used for the computation of the Italian IBE index, namely Plecoptera, Ephemeroptera, Trichoptera, Gammaridae and Palaemonidae, Asellidae, Oligochaeta, the genus Leuctra, Baetidae and Caenidae.

Table 4.3.1 Environmental variables collected at each sampling site.

Environmental predictive variables	
elevation (m)	boulders (surface, %)
distance from source (km)	rocks (surface, %)
gradient (%)	pebbles (surface, %)
watershed drainage area (km^2)	gravel (surface, %)
water flow (score, 1-5)	sand (surface, %)
	silt and clay (surface, %)

The records available for both predictive and faunistic variables were divided into three subsets (training, validation and test). The training subset included 50% of the records (n=77), while the validation and test subsets contained 25% of the records each (n=38). The three subsets were defined according to a stratified procedure, using elevation as the stratification criterion. Therefore, each subset includes samples from sites at different elevations. Faunistic information was exploited at its simplest (and most reliable) level, i.e. as binary (presence/absence) data. All predictive variables, that include heterogeneous quantitative and semiquantitative environmental variables, were normalized into the [0,1] interval.

The composition of the benthic macroinvertebrate fauna was modelled using feed-forward multilayer perceptrons. The number of nodes in the hidden layer was defined after empirical tests and the structures of the ANNs that provided the best results are shown in Table 4.3.2. The validation subset was used to compute the mean square error (MSE) of the ANN after each epoch, whereas the test set was used to test the performance of the ANN after completion of the training procedure.

The learning procedure was iterated over 100 000 epochs, restarting the learning procedure each time the validation began to increase, and keeping the set of synaptic weights that provided the minimum validation error. In order to prevent overtraining, only a random subset of the training patterns (38 patterns) was submitted to the ANN at each training epoch, and white noise in the [-0.01,0.01] range was added to each input value at each epoch. Sigmoid activation functions were used in all the nodes of the hidden and output layers of the ANN, whereas the error back-propagation algorithm was selected for adjusting the ANN weights during the training procedure. The learning rate and the momentum were constant and set respectively to 0.90 and 0.10.

Table 4.3.2 Four different model outputs, corresponding to different modelling strategies were selected. The optimal ANN structure for each modelling strategy was defined after empirical tests.

MODEL OUTPUTS	MODELLING STRATEGY	ANN STRUCTURE
Only taxa present in more than 5% samples	1 model 65 outputs	11-19-65
Only taxa present in more than 5% samples	65 models 1 output each	11-5-1
Only taxa present in more than 20% samples	1 model 19 outputs	11-5-19
Only taxa involved in IBE index (Ghetti 1997) computation	1 model 8 outputs	11-14-8

The continuous ANN outputs, representing the probability of presence in a given site for each modelled taxon, were converted back to binary presence/absence estimates using a threshold function set to 0.5. The percentage of Correctly Classified Instances (CCI) was then computed for each modelling strategy and for each taxon, but a more reliable method for evaluating the accuracy of the models was needed. Therefore, the K statistic (Cohen 1960, Kraemer 1982) was also computed. In particular, this method tests the null hypothesis of independence between the modelled presence and absence data and the observed data. Finally, the modelling strategies were compared by computing Jaccard's dissimilarities (Jaccard 1908) between observed and modelled patterns (i.e. samples).

Results

Only data belonging to the independent test set, which was not used during the training and validation phases, were used for evaluating the performance and the accuracy of the different models. In the case of Strategy C the effects of the reduction of the number of modelled taxa from 65 to 19 were analysed by comparing the first Principal Coordinates (PCoo1) obtained from PCOs performed on the two data sets. The overall agreement between the two cases was very good, as shown in Fig. 4.3.1, and Spearman's rank correlation was highly significant ($r=0.905$, $p<0.01$). The Mantel test confirmed this result, as the null hypothesis of independence between the dissimilarity matrices was rejected ($r=0.67$, $p<0.01$). The results about the performances of the models trained according to the four different strategies are shown in Tables 4.3.3 to 6, in which the percentage of CCI, the values in the four confusion matrix cells and the K statistic are reported. Only taxa that were associated to significant K statistics, i.e. predicted by the ANN models in a way that was significantly different from random, have been included in the tables.

Table 4.3.3 Strategy A modelling results. Only 6 out of 65 taxa, which are associated to significant K statistics are shown.

	Overall CCI: 83.8%					
Strategy A	CCI %	1-1	1-0	0-1	0-0	K
Baetis	76.32	19	4	5	10	0.50
Simuliidae	73.68	12	2	8	16	0.48
Elmidae	81.58	5	5	2	26	0.47
Hydropsychidae	78.95	6	5	3	24	0.46
Rhyacophilidae	76.32	5	7	2	24	0.38
Lumbricidae	68.42	7	8	4	19	0.31

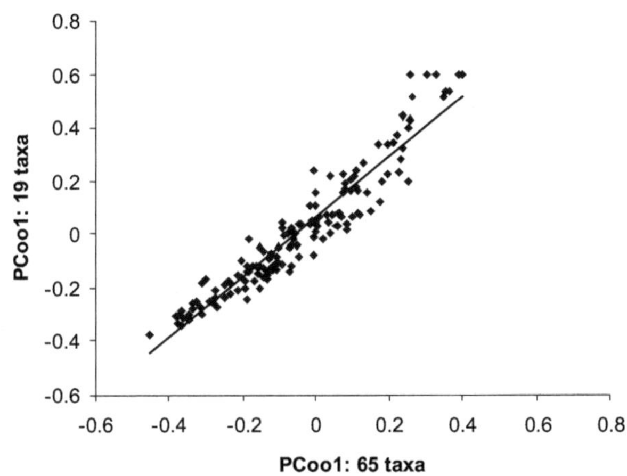

Figure 4.3.1 The overall structure of the data set including taxa that were present in more than 20% of the samples (19 taxa) was compared to that of the data set including taxa that were present in more than 5% of the samples (65 taxa). The information contained in the two data sets was similar, as shown by the agreement between the first Principal Coordinates obtained from Jaccard's dissimilarity matrices (Spearman's $r=0.905$, $p<0.01$).

Table 4.3.4 Strategy B modelling results. Only 10 out of 65 taxa, which are associated to significant K statistics are shown.

Overall CCI: 80.6%						
Strategy B	CCI %	1-1	1-0	0-1	0-0	K
Dinocras	89.47	5	1	3	29	0.65
Simuliidae	78.95	12	2	6	18	0.57
Rhyacophilidae	76.32	8	4	5	21	0.46
Baetis	73.68	22	1	9	6	0.39
Hydropsychidae	65.79	11	0	13	14	0.38
Ephemerella	71.05	7	5	6	20	0.34
Ceratopogonidae	68.42	8	6	6	18	0.32
Onychogomphus	84.21	3	3	3	29	0.41
Limoniidae	78.95	4	5	3	26	0.37
Limnephilidae	76.32	4	1	8	25	0.35

Table 4.3.5 Strategy C modelling results. Only 7 out of 19 taxa, which are associated to significant K statistics are shown.

Overall CCI: 68.3%						
Strategy C	CCI %	1-1	1-0	0-1	0-0	K
Baetis	73.7	19	4	6	9	0.44
Hydropsychidae	71.1	10	1	10	17	0.43
Elmidae	73.7	6	4	6	22	0.36
Gammaridae	68.4	9	9	3	17	0.36
Leuctra	76.3	4	8	1	25	0.35
Lumbricidae	68.4	8	7	5	18	0.32
Ceratopogonidae	68.4	8	6	6	18	0.32

Table 4.3.6 Strategy D modelling results. Only 3 out of 8 taxa, which are associated to significant K statistics are shown.

Overall CCI: 73.3%						
Strategy D	CCI %	1-1	1-0	0-1	0-0	K
Plecoptera	86.84	8	1	4	25	0.67
Ephemeroptera	73.68	23	7	3	5	0.33
Trichoptera	68.42	23	12	0	3	0.23

The comparisons between the results of the different modelling strategies, although based on a small number of efficiently predicted taxa, provided some useful hints. In particular, the comparison between Strategies A and B (Tables 4.3.3, 4.3.4) showed that the predictions were slightly more accurate when a set of models, one for each taxon, was trained instead of a single model simultaneously predicting all the taxa. In fact, with Strategy B not only was the number of taxa efficiently predicted larger (10 instead of 6 out of

65), but the average value of the K statistics for the predicted taxa was also slightly larger. This evidence, however, is not in agreement with the results for other groups of organisms, like fishes or diatoms (see sections 3.8 and 5.8), and is probably related to the smaller spatial scale at which the benthic macroinvertebrates respond to environmental conditions. Four taxa, namely Simuliidae, Rhyacophilidae, Hydropsychidae and the genus Baetis, were efficiently predicted both by Strategy A and by Strategy B, while two taxa (Elmidae and Lumbricidae) were efficiently predicted only by Strategy A, i.e. by using a single model for predicting all the species. Since information about interspecific interactions can only be embedded into this kind of model, it is possible that the success in modelling Elmidae and Lumbricidae depends on consistent association with other taxa or on the role that biotic interactions play in determining their distribution.

In Strategy C, a single model was trained for predicting the 19 taxa that were present in more than 20% of the samples. The accuracy of the predictions was not very different from the previous cases, as seven taxa had K statistics significantly different from zero (Table 4.3.5). Four of these taxa (Hydropsychidae, Elmidae, Lumbricidae and the genus Baetis) were also included among those that were efficiently predicted by the Strategy A model.

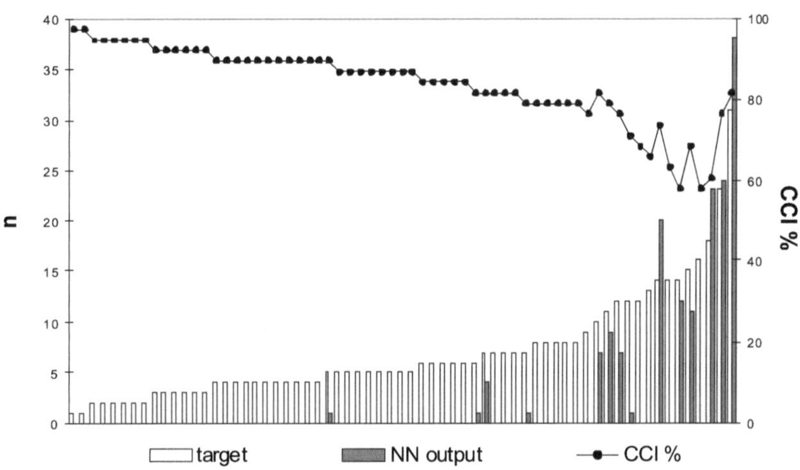

Figure 4.3.2 Strategy A: comparison between 65 ANN outputs (grey bars) and targets (white bars). The percentage of CCI is also shown (solid line with black dots).

Finally, eight taxa were modelled according to Strategy D (Table 4.3.6). In particular, these taxa are the ones that are routinely used for computing the IBE index. Obviously, these taxa have been considered for the biotic index because they have distinct ecological characteristics, and this is also the reason why they were selected as targets for ANN modelling. Three out of eight taxa were efficiently predicted by the ANN model, namely Plecoptera, Ephemeroptera and Trichoptera, but it is important to point out that these taxa are certainly the most sensitive to disturbance and pollution.

The difference between the apparently high CCI percentages, ranging from 83.8% to 68.6%, and the limited number of taxa that can be reliably predicted by the ANN models needs some explanation. In Figs. 4.3.2 to 5 the observed (target, white bar) and modelled data (ANN output, grey bar) are ranked according to the observed frequency of the taxa in the test set (n=38). It is obvious that the two bars are similar in height when the predicted values closely approximate the observed ones. The bars are not labelled to avoid clutter and because distinguishing each modelled taxon is not relevant in this case. The CCI percentage (solid line with black circles) was also plotted for each taxon.

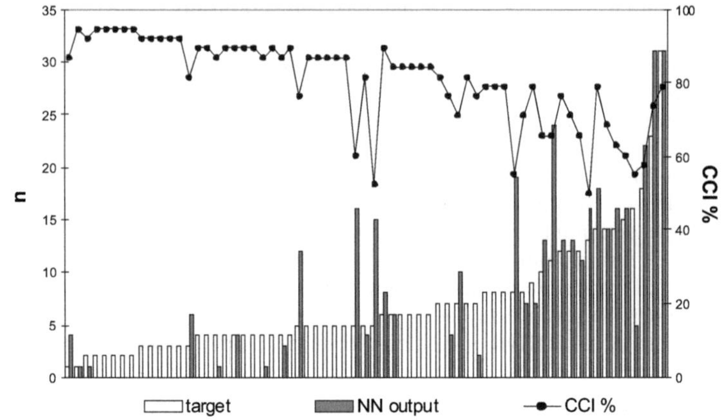

Figure 4.3.3 Strategy B: comparison between 65 ANN outputs (grey bars) and targets (white bars). The percentage of CCI is also shown (solid line with black dots).

When the number of taxa to be modelled was large (Strategies A and B, 65 taxa), the CCI percentage tended to be inversely correlated to the taxon frequency. This inverse relationship is a clear symptom of model malfunction, as the predictive ability of a model should not be related to the frequency of the taxa to be predicted. In fact, the models mainly failed in predicting the rarest taxa, and this bias was caused by the tendency of the ANNs to output only absence predictions when those taxa were considered.

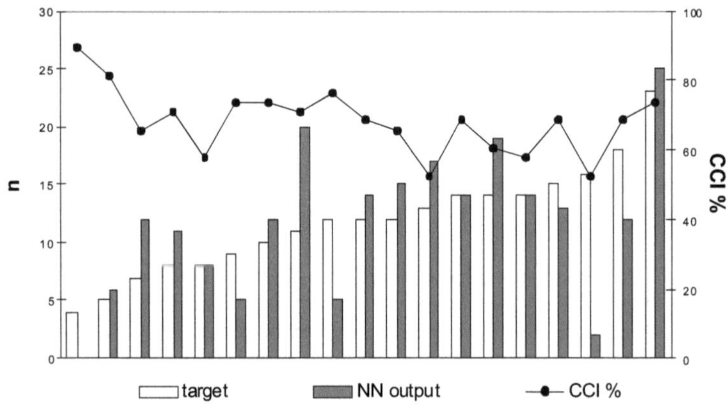

Figure 4.3.4 Strategy C: comparison between 19 ANN outputs (grey bars) and targets (white bars). The percentage of CCI is also shown (solid line with black dots).

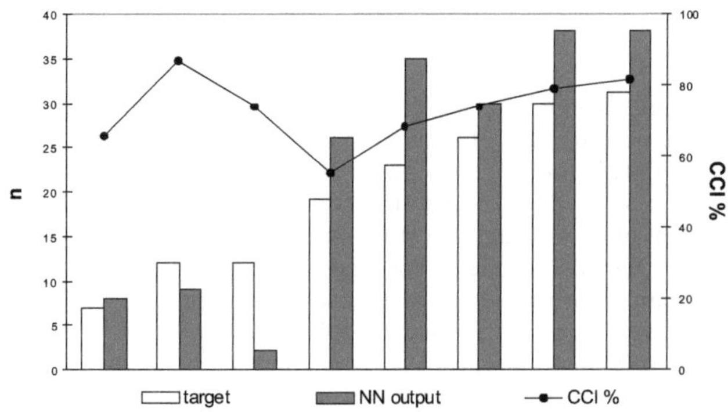

Figure 4.3.5 Strategy D: comparison between 8 ANN outputs (grey bars) and targets (white bars). The percentage of CCI is also shown (solid line with black circles).

In the case of Strategies C and D, in which only frequent taxa have been modelled, the inverse relationship between CCI percentage and taxon frequency was not observed, and the overall agreement between predicted and observed presence data was better than in the case of Strategies A and B. Therefore, these strategies were more effective in providing unbiased predictions about community structure, even though they obviously traded resolution for accuracy.

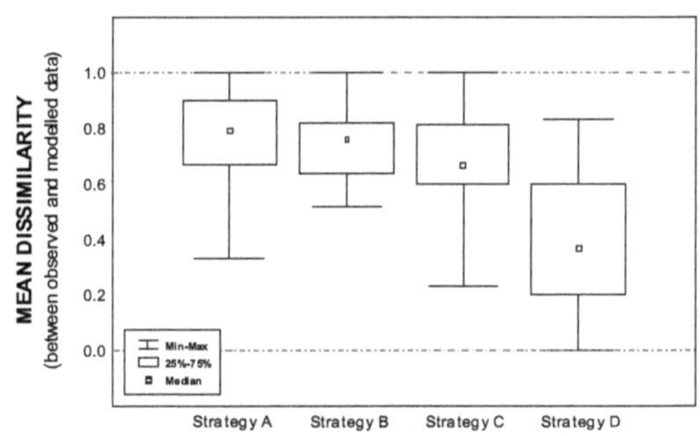

Figure 4.3.6 Comparison of the mean dissimilarities between observed and modelled samples for the four selected ANN training strategies.

Finally, observed and modelled data were compared using Jaccard's dissimilarities (Jaccard 1908) as a criterion for summarizing their resemblance. In Fig. 4.3.6 the distribution of these dissimilarities is shown, and it is evident that the closest match between modelled and observed data was obtained when the eight taxa that are used in the IBE index computations (Strategy D) were taken into account. In fact, the three lower quartiles in the dissimilarity distribution for the latter training strategy do not extend beyond the lowest quartile for the other training strategies (A, B and C). In other words, the similarity relationships that describe the structure of the test data set were closely reproduced by the ANN model when only a small number of ecologically significant taxa were selected as ANN outputs.

Conclusions

The benthic macroinvertebrate data set that was available for our study was certainly too small to support the development of accurate models. However, it provided a good opportunity for testing different training strategies and collecting useful hints for further developments. As in the case of other groups of organisms (see chapters 3 and 5), rare taxa (as well as very frequent ones, although the latter case is less likely to occur) could not be accurately predicted by the ANN models, independently of the training strategy. In fact, ANN models tend to "learn" that predicting only absence of rare taxa is the best solution for minimizing errors, even though this practice is obviously not appropriate for a real model. Obviously, the only solution to this problem would be a larger data set, but the way data are collected also plays a major role. In particular, more information is needed to model taxa that are insufficiently frequent. This goal can be attained, for instance, by planning the sampling activities at different spatial scales, thus allowing the collection of information about widely distributed taxa as well as about taxa that are only found in limited areas. It is obvious that a homogeneous spatial allocation of the sampling effort, although very convenient from a practical point of view, is not the best practice in this case. On the contrary, a multi-scale approach is needed, in which part of the samples are collected according to a ecoregional systematic sampling design, while other samples are collected in sub-areas where local maxima in beta diversity are detected. This way more information about the relationships between environmental variables and spatial distribution of benthic macroinvertebrates would probably be available.

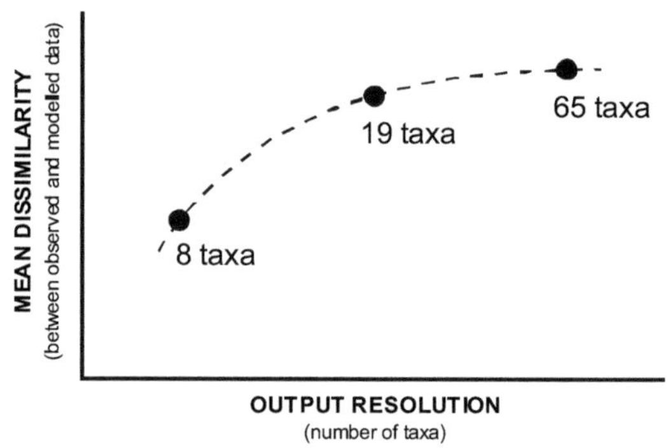

Figure 4.3.7 Three training strategies involving a single model for simultaneously predicting all the taxa are qualitatively compared. The comparison criterion is the mean dissimilarity between observed and predicted samples. The optimal efficiency of the modelling approach, corresponding to the minimum mean dissimilarity, was observed in the case of the modelling strategy based on the smallest set of taxa.

As for the different training strategies that have been tested, the macroinvertebrate fauna was predicted more efficiently when a set of single-taxon models was trained instead of a single model with multiple outputs. This result was not in agreement with previous findings obtained for other organisms (see chapters 3 & 5 in this book). Given the limited size of our data set, it is not easy to figure out whether this is a particular characteristic of benthic macroinvertebrate fauna or not. However, it is certainly possible that the lack of efficiency of the single model approach was somehow related to the complexity of underlying interspecific associations or interactions that were not adequately incorporated into a single ANN model.

Finally, the best training strategy among the ones we tested was based on very broad taxonomical units, namely on the taxa that are routinely used in the Italian IBE index computation. In particular, this approach was the one that gave the closest approximaiton of the observed structure of the dissimilarities among the samples in the test set. This result is not surprising, because the ecological characteristics of the taxa considered in the IBE index are certainly well defined. Therefore, they represent entities that are probably easier to model than others that are less closely related to the environmental variables. This result can be very useful in other ecoregions, where species that have been selected for other biotic indices could probably play a similar role in defining the structure of the macroinvertebrate assemblage. The different efficiencies, measured in terms of mean dissimilarity between observed and predicted data, of the three strategies involving a single model for the prediction of all the species is qualitatively shown in Fig. 4.3.7. It is obvious that the output resolution, i.e. the potential accuracy of the model, can be expressed as the number of modelled taxa, although the taxonomic level of the latter also plays a role. According to our results, the mean dissimilarity between observed and modelled data tends to increase with the output resolution, i.e. with the number of modelled taxa. Therefore, even though only a few cases have been considered in our study, our results support the hypothesis that the taxa to be modelled should be limited to the minimum set that provides the relevant information for correctly reproducing the relationships among observed samples.

4.4 Predicting Dutch macroinvertebrate species richness and functional feeding groups using five modelling techniques*

Gevrey M†, Park YS, Verdonschot PFM, Lek S

Introduction

When establishing a quantitative model to predict macroinvertebrate communities from environmental variables, the variety and complexity of variables often make the process of selecting a modelling method difficult. The most popular prediction method in ecology is multiple linear regression (Holler et al. 1993, Chhetri and Fowler 1996, Green 1996, Quensen and Woodruff 1997, Kolozsvary and Swihart 1999). In spite of numerous qualities, this method has two major weaknesses. Firstly, the parametric approaches involved assume distributions of relationships between data that may or not may hold. Secondly, the assumption of the linearity of the data is often questionable. In recent years, considerable attention has been given to the development of techniques for exploring data sets. New computational methods either overcome the parametric assumption, such as Partial Least Squares (PLS, Wold et al. 1983), or identifying non-linear relationships between the data, such as General Additive Models (GAM, Hastie and Tibshirani 1986, 1990), Regression Trees (RT, Breiman et al. 1984) and Artificial Neural Networks (ANNs, Rumelhart and Mc Clelland 1986). The predictive capacity of Multiple Linear Regression (MLR) was compared with (i) ANNs (Lek et al. 1996b, Brey et al. 1996, Paruelo and Tomasel 1997, Brasquet et al. 1999, Kemper and Sommer 2002), (ii) RT (Rejwan et al. 1999, Boone and Krohn 2000), (iii) PLS (Sanz et al. 1999, Schmilovitch et al. 2000, Dane et al. 2001, Delalieux et al. 2002), and (iv) GAM (Ette and Ludden 1996, Brosse and Lek 2000a). Several studies have compared different modelling techniques; to predict vegetation types (Cairns 2001), to fit the biological structural activity relationship in microbiology (Ramos-Nino et al. 1997), to model fish microhabitats (Brosse and Lek 2000b), to model fish species distributions (Olden and Jackson 2002), to predict the abundance of aquatic insects (Wagner et al. 2000a,b), to develop quantitative inference models in paleolimnology (Racca et al. 2001), to predict forest characteristics (Moisen and Frescino 2002), and to capture ozone behaviour (Gardner and Dorling 2000).

Species richness (SR) is an integrative descriptor of the community, as it is influenced by changes of natural environmental variables as well as anthropogenic disturbances (Rosenberg and Resh 1993). Therefore, it is commonly used as an ecological indicator for ecosystem assessments. The functional feeding groups (FFGs) of benthic macroinvertebrates are guilds of invertebrate taxa that obtain food in similar ways, regardless of taxonomic affinities. Therefore, they can represent a taxonomicaly heterogeneous assemblage of benthic fauna as well as a variety of disturbances of their habitats. Moreover, they reflect

* Funding for this research was provided by the EU project PAEQANN (N° EVK1-CT1999-00026).

† Corresponding: gevrey@cict.fr

the food resources available in a given area, therefore their distributions respond mostly to disturbances that alter the food base of the system (e.g., Hershey et al. 1988, Hart and Robinson 1990). The proportion of different groups may change in response to disturbances that affect the food base of the system, thereby offering a means of assessing disruption of ecosystem function. Therefore, the percentages of FFGs have commonly contributed as indicators of rapid bioassessment (Resh and Jackson 1993, Barbour et al. 1999). Predicting SR and FFGs is valuable for aquatic ecosystem management.

The primary objective of this work was to compare several recently developed techniques to predict a simple community index, SR. The second objective was to determine the relative strength of these methods while increasing the complexity of the variables i.e. in the prediction of FFGs. The last objective was to compare their ability to select those environmental variables that best predict the macroinvertebrate SR and FFGs.

Materials and Methods

Data description

We used the dataset compiled and presented by Verdonschot (1990, 1994). Samples came from 664 sites situated in the province of Overijssel (The Netherlands) including species abundances and a number of environmental variables (Table 4.4.1). The sampling dates were spread over the four seasons as well as over several years (1981 up to and including 1985). Six hundred fifty sites were used for our study, as fourteen sites were discarded due to missing values or other inconsistencies.

Table 4.4.1 Environmental variables used in the models.

Parameter	Definition
pH	Acidity
Ca	calcium Ca^{2+}
DO	dissolved oxygen
NO_3	nitrate NO_3^-
NH_4	Ammonium
conductivity	Conductivity
slope	Slope
depth	Depth
width	Width
temp	water temperature
submerg	% sampled habitat of submerged vegetation
emerg	% sampled habitat of emergent vegetation
float	% sampled habitat of floating vegetation
velocity	current velocity
season	Season
silt	% sampled habitat of silt substrate
bank	% sampled habitat of bank vegetation
gravel	% sampled habitat of gravel substrate

The sampling objective was to capture the majority of the species and their relative abundances present at a given site. At each site, major habitats were selected over a 10 to 30 m long stretch of the water body and were sampled with the same sampling effort.

Macroinvertebrate samples were taken to the laboratory, sorted by eye, counted and identified to species level. From 853 macroinvertebrate species, we calculated the SR to be used as response variable. Furthermore, among six existing FFGs, the species richness of four FFGs (filterers, predators, scrapers and shredders) were calculated following Resh and Jackson (1993) and Barbour et al. (1996, 1999), because these four groups are used for the rapid bioassessment of aquatic ecosystems. Finally, the response variables considered in our model correspond to all species richness and number of species in each of 4 FFGs.

Table 4.4.2 Correlation matrix of the 18 environmental variables used in the models. A dark colour indicates strong linearity between these variables, whereas a light colour indicates weak linearity between variables. The abbreviations for environment variables are given in Table 4.4.1.

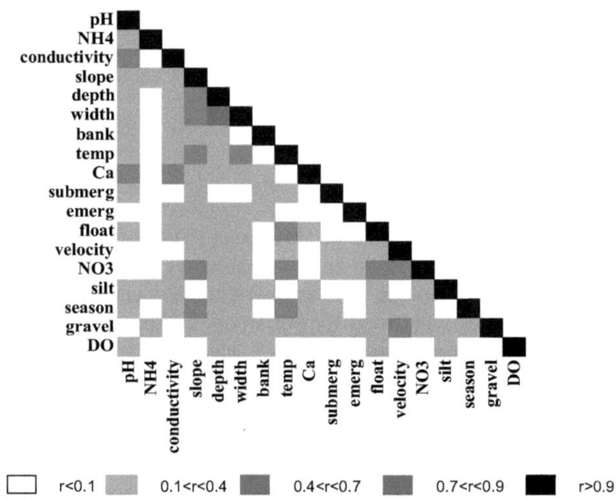

For predictor variables, many abiotic and biotic variables were collected at each sampling site. Some were measured directly in the field (width, depth, surface area, temperature, transparency, percentage of vegetation cover, percentage of sampled habitat), and of these, some (such as regulation, substratum, bank shape) were classified into two or several classes according to their properties. Field instruments were used to measure oxygen, electrical conductivity, stream velocity and pH. Surface water samples were taken to determine chemical variables (PO_4^{3-}, Cl^-, K^+, NO_3^-...). Other parameters, like land-use, bottom composition, and distance from source, were gathered from additional sources (data from water district managers, maps). From the 90 environmental variables available in this database, 18 environmental variables were selected on the basis of sensitivity analysis performed on a preliminary ANN model using a backpropagation learning algorithm, showing high contributions to the explanation of species richness (Table 4.4.1). These 18 variables showed low correlation coefficients for most of the variables (less than 0.4) (Table 4.4.2). In order to perform the test of the models, i.e., to validate the models, these sites were randomly split into two data sets, 500 samples for training to calibrate the models and 150 for testing.

Modelling

The five modelling techniques used in this study will be described briefly (for detail, see chapter 2). Modelling and analyses were conducted in S-Plus 2000 (Mathsoft 1997) for the

MLR, GAM and RT, Matlab 6.0 (MathWorks 1998) for the ANN and Simca-P 9.0 (Umetrics 2001) for the PLS.

Multiple linear regression (MLR)

MLR is a powerful and useful technique to determine the form and strength of a linear relationship between variables. It is used to predict the values of the dependent variable (y) for given values of the n independent variables (x1, x2,...,xn). MLR is a direct extension of simple linear regression. The model can be written in the following form (Walker and Lev 1969) for object i:

$$y_i = a_0 + a_1 x_{i1} + a_2 x_{i2} + ... + a_n x_{in} + \varepsilon_i \quad (4.4.1)$$

The best-fit line is made up of a separate 'slope' for each of the independent variables. The second component of the output is the intercept (a_0). This is the predicted value of y when all independent variables are equal to zero.

Generalized additive models (GAM)

This method (Hastie and Tibshirani 1986, 1990) extended the traditional linear statistical model. It can be applied in any setting where a linear or generalised linear model is typically used. These settings include standard continuous response regression, categorical or ordered categorical response data, count data, survival data and time series data. GAM allows non-parametric functions to be estimated from the data using smoothing operations. In this study, we used a cubic smooth spline function to determine the non-parametric estimation of the species-environment relationships.

Regression trees (RT)

Tree-based methods (Breiman et al. 1984) involve dividing the observations into groups that differ with respect to the variable of interest. A tree-based procedure automatically chooses the grouping that results in homogeneous groups that have the largest difference in proportion of the variable of interest. The tree-based method first divides the observations into two groups. The next step is to subdivide each of the groups based on another characteristic. The process of subdividing is separate for each of the groups. This is an elegant way of handling interactions that can become complicated in traditional linear models. When the process of subdivision is complete, the result is a classification rule that can be viewed as a tree. For each of the subdivisions, the proportion of the variable of interest can be used to predict the effect of that variable. The structure of the tree provides insight into the characteristics that are relevant. RT does not have to conform to the same distributional restrictions as classical statistics methods and there is no assumption of a linear model. This would generate a tree with a lot of branches and as many terminal segments (leaves) as there are cases. Normally, some "stopping rule" is applied before we arrive at this extreme condition. Inevitably this means "impure partitions" occur, but this is necessary to balance accuracy against generality. A tree which produces a perfect classification of training data would probably perform poorly with new data.

Partial least square (PLS)

PLS (Wold et al. 1983) is a method for constructing predictive models when the factors are many and highly co-linear. The general idea of PLS is to try to extract from many factors a few underlying or latent factors, accounting for as much of the manifest factor variation as possible while modelling the responses well. In the sample, latent factors and latent responses are extracted from the independent (X) and dependent variables (Y), respectively. The extracted latent factors are used to predict the extracted latent responses. Next, the predicted latent responses are used to predict the dependent variables. Last, inferences are made from the sample to the population.

Artificial neural network (ANN)

ANN (Rumelhart and McClelland 1986) is a computational tool, derived from a simplified concept of the brain which enables a nonlinear relationship between a dependent and some independent variables to be determined. A neural network contains nodes, called neurons, which are interconnected in a net-like structure generally composed of three layers: one input layer, one output layer and one or many intermediate (hidden) layer (s). The required number of hidden neurons is optimized by an iterative process. The degree of influence between interconnected neurons is represented by numerical weights called connection weights. The overall behaviour of the system is modified by adjusting the connection weight values through the repeated application of the back-propagation algorithm. ANN training is terminated when the error function, which measures the difference between calculated and desired output values, is minimized. In this study, we used only one hidden layer. The input layer contained 18 neurons corresponding to the 18 environmental variables and one neuron was used in the output layer. Several models were then used, i.e. as many as the variables to be predicted. The ANN models were optimised with 10 neurons in the hidden layer and nearly 500 iterations.

Evaluation of environmental variable contribution

Depending on the modelling method used, different ways to determine the contribution of the environmental variables were applied: (i) with MLR the influence of each variable was roughly assessed by checking the final values of regression coefficients; (ii) with GAM, the "Analysis of Deviance" provided the influence of the variables. For each smoothing effect in the model, this analysis gave a chi-test that compared the deviance between the full model and the model without each respective variable to a significance level ($P<0.05$); (iii) with PLS, the variable importance in the projection (VIP) was obtained that reflected the importance of terms in the model both with respect to Y, i.e. its correlation to all the responses, and with respect to X (the projection) with designated data, i.e. close to orthogonal X. The VIP values mainly reflect the correlation of the terms to all the responses. VIP values were computed, by default, from all extracted components; (iv) the most important variables in the RT model were those selected by the pruning method, i.e. those that composed the pruned tree; the pruning is a standard approach to fitting RT models: an overly large tree is fitted and then pruning is used to simplify the tree; (v) with ANN, the PaD algorithm (Gevrey et al. 2003) was applied to the model to classify the environmental variables by order of importance according to a calculation on the partial derivatives.

Comparison of model performances

The predictive performances of the models were evaluated through independent estimates from test sets of the global root mean squared error (RMSE) and the correlation coefficient (r) between observed and predicted values. RMSE indicates the average error in the analysis. The correlation coefficient indicates the quality of fit of all the data to a model.

Results

Established models MLR, PLS, GAM, ANN and RT were used to predict the same data cases. The correlation coefficient and errors obtained using learning and testing sets for each modelling technique are illustrated in Figures 4.4.1 and 4.4.2.

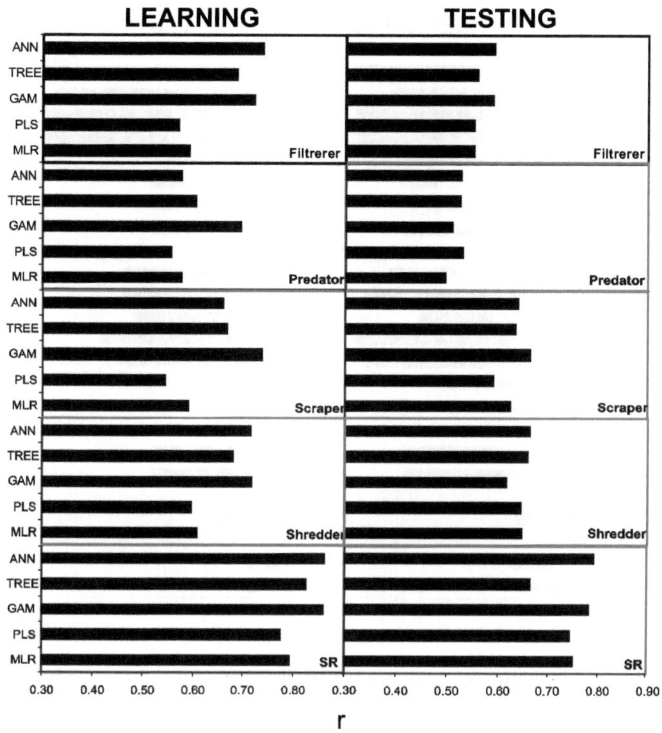

Figure 4.4.1 Predicting power (correlation coefficient; r) of each of the 5 models (ANN, TREE, GAM, PLS, MLR) used per 4 FFG (filterer, predator, scraper, shredder) and for the species richness (SR), on the right side, the testing set and on the left side, the learning set.

Comparison of methods

The results suggest that all five models often perform competitively for RMSE and r, but occasional erratic behaviour can be anticipated. GAM and ANN performed marginally better using the learning data set, but ANN performed marginally better using the testing data set. In terms of coefficient r, as shown in Fig. 4.4.1, the MLR and PLS models clearly exhibited similar levels of performance with the learning data sets, and also with the testing data sets. The values of r were generally lower than those obtained with the other models, and the RMSE values are higher (Fig. 4.4.2). This may be due to the non-linear relationship between the environmental variables and the SR or the FFG. The RT model had a medium behaviour with respect to the learning data set as well as to the testing data set. GAM was also intermediate with respect to the testing data set. Using the training data set the maximum r was 0.736 with the FFG, and 0.862 with SR. When using the testing data set, the FFG maximum r was 0.666 and 0.792 with SR. In the case of the filterers, the RMSE values obtained with the GAM and the ANN models were never higher than 0.1242 using the learning and the testing data set.

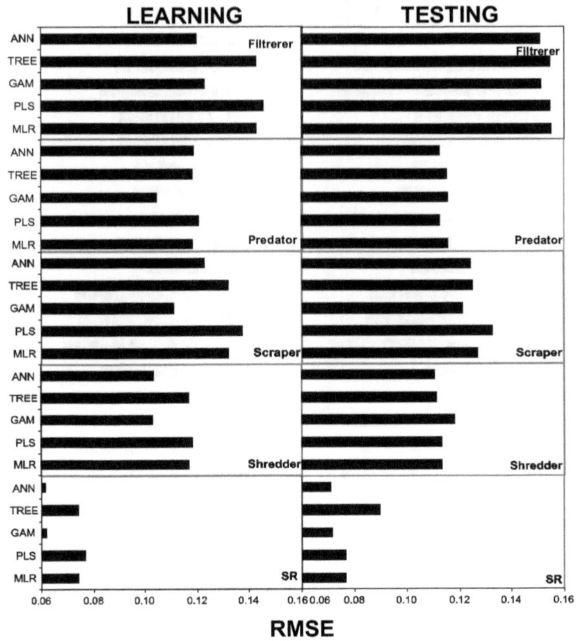

Figure 4.4.2 Root mean square error (RMSE) values of each of the 5 models (ANN, TREE, GAM, PLS, MLR) used per 4 FFGs (filterer, predator, scraper, shredder) and for the species richness (SR), on the right side, the testing set and on the left side, the learning set.

Table 4.3.3 The contribution of each environmental variable (grey colour) to all 5 models for each response variable (SR and 4 FFGs). Numbers (last column) refer to the number of models that selected the respective variable.

	Filtrerer					
	MLR	PLS	GAM	TREE	ANN	
pH				■	■	2
Ca			■	■	■	3
DO	■	■		■	■	4
NO3	■	■	■		■	4
NH4				■		1
conductivity			■	■	■	3
slope		■	■	■	■	4
depth		■				1
width	■	■	■	■	■	5
temp		■			■	2
submerg						0
emerg						0
float		■				1
velocity				■	■	2
season						0
silt	■			■		2
bank						0
gravel						0

	Predator					
	MLR	PLS	GAM	TREE	ANN	
pH	■	■	■	■	■	5
Ca	■	■	■	■		4
DO				■		1
NO3	■	■			■	3
NH4	■				■	2
conductivity		■		■	■	3
slope			■	■		2
depth			■			1
width			■			1
temp				■		1
submerg	■	■	■		■	4
emerg	■					1
float	■			■	■	3
velocity		■	■			2
season	■				■	2
silt						0
bank	■				■	2
gravel						0

	Scraper					
	MLR	PLS	GAM	TREE	ANN	
pH	■	■	■	■	■	5
Ca	■	■	■	■	■	5
DO	■					1
NO3						0
NH4						0
conductivity		■		■		2
slope	■	■		■	■	4
depth	■	■	■	■	■	5
width		■	■	■		3
temp						0
submerg	■		■	■		3
emerg				■		1
float						0
velocity						0
season		■				1
silt						0
bank						0
gravel			■			1

	Shredder					
	MLR	PLS	GAM	TREE	ANN	
pH	■		■	■	■	4
Ca						0
DO			■			1
NO3	■	■	■	■	■	5
NH4					■	1
conductivity				■		1
slope		■	■	■	■	4
depth		■		■		2
width	■	■	■	■	■	5
temp		■		■		2
submerg						0
emerg				■		1
float						0
velocity				■		1
season						0
silt	■			■		2
bank			■			1
gravel	■	■			■	3

	SR					
	MLR	PLS	GAM	TREE	ANN	
pH	■	■		■	■	4
Ca		■	■	■		3
DO			■	■		2
NO3	■	■	■	■		4
NH4	■		■	■	■	4
conductivity	■	■			■	3
slope	■	■	■	■	■	5
depth		■	■	■		3
width		■	■	■		3
temp	■	■				2
submerg	■			■	■	3
emerg	■			■	■	3
float	■	■			■	3
velocity						0
season	■	■				2
silt						0
bank	■			■	■	3
gravel						0

Comparison of predicted output

There was a large difference between the results obtained in predicting the FFGs in comparison to the SRs (Fig. 4.4.2). For all five models the prediction of the SRs was better than for the FFG. The ANN and GAM models performed best whatever the predicted variables.

For the training data set the maximum value of r was 0.736 for FFG prediction but the mean values was 0.642, and for SR prediction the maximum value was 0.823 and the mean 0.862. For the testing data set, the maximum value obtained when the FFG's were predicted was 0.666 and the mean was 0.592, as for SR the maximum value predicted was 0.792 and the mean 0.747. It was easier to predict SR than the FFGs independently. In fact, there was more information included in the SR in comparison to the FFGs, where the information is divided over four groups.

Comparison of variable contribution

Table 4.4.3 shows that the important environmental variables varied according to the models used. Some variables were shown to be important by all or four out of five models. These environmental variables can be seen as the most important variables. For the group of filterers the most important variables were width, DO, NO_3 and slope. For the group of predators, these were pH, Ca and submerged vegetation. Ca, pH, depth, and slope were the most important variables for the scrapers and width, NO_3, pH, and slope for the shredders. In general, slope was an important variable for predicting the filterers, scrapers and shredders while pH commonly explained the predators, scrapers and shredders. Width and NO_3 were common predictors for both filterers and shredders and Ca was the predictor common to explain the predators and scrapers. When the SR was predicted, the most important variables were pH, NO_3, NH_4 and slope.

Discussion and conclusion

Regression models play an important role in data analysis by providing prediction rules, and in identifying important predictor variables. The simple linear models, however, often fail in real life because responses are not linear. MLR fails if high correlations exist between input variables. Although MLR is the most commonly applied technique, it is not the most suitable when the data set contains col-inear variables, which is often the case for environmental data (Delalieux et al. 2002).

PLS, which deals with linear relationships, offers the possibility to combine co-linear variables in the X-matrix and to decide which variables contribute to explain the variance of the Y-matrix. It is also robust to atypical observations (outliers) and some statistical packages (such as simca-P) even tolerate missing values up to about 10-20% of the independent variables without seriously affecting the outcome, unless the data are missing in some systematic way (Eriksson et al. 1995, Lindgren et al. 1994). The PLS method is, in contrast to many other regression methods, suitable for data sets with fewer observations than variables and a high degree of intercorrelation between the independent variables (Delalieux et al. 2002).

GAM goes one stage further. Rather than using predefined functions as approximations to the data being modelled, the data themselves dictate the form of the function through spline smoothing. The GAM approach has several advantages in ecology. Firstly, by retaining the familiar regression structure, model building is relatively straightforward and the contributions of different predictors may be assessed. Secondly, by providing a close fit to the original data, GAM may produce models simpler than an equivalent GLM with several polynomial terms (Suarez-Seoane et al. 2002). However, one must be extra cautious not to over-fit the data, i.e., apply an overly complex model (with many degrees of freedom) to data so as to produce a good fit that likely will not replicate in subsequent validation studies

Methods, like ANN and RT, have the advantage that they can handle complex relations among data. Their drawback is that they are highly complex in regard to their algorithmic basis and therefore require substantial expertise and human interaction or interpretation (or both). RT when compared to ANN is a much more interpretable model. However, the main advantage provided by ANN is the ability to represent any smooth measurable functional relationship between one or more predictor and predicted variables.

In conclusion, this paper showed the feasibility of using environmental variables to determine SR or FFGs in aquatic macroinvertebrates. In order to find out which model is best, several methods were compared. Major differences between modelling techniques exist, and are critical in deciding which procedure to use. Each of the methods examined here have different advantages but also drawbacks. However, best results, root mean squared errors and correlation coefficient, were obtained with ANN and GAM. RT performs better in comparison to PLS and MLR. The latter are not the most suitable methods with respect to ecological data as relationships between environment and biota are non-linear.

Predictive modelling of macroinvertebrate communities provides an important tool for evaluating species-environment interactions. The choice of the best model to use has to be made by people who have a thorough knowledge of the properties of the model as well as of the data.

4.5 Comparison of clustering and ordination methods implemented to the full and partial data of benthic macroinvertebrate communities in streams and channels*

Nijboer RC†, Park YS, Lek S, Verdonschot PFM

Introduction

We analysed benthic macroinvertebrate data from streams and from channels with three techniques: the self-organizing map (SOM), non-hierarchical clustering (NHC) and canonical correspondence analysis (CCA). These techniques are often used in bioassessment. Assessment of the integrity of the biological elements of surface waters is an important aspect in water management (Barbour et al. 2000). To restore surface waters, managers need techniques to identify the community present and to predict which community they can expect if the environment changes either due to degradation or to restoration measures (Verdonschot and Nijboer 2000). Therefore, understanding community patterns is fundamental for ecosystem management. Benthic macroinvertebrates are recognised as one of the most reliable biological indicator groups in aquatic ecosystems (Hellawell 1986). They play a key role in food web dynamics, linking producers and top carnivores, and a number of species show clear responses to environmental variables. Their spatial sedentariness and intermediate life span, from several months to several years, make macroinvertebrates ideal as an integrative and continuous indicator group of water quality (e.g., Sládeček 1979, Hellawell 1986). Many useful biological indices use benthic macroinvertebrates (reviews of biological indices and metrics: Metcalfe 1989, Resh and Jackson 1993, Verdonschot 2000). For a long time, biological assessment of water quality had been uni-dimensional (Cairns and Pratt 1993) and focussed, for example, on organic pollution. However, Karr (1991a) stressed that species can react to a complex of factors and that they can also influence each other, for example by competition, thus the use of one species as indicator has its shortcomings. During the last decades, ecological assessment systems have been developed. This was stimulated by the development of integrated ecological indices (Karr et al. 1986, EPA 1988). In ecological assessment the overall environment is added to the biological component (Odum 1971), and the combination of species composition and environmental variables is used to assess the quality of surface waters (Verdonschot 1990). In these ecological assessment systems often a group of organisms, or even a whole community is used as bioindicator e.g., in RIVPACS (Wright et al. 1993b, Wright 2000) and EKOO (Verdonschot and Nijboer 2000). Multivariate techniques such as principal component analysis, cluster analysis, and correspondence analysis have been used to understand these ecological data, to extract communities, and to relate these to the environment (e.g., Gauch 1982, Jongman et al. 1995, Ludwig and Reynolds 1988, Legendre and Legendre 1998). Assessment sys-

* Funded by PAEQANN project (EVK1-CT1999-00026). The authors thank Cajo ter Braak for giving useful comments on an earlier version of the manuscript.

† Corresponding: rebi.nijboer@wur.nl

tems, such as EKOO and RIVPACS, are based on a stepwise progression of clustering and ordination. The basic unit is the community which is interpreted using cluster analysis.

Recently, artificial neural networks (ANNs) have been used for classifying groups (e.g., Chon et al. 1996) and patterning relationships between variables (Lek et al. 1996a). The ANN proved to be a versatile tool for dealing with problems in the extraction of information out of complex and non-linear data (Hoang et al. 2001), and could be effectively applied in classification and association (Lek and Guegan 2000). ANNs have been successfully applied to classify communities and to predict species distribution, communities and community parameters such as diversity (Lek and Guegan 2000, Recknagel 2003). Other authors have related community characteristics to environmental variables (e.g., Lek et al. 1996a, Recknagel et al. 1997). Among ANN techniques, the Self-Organizing Map (SOM) which is based on an unsupervised learning algorithm is often used to analyse the community structure. In several applications, e.g., Chon et al. (1996, 2000a, b, c) and Park et al. (2001a, 2003a), the SOM was successfully used to pattern benthic macroinvertebrate communities. However, analysing community patterns is difficult because the data sets are nonlinear and composed of many species varying over different locations and time and with a different distribution and density. At one site, only part of the community present is collected at any given moment. Therefore, each sample contains some information about the community but none are complete. Analysing large data sets is always an interpretation of the real situation; it is difficult to make the community structure apparent (Giraudel and Lek 2001).

In classification, sampling sites are clustered to reduce the variability and complexity of ecosystems and to make the results more useful in water management. Various techniques can give insight into the structure of communities, but the results can differ between techniques or within a technique depending on the choices that are made, e.g., the number of clusters and the basic algorithm. If the data set is small and shows clear gradients, different techniques show similar results (Giraudel and Lek 2001). But if the gradients in the data are less clear, different techniques might result in different community structures. Furthermore, classification and ordination results could differ if some species are excluded from analysis. This is frequently the case because rare species can add noise to the analyses (Gauch 1982, Marchant 2002) or because processing complete macroinvertebrate samples implies high costs. For management purposes, it is more effective if the number of species required as inputs in an assessment system can be reduced. In addition, results of clustering data sets with over 500 species are hard to analyse and interpret. Many researchers have reduced the species data before using ANN to pattern the community. Chon et al. (2000a) for example, summed the densities of taxa to seven selected taxa. Park et al. (2003a) expressed the biotic data as EPTC (Ephemeroptera, Plecoptera, Trichoptera, and Coleoptera) richness. Hoang et al. (2001) used only presence/absence of the 37 most common taxa. Walley and Fontama (1998) used the Biological Monitoring Working Party (BMWP) score or number of families as a biological index. However, these researchers did not compare the results obtained by analysing only part of the data with those obtained analysing the complete data set.

The aims of our study were: (1) to investigate the differences between classification with an unsupervised artificial neural network and with a classical clustering technique using two large data sets, (2) to test the stability of the classification results from both techniques, if only part of the taxon data is used, (3) to compare the gradients in environmental variables resulting from an artificial neural network with those resulting from a classical ordination technique, and (4) to study the ordination results using the complete and reduced data sets.

Materials and methods

Data

Comparison of techniques could yield different results between data sets, depending on the total number of sites or taxa, the number of taxa per site, the distribution of species over the sites or the distribution of individuals among the taxa. Therefore, two different data sets, both including samples taken in different seasons were used in this study. The first data set contained 563 samples from streams and included 767 macroinvertebrate taxa. The second data set was composed of 408 samples from small channels, less than 15 m wide, used for drainage in agricultural areas. In this data set 695 macroinvertebrate taxa were present. The benthic macroinvertebrate data in both data sets were collected by water district managers all over the Netherlands by using a standard sampling protocol (Verdonschot 1990). In summary, a 5 m surface sample was taken with a pond dip net. The sample was divided over the dominant habitats representative for the stretch of stream or the channel. In both data sets animals were identified to species level where possible. The abundances were log-transformed (ln(x+1)) to normalise the data distribution.

Water district managers measured environmental variables. The variables measured and the methods used differed between water managers. Therefore, in a first step, variables that were measured in less than 90% of the samples were removed from the data. In a second step, environmental variables were selected from the total data set based on their importance, which appeared from the first ordination results and expert opinion. Different variables were selected for streams and channels. For all variables annual means were used. To estimate lacking values for environmental variables (less than 10% of the sites per variable) average values were used from samples that were classified in the same cluster using non-hierarchical clustering. Finally, the stream data included 19 environmental variables, the channel data 22 variables.

Self-organizing map (SOM)

We used the SOM to pattern and classify species communities and to relate them to environmental variables. The SOM is an unsupervised learning algorithm for clustering, visualisation, and abstraction. The SOM is used to represent the data set in another, more usable form; it is an approximation of the probability density function of the input data (Kohonen 2001). The SOM consists of two different units (i.e. computational units) of input and output layers, connected by the computational weight vectors (i.e. connection intensities). To train the SOM, initially the community data with species densities were subjected to the SOM as inputs. When the input vector x is sent through the network, each neuron k of the network computes the summed distance between weight vector w and input vector x. The output layer consists of N output neurons which usually constitute a two dimensional grid giving better visualisation. The form of the output layer is a hexagonal lattice, because it does not favour horizontal or vertical directions as much as the rectangular array (Kohonen 2001). Among all output neurons, the best matching unit (BMU) which has minimum distance between weight and input vectors becomes the winner. For the BMU and its neighbouring neurons, the new weight vectors (w) are updated by the SOM learning rule. Training is usually done in two phases: first rough training for ordering with a large neighbourhood radius, and then fine tuning with a small radius. This results in training the network to classify the input vectors by the weight vectors they are closest to. The detailed algorithm of the SOM is described by Kohonen (1989, 2001) together with theoretical considerations, and Chon et al. (1996) and Park et al. (2003a) included ecological applications.

The number of output neurons (map size) affects the resolution of patterns resulting from the SOM. Therefore, the map size is an important parameter. To find the optimum map size, we trained the SOM with different map sizes ranging from 15 to 200 output units. Finally, we chose 40 (i.e., 8x5) as number of SOM output neurons on the 2D hexagonal lattice based on our experience and reasonable ecological meaning with each technique. The learning process of the SOM was carried out using the Matlab SOM Tool Box (Alhoniemi et al. 1999, MathWorks 1998). After training the SOM, hierarchical cluster analysis using Ward's linkage method was conducted to find clusters on the units of the SOM map according to their similarities. The clusters were characterised using expert judgement and literature about the indicative taxa (Mol 1984, Gittenberger et al. 1998, Drost et al. 1992, Geijskes and van Tol 1983, Bos and Wasscher 1997, Smit and van der Hammen 2000). To analyse the relationship between biological and environmental variables, the mean values of environmental variables were visualised on the SOM map. To do this, we calculated the mean value of each environmental variable in each output unit of the trained SOM (Park et al. 2003a) and represented these means by a grey scale.

Non-hierarchical clustering (NHC)

The same data sets were clustered by means of the NHC using the program FLEXCLUS (van Tongeren 1986). The strategy is based on an initial, non-hierarchical clustering, following the algorithm of Sørensen (1948) for a site-by-site matrix based on the similarity ratio, using species abundances. During this initial clustering, sites are fused according to single linkage but a fusion is skipped when two sites with a lower resemblance to each other than a specified threshold would become members of the same cluster. The value of the threshold depends on the number of sites clustered and the cluster homogeneity. The homogeneity of a cluster is defined as the average resemblance (based on the similarity ratio) of the sites of this cluster to its centroid. The initial clustering is optimised by relocative centroid sorting. Large and/or heterogeneous clusters are divided, small and/or comparable clusters (with a high resemblance) are fused, and then sites are relocated. During the relocation procedure, each site is compared to each cluster (as it was before relocation of any site) and, if necessary, moved to the cluster to which its resemblance is highest. Before a site is compared to its own cluster, the respective site is removed from that cluster and the new cluster centroid is computed.

To make both techniques (SOM and NHC) comparable, the number of resulting clusters should be the same. In the SOM modelling procedure 40 output units were chosen which were later grouped into 19 groups for streams and 23 groups for channels. The same numbers of clusters were chosen in FLEXCLUS. Therefore, some runs were done to explore the correct threshold value to obtain this number of clusters. After the initial clustering 50 relocation cycles were carried out. This was sufficient to result in stable clusters.

Comparison of SOM with NHC classifications

Distribution of the sites over the clusters

First, the distributions of the sites over SOM and NHC clusters were compared. This was done by constructing a matrix, with SOM clusters in the rows and NHC clusters in the columns. In the matrix cells, the number of sites that occurred in the respective combination of SOM and NHC clusters was given. For each row and column the total number of sites and the maximum number of sites were calculated. The total number of sites minus the maximum number of sites was considered as the deviation. The total deviation was calculated for all rows together as well as all columns together. Finally, the percentage was calculated by dividing the total deviation by the total number of sites. Two types of error were calculated, (1) type a error calculated over the rows, here, this signifies the percentage of sites that were combined in a cluster in the SOM but spread over other clusters using NHC, and (2) type b error calculated over the columns, in this case it is the percentage of sites that were combined in a cluster in NHC but spread over other clusters in the SOM.

Cluster characteristics

The stability of a classification can be expressed by the mean isolation value over all clusters. The mean isolation value is calculated for each cluster by dividing the homogeneity of a cluster by the resemblance of a cluster to the most similar cluster. The homogeneity is the average similarity between all combinations of two sites in the cluster. The resemblance is the similarity of the sites within one cluster to the sites of the most similar cluster. If the isolation is higher than 1, the homogeneity of the cluster is higher than the resemblance to the most similar cluster. In these calculations, the similarity measure used is the similarity ratio. For all NHC classifications minimum, average, and maximum homogeneity, resemblance and isolation were calculated. For the site groups resulting from the SOM the same values were calculated using the FLEXCLUS program. Therefore, the classification resulting from the SOM modelling was introduced as a fixed classification into FLEXCLUS. An additional cluster characteristic is the distribution of the sites over the clusters. We used the number of clusters with only one site and the maximum number of sites within a cluster.

Typifying taxa

For each cluster a set of indicators was established by using a calculation of typifying weights. A typifying weight represents the indicative value of a species for a cluster. The typifying weight for a single taxon differs between clusters. The weights for all species per cenotype were calculated using the program NODES (Verdonschot 1990). The clusters and the sites with the abundances of the taxa were used as input. In NODES the typifying weight of a taxon was calculated per cluster by combining the formulae of constancy, fidelity, and concentration of abundance (Boesch 1977, Verdonschot 1984). The greater the weight, the more characteristic the taxon is for a cluster. For example, if a species occurs within one cluster with a high frequency of occurrence and high abundance and it does not occur in any of the other clusters, the typifying weight of the taxon for that cluster is extremely high and low for all other clusters. If a taxon occurs in all clusters in about the same frequency and with similar abundances, the typifying weight of that taxon is low for all clusters. The weights vary from one to twelve (Verdonschot 1990). The taxa can be divided into four indicator groups: indifferent taxa (weights 1-3), slightly typifying taxa (weights 4-6), moderately typifying taxa (weights 7-9), and highly typifying taxa (weights 10-12).

To compare the results of classification between SOM and NHC, the number of highly typifying taxa (all taxa that have a typifying weight > 10 for one or more clusters) was calculated for classifications with 40 and 19/23 clusters (19 clusters for stream data, 23 clusters for channel data), respectively. The overlap of highly indicative taxa between SOM and NHC classification was further determined by calculating the mean abundance and frequency of occurrence for three groups of taxa: (1) only typifying in SOM, (2) only typifying in NHC, and (3) typifying in both classifications, using stream data and 40 clusters.

Classification with reduced taxa data

We used three methods for taxa reduction. Firstly, we considered the frequency of occurrence and abundance of taxa in the samples from the different sites. We excluded the species that occurred at less than 1% of the sites and had an average abundance of less than 8. The species that remained in the data were called 'dominant species'. Secondly, we considered taxonomic groups according to the ecological indicative value. We used species' abundances of five taxonomic groups for streams as well as for channels. These groups include a variety of species, which we thought to be ecologically indicative for the respective water types. For streams we included Ephemeroptera, Chironomidae, Gastropoda, Oligochaeta, and Crustacea. For channels we included Chironomidae, Coleoptera, Crustacea, Heteroptera, and Gastropoda. The third method of taxon reduction was done after the first SOM including all taxa. We selected indicator taxa (in this study defined as taxa that are indicative for a certain SOM unit). After training the SOM, we considered the weights of the SOM as occurrence probability ranging between 0 and 100% in each SOM unit. In the weight matrix, high values represent a high probability of occurrence in the corresponding neuron, while low values represent a low probability of occurrence. For example, a species with maximum 100% in a certain SOM unit displayed, we can observe this species with relatively high abundances in most of the sampling sites assigned to the unit. Using these values we selected species that had a maximum probability > 5% (in one or more of the output units the probability is > 5%). We considered these selected species as 'indicator species' in this study.

Table 4.5.1 The number of taxa in each of the data sets. The numbers in parentheses are the percentages of the number of taxa in the complete data set.

Water type	Number of species in complete data set	Indicator taxa	Dominant taxa	5 taxonomic groups
Streams	767 (100)	270 (35.2)	214 (27.9)	255 (33.2)
Channels	695 (100)	310 (44.6)	241 (34.7)	391 (56.3)

The reduction of species resulted in three new data sets which included about 28-56% of the species in the complete data sets (Table 4.5.1). With the smaller data sets SOM and NHC were repeated, using the same methods. The results of the complete data sets were compared with the results using part of the data by comparing (1) site distribution over the clusters and (2) measures for classification consistency.

Comparison of SOM with Canonical Correspondence Analysis (CCA)

The results of the SOM were compared with ordination results to study the role of the environmental variables in the data. Within the SOM the most important environmental variables were selected using the distribution of variables on the map. Used in this way, the SOM is a method for relating community data to environmental data. The state of the art method to this aim is CCA (ter Braak and Verdonschot 1995). This is the reason why we compared the results of the SOM with CCA which was carried out with the same data using the program CANOCO (ter Braak and Šmilauer 2002). Before analysis, the data were transformed. Species data were transformed into Preston classes (Preston 1962, Verdonschot 1990) and environmental variables except for pH were log transformed (ln(x+1)).

Ordination with reduced taxa data

Ordination results were compared between the complete dataset, the data with indicator taxa, with dominant taxa and with five taxonomic groups. To judge the strength of the ordination, the eigenvalues of the first four axes were used. Both measures illustrate the variance in the data that is explained by the ordination axes. The first measure includes all variation, the second one only the variation that is explained by the environmental variables. Forward selection was used to compare the order of importance of the resulting environmental variables between the results using the different datasets. The conditional effects of the variables were used. This means that first the most important variable was chosen, followed by the second variable that explains most of the remaining variation, and so on.

Results and discussion

Patterning sites using the SOM

Channels

Using the data matrix of channels to train the SOM of 40 output units (5 x 8 map), we obtained the distribution pattern of the channel communities. Clustering the 40 units of the SOM with the help of the dendrogram with Ward's algorithm resulted in 23 clusters, which are illustrated in Fig. 4.5.1.

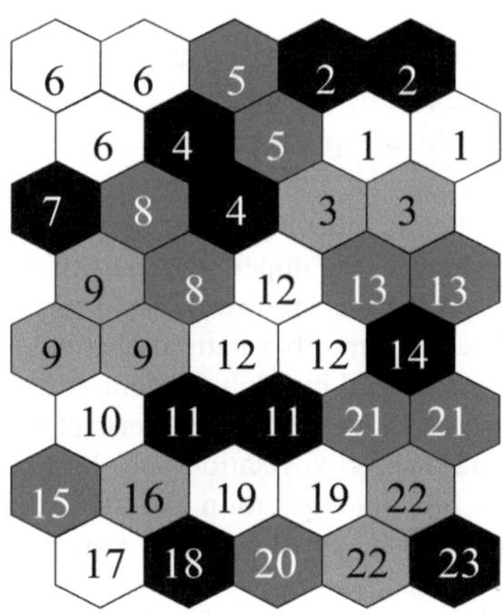

Figure 4.5.1 Classification of the SOM map for channels. Cells with the same number belong to the same cluster. The numbers of these clusters refer to Table 4.5.2. Different gray scales were used to indicate differences of clusters.

After training the SOM with community data, the mean values of the environmental variables were visualised on the trained SOM map. This technique is useful to identify the associations between environmental variables and communities. On the SOM map, a clear gradient in the distribution of a variable represents a high contribution to the classification (Fig. 4.5.2). The results of the SOM for channel data show that some variables have a very restricted distribution over the map (only few cells are darkly coloured) while others had more similar values over all the cells of the SOM (these variables are lightly coloured in many cells). In the right upper part of the SOM map the brackish channels occur. These channels have high chloride levels and high conductivity. Most of them are situated along the coast and their soil mainly consists of clay. In these channels emergent vegetation is dominant. Most of the channels that are wide and deep also occur in the right upper part of the SOM map.

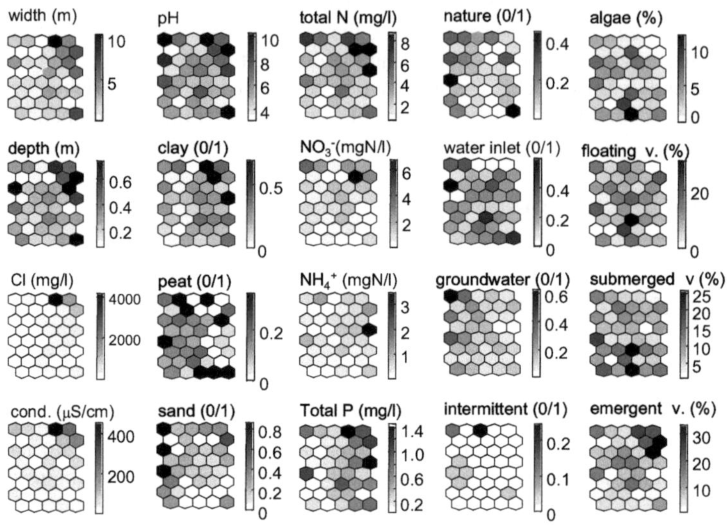

Figure 4.5.2 The distribution of environmental variables over the SOM map for channels, the level of the variables is indicated by a grey scale, ranging from light grey for low values to black for the highest values. (cond.=specific conductivity, v.=vegetation cover).

Eutrophication appeared to be the most important degradation factor. The nutrient levels (TN, TP, NO_3^- and NH_4^+) are high in the upper right part of the map. In the left lower part of the SOM map, nutrient concentrations are lower. The soil of channels that are situated in nature reserves mostly consisted of sand, but natural channels on peat also occurred. Clay channels were in most cases influenced by high nutrient or chloride levels. The vegetation types did not show a clear pattern except for the emergent vegetation that was correlated with the brackish channels. Submerged vegetation had the highest percentage coverage in the lower part of the map, but floating vegetation and algae did not show any gradient. Intermittent channels were characteristic for the middle unit in the upper line in which very shallow channels occur.

For some groups the indications given by the characteristic taxa (Table 4.5.2) were clearly linked to the extent of the values for the related environmental variables (Fig. 4.5.2). For example, the brackish channels (high chloride level and high specific conductivity) were inhabited by typical brackish water species. The group with submerged vegetation had species that live in vegetation. The left upper group was characterised by Tubificidae, indicating organic pollution, which corresponds with the distribution of the environmental variables. However, differences between groups in the middle of the map and differences between neighbouring groups were often less clear and only small. These are the channels that have no extreme characteristics and a more overlapping species composition. The indicator species are common species that can occur in many channel types and are not indicative of a specific environment. Some species indicated slowly flowing water, e.g., in groups 9 and

5, but this variable was not included in the channel data and therefore not illustrated on the SOM map (Fig. 4.5.2).

Table 4.5.2 Characterisation of the communities in the 23 channel groups identified by the SOM.

Group	Characteristic taxa	Characterisation
1	*Chironomus sp., Gammarus duebeni, Sigara lateralis*	slightly brackish
2	*Gammarus zaddachi, Palaemonetes varians, Nereis diversicolor*	Brackish
3	*Dytiscus circumcinctus, Culex sp., Lestes viridis*	Temporary
4	*Ilyodrilus templetoni, Dero digitata, Spirosperma ferox*	sand, no vegetation
5	*Limnodrilus claparedeianus, Macropelopia sp., Gammarus roeselii*	organic soil, slowly flowing water
6	*Tubificidae juvenile with hairchaetae, Tubificidae juvenile without hairchaetae, Psectrotanypus varius*	Saprobic
7	*Limnephilus lunatus, Arrenurus virens, Zavrelimyia sp.*	dense vegetation
8	*Arrenurus securiformis, Gyrinus marinus, Ablabesmyia monilis*	Peat
9	*Pisidium sp. Clanotanypus nervosus, Caenis horaria*	sand, slowly flowing water
10	*Sialis lutaria, Tanytarsus sp., Polypedilum nubeculosum*	vegetation, organic soil
11	*Sphaerium corneum, Athripsodes aterrimus, Hygrotus versicolor*	vegetation, oxygen rich
12	*Trianodes bicolor, Holocentropus picicornis, Limnesia maculata*	vegetation, oxygen rich
13	*Piscicola geometra, Rhantus frontalis, Cricotopus gr. intersectus*	Wide
14	*Unionicola crassipes, Hydrovatus cuspidatus, Ablabesmyia longistyla*	wide, eutrophic
15	*Planobarius corneus, Musculium lacustre, Haliplus heydeni*	hypertrophic, filamentous algae, vegetation
16	*Polycelis sp., Xenopelopia sp, Laccophilus hyalinus*	wide with organic soil
17	*Asellus aquaticus, Bithynia tentaculata, Planorbis planorbis*	eutrophied, low oxygen level
18	*Stagnicola palustris, Ceratopogonidae, Lymnaea stagnalis*	oxygen poor, covered by *Lemna sp.*
19	*Gyraulus albus, Arrenurus crassicaudatus, Haliplus immaculatus*	vegetation, low oxygen level
20	*Arrenurus globator, Haliplus sp., Ilyocoris cimicoides*	vegetation, nutrient rich, organic pollution
21	*Ischnura elegans, Arrenurus latus, Arrenurus sinuator*	vegetation, moderately polluted
22	*Theromyzon tessulatum, Glossiphonia heteroclita, Noterus clavicornis*	large channels, moderately polluted, open, few vegetation
23	*Cricotopus gr. sylvestris, Radix peregra, Sigara striata*	hypertrophic channels

Streams

Using the data matrix of streams to train the SOM of 40 output units, i.e. 5x8 map, we obtained the distribution pattern of the stream communities. Clustering the 40 units of the SOM with the help of the dendrogram with Ward's algorithm resulted in 19 clusters, which are illustrated in Fig. 4.5.3. When mean values of the environmental variables were visualised on the SOM map trained with community data of streams, three gradients appeared to be important (Fig. 4.5.4), a dimensional gradient, a nutrient/organic pollution gradient and a morphological alteration gradient. Wide and deep streams are situated in the right upper part of the SOM map versus small shallow streams at the lower and left part of the map. Many of the smaller streams were still in a natural state, meandering, having a natural profile and being situated in nature reserves (often forests). On the contrary, streams in the upper part of the SOM map were influenced by human impact, they were normalised and canalised. The deeper and wider streams often contained weirs, current velocity was low and silt had deposited. Organic pollution played a role in the left part of the SOM map (N-Kjeldahl, total phosphorus (TP), and NH_4^+ contents were high). The nitrate (NO_3^-) concentration was high in the lower units of the map. The oxygen content was high in the smaller streams with high current velocities.

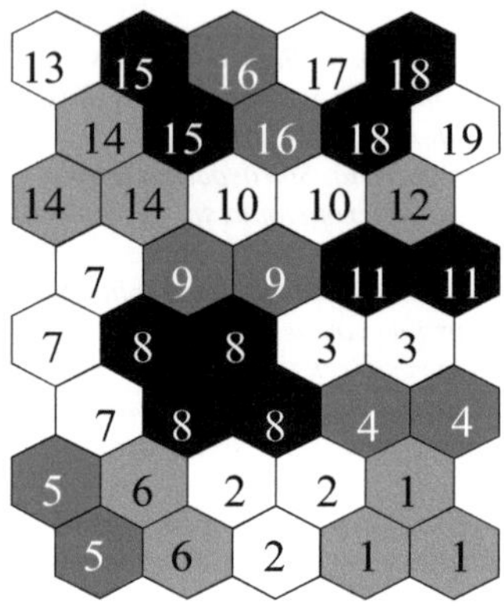

Figure 4.5.3 Classification of the SOM map for streams. Cells with the same number belong to the same cluster. The numbers of these clusters refer to Table 4.5.3.

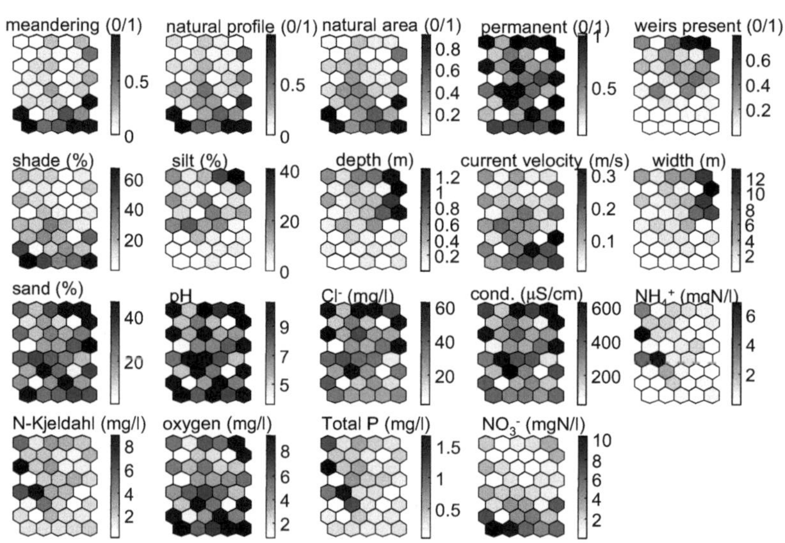

Figure 4.5.4 The distribution of environmental variables over the SOM for streams, the level of the variables is indicated by a grey scale, ranging from light grey for low values to black for the highest values (cond.=specific conductivity).

The ecological indications of the characteristic taxa mainly confirmed the distribution of the environmental variables over the SOM for streams (Table 4.5.3). In the lower part of the map the species of fast flowing upper courses occurred and in the upper part species that indicate normalised and often larger streams were found. However, sometimes the species indicated a different environment that could be interpreted from the SOM. For example, the species of groups 9 and 10 indicate middle-lower courses and upper courses, respectively. From the map, it appears that the dimensions should be the other way around. Probably, other variables played a role, such as normalisation. Another example is group 16, which seems to be polluted, considering the indicator species. This is not shown by the nutrient concentrations on the SOM, only chloride concentration and conductivity are high. Variables, such as the presence of vegetation, are indicated by the species but were not included in the data analyses. This sometimes explains the differences between interpretation from the species and interpretation from the environmental variables.

Table 4.5.3 Characterisation of the communities in the 19 stream groups identified by the SOM.

Group	Characteristic species	Characterisation
1	*Gammarus fossarum, Baetis vernus, Hydropsyche angustipennis*	upper-middle course, fast flowing
2	*Gammarus pulex, Sericostoma personatum, Velia caprai*	undisturbed small upper course
3	*Anabolia nervosa, Gammarus roeselii, Mystacides nigra*	normalised lower course, with vegetation
4	*Hygrobates nigromaculatus, Platambus maculatus, Lebertia inaequalis*	undisturbed upper course
5	*Nemoura cinerea, Glphpell, Zavrelimyia sp.*	intermittent small upper course
6	*Elodes minuta, Plectrocnemia conspersa, Brillia modesta*	Springbrook
7	*Micropsectra sp., Conchapelopia sp., Prodiamesa olivacea*	undisturbed, slowly flowing small upper course
8	*Tubificidae juvenile with hairchaetae, Tubificidae juvenile without hairchaetae, Limnodrilus hoffmeisteri*	stream with organic pollution
9	*Limnodrilus claparedeianus, Cryptochironomus sp., Neumania deltoides*	normalised, slowly flowing middle-lower course
10	*Stylaria lacustris, Limnesia koenikei, Armiger crista*	normalised upper course
11	*Micronecta sp., Caenis luctuosa, Cyrnus trimaculatus*	normalised middle course
12	*Mideopsis orbicularis, Caenis horaria, Ischnura elegans*	sand with silt, slowly flowing
13	*Procladius sp., Pisidium sp., Cricotopus gr. Sylvestris*	slowly flowing hypertrophic upper-middle course
14	*Chironomus sp., Psectrotanypus varians, Radix ovata*	polluted slowly flowing stream with vegetation
15	*Asellus aquaticus, Helobdella stagnalis, Cloeon dipterum*	normalised lower course
16	*Anisus vortex, Bithynia tentaculata, Sigara striata*	slowly flowing, polluted, low oxygen level
17	*Laccophilus hyalinus, Hygrotus versicolor, Bithynia leachi*	slowly flowing, polluted, moderate oxygen level
18	*Arrenurus globator, Haliplus immaculata, Limnesia undulata*	middle-lower course, vegetation, very low current velocity
19	*Arrenurus crassicaudatus, Limnesia maculata, Molanna angustata*	normalised middle-lower course with vegetation

Comparison between SOM and NHC classifications

Distribution of the sites over the clusters

The distribution of the sites over the clusters strongly differed between the SOM and the FLEXLCUS classification with the same number of clusters (Table 4.5.4). About half of the sites were put in different clusters. This goes for the classification in 40 clusters as well as for the classification in 19/23 clusters (23 channel clusters and 19 stream clusters found on the SOM map). This result was not found by Aguilera et al. (2001) who compared the Kohonen Neural Network classification of coastal waters in four groups with numerical classification. Probably the small number of groups and the fact that both of their classifications were based on Euclidean distance caused the results to be quite similar. Chon et al. (1996) observed that classification with the Kohonen Network and classification with clustering based on average linkage between groups (Norusis 1986), showed similar results. However, this was done for a small very distinct data set with 10 sampling sites and 8 tree species. Apparently, if larger complex data sets are used, the results of different techniques are less similar.

Table 4.5.4 Comparison of the site distribution over the clusters using SOM and NHC.

	Streams 40 clusters (% of sites)	**Channels 40 clusters (% of sites)**	**Streams 19 clusters (% of sites)**	**Channels 23 clusters (% of sites)**
type a deviation*	43	44	37	48
type b deviation**	52	53	48	53

* % of sites that was in one group using SOM, but put in different groups using NHC.
** % of sites that originates from different groups using SOM and is included in the same group using NHC.

Table 4.5.5 Cluster characteristics for the SOM classification and the NHC classification for 2 water types: STR, stream and CH, channel (all calculations were done excluding clusters of 1 site).

			Homogeneity			**Resemblance**			**Isolation**			**No. of sites**	
Water type	**Tech-nique**	**No. of Clus-ters**	**min**	**mean**	**max**	**min**	**mean**	**max**	**min**	**mean**	**max**	**min**	**max**
STR	NHC	40	0.41	0.50	0.66	0.27	0.52	0.76	0.70	1.03	2.41	1	71
STR	SOM	40	0.22	0.42	0.60	0.51	0.69	0.87	0.32	0.61	0.93	2	45
STR	NHC	19	0.32	0.47	0.60	0.27	0.46	0.57	0.73	1.07	1.98	1	100
STR	SOM	19	0.25	0.44	0.58	0.57	0.71	0.87	0.41	0.61	0.83	8	71
CH	NHC	40	0.39	0.53	0.63	0.36	0.53	0.72	0.70	1.01	1.59	1	79
CH	SOM	40	0.25	0.46	0.58	0.38	0.70	0.82	0.46	0.66	1.41	3	23
CH	NHC	23	0.39	0.52	0.63	0.29	0.52	0.69	0.71	1.03	1.58	1	96
CH	SOM	23	0.30	0.46	0.58	0.41	0.73	0.86	0.48	0.64	1.23	3	39

Cluster characteristics

Using the SOM, a less uneven distribution of the sites over the 40 clusters was resulted (the maximum number of sites in a cluster was 45 for streams and 23 for channels, Table 4.5.5).

The classification in 40 clusters resulting from NHC had a much higher maximum number of sites within a cluster, 71 and 79 sites for streams and channels, respectively. Using SOM there were no clusters composed of a single site. Using NHC there were 10 clusters consisting of one site for streams as well as for channels. Classification in 19/23 clusters resulted in higher maximum numbers of sites within the clusters. Again, using SOM the sites were more evenly distributed over the clusters.

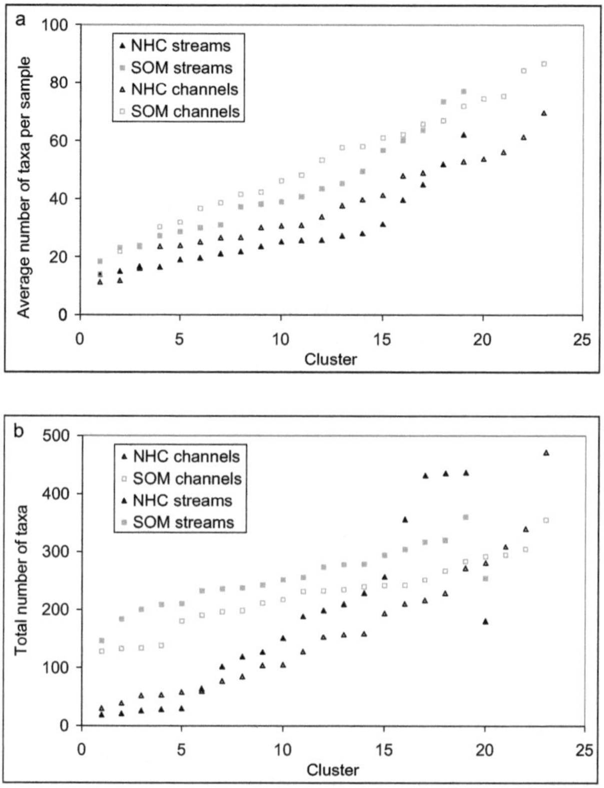

Figure 4.5.5 The average number of taxa per site (a) and the total number of taxa per cluster (b) for stream and channel clusters (19/23 clusters) using SOM and NHC.

Isolation values were similar for stream and channel data. Both, using 40 or 19/23 clusters, minimum, maximum, and mean isolation were lower for the SOM classification compared to NHC classification. This was due to lower homogeneity and higher resemblance for the SOM classification. Thus, NHC resulted in a classification of more distinct clusters. This could partly be explained by the fact that there were fewer clusters using NHC if the single sites were not included and that one or two very large clusters were made. To optimise the comparison, the classification in 40 clusters using NHC was repeated for streams, thereby deleting the 10 single sites from the data set. Again "single-site clusters" were formed. The isolation values for the complete classification were similar. This confirmed that the excluded sites were not outliers, otherwise the isolation value would have improved. Apparently, NHC always finds sites that are more different from all the others than the differences between groups of sites. Another explanation for the higher isolation values for the NHC is that the isolation is calculated using the similarity ratio which is also used to cluster the sites with this technique. It is therefore recommended to develop an evaluation measure for the SOM and then test SOM and NHC with this measure.

The average number of taxa per sample and the total number of taxa in the clusters were higher in the SOM classification compared to the NHC classification (Fig. 4.5.5a, b) using 19/23 as well as 40 (results not shown) clusters. In NHC, the "single site clusters" had very low numbers of taxa, which was probably the reason for the separation of these sites. The SOM classification shows a steeper line for the average number of taxa per sample, which means that this technique uses taxa richness to cluster the sites more than NHC does. This can explain the many successful applications of the SOM in combination with the multi-layer perceptron and the backpropagation algorithm for the prediction of taxon richness and diversity (e.g., Park et al. 2003a). The fact that the SOM clusters contained larger numbers of taxa could be an explanation for the lower isolation values compared to NHC results.

In four stream clusters and in three channel clusters the number of species was higher in the NHC classification compared to the SOM classification. These were the clusters with a high number of sites. Because the other clusters had fewer sites than most of the SOM clusters, the number of species in the other clusters was lower. Again, the distribution of the numbers of sites over the clusters seems to be important in evaluation of cluster characteristics. Apparently, NHC made one or two large clusters which included sites with high numbers of taxa. The remaining clusters were small and consisted of samples in which the taxa that occurred in the large clusters were lacking.

Using 40 clusters the pattern was the same, only the differences between NHC and SOM were slightly smaller. Using NHC the average number of taxa per cluster was 125 for channels and 127 for streams. With SOM the average values were 171 and 168 for channels and streams, respectively. In both classifications the number of taxa increased by using only 19/23 groups. This indicates that sites that included partly different taxa were put together. This is possible if a number of dominant taxa have high densities and therefore the similarity between the clusters is high.

Fig. 4.5.5a also shows that the average number of taxa in the samples in both classifications is higher in the channel data compared to the stream data, although the total number of taxa in the clusters was less for channel data (Fig. 4.5.5b) and the overall number of taxa was less in channels (695 taxa) than in streams (767 taxa). This indicates that many taxa in channels are more widespread. In streams, taxa are more restricted to certain sites, which results in a lower number of taxa in the samples, but a higher total number of taxa. However, this was not reflected in higher isolation values for the stream clusters (Table 4.5.5).

The choice of the classification technique and the options within the technique are important and strongly influence the results. Mangiameli et al. (1996) concluded that the SOM was superior to seven hierarchical clustering algorithms tested. They observed that the SOM classification is robust across all kind of data with different imperfections, such as outliers. In our study isolation values were higher for the NHC classification. However, it is not possible to state that one technique is better than the other because they focus on different characteristics of the data. The suitability of the classification depends directly on the application goal. The SOM technique results in a more uniform distribution of the sites over the clusters. This results in higher resemblance and thus lower isolation values. But, on the other hand, uniform distribution of the sites over the clusters can be a great advantage in the development of models for prediction of communities. The skew distribution of the NHC results is probably closer to reality. In the data sets there are large groups of sites that are quite similar and originate from eutrophied channels or normalised streams, while more extreme or undisturbed situations occurred less frequently and therefore made smaller clusters. But, the separation of single sites is a problem in this technique, especially if the sites are not real outliers, but just the endpoints of large gradients.

It is hard to choose the right number of clusters within a technique. Probably there is not just one best option. Both, using 40 and 19/23 clusters resulted in similar isolation values, meaning that the classifications were evenly distinct. Therefore, one should clearly keep the application of the classification in mind, and decide on the number of clusters using cluster tables in which the grouping of the sites is visualised.

Table 4.5.6 Number of sites in which indicator species occur and average abundance for three groups of indicator species, (1) only indicative in NHC, (2) only indicative in SOM, and (3) indicative in both techniques (stream data divided into 40 clusters).

	NHC		SOM		SOM and NHC	
	No. of sites	**Average abundance**	**No. of sites**	**Average abundance**	**No. of sites**	**Average abundance**
average	28	15	41	21	50	34
minimum	1	1	2	1	3	1.5
maximum	185	268	185	203	171	690
10 percentile	3	1.5	7	1.9	13	2.8
90 percentile	70	45	107	34	98	55

Typifying taxa

The number of highly typifying taxa was higher in the NHC classification using 40 clusters, (235 for streams and 289 for channels) than using SOM (217 in streams and 214 in channels). About half of the highly typifying taxa overlapped between NHC and SOM results (126 for streams and 148 for channels). Using 19/23 clusters the numbers of highly typifying taxa decreased for both techniques and both data sets. Probably, the fusion of the sites into fewer clusters made the number of taxa per cluster increase and fewer taxa became characteristic for a cluster. SOM classification had the same number of typifying taxa as the NHC classification for the channels and a higher number for the streams (131 for SOM and 91 for NHC). Thus, there was no general trend either for the number of typifying taxa resulting from a classification technique or for the number of typifying taxa in relation to water type (streams or channels). Using 19/23 clusters only 40 taxa overlapped between both techniques for streams and 36 taxa for channels. The results were thus similar between streams and channels. The number of overlapping typifying taxa was small. This means that the classifications were based on different assemblages of taxa and sites confirming the results mentioned earlier in this paragraph. Typifying taxa which occurred in both classifications were taxa with high abundances and occurrence in many sites (Table 4.5.6). Taxa that were only typifying in the SOM classification also had relatively high frequency and abundance, while the ones that were highly typifying in the NHC classification were less abundant and less frequent. This also means that NHC classifies on less widely distributed taxa than SOM does.

Classification with reduced taxa data

Distribution of the sites over the clusters

Using only part of the taxa data resulted in differences in site distribution over the 40 as well as the 19/23 output units for both stream and channel data. Using the SOM with indicator taxa the results were quite similar to those using all species, the deviation was 20% for streams and 12% or 13% for channels using 23 groups or 40 output units, respectively

(Table 4.5.7). This was expected because the indicator taxa were those taxa with a high maximum probability in the initial SOM including all taxa.

Table 4.5.7 Comparison of the distribution of the sites over the 40 and 19/23 clusters using different parts of the taxa data.

Data set	Deviation	Streams		Channels	
		40 output units	19 groups	40 output units	23 groups
		SOM			
Indicator taxa	type a *	20	20	13	12
	type b **	20	21	13	12
Dominant taxa	type a	44	37	33	31
	type b	44	40	35	29
5 taxonomic groups	type a	54	45	52	52
	type b	55	49	57	56
		NHC			
Indicator taxa	type a	16	15	15	17
	type b	15	20	19	25
Dominant taxa	type a	17	12	19	17
	type b	15	13	17	30
5 taxonomic groups	type a	25	24	31	25
	type b	27	22	34	38

* % of sites that was in one group using all taxa, but put in different groups using the reduced taxa data.
** % of sites that originates from different groups using all taxa and is included in the same group using reduced taxa data.

For dominant species the errors were about 30%. Using 5 taxonomic groups, the distribution of the sites over the groups was very different from using the complete data set for streams as well as for channels, although the number of taxa in these data sets was higher than for dominant taxa. In most cases more than half of the sites were put in a different group. This means that using part of the data including rare as well as common dominant species gives worse results than using part of the data in which only dominant taxa occur. This is explained to some extent by the fact that most of the indicator taxa, which were of importance for the SOM classification were also dominant (148 and 207 taxa overlapped between indicator and dominant taxa for streams and channels respectively). The overlap between dominant or indicator taxa and the taxa of the 5 taxonomic groups was smaller. Thus, dominance plays a role in SOM classification, but not a large role, otherwise the errors for using only dominant taxa would have been smaller. Using larger groups by clustering the 40 output units into 19 and 23 groups for streams and channels respectively, the deviation, if using reduced taxon data, was similar. Using NHC type a as well as type b, deviations were both smaller in all cases than they were in the SOM results (Table 4.5.7). Only the NHC classification with the indicator taxa resulted in larger type a as well as type b deviations in comparison to the SOM results. This could be explained by the fact that the indicator taxa were chosen from the first SOM training and not from the NHC results. Using 5 taxonomic groups in the channel data resulted in the largest deviations of 34% and 38% for type a and type b deviations, respectively. This was similar to the SOM results.

Because the errors were lower for dominant taxa, the NHC seems to cluster sites with similar abundances of dominant taxa.

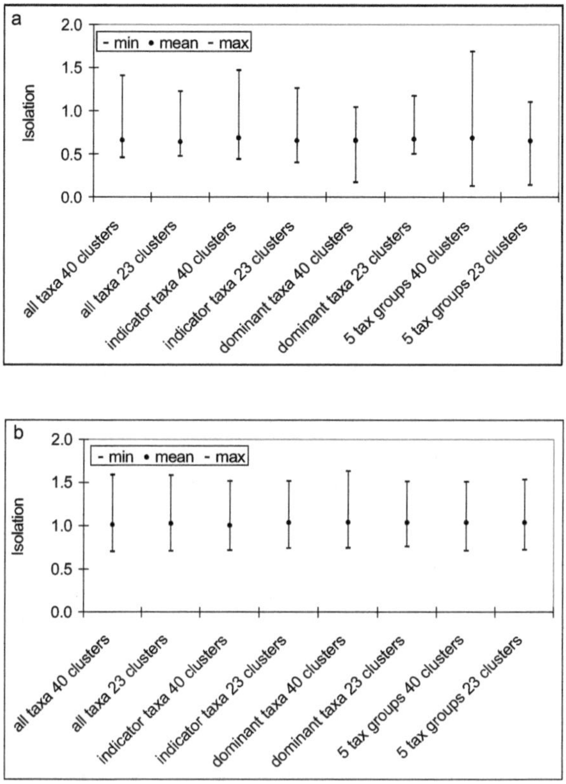

Figure 4.5.6 Minimum, mean and maximum isolation values for the classification in 40 and 23 site groups using SOM (a) and NHC (b) for channel data.

Cluster characteristics

Although the distribution of the sites over the clusters differed strongly between complete and reduced data using the SOM, the isolation values of the resulting classification for channels changed only slightly (Fig. 4.5.6a). This means the classification results were not worse, they were only different. Also the reduction of the number of clusters from 40 to 19/23 did not result in different isolation values. Similar results were found for streams, isolation values varying between 0.61-0.68. Other researchers have reduced the number of taxa before analysis. Chon et al. (2000a), for example, summed the species density into 7 selected taxa of high taxonomic level to avoid noise, caused by species with low densities. They assumed that this might contribute to stabilising the process. However, we did not observe a change of the isolation values of the clusters, which indicates that reducing the number of taxa did not improve nor worsen the classification. Probably, this was caused by the fact that in our study taxa were deleted from the data. Chon et al. (2000a) put many taxa into one by adding their densities. In addition, they used taxon richness within the seven taxa and density of the taxa as input variables. This leads to a small but distinct data set with large gradients in densities and number of taxa.

Similar results were retrieved from the NHC classification of channel sites (Fig. 4.5.6b). For this technique, the results were less surprising, because the changes in distribution of sites over the clusters were much smaller than for the SOM results. The NHC classification had higher isolation values as was already observed in paragraph 3.2. Also the NHC classification kept similar isolation values for the resulting clusters if another number of clusters was made or if reduced taxon data were used. Thus, a different classification of the samples did not result in higher or lower isolation values. Similar results were observed for the stream data: isolation values ranged from 1.03 to 1.18.

In conclusion, for SOM only indicator taxa give similar results compared with using all taxa, and for NHC indicator taxa as well as dominant taxa were useful. Indicator taxa are to be selected after a first classification analysis and are therefore not useful to reduce sampling and sorting costs. Dominant taxa could be useful if NHC is used. Excluding the rare taxa from data analyses could save costs but still there is a difference of 12% to 25% with the classification that included all taxa. Using only 5 taxonomic groups resulted in a completely different classification compared to using all taxa. Still, the question remains whether the community is better or worse described by using all species data or only a selection. Schleiter et al. (1999) concluded that dimension-reducing pre-processing of the data in which the most indicative species are selected caused an increase of the generalisation performance of ANNs and a considerable reduction of the calculation effort. Many rare species are unlikely to be detected by sampling and, even when detected, the estimated abundances of such species are unreliable (Manté et al. 1995). However, this was not confirmed by our study because the isolation values were not lower if all or a selection of taxa was used. Moreover, rare species appeared to be indicative for unimpacted sites and specific habitats (Nijboer and Schmidt-Kloiber 2004).

Comparison between SOM and CCA

Comparing the SOM and CCA results for channels, Figs. 4.5.2, 4.5.7, and 4.5.8 show that the main gradients were similar for both techniques. The ordination diagram (Fig. 4.5.7) shows that chloride and conductivity explained the largest part of the variation in the data (these variables have the longest arrows). On the SOM, the distribution of these variables is restricted to limited areas but the differences between these areas and the others are large (Fig. 4.5.2). Figure 4.5.8 summarises the distribution of environmental variables on the community data set used to train the SOM in Figures 4.5.2 and 4.5.4.

The other variables that are restricted to limited areas on the SOM are: intermittency, NH_4^+ and NO_3^-. In the ordination diagram these variables have shorter arrows than chloride and conductivity, thus they were of less importance. The arrow for intermittency is short, probably because it concerned only few sites and thus these variables had less influence on the total analysis. However, in SOM this variable was of more importance, but only in a limited number of output units. The same goes for emergent vegetation, which was of no importance in CCA but characteristic in a small area on the SOM.

Total phosphorus and width are also restricted to a limited area on the SOM map, but the number of sites in this area was higher, which explains the longer arrows in the ordination diagram. The gradient of these variables was longer. The variables that are widely spread on the SOM map have only moderately sized arrows in the ordination diagram. These variables had a smaller gradient, the differences between the sites were smaller. This is visible in the results of both techniques.

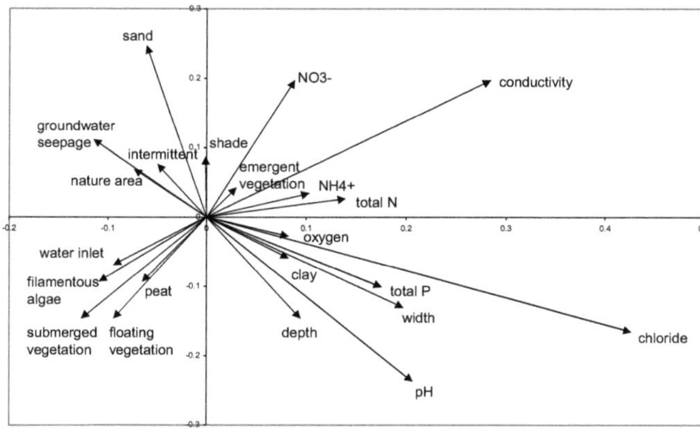

Figure 4.5.7 Ordination diagram resulting from Canonical Correspondence Analysis for channels.

Channels	Streams
chloride, conductivity, intermittent, NH_4^+, NO_3^-	NH_4^+, N-Kjeldahl, total phosphorus
width, total phosphorus, emergent vegetation,	silt, dam, depth, width
sand, nature area, ground water seepage	NO_3^-, shade, meandering, natural profile, natural area, current velocity
floating vegetation, filamentous algae, submerged vegetation, peat	Oxygen, sand
water inlet, depth, pH, total nitrogen, clay,	chloride, conductivity, permanent, pH

distribution on the SOM restricted to limited areas

widely spread over the SOM

Figure 4.5.8 Characteristics of the environmental variables according to their distribution on the SOM. Variables in one group have a similar distribution pattern on the SOM.

For streams, the patterns were even more similar (Figs. 4.5.4, 4.5.8, 4.5.9). The variables NH_4^+, N-Kjeldahl, and total phosphorus are positioned together in the ordination diagram, represented by long arrows, correlated with the second axis, and also in one group, restricted to a limited area on the SOM map. The second group, consisting of silt, presence of dams, depth, and width is also recognisable as a group in the ordination diagram. These variables explained most of the variation on the first axis. They are distributed over a larger area on the SOM map than the first group but from the ordination diagram, it appears that they had long gradients in the data and were therefore important. The last group of variables that are widely spread over the SOM show very small arrows on the ordination diagram. This means that these variables had values that did not differ much between sites. Their influence was only small. The middle group in Fig. 4.5.9 included variables that have long arrows in the ordination diagram, but pointing towards the left lower part of the diagram. These variables were important in about half of the SOM map. The ordination diagram shows that they represent large gradients.

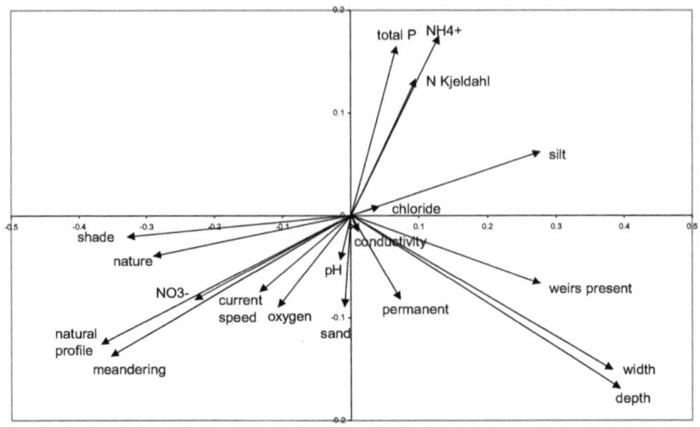

Figure 4.5.9 Ordination diagram resulting from the Canonical Correspondence Analysis for streams.

Although linking environmental variables to the SOM is an indirect technique and analysing the relationships between environment and species data in the CCA is a direct technique, both techniques showed similar results concerning the main gradients in the data. However, this should be tested with data sets including only smaller gradients. The CCA focused more on the length and the direction of gradients while the SOM focused more on the distribution of variables over the clusters. Giraudel and Lek (2001) also concluded that it is not possible to control the direction of the gradients with the SOM. Many researchers used ordination in combination with classification to relate communities to environmental variables (e.g., Verdonschot 1990). But, the availability of both, classification and relating environmental variables to the clusters within one technique could be an advantage, especially, if a model to predict communities from environmental variables is going to be the next step.

Ordination with reduced taxon data

The sum of the eigenvalues of the first four axes decreased if only parts of the taxon data were used. This means that the gradient in the data became smaller. This trend was observed for streams as well as for channels (Table 4.5.8). This result was expected because as more taxa were excluded from the data, variation decreased. The results confirm that with deleting taxa, information is deleted that is not represented by other taxa. For channels, the deviation in the sum of eigenvalues was small for indicator taxa, larger for dominant taxa and largest for taxa from five taxonomic groups. In streams, dominant taxa showed the smallest deviation followed by indicator taxa and taxa from five taxonomic groups. This indicates that dominant species are more important in the stream data than they are in the channel data, although the stream data set with only dominant taxa is the stream dataset with the lowest number of taxa.

The effect on the order of importance of the environmental variables, as resulting from forward selection, appeared to be small (Table 4.5.9). The results did vary between data sets but only for the variables of minor importance (all variables were significant, but the amount of variation in the data they explained differed). The most important variables were the same ones or were only exchanged with the next or previous variable in order of importance. Using only dominant taxa in stream data resulted in an exchange of shade (the fourth variable) with nitrate (the eleventh variable). Probably, the taxa that were deleted were related to shaded waters. However, this change in major variables, using dominant taxa did result in the smallest deviation of eigenvalues compared to the other partial data sets (Table 4.5.8).

The deviation of the order of environmental variables, compared to when all taxon data were used, was highest when only 5 taxonomic groups were used, followed by the use of dominant species in both stream and channel data (Table 4.5.9). Using indicator species resulted in the most similar order of importance of the environmental variables. For streams this order differed from the order of the extent of deviation of the eigenvalues. This can be explained by the fact that using only dominant taxa, another variable becomes more important while the extent of variation in the data remains similar.

Conclusions

Analysing community patterns appeared to be difficult and not objective. There are many techniques that could be used and the two examples in this study showed that different results are obtained with these techniques. A large percentage (50%) of the sites was clustered with other sites if non-hierarchical clustering was used instead of a SOM or the other way around.

The differences depend on the community characteristics on which the technique focuses. One technique is not always better than the other, the most appropriate technique should be chosen depending on the goal of the study and the application of the classification. The SOM appeared to cluster sites with similar numbers of taxa and similar densities for the most dominant taxa. The NHC clustered on the similarity between all species, which the abundance plays a major role. The number of taxa was of minor importance. Stream and channel data showed similar results, although the number of taxa per site was higher for channels, while the total number of taxa was lower.

The number of clusters that should be included in the classification can be chosen within the classification techniques. However, it is very hard to interpret which number of clusters is the most appropriate. It is useful to try classifications with different numbers of clusters and compare the isolation values but this will not automatically lead to the best solution. In this study the mean isolation value of the clusters appeared to be quite similar between the classifications of 40 and of 19/23 clusters. Therefore, it might be better to relate the number of clusters to the application goal of the classification.

Table 4.5.8 Eigenvalues of first, second, third and fourth ordination axes and sum of these four eigenvalues for complete and partial data sets as a result from Canonical Correspondence Analysis.

data set	1st axis	2nd axis	3rd axis	4th axis	sum of axes 1-4
streams all taxa	0.344	0.133	0.082	0.074	0.633
streams indicator taxa	0.339	0.130	0.076	0.064	0.609
streams dominant taxa	0.349	0.134	0.074	0.072	0.629
streams 5 tax groups	0.284	0.126	0.085	0.060	0.555
channels all taxa	0.257	0.146	0.085	0.068	0.556
channels indicator taxa	0.255	0.144	0.082	0.066	0.547
channels dominant taxa	0.250	0.134	0.071	0.061	0.516
channels 5 tax groups	0.260	0.090	0.070	0.062	0.482

Table 4.5.9 Results of forward selection in Canonical Correspondence Analysis. The columns with 'all taxa' show the order of importance of the environmental variables from high importance to low importance. The columns of the other data sets show the deviation, caused by using all taxa, in the number of positions in order of importance (a positive number means the variable has become higher in order of importance, a negative number indicates a lower importance). The total deviance is the total number of exchanged positions.

Channels				Streams			
all taxa	indicator taxa	dominant taxa	5 tax. groups	all taxa	indica- tor taxa	dominant taxa	5 tax. groups
Chloride	0	0	0	depth	0	0	0
Conductivity	0	0	0	natural profile	0	0	0
Width	0	0	0	NH_4^+	0	0	0
Sand	0	0	0	shade	0	-6	0
NH_4^+	0	0	0	width	0	0	0
pH	0	0	-1	oxygen	0	0	-6
natural area	0	0	+1	current speed	0	0	-4
floating vegetation	0	0	0	chloride	0	-1	0
water inlet	-1	0	-2	permanent	0	+1	-1
Peat	+1	0	-8	pH	-1	-1	+1
Depth	0	0	-2	NO3-	+1	+7	+4
Total phosphate	0	-1	+3	conductivity	0	0	+6
NO_3^-	0	+1	+3	silt	0	0	0
Clay	-1	-1	+2	natural area	0	0	-1
Submerged vegetation	+1	+1	-1	meandering	0	-1	-1
Shade	0	0	-3	sand	0	+1	+2
Emergent vegetation	0	0	+2	total phosphate	0	0	-1
Intermittent	-2	-4	+4	N Kjeldahl	-1	0	+1
Groundwater seepage	+1	+1	+2	dams	+1	0	0
Filamentous algae	+1	+1	-1				
Total nitrogen	0	+1	+1				
Oxygen	0	+1	0				
Total deviation	4	6	18	total deviation	2	9	14

Reducing the taxon data resulted, with both techniques, in another distribution of the sites over the clusters for both streams and channels. For the SOM the classification changed more than for the NHC, thus the results again depended on the technique that was used. The classifications had similar values for isolation, thus they were not worse than if the complete data were used. If reduction of the data is desirable one should at least compare the differences with the classification of the complete data to evaluate the suitability of using reduced taxon data. Therefore, it does not yet add to cost effectiveness in water management. The relations with the environmental variables were quite comparable between SOM and CCA. Both techniques could be used together to get the most information out of the data. The main gradients were the same. The advantage of the SOM is that the environmental variables can be related to the clusters on the SOM. The CCA however, is more suitable for showing the length and direction of the gradients. Reducing taxon data reduced the amount of variation in the data but this reduction was only represented by less important environmental variables.

4.6 Prediction of macroinvertebrate diversity of freshwater bodies by adaptive learning algorithms*

Park YS†, Verdonschot PFM, Chon TS, Gevrey M, Lek S

Introduction

The natural distribution of organisms is determined primarily by their environmental requirements (Huntley 1999). Thus, understanding community patterns is important to manage target ecosystems. Especially in aquatic ecosystems, communities of benthic macroinvertebrates are important to monitor changes of the target system. Benthic macroinvertebrates constitute a heterogeneous assemblage of animal phyla and consequently it is probable that some members will respond to stresses placed upon them (Hynes 1960, Hawkes 1979). Many are sedentary, which assists in detecting the precise location of pollutant sources, and some have relatively long life histories. They provide both a facility for examining temporal changes and integrating the effects of prolonged exposure to intermittent discharges or variable concentrations of pollutants (Hellawell 1986). Therefore, it is promising to characterize the changes occurring in communities to assess target ecosystems exposed to environmental disturbances.

Species richness is an integrative descriptor of the community (Lenat 1988), as it is influenced by a large number of natural environmental factors as well as anthropogenic disturbances (Cummins 1979, Rosenberg and Resh 1993). The disturbances of environmental factors may lead to spatial discontinuities of predictable gradients and losses of taxa (Ward and Stanford 1979). Species richness is known to be sensitive to environment changes in stream ecosystems (Resh and Jackson 1993), and is used as a biological indicator of disturbance. As with species richness, diversity indices decrease under increasing disturbance and stress on the ecosystem. The Shannon-Weaver diversity index (Shannon and Weaver 1949) is commonly used to describe the diversity of a particular community. The index is a function of both the number of species in a sample and the distribution of individuals among those species (Klemm et al. 1990). The diversity index is often used as an ecological indicator for the assessments of ecosystems (Bahls et al. 1992).

Development of methods for patterning spatial and/or temporal changes in communities has currently become an important issue in ecosystem management. The River Invertebrate Prediction And Classification System (RIVPACS) was developed to assess water quality. The RIVPACS and its derivates belong to the first integrated ecological assessment analysis techniques (Wright et al. 1993b, Norris 1995). The models are based on a stepwise progression of multivariate and univariate analyses (Barbour et al. 1999). With nonlinear and complex ecological data, however, nonlinear analysing methods should be preferred (Blayo and Demartines 1991). An artificial neural network is a versatile tool for dealing with problems to extract information out of complex, nonlinear data, and could be effectively applicable to classification and association (Lek and Guégan 2000, Recknagel 2003).

* This work was supported by the EU project PAEQANN (EVK1-CT1999-00026).

† Corresponding: park@cict.fr

In ecological modelling, artificial neural networks are more and more used for data organization and classifying groups (Chon et al. 1996, Park et al. 2001a), patterning complex relationships between variables (Lek et al. 1996a, Scardi 2000), and predicting population development (Tan and Smeins 1996, Stankovski et al. 1998). Most of these models used two popular artificial neural networks: a multiplayer perceptron using backpropagation algorithm (Rumelhart et al. 1986b) and a Kohonen's Self-Organizing Map (Kohonen 1982). In the study of the benthic macroinvertebrates, in particular, a SOM has been used for patterning communities (Chon et al. 1996, 2000a, Park et al. 2001a, 2003a), for water quality assessments (Walley et al. 2000, Aguilera et al. 2001), and for prediction of population and communities (Céréghino et al. 2001, Obach et al. 2001). In addition, the MLP has been applied to the prediction of community parameters and species composition (Chon et al. 2001, Park et al. 2001a, 2003a), and bioassessment of water quality (Schleiter et al. 1999). The networks are mainly used to predict target values or to classify input vectors in a model. It is not easy to conduct both classification and prediction in such networks at the same time. However, patterning and predicting could effectively be carried out in a network. One example is a counterpropagation network (Hecht-Nielsen 1987), which consists of unsupervised and supervised learning algorithms. It classifies input vectors and predicts output values. This study aims to apply a counterpropagation network for patterning and for predicting the ecological data consisting of benthic macroinvertebrate communities and environmental variables. It could be a useful tool in managing aquatic ecosystems according to the EU Water Framework Directive (European Parliament 2000).

Materials and methods

Ecological data

To implement the CPN, benthic macroinvertebrate communities and the corresponding environmental variables were used. The data sets were extracted from the EKOO database in the Netherlands (Verdonschot and Nijboer 2000). The data were collected at 664 sites (Fig. 4.6.1) of 23 different water types (Table 4.6.1) in the province Overijssel, The Netherlands. The EKOO studied the relative abundances of macroinvertebrates at given sites according to the characteristics of the environmental variables. At each sampling site, the major habitats were selected over a 10- to 30-m long stretch of the water body and were sampled with the same sampling effort. The sampling effort was thus standardised for each site. A total of 854 species were recorded, Chironomidae, Coleoptera, and Oligochaeta being the most abundant taxa in the dataset. From the community matrix, two community indices; species richness (SR; number of species at each sampling site) and diversity index of Shannon-Weaver (SH) were extracted to evaluate the benthic macroinvertebrate community structure at each sampling site. The mean species richness was 54.46 (±0.94 SE) ranging from 2 to 132, and mean diversity index was 5.29 (±0.03 SE) ranging from 0.49 to 6.77.

Ninety environmental variables (34 quantitative variables in Table 4.6.2 and 56 qualitative variables in Table 4.6.3) were also measured at sampling sites. Qualitative variables were evaluated 0 or 1 according to their characteristics in the sampling areas. We used two different data sets based on quantitative and qualitative environmental variables. They were separately used to predict species richness (SR) and Shannon diversity index (SH) of macroinvertebrate communities. The prediction abilities of each data set were compared. Out of 664 data sets 500 were used to train the network, while the remaining sets (164) were applied to test the feasibility of the trained network in each data set.

Table 4.6.1 Water types of sampling sites and number of samples collected in each habitat.

Acronym	Water type	No samples
BB	Lower courses	24
BK	Springs sources	21
BO	Upper courses	63
BP	Remaining stream pools	17
BR	Springs	22
BV	Spring ponds	1
DW	Temporary water	25
KA	Canals	35
KB	Regulated small rivers	34
KO	Deep ponds	27
LS	Peat ditches	29
ML	Middle courses	29
MM	Small lakes	24
PE	Peat pits	26
PO	Shallow pools	24
RM	Large lakes	10
RR	Rivers	33
SB	Regulated streams	24
SG	Spring gutter	1
SL	Ditches	97
VA	Peat canals	42
VE	Moorland pools	32
ZW	Sand and clay pits	24

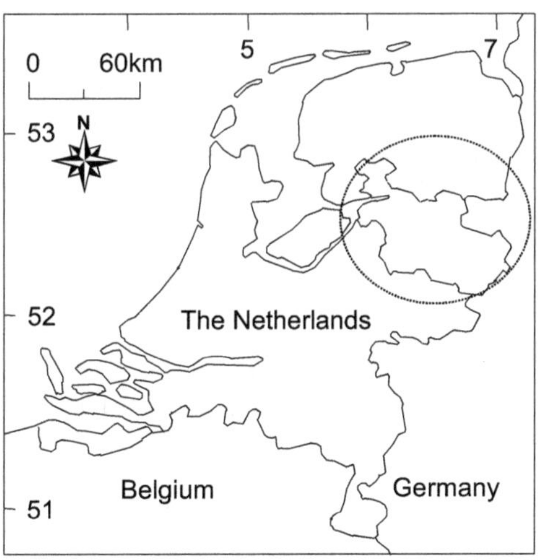

Figure 4.6.1 Study area in the province of Overijssel, The Netherlands.

The qualitative variables were coded in 0 (absence) and 1 (presence) and used in the model without any transformation. However, the input data – both quantitative environmental variables and biological attributes (SR and SH) – were proportionally scaled between 0 and 1 in the range of the minimum and maximum values. Environmental variables showing high variations were log-transformed before normalization to delimitate variations. In log-transformation, value 1 was added to avoid the problem of log of zero. In the modelling process, sampling sites were classified using environmental variables in the SOM layer, and then the network was applied to predict SR and SH in the Grossberg layer according to the output signal of the SOM.

Modelling procedure

The CPN proposed by Hecht-Nielsen (1987), is a combined network of the two ANN: the Kohonen SOM (Kohonen 1982) and the Grossberg outstar (Grossberg 1982). The name counterpropagation is derived from the initial presentation of this network as a five-layered network with data flowing inward from both sides. There is literally a counterflow of data through the network (Fig. 4.6.2). The network is designed to approximate a continuous function f:$A \subset R^m \rightarrow B \subset R^n$ defined on a dataset A, and serves as a statistically optimal self-programming lookup-table (Hecht-Nielsen 1987).

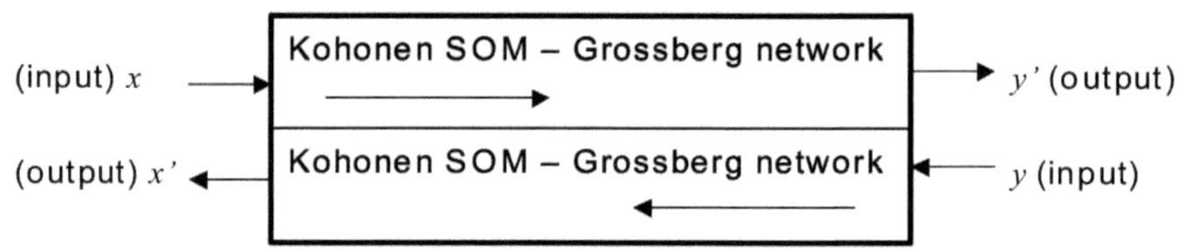

Figure 4.6.2 A Schematic diagram of full counterpropagation network.

The full network works best if the inverse function f^{-1} exists and is continuous. It is assumed that the **x** and **y** vectors are drawn from A and B, respectively. During the training, (**x**,**y**) of f (where **y**=f(**x**)) are presented to the network from both sides. These **x** and **y** vectors then propagate through the network in a counterflow manner to yield output vectors **x'** and **y'** which are intended to be approximations of **x** and **y**, respectively (Hecht-Nielsen 1987, Lin and Lee 1996). Recently, these characteristics were successfully applied for patterning hierarchical relationships among taxonomic groups of benthic macroinvertebrate communities (Park et al. 2001b). However, the structure of the five-layer network is complex, and can be considerably simplified without loss of accuracy (Hecht-Nielsen 1987, 1990, Lin and Lee 1996). In this study we used a forward-only CPN composed of three layers which is a specific type of CPN without counterflow, only flowing from input to desired output (Fig. 4.6.3).

In the modelling process, initially the data vectors **x** (explanatory variables) and **y** (dependent variables) are sent to the SOM and the Grossberg layers, respectively. Then, the weights are updated for a given set of data vectors **x** and **y**. For the CPN this occurs in two phases. First, the SOM layer is trained. It works like a hidden layer of a multiplayer perceptron. At the SOM layer, when the input vector **x** is sent through the network, each neuron k of the network computes the distance between the weight vector **v** and the input vector **x**. Among all N output neurons in two dimensions, the best matching unit (BMU) of minimum distance becomes the winner (e.g., neuron q). The BMU and its neighbouring neurons are allowed to learn by changing their weights so as to further reduce the distance between weight and input vectors as follows:

$$v_{jk}(t+1) = v_{jk}(t) + \eta(t)(x_j - v_{jk}(t))z_k \quad (4.6.1)$$

where v_{jk} is the weight between neuron j of the input layer and neuron k of the SOM layer, z_k is assigned 1 for the winning (and its neighbouring) neuron(s) while it is assigned 0 for the other neurons, and $\eta(t)$ denotes the fractional increment of the correction ($0 < \eta(t) < 1$). The radius that defines the neighbourhood is usually set to a larger value early in the learning process, and is gradually reduced as convergence is reached. This results in training the layer to classify the input vectors by the weight vector **v** they are closest to.

Table 4.6.2 Thirty-four quantitative environmental variables used in the model. The groups (A-I) are based on the similarity of distribution patterns of each variable on the trained SOM map in Fig. 4.6.4.

Variables	Acronyms	Unit	Mean (SE)	Groups
Percentage cover emergent vegetation	BOVE%	%	6.77 (0.54)	A
Percentage cover floating vegetation	DRIJ%	%	11.67 (0.89)	A
Percentage cover floating algae	FLAL%	%	3.82 (0.56)	D
Percentage sampled habitat: emergent vegetation	MMBO%	%	16.16 (0.86)	I
Percentage sampled habitat: detritus	MMDE%	%	9.01 (0.69)	H
Percentage sampled habitat: floating vegetation	MMDR%	%	12.96 (0.79)	I
Percentage sampled habitat: gravel	MMGR%	%	1.36 (0.20)	H
Percentage sampled habitat: clay	MMKL%	%	0.51 (0.14)	I
Percentage sampled habitat: bank	MMOE%	%	18.24 (0.91)	I
Percentage sampled habitat: submerged vegetation	MMON%	%	12.05 (0.76)	D
Percentage sampled habitat: silt	MMSL%	%	15.67 (0.73)	F
Percentage sampled habitat: stones	MMST%	%	0.72 (0.13)	H
Percentage sampled habitat: peat	MMVE%	%	2.20 (0.26)	D
Percentage sampled habitat: sand	MMZA%	%	10.51 (0.65)	H
Dissolved oxygen percent saturation	O2%	%	90.70 (1.66)	G
Percentage cover bank vegetation	OEVE%	%	6.10 (0.57)	I
Percentage cover submerged vegetation	ONDE%	%	11.23 (0.92)	D
Percentage cover all vegetation	TOTB%	%	33.14 (1.38)	A
Width of stream	WIDTH	m	64.24 (18.16)	E
Ratio width/depth	WD/DP		28.51 (4.54)	C
Calcium	Ca++	mg/l	51.21 (1.01)	B
Chloride	Cl-	mg/l	52.79 (1.98)	B
Depth	DEPTH	m	1.13 (0.06)	E
Silt thickness	DSAPR	m	0.11 (0.01)	I
Electric conductivity	ECOND	uS/cm	427.95 (9.18)	B
Ammonium	NH4+	mgN/l	1.46 (0.14)	F
Nitrate	NO3-	mgN/l	3.87 (0.32)	H
Oxygen concentration	O2	mg/l	9.71 (0.16)	G
Ortho-phosphate	O-P	mgP/l	0.29 (0.03)	F
Acidity	pH		7.13 (0.04)	B
Flow velocity	VELOC	m/s	0.07 (0.01)	H
Water temperature	TEMP	°C	13.26 (0.24)	C
Total-phosphate	T-P	mgP/l	0.51 (0.05)	F
Slope	VERVA	m/km	5.91 (0.81)	H

Once the SOM layer is trained, the Grossberg layer (Grossberg 1969), the output layer of the CPN, can be trained. This is done in supervised mode according to the following procedure. An input vector **x** is applied to the CPN, the output of the SOM layer is established, and the Grossberg layer outputs are calculated. In this process, the Grossberg layer receives z vector signals from the SOM layer. If the difference between the desired output and the calculated output of the Grossberg layer is greater than the acceptable error, the weights are updated using the following Grossberg outstar learning rule:

$$w_{ik} = w_{ik} + \beta(y_i - w_{ik})z_k \tag{4.6.2}$$

where w_{ik} is weight between neuron k of the SOM layer and neuron i of the Grossberg layer, β is the learning rate, and z_k is assigned to 1 for neuron q (BMU) while set to 0 for all other neurons of the SOM layer. The weights correspond to the averages of the desired outputs **y** associated to the inputs **x** according to the equiprobability of the winning neurons of the SOM layer. By repeating this process until the weight differences become sufficiently small, the relationship of the two variable sets is preserved in the weights of the network. Finally the trained CPN actually functions as a statistically optimal self-programming look-up table.

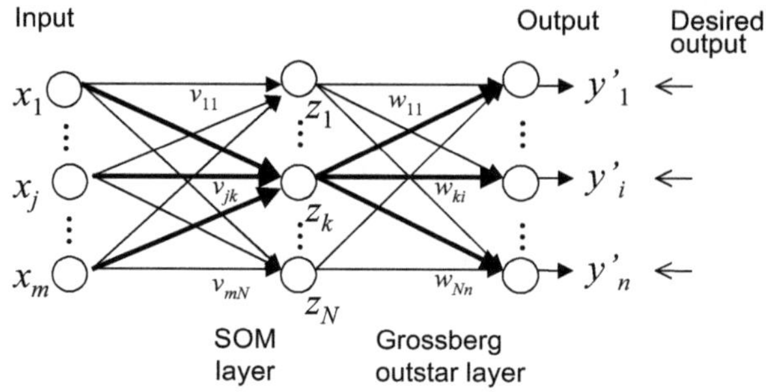

Figure 4.6.3 Schematic diagram of a forward-only counterpropagation network.

The SOM layer of the CPN works as a hidden layer in a multiplayer perceptron. Accordingly, a number of neurons of the SOM layer affect a predictability of the CPN. Therefore, to find the optimum neuron size of the SOM layer we trained the network with different map sizes, and chose 63 (9 × 7) and 56 (8 × 7) neurons for the SOM layer on the 2D hexagonal lattice for quantitative dataset and qualitative dataset, respectively. After training the CPN in this study, a unified-matrix algorithm (U-matrix, Ultsch 1993) was applied to detect the cluster boundaries on the map of the SOM layer. The algorithm is commonly used to show clusters on the SOM showing distances between neurons. High values of the U-matrix indicate cluster boundaries.

Relationships between biological and environmental variables

The weight vectors of the SOM layer tend to approximate the probability density function of the input vector (Kohonen 2001). The visualization of these vectors according to different input variables is an efficient way to understand the contribution of each input variable to the clusters on the trained SOM map. Therefore, the values calculated for each input variable during the learning process were visualized on the trained SOM map with a grey scale to represent the relationships between the input variables and the clusters of the trained SOM.

Furthermore, to understand the relationships between input (environmental) variables and output (biological) variables, mean values of output variables were calculated in corresponding units of the trained SOM map. If the output neuron is not occupied with any input vectors, the vacant neurons are replaced by the mean value of the neighbouring neurons. These mean values of environmental variables assigned to the SOM map were visualised on the grey scale, and then compared with maps of sampling sites as well as with biological attributes. Furthermore, to compare the relationships between environmental variables, they were classified into several groups based on their distribution patterns on the trained SOM map with weight vectors of the trained SOM. In addition, correlation coefficients were calculated within each group using the weight vectors.

Multiplayer perceptron with backpropagation algorithm (BP)

The multiplayer perceptron with backpropagation algorithm was also applied to the EKOO database to predict species richness and diversity index with quantitative environmental variables, and then the performance of the models was compared with those of the CPN. The BP is most popular and used more than other neural network types in various fields of investigation. It is an interactive algorithm designed to minimize the mean square error between the computed output of the network and the desired output. The network normally consists of three layers: input, hidden, and output layers. It requires input vectors in the input layer, as well as target (or desired) values in the output layer corresponding to each input vector. In this study the network consists of 34 neurons in the input layer, 10 neurons in the hidden layer, and two neurons in the output layer. The learning algorithm of the BP is very popular and common, and the detailed description will not be given here. Procedures of the learning rules can be found in Rumelhart et al. (1986b), Kung (1993), and Lek and Guégan (2000). After the learning process, the datasets not included in the training process were applied to test the reliability of the trained BP, and then the prediction power was compared with that of the CPN.

Results

Quantitative environmental data set

Patterning samples with environmental variables

The CPN patterned the input vectors in the SOM layer, and a U-matrix method clustered the trained SOM map. The results showed five clusters (I-V) of sampling sites grouped according to environmental gradients, and two subclusters Va and Vb were observed in cluster V (Fig. 4.6.4). Each cluster was mainly associated with the characteristics of the water types. For instance, cluster I mainly consisted of sites of moorland pools (VE), cluster II of ditches (SL), cluster III of stagnant water bodies (VA, PE, PO, and KA), cluster IV of large rivers and lakes (RR, RM, KA, and ZW) and ditches (SL). Finally, clusters Va and Vb were characterized respectively by springs and upper watercourses (BK, BO and BR), and by intermittent or regulated streams (BP, DW and SB). These distribution patterns show the characteristics of natural key conditions. The sampling sites located on the left areas of the SOM map were mainly from unregulated water systems, whereas sites on the right were from regulated areas (Fig. 4.6.4). The water types are listed in Table 4.6.1, and their characteristics are described by Verdonschot (1990).

Fig. 4.6.5 displays the contribution of each input variable for the classification of sampling sites on the trained SOM map. Dark areas represent high values, while light ones display low values. Acronyms of environmental variables are shown in Table 4.6.2. Each variable displayed a high gradient distribution on the trained SOM map. Nine groups were observed among the input variables according to their distribution similarities: the first eight groups (A-H) showed high correlations among environmental variables within each group (mean correlation coefficient r=0.74 (±0.06 SE) ~ 0.93 (±0.06 SE)), whereas the last group (I), which does not belong to any other group, showed relatively low correlations (mean correlation coefficient r=0.34 (±0.04 SE)) among variables within the group.

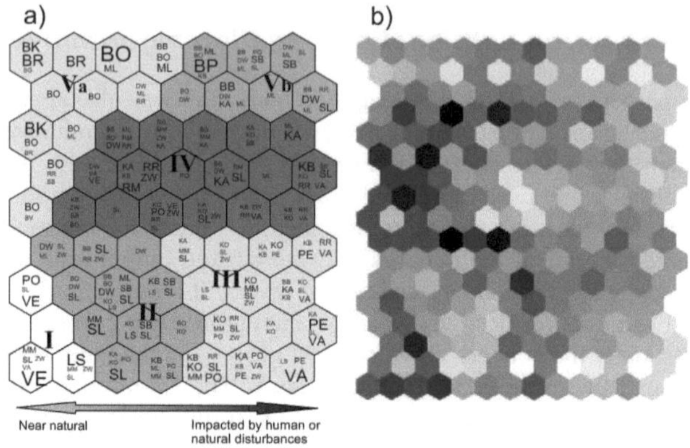

Figure 4.6.4 Classification of sampling sites with quantitative environmental variables using the SOM. The U-matrix algorithm was applied to cluster the SOM map. The Latin numbers (I-V) represent different clusters. The acronyms in the hexagonal neurons represent different habitats, and are shown in Table 4.6.1. The size of the acronym font is proportional to the number of sampling sites in the same habitats in the range of 1-18 sampling unites.

The groups of variables show different aspects of the environment. For example, group B was related to electric conductivity and group F was characterised by inorganic nutrients (NH_4^+, T-P, and O-P). The groups also showed different local habitat characteristics. Groups A and D were concerned with the percentages of vegetation cover, whereas group H typically represents the characteristics of upper water course habitats showing high percentages of detritus, stones, sand, and gravel with high current velocities and strong slopes. The morphological characters of streams (width and depth) were grouped together in the group E.

The next step is to compare the relationship between clusters of sampling sites and groups of environmental variables. Clusters I and II were related to low values of group B and high values of group D, and cluster III was represented by high values of groups D and G, and low values of group H (Fig. 4.6.4). Similarly, cluster IV displayed high values of groups B and E and variables MMBO%, MMKL%, and MMOE% of group I, and sub clusters Va and Vb were strongly related with high values of groups H and F, respectively. Furthermore, the sampling sites in the left areas of the SOM map (Clusters I, II, Va) mainly display natural water systems, while the sites in the right areas (Cluster III, IV, Vb) reveal either natural or anthropogenically disturbed aquatic systems. Overall, the Fig. shows that sites of clusters I and II in the lower areas of the SOM map are not disturbed and contain well developed vegetation, whereas the sites of cluster Vb in the upper area are disturbed by regulation and nutrients (e.g., nitrate, ammonium, ortho-phosphate, and total phosphate) which are presumably due to increased amounts of dissolved ions entering the water through agricultural activities.

Relationship between environmental variables and community indices

To evaluate the relationships between environmental variables and diversity indices (SR and SH), the mean values of SR and SH were visualized on the trained SOM map in grey scale (Fig. 4.6.6). The results showed that SH and SR were higher in the lower right areas of the SOM map than in the upper left areas. The low values were characterised by group H representing high percentages of stone, gravel, sand, and detritus in substrates and water types of springs and upper courses (cluster Va), and by group F showing high concentration of nutrients or sites representing intermittent and regulated water systems (cluster Vb). SR and SH were also related to dissolved oxygen (group G).

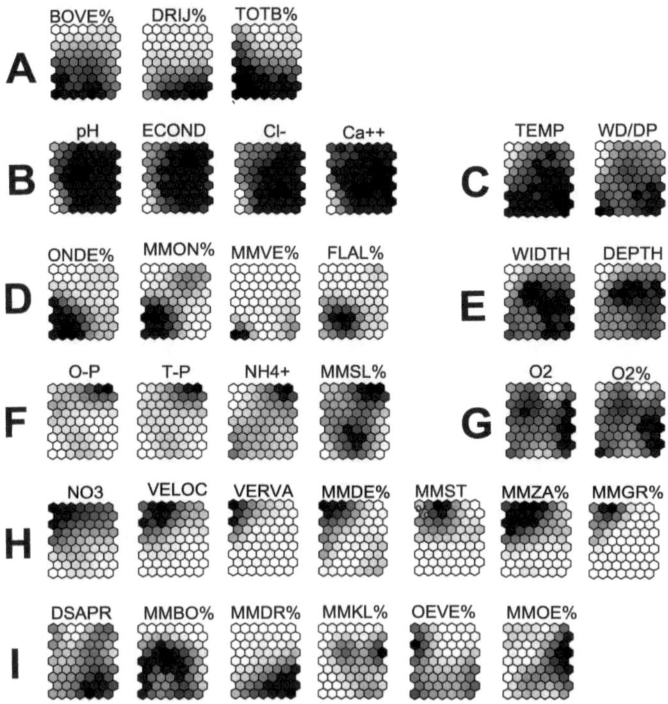

Figure 4.6.5 Component planes displaying the contribution of each quantitative environmental variable to classification of sampling sites. Based on the distribution pattern on the SOM map, nine groups (A-I) were identified. The names of the environmental variables are given in Table 4.6.2. Dark represents high values, whereas light is for low values.

Predicting community indices

The trained CPN serves as a 'look-up table' for finding the corresponding values between the input and output variables. The Grossberg layer of the trained network showed high predictability in the learning process (Fig. 4.6.7a, b). Correlation coefficients between observed and estimated values were 0.90 ($P<0.01$) for both SH and SR. In both cases overestimations were observed at low values, while under-estimations were observed at high values. This is caused by the structural characteristics of the data. There are few cases with low values in both SH and SR. The residuals between observed and estimated values averaged 0.000 for both SH and SR, and their standard deviations were 0.044 and 0.065 for SH and SR, respectively. The distribution of error values showed that most error values lie around zero. The data not used in the learning process were applied to test the feasibility of the trained network.

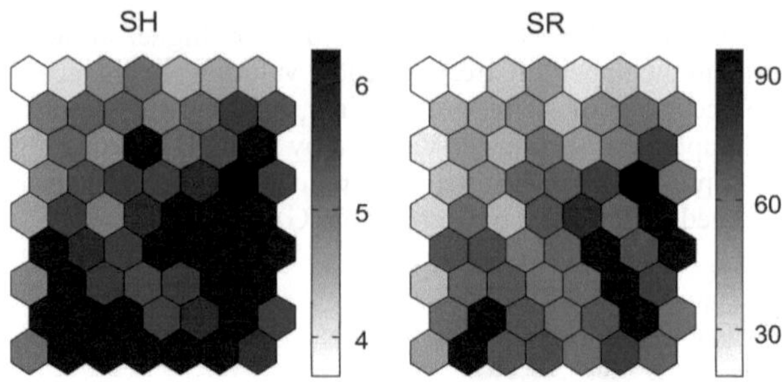

Figure 4.6.6 Distribution of Shannon's diversity index (SH) and species richness (SR) on the SOM map trained with quantitative environmental variables. Dark represents high values, whereas light displays low values.

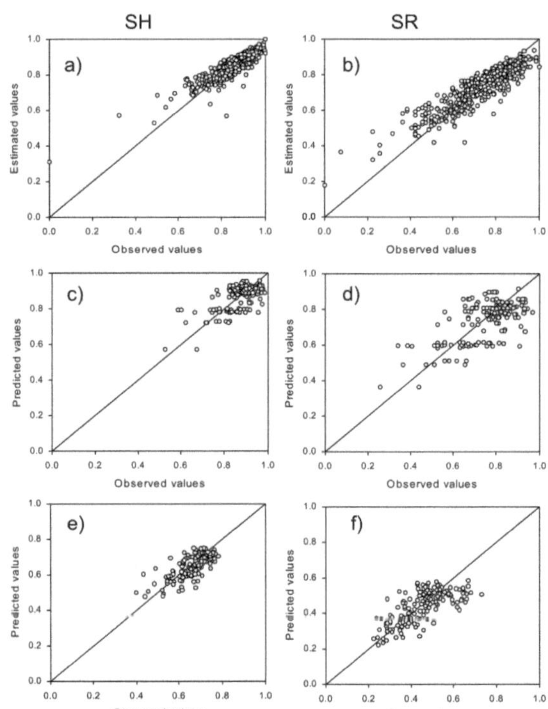

Figure 4.6.7 The results of the model to predict diversity index (SH) and species richness (SR) with 34 quantitative environmental variables. Learning results of the model for SH (a) and SR (b) and results of the model tested by the data set not used in the learning process (c and d for SH and SR, respectively). SH (e) and SR (f) were also predicted by with the multiplayer perceptron with quantitative environmental variables.

The results showed a high predictability of the network. The correlation coefficients between observed and predicted values were 0.70 and 0.67 for SH and SR, respectively ($P<0.001$) (Figs. 4.6.7c, d). The residuals between observed and predicted values were located around zero showing averages of 0.009 (±0.059 SD) and 0.014 (±0.103 SD) for SH and SR, respectively. A majority of frequencies of the error terms also appeared around zero. Thus, the results showed that the trained CPN corresponded well to the reality of SH and SR.

Qualitative environmental data

Patterning samples with environmental variables

The CPN was also applied to predict SH and SR with qualitative environmental data. At the first step the SOM layer classified sampling sites into three major clusters (I, II and III) according to the gradient of environmental variables (Fig. 4.6.8). The SOM effectively classified different water types. Cluster I was characterized by the streams BK, BO and BR which are water types of springs and/or upper water courses, and related to cluster Va of the quantitative dataset (Fig. 4.6.4). Cluster II represents water types of ponds, pools, and/or lakes characterized by the KO, MM, and VE, and cluster III represents those of ditches and/or canals characterized by LS, SL and VA.

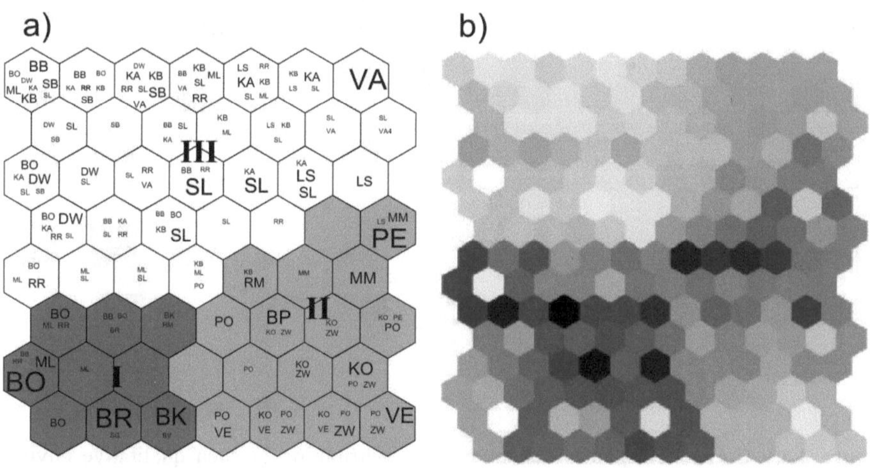

Figure 4.6.8 Classification of sampling sites with qualitative environmental variables using the SOM. The U-matrix algorithm was applied to cluster the SOM map. The Latin numbers (I-III) represent different clusters. The acronyms are explained in Fig. 4.6.4. The size of the acronyms font is proportional to the number of sampling sites in the same habitats in the range of 1-20 sampling unites.

These clusters were explained by the distribution of each component (environmental variable). Fig. 4.6.9 shows the distribution of environmental variables calculated on the trained SOM map. The darker the intensity, the higher the probability of each variable. Environmental variables showed eight groups according to their distribution patterns on the map. In groups they showed high correlation between variables, although they were qualitative (presence/absence) data. The qualitative dataset given to the SOM as input is converted into a quantitative dataset as a probability of occurrence at each sampling site (or units of the SOM) through the learning process of the SOM. The component map displays the probability of occurrence of each variable in each SOM map unit. Based on this component plane, cluster I was effectively explained by high values of groups A and B, and low values of group L, cluster II by group H, and cluster III was characterised by high values of groups E, J, and F, and low values of groups A, B, H, I and J. Fig. 4.6.9 also shows some variables such as groups A, B, E, and H are very specifically related to water types.

In considering environmental status, the upper areas of the SOM map concern environments disturbed and modified by physical and chemical factors, whereas the lower areas concerned relatively less disturbed and near natural environments. These characteristics also agree with the clusters of sampling sites. The acronyms of environmental variables are shown in Table 4.6.3. Cluster I on the lower left areas of the SOM map is concerned with springs and upper courses representing relatively clean and unmodified environments, whereas cluster II on the upper areas results mainly from ditches and canals polluted and impacted by human activities.

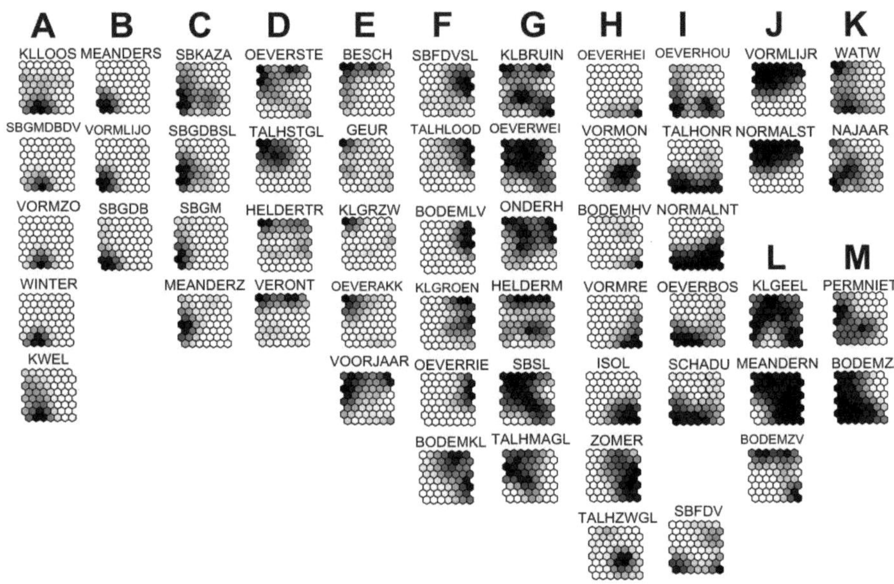

Figure 4.6.9 Component planes displaying the contribution of each qualitative environmental variable to classification of sampling sites. Based on the distribution pattern on the SOM map, 12 groups (A-L) were identified. The names of the environmental variables are given in Table 4.6.3. Dark represents high values, whereas light is for low values.

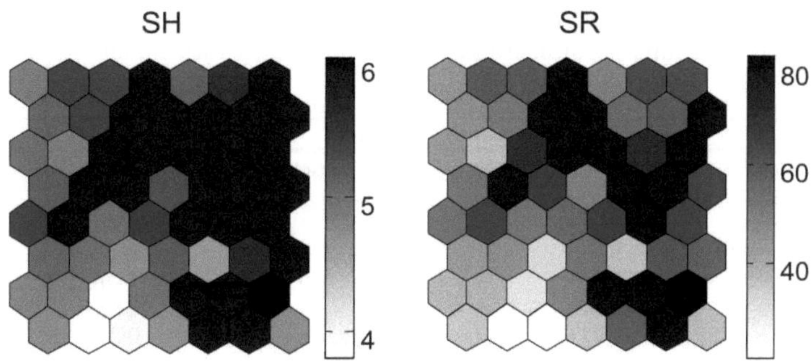

Figure 4.6.10 Distribution of Shannon's diversity index (SH) and species richness (SR) on the SOM map trained with qualitative environmental variables. Dark represents high values, whereas light displays low values.

SH and SR could also be explained by the relationships with clusters and environmental variables. Their mean values in each unit were visualized on the trained SOM (Fig. 4.6.10). Dark represents high values, whereas light is for low values. Both SH and SR showed high values in the upper areas and in the right areas of the SOM map which belong to clusters II and III, while the values were relatively low in the lower left areas which are assigned to cluster I. Cluster I is characterized by the water types of springs and upper watercourses. The low values of SH and SR were related to high probabilities of colourless water, very irregular shape of sampling area, substrate course with material, detritus, leaves, and peat, seepage, and winter in group A, and irregular linear shape of sampling areas, strong meandering, substrate course with detritus and leaves in group B (Fig. 4.6.9, Table 4.6.3). Groups A and B are mainly represented by a high frequency of irregular shapes of sampling areas, substrate course with detritus and leaves, and season winter. And the low species richness and diversity index are also related to low probability of yellow colour of water and soil type of sand and peat in the group L (Fig. 4.6.9, Table 4.6.3). These characteristics are mainly observed in the upper watercourses.

Predicting community indices

The Grossberg layer showed high predictability of SH and SR4 with qualitative environmental variables ($r=0.90$, $P<0.001$ for both SH and SR) (Fig. 4.6.11a, b) like that of quantitative data. Here again, overestimations were observed at low values, while underestimations occurred at high values. However, the new data sets not used in the learning process showed very low correlations between observed and predicted values (Figs. 4.6.11c, d). Therefore, this model was not successful.

In results of learning process, the residuals between observed and predicted values were centred near zero showing averages of 0.000 (±0.045) and 0.000 (±0.067) for SH and SR respectively. However, the distribution of the residuals did not show normality ($P<0.05$). The results of the testing phase also displayed similar patterns compared with those of the learning process.

Table 4.6.3 Fifty-six qualitative environmental variables used in the model. The groups (A-L) are based on the similarity of distribution patterns of each variable on the trained SOM map in Fig. 4.6.8.

Variables	Acronyms	Groups	Variables	Acronym	Groups
Autumn	NAJAAR	K	Profile slope: irregular	TALHONR	I
Bank type: Field	OEVERAKK	E	Regular shape	VORMRE	H
Bank type: Forest	OEVERBOS	I	Seepage	KWEL	B
Bank type: Heathland	OEVERHEI	H	Shadow	SCHADU	I
Bank type: Pasture	OEVERWEI	G	Shape very irregular	VORMZO	A
Bank type: Reeds	OEVERRIE	F	Smell	GEUR	E
Bank type: Urban	OEVERSTE	D	Soil type: clay	BODEMKL	F
Bank type: Wooded bank	OEVERHOU	I	Soil type: peat	BODEMLV	F
Color of the water grey-black	KLGRZW	E	Soil type: peat in fenland	BODEMHV	H
Colour brown	KLBRUIN	G	Soil type: sand	BODEMZA	M
Colour green	KLGROEN	F	Soil type: sand/peat	BODEMZV	L
Colour yellow	KLGEEL	L	Spring	VOORJAAR	E
Colourless	KLLOOS	A	Strong canalisation	NORMALST	J
Irregular shape	VORMON	H	Substrate coarse detritus/leaves	SBGDB	B
Isolation	ISOL	H	Substrate coarse detritus/leaves/silt	SBGDBSL	C
Linear shape irregular	VORMLIJO	B	Substrate coarse material	SBGM	C
Linear shape regular	VORMLIJR	J	Substrate coarse material/detritus/leaves/peat	SBGMDBDV	A
Maintenance	ONDERH	G	Substrate fine detritus/peat	SBFDV	I
Meandering none	MEANDERN	L	Substrate fine detritus/peat/silt	SBFDVSL	F
Meandering strong	MEANDERS	B	Substrate sand	SBKAZA	C
Meandering weak	MEANDERZ	C	Substrate silt	SBSL	G
No canalisation	NORMALNT	I	Summer	ZOMER	H
Pollution	VERONT	D	Temporary	PERMNIET	M
Profile consolidation	BESCH	E	Transparency: clear	HELDERH	I
Profile slope: 0 – 30 degrees	TALHZWGL	H	Transparency: slightly turbid	HELDERM	G
Profile slope: 30 - 45 degrees	TALHMAGL	G	Transparency: turbid	HELDERTR	D
Profile slope: 45 - 75 degrees	TALHSTGL	D	Water level fluctuation	WATW	K
Profile slope: 75 - 90 degrees	TALHLOOD	F	Winter	WINTER	A

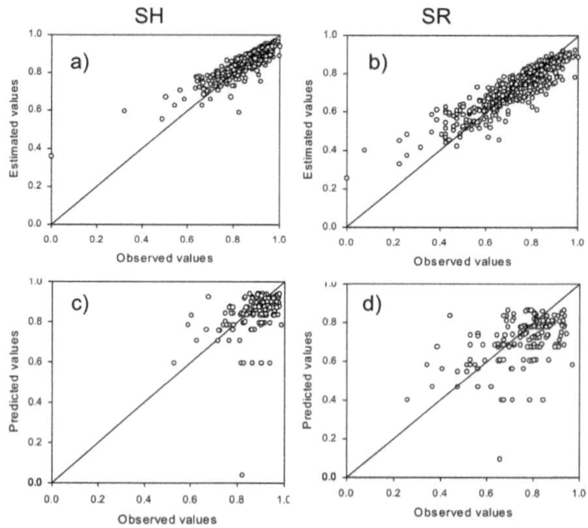

Figure 4.6.11 The results of the model to predict diversity index (SH) and species richness (SR) with 56 qualitative environmental variables. Learning results of the model for SH (a) and SR (b) and results of the model tested by the data set not used in the learning process (c and d for SH and SR, respectively).

Comparison with a multiplayer perceptron

The multiplyer perceptron with the backpropagation algorithm was also trained to predict SH and SR with quantitative environmental variables. The trained results showed high correlation coefficients between observed values and values estimated by the model ($r=0.87$ and 89 for SH and SR, respectively, $P<0.01$). And the test results with new datasets not used in the training phase showed SH and SR were well predicted with high correlation coefficients between desired values and predicted values by the model ($r=0.75$ and 0.72 for SH and SR, respectively, $P<0.01$) (Figs. 4.6.7e, f). The residuals between observed and predicted values were centred near zero showing averages of 0.015 (±0.079) and 0.011 (±0.050) for SH and SR respectively.

Discussion and conclusion

The CPN was implemented to pattern sampling sites and to predict SH and SR with the environmental variables available in this study. In the first step, the network classified sampling sites into five clusters based on environmental variables using the SOM algorithm, and afterwards the diversity indices (SR and SH) were predicted in the output layer of the network. Thus, the CPN proves to be a general approach to explain the variation of ecological data in two steps: ordination methods to summarize the variability of the data as a first step, and exploration for possible relationships between biological and environmental variables as a second step (Jongman et al. 1995).

The CPN shows different performance to predict SH and SR with quantitative environmental variables and qualitative variables. In both cases, the sampling sites were effectively classified according to environmental gradients. The clusters were mainly related to the water types of the sampling sites, and the coincidence also observed between clusters of the quantitative dataset and the qualitative dataset representing the classification of sampling sites was similar in overall terms. However, in considering the predictability of SH and SR, two models for quantitative dataset and qualitative datasets showed different performances. In both the quantitative dataset and the qualitative dataset, the training results showed high predictabilities with correlation coefficients higher than 0.90. However, the testing results showed differences between each other: displaying the model of the quantitative dataset has higher performance than that of the qualitative dataset. These results demonstrate that quantitative variables are preferred in CPN modeling. . These characteristics are commonly observed in ecological studies. A quantitative dataset has much more information than a binary dataset. Therefore, it can represent its system more reasonably. However, it is very difficult to explore these datasets if quantitative variables have hig levels of background noise or if there are many variables showing very low occurrence frequency and/or very low values. Although qualitative variables did not show high predictability, they may be important factors in the modelling techniques, depending on the model structure

The SOM layer showed the ability to produce a classification of input vectors as well as visualization of relationships among input variables in their contribution to the classification. The analysis using visualization of component planes is comparable to principal component analysis (PCA), but more directly describes the discriminatory power of the input variables in the mapping procedure (Kohonen 2001). The clear distribution of a variable along a gradient represents a high contribution to the classification of input vectors. In this study, the sampling sites were classified into five clusters, and input variables were divided into nine groups. Each cluster was explained very well by environmental groups (Figs. 4.6.4, 4.6.5, 4.6.9, 4.6.10). Furthermore, by overlapping the distribution of both input variables and mean values of diversity indices on the SOM map, the relationships between explanatory (input) variables and dependent (output) variables could be analysed. When there are strong relationships between input variables and output variables, the component planes show clear gradients and similar patterns of their distribution on the trained SOM map. In this study, this approach showed that species richness and diversity indices were strongly related to the concentrations of nutrients, dissolved oxygen, and percentages of vegetation cover in quantitative datasets (Figs. 4.6.5, 4.6.6), and the shape of sampling areas, substrate types, and sampling time in qualitative datasets (Figs. 4.6.8, 4.6.9). Similarly, the composition of the communities was also influenced by different water types. The diversity indices were lower in springs and upper courses and disturbed aquatic systems, whereas they were higher in natural areas (Figs. 4.6.4-6, 4.6.8, 4.6.9). These characteristics, low biodiversity in springs and upper courses, are generally observed in stream ecosystems, and follow the river continuum concept (Vannote et al. 1980). However, it is necessary to quantify the distribution gradient of each variable as well as the relationships between biological and environmental variables.

The structure of the CPN is similar to a combination of two networks; SOM and multiplayer perceptron with BP. Especially when prediction output values are considered, the CPN is related to the BP. The BP generally presents some advantages in handling more classes and complex data, although there is still debate on this point (Ruiz and Srinivasan 1997, Ellingsen 1994). In contrast, the counterpropagation network is more effective in noise sensitivity, can handle more data per class, and performs well without being influenced by the increase in data size. When comparing the prediction power between the CPN and the BP, the BP showed relatively higher predictability for SH and SR than the CPN in this study. This agrees with Ruiz and Srinivasan (1997). However, it is difficult to decide which algorithm should be better suited for patterning communities at the present time, because information extraction and noise sensitivity are equally important in adaptive learning processes with ecological data (Ruiz and Srinivasan 1997, Park et al. 2001b). The rare species are generally considered as noise in most quantitative ecological analysis. However, if the model is too sensitive to noise, it cannot manage variables showing low occurrence and/or low values. For example, in this case, the rare species which are important as good water quality indicators are considered as noise, so important information for the assessment of the water quality is lost. Therefore, the level of noise sensitivity should be adjusted according to the objectives of the model.

In patterning community dynamics using the CPN, Park et al. (2001b) reported that predicted values of the CPN showed an averaging effect in that dominant taxonomic groups appeared more consistently while groups with low densities tended to disappear. The averaging effect would be advantageous in some cases. In the ecological community, for example, if rare species are accidentally introduced at low densities, they would be considered as noise in the data sets. Then the averaging effect in the patterning would be more effective, and not mislead the assessment of the water quality due to the presence of accidental taxa.

In ecological modelling, the CPN can be applied from two points of view based on its structure. First, we can consider the possibility of a full counterpropagation structure. In this case the network uses counterflow of information in both directions. Thus, we can predict both input and output variables from each other i.e. we can predict biological attributes from environmental variables, and we can also predict environmental variables from biological attributes.

Another possibility is the use of a simplified forward-only network. Here, we can use the characteristics of the network which combines unsupervised and supervised learning algorithms. Thus, the network can pattern sampling sites, interpret relationships among environmental variables, biological attributes and sampling sites as well as relationships among input variables, and predict biological attributes for the assessment of target ecosystems in a network. Thus, the CPN can be used as a tool for assessing ecological status and predicting the water quality of target ecosystems.

The structure of the normal community may be changed by perturbations in the environment and the degree of change in community structure may be used to assess the intensity of the environmental stress (Hellawell 1986). Species richness is a function of the stability of the environment (Legendre and Legendre 1998). A stable environment contains more species and more niches, because a more stable environment involves a higher degree of organization and complexity of the food web (Margalef 1958). The niche of a species is the set of environmental conditions that the species does not share with any other sympatric species, so species richness is related to the number of niches (Hutchinson 1957). The diversity index further accommodates the evenness concepts in addition to the taxon richness, and represents heterogeneity of species composition, characterizing the ecological status of communities at a given site and a given time (Hellawell 1986). Based on these facts, species richness and diversity indices are frequently used as biological indicators of target ecosystems in combination. It is worth predicting these indices with their explanatory variables, and they can be used as a tool for the assessment of disturbances in a given ecosystem. In conclusion, the CPN was successfully implemented for patterning and predicting ecological data showing its applicability as a tool for assessing the ecological status and for predicting the water quality of target ecosystems.

4.7 Hierarchical patterning of benthic macroinvertebrate communities using unsupervised artificial neural networks*

Park YS†, Kwak IS, Lek S, Chon TS

Introduction

Patterning communities is essential to reveal the ecological states of the target ecosystem effectively and consistently. Especially in aquatic ecosystems, the composition of residential communities rapidly varies in response to various impacts of natural and anthropogenic perturbations such as flooding and pollution (Hawkes 1979, Hellawell 1986, Spellerberg 1991). Particular attention has been recently focussed on properly assessing changes in water quality through community patterning. As well documented, field community data are nonlinear and complex because they involve many species, fluctuating greatly depending upon numerous effects of endogenous (e.g., physiological development, life cycle, etc.) and exogenous factors (e.g., precipitation, pollution, etc.) (Jongman et al. 1995, Legendre and Legendre 1998). A complex system like the responses of communities to their environments usually develops a hierarchical structure (Allen and Starr 1982, O'Neill et al. 1986); in particular, benthic macroinvertebrates in streams clearly develop taxonomic and functional hierarchies that are essential to establish organization in communities (Cummins et al. 1973, Cummins 1974). Additionally, habitats of benthic macroinvertebrates in streams are also classified hierarchically, taking into account the fact that variables are revealed differently across different space and time scales on which a system is viewed (Frissell et al. 1986, Minshall 1988). Since a hierarchical nature is an essential part of stream ecosystems, the determination of the appropriate methods of examination has been a key concept in investigating aquatic ecosystems (Minshall 1993). Consequently, the hierarchical classification approaches could provide in-depth and comprehensive understanding of community organization and water quality in the target ecosystem.

Assessment of water quality and prediction of community dynamics in streams are essential for diagnosing ecosystem health and for providing policies of sustainable management of stream ecosystems. Especially benthic macroinvertebrate communities are effective in indicating water quality and could effectively reveal ecological states of the target aquatic ecosystem. They constitute a heterogeneous assemblage of animal phyla, and consequently it is probable that some members will always respond to stresses placed upon them (Hynes 1960, Hawkes 1979, Hellawell 1986). Communities have been analyzed by conventional multivariate statistical methods (Ludwig and Reynolds 1988, Jongman et al. 1995, Legendre and Legendre 1998), however they are limited in extracting information effectively out of complex data. As an alternative tool to deal with this problem of complexity in ecological data, artificial neural networks (ANNs) have been utilized for patterning

* Part of grant from the Korea Science & Engineering Foundation (N° R01-2001-000-00087-0) and the EU PAEQANN project (N° EVK1-CT1999-00026).

† Corresponding: park@cict.fr

communities. The ANNs are well known for their ability to extract information from nonlinear and complex systems, and have been well applied to the study of secosystems (Lek and Guégan 2000, Recknagel 2003). Among the ANN techniques, Kohonen's Self-Organizing Map (SOM) (Kohonen 1982, 1989, 2001) is the most popular unsupervised learning algorithm. In aquatic ecosystems, the SOM has been used for classifying communities (Chon et al. 1996, Foody 1999, Park et al. 2001a, 2003a), for water quality assessments (Walley et al. 2000, Aguilera et al. 2001), and for population and community predictions (Céréghino et al. 2001, Obach et al. 2001). The classification by the SOM, however, has the problem of objectivity in finding similarities among the map units (Chon et al. 1996). When the groups are located far apart on the map, it is difficult to judge to what extent they are similar. Furthermore, due to randomness in iterative calculations and variability in determining parameters in the learning process of the network, the grouping presents a slightly different conformation after each training task.

Thus, to effectively define clusters among units of the SOM map, differentiation in the degree of clustering per se is additionally required. To divide the map into certain sub-areas, the unified-matrix algorithm (Ultsch and Siemon 1990, Ultsch 1993) is currently the most often used. However, it is not an easy task to efficiently reveal different degrees of clustering based on this distance matrix. In this study, we propose a combinational method for successively clustering communities through self-organization. The model developed was further evaluated with new data sets to detect the effect of sub-groupings on community development.

Materials and methods

Ecological data

The data of benthic macroinvertebrate communities were obtained from the database of the Laboratory of Ecology and Behavior System, Pusan National University, Korea. The communities were seasonally sampled at the study sites in the Suyong (SY), Cheolma (CM), Hoedong (HD), and Soktae (ST) streams in the Suyong River in Korea (Fig. 4.7.1) in October, 1989, and in January, May, and August, 1990. The Suyong River is a forth order river, 28.5 km in length with a catchment area of 199.5 km^2, passing through Pusan city, which has a population of more than four million people. Two tributaries CM and SY flow through agricultural areas down to the Hoedong reservoir. The HD, which is located in the lower area of the reservoir, is characterized by abundant filamentous algae and low current velocity, but has a great variation in discharge rates due to the water drained from the reservoir. The ST runs through the populated residential area, being disturbed with heavy pollution caused by organic matter in domestic sewage (Kwon and Chon 1991, Kang et al. 1995).

The benthic macroinvertebrates were collected using a Surber sampler. The dataset consisted of 76 samples from 19 sites in four seasons. In the study sites, 84 species were recorded and the communities collected were dominated by Chironomidae, Tubificidae, Erpobdellidae, Hydropsychidae and Baetidae. The dominant families occurred differently according to the impact of environment at the sampling sites (Kwon and Chon 1991, 1993, Kang et al. 1995). A wide range of levels of organic pollution was observed in the study area (Table 4.7.1). The enrichment states ranged from oligosaprobity to polysaprobity going downstream. The general characteristics of communities and ecological assessment on the Suyong River has been reported in Kwon and Chon (1991, 1993), Kang et al. (1995) and Chon et al. (2000a, b, c).

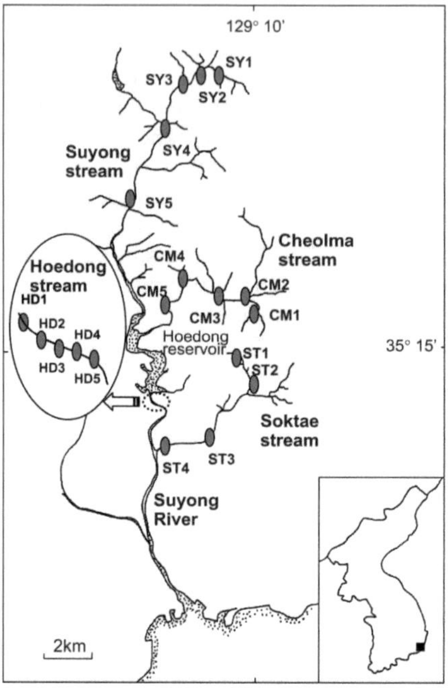

Figure 4.7.1 Study sites. Water quality of sampling sites is given in Table 4.7.1.

The densities of 84 species in total were provided as input to the network (number of computation units in the input layer: 84). The input data were transformed by natural logarithm in order to reduce variations in densities. To avoid any problems of logarithm zeros, the number one was added to the density of each species. Subsequently the transformed data were scaled proportionally between 0 and 1 in the range of the minimum and maximum density for each species to give same weights (or importance) to each species. After the learning processes of the SOM and the ART, a new dataset, which was seasonally sampled at SY2 from 1993 to 1995, was provided to the network to test the trained model.

Modelling process

In order to pattern community data at different scales, the two-step classification process using artificial neural networks was applied (Fig. 4.7.2). First, the densities in different taxa collected from benthic macroinvertebrate communities were fed into the SOM (Kohonen 2001). Weight vectors (i.e., connection intensities) of the SOM, containing the conformational characteristics of grouping in communities, were subsequently fed into another type of SOM network, the ART (Carpenter and Grossberg 1987), to find clusters in the units of the SOM map. By setting different levels of dissimilarity threshold for grouping, the clustering was further conducted in the hierarchical patterns by the ART (Lin and Lee 1996).

The SOM approximates the probability density function of input data through an unsupervised learning algorithm, and is an effective method for clustering, visualization and abstraction of complex data (Kohonen 2001). In the learning process, initially the weight vectors, which are connectivity intensities between input and output layers, are randomly assigned small values (Fig. 4.7.3). When the input vector x is sent through the network, the distance between the weight vector w and the input vector x is calculated by Euclidean distance ||x-w||. The output layer consisted of N output neurons (i.e. computational units) on a 2D hexagonal lattice (Kohonen 2001). Among all N output neurons, the best matching unit (BMU), which has the minimum distance between weight and input vectors, becomes the winner. The weight, w_{ij}, of the network is updated as follows:

$$w_{ij}(t+1) = w_{ij}(t) + \eta(t)Nr(t,r)[x_j - w_{ij}(t)] \tag{4.7.1}$$

where η(t) denotes the fractional increment of the correction, and Nr(t, r) is a predefined neighbourhood function determining the radius from the BMU in the map. The neighbourhood radius is usually set to a larger value early in the learning process, and is gradually reduced as convergence is reached. This results in training the network to classify the input vectors according to changes in the weight vectors they are closest to. The detailed algorithm of the SOM can be found in Kohonen (2001) for theoretical considerations, and Chon et al. (1996), Giraudel and Lek (2001), and Park et al. (2003a) for ecological applications.

Table 4.7.1 Mean BOD and overall saprobity status of sampling sites. Saprobity was evaluated based on the BOD and biotic indices (Trent biotic index, Biotic score, and Shannon diversity index).

Sampling site		BOD (ppm)	Saprobity
Suyong	SY1	2.7 (1.4)	β-mesosaprobity
	SY2	2.5 (1.5)	Oligosaprobity
	SY3	2.3 (1.1)	Oligosaprobity
	SY4	1.6 (0.3)	Oligosaprobity
	SY5	2.6 (2.9)	Oligosaprobity
Cheolma	CM1	2.6 (1.6)	Oligosaprobity
	CM2	1.8 (1.3)	Oligosaprobity
	CM3	2.1 (1.6)	Oligosaprobity
	CM4	1.9 (0.8)	Oligosaprobity
	CM5	2.0 (1.5)	Oligosaprobity
Hoedong	HD1	4.4 (1.0)	Oligosaprobity
	HD2	3.9 (1.3)	Oligosaprobity
	HD3	8.1 (3.7)	Oligo-β-mesosaprobity
	HD4	8.3 (1.6)	α-mesosaprobity
	HD5	9.2 (1.7)	α-mesosaprobity
Soktae	ST1	2.6 (1.2)	Oligosaprobity
	ST2	17.8 (11.7)	α-mesosaprobity
	ST3	40.9 (29.2)	Polysaprobity
	ST4	55.5 (23.4)	Polysaprobity

Self-organizing map (SOM)

The size (number of output units) of the SOM map is important to detect the deviation of the data. If the map size is too small, it might not explain some important differences that should be detected. Conversely, if the map is too big, the differences are too small (Wilppu 1997). The map size is especially important to accommomodate hierarchical levels in community grouping. Thus, we trained the SOM with different map sizes, and chose the optimum map size based on the minimum values of quantization and topographic errors. The quantization error is the average distance between each input vector and its BMU and is used to measure map resolution (Kohonen 2001). The topographic error represents the accuracy of the mapping in preserving topology, because the error value is calculated from the proportion of all data vectors for which first and second BMUs are not adjacent for measuring topology preservation (Kiviluoto 1996).

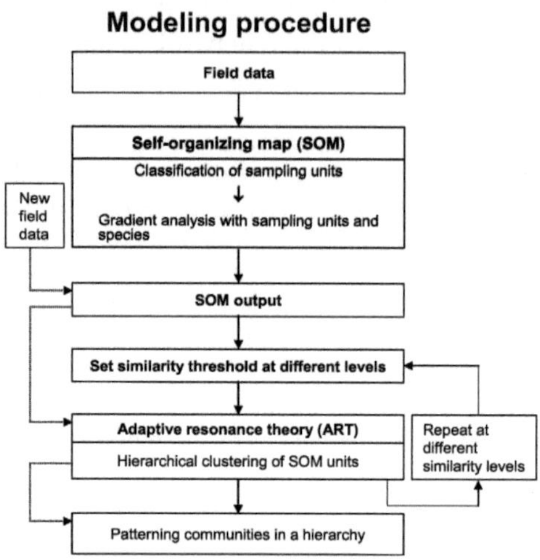

Figure 4.7.2 Flowchart of the modelling procedure in hierarchical patterning of community data by the combined use of artificial neural networks. The solid lines represent the learning process of the network, whereas the dotted lines indicate the evaluation process for new field data not used in the learning process.

Adaptive resonance theory (ART)

The ART is able to carry out stable self-organization of datasets for an arbitrary number of input vectors (Carpenter and Grossberg 1987). The fundamental characteristic of the ART lies in its ability to dynamically self-adjust its output size depending on the complexity of the network (Baraldi and Alpaydin 1998). The algorithm selects the first input as the exemplar for the first cluster, and the new input is clustered with the first if the distance to the first is less than a given threshold. Otherwise it is the exemplar for a new cluster. As the threshold value increases, the group size accordingly increases. This provides a basis for organizing the input data in a hierarchical manner correspondingly to the threshold level. In this study the weight vectors produced from the SOM were then fed into the ART (Figs. 4.7.2, 4.7.3).

A modified algorithm in ART (Pao 1989) was used in this study. Bottom up weights $b_{jk}(1)$ between input neuron j and output neuron k are initialized with some small numbers. After the input value y_j, which is the weight of each output neuron of the SOM, has been fed into the network, the distance $d_k(t)$ is measured for the degree of dissimilarity between input and weight values for each output neuron k, and used as a criterion for grouping input data through training by ART.

As each new input vector is sent into the network, the distance is calculated and the output neuron k which is closest is selected as k*. If $d_{k*}(t)$ is smaller than ρ, which is a threshold parameter for determining vigilance, the input vector is assigned to output neuron k*. The weight of neuron k*, $b_{k*j}(t)$, is then updated as follows:

$$b_{k^*j}(t+1) = \frac{c}{c+1} b_{k^*j}(t) + \frac{1}{c+1} y_j \tag{4.7.2}$$

where c is the number of sample units classified to node k*. If $d_{k*j}(t)$ is larger than the input is assigned to new output neuron. This means that the input vector entered is 'patterned' (or classified) as a new pattern (or cluster), not belonging to one of the previously existing patterns. Then, its weight $b_{k*j}(t)$ is newly assigned with the input vector. These weights produced by the ART preserve the conformational characteristics of the input data for grouping, and hence the associations among the communities were projected into the space defined in the ART (Zurada 1992). For hierarchical clustering in this study, initially the learning process was begun with whole map sizes. When clusters were found, these input vectors were replaced with their corresponding weights from the ART (Fig. 4.7.2).

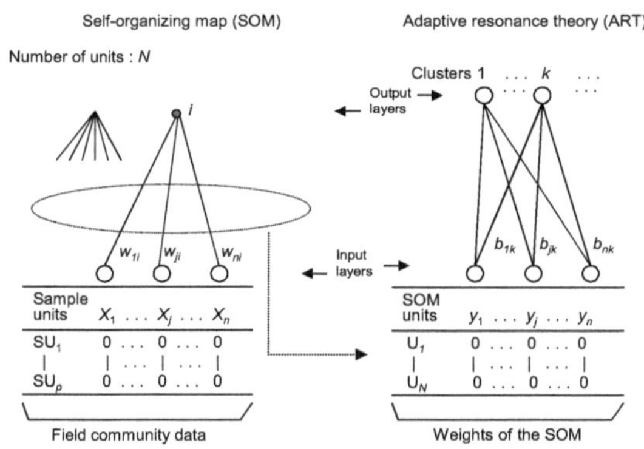

Figure 4.7.3 A schematic diagram of the Self-Organizing Map (SOM) and adaptive resonance theory (ART). The reference (weight) vectors of the SOM are given to the ART as input vectors. The number of output neurons (clusters) is not fixed in the learning process. The number of clusters increased according to the decrease of dissimilarity threshold values.

Results

Classifying communities

We chose 63 (9 × 7) output units of the SOM on the 2-D hexagonal lattice based on two different indices (quantization and topographic errors) for determining map size (Table 4.7.2), showing lower errors in both indices. The errors for quantization and topography were 1.256 and <0.00, respectively. The topographic error indicated that the first and second BMUs of all input vectors were adjacent hexagons showing smooth training in the SOM in terms of topology.

The trained SOM classified samples according to the gradient of community composition (Fig. 4.7.4). The acronyms in each unit of the SOM map stand for samples. The first three letters in the acronyms indicate the study sites (Fig. 4.7.1) while the last three represent the sampling season: SPR; spring, SUM; summer, AUT; autumn, and WIN; winter. (e.g., ST1SPR; samples at ST1 in spring). The grouping was firstly arranged according to the geographical distribution of the sample sites, e.g., ST, HD, CM, and SY. For example, samples collected from CM are mostly located in the lower areas of the SOM, while those belonging to HD are more concentrated in the upper right areas. Furthermore, temporal variations in different seasons were also observed locally. For instance, samples in summer at SY1-5 were grouped together in the same unit or were located near each other.

Table 4.7.2 Changes of quantization error and topographic errors at different SOM map sizes.

Map size	10	20	30	40	54	63	70	80
Quantization error	1.561	1.476	1.382	1.355	1.290	1.256	1.242	1.234
Topographic error	0.000	0.013	0.000	0.000	0.000	0.000	0.001	0.000

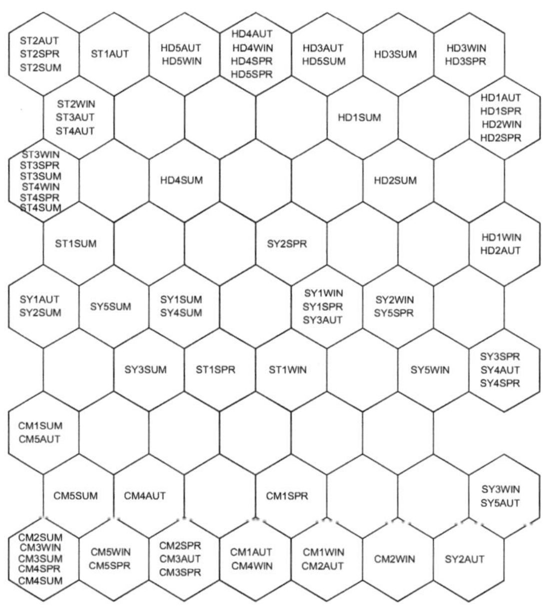

Figure 4.7.4 Classification of sampling units by the trained SOM. Acronyms in units stand for samples: the first three letters represent sampling sites (see Fig. 4.7.1), and the last three indicate the sampling season: SPR; spring, SUM; summer, AUT; autumn, and WIN; winter.

The grouping on the map also revealed the impact of pollution, and was comparable to the pollution states of the sampling sites displayed in Table 4.7.1. The upper areas of the trained map represented polluted sampling sites, whereas the lower areas showed relatively clean sites. For example, sites ST2-4 were strongly concentrated in the upper left corner of the SOM (e.g., nodes (1,1), (2,1) and (2,3)). The sites were heavily polluted from domestic sewage, showing polysaprobity (Table 4.7.1). Moreover, the samples from the less polluted site, ST1, with oligosaprobity, were not grouped with sites ST2-4 and were more widely scattered in the upper left area of the map (Table 4.7.1). Similarly, sites HD4-5 were heavily affected by domestic and industrial waste and were located close to the nodes of ST2-4 in the upper left areas of the SOM map, whereas the less-polluted sites HD1-2 were placed more loosely in the upper right area of the map. Site HD3 was relatively clean but occasionally disturbed by domestic waste. The samples of HD3 fell on the boundary between the areas of HD4-5 and HD1-2. The clean sites from CM with oligosaprobity (Table 4.7.1) were mostly located in the lower areas of the SOM. The sampling sites of SY were relatively clean ranging from oligosaprobity to ß-mesosaprobity (Table 4.7.1), and communities from these sample sites were diverse. Correspondingly, the samples occupied a wide range in the middle areas of the map.

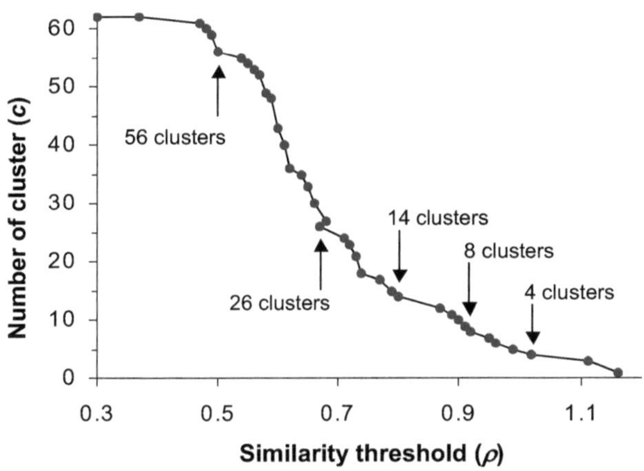

Figure 4.7.5 Relationships between number of clusters in the SOM units and dissimilarity thresholds in the ART. The network was trained at different dissimilarity threshold levels to find clusters of the SOM units.

Overall, the groups of communities on the SOM map efficiently revealed the impact of natural and anthropogenic factors. The samples were grouped according to the streams. At the same time, they were also arranged by different levels of pollution: the lower areas of the map had the less polluted sites while the upper left areas had the highly polluted sites (Fig. 4.7.4). The results of the SOM also generally confirmed the groupings revealed by previous studies: the communities were classified based on the streams and degree of pollution (e.g., Chon et al. 1996, 2000a, b).

Hierarchical classification of SOM units

Although the samples were grouped on the SOM map, it is still difficult to recognize the differences in similarities among the units of the SOM map. In this study we produced additional clusters in hierarchical levels by implementing the ART as mentioned previously. For this purpose, the weights of the trained SOM were given to the ART as input and the units of the SOM map were further grouped in different scales by adjusting dissimilarity threshold levels in the ART. The number of input neurons was 84; each neuron corresponding to one species used in the SOM.

As the dissimilarity threshold was increased, the number of groups decreased correspondingly (Fig. 4.7.5), revealing higher levels in the hierarchy. The optimal number of clusters can be determined based on the relationships between the number of clusters (c) and the threshold values (ρ): the decrease in the number of clusters was stabilized (i.e., $dc/d\rho=0$ and $c\neq1$) at various points of the dissimilarity threshold values as they were gradually increased. In this study three stabilized levels of dissimilarity were mainly observed in higher ranges (Fig. 4.7.5): fourteen subclusters in the range 0.79~0.87 of similarity threshold, eight medium-sized clusters in the range 0.91~0.95, and four large clusters in the highest level in the range 0.99~1.11. Small groupings with 26 and 56 clusters were also observed at lower similarity thresholds (Fig. 4.7.5), however the groups at these low levels were too numerous to be differentiated in the present study and are not considered here.

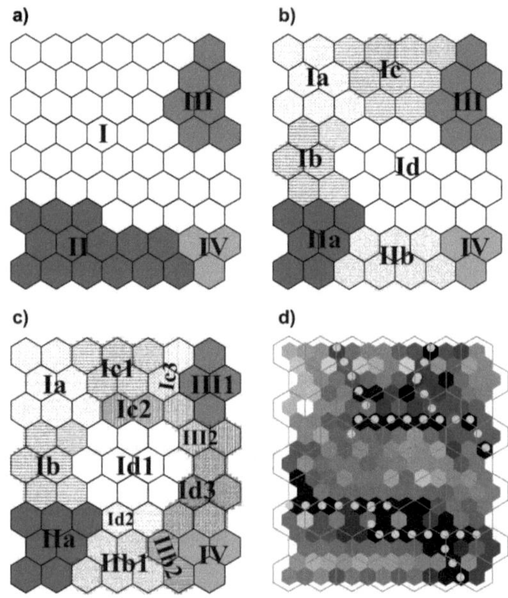

Figure 4.7.6 Hierarchical clustering of the trained SOM map by the ART and clustering by the U-matrix. The different clusters are displayed with characters. From four large clusters (a), the clusters were divided into subclusters based on the corresponding similarity levels (b, c). The symbols represent the classification directions of each cluster. I-IV stand for the four large clusters (a), a-d is in the eight medium-sized clusters (b), and 1-3 is in the 14 small clusters (c). The clusters determined by the U-matrix are indicated with white dotted lines based on the grey scale of U-matrix distance (d).

Fig. 4.7.6 shows the hierarchical clustering projected on the SOM map. With the low number of clusters at the higher dissimilarity level (Fig. 4.7.6a), the samples fell into four clusters (I - IV): the groups of ST-HD-SY (cluster I), CM (cluster II), HD (cluster III), and SY (cluster IV) (Fig. 4.7.6a). The groupings generally reflected the differences in streams and pollution. The ST-HD-SY group (cluster I) accommodated a wide range of the sample sites with the polluted sites in the upper left areas and the intermediately polluted sites SY in the middle area of the SOM. The communities in SY especially were diverse, and were divided into two groups with different levels of pollution, SY (cluster IV) and ST-HD-SY (cluster I). The sample sites in HD also showed different levels of pollution. Sites HD4-5 were located in cluster I (ST-HD-SY group), showing α-mesosaprobity, while the less-polluted sites HD1-3 below the Heodong reservoir joined cluster III in β-mesosaprobity. This indicated that communities collected from different sample sites in HD were diverse. The relatively clean sites from CM (cluster II) occupied the lower areas of the SOM map.

The eight medium-sized clusters (Ia-IV) in the lower hierarchical level were also classified based on the pollution status and the morphology of the streams (Fig. 4.7.6b). Large cluster I was divided into four subgroups Ia, Ib, Ic and Id according to pollution levels. Cluster Ia showed groups of the highly polluted sites, including ST3-4 with polysaprobity. Cluster Ic mostly accommodated HD4-5 with α-mesosaprobity, and the remaining groups showed relatively lower levels of saprobity. Unlike cluster I, the communities collected from the same stream in cluster II were separated based on the stream gradient: upstream CM1-2 (cluster IIb) and downstream CM4-5 (cluster IIa). Clusters III and IV, which had smaller areas in the SOM, were not divided in this case (Fig. 4.7.6b).

At the lowest hierarchical level with fourteen clusters (Fig. 4.7.6c), the sample sites were further divided based on various factors including location of sample sites, season and environmental impact. For example, the communities in cluster Id in the middle areas of the map were further sub-subgrouped according to the locations of SY1-2 and SY4, whereas some communities in subcluster Ic were separated based on season (e.g., Ic3). The sample sites in cluster III were divided into two sub-subclusters III1 and III2. Sub-subcluster III1 included the samples collected in spring and the samples collected in HD3 more selectively. The samples in cluster IV were not sub-subgrouped at the lowest hierarchical level.

Comparison with U-matrix method

The unified-matrix algorithm (U-matrix) was applied to the same results of the SOM to be compared with the performance of grouping by the ART (Fig. 4.7.6d). The U-matrix calculates distances between neighbouring map units, and displays the cluster structures of the map as mentioned before. High values of the U-matrix indicate cluster boundaries, while low values reveal clusters themselves, which can be visualised using grey shades. A dark shade reveals large differences between the map units, whereas a light shade represents map units similar in relative terms. Through the U-matrix the units of the SOM map were also grouped similarly to the clustering by the ART at the highest hierarchical level in general (Fig. 4.7.6a). The boundary at the lower area between cluster I and clusters II-IV occurred on the U-matrix, and the boundary between cluster I and cluster III at the upper right area of the map was also observed on the U-matrix. In the upper left areas, however, the border line was additionally formed in the U-matrix (Fig. 4.7.6d) while, in the corresponding areas in cluster I on the ART (Fig. 4.7.6a), the border line was not observed at the highest hierarchical level. At the subcluster level (Fig. 4.7.6b), however, cluster Ic was matched to the boundary appearing in the U-matrix. However, clusters were frequently not apparent on the U-matrix. For example, the sample sites of ST were not clearly differentiated from those of SY on the U-matrix (Fig. 4.7.6d), although community compositions were different between ST and SY as mentioned above. The ART showed distinctive clusters between ST and SY at the sub- and sub-sub-grouping levels (Fig. 4.7.6b, c). This indicated that the boundaries defined by the ART could differentially contribute to those formed from the U-matrix. Further investigation is required to determine the extent to which the boundary formation could be defined in the hierarchical concept in relation to groupings in the U-matrix.

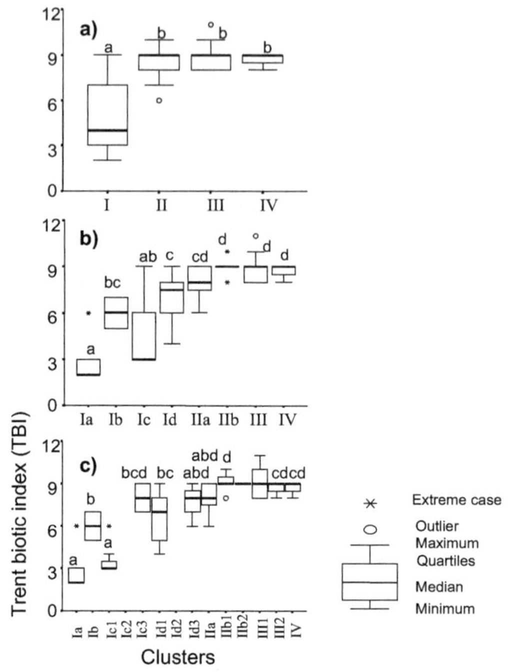

Figure 4.7.7 Box plots of Trent Biotic Index (TBI) to assess water quality at different clusters of different hierarchical similarity levels. a) four clusters, b) eight clusters, and c) 14 clusters. Cluster numbers correspond to the clusters defined in Fig. 4.7.4. The same characters on the box-plot indicate no significant difference at the 5% level of confidence by using Tukey's multiple comparison test.

Variations of biotic index in different clusters

The values of the Trent biotic index (TBI) were calculated and compared among the communities in different clusters on the SOM map (Fig. 4.7.7). The hierarchical grouping clearly corresponded to the pollution levels at the sampling sites. The different characters on the abscissa give a significant representation of different groups based on Tukey's multiple comparison test with a 0.05 significance level. Changes in water quality index were effectively reflected in the hierarchical classification. As the communities were grouped at finer levels, local variations existed in BOD values. When the clusters showed high variations, they were divided into subclusters, representing different levels of pollution with highly polluted cluster (Ia) and moderately polluted clusters (Ib, Ic and Id) (Fig. 4.7.7b). In a similar manner cluster II was divided into subclusters showing moderately polluted (IIa) and relatively clean (IIb) clusters, whereas clusters III and IV, which showed homogeneity in water quality, were not subclustered at this hierarchical level (Fig. 4.7.7b). This process of cluster division according to pollution level was further observed in sub-sub-clusters (Fig. 4.7.7c). For instance, sub-subcluster Ic1 was differently grouped from Ic3. This was due to different levels of pollution: the samples in Ic1 reflected high pollution while those in Ic3 represented a lower level of pollution (Table 4.7.1). These results showed how effectively the hierarchical grouping system would contribute to investigating water quality in relation to community composition.

Evaluating new samples and seasonal patterns

Once the networks have been trained with input data, new data not used in the learning process could be recognized on the SOM; they may be classified either as one of the already determined patterns or as a new pattern at the corresponding hierarchical levels. The patterns of new data were then determined by the trained network. Through recognition of a series of input data on the hierarchical map, diagnostics of community changes were possible. Fig. 4.7.8 shows an example of detecting changes in community development in different seasons. The data were seasonally collected at SY2 for two years from November 1993 to November 1995. The numbers (1-9) in the units of the SOM map stand for sampling season of collection from autumn 1993 to autumn 1995 sequentially (e.g., 1 for autumn in 1993, 2 for winter in 1993, etc.). The tracks of the recognized samples represented development of communities in different seasons. In the highest hierarchical level on the SOM (Fig. 4.7.8a), all the samples belonged to cluster I. Consequently community changes were not clearly distinguished in the highest level.

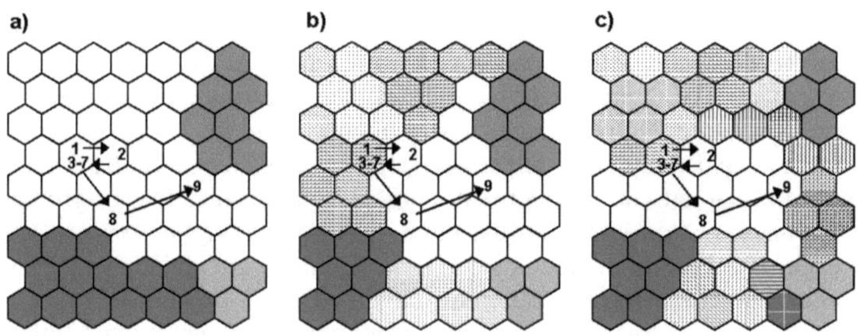

Figure 4.7.8 Evaluation of new communities not used in the learning process at different similarity levels. The data were seasonally collected at SY2 on the Suyong stream for two years from November 1993 to November 1995. The numbers stand for samples from autumn 1993 to autumn 1995 sequentially. a) four clusters, b) eight clusters, and c) 14 clusters.

In contrast, the tracks crossed between clusters Ib and Id1 in the next high level of hierarchy (Fig. 4.7.8b). The changes in the clusters were shown between '2 and other samples of 1 and 3-7' and again between '3-7 and 8-9'. This indicated that communities in cluster Ib jumped from to cluster Id1 in winter 1993 and changed back to cluster Id1 in summer 1995 (Fig. 4.7.8b). This demonstrated that the map could detect different degrees of changes in community development as time progressed. Correspondingly, the change in clusters reflected different water quality. This change was in accord with field observations. The motorway construction and the restoration project of the river initiated in the early 1990s were completed sometime early in 1995, and water quality was correspondingly improved. However, the jump in the community development was not further detected with the 14 subclusters (Fig. 4.7.8c).

Discussion and conclusion

The effectiveness in classification of communities using artificial neural networks (ANNs) has been discussed by Chon et al. (1996) and Park et al. (2003a). ANNs efficiently extract information from complex and nonlinear ecological data, and consequently provide a comprehensive understanding of multivariate datasets in a reduced dimension. In this study different degrees of groupings were further elucidated by combinational implementation of the SOM and the ART.

Through the SOM, communities were classified mainly based on the sampling sites in different streams and environmental impacts (Fig. 4.7.4). When the ART was further applied to the groupings in the SOM, hierarchical clusters were formed at different dissimilarity levels (Fig. 4.7.6a-c). The large clusters were successively divided into subclusters, differentially revealing characteristics of topography, environmental disturbances and temporal variations in the samples collected.

Through these hierarchical classification processes, samples were effectively assessed in different levels of association. For instance, the upstream (HD1-2) and downstream (HD4-5) samples in HD were grouped differently: in cluster III and in cluster I respectively at the highest hierarchical level (Fig. 4.7.6a), indicating that the communities of these two areas were very different although they are located close to each other in the same stream. This difference of grouping was caused by organic pollution. Sites HD4-5 were heavily polluted, being affected by domestic sewage and waste from a junkyard located near the sampling areas, whereas sites HD1-2 were relatively clean. These characteristics were further displayed in the community composition. For example, *Asellus hilgendorfi* and *Cardina dentriculata* were exclusively collected at sites HD1-2. *Asellus* is an indicator species of ß-mesosaprobity and habitats in a moderately enriched region of rich alga growth (Wiederholm 1984, Kwon and Chon 1991). Furthermore, macrophytes *Hydrilla verticillata* and *Potamagetum criptus* and the filamentous algae *Oedogonium* were abundant, serving as a major energy source for the macroinvertebrates. At sites HD4-5, in contrast, species richness was low while the species tolerant to pollution were dominant, including *Limnodrilus hoffmeisteri* and *Chironomus* sp. In this case domestic organic waste was the main food supply for the tolerant species. Samples in HD3 were mostly located at boundary areas between HD1-2 and HD4-5 on the SOM (Fig. 4.7.4). The field data also confirmed this: community compositions were intermediate between HD1-2 and HD4-5 sites. The BOD values also lay in the middle range although the saprobity level of HD3 was similar to that for HD1-2 (Table 4.7.1).

The benthic communities in two streams, SY and CM, were patterned differently on the hierarchical SOM, although their BOD values were similar (2.35 ppm and 2.06 ppm in SY and CM, respectively, t-test, $P>0.05$). Community composition showed large differences between these streams. Kwon and Chon (1991) reported that variations in community composition were diverse and Chironomidae occurred more abundantly in SY than in CM. These characteristics of the streams SY and CM were also clearly identified in the hierarchical patterning of communities on the SOM map: communities in SY and CM occupied different regions and the conformations of the groups were different at different hierarchical levels (Fig. 4.7.6a-c). These results demonstrate that community grouping could reveal an extra dimension of ecological assessment of polluted communities, which chemical measurements such as BOD can not convey sufficiently.

It is helpful to understand communities in the aspect of hierarchical organization. A complex system like a community usually develops hierarchical organization (Allen and Starr 1982, O'Neill et al. 1986, Urban et al. 1987, Allan and Hoekstra 1992). Benthic macroinvertebrate communities in streams usually have clear taxonomic and functional hierarchies, and these are essential to verify organizational characteristics in communities

(Cummins et al. 1973, Cummins 1974). The study of complex systems emphasizes the importance of scale (O'Neill 1989, Levin 1992), and developments in hierarchy theory demonstrate how processes and constraints change across the scales (Allen and Starr 1982, O'Neill et al. 1986). The hierarchical aspects were effectively revealed by combinational implementation of ANNs in unsupervised learning as demonstrated in this study. To obtain more detailed information on hierarchical grouping in communities, however, grouping of 'taxa', in addition to grouping of 'communities', would also be required. For the present study, however, we only showed grouping in communities. Hierarchical grouping of taxa has many additional points to be covered (e.g., functional groups, taxonomic differentiation, etc.).

To cluster the units of the SOM map, the U-matrix algorithm is conventionally used these days. The matrix gives a picture of the topology of the unit-layer and therefore also of the topology of the input space (Ultsch 1993). Sometimes, however, it is not an easy task to detect clear boundaries on the grey-scale map of the U-matrix, although there are distinct differences in ecology and in environmental impact as shown in this study. Vesanto and Alhoniemi (2000) demonstrated that the U-matrix did not present clusters in their dataset, while an agglomerative clustering method showed clear clusters. In ecological studies, a fuzzy c-means clustering method (FCM; Giraudel et al. 2000) and a k-means algorithm (Park et al. 2003a) have also been used to cluster the units of the trained SOM map. Different methods to group the map have both strengths and weaknesses according to their clustering algorithms (Jongman et al. 1995, Legendre and Legendre 1998, Vesanto and Alhoniemi 2000). In this study we propose a model using two artificial neural networks. However, it is a difficult task to compare the various methods with respect to their grouping efficiency. For the present study we initially proposed the model and showed its efficiency in comparison with the U-matrix. In the further study, the results should be compared with results from other methods.

Chon et al. (2000a), in their implementation of the SOM and the ART for grouping of communities, used the ART initially before the SOM to train the similar datasets, reporting that the ART appeared to group the community data more efficiently with a low level of noise. They used only seven taxa at the family and class levels; the species and genus data were summed and consequently there were relatively few zeros in the densities. In this study, however, densities of 84 species were directly used as input data and consequently included many zeros (i.e., high level of noise) in the community data. As discussed in Chon et al. (2000a) the SOM appeared to be slightly more efficient in grouping as the data have a relatively high level of noise while the ART appeared to be more sensitive to noise in this type of ecological data. Additionally, the visual graphics of the two-dimensional map obtained by the SOM were needed as a basic framework for accommodating hierarchical clustering visually. Based on this reasoning we used the SOM initially before the ART in this study.

The SOM is an effective learning algorithm for the visualization and abstraction of high-dimensional data in low dimensions. It converts the nonlinear statistical relationships between high-dimensional data into simple geometric relationships of their virtual units on a low-dimensional display, usually a two-dimensional lattice (Kohonen 2001). However, the visualization of the trained SOM only reveals qualitative information. By implementing two-level classifications, the distances between the groups in the SOM map can be efficiently defined through hierarchical levels as shown in this study.

In conclusion, the combined use of unsupervised learning algorithms with the SOM and the ART efficiently assesses groupings in a hierarchical pattern in community data and reveals differential effects of community composition and environmental factors such as pollution and season. This hierarchical patterning could extract community features to lead to a comprehensive understanding of community variations in spatial and time domains. Accordingly, hierarchical clustering would be a helpful approach for understanding complex ecological data as well as for providing information for management of polluted aquatic ecosystems on different scales.

4.8 Species spatial distribution and richness of stream insects in south-western France using artificial neural networks with potential use for biosurveillance*

Compin A†, Park YS, Lek S, Céréghino R

Introduction

The major goal of the PAEQANN European project is to provide tools to aquatic ecosystem managers, by using aquatic communities as ecological indicators and ANNs as modelling techniques. In lotic ecosystems the species composition of benthic communities depends on the diversity and stability of the stream habitats (Cummins 1979, Ward and Stanford 1979) which provide the possibilities of development (Malmqvist and Otto 1987). Therefore, benthic invertebrates are widely used as indicators of short- and long-term environmental changes in running waters (Hellawell 1978, Lenat 1988, Smith et al. 1999, Hawkins et al. 2000). Because they are ubiquitous, basically sedentary, with a large number of species, and strongly influenced by many natural and/or anthropogenic disturbances, aquatic invertebrates are by far the most commonly used indicators for the assessment of freshwater ecosystem quality (Rosenberg and Resh 1993). However, the very high diversity of aquatic invertebrates – 70% of the overall animal species recorded in European continental waters (Illies 1978) – and the difficulty to obtain specific identifications make quantitative approaches using macroinvertebrates unsuitable for the assessment of long term or large-scale changes in water quality. In the Adour-Garonne drainage basin (SW France) these quantitative studies have often been restricted to a single valley or range of mountains (Décamps 1968, Vinçon and Thomas 1987, Vinçon and Clergue 1988, Giudicelli et al. 2000), and were usually based on a single taxonomic group (e.g., one insect order).

An important development for water management is the generation of practical tools which provide accurate biological assessments of river conditions without requiring a high level of expertise, effort and time for their users. These "rapid assessment" aproaches are designed to fulfil two objectives (Resh and Jackson 1993). First, reducing the effort (and cost) in sampling, sorting, and identification procedures. This can be achieved for exemple by considering only a fraction of the macroinvertebrates collected. A second objective is to summarize the results of site surveys by using single-score measures that can be understood by non-specialists.

Species richness is such a measure, and is commonly used as an integrative descriptor of the community (Lenat 1988). It is influenced by a large number of environmental factors which can determine gradients in stream species richness (Vannote et al. 1980, Minshall et al. 1985) and it is also strongly influenced by natural and/or anthropogenic disturbances (Rosenberg and Resh 1993), which may lead to spatial discontinuities of these predictable gradients (Ward and Stanford 1979, 1983) and losses of taxa (Brittain and Saltveit 1989).

* Funded by the EU PAEQANN project (EVK1-CT1999-00026). The authors thank also to the French Water Agency (AEAG) for supporting the database construction.

† Corresponding: compin@cict.fr

Resh and Jackson (1993) observed that species richness measures were sensitive to the impact of human activities on stream ecosystems, and this was particularly true of some aquatic insects, e.g., Ephemeroptera, Plecoptera or Trichoptera (EPT), which can be considered as good biological indicators of disturbance in streams. Thus, the species richness of a restricted number of selected taxonomic groups is a good descriptor of the influence of disturbance upon the biota (Lenat 1988).

An a priori framework for developing biological indicators is a stream classification based on macroinvertebrates, to characterize how ecosystems differ in terms of species assemblage. An interest of such classifications is that the stability of species assemblages may be used to define representative and/or reference sites for biological surveillance (Hughes et al. 1986), as any structural change in population features can indicate environmental changes in streams from a given region or a longitudinal section. At a large geographic scale, such stream classifications detecting several sub-regions associated to their characteristic macroinvertebrate assemblages are basically necessary to calibrate biological indicator measures.

Using macroinvertebrates, we deal with ecological data that are bulky, nonlinear and complex, showing noise, redundancy, internal relations and outliers (Gauch 1982, Jongman et al. 1995). Great changes can also appear in variables, and complex interactions can occur between explanatory and response variables (Jongman et al. 1995). Traditionally, conventional multivariate analyses have been applied to solve these problems (Bunn et al. 1986, Ludwig and Reynolds 1988, Legendre and Legendre 1998). With these nonlinear and complex ecological data, however, nonlinear analysing methods should be preferred (Blayo and Demartines 1991). One of these methods is artificial neural networks (ANNs), which are versatile tools to extract information out of complex data, and which could be effectively applicable to classification and association.

This paper describes how ANN methods can be used: i) to contribute to the understanding of large-scale geographic patterns in aquatic macroinvertebrate assemblages; ii) to obtain taxa richness predictions, with simple environmental attributes as input variables; iii) to replace or complement existing tools for water quality biosurveillance and management (Fig. 4.8.1). The results of recent studies, which focused on macroinvertebrates from four orders of aquatic insects (EPTC) in the Adour-Garonne stream system (South-Western France) are used to highlight the concepts.

Methods

Field data

The Adour-Garonne stream system (South-Western France) has a 116 000 km^2 drainage basin. It contains 120 000 km of permanent running waters flowing from the Pyrénées mountains and Massif Central mountains to the Atlantic Ocean. This basin has a large human population (6 million inhabitants) distributed in urban and agricultural areas, and is potentially and effectively affected by anthropogenic disturbances. From our laboratory database, we selected sampling sites ranging from 10 to 2500 m a.s.l. (Fig. 4.8.2), representing high mountain to plain or coastal areas. Samples were taken from 1988 to 1998. Each site was sampled at two periods (summer and winter). The species lists (detailed in Céréghino et al. 2001) were used to model the species' distributions and characteristic species assemblages, and the species richness was calculated as the sum of recorded species among the two periods. Samples were taken from the various substratum types: sand (< 2 mm), gravel (2-20 mm), pebbles (20-200 mm) and cobbles (>200 mm), using a standard Surber sampler (sampling area 0.1 m^2, mesh size 0.3mm). They were distributed in proportion to the relative abundance of these substrata. Depending on the heterogeneity (or homogeneity) of bed-paving substrate, 5 – 8 sample-units were thus taken from the various substratum types. All samples were taken in the part of the channel that is always covered by water.

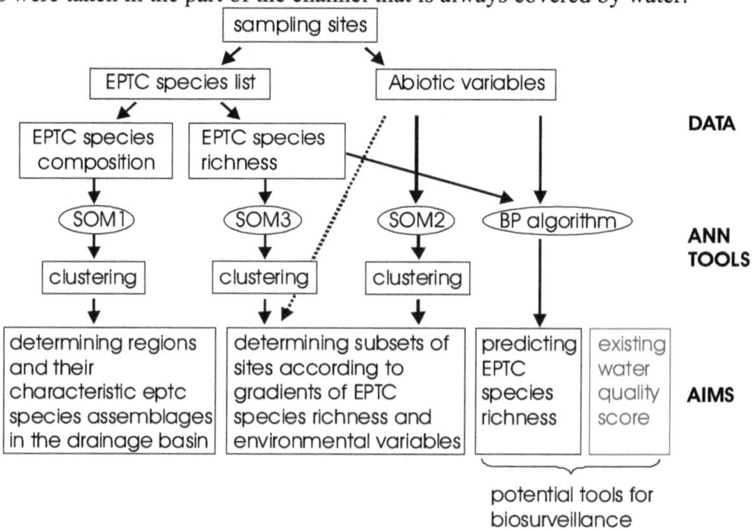

Figure 4.8.1 Overview of the aims, material and methods of the study. ANN = Artificial Neural Networks; EPTC = Ephemeroptera, Plecoptera, Trichoptera, Coleoptera; SOM = Self-Organizing Map; BP = Back Propagation.

We focused on EPTC species, aquatic insects commonly identified to the species level in freshwater studies, and thus we added Coleoptera to the standard EPT index. Indeed, Coleoptera are major components of stream invertebrate communities (Cayrou et al. 2000), and contain sensitive taxa particularly in the family Elmidae (this family being taken into account in the calculation of the IBGN water quality index in French rivers). Barbour et al. (1996) found that both the number of Coleoptera and EPT taxa decreased with increasing disturbance, and we thus suggest that considering the four insect orders should enhance the accuracy of water quality assessments. Each site was characterised by abiotic variables: elevation, stream order, slope, distance from the source, and maximum water temperature. These variables were chosen for two reasons: i) they relate the location of sampling sites within the stream system without a priori consideration of any disturbance, and ii) they are easy to collect using a map and a min-max thermometer.

Models processing

The two datasets and ANN methods we used in the different studies presented in this paper are grouped in the Table 4.8.1. The data were first processed using Self-Organizing Maps (SOM), an unsupervised neural network algorithm (Kohonen 1982). Three different SOMs were calculated in order to summarize the variability of the data and to cluster the sampling sites according to different input variables: EPTC species (SOM1); E, P, T, and C species richness (SOM2) and environmental variables (SOM3). The SOM performs a non-linear projection of the data space onto a two-dimensional space. A detailed description of the

SOM methodology was given in Céréghino et al. (2001). This network consists of two layers of neurons: the input layer is composed of neurons connected to the sampling sites (one per sampling site), the output layer is made up of neurons organized on an array with rows and columns laid out on a hexagonal lattice. In the output layer, the neurons act as virtual sites and approximate the probability density function of the input data. The training was broken down into two phases with a specific number of iterations: first ordering with a large neibourhood radius and then fine tuning with a small radius. The input variables and characteristics of the datasets are given in the Table 4.8.1. A k-means algorithm or the unified-matrix (U-matrix) approach was then applied to detect the cluster boundaries on the trained maps. With the k-means algorithm method, the retained number of clusters was justified according to the minimum Davies-Bouldin index (Ultsch 1993). Correlation coefficients between E, P, T, and C richnesses were assessed for observed data (field) and predicted data (i.e. weights of output neurons of the trained SOM), in order to establish relationships among biotic variables.

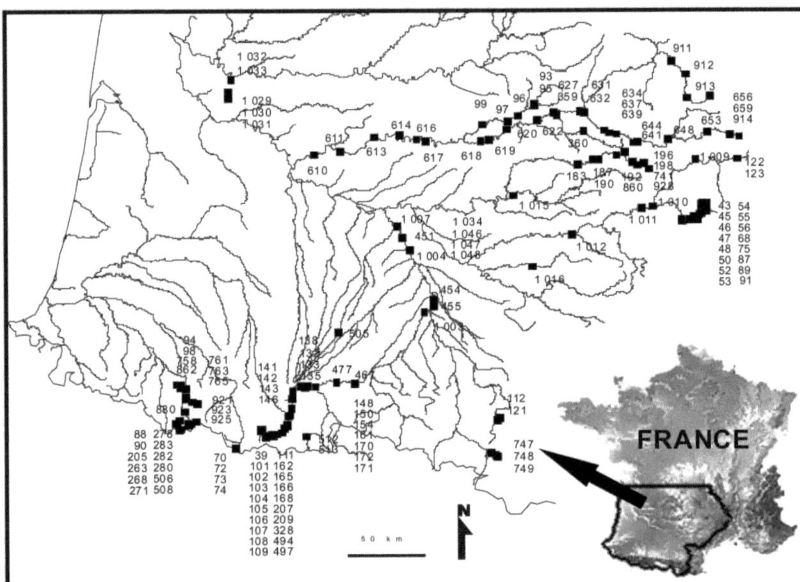

Figure 4.8.2 Map of the Adour-Garonne stream system, and location of the sampling sites.

As a second step, a multi-layer perceptron (MLP) using the backpropagation (BP) algorithm (Rumelhart et al. 1986b) was used to predict the EPTC richness (output variable) from environmental data (four input variables, see Table 4.8.1). To find the optimum number of hidden neurons of the MLP model, we trained models with different numbers of hidden neurons (from 3 to 10), and finally chose five neurons in the hidden layer as showing the best performance. Therefore, we used a 4-5-1 structure for the MLP model. The learning and momentum coefficients were 0.75 and 0.95, respectively. Out of 155 sampling sites, 130 were randomly selected and used to train the network, whereas the remaining 25 sites were used to test the trained MLP. During the learning process, the values of the error between estimated and observed values were calculated and the training was stopped when error values gradually increased for several learning iterations, to avoid overfitting. After the learning process, correlation coefficients between observed and estimated values were calculated for both learning and testing datasets to verify the calculated for both learning and testing datasets to verify the predictability of the network. A sensitivity analysis, i.e., a method to study the behaviour of a model (Scardi and Harding 1999), was conducted to determine the contribution of each input variable on the values of the output variable of the model.

Table 4.8.1 Characteristics of the two datasets and ANN methods used in the studies presented in this paper. E=Ephemeroptera, P=Plecoptera, T=Trichoptera, C=Coleoptera.

Aim of the study	**Representing and clustering the distributions of EPTC sampling sites according to the input variables**			**Predicting EPTC Richness**
Number of sampling sites	252 Dataset 1	155	155 Dataset 2	155
Input variables	283 EPTC species	Elevation Stream order Distance from source Maximum water temperature	E sr* P sr* T sr* C sr*	Elevation Stream order Distance from source Maximum water temperature
ANN method	SOM (SOM1)	SOM (SOM2)	SOM (SOM3)	MLP
Number of nodes in the input layer	283	4	4	
Number of iterations Ordering/Tuning	2000/75000	3000/7000	3000/7000	
Number of units in output layer (Rows/Columns)	150 (10/15)	140 (14/10)	140 (14/10)	
Clustering method	U-Matrix	K-means	K-means	

* species richness

Results

Spatial distribution patterns of EPTC species

The non-linear projection of presence – absence data in a two dimensional space (Fig. 4.8.3) allowed us to cluster our sites according to the similarity of their species composition. Four major clusters (i.e. "regions" 1, 2, 3 and 4, see Fig. 4.8.3) could be identified on the SOM (SOM 1, see Table 4.8.1).

Regions 1 and 2 were formed of 2 and 5 sub-regions respectively. Then, these regions were plotted on a geographic map of the Adour-Garonne drainage basin, in order to make interpretations (Fig. 4.8.4). Region 1 encompassed sites from the Massif Central Mountains (eastern part of the drainage basin) above 500 m a.s.l., with 2 sub-regions corresponding to the River Lot (1a) and the River Tarn (1b) stream systems respectively. All sites from region 2 belonged to the Pyrenees mountains (South part of the drainage basin), and were partitioned into 5 sub-groups. Three sub-groups corresponded to catchment areas of large rivers: Rivers Lez and Garonne from 800 to 500 m a.s.l. (2a and 2e), Gave d'Ossau (2b), Neste d'Aure (2c).

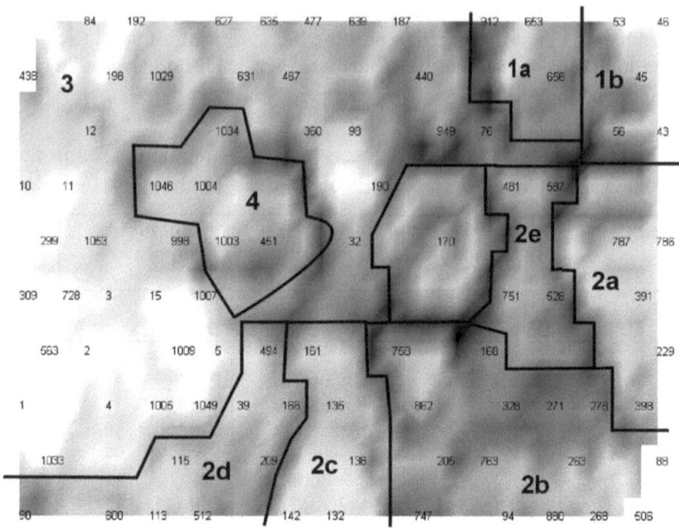

Figure 4.8.3 Distribution of the sampling sites on the Self-Organizing Map (SOM1). Numbers correspond to the code of the 252 sampling sites. In order to lighten the Fig., hidden points are not represented. 1a – 4 (bold) are the regions or clusters of the map (see text).

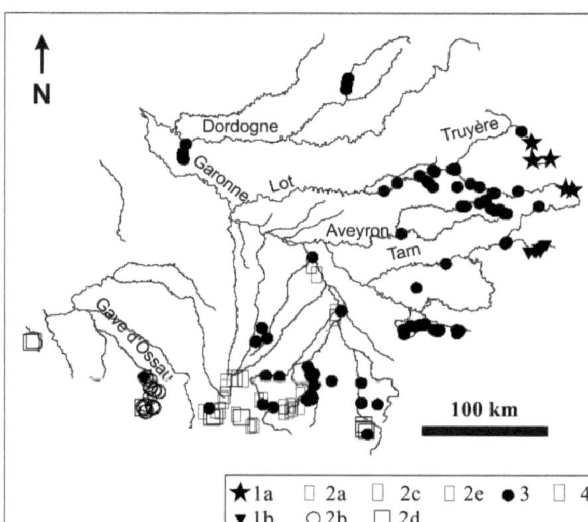

Figure 4.8.4 Distribution of sampling sites in the Adour- Garonne stream system and correspondence with their position on the Kohonen map. The legend is explained in the text.

All sites from sub-group 2d were Pyrenean springs. Sites from singular stream environments were clearly segregated by the SOM algorithm, e.g., sites 170 – 172 belonged to a watercress bed (Fig. 4.8.4), and were not considered in the interpretation of the map. Region 3 clearly represented piedmont zones from the Adour-Garonne drainage basin, and included sites from both Massif Central and Pyrenees rivers. Finally, region 4 corresponded to the Toulouse city agglomeration. The distribution of each of the 283 species (one map per species) was visualised in the Kohonen map (Fig. 4.8.5).

To summarise EPTC assemblages characterising each region, we recorded the presence of each species in a table, where we also indicated the probabilities of occurrence, calculated as [number of sites where the species was recorded / total number of sites defining the region]. EPTC richness (Fig. 4.8.6a) ranged from 45 to 159 species according to the region considered. Richness values were the lowest in springs (2d), in the agricultural Garonne region (2e), and in the urbanised Toulouse region (4).

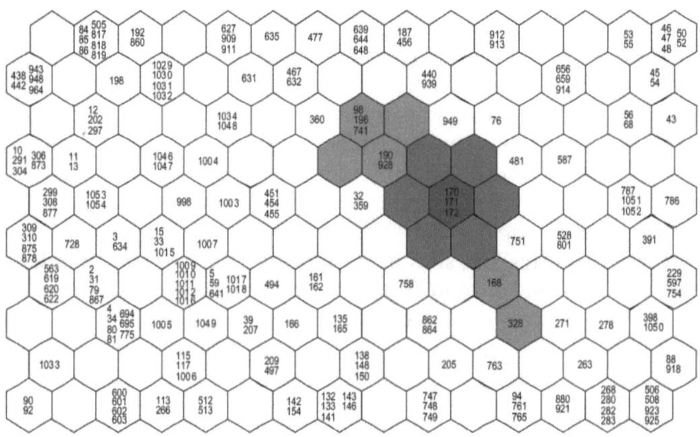

Figure 4.8.5 Example of the representation of a species distribution on the Kohonen map (*Agapetus fuscipes*, Trichoptera). The darkness of shading indicates the relative influence of the species considered upon the classification of sites.

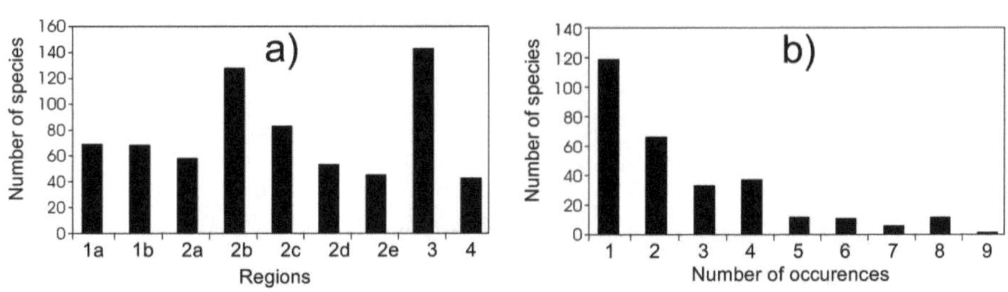

Figure 4.8.6 Species richness patterns derived from the SOM analysis. (a) number of species per identified regions; and (b) number of species occuring in 1-9 regions.

Higher richness was observed in both piedmont and mountain regions. We also plotted the number of species occurring in 1 to 9 regions (Fig. 4.8.6b). Most species occurred in only one (121 species) or two (63 species) regions. They therefore had the strongest influence upon the stream classification, and should require particular attention as indicator species. 31 and 34 species appeared respectively in 3 and 4 regions, and 4 to 10 species appeared in 5 to 8 regions. Finally, only one species – *Baetis rhodani* (Ephemeroptera) - occurred in all regions. Three main spatial distribution patterns could be identified: i) local distribution, i.e. species occuring in a restricted geographic and/or altitudinal area (e.g., *Baetis buceratus*), ii) longitudinal zonation, i.e. species occuring in different geographic areas, but within a characteristic altitudinal range (e.g., *Brachyptera seticornis*), and iii) regional distribution, i.e. widespread species (e.g., *Baetis rhodani*). Any species associations can also be pointed out by overlapping the representations of several species distributions on the Kohonen map.

Distribution of sampling sites according to EPTC richness and environmental variables

After training the SOM with environmental variables (SOM2, Table 4.8.1), the k-means algorithm helped to derive four clear clusters (A, B, C and D) based on the minimum Davies-Bouldin index (DBI = 0.91) (Table 4.8.2). Thus, sampling sites were clustered into four subsets (Fig. 4.8.7) according to a gradient of stream order and elevation. The abscissa on the SOM was explained by the gradient of elevation (from low (left) to high (right)), whereas the ordinate of the map represented the stream order and the distance from source (from low (top) to high (bottom)). Stream order was significantly correlated with the distance from source ($r= 0.82$, $p<0.01$), and elevation was correlated negatively to the maximum water temperature ($r= -0.80$, $p<0.01$). Sites in cluster A were at low elevations (< 400 m) and high stream order (5th-7th), sites in cluster C were at low stream order (1st-2nd) and low elevations (< 500 m), and sites in cluster D were at low stream order (1st-2nd) and high elevations (> 1300 m), whereas sites in cluster B were at intermediate stream order (3rd-4th) and elevations (500-1200 m). Thus, we can consider that clusters reflected the longitudinal location of sampling sites, chiefly with respect to stream order and elevation.

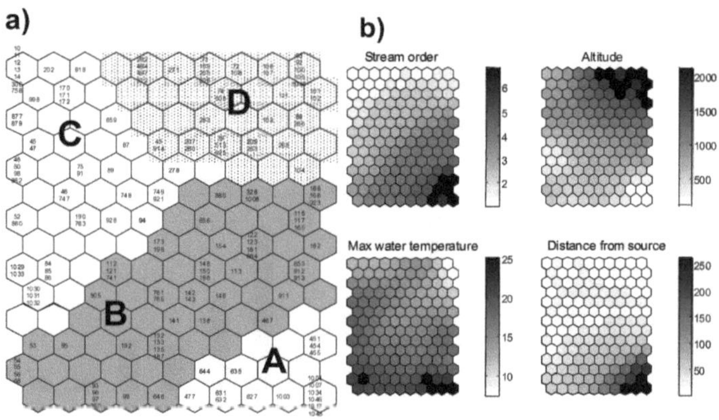

Figure 4.8.7 (a) Distribution of sampling sites on the Self-Organising-Map according to the four environmental variables (SOM2), and clustering of the trained SOM. Codes correspond to sampling sites (e.g., 765, see also Fig. 4.8.1). Grey shades were used to visualize clusters A – D derived from the k-means algorithm. Sites which are neighbours within clusters are expected to have similar features. (b) Gradient analysis of each environmental variable on the trained SOM, with visualization in grey scale. The mean value of each variable was calculated in each output neuron of the above-trained SOM, dark represents high values while light is low. (Altitude: m a.s.l., maximum water temperature: °C, distance from the source: km).

The SOM trained with EPTC richness (SOM3, Table 4.8.1) was also divided into four subsets based on the minimum Davies-Bouldin index (DBI = 0.91) (Table 4.8.2). Thus, sampling sites were distributed into four clusters (I, II, III and IV) according to a gradient of EPTC species richness (Fig. 4.8.8). Moreover, the comparison of distributions of sites in Figs. 4.8.7 and 4.8.8 helped to relate EPTC richness to environmental conditions within the stream system. Bottom areas of the SOM had the highest EPTC richness, whereas top areas showed low richness.

Table 4.8.2 Davies-Bouldin index (DBI) of k-means clustering at different numbers of clusters on the trained Self-Organizing Maps for environmental (SOM2) and biological (SOM3) data sets. The retained number of clusters was justified according to the minimum DBI.

	Number of clusters					
Input data set	2	3	4	5	6	7
Environmental variables (SOM2)	1.07	0.97	0.91	1.09	1.01	1.02
EPTC Richness (SOM3)	0.97	1.10	0.91	1.31	1.10	1.14

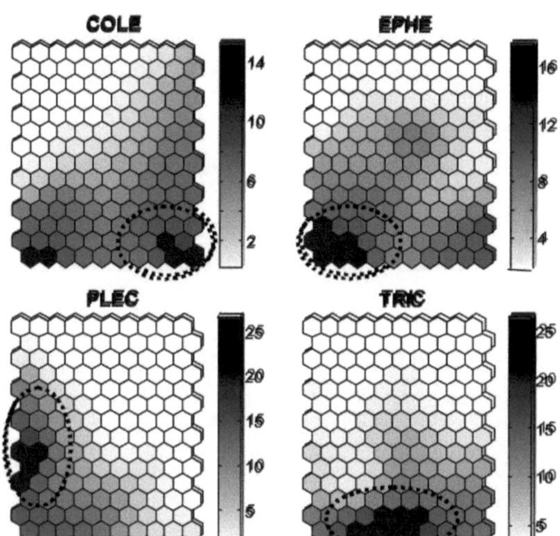

Figure 4.8.8 (a) Distribution of sampling sites on the Self-Organizing Map according to EPTC species richness (SOM3), and clustering of the trained SOM. Codes in each unit of the map represent sampling sites (Fig. 4.8.1), grey shades were used to isolate clusters I – IV derived from the k-means algorithm. (b) Gradient analysis of species richness for each insect order on the trained SOM, with visualization in grey scale. Dark represents high richness values, while light means low richness values.

Clusters I and IV were mainly classified by the overall EPTC richness. Sites in cluster I had high EPTC richness, and chiefly belonged to 3rd – 4th order streams: 70% were also associated to cluster B in Fig. 4.8.7. Most sites (80%) in cluster IV were previously associated to clusters C (40%) and B (40%). They had the lowest EPTC species richness and were primarily located at low stream order (1st – 2nd) Clusters II and III were separated by the richness of Plecoptera. Sites in cluster II had high Plecoptera richness with moderate richness for other insects (Ossau valley in the Pyrenees, stream order 1-2, 70% of these sites thus belonged to cluster D). Sites in cluster III had low numbers of Plecoptera species, along with moderate richness for other taxa. They were located in the intermediate part of the altitudinal gradient (piedmont zones). Indeed, they were chiefly assigned to clusters B (45%) and C (30%) in Fig. 4.8.7. Thus, the overall EPTC richness gradient identified from Fig. 4.8.5 resulted from different richness distribution patterns, which were characteristic for each insect order. When the distribution of species richness was examined for each insect order on the trained SOM (Fig. 4.8.8), the map units on the bottom of the SOM showed highest richness values for Ephemeroptera, Trichoptera and Coleoptera. The units on the left bottom corner showed the highest richness for Plecoptera, whereas the units on the right top corner corresponded to the lowest richness values.

There was a high coincidence between observed (i.e. field data) and predicted (i.e. from the output neurons of the SOM) species richness in each taxonomic group (Fig. 4.8.9). Bar charts represent the histograms of observed and predicted values of each taxonomic group, and the correlation coefficient were highly significant ($r>0.74$, $p<0.01$) except for Coleoptera ($r=0.45$, $p>0.1$). The scattergrams on the right upper corner of Fig. 4.8.9 show the relationships among taxonomic groups in observed data, while the charts on the left bottom corner show the relationships for predicted data. Correlation coefficients were higher in predicted than in observed data. Species richness relationships were highly significant for both observed and predicted data among Ephemeroptera, Trichoptera and Coleoptera, but the correlation was relatively low when Plecoptera were plotted against other insect orders.

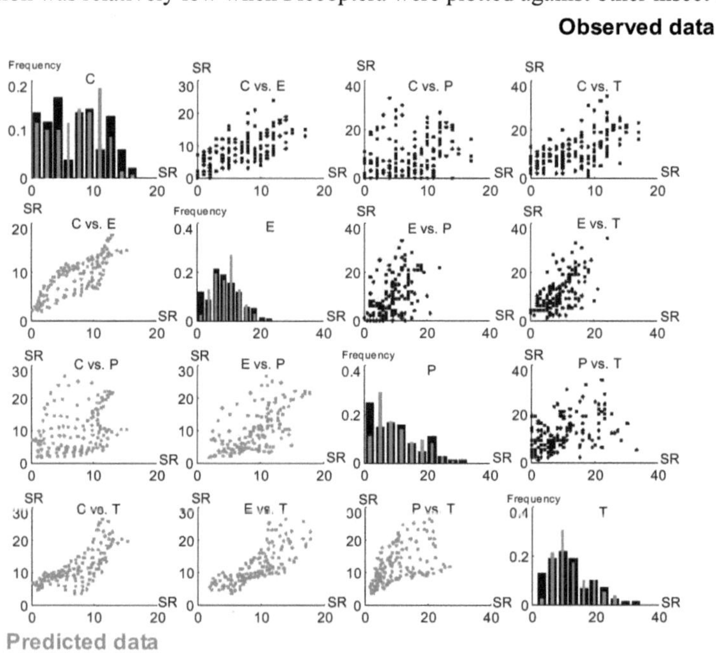

Figure 4.8.9 Observed (black bars) and predicted (grey bars) frequencies of species richness in each insect order (panels on the downward diagonal line), and species richness relationships between insect orders in observed (black dots) and predicted (grey dots) data. E= Ephemeroptera, P= Plecoptera, T= Trichoptera, C= Coleoptera; SR= Species Richness.

Relationships between biological and environmental variables

The SOM has shown its high performance for visualization and abstraction for our nonlinear and complex ecological data. However, it was not easy to include environmental variables in the SOM trained with biological variables. Thus, we suggest a method to introduce (or include) environmental variables into the SOM map trained with biological variables (SOM3), in order to understand their effects on biological variables and on the classification of sampling sites in the trained SOM. To this end, the mean value of each environmental variable was calculated in each output neuron of the trained SOM, then each variable was visualized on the trained SOM map (Fig. 4.8.10). Dark represents high values, while light represents low values. The areas with the highest values were marked with a circle for each variable. Environmental variables showed gradient distributions on the SOM map. Stream order increased from the left to the right side of the map. Elevation was the highest in the upper left area, and showed the clearest gradient among environmental variables. Distance from the source was lower in left areas, and higher in right areas of the SOM map. Maximum water temperature did not show a clear gradient in its distribution on the map. These results revealed that elevation was the most important factor in patterning sampling sites according to EPTC richness, while the effect of the maximum water temperature was the lowest.

At this point, we have three types of parameters (sampling sites, biological and environmental data) on the trained SOM map. Using these data, we superimposed each parameter on the same SOM map (Fig. 4.8.11). We can compare the relationships among clusters (and/or sampling sites), EPTC richness, and environmental variables. Sampling sites on the lower area of the SOM map have the highest species richness (Fig. 4.8.10).

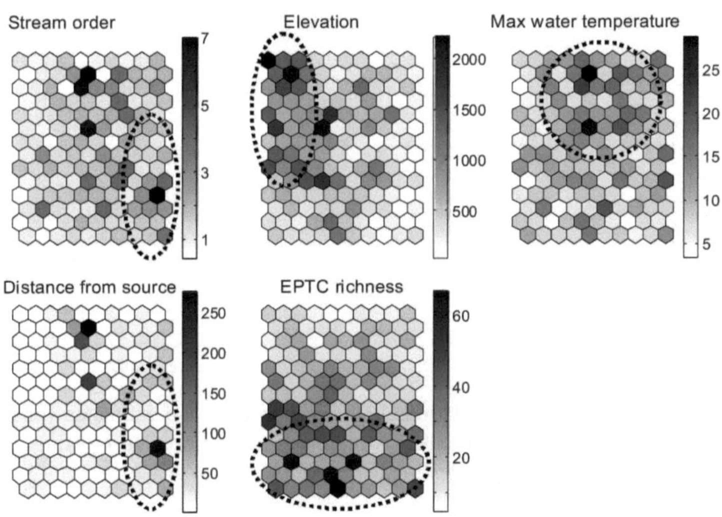

Figure 4.8.10 Visualization of environmental variables and overall EPTC richness on the trained SOM map (SOM3). The mean value of each variable was calculated in each output neuron of the trained SOM. Dark represents a high value, while light is low. The areas with the highest values are marked with a dotted circle.

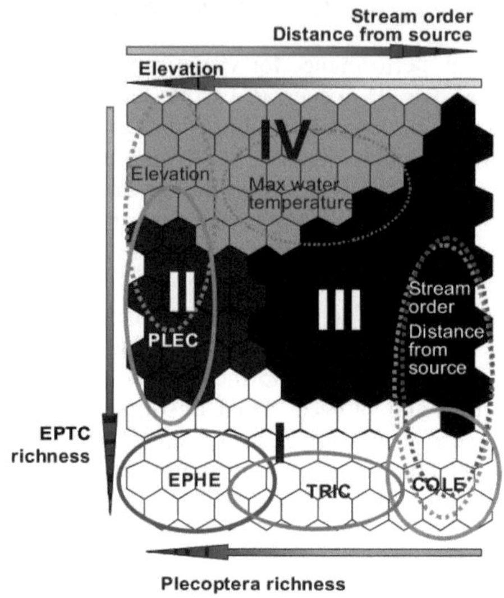

Figure 4.8.11 Comparison of relationships among clusters (and/or sampling sites), EPTC richness, and environmental variables. Each parameter from Figs. 4.8.7 and 4.8.9 is overlaid on the trained SOM map. COLE = Coleoptera, EPHE = Ephemeroptera, PLEC = Plecoptera, and TRIC = Trichoptera.

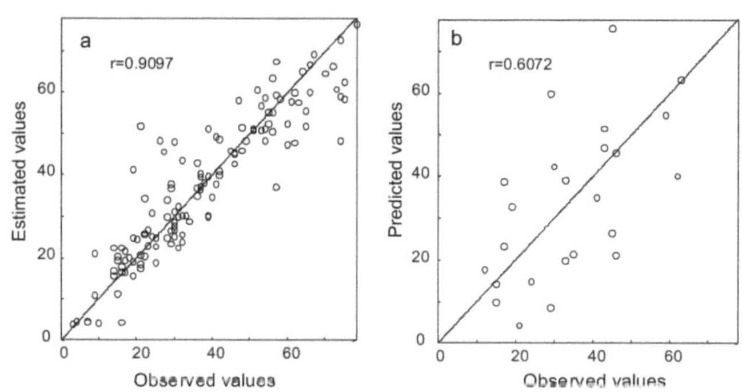

Figure 4.8.12 Scatter plots of correlations between observed and estimated (or predicted) values by the trained BP. The diagonal lines represent perfect prediction values (predicted and observed values). a) learning, b) testing.

Prediction of the taxonomic richness

The MLP was applied to predict EPTC richness as an output variable, using the four environmental variables as input. The convergence of the learning process was generally reached after 3,000 iterations under mean error terms of 0.01. The trained BP showed accuracy in predicting the overall EPTC richness on the basis of the environmental variables (r^2=0.83, p<0.001 and r^2=0.37, p<0.01 for training and test data sets, respectively) (Fig.

4.8.12). There was, however, an underestimation of some high EPTC richness and an over-estimation of around 30 EPTC richness values.

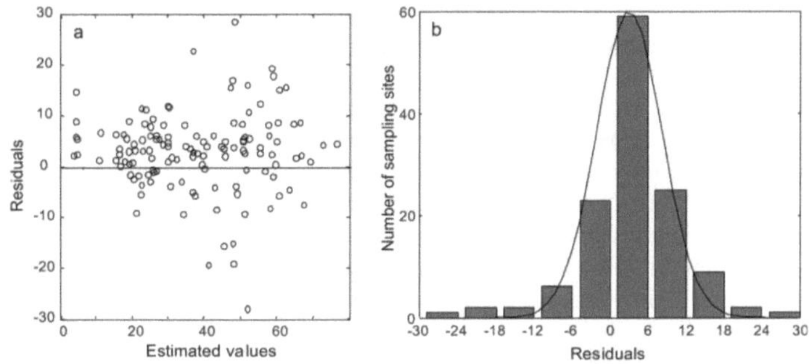

Figure 4.8.13 Relationships between residuals and estimated values (a), and histogram of residuals (b).

The residuals were well distributed near the horizontal line representing the residual mean (r=-0.01, p>0.5) (Fig. 4.8.13a). The histogram of residuals revealed that most values were centred near zero. To test the normality of model residuals, the statistical test of Lilliefors (1967) was applied. The test did not reject the null hypothesis that the residuals are normally distributed (p=0.2) (Fig. 4.8.13b). The relationships showed no obvious sign of dependence of residuals, showing that the BP fitted the data with no bias.

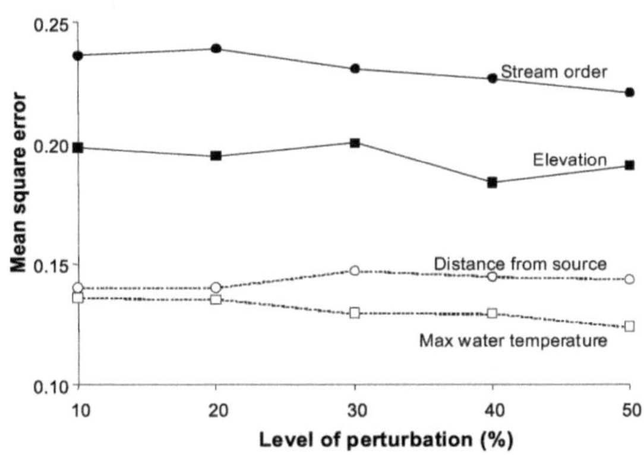

Figure 4.8.14 Sensitivity analysis of the BP. Mean square error values were measured at different levels of perturbation of the input variables.

Sensitivity analysis was carried out to evaluate the effect of small changes in each input on the neural network output. This was achieved by adding a random variation to each input variable of the network (Scardi and Harding 1999). To measure a response in output values, mean square errors were calculated at different levels of the perturbation of input variables (from 20% to 100% of the input range). The sensitivity analysis showed that elevation and stream order provided the highest contributions among the four input variables when predicting EPTC richness, whereas maximum water temperature provided the lowest contribution (Fig. 4.8.14). This is in agreement with the results of the ordination based on the trained SOM (Figs. 4.8.8, 4.8.10, 4.8.11).

Discussion

Stream classifications and modelling techniques

Whatever the modelling technique, stream ecologists use classification and ordination to characterise how ecosystems differ in terms of biotic (e.g., species assemblages, species richness) and/or abiotic (environmental variables) attributes (e.g., Giudicelli et al. 2000, van Sickle and Hughes 2000, Dethier and Castella 2002). By knowing what the ecosystem should be like in a given geographic zone, they also can determine the degree to which human activity has altered it (Hawkins et al. 2000). During the last decade, such approaches to river bioassessment using macroinvertebrates were thus developed in Europe (e.g., RIVPACS system), Australia (AUSRIVAS system), and North-America (BEAST system) (reviewed in Wright et al. 2000). Most of these techniques are based on multivariate analyses, and basically use classifications of reference sites from rivers of high biological quality to provide site-specific predictions of the macroinvertebrate fauna to be expected under undisturbed conditions, using a small set of environmental characteristics. In this context, our study is an attempt to use artificial intelligence techniques for biota prediction in river bioassessment.

Ordination and cluster analyses are frequently used in the early exploratory phase of ecological investigations as their results may suggest relationships that should be studied in more detail in subsequent research (Jongman et al. 1995), whereas regression analyses may be helpful to study more specific questions in the later phases of research. This analysis procedure (ordination and/or cluster analysis first, then regression analysis) was used in this study. During the learning process of the SOM, neurons that are topographically close in the array will activate each other to learn something from the same input vector. This results in a smoothing effect on the weight vectors of neurons (Kohonen 2001). Thus, these weight vectors tend to approximate the probability density function of the input vector. Therefore, the visualization of elements of these vectors for different input variables is convenient to understand the contribution of each input variable with respect to the clusters on the trained SOM. Although the SOM visualization is an indirect gradient analysis like a Principal Component Analysis (Kohonen 2001), SOM can be used as an analysis tool to bring out relationships between sampling sites, environmental variables, and biological variables. Thus, this approach is much more practical to analyse the relationships between variables than general indirect gradient analysis. Giraudel and Lek (2001) compared the SOM algorithm to conventional ordination techniques for ecological community ordination, and concluded that the SOM is fully usable in ecology and can complement classical techniques by offering a non-linear approach to modelling and a method for visualising data and for achieving community ordination.

Species assemblages

We visualised the spatial distribution of each of the 283 species considered, and we could therefore derive negative or positive species association. Numerous site-specific data (i.e. local scale biodiversity) were compiled in order to derive spatial distribution patterns of stream macroinvertebrates at the regional scale. As a first step before further investigations, our analysis revealed several EPTC geographical zones. Such a regional classification of stream ecosystems provides a useful framework for studying and managing streams in different geographic areas (Witthier et al. 1988). The number of species characterising each region ranged from 43 to 147, underlining the expected longitudinal (e.g., high or low mountain, piedmont) and geographical (Pyrenees, Massif Central) differences (Culp and Davies 1982). These results also support the idea that biodiversity depends on the environmental heterogeneity (Ward and Stanford 1983), and is both reduced by environmental constancy (e.g., springs) and under severely fluctuating conditions (e.g., severe flow fluctuations due to hydropower generation). In these conditions, SOM may help to identify disturbed sites at the regional scale. Any modification of the species composition will create a faunal discontinuity that will be visualised in the self-organised map by the unexpected position of the considered site regarding to its geographical position. In our map, the most striking example of such a faunal discontinuity could be the segregation of the Toulouse City neighbourhoods.

Finally, this study and similar works (e.g., Frissel et al. 1986, Hughes et al. 1986, Omernik 1987) provide an explicit scheme of the implicit knowledge that stream ecologists already have. Biotic features of streams within the same region and/or longitudinal section tend to be similar, and those characteristics tend to differ when streams belong to more distinct areas. Any tool able to provide a stream classification is therefore of obvious value to both resource managers and researchers to assess spatial and temporal variability.

EPTC species richness

E, P, T, and C occurred in all the streams considered. Ephemeroptera, Trichoptera and Coleoptera were widespread from mountain to plain areas, and species richness relationships between these three insect orders were highly significant, for both observed and predicted data. Plecoptera were rather located in the upper mountainous sections of the stream system, and species richness relationships between Plecoptera and Ephemeroptera, Trichoptera or Coleoptera were non-significant. According to Gaston (1996), concordant spatial patterns in species richness among different taxa may result from: i) random mechanisms; ii) biotic interactions among different taxa; iii) common environmental determinants; or iv) spatial covariance in different environmental factors that independently account for diversity variation in different taxa. If local systems were compared, it is likely that a high degree of concordance would be generated through biotic factors (Paszkowski and Tonn 2000). However, at broader spatial scales such as the Adour-Garonne stream system, significant correlations among aquatic insect species richness is almost certainly due to similar responses by different taxa to environmental conditions rather than to biotic interactions (Heino 2002).

Stream classifications: species composition or species richness?

We derived a classification of streams from the Adour-Garonne drainage basin, using the similarity of their species compositions. The biogeographic model obtained thus referred to a regional-specific fauna. The stream classification based on species richness referred to stream order, elevation, and distance from the source (and to a lesser degree to water temperature), i.e. the downstream location of sampling sites within a stream system, which rather fitted with a broader typological approach. Thus, it was not surprising that SOMs with species compositions and species richness provided different stream classifications. The former clearly segregated different geological areas (Pyrenees in the Southern part of the drainage basin, Massif Central Mountains in the Eastern part, and alluvial plain areas), whereas the latter provided a model of longitudinal gradients in species richness (two communities from different geographic areas could differ in terms of species composition, but they could have similar number of EPTC species when sites were located at comparable elevations and stream orders). The sensitivity analysis of the BP algorithm showed that elevation and stream order made the largest contributions to the predictive model. Although many criteria have been proposed for classifying running waters, few typological systems have had more than local acceptance (Pennak 1971). Therefore, our results support the idea that the most "universal" classification systems remain those proposed i) by Illies (1961) and Illies and Botosaneanu (1963) which recognise eight zones within a single drainage system, ranging from zone I (springs or "eucrenon") to zone VIII (brackish zone or "hypopotamon"), and ii) by Vannote et al. (1980), i.e. the River Continuum Concept which implies a classification based on stream size and location within a stream system. Despite geographic differences in species assemblages, these two models have the broadest validity to describe spatial patterns of community organisation and diversity. Of course, the fauna of a stream system must be investigated before it can be categorised within the expected zones, and this is always a time- and money-consuming process. However, scientific studies and large-scale surveys of stream ecosystems have led to the development of extensive databases, particularly in Europe and in the United States. These data can be used to derive a [sites x species] matrix, then data may be analysed using a-posteriori inductive approaches (Whittier et al. 1988, Cayrou et al. 2000).

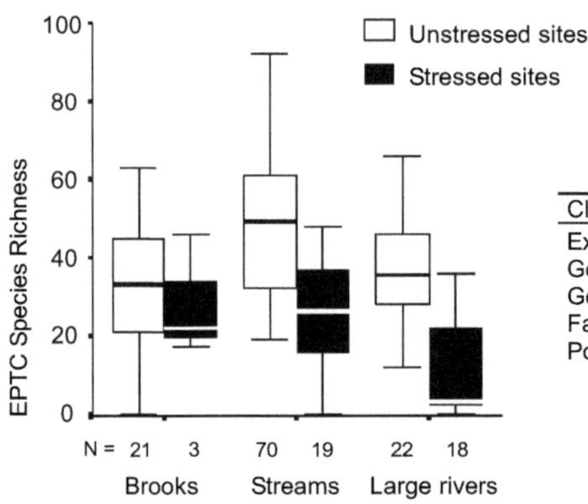

Class	Streams	Large rivers
Excellent	>50	>35
Good	42-50	25-35
Good-fair	34-41	14-24
Fair	25-33	3-13
Poor	0-24	0-2

Figure 4.8.15 EPTC richness distributions for brooks, streams and large river clusters at unstressed and stressed sites. The table is derived from the Fig. and indicates the values of the five water quality classifications for streams and large rivers. Upper classes were defined as species richness values above the median species richness at unstressed sites, and lower classes were defined as species richness values below the median species richness at stressed sites. The other classes (Good, Good-Fair and Fair) were then defined by dividing the remaining species richness ranges into three equal groupings.

Predicting the species richness

Using MLP, EPTC richness was predicted with environmental variables, and we evaluated the importance of each variable to estimate the richness. It is recognized that MLP is able to make better predictions than regression models (Lek et al. 1996b, Paruelo and Tomasel 1997), and a sensitivity analysis is applied to explain the contribution of input variables to output variables. Recently, Gevrey et al. (2003) reviewed several methods proposed for the sensitivity analysis of ANN in ecological applications. However, sometimes it is not sufficient to explain the relationships between explanatory and response variables in terms of causality. In this case the ordination approach could be helpful to explain the relationships between input and output variables. Predicting the species composition of the EPTC community with environmental variables as input data remained an impossible task for us, when about 300 species had to be dealt with. However, the final objective of such studies is to provide methods for rapid assessments of water quality and for water framework directives, and species richness of aquatic insects is commonly recognised as a good biological indicator of disturbance in streams (Rosenberg and Resh 1993). Thus, if we can predict what the richness should be like under undisturbed conditions in a given area, we can provide explicit spatial distribution schemes which may be useful for further studies, and in stream management.

A potential application in biosurveillance

We recently calibrated for the Adour-Garonne basin a water quality score, based on EPTC species richness measures (Compin and Cereghino 2003). Principal Components Analysis (PCA) was performed for five abiotic variables (elevation, stream order, slope, distance from the source, and maximum water temperature) recorded at 113 unstressed sites. The coordinates of the sites on the most significant axes of the PCA were then used to classify sites into three clusters corresponding to brooks, streams and large rivers using an agglomerative clustering technique. Significant differences in EPTC species richness distribution between the clusters and between stressed and unstressed sampling sites for each cluster were tested using Mann and Whitney non-parametric tests.

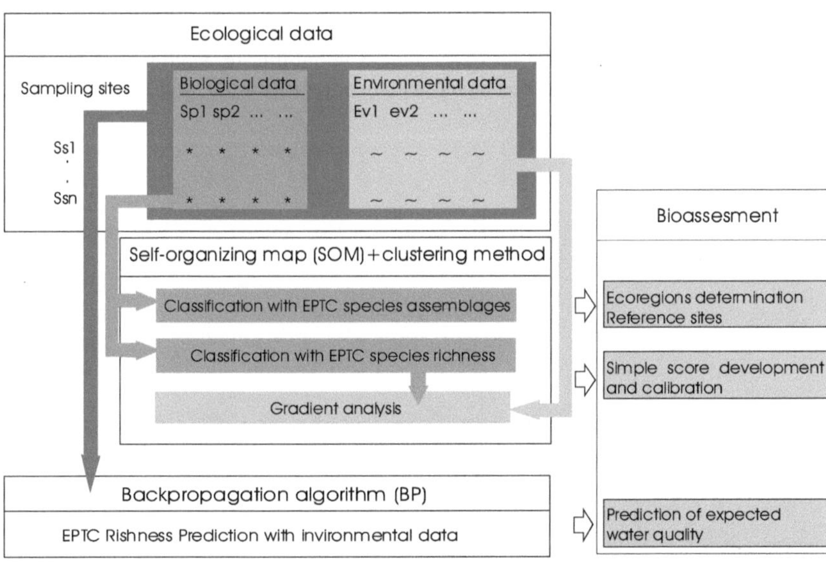

Figure 4.8.16 Synthesis of the ANN tools cited in this chapter and examples of their potential application to water quality assessment.

The clusters grouped sampling sites having significantly comparable richness, whereas between-regions differences in species richness distributions were significantly different. Therefore, species richness to disturbance relationships was assessed differently according to the region considered. For each region, five richness classes were determined, using box-plots of species richness distribution among disturbed and undisturbed sampling sites (Fig. 4.8.15). As we were able to provide quite good prediction of the species richness expected under natural conditions in a given area, we may provide a valuable tool for the design of management actions.

Conclusion

Two different neural networks have been applied to suggest practical approaches for understanding ecological data (Fig. 4.8.16). The SOM showed high performance for visualization and abstraction of ecological data. The trained SOM efficiently classified sampling sites according to different input variables, and displayed a distribution of each component (input variable). The component planes helped to interpret the contribution of each component to the classification. Additionally, by introducing new variables (i.e. environmental variables) not used in its training phase, the SOM showed high performance in analyzing the relationships among sampling sites, biological variables and environmental variables. This method could be used to extract relationships between sampling sites, communities, and environmental variables, although the algorithm is theoretically an indirect gradient analysis. However, it remains necessary to quantify the relationships among variables.

After understanding the relationships between biological and environmental variables using the SOM, the BP, used as a nonlinear predictor, showed accuracy in predicting EPTC richness on the basis of a set of four environmental variables. Thus, this prediction could be valuable in the assessment of disturbances in given areas.

Finally, approaches using two different ANNs (first understanding data sets using visualization and abstraction methods with SOM and second prediction for target variables with BP) showed that they could take into account the variability of ecological data efficiently. Therefore, this procedure could be chosen when ecological modelling is applied to nonlinear and complex ecological data.

4.9 Patterning community changes in benthic macroinvertebrates in a polluted stream by using artificial neural networks*

Kwak IS, Song MY, Park YS, Liu G, Kim SH, Cho HD, Cha EY, Chon TS†

Introduction

Data for community dynamics are complex and difficult to analyze, since communities consist of many species varying in a non-linear fashion in spatial and temporal domains. However, investigation of community changes in disturbed aquatic ecosystems is critical for diagnosing temporal community responses to stressful sources and for establishing sustainable management policies to solve the problems of polluted aquatic systems. Although there have been numerous accounts of conventional multivariate analyses on community patterning through clustering and ordination (e.g., Bunn et al. 1986, Legendre and Legendre 1987, Ludwig and Reynolds 1988, Quinn et al. 1991) or on community-environment relationships (e.g., van Dobben and ter Braak 1998), the conventional statistical methods are limited in the sense that they are mainly applicable to linear data and are less flexible for handling ecological data with missing values, noise, etc. Additionally, the studies have mostly been carried out on static community patterns from single samplings.

Artificial neural networks (ANNs) are an alternative tool for solving the problem of complexity residing in community data. They are problem oriented and are adaptively flexible for applications (Lippmann 1987, Zurada 1992, Haykin 1994). The multi-layer perceptron has been extensively used in prediction of communities by revealing complex relationships between communities and environmental factors, such as algal bloom (e.g., Recknagel et al. 1997) and establishment of grasslands (e.g., Tan and Smeins 1996). Additionally, temporal networks have been developed to predict community dynamics in a time-delayed manner. The partially and fully connected recurrent ANNs have been utilized to predict short-term community changes (Chon et al. 2000c, 2001). In this case, however, the models were only used for predicting the occurrence of communities. Grouping of community changes has rarely been conducted using artificial neural networks.

Recently, however, patterning of community changes has been focused on ecological water quality assessment. Successful management of aquatic ecosystems requires better understanding of the patterns of community development, i.e., either progression of pollution or recovery from the stressful agents. In conventional methods, however, not many studies have focused on grouping of community changes in the temporal domain per se. Mostly communities were clustered in static terms. Since Legendre et al. (1985) and Legendre (1987) discussed chronological clustering in multivariate datasets to represent the succession of species within a community by using ordination and segmentation techniques, specific results concentrating on either the methods for clustering community changes or

* Supported by a grant N° R01-2001-000-00087-0 from Korea Science & Engineering Foundation

† Corresponding: tschon@pusan.ac.kr

groupings from field data have not been reported. Similarly, in ANNs, not many studies have been carried out on grouping of the temporal development of communities. By implementing the Self-Organizing Map (SOM) (Kohonen 1989), groupings on communities have been conducted for clustering and ordination on static community patterns (e.g., Chon et al. 1996, Foody 1999, Kwak et al. 2000). Grouping of community changes is not an easy task since patterning of temporal developments has the problem of the unlimited increase in the number of variables as the sampling periods are increased. Chon et al. (2000a) recently grouped community changes by the combined use of unsupervised ANNs, the Adaptive Resonance Theory (ART) plus the SOM. In this case, however, only noise was filtered through the ART, and dimension reduction from the original datasets was not carried out. Additionally, community changes were grouped in a relatively short period of less than six months. In this study, we further address the feasibility of the SOMs in dimension reduction of sequential datasets of community changes and grouping of community changes over a longer period. The proposed model could be used for detecting community changes commonly-occurring in the survey area.

Methods

Implementation of the SOM

Since communities consist of multiple taxa, the dimension would be greatly increased if community changes were patterned sequentially in a given period. In order to solve this problem of increase in dimension, the SOMs were carried out in two processes in this study: i) initial training of the SOM with the one-time (1-month in this case) samples, and ii) secondary training of the SOM with coordinates of the 1-month SOM for a longer period (e.g., 3 months, 12 months, etc.; Fig. 4.9.1).

In the SOM, M neurons are used for the output layer, which could be empirically determined based on efficiency of convergence and discrimination capability among the patterned nodes. In this study, 81 (9x9) nodes were used on the two-dimensional array. The weights between node j of the output layer and node i of the input layer in the SOM were represented as $w_{ij}(t)$ at iteration t. When the input data x_i (densities in taxa i) with N taxa were sent through the network, each output neuron, j, computed the Euclidean distance between input vector and weights. The neuron with the weight vector which has the shortest distance to the input vector was chosen to be the winning neuron. The winning neuron (and possibly its neighbouring neurons) was allowed to learn by changing the weights in the manner to further reduce the distance between the weight and the input vector as shown below:

$$w_{ij}(t+1) = w_{ij}(t) + \eta(t)(x_i - w_{ij}(t))Z_j \qquad (4.9.1)$$

where Z_j is assigned 1 for the winning and its neighbour neuron(s) while it is assigned 0 for the rest neurons, and $\eta(t)$ (e.g., 0.1) denotes the fractional increment of the correction. For details of the algorithm in the SOM, see Kohonen (1989), Chon et al. (1996), and Park et al. (2003a).

After initial training of the SOM with the 1-month samples, the coordinate data (i.e., locations of the nodes) on the SOM were provided as new input data to the SOM for a longer period (Fig. 4.9.1). Since 1-month sampling data was accordingly extracted from the coordinate data on the SOM, we assumed that the tracks of the x, y coordinates of the patterned nodes on the SOM would convey information on community changes. In other words, the tracks of the nodes following the months progressively (e.g., April, May, etc) would solely represent temporal changes in communities during the specified period (Chon et al. 1996). If the one-month coordinate data could be appended for a longer period in a sequence (i.e., the tracking of nodes on the SOM), the merged data would represent changes in community patterns. In this study, the sequence of coordinates on the x- and y-axis served as input data for training with the SOM (Fig. 4.9.1). For each sample unit, *T*x2 (x and y locations) coordinates were used for an input period of *T* (i.e., months). As *T*=12 (i.e., 12 months), for instance, 24 (=12x2) variables were provided to the SOM for each sample unit in total. Since the number of taxa, N, was not required as input nodes for training the SOM in this study, substantial reduction in dimensions was achieved as the coordinate data were sequentially merged in different periods.

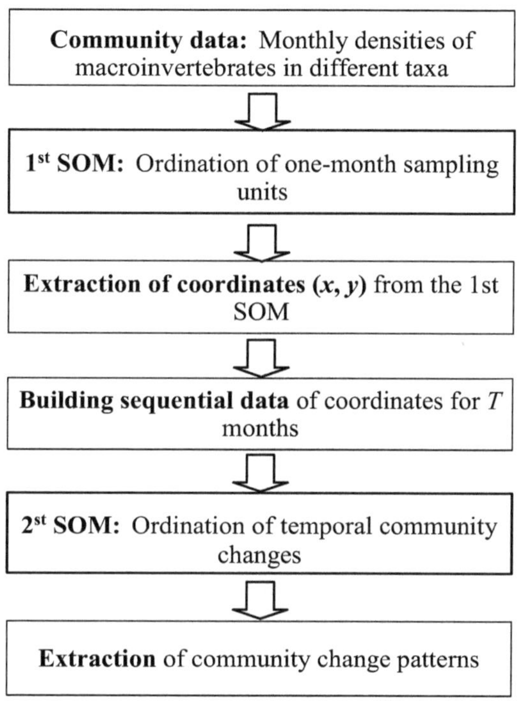

Figure 4.9.1 Schematic diagram of combinational implementation of the SOMs for grouping community changes.

Grouping of community changes was conducted with the specified sampling periods starting with different months. For the seasonal and yearly groupings, the season or the year to which the starting month belonged was used for naming the sampling period for training. If April 1996 is the beginning of the sampling period, for instance, it was considered as 'Spring' for the three-month samples and '96 for the 12-month samples. In other words, the community changes sampled were named based on the starting months.

Field data

For four years from 1996 to 1999, benthic macroinvertebrates were collected monthly at 11 sample sites located at different habitats within 200m of the Yangjae stream, a tributary of the Han River, Korea (Fig. 4.9.2), which was heavily polluted with organic matter from the Seoul metropolitan area. The Yangjae stream, however, has slightly recovered due to the restoration efforts by the local government. For instance, *Baetis sp.* (Ephemeroptera), an indicator taxa for partial water recovery (Hellawell 1986), was seldom observed in the early part of the survey period, but has been collected since June 1997.

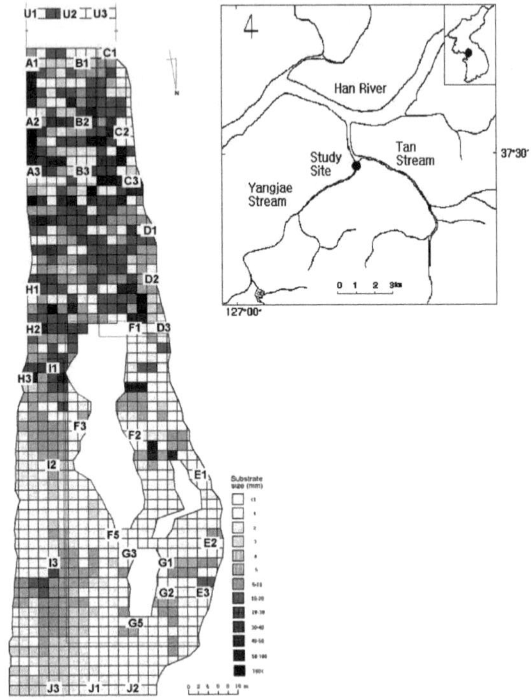

Figure 4.9.2 Location of the sample sites for collecting benthic macroinvertebrates at the Hakyeoul Reach in the Yangjae Stream, Han River, Korea.

The surveyed area mainly consisted of two sections (Fig. 4.9.2): the upstream straight section with larger substrate size and the downstream curved section at the bottom right corner with smaller substrate size. Three to four replicates were collected at each sample site. During the survey period, 3 phyla, 4 Classes, 10 orders, 26 families and 39 species were collected. However, only a few taxa were highly dominant at the surveyed area during the study period. Chironomidae, dominant with *Chironumus* sp., and Oligochaeta, mostly consisting of *Limnodrilus hoffmeisteri* (Tubificidae), were extremely abundant at the sampling sites. This was in accordance with the previous reports that only a limited number of tolerant taxa (e.g., *Chironomus* sp., Tubificidae, etc.) are strongly dominant in aquatic systems polluted with organic matter (e.g., James and Evison 1979, Hellawell 1986, Rosenberg and Resh 1993). Other taxa except *Chironumus* sp. and Tubificidae were mostly collected at very low densities with a few exceptions. Considering the limited number of abundant taxa, we used three representative taxa, the two dominant *Chironomous* sp. and Oligochaeta, and an indicator of slight water recovery, *Baetis* sp., as input data for training with the SOM in this study. In general, densities of benthic macroinvertebrates were greatly affected by flooding and low temperature at the survey area. In the monsoon climate in Korea, flooding with extremely high precipitation regularly occurs in early summer, while drought and severely low temperatures occur in winter.

a)

	0	1	2	3	4	5	6	7	8
0	F43, F44, J41 J42, **G42**	U29, U41, E29 I19, J29, **G43**	C35, J44	H19		U38, C16, H38 I18, J30, **B18** **B32, G31**	U16, U31, U42 C42	F31, H30, H31 I30, I31, **B31**	A31, A39, C39 D39, F39, H39 **I39, B39**
1	C32, C38, D29 D41, E41, H29 H43, I29,J5, J6 J7, J43, **B29**	C31, I41	I32, I44, I45 J32	A8, A30, E46 J45		C30, I28, **G30** **G44**	A16, E31, F30 F42, H28	U39, E39, I42 J39	U27, C27, D16
2	U44, C41, I43 J19, J38, G41	C29, F29, F41 H6, I35, **G47**	U32, C43, E30 E44, I8, J31 **B42, G45**	U34, C46, C47 D43, F36, J46 J47, **G46**	U30, U43, A40 C36, F35, F45 I36, I47	U8, U35, U36 A42, C8, D30 F46, I37, I40 **B46**	U19, A28, F40 I17	U17, U18, U26 D42, E42, I26 **B9, B17**	A17, A18, A26 A27, C28, I27 **B26, B27, G39**
3	A29, E43, J16 **B41, G35**	A41, E34, J10 J37	U33, C34, C37 C45, D34, H46 I38, J34, J36 **G29**	C33, E33, F33 F34, H32, J25 J28, **B8, B30**	U46, U47, A32 A33, A34, C44 D8, D35, D36 D37, H8, H35 H40, **B28, B34** **B40, B43**	A35, A43, A46 E19, E28, H3 I46, **B35, G8**		C18, E23, F16 H42, **B20**	U28, D17, D18
4		J35	U3, D31, D32 D33, D45, H41 I33, J33, **B10** **B33, B38, G32** **G33, G34, G36**	U45, D46, F32 F47, H33, H34 H44, I22, J48 **B45**	E8, E32, E45 H36, H37	U40, A3, D44 E25, I15, J1 J26, **B19, B36** **B47**	A15, A19, A47	**B1**	D27, E18
5	U5, U14, D19 D38, I10, I11 I25, **B7**	F22, I21, J12 J22, **B16**	U25, E47, I20 **G11**	A45, E35, E36 F37, H45, J18	A25, D40, F38 J8, **B37**	C26, F8, I1 J21, **G28**	U22, C20, H1 I48, J17, J40	C40, D7, E3 H17, H22	A1, A2, A36 C2, C19, I2 J2, **B2, G19**
6	C25, D25, I12 I24, J14, **B44**	E48, **B12, B25** **G12**	U7, I6, J11, J15 **B15, B48, G48**	U37, U48, E22 F48, H25, H26 H48, I34, J9 **B11, B22, G25**	U12, U13, U20 A38, C22, F1 F24, I7, I9, **B4** **B21**	U10, E37, E38 E40, F15, H7 H18	U11, I3	U1, A20, A22 C1, C7, D12 E1, F19, F26 J3,**G18, G40**	U2, C17, D2 E2, F2, H2, **G2**
7	A5, H16, I16 **B5, B13, B14** **G13**	U4, A37, C4 C5, C48, D6 F11, J13, **G6** **G22**	D47, D48, H47 J24, **B3, G10** **G37**	A10, D1, D13 F21, F23, J20 J23	A14, C23, C24 F7, F25, H11 **B24, G38**	U15, A12, A23 C10, C11, E10 E24, F14, H20 H24, **G24**	U9, U21, U24 A7, A21, A24 C15, C21, D4 D10, H21, **G3** **G14**	D14, E7, E15 H4, H23, J27	D11, D15, D26 D28, E17, F3 F18, F28, H15 **G1, G16, G26**
8	A6, A44, A48 D5, D22, E4 E5, E6, F4 H5, I5, I13 I14, J4, **B6** **G4, G5, G7** **G23**	A4, D23, F5 F13, **G20**	U6, C6, C14 E13, H13, I4	U23, A13, C3 D3, H10, H12 I23, **B23, G21**	A9, C13, D21 D24, E14, F6 F10, F12, F20 **G9**	A11, D20, E11 E16, E20, E21 F9, H9	C9, E9, E12	C12, D9, F27 H27	E26, E27, F17 **G17, G27**

straight region curved region

b)

	0	1	2	3	4	5	6	7	8
0									SU
1	SU								
2			AU						
3			WI		WI				
4				WI					
5									SP
6				SP					
7							WI		SU
8	SU								

c)

	0	1	2	3	4	5	6	7	8
0		98, 99				97, 98, 99		98	99
1	96, 98, 99								
2	99	98, 99	98, 99	99, 00	99	96, 99, 00		97, 98, 99	97, 98
3			98, 99	98	96, 98, 99, 00	96, 99, 00			
4			98, 99	98, 99, 00		96, 97, 98, 99			
5	96, 97			97					96, 97
6	97, 98		96, 97, 00	98, 99, 00	96, 97, 98	97, 99		96, 97, 98	96
7	96, 97	96, 97	00	97, 98	97, 98	97, 98	96, 97, 98		96, 97, 98
8	96, 97, 98		96, 97	96, 97, 98	96, 97	96, 97			

Figure 4.9.3 The map trained by the SOM for grouping 1-month community data. The letters in the Fig. indicate the name of the sample sites (Fig. 4.9.2). The sample sites B and G in bold characters were used for validation. The numbers following the letters indicate the starting month of the sampling period. For instance, 1, 13, 25, and 37 represent April in '96, '97, '98, and '99 respectively. The increment of the number represents the following months (e.g., 2; May, 3; June, 4; July, etc)). (a) grouping of communities, (b) seasonal grouping, and (c) yearly grouping.

For training with the models, the input datasets (densities of the selected taxa) were transformed by natural logarithm to achieve normal distribution and to emphasize the differences in low densities. Subsequently, the transformed data were proportionally normalized between 0.01 and 0.99 in the range of the maximum and minimum densities for each taxon collected during the survey period. Coordinates (x, y) defined in the first SOM were also proportionally rescaled between 0.01 and 0.99 in the range of the maximum and minimum values for training with the secondary SOM. The data collected from the 9 sampling sites for 4 years were used for training, while the data collected at the other two sites, B and G, were used for validation.

Results

Grouping of one-month samples

Overall, grouping of 1-month samples by the SOM reflected temporal and spatial variation of communities. This was generally in accordance with patterns observed in the previous results on community grouping (Chon et al. 1996, 2002, Kwak et. al. 2000). Fig. 4.9.3a shows grouping of communities based on 1-month field data. Temporal patterns were observable (Figs. 4.9.3b, c). By counting the nodes grouped with more than six samples, seasonal groupings appeared strongly. In accordance with low temperatures in winter, communities were grouped on 4 nodes (e.g., (2 (column), 3 (row)) and (4,3)) (Fig. 4.9.3b). Additionally, communities affected by flooding in summer were placed on 3 nodes (e.g., (0,1) and (8,7)). Community grouping in spring was also observable on 2 nodes (e.g., (3,6) and (8,5)), while grouping in autumn was found only on 1 node (e.g., (2,2)).

Additionally, communities were also grouped on a yearly basis (Fig. 4.9.3c). Two-year groupings for the year of 1996-1997 abundantly appeared on 7 nodes (e.g., (0,5) and (4,8)), and groupings for 1997-1998 and 1998-1999 were also observed on different nodes (Fig. 4.9.3c). However, the other two-year combinations were not clearly observed. For instance, the group of 1997-1999 was only observed on 1 node (5,6). The three-year and four-year groupings were also occasionally observed on the SOM, but they were not abundant on the map compared with the two-year groups. In the three- or four-year groups, communities mostly show low densities, reflecting flooding effects in August in summer (e.g., node (0,1), Fig. 4.9.3a).

The sample sites were also spatially grouped according to the curved and straight sections of the sampling sites (Fig. 4.9.3a). Communities sampled in the curved section were grouped on 4 nodes (e.g., (0,8) and (8,7)), while communities collected at the straight section appeared on 7 nodes (e.g., (4,3) and (8,2)). In the curved section (e.g., E, F and G) (Fig. 4.9.2), sediments were strongly collected and the burrowing type of organisms such as Oligochaeta and *Chironomus* sp. occurred abundantly. In the straight section (e.g., U, A and B), in contrast, the amount of sediments was low, and diverse communities including *Baetis sp.* were collected more abundantly. However, spatial grouping was, in general, weaker than temporal grouping (Figs. 4.9.3b,c).

The data from sites (B and G) used for validation were located on the nodes with similar community compositions as shown in Fig. 4.9.3a (e.g., communities shown in bold characters on the map; nodes (0,8), (4,3) and (8,7)). For the purpose of overall comprehension on community dynamics at the surveyed area, they were presented with the trained sample sites on the same figure.

Community changes in three months

For patterning temporal changes in communities, the 3-month SOM was obtained through training with the coordinate data by the secondary SOM (Fig. 4.9.1). Similar to the case of the 1-month SOM (Fig. 4.9.3), changes of communities were accordingly grouped spatially and temporally (Fig. 4.9.4). The validation data for sample sites B and G were in general matched to the nodes grouped with similar community compositions (e.g., the communities shown in bold characters on nodes (0,0) and (7,8) in Fig. 4.9.4a). Compared with the 1-month SOM (Fig. 4.9.3), however, differences were also observed in the 3-month SOM. Seasonal groupings were more apparent with a larger number of communities in groups: spring on 3 nodes (e.g., (0,8) and (1,6)), summer on 7 nodes (e.g., (0,0) and (8,4)), autumn on 3 nodes (e.g., (1,4) and (2,5)), and winter on 4 nodes (e.g., (0,6) and (3,2)) (Fig. 4.9.4b). Groups in summer were more strongly observed compared with the 1-month SOM, reflecting flooding effects on benthic communities. The increase in groupings in the 3-month SOM is understandable, since four seasons appear clearly in Korea and community changes in 3 months are consequently well matched to seasonal variation.

a)

	0	1	2	3	4	5	6	7	8
0	U16, U17, U26 A16, A17, A26 C16, D16, I26 **B18, B26**	U15, U37, A15 A25, C26, D15 F38, H37	A29, F29, H29 I29	C38, J38	U41, C41, D29 D41, E41, I41 J29, **G43**	U38, A30, F30 H38, **B17**	H30, I17, I30 **B31**	U27, A27, C27 D17, F39, I27 I39, **B27**	E27, F27, H17 H27, J17, J27 **G27**
1	U14, A37, C25 F37, **B7, G37**	U25, I16, I25 **B16, B25**	A38, D38, I38 **B30, B38**	U34, A41, E30 I35, **G29**	C35, I45	C29, F41, H41	U30, I36, **B8** **G30**	U42, A31, A39 E31, F31, H31 H39, **B39, G31**	U39, C39, D27 D39, E39, J39 **G39**
2	A6, C6, C14 D6, D25, E6 H16, H26	U24, A14, A24 E21, E37, F14	U7, A45, I46 **B45**	U45, A32, A33 D35, F32, H32 H33, H34, H35 **B33, B34, B35**	U33, C45, D34 E44, F33, F34 H36	U29, E29, F44 **B29, G42**	C34, I44, J44 J45	C30, C36, F42 J30	U31, C42, D42 I18, I31, **B32** **G44**
3	D14, E16, E26 H1, H20, J26 **G26**	E22, E38, I1 **B19, G38**	A7, C7, C15 D7, E7, E18 H2, H7, H18	U18, U35, U46 A42, C46, D36 E45, F35, F46 **B46**	C43, D43, F45 J46	U32, C32, J31 J32, **G45**	F43, I43, J5 J41, J42, J43 **G41**	U43, D30, D44 E33, **B9, B28** **B40, B43, G46**	U28, A28, C28 E28, E42, H42 I42, J40, **G40**
4	A1, C17, E1 E17	A18, A19, C18 C19, E23, F16 J1	A34, J25	F36, I32	U44, E34, H6 H43, J7, J16	C31, E43, J6 **B41**	J36	C33	U40, C40, D28 D40, E40, F28 F40, H28, I28 I40, J28
5	A20, D9, D10 D26, F17, F26 **G1, G16, G17**	D13, E24, E25 F1, F15, F25 H21, **G15, G25**	U8, U19, C8 D8, E8, E19 F8, H8, **B20** **G8**	A8, H19, J19	E46, I8, I19 J10, **B42**	H44, I6, I33 J8, **B36**	D31, D32, E32 J33, J34, J35 **G32, G33, G34**	C44, D33, I34 **G28**	U36, A40, C37 D37, H40, I37 J37, **B37**
6	U9, C9, C10 C11, C21, E9 E10, E14, F7	U22, A22, D2 D12, E15, H15 **G14**	U20, A10, C22 E36, F6, F13 F23, F24, **G13**	U5, D19, D46 F22, H46, J22 **G36**	D45, I21, **G7** **G35**	F21, H45, J9 J21	J4, J14, **B6**	I20, J24, **B44**	I15, I24, **B15** **B24**
7	D1, D3, H14	U1, U10, U11 U21, A21, A23 C1, C20, D11 F18, H22, H23	A9, A12, E12 F10, F12, H9 H10, H11, H12 H24, **B21, B22**	U12, U23, C23 D23, F20, J20 **B3, B23, G9** **G23**	U3, I22, J12 **B10, G11**	I10, I11, I14 J11, **B14, G10**	H5, I7, J18	U13, C24, E35 J15	U6, A13, A44 J23, **G6**
8	A2, A35, A46 E2, F2, H3 I2, J2, **B1** **G2, G18**	A11, C12, D20 E11, E20, F9 F19	U2, A3, A36 C2, D4, D18 E3, F3, I3 **B2, G3, G19**	A43, H4, J3	I9, I23, **B11**	A4, C3, D21 E4, F11, I12 **B4, B5, B12** **G4, G5, G20** G21	U4, C4, I4 J13, **B13**	A5, C5, D5 D22, E5, F4 F5, I5, I13 **G12, G22**	C13, D24, E13 H13, H25, **G24**

straight region   curved region

b)

	0	1	2	3	4	5	6	7	8
0	SU				SU			SU	SU
1									SU
2				WI					
3			AU	WI					
4		AU							SU
5		SP	AU						
6	WI	SP							
7		WI							
8	SP							SU	

c)

	0	1	2	3	4	5	6	7	8
0	97, 98	97, 98, 99			98, 99			98, 99	97, 98
1	99							98, 99	99
2	96, 97, 98	97, 98		98, 99	98, 99				98, 99
3	97, 98		96, 97	99, 00			99	98, 99	98, 99
4		97			96, 99				98, 99
5	96, 97, 98	97, 98	96, 97				98		99
6	96, 97	97, 98	97, 98						
7		96, 97, 98	97	96, 97, 98		97			
8	96	96, 97	96			96, 97		96, 97, 98	97, 98

Figure 4.9.4 The map trained by the SOM for grouping community changes based on coordinate data for 3 months. The coordinate data were extracted from the SOM map trained with community density data as shown in Fig. 4.9.3. Descriptions of symbols are explained in Fig. 4.9.3. (a) grouping of communities, (b) seasonal grouping, and (c) yearly grouping.

Yearly groupings were also observed on the 3-month coordinate SOM (Fig. 4.9.4c). The two-year groupings were prevalent on the SOM as shown on the 1-month SOM (Fig. 4.9.3b), being similarly dominated by 1997-1998 and 1998-1999. The three-year groupings (nodes (0,2) and (0,5)) were additionally observed (Fig. 4.9.4c). However, the three- and four-year groupings were not strongly observed on the SOM based on the 3-month coordinate data.

Spatial groupings tended to be stronger in the 3-month SOM (Fig. 4.9.4a) compared with the results from the 1-month SOM (Fig. 4.9.3a): the straight section on 7 nodes (e.g., (0,0), (1,7) and (3,2)) and the curved section on 6 nodes (e.g., (1,5), (2,6) and (5,8)). Communities shown in groups were in general similar to those shown in the 1-month SOM (Fig. 4.9.3a).

The SOM results based on the coordinate data (Fig. 4.9.4a) were compared with the SOM based on the 3-month density data (Fig. 4.9.5a). The general conformation was similar to the SOM trained with the coordinate data in temporal and spatial domains. The field data from sites B and G for the validation were also accordingly recognized on the nodes with similar community compositions (e.g., communities shown in bold characters on the nodes (3,0) and (4,7) in Fig. 4.9.5a). Similar to groupings observed on the SOM based on 3-month coordinate data (Fig. 4.9.4b), seasonal groupings were apparent on the SOM based on the 3-month field data (Fig. 4.9.5b): summer flooding on 4 nodes (e.g., (0,8) and (8,6)), cold temperature in winter on 3 nodes (e.g., (4,1) and (5,6)), spring on 4 nodes (e.g., (0,0) and (3,2)), and autumn on 2 nodes (e.g., (2,8) and (4,0)). Yearly groupings (Fig. 4.9.5c) were also similar to the SOM based on the 3-month coordinate data (Fig. 4.9.4c), being dominated with the 2-year groupings (1996-1997 and 1998-1999), and occasionally with the 3-year groupings.

Community changes in 12 months

The grouping was further carried out over a longer duration of 12 months (Fig. 4.9.6a). Over the 12-month period, community groupings were also formed accordingly. The patterns of community changes were characterized with density decrease occurring in the flooding period in summer. The number of patterned groups, however, decreased in general. In summer 5 nodes (e.g., node (1,8)) were grouped, while 4 nodes (e.g., node (4,5)) were patterned in winter (Fig. 4.9.6b). As stated above, grouping with community changes was conducted in specified sampling periods starting with different months. For the yearly grouping, the year to which the starting month belonged was used for indicating the name of the samples of community changes. For patterning 'seasonal' variation in the yearly grouping, for instance, the season of the starting month in the sampling period was used for representing the sampled data, although the yearly sampling period covered the entire four seasons. If April 1996 is the beginning of the sampling period for the yearly data, for instance, it was considered as 'spring' in the seasonal grouping as stated above.

Yearly grouping was also observed (Fig. 4.9.6c), however grouping was not clearly formed compared with the 3-month SOM (Fig. 4.9.5c). The communities were grouped in 1998, 1996 and 1997, while communities in 1999 did not form strong groups. While 1998 groups were placed at the left edge of map, 1996 groups occurred at the right edge of the map. Two-year groups, however, were observed only in a few cases. This indicated that yearly groupings were not frequently observable in the 12-month SOM compared with the 3-month SOM (Fig. 4.9.5c).

Although weaker than temporal groupings, spatial groupings were also observed in the 12-month SOM (Fig. 4.9.6a). The samples sites in the straight section (e.g., U and A) were grouped on nodes (1,4), (4,5) and (8,8), while the curved sites (e.g., E and F) occurred on nodes (1,0), (6,3), (8,5) and (8,6). This implies that community changes at different habitats could be identified on a yearly basis. We also checked the SOM trained with the 12-month field data (Fig. 4.9.7) to evaluate the results from the SOM based on the coordinate data (Fig. 4.9.6). Community groupings appeared similarly according to temporal and spatial variations in both the coordinate data and the field data for 12 months. The number of groups in winter was slightly increased to 8 nodes (e.g., (5,0) and (8,0)), while the number of groups in autumn was slightly decreased to 4 nodes (e.g., (1,1) and (6,2)) (Fig. 4.9.7b) on the SOM based on field data. The number of groups in spring and summer were the same in both SOMs (Figs. 4.9.6b and 4.9.7b). Yearly grouping was also similar to the SOM based on coordinate data, appearing abundantly in 1998 and 1996.

a)

	0	1	2	3	4	5	6	7	8
0	U38, A30, A38 C38, D38, E38 F38, H38, I38 J38, **B38**	C17, D26, E17	A1, E1, H1 I1, J1, **G1**	A20, D9, D10 E15, E26, F17 F26, **G16, G17** **G26**	U8, A8, C8 D8, E8, E19 H8, **G8**	U10, U20, A19 A21, A23, C19 C20, E23, H20 H22, J26	U1, U21, A35 C1, D1, F18 **B1, G18**	D27, E2, E27 F2, F27	A42, A46, H3 J2, **G2**
1	U17, U26, A17 A26, D15, I26 **B26**	E22, **B19, G38**	U18, U19, A18 C18, D16	E25, F1, F16 F25, **G15, G25**	U9, A22, C9 E9, E10, E14 E21, F14, H21	U11, C11, C21 D11, E20, F8 F19, H23, **G27**	C12, D17, E11 H14	U2, A2, C2 I2	H27, H41, J17 J27, **G40**
2	U15, U16, A15 A16, A25, C26 **B25**	C16, D25, E16	C7, D7, E7 E18, H2, H7	C10, D13, E13 E24, F15, F24 H13, H25, J25 G14, G24	A10, A13, C22 E35, E36, F6 F7, F23	C14, F5, F13 **G13**	U12, A11, F9 F11, F20, H10 H11, **B3, B23**	C23, D3, D20 I3, **G21**	D4, D21, E3 F3, H4, J3 **B4, G3**
3	U25, U37, A37 C25, F29, F37 I25	U14, A14, A24 D14, E37, H26 **G37**	C5, C15, D24	A9, A12, C13 E12, F10, F12 H9	D19, F22, H12 J22, **B22, G36**	U23, J20, J23 **B14, G9**	U3, A3, D46 I9, I22, J11 **B11, G11**	U4, C3, C4 D23, I23, J12 **B12, G20**	A4, E4, I11 I12, J4, J5
4	A29, C37, D37 H37, I37, **B37**	I40, **B30**	U24, H40, I15 I24, J37, **B7** **B15, B24**	U22, A36, D2 D12, D18, H15 B2, B17, G19	F21, H24, J21 **B20, B21**	D45, H45, H46 I20, I21, J9 **B10, G10, G35**	H44, I6, I8 I33, J8	I10, J14, J36	I14, **G5, G7**
5	H28, I28	U40, C40, F28 F40, **G28**	A43, D28, D40 E40, H42	U13, C24, I7 J15, J18, **G12**	U7, A7, J24	U36, D36, E32 H36, **B16**	D31, D32, J10 J33, J34, J35 **G32, G33, G34**	I43, **G41**	U5, I4, J13 **B5, B13, G6** **G22**
6	F42, J40	U28, A28, C28 E42, I42, **G44**	U43, E28, I18 J28, **B28**	E45, H18	A40, C33, C44 D33, D44, E33 F32, F33, H33 I34, **B40, B43**	U35, U46, D30 D35, D43, F35 F36, F45, F46 H32, H35, I36 **B35, B36, B46**	C43, C46, E46 I19, I32, J32 **B42, G46**	U44, E34, E43 J19, **B6, B41** **G31**	A5, D5, D22 E5, F4, H5 I5, I13, **G4**
7	U27, U39, E39 I27, J39, **B27** **G39**	C30, C36, I17	U30, J29, J30 **G29**	C35, C45, J31 J44, J45, J46 **G45**	U34, U45, A32 A33, A34, A41 A45, H34, I46 **B33, B34, B45**	D34, E30, **B8**	H19	U29, C32, D29 E29, F44, H43 J43, **B29, G42** **G43**	C31, J6
8	A27, A31, A39 C39, D39, F39 H39, I39, **B39**	C27, F30, H30 I30, **B31**	U31, U42, C42 D42, E31, F31 H31, I31, **B9** **B32, G30**	E44, H17	U32, U33, C34 F34, I35, I44 I45	U6, C6, D6 H6, H16, J16 **B18, B44**	A6, A44, D41 E6, E41, I16 J7, **G23**	U41, C29, C41 F41, H29, I29 I41	F43, J41, J42

straight region   curved region

b)

	0	1	2	3	4	5	6	7	8
0	SP				AU				
1					WI				
2				SP					
3	SP	SP							
4									
5									
6					WI	WI			SU
7					WI				
8	SU		AU			SU		SU	

c)

	0	1	2	3	4	5	6	7	8
0	99		96	97, 98	96	97, 98	96, 97		
1	97, 98			97, 98	96, 97	97, 98			
2	97, 98		96	97, 98	96, 97, 95, 99		96, 97		96
3	98, 99	97, 98, 99		96, 97	97, 98		96, 97	96, 97	96, 97
4	99		97, 98, 99	96, 97		97, 99			
5				97			98, 99		96, 97
6		98, 99			98, 99	98, 99	98, 99, 00	99	96
7	98, 99			99	98, 99			98, 99	
8	98, 99		98, 99		98, 99	96, 97	96, 97, 99	98, 99	

Figure 4.9.5 The map trained by the SOM for grouping community changes based on field data for 3 months. Community density data were used as input data. Descriptions of symbols are explained in Fig. 4.9.3. (a) grouping of communities, (b) seasonal grouping, and (c) yearly grouping.

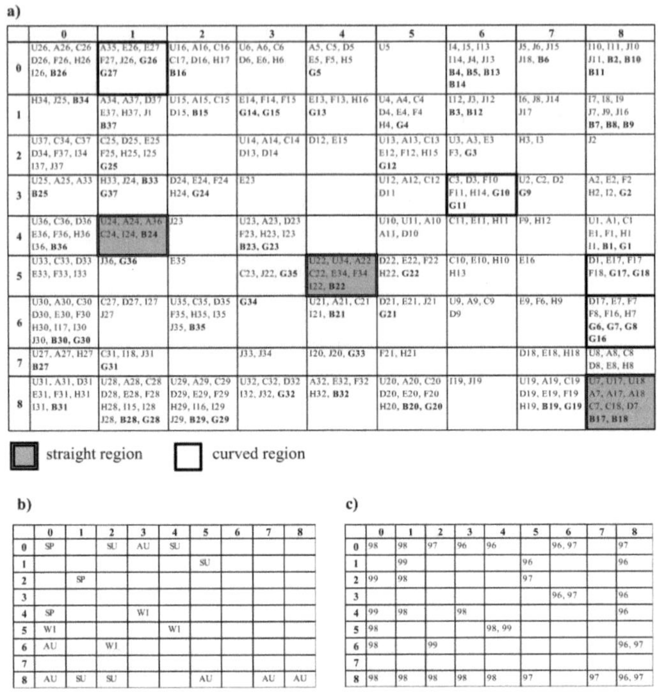

Figure 4.9.6 The map trained by the SOM for grouping community changes based on coordinate data for 12 months. The coordinate data were extracted from the SOM map trained with community density data as shown in Fig. 4.9.3. Descriptions of symbols are explained in Fig. 4.9.3. (a) grouping of communities, (b) seasonal grouping, and (c) yearly grouping.

a)

	0	1	2	3	4	5	6	7	8
0	D17, E8, E17 F8, F17, H11 **G7, G8, G17**	C7, D16, E7 F7, H7	U7, U16, U17 A7, A16, A17 C17, **B17**	U8, A8, C8 D7, D8	U9, U10, U19 A9, A19, C19 D9	U23, A23, E23 F23, H23, I23 **B23**	U22, A22, E22 F22, H22, I22 **B22**	U21, A21, C21 D21, E21, F21 H21, I21, J21 **B21, G21**	U33, A33, C33 D33, E33, F33 H33, I33, **B33**
1	E15, E16, F15 F16, **G15, G16**	A6, C6, D6 E6, F6, H6 **G6**	U6, C16, F14 H17, **B16**	D11, E14, H14	U11, A10, D10	A32, F24, **B32**		D22, J22, **G22**	J33, J34, **G34**
2	C15, D12, D13 D14, D15,**G14**	U15, A15	U14, A14, C14	U12, A11, A12	C11, D20, E11 E20, F11, F20 **G10, G11, G20**	U20, A20, C20 H20, I20, **B20**	U32, C32, D32 E32, F32, H32 I32, J32, **G32**		U34, A34, C22 C34, D34, E34 F34, H34, I34 **B34**
3	E13, F13, **G12** **G13**	U13, A13, C13 H16	C12, E12, F12 H15	C9, E9, E18 F18, **G18**	H9	J20, **G33**		D25, E25, J25 **G25**	U25, A25, F25 H25, I25, **B25**
4	U5, A5, C5 D5, E5, F4 F5, H5, **G5**	U4, A4, C4 D4, E4, H4	C2, C10, D2 E10, F10, H10 H13, **G9**	D1, D18, F9 H12	H1, J15, **B14**	I16, J16	D26, E26, F26 J26, **G26**	U26, A26, C26 H26, I26, **B26**	U37, A37, C25 C37, D37, E37 F37, H37, I37 J37, **B37, G37**
5	U18, A18, C18 H18, **B18**	H8, J18	U3, A3, C3 D3, E3, F3 H3, **G3**	U1, U2, A1 A2, C1, E1 E2, F1, F2 H2	I1, J1, **B1** **G1**	I10, J10	J7, J13, **B7** **B12,B13**		
6	U30, A30, C30 D30, E30, F30 H30, I30, **B30** **G30**	I18, J30	D19, E19, F19 J19, **G19**	**B19**	I2, I8, J2 J3, **B2, B3** **G2**	I11, J4, J11 **B4, G4**	I4, I5, I12 J5, J6, **B5** **B6**	D23, I13, J23 **G23**	U35, A35, C23 C35, D35, E35 F35, H35, I35 J35, **B35, G35**
7	I17	**G31**	H19, I19, J31	**B15**	I7, I9, I15 J9, **B9**	I6, J8, J12 J17, **B8, B10** **B11**	I3, I14, J14	D24, E24, J24 **G24**	H24, J36, **G36**
8	U29, A29, C29 D29, E29, F29 H29, I29, J29 **B29, G29**		U31, A31, C31 D31, E31, F31 H31, I31, **B31**	J28	U28, A28, C28 D28, E28, F28 H28, I28, **B28** **G28**	D27, E27, F27 J27, **G27**	U27, A27, C27 H27, I27, **B27**	C36, D36, E36 H36, I36	U24, U36, A24 A36, C24, F36 I24, **B24, B36**

straight region   curved region

b)

	0	1	2	3	4	5	6	7	8
0						WI	WI	WI	WI
1		AU							
2							AU		WI
3									
4	SU	SU							SP
5			SU	SP					
6	AU								WI
7									
8	SU		AU		SU				SP

c)

	0	1	2	3	4	5	6	7	8
0	96, 97		97		96, 97	98	98	97	98
1	97	96							
2	97				96, 97	97	98		99
3									98
4	96	96	96					98	99
5			96	96					
6	98				96		96		99
7						96, 97			
8	98		98		98		98		98, 99

Figure 4.9.7 The map trained by the SOM for grouping community changes based on field data for 12 months. Community density data were used as input data. Descriptions of symbols are explained in Fig. 4.9.3. (a) grouping of communities, (b) seasonal grouping, and (c) yearly grouping.

Patterns in community changes

By combinational application of the SOMs, it was possible to identify the most commonly observed community changes in the surveyed area. In the groups of the 3-month SOM based on the coordinate data, for instance, a typical community change after flooding was observed on node (8,1) in Fig. 4.9.4a. Actual data for community changes corresponding to node (8.1) are shown in Fig. 4.9.8a. The densities in *Chironomus, Baetis* and Oligochaeta were initially high but rapidly dropped in the third month. The abrupt decrease in densities was frequently observed after strong flooding in streams in Korea in summer. Similar community changes were also observed on the SOM based on 3-month field data (Fig.

4.9.8b). Other grouping patterns in community changes over 3-months were observable on the SOM, and they mostly occurred also on the SOM based on field data correspondingly.

Typical community changes were also identifiable on the 12-month SOM. For instance, a strong one-year group starting from August was observed on the SOM in Fig. 4.9.9, both from the coordinate data (node (2,8), Fig. 4.9.9a) and from the field data (node (0,8), Fig. 4.9.9b). In this group, densities in Oligochaeta were initially low, but gradually increased in the later period. Densities in *Chironomus* sp., however, were in the intermediate range with some fluctuation. Additionally, *Baetis* densities were initially low but increased qtrongly in the last part of the 12-month period. This grouping characteristically represented the pattern of community dynamics that would be frequently expected in the survey area.

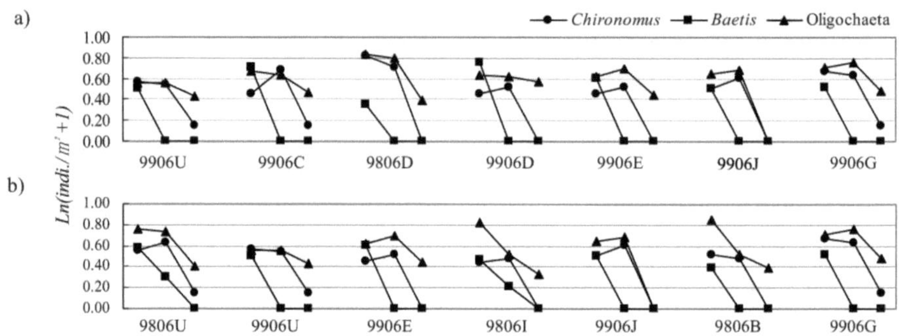

Figure 4.9.8 A pattern of community changes for 3 months. The name of the sample sites indicates "year (two digits)"-"month (two digits)"-"sample site". (a) community changes appearing on node (8,1) of the SOM in Fig. 4.9.4a defined with coordinate data, and (b) community changes on node (0,7) of the SOM in Fig. 4.9.5a defined with field data.

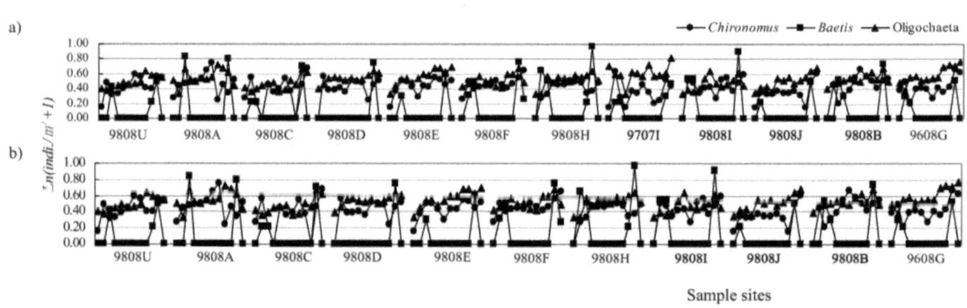

Figure 4.9.9 A pattern of community changes for 12 months. The symbols for the sample sites are explained in Fig. 4.9.8. (a) community changes on node (2, 8) of the SOM in Fig. 4.9.6a defined with coordinate data, and (b) community changes on node (0, 8) of the SOM in Fig. 4.9.7a defined with field data.

Discussion and conclusion

In this study the combinational use of the SOM was applied to grouping community changes. We demonstrated that the model developed was capable of identifying community changes commonly occurring in the survey area. Dimension reduction was achieved by using the coordinate data from the SOM, and datasets for up to a one-year period were efficiently grouped. Grouping of community changes on a large scale is useful for acquiring a comprehensive view of the progressive or regressive development of aquatic ecosystems in stressful situations and is also urgently required for establishing policies for sustainable management of aquatic ecosystems. The model proposed could be used a a novel tool to detect the representative community changes observed in the survey area.

Although there have been numerous accounts concerning the classification of communities through conventional multivariate analyses in ecology as stated above (e.g., Bunn et al. 1986, Legendre and Legendre 1987, Ludwig and Reynolds 1988, Quin et al. 1991), not many studies have been focused on grouping of community changes per se. Similarly grouping in community changes has been rarely reported in artificial neural networks (Chon et al. 2000a, b), although there have been many studies on prediction of community dynamics using ANNs. In classification, communities were mostly grouped in static terms, not in dynamic terms in ANNs, although the data might have been collected sequentially (see Lek and Guegan (2000) and Recknagel (2003) for implementation of ANNs).

Dimension reduction is a major requirement especially when the data consist of a higher dimension than that tolerated. Generally ecological community data consists of many species at numerous sampling sites over a long period. Dimension reduction is frequently considered as a preprocessing step for data analysis in ecology. This study demonstrated that dimension reduction could be efficiently carried out with the combinational use of the SOMs. This advantage would be more appreciated if the sampling period was lengthened.

A combination of the ART plus the SOM has been used for grouping community changes (Chon et al. 2000a). In this case the number of input nodes was dependent upon the number of taxa times the input period. In this study, in contrast, the number of input nodes was constant with 2 (i.e., x, y coordinate) times input period, while the number of taxa is included in input data for the model of the ART plus the SOM (Chon et al. 2000a). The model developed in this study shows an advantage of dimension reduction. In this case, however, only three taxa were used for revealing the impact of pollution more strongly in community response and for the simplicity of modeling. Since non-linearity would be more complex and difficult for analysis as the number of variables is increased, further investigation would be necessary into its application to the datasets with a higher number of taxa, for instance, community changes in clean aquatic ecosystems. Additionally, since the secondary SOM is heavily dependent upon the initial coordinate data obtained by the first SOM (Fig. 4.9.1), it is important to have the appropriate coordinate datasets from the first SOM, and it is desirable to conduct a sufficient number of training sessions with the first SOM to get the most representative community groupings collected at the survey area.

In summary, the combinational execution of the SOMs was demonstrated in grouping community changes in periods of up to 12 months through dimension reduction of input data. It was possible to find seasonal and yearly community changes that are typically expected in polluted streams in Korea. The combinational use of the SOM could be useful for mining large-scale time series data to find representative community changes occurring in the survey area.

4.10 Patterning, predicting stream macroinvertebrate assemblages in Victoria (Australia) using artificial neural networks and genetic algorithms*

Horrigan N†, Bobbin J, Recknagel F, Metzeling L

Introduction

Macroinvertebrate assemblages are widely used for biomonitoring of stream ecosystems. Several modern assessment concepts and approaches have been desribed. The so-called referential approach (Parsons and Norris 1996, Marchant et al. 1999, Smith et al. 1999) is based on the comparison of macroinvertebrate communities between potentially impacted sites and reference sites considered to be pristine. Knowing the relationships between environmental variables and macroinvertebrate occurrence at reference sites, it is possible to predict species or taxa, which should occur at the remaining sites in the absence of anthropogenic stress. The ratio of observed/expected (O/E) families is used as a measure for site-specific ecological conditions.

Statistical and computational techniques have been successfully integrated into the referential approach facilitating stream site classification and prediction of macroinvertebrate assemblages. Classification or grouping of macroinvertebrates into assemblages is sometimes criticized as an arbitrary procedure as they are usually distributed in continuous gradients rather than well defined separate groups (Chessman 1999). However in order to deal with large numbers of macroinvertebrate taxa it is often crucial to consider groups instead of individual taxa provided appropriate classification techniques are available.

Widely used statistical methods for data classification and ordination are cluster and principal component analysis. Both methods have shortcomings in coping with heterogeneous and nonlinear data, and results can be confounded by outliers and missing data. Artificial neural network (ANN) based classification techniques such as Kohonen or Self-Organizing Maps (SOM) may help to overcome these shortcomings. A number of ecological case studies have shown that SOM are an efficient classification tool (Chon et al. 1996, 2003, Cereghino et al. 2001, Park et al. 2001a, 2003a, Brosse et al. 2001, Giraudel and Lek 2001).

ANN as well as genetic algorithms (GA) prove to be appropriate for the prediction of macroinvertebrate and fish assemblages in streams. Multi-layer perceptron ANN were successfully applied to predict the occurrence of stream macroinvertebrates from environmental variables (Walley and Fontama 1998, Schleiter et al. 1999, Pudmenzky et al. 1998, Hoang et al. 2001). GA were used to predict fish distribution from physical characteristics of streams (d'Angelo et al. 1995) and to select input variables of classification tree models predicting benthic macroinvertebrate communities in Belgian watercourses (Goethals et al. 2003).

---

* The authors are grateful to the EPA Victoria for providing the database.

† Corresponding: nelli.horrigan@adelaide.edu.au

Even though ANN have clearly demonstrated their potential for ecological applications in terms of classification and prediction they store learned models in a highly distributed manner by means of connection weights, which bear little resemblance to human understanding of rules or concepts. By contrast, GA can be used for knowledge discovery by deriving predictive models or rule sets, which can easily be understood (Recknagel 2001). Recknagel et al. (2002) compared applications of ANN and GA in terms of forecasting and understanding of algal blooms in Lake Kasumigaura (Japan). It was demonstrated that models explicitly synthesized by GA not only performed better in seven-days-ahead predictions of algal blooms than ANN models, but provided more transparency for explanation as well. The present paper demonstrates the use of both ANN and GA for the classification and prediction of macroinvertebrate spatial assemblages in the stream system of Victoria (Australia). The stream database contains abundances of macroinvertebrates in conjunction with environmental and stream habitat characteristics. Both ANN and GA are applied in order to best compromise: (i) the discovery and explanation of patterns of macroinvertebrate occurrence within the Victorian landscape, and (ii) the prediction of these patterns from environmental variables. The predictive and explanatory performance of both ANN and GA will also be compared.

Material and methods

Data

The stream database for this study was provided by the Victorian Environment Protection Authority, Australia. It contained abundances of 128 macroinvertebrate families sampled at 407 stream sites between March 1990 and November 1998. The sampling sites were chosen in order to represent the main types of rivers in each of the 25 drainage basins defined by the Australian Water Resources Council (AWRC). Most sites were sampled on four occasions in spring and autumn over the two consecutive years and seasonal habitat data for single sites were combined (Marchant et al. 1999). At each site, two habitats were sampled separately: the main-channel (often a riffle) and the bank or edge of the channel. In order to simplify clustering, only the database including edge habitats was used for this study. A sample consisted of a macroinvertebrate collection over a 10m transect for each habitat using a D-frame hand net (0.25 mm mesh), followed by 30 minutes picking of live specimens. Macroinvertebrates were preserved in 70% ethanol and identified to family level. Specimens of Oligochaeta, Hydracarina, and Nematoda were not identified further (Marchant et al. 1999). Only environmental variables presumably not affected by human activity (natural variables) were used for this study. The variables distance from source, slope, altitude, catchment area, width, alkalinity, macrophyte taxa and macrophyte abundance category were log-transformed.

Modelling techniques

Three modeling techniques (SOM, MLP and GA) were applied in order to pattern, predict and explain occurrences of macroinvertebrate assemblages based on 19 environmental variables utilising all environmental and macroinvertebrate data of the 407 stream sites. The conceptual framework for this study is shown on Fig. 4.10.1. Neural Solutions 4.0 and Matlab 5.3 software with the SOM Toolbox developed by The Laboratory of Computer and Information Science (CIS) at Helsinki University of Technology was used for this study. The underlying GA was designed and implemented in C++.

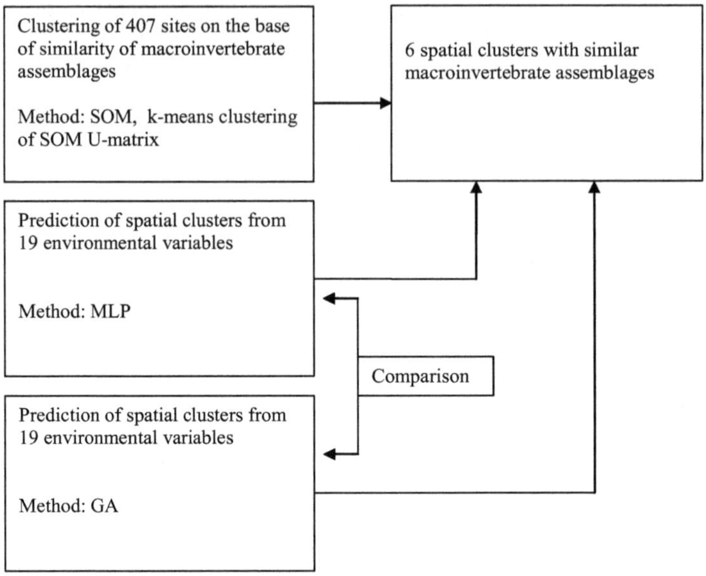

Figure 4.10.1 The conceptual framework for the study

Patterning of macroinvertebrate assemblages by SOM

The abundance pattern of 128 macroinvertebrate taxa was examined using a SOM with a 9x11 topology. The resulting U-matrix was partitioned by k-means algorithm into 6 clusters, which were used for the further analysis. In order to relate environmental variables to the above mentioned clusters, we built another SOM with 19 environmental variables and the number of the cluster as a separate variable 20 variables in total, all data normalized between 1 and 0. We used component planes for this analysis, which visualise continuous values of each variable used for building SOM on the same spatial scale as all the other contributing variables. Comparison of component planes allows the identification of common trends and correlations between variables, which might not be easily detectable by statistical methods. Although SOMs have been widely applied for the analysis of various ecological problems, analysis of component planes has not been extensively studied so far. Park et al. (2003c) applied it to study the contribution of each environmental variable to the classification of sampling sites from different water bodies in The Netherlands.

Prediction of the assemblage types by ANN

In order to predict the types of macroinvertebrate assemblages by means of a multilayer perceptron, a 25x407 data matrix was created. It considered the 19 environmental variables as inputs and the 6 spatial groups derived from SOM for each of the sites as outputs. All these data were normalized into the range between 0 and 1. The MLP contained 19 neurons in the input layer, 10 neurons in the hidden layer and 6 neurons in the output layer. The sigmoid function was used as transfer. The database was randomly subdivided into a training subset (65% of the data), a cross-validation subset (10%) and a testing subset (25%). The accuracy of MLP reported in this paper is obtained from the simulation on the testing subset which was not used for training purposes. The optimum training error was achieved by 1500 iterations.

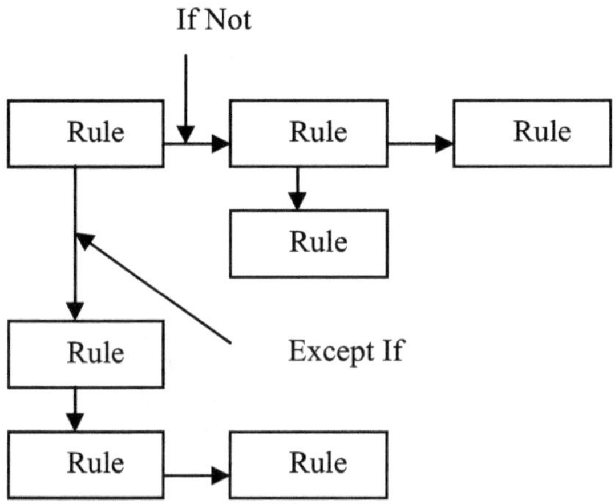

Figure 4.10.2 Structure of an evolved rule tree

Prediction of the assemblage types by GA

A GA consists of a population of individuals where each individual represents a model. Individuals are modified by mutation and crossover and the best individuals are selected to form a new population. Each new population is called a generation. In the context of the present paper a GA is used to evolve associations between physical and chemical properties of streams (attributes) and spatial clusters derived from partitioning of SOM U-matrix (outputs) based on similarities of macroinvertebrate assemblages. Attributes are associated with outputs by means of a classifier or rule.

Rules are combined to the rule sets by using a ripple-down structure shown in Fig. 4.10.2. When a rule is true any consecutive horizontal rule is immediately tested. If a rule is not true then the consecutive vertical rule is tested. Horizontal arrows in Fig. 4.10.2 represent exceptions to the rule to their left, and vertical arrows point to the rule to be tested if the current rule is not true. The last rule found to be true has its action implemented. If no true rule is found then the evolved default action is performed. Rule D in Fig. 4.10.2 would have its action performed if and only if rule A is true, rule B is not true and rule D is true. The approach used by the GA facilitates gradual evolution of the model by allowing mutation processes to slightly modify the model behaviour with exceptions to current rules. Information contained in the rules is represented symbolically, where the symbols are associated with values in a parameter vector that is co-evolved alongside the rulesets. Each individual in the population is a complete ruleset. During each generation the structure of the ruleset is evolved by means of discrete operators (addition, subtraction and modification of the rules), and the parameters which define the values on the rules are modified by means of a self adaptive evolutionary algorithm (Schwefel 1995, Baeck 1996).

Results

Patterning of macroinvertebrate assemblages by SOM

The resulting SOM U-matrix of this data is shown in Fig. 4.10.3a and the six clusters or groups resulting from partitioning by K-means algorithm in Fig. 4.10.3b. The six clusters or groups roughly correspond to the five ecological zones defined by the EPA Victoria (Fig. 4.10.4). Group 1 corresponds to Forest A, Group 2 to Cleared Hills and Coastal Plains, Groups 3 and 4 to Forest B, Group 5 to Highlands and Group 6 to Murray and Western Plains.

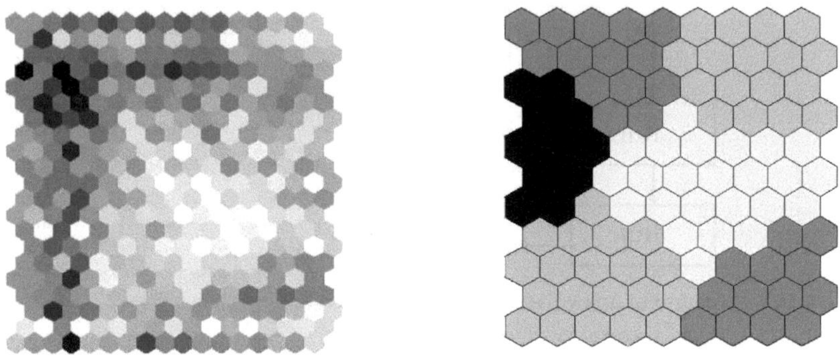

a b

Figure 4.10.3 SOM outputs: (a) U-matrix, (b) Partitioning into 6 clusters by the K-means algorithm.

Fig. 4.10.5a shows a distribution of 6 groups or clusters on SOM grid (see Vesanto et al. 2000 for explanation of hit diagrams). Fig. 4.10.5b shows selected component planes which appear to have some patterns corresponding to the distribution of 6 groups on the same scale. Alkalinity is particularly low in the area corresponding to group 1, and particularly high in groups 6 and 3. The number of macrophyte taxa is comparatively low at groups 1 and 2, and comparatively high at all other groups. The distribution of the macrophyte category amongst clusters is quite patchy, but distinctively different at groups 1 and 6. Distribution of values for vegetation category does not follow horizontal gradient characteristic for distribution of clusters at first sight, but it might be an important variable to distinguish between groups 1 and 6, and to some extent between groups 3 and 4. Sites belonging to groups 1 and 6 clearly differ in relation to slope, with high values for this variable at group 1 and low values at group 6. Group 1 can also be characterized by relatively high altitude, although for other groups altitude does not seem to fall into any distinctive pattern. Other variables do not appear to have any distinctive pattern corresponding to the distribution of clusters (groups) and will not be considered here.

Prediction of the assemblage groups by ANN and GA

The average percentage of correct predictions by the ANN of the six assemblage groups of macroinvertebrates as discovered by the SOM was 88.56% while the average percentage of correct predictions by GA was 77.1%. The mean squared error (MSE) and the percentage of correct predictions (PCP) for each group by the MLP and the GA is shown in Table 4.10.1.

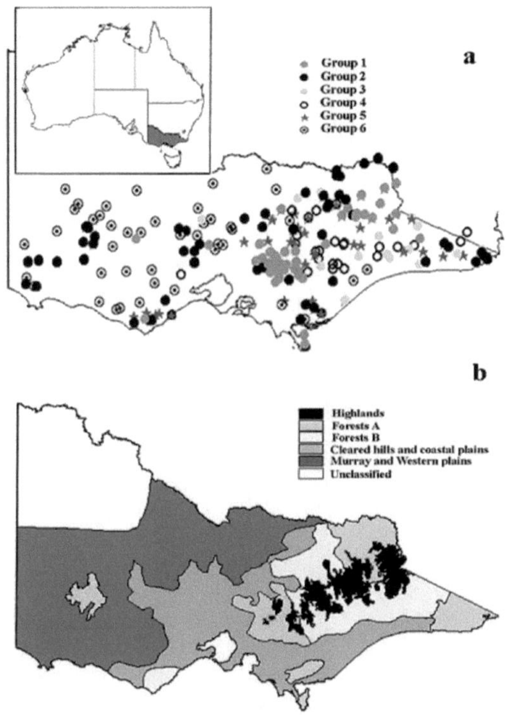

Figure 4.10.4 a) Distribution of macroinvertebrate groups resulting from SOM (sites belonging to the same group have the same marker) b) biological regions in Victoria based on benthic macroinvertebrates (Metzeling et al. 2001).

Table 4.10.1 Mean square error (MSE) and percentage of correct predictions (PCP) by applications of ANN and GA for each of 6 groups.

	Group 1	Group 2	Group 3	Group 4	Group 5	Group 6
MSE (ANN)	0.069	0.132	0.054	0.15	0.11	0.03
PCP (ANN) %	91.17	82.35	94.11	83.33	84.31	96.07
PCP (GA) %	93.42	96.97	84.44	89.47	53.17	80.90

GA rules

For the sake of space we consider here ruleset for group 1 only (Table 4.10.2). All the values considered by the GA ruleset fall into the minimum and maximum range within SOM cluster (group). However, taking into consideration that group 1 is heterogeneous and spatially scattered (Fig. 4.10.4) the range and averaged values for the predictor variables can give only a very approximate and rough idea about combinations of variables contributing to the occurrence of this particular macroinvertebrate assemblage. On the contrary, rules resulting from application of GA give more detailed and directional descriptions of the physical variables, allowing for spatial heterogeneity. For example, the average value of the variable latitude is –37.52. In the ruleset it takes two directions: latitude is not between -38.2 and -37.4 (latitude >-38.25) and latitude is between -37.4 and -36.4. This is the case for the other variables as well. Vegetation category has average value of 3.52, in the ruleset its two directions are: vegetation category >2.07 and vegetation category < 2.07. The same stands for shade and alkalinity. Variables altitude and macrophyte category have only one direction each in agreement with averaged values for the cluster.

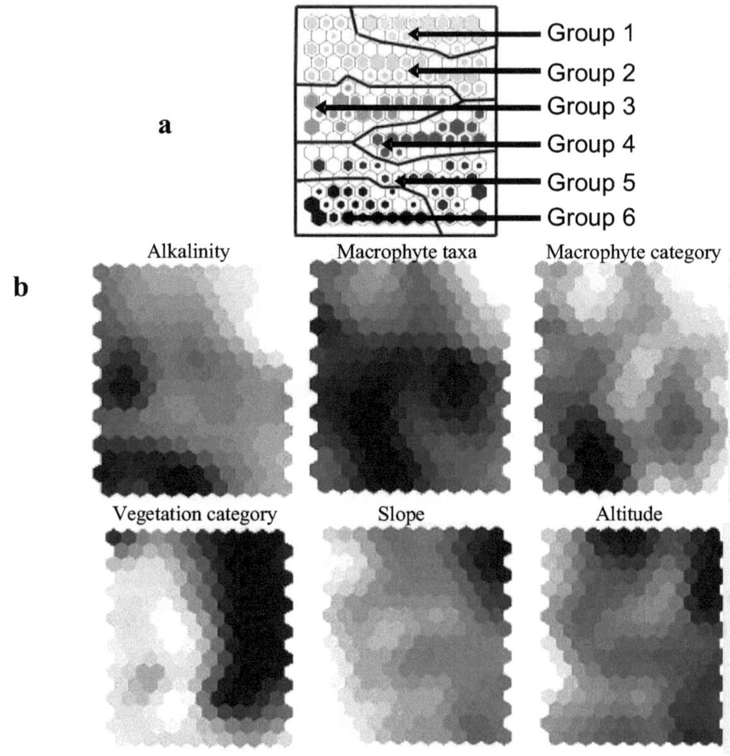

Figure 4.10.5 a) SOM hit diagram showing distribution of 6 groups (clusters) on SOM grid b) selected SOM component planes for environmental variables (all data normalized between 0 and 1, darker shades correspond to higher values).

Discussion and conclusion

In this study we applied and compared three machine learning methods: Self-Organizing Maps neural network, Multilayer Perceptron neural network and Genetic Algorithm. The SOM neural network with K-means algorithm applied has been able to produce a meaningful clustering largely in accordance with previously defined bioregions of the state of Victoria. Traditionally in ecological applications neural network models are used for the prediction of taxa occurrence or abundance from a set of environmental variables.

Table 4.10.2 Characterisation of the macroinvertebrate assemblage group 1 by means of environmental variables in terms of descriptive statistics and ruleset from GA.

	VARIABLE	MEAN (MIN/MAX)	VARIABLE	MEAN (MIN/MAX)
Descriptive Statistics of Group 1 by SOM	LATITUDE	-37.52 (-39.072/ -36.36)	WIDTH	0.61 (-0.20/ 1.48)
	LONGITUDE	146.25 (143.28/148.47)	BEDROCK%	4.95 (0.00 / 60.00)
	REACH PHI	-3.10 (-7.22 / 4.03)	BOULDER %	13.81 (0.00 / 65.00)
	SUBSTRATE HETEROGENITY	2.68 (1.25 / 5.00)	COBBLE %	26.70 (0.00 /70.00)
	VEGETATION CATEGORY	3.52 (2.00 / 4.00)	PEBBLE %	12.07 (0.00 / 50.00)
	SHADE	3.07 (1.00 / 5.00)	GRAVEL %	11.47 (0.00 / 42.50)
	DISTANCE FROM SOURCE	0.89 (-0.70/ 2.00)	ALKALINITY	0.96 (0.57/ 1.71)
	SLOPE	1.19 (-0.30/ 2.52)	MACROPHYTE TAXA	0.25 (0.00 / 0.88)
	ALTITUDE	2.63 (1.30/ 3.23)	MACROPHYTE CATEGORY	0.06 (0.00 / 0.60)
	CATCHMENT AREA	1.51 (-0.10 / 3.23)		
Rule Set for Group 1 by GA	IF 0.51 < CATCHMENT AREA < 1.02 OR IF MACROPHYTE CATEGORY < 0.13 AND VEGETATION CATEGORY > 2.07 AND ALTITUDE > 1.88 AND SHADE < 0.323 OR SHADE > 1.17 AND LATITUDE > -38.2 OR LAT < -37.4 OR IF MACROPHYTE CATEGORY < 0.13 AND VEGETATION CATEGORY < 2.07 AND LATITUDE > -38.25 AND ALTITUDE > 1.88 AND ALKALINITY < 1.08 AND -37.4 < LATITUDE < -36.4 OR IF MACROPHYTE CATEGORY < 0.13 AND VEGETATION CATEGORY > 2.07 AND ALTITUDE > 1.88 AND 0.323 > SHADE > 1.17 AND -37.4 < LATITUDE< -36.4 AND 1.09 < ALKALINITY < 1.39 OR IF ALKALINITY > 1.087 AND VEGETATION CATEGORY > 2.07 AND -3.38 < REACH PHI < -2.01 THAN ASSEMBLAGE 1 ELSE ASSEMBLAGES 2 TO 6			

In this study we explored the question of whether it is possible to predict occurrence of a unit larger than the separate taxa, in our case defined as the pattern in abundance and co-occurrence of all taxa recorded. Although, separating these patterns as distinct clusters or groups might be artificial, it appears that both MLP and GA are well capable of predicting these groups from the set of environmental variables. Contrary to the previous findings of Recknagel et al. (2002) showing that in time-series, the case predictive power of GA was higher that that of ANN, in our case ANN outperformed GA by approximately 10%, although both methods were able to meet the commonly acceptable 70% threshold of correct predictions.

The sites used for the analysis were reference sites, presumably least affected by agricultural practices or urban developments. We assume that in this case it should be easy to explain patterns in abundance and co-occurence of various macroinvertebrate families together, by a range of environmental variables. We tried to do this by examining SOM component planes and GA rulesets for the environmental variables likely to be important in distinguishing group 1 from the other groups. In terms of explanatory power GA is commonly considered as offering more transparency by the generation of rules providing the directional explanation for environmental heterogeneity, while neural networks are considered to be a "black" or "grey" box technique. SOM component planes provided an easy and highly visual way to assess relationships between variables and to suggest which ones are likely to be of importance in shaping macroinvertebrate assemblages within each group. Although this approach can be criticized as being, to a certain extent, qualitative and intuitive, we suggest that it still can be valuable when quick and visual assessment of data is needed. GA rules provide a qualitative approach but are not easy to follow and understand, it might be useful where more in-depth assessment is needed.

The prediction of defined macroinvertebrate assemblages instead of separate taxa can be used as an extention of the referential approach outlined in the introduction. If the type of macroinvertebrate assemblage predicted under particular environmental conditions does not match that actually found, it might be then compared against other possible assemblages indicative for various stresses such as increased salinity and others.

In conclusion, this study demonstrated that ANN and GA provide different approaches to the problem and neither of them were clearly favored in the context. Both methods were able to predict spatial groups of macroinvenvertebrates from environmental variables with high efficiency and provide an explanation from slightly different angles. We recommend the use of the both methods in combination for achieving the most accurate predictions and the highest explanatory power.

5 Diatom and other algal assemblages

Editor: Descy JP*

5.1 Introduction

From low-order streams to lowland rivers, micro-algae are at the basis of the food web and are key elements in ecological and biogeochemical processes. Modelling the composition of algal assemblages in river systems has been a challenge rarely addressed by freshwater ecologists. Indeed, most algal models – essentially phytoplankton models – have been designed to simulate total algal biomass, which seems sufficient to take into account the functional role of micro-algae in ecosystems. However, micro-algae form high diversity assemblages, which are sensitive to disturbance and stresses, and therefore can be excellent indicators of environmental change, at different scales in space and time. Moreover, some key ecological processes which are species- or class-specific cannot be represented satisfactorily by chlorophyll a models: for instance simulating variations of silicon in lowland rivers obviously involves modelling planktonic diatom growth and losses. As all diatoms do not respond in the same way to environmental conditions, have different growth rates and have different settling rate and sensitivity to grazing, predicting Si concentration may quickly become challenging and certainly require modelling different diatom species or categories. Furthermore, simulating phytoplankton dynamics in the extremely variable environment of lowland rivers involves development of non stationary models which take into account variations of the forcing variables (light, discharge) in time and space. Not surprisingly, few research teams have developed such complex models (Billen et al. 1994, Bormans and Webster 1999, Thébaut and Qotbi 1999, Everbecq et al. 2001, Schöl et al. 2002, Descy et al. 2003).

However, process-based, dynamic simulation model clearly have limits. For instance, they can deal with only a few functional groups (see e.g. Eliott et al. 1999), and certainly not with a high number of species. This capacity is further limited by the lack of knowledge on eco-physiological data for most phytoplankton species, and when it comes to benthic micro-algae, eco-physiological data are almost totally missing.

Therefore, there is a need for an alternative approach for predicting the composition of algal assemblages at the species level, and it may be provided by the use of data-based models using machine-learning techniques. In this area a great deal of pioneer work has been carried out by Recknagel et co-authors in a series of applications dealing with the prediction of cyanobacteria blooms, mostly in lakes (Recknagel et al. 1997, Recknagel 2003). These applications have demonstrated the ability of Artificial Neural Networks (ANN) and related algorithms to predict abundance of single algal species in freshwaters, even if eco-physiological knowledge is scarce. Sensitivity analyses of the models may help identifying key variables which determine species distribution, and new techniques which allow for adaptation of taxa, can be used for allowing flexibility of the species response to environmental factors.

In the following chapter, we present applications of machine-learning techniques used for classification and prediction of mostly riverine micro-algal assemblages. Two papers

* Correspondence: jpdescy@fundp.ac.be

are devoted to prediction of planktonic cyanobacteria. First, an application of CBR (Case-Based Reasoning) is developed for prediction of chlorophyll a and *Microcystis* blooms in Lake Kasumigaura. The author (P.A. Whigham) explores in great detail the possibilities of this machine learning technique when trained on time series ecological data sets. The second paper, by G.-J. Joo and K.-S. Jeong, investigates the potential of SOM (Self Organising Maps) for understanding the changes in the communities of cyanobacteria which form blooms in conditions of low discharge in the heavily regulated Nakdong River, Korea. This analysis nicely shows the effects of environmental variables on the changes of dominance among several genera of cyanobacteria.

The following papers all address modeling and prediction of benthic diatom assemblages in rivers, at different scales, with two main objectives: classification of these assemblages as related to environmental gradients, and prediction of community structure. All studies were carried out in the framework of the EC-5thFP PAEQANN project, devoted to the use of ANN for predicting aquatic communities in fresh waters. In several of the papers presented here, diatoms are identified and key elements of the river benthos, useful for stream classification and for assessment of ecological status of rivers. All studies involved careful collection of environmental data and biological data in France, Belgium, Luxemburg and Austria, structured into partial or global data bases (Gosselain et al. 2004). The data were processed with different techniques, in different river basins or across several basins, in order to identify diatom assemblages which may correspond to reference and altered conditions in these basins. Indeed, benthic diatoms are known to respond to water quality changes driven by natural factors such as those depending of geochemistry, as well as to alterations from anthropogenic pressures. Hopefully, the papers presented here are significant contributions to stream classification, and will be useful in water quality management. Moreover, the results have also contributed to identify possible taxonomic confusions and to assess their consequences, but the main fundamental input may be in improvements to the knowledge in benthic diatoms ecology. Indeed, the results have allowed confirming the auto-ecology of many species, have provided clues to the auto-ecology of less well-known diatoms, and have enabled characterization of assemblages across several river basins of the European continent.

5.2 Applying case-based reasoning to explore freshwater phytoplankton dynamics*

Whigham PA†

Introduction

The prediction and explanation of algal abundance and succession is of major interest to freshwater ecologists. Freshwater data often exhibit characteristics such as non-linearity, a non-normal distribution, complex relationships, sampling and scale dependence, noise and non-independence of observations (Gillman and Hails 1997, Mastrorillo et al. 1997b) that make analysis and modelling difficult. In practice these issues are often ignored and classical statistical analyses are relied upon despite their underlying assumptions being violated (Fielding 1999). Recent work has focussed on the use of neural networks, recurrent neural networks, rule-based induction systems, and evolutionary techniques to produce models that predict the value of various freshwater variables. Within the machine learning field, artificial neural networks have received much attention (Lek et al. 1996b, Tan and Smeins 1996, Mastrorillo et al. 1997b, Brosse et al. 1999b, Scardi 2001, Wilson and Recknagel 2001), though genetic algorithms (d'Angelo et al. 1995, Bobbin and Recknagel 1999, Whigham and Recknagel 2001a), classification and decision trees (Stankovski et al. 1998, Whigham 2000) and cellular automata (Silverton et al. 1992, Dunkerley 1997, Dunkerley 1999) have also been used to examine ecological systems. However, there are few studies employing case-based reasoning (CBR) in an ecological context (van Den Brink et al. 2002, Szabados et al. 2003), although they have been demonstrated to perform well in comparative studies with other machine learning techniques (MacGillivray 2000).

This paper will examine the use of CBR in the domain of time-series ecological models, using a freshwater algal system as a case study. The contribution of this work is to demonstrate that a time-series interpretation of CBR can be used to consider similar sequences of behaviour (Nakhaeizadeh 1993), and that identifying these sequences offers different information from standard methods of model construction. The paper is organised as follows: The next section introduces case-based reasoning as a predictive technique; the following section applies a tim pwhigham@infoscience.otago.ac.nz e-series based CBR system to the prediction of chlorophyll-*a* and *Microcystis* based on a set of water quality variables, that highlight the issues and limitations involved in applying CBR to time-series data. The discussion is on future research directions.

* The author would like to thank Mr. Alec Holt for discussions and background information on case-based reasoning, the anonymous reviewers for clear and useful directions, and Egbert Van Nes for supplying supporting literature.

† Correspondence: pwhigham@infoscience.otago.ac.nz

Case-based reasoning

Case-based reasoning (CBR) is a concept matching technique that was developed in the late seventies (Schank and Abelson 1977). CBR uses previous experience, represented as a set of cases, to predict the result of new, unlabelled, cases that are presented. A fundamental notion with CBR is the concept of similarity, which is used to select from the case base the most appropriate case to match against the current new observation.

The retrieval and adaptation of old cases and the retention of new cases require a process (Klolodner 1993, Watson 1997). This process is defined as the CBR cycle and has four main components:

1. Retrieve the most similar case(s): When the CBR system is given a new problem, the case base is searched for similar cases, defined by a some metric that determines the 'distance' between the new problem and each case from the case base. The solution is based on the most similar (nearest neighbour) case, or some weighted combination of a set of similar cases. Typical metrics are the Euclidean and absolute distance between cases, where each attribute is treated as an orthogonal axis in n-dimensional space.
2. Reuse the information and knowledge in that case(s) to solve the problem: Use the solution of these similar case(s) as the solution to the current problem or as a guide to creating a new solution.
3. Revise the proposed solution as necessary: Since the new problem may differ from those retrieved from the case base, the facility is available for the adaptation of the retrieved solution based upon other information sources (i.e. user experience, literature, other model results) to suit the current problem. The suggested solutions can therefore be modified if desired by the user.
4. Retain the new solution as part of a new case: Any new problem/solution pair can be a new case. This new case is stored in the case base and is available for future problems. This allows the case base to grow with use, and effectively to learn from its experiences.

In the context of this study, the revision and retention of new cases is not considered, however a method (see below) for generating new case values, based on a weighted interpolation procedure, indicates some possible directions for future research.

A key component of CBR is the method of computing similarity between objects. There are many different ways of computing similarity, including (Bridge 1998):

- the *feature-based approach*, where the similarity is based on feature commonality and difference;
- the *geometric approach*, where each case is represented by n features, defining a location in n-dimensional space. Similarity is then based on the distance between objects in this space; and
- the *structural approach*, where objects are represented by a graph structure (nodes, links), and similarity is based on the difference between these graph structures.

The flexibility of CBR is based on the fact that objects may be complex, have mixed attribute types, and not necessarily fit within a standard format. Hence, there are few CBR applications in standard time-series modelling since this domain is well covered by other statistical and machine learning techniques. The purpose of this paper is to explore the properties of CBR in an ecological time-series context and to consider what different types of information can be obtained from the use of CBR compared with other common methods. The hypothesis is that CBR should be considered as an additional tool for the study of freshwater dynamics and the analysis of complex ecosystems.

Time series CBR

CBR has been successful in domains where there are independent cases, and where an appropriate result from the CBR matching is to give a list of similar cases, ordered by the similarity metric. Of course, in a time-series domain there will be many cases that require matching, and a single time-series solution would seem an appropriate response. Previous research in CBR with temporal data include weather prediction (Roydhouse and Jone 1995), process control (Brann et al. 1995), robotics and planning (Ram and Santamaria 1997), however the application of CBR to multivariate time series systems has been little used, probably because there are many other techniques that can be applied to this type of data, and CBR is most often applied where there are mixed data types, and it is difficult to formulate a model.

Reasoning about dynamic systems is difficult because the context of any event needs to be taken into account when assessing similarity. This is particularly difficult with CBR where typically each individual case is registered in the system, without a context of other cases. Additionally, the concept of similarity between time series is a complex problem that is often computationally expensive. Some open issues when applying CBR in the temporal domain have been identified as (Ram and Santamaria 1997):

- How should continuous cases be represented?
- When do cases start and end (in the temporal sense)?
- When are two experiences different enough to warrant consideration as independent cases?
- What is the scope of a single case?

Additionally, in the scope of modelling freshwater systems, the issues include:

- How are predicted models to be presented?
- What can be learned regarding the behaviour of the system through case similarity?

Case study example

The dataset used for this study is a weekly-interpolated time series of water quality variables, zooplankton and cholorophyll-*a* for Lake Kasumigaura, in South-Eastern Japan (Table 5.2.1). The data has been studied previously with some success, based on using neural networks (Recknagel et al. 1998), genetic programming equation and rule models (Bobbin and Recknagel 1999, Whigham and Recknagel 1999), and optimised difference equations (Whigham and Recknagel 2001a). The CBR approach will be applied to the modelling and understanding of chlorophyll-*a* dynamics for this system.

Data characteristics of Lake Kasumigaura

Lake Kasumigaura is a large, shallow water body where no thermal stratification occurs. The lake has high external and internal nutrient loadings and therefore primary productivity is high. Algal succession changes species abundance year by year, therefore making it very difficult to predict algal blooms or develop causal models of algal behaviour. Kasumigaura is dominated by harmful blue-green algal species such as *Microcystis spp*, *Oscillatoria* and *Anabaena flos aquae*.

Predicting Chl-a concentration

The CBR system used for this study is one component of a time series toolbox, TSToolBox (Keukelaar 2002). The CBR component was created for this study since available commercial CBR systems would not easily integrate the time-series nature of the data, nor would they give explicit and appropriate feedback for the modelling (i.e. time-series plots) automatically. The initial similarity measure used between any two cases was based on the geometric distance in n-dimensional space, based on the selected independent variables. Since each variable has a different range of values, those variables with a larger absolute value will dominate the distance measure.

Table 5.2.1. Factors measured in Lake Kasumigaura with the weekly time series data.

Measured Factor	Average (± SD)	Units
Orthophosphate	14.14 (± 25)	μg/l
Nitrate	520.56 (± 503)	μg/l
Secchi depth	85.43 (± 44)	Cm
Dissolved oxygen	11.2 (± 2)	mg/l
pH	8.74 (± 0.5)	-
Water temperature	16.36 (± 7)	°C
Rotifera	229.2 (± 293)	ind/l
Cladocera	169.9 (± 221)	ind/l
Copepoda	156.4 (± 83)	ind/l
Chlorophyll-a	74.43 (± 42)	μg/l
Microcystis	38563 (± 95202)	cells/ml

All variables were therefore normalised between a range of 0 and 1 to ensure that the range of values did not distort the relative contribution of each variable when measuring similarity between cases. Each variable was also given a weighting between 0 and 1, although initially these were all set to 1. The weights are used to indicate the significance of each variable. Later sections will explore optimising these weights as a method to investigate the significance of each variable in prediction. The root mean square error (RMSE) was used to measure the accuracy of the resulting time series predictions. Hence, for n independent variables, each with weight W_n, the similarity distance measure D_{ij} between two cases, i and j, with values for each independent variable n of V_{in} and V_{jn} , is

$$D_{ij} = \sqrt{\frac{\sum_{n} W_n \left(V_{in} - V_{jn}\right)^2}{n}} \tag{5.2.1}$$

Examining the chlorophyll-*a* concentration for the 10 years (Fig. 5.2.1) it is clear that the years 1984-86 were significantly more productive than later years. Note that CBR cannot extrapolate values, since the current case set is used to give the value of the predicted, most similar, unknown case. Therefore, if the previous time series does not have examples of the extremes of the dataset then these values cannot be predicted.

Performance when no temporal relationships are considered

To illustrate the CBR performance, the case based was set to 1984-1985, and the system used to predict the chlorophyll-*a* behaviour for 1987-1988, using nearest neighbour similarity to match the cases for each weekly value. All water quality and zooplankton variables were used for the similarity match, with equal weighting. The resulting prediction had a RMSE of 0.122 with the predicted values shown in Figure 5.2.2.

The main points to illustrate with Figure 5.2.2 are that the prediction is ragged, fairly inaccurate and uses very few cases to match the two years of data. However, the overall behaviour, in terms of the rise and fall of the chlorophyll-*a*, is represented to some extent. The first improvement to this prediction is to allow the system to automatically optimise the weights for each independent variable to minimise the error of the predicted concentration.

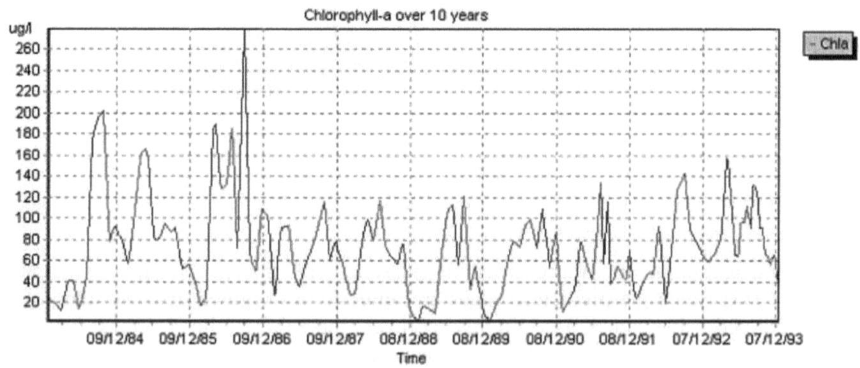

Fig. 5.2.1. Chlorophyll-*a* concentration in Lake Kasumigaura for 1984-1993.

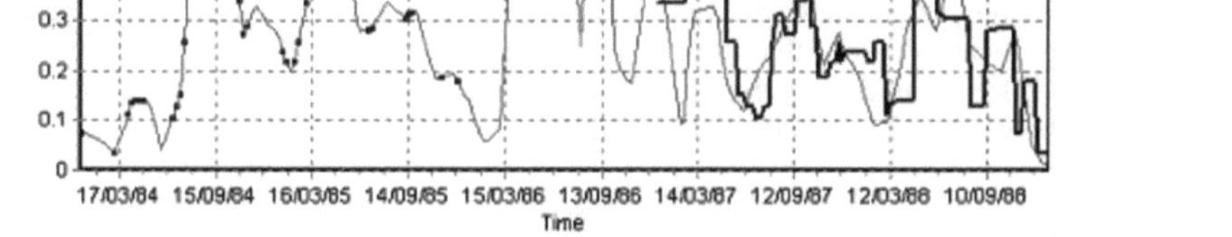

Fig. 5.2.2. Initial prediction of chlorophyll-a (expressed as a fraction of the measured maximal concentration) in Lake Kasumigaura for 1987-88, based on 1984-85 cases – RMSE = 0.12.

The TSToolBox uses an evolutionary algorithm, based on a simple Genetic Algorithm (GA) (Goldberg and Holland 1988) to find the most suitable weights (Jarmulak et al. 2000), although other optimisation techniques, such as Simplex and Powell, could also have been used.

The parameters were set at a population size of 100, for 50 generations, using a mutation probability of 5% and crossover probability of 90%. A generational population was used, with 5-member tournament selection to select parent individuals for the next generation. This optimisation of the weights for each variable can be interpreted as indicating the independent variables that are most important in selecting the case conditions. The resulting prediction after optimisation (Fig. 5.2.3) had a RMSE of 0.073, and overall the dominant variables were *orthophosphate* (0.97), *Secchi depth* (0.78), *Rotifera* (0.6) and *pH* (0.52).

This implies that the phosphate and turbidity levels are major factors in the production of chlorophyll-*a*, and that *Rotifera* levels are correlated with chlorophyll-*a* levels.

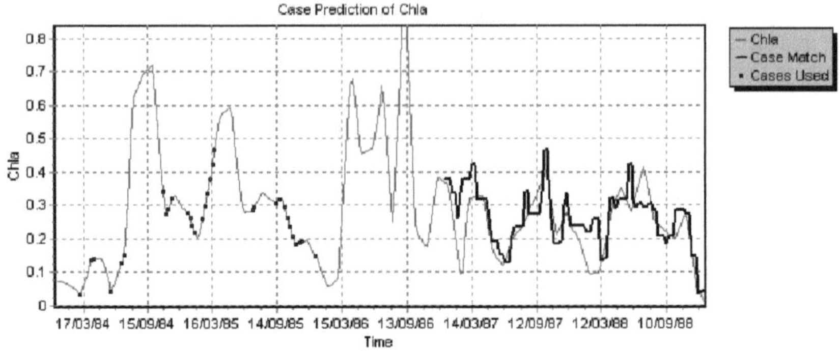

Fig. 5.2.3. Best prediction of chlorophyll-a (expressed as a fraction of the measured maximal concentration) in Lake Kasumigaura using evolved weights – RMSE = 0.07.

A second technique for improvement is possible with this simple, non-temporal approach to CBR. Since the predicted values are numeric, a weighted distance interpolation for the predicted value can be used. In this approach, the value of the N nearest neighbours using the similarity distance neighbourhood can be averaged, where the distance from the predicted point is used to inversely weight the contribution of the final predicted value. This approach is the same as an inverse distance weighting interpolation procedure, commonly used to interpolate spatial data (Burrough and McDonnell 1998). Using equal weights for all independent variables, and a neighbourhood of 5 (i.e. 5 nearest matches are interpolated) the resulting RMSE for 1987-88 is 0.11, and with optimised weights, based on the previous settings, the RMSE is 0.06. This optimised weight prediction is shown in Figure 5.4, and demonstrates that the interpolation has a smoother and more realistic interpretation of the cases. The dominant weighted variables were Secchi depth (0.77), pH (0.64), Rotifera (0.43) and orthophosphate (0.37). Although the ordering of significance is different from the previous experiment, the first 4 variables are the same. The accuracy of this prediction is significantly better than multivariate linear regression and evolved equations using the TSToolBox evolutionary equation discovery system, although due to the method of training for the weighted variables it is difficult to give a direct comparison.

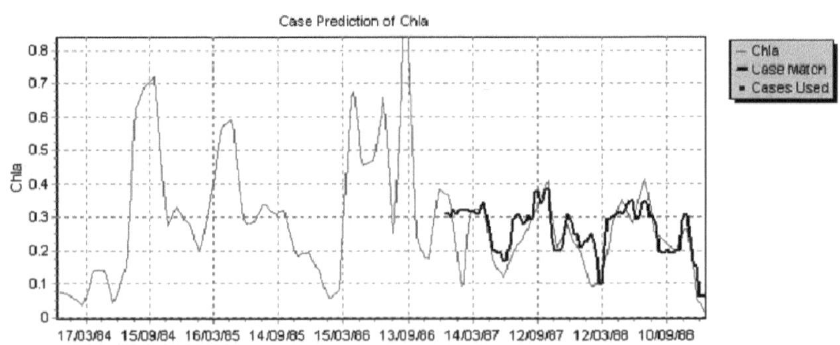

Fig. 5.2.4. Nearest Neighbour interpolation (5) for prediction of chlorophyll-a (expressed as a fraction of the measured maximal concentration) in Lake Kasumigaura for 1987-88 with RMSE = 0.06.

Introduction of temporal relationships to case matching

The previous examples treat each case selection independently from one time step to the next, and therefore do not incorporate the temporal shape of the cases or the temporal correlations that exist in the data. The first extension to the previous approach is to extend the similarity measure to use matching sequences of the independent data. The simplest approach is to search for a sequence length of p points that best match the current point, by extending Equation (5.2.1) to give a sequence similarity instance between two sequences. Given D^t_{ij} is the distance metric between two cases i and j at time t, the sequence similarity distance for p points at time t, S^t_p is defined as:

$$S_p^t = \sum_{x=0}^{p} D_{ij}^{t-x} \tag{5.2.2}$$

This sequence matching allows the case context to be incorporated into the selection of the most appropriate situation, and should therefore help to select an appropriate case match if the time scale and processes described by the data have a temporal context. A study of the improvement or otherwise of the prediction as p increases for S^t_p should support an understanding of the temporal context of the system. Note, however, that a lack of improvement in prediction as p increases does not necessarily imply that the behaviour is best described without temporal context.

Using a sequence of length 1 (i.e. current plus previous point) for matching, the RMSE was higher than the non-temporal measure (although not significantly), and the dominant variables were *orthophosphate* (1.0), *Secchi depth* (0.74), *pH* (0.68) and *dissolved oxygen* (0.56) (Fig. 5.2.5). Extending this to a sequence length of 2 (i.e. current day plus matching sequence 2 weeks in past) the RMSE = 0.092, and the dominant variables were *pH* (1.0), *Secchi depth* (0.85), *Rotifera* (0.84) and *orthophosphate* (0.77). Although the predicted accuracy was less this result further supports the basic variables of *pH*, *Secchi depth* and *orthophosphate* are drivers of the system. Additionally, there is some support to suggest that *Rotifera* could be used as an indicator for bloom conditions. Extending the temporal sequence matching to 3 weeks in the past further reduced the accuracy of the model (RMSE = 0.096), indicating that information from 3 weeks previously does not help understand the current context. This is supported by the correlation coefficient, that is approx. 0.9 for 1 week in the past for chlorophyll-*a*, but has dropped to approximately 0.7 by 3 weeks.

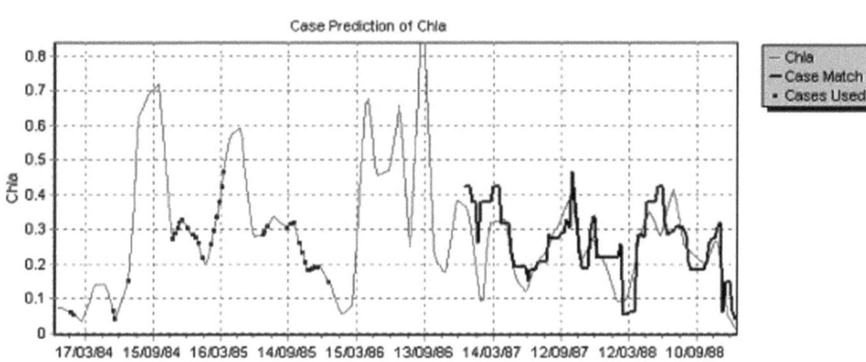

Fig. 5.2.5. Prediction of chlorophyll-a (expressed as a fraction of the measured maximal concentration) in Lake Kasumigaura. Sequence matching using 1 time step in the past. RMSE = 0.07.

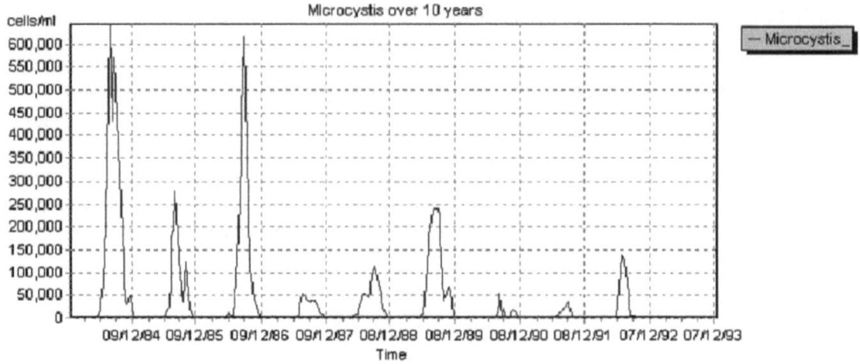

Fig. 5.2.6. Changes in cell concentration of *Microcystis* in Lake Kasumigaura from 1984-1993.

What conditions lead to increased Microcystis?

Extracting and describing the general conditions that cause an increase in *Microcystis* is a more difficult example to demonstrate the utility of a CBR approach. The prediction of *Microcystis* from this dataset (Fig. 5.2.6) has been previously studied (Recknagel et al. 2002) where the timing and magnitude of the bloom in 1986 and the non-occurrence in 1993 were successfully explained by only one complex rule-based model. Since this is a difficult problem it is informative to use CBR to explore why the non-occurrence of *Microcystis* occurred in 1993. Using equal weights for all variables, nearest neighbour for similarity and the cases from 1984-85 to predict 1986, the RMSE is 0.12, as shown in Figure 5.2.7. This result could not be improved using a sequence match between 1 and 5 weeks, and in fact the best sequence match occurred with 3 weeks and a RMSE of 0.16.

When the weights of each variable were optimised the dominant variables were *Secchi depth* (0.98) and *water temperature* (0.97). Using 1984-85 to predict 1993, the system predicted a *Microcystis* bloom that was not recorded, although the peak was the least of all years predicted. Clearly conditions for the non-bloom were different from these early years. Using all years until 1993 as the case base and optimising the weights for the variable selection, the result was only a small bloom prediction that was essentially matched with the year 1990 and had a RMSE of 0.018.

This result, shown in Figure 5.2.8, had dominant variables of *dissolved oxygen* (1.0), *Rotifera* (0.99), *Cladocera* (0.97) and *Secchi depth* (0.96). By examining the data for this period the lack of a *Microcystis* bloom can be accounted for by relatively high turbidity and relatively low dissolved oxygen. The *zooplankton* numbers were consistent for each of the years in the 1990's and therefore do not indicate a dominant influence on algal cell counts.

This paper will endeavour to address these issues through a case study, commencing with a simple application of CBR that does not take the temporal nature of the data into account, and by systematic additions to the system to demonstrate the use of CBR in the domain of multivariate time series models.

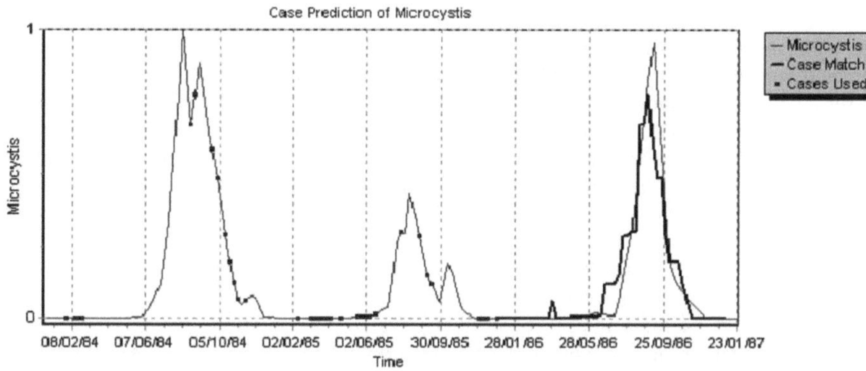

Fig. 5.2.7. Prediction of Microcystis cell abundance (expressed as a fraction of the measured maximum) in Lake Kasumigaura using optimised nearest neighbour, RMSE = 0.08.

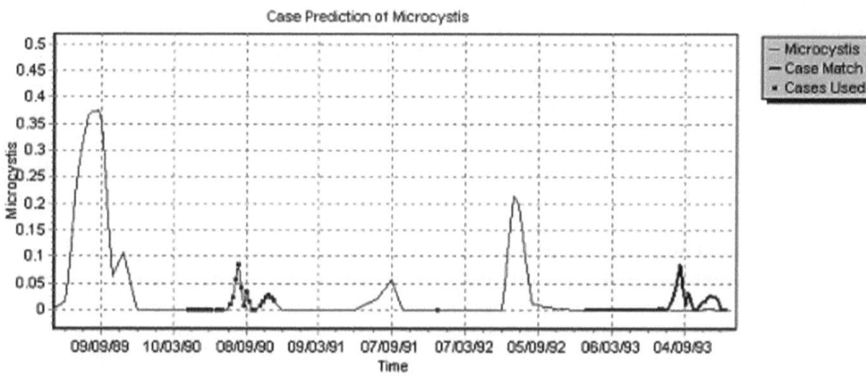

Fig. 5.2.8. Prediction of Microcystis cell abundance (expressed as a fraction of the measured maximum) in Lake Kasumigaura for 1993 – note case selection from 1990.

Discussion and conclusion

The previous examples have demonstrated that CBR approaches to ecological time-series warrant further research. Although CBR does not construct an explicit model, various characteristics of the system being examined can be extracted from the optimisation of weights and the use of sequences to determine time behaviour. Additionally, CBR allows mixed variable types and can incrementally use new observations without having to relearn a model. A criticism of the work described here is that the described relationships between the variables driving chlorophyll-*a* production were neither new nor unexpected – of course they confirmed previous understanding of freshwater system dynamics but did not present any new insight to the process. This is true, however a number of descriptions from the CBR system are different from other methods, and offer additional information regarding system behaviour. For example, being able to find analogous past time sequences to current situations is a powerful method of comparing possible management options (What did we do in the past? What were the results? What other times were similar? What range of responses has the system demonstrated in similar previous situations?). This type of relationship has been examined in this paper. Extending this concept, the notion of prototypical sequences could be developed. These prototypes would describe the common patterns exhibited by the system under certain conditions. The use of prototype descriptions has been previously studied in the medical domain (Schmidt and Gierl 2000), where these prototypes describe short, medium and long term trends. These prototypes were used to guide the retrieval process and to decrease the amount of stored cases, by removing redundant cases. For our purposes, a prototype would describe the basic independent variable time-series patterns that represent conditions that describe the majority of the systems behaviour. These prototypes could then be used as explanatory models. Current research is focussed on how to construct these prototypes for the generalised behaviour of various temporal conditions of freshwater systems.

A second criticism of CBR is the problem of extrapolation beyond the currently observed set of conditions and responses. For conditions such as global warming it is desirable to be able to examine the response of the system as conditions move outside of the range of observed variations. This is clearly one area where the basic case-based approach is weak, and would need to be supplemented with some form of hybrid model. One possibility would be to use the CBR system as a gating algorithm to switch between different models that have been trained for various 'states' of the ecosystem. These states would probably have been identified previously as prototypes. This approach would be similar to methods such as the mixture of expert's model framework, while still maintaining some desirable properties of the CBR system.

A recent successful use of CBR in an ecological context has been demonstrated with the Perpest model (van Den Brink et al. 2002). This work differs from that described here since the Perpest model does not explicitly use the context of time in the measurement of similarity. However, the work showed in detail how CBR can be used to produce complex descriptions of ecological response. The work described here further supports the utility of CBR for ecological problems, and argues that the explicit representation of time within the similarity framework will allow other useful descriptions and models to be produced.

Other extensions to the current work include exploring similarity measures based on previous research in time modelling (Bollabas et al. 1997) that allow more accurate similarity matches between cases, and between sequences. Additionally, since the number of cases with time series data is large, and research is required to determine appropriate indexing structures to allow fast matching between cases and sequences.

CBR clearly has a role to play in understanding the dynamics of freshwater systems, both in terms of prediction and as a model that allows direct comparisons with behaviour from the past. CBR is not a replacement for other models such as regression and artificial neural networks, however it has different properties that should be viewed as a complement to other time-based ecological descriptions and models.

5.3 Modelling community changes of cyanobacteria in a flow regulated river (the lower Nakdong River, S. Korea) by means of a Self-Organizing Map (SOM)*

Joo GJ, Jeong KS†

Introduction

The study of ecosystem dynamics requires a comprehensive approach that can generalize and synthesize hypotheses and existing knowledge. It is not possible to deal with the ecosystems in a simplistic fashion (Wallace et al. 1991), because they contain some properties that cannot be seen from each individual scale (Odum 1983). This may prevent researchers from readily gathering fruitful information within experimental or survey datasets. Analyzing and modelling ecological systems must also take into consideration the multi-modality of data, non-linearity, many zeros, multi-colinearity, and other factors (Fielding 1999). Mechanistic models have played a particular important role in helping to synthesize results of ecosystem research and make predictions about future ecosystem behavior.

One focal area for research and modelling in freshwater ecosystems has been the blue-green algae (cyanobacteria). They have been studied for several decades, and have both an interesting ecology and implications for water quality management (Shapiro 1984). The accelerated eutrophication of aquatic ecosystems is recognized as a global problem leading to recurrent harmful algal blooms in lakes and rivers. Diverse experimental approaches have been used in an attempt to reveal the proximal cause of blooms.

Cyanobacterial blooms have caused serious problems for natural ecosystems and the human society. Toxicity of water as well as malfunction of water purification systems are typical problems for drinking water (Jang 2002). Cyanobacteria have competitive advantages that enable them to dominate the final phase of phytoplankton succession in lakes (Reynolds 1984). Shapiro (1990) explained these advantages over other phytoplankton from the perspective of various limnological attributes.

Blue-green algal blooms in real systems appear to be the consequence of complex synchronization among various environmental parameters. Blue-green proliferation in river ecosystems has been well documented in Paerl and Bowles (1987), Köhler (1993) and Sherman et al. (1998). Due to the elongated retention time and accelerated eutrophication caused by regulation of water flow, severe bloom events can be observed in rivers with dams and other restrictions to water flow (Ha et al. 1998). There are various examples of phytoplankton models that take into consideration the blue-greens (Kamp-Nielson 1978, Reynolds 1984, Sommer et al. 1986, Kromcamp and Walsby 1990). These early models are somewhat successful in accounting for algal dynamics, but do not explain all observed variations. In recent, more sophisticated techniques, Machine Learning (ML) such as Arti-

* Authors are grateful to Dr. Karl E. Havens for critical revision of English and ecological logic and to Prof. Friedrich A. Recknagel for his warm and continuous attention and comments. This study is granted from the Institute of Environmental Technology and Industry (Project No. R12-1996-015-00035-0).

† Correspondence: pow0606@hanafos.com

ficial Neural Networks (ANNs) have been used to improve the predictive understanding of bloom dynamics (Lek et al. 1996b, Chon et al. 1996, Recknagel et al. 1997).

Artificial Neural Network is known as a good pattern recognizer as well as predictor (Chon et al. 2000c, Recknagel and Wilson 2000, Jeong et al. 2001). This computerized algorithm can satisfy the necessity for dealing with the non-linear nature of ecological datasets, due to the model's characteristics of architecture (Lin and Lee 1996, Lek et al. 2000). Pascual and Ellner (2000) suggested neural network as ecological model of linking environments and biological characters, and Roadknight et al. (1997) showed the ability of neural network for environmental data interpretation. Other researches have encouraged the application of neural networks to ecological modelling (e.g. Karul et al. 1998, Wen and Lee 1998, Moatar et al. 1999, Özesmi and Özesmi 1999, Spitz and Lek 1999, Schleiter et al. 1999, Scardi 2001). The complicated patterns of blue-green community change can be effectively dealt with by data-driven inductive models, such as Self-Organizing Maps (SOMs). Furthermore, water flow regulation, which gives rise to the unusual physical-chemical conditions favoring blooms in regulated rivers, can be considered in the model algorithm, which is more difficult in mechanistic models. Self-Organizing Maps have previously been used to predict community-level changes of macro-invertebrates and fish in stream ecosystems (Chon et al. 1996, Giraudel et al. 2000).

The lower Nakdong River is a good example of regulated river (a river-reservoir hybrid) that experiences blue-green algal blooms. Eutrophication has rapidly progressed since the construction of an estuarine dam in 1987 (Joo et al. 1997). The river has reservoir-like ecological patterns, and plankton dynamics are different from those of natural rivers (Ha et al. 1998, Ha et al. 1999, Kim et al. 1998, Kim et al. 2001). There have been previous efforts to model water quality (Song et al. 1993, Cho et al. 1996, Shin et al. 1998), but ecological attributes are not as well known. In one of the rare ecological modelling studies, Jeong et al. (2003a, b) derived population-based time-series ecological models by means of recurrent neural networks for two bloom-forming algal species. Even though their results were promising, the only cyanobacterial taxon considered was *Microcystis aeruginosa*; changes in the composition of the phytoplankton community during blooms were not considered.

In this study, community changes of cyanobacteria in the lower Nakdong River were modelled using the SOM algorithm. It involves a non-linear clustering method that can suitably reflect the nature of ecological data in the system. The environmental variables which are associated with groups identified by the SOM plane can be assumed as preferred conditions for the algal species dynamics. By considering the results in the context of information from Jeong et al. (2001, 2003a), the characteristics of blue-green community dynamics in flow-regulated systems can be recognized.

Materials and methods

Description of the study site

The Korean Peninsula is situated in the far-eastern part of Asia, and the Nakdong River basin lies in the southeastern part of South Korea (35° ~ 37° N, 127° ~ 129° E) (Fig. 5.3.1). South Korea experiences four distinct seasons, with a hot summer (June to August) and a cold winter (December to February). A short rainy season with concentrated precipitation in summer is one of important climate characters of S. Korea. A monsoon (late June to late July) and several typhoons typically occur in the summer, and > 60% of total annual rainfall (ca. 1,200 mm) is concentrated in this season (Park 1998). The rainfall also is differently distributed across the basin. The annual mean water temperature at the study site was

13.7°C. The mean water temperature was 2.2°C during the coldest month (January), and 25.9°C in the warmest (August).

The main channel of the river is 526 km long, and the catchment area occupies about 25% (23,817 km^2) of the whole country. The study site (Mulgum) is situated on 27.4 km upstream of the estuarine dam at the river mouth, and has a maximum water depth of ca. 11 m; mean water depth is ca. 4 m; river width is 250-300 m. The average retention time was about 3 days at the study site (Song et al. 1993, Jeong et al. 2001), and it varied from 6 hours to 10 days according to the rainfall frequency (concentrated rainfall events caused shorter retention time while dry winter to spring it marked longer one). The discharge ranges from about 200 to 10,000 $m^3 s^{-1}$.

The Nakdong River is a representative flow-regulated river ecosystem (Joo et al. 1997), and the phytoplankton succession is similar to that of lakes and reservoirs. There is even a "clear water phase" attributed to a sharp increase of zooplankton grazing during spring and autumn (Kim et al. 1998).

Over 10 million people depend on the river for their drinking, agricultural, and industrial water supply. The Nakdong River has 4 multi-purpose dams and an estuarine dam. Physical alterations of the river, industrialization, and urbanization have accelerated eutrophication of the lower part of the river (Kim et al. 1998).

Data collection and analysis

Limnological parameters were collected over a five-year period (1994-1998). Precipitation data were obtained from 5 representative meteorological stations within the Nakdong River basin (Andong, Taegu, Hapchun, Jinju, and Miryang). River flow data were obtained from the Flood Control Center. Data for irradiance, wind velocity, and evaporation were collected from the Busan Local Meteorological Station, which is closest to the study site. Three-day-averaged (discharge, wind velocity) or summed (irradiance, evaporation and rainfall) values of those parameters were used, in association with the limnological sampling dates.

River water samples were weekly collected at a depth of 0.5 m at the study site around at noon. The previous study reported that the cyanobacterial community almost evenly distributed during the daytime, whereas they tended to be accumulated at the surface layer during night when wind velocity was low (Ha et al. 2000). The following water quality parameters were measured: water temperature, Secchi transparency, pH, turbidity, concentrations of dissolved oxygen (DO), nitrate (NO_3^--N), ammonia (NH_4^+-N), phosphate (PO_4^{3-}-P), dissolved silica (SiO_2), chlorophyll *a* (chl. *a*), phytoplankton biovolume, and zooplankton abundance. Water temperature and DO (dissolved oxygen) were determined with an YSI DO meter (Model 58); transparencies were determined using a 20-cm Secchi disc; pH was measured with an Orion pH meter (Model 250A); turbidity (NTU) was measured with a Turbidimeter (Model 11052). Water samples were filtered through Whatman GF/C filters to determine the soluble nutrient concentrations, and the filtrates were frozen and analyzed by a QuikChem Automated Ion Analyzer (NO_3^--N, No. 10-107-04-1-O; NH_4^+-N, No. 10-107-06-1-B; PO_4^{3-}-P, No. 10-115-01-1-B; SiO_2, No. 10-114-27-1-A). Chlorophyll *a* concentration was measured with a spectrophotometer (using filtrates on pore size 0.45 μ MFS membrane filter), using extraction methods described by Wetzel and Likens (1991).

Identification of phytoplankton species was conducted with a Nikon light microscope (×1,000), following Foged (1978), Cassie (1989) and Round et al. (1990). Phytoplankton samples were preserved with Lugol's solution when collected, and enumerated using an inverted microscope (ZEISS, ×400) by the Utermöhl sedimentation method (1958). Biovolumes of individual species were estimated from mean cell dimensions and the cellular shape of each species as described in Wetzel and Likens (1991). Mean cell biovolumes were based on individual cell volume calculations of 10 to 25 cells.

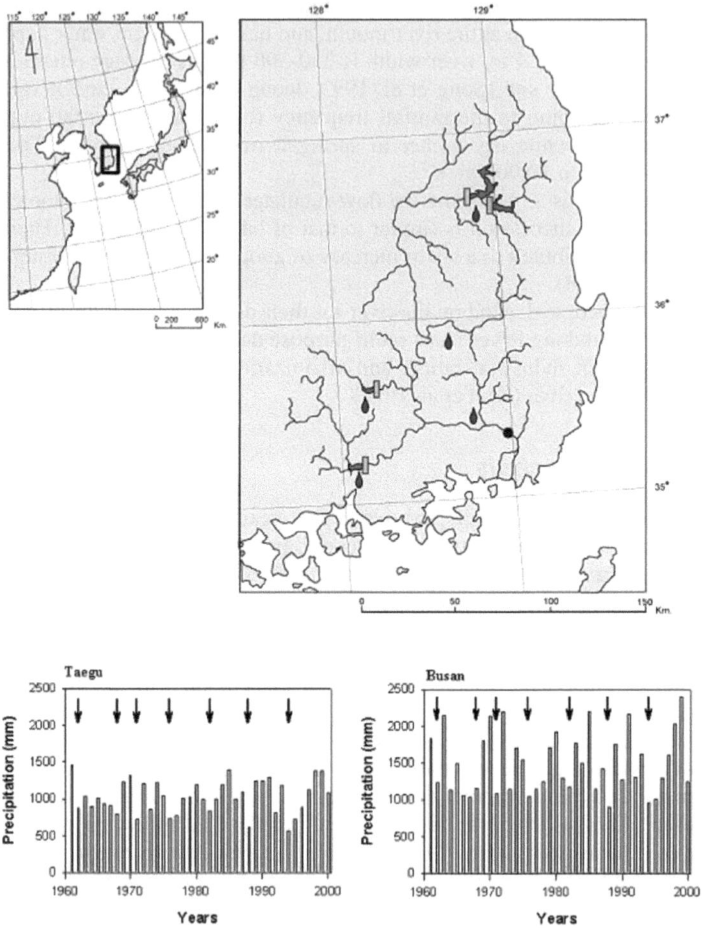

Fig. 5.3.1. Map of the Nakdong River and the long term rainfall regime. In the map of the river basin (A): representative meteorological stations (grey droplets), the sampling station (Mulgum) (black circle) and multipurpose dams (grey bars) are shown. The long term rainfall regime in two representative stations is shown in B (arrows indicate dry years).

Zooplankton was collected from a depth of 0.5 m using a 3.2 L Van Dorn water sampler until a total of 8 L of water was obtained. Water samples were filtered through a 35-μm-mesh net, and the retained zooplankton was preserved with 10% formaldehyde (final concentration: 4%). Meso- to macro-zooplankton (almost exclusively Cladocera and Copepoda, respectively) were counted with an inverted microscope at ×25-50 magnification. Micro-zooplankton (mostly Rotifera) were enumerated with the inverted microscope at × 100-400 magnification. Zooplankton taxa were identified to the genus or species level (except for juvenile Copepoda) using Koste (1978), Smirnov and Timms (1983), and Einsle (1993).

Self-Organizing Map for blue-greens

A Self-Organizing Map (SOM; Kohonen 1982) is capable of reducing data dimensionality (Lin and Lee 1996). It is a competitive network system in which neurons (processing elements) in Euclidean map space compete with each other. To identify pattern as well as to cluster the blue-green algal dynamics, a two-dimensional Kohonen network was adapted.

A characteristic property of SOM distinguished from other clustering methods (e.g. Principal Component Analysis, Correspondence Analysis, Cluster Analysis etc) is that it is possible to use various distance metric for similarity. It is known that Euclidean distance is sometimes not satisfactory for species data due to double-zero problems (Legendre and Legendre 1998). Giraudel and Lek (2001) summarized the possibility of using different distance metric. They argued when SOM used another distance metric to avoid this problem, the learning equation had to be adapted in order to be compatible with the chosen distance. In addition, Walley and O'Connor (2001) argued that SOM based on Euclidean distance can cause problems on biological and environmental data on "ordinal scale" (usual data scale of biological species is ratio or interval scale). Even though this point is important, many of ecological applications of SOM were successful on the basis of Euclidean distance for macroinvertebrates (Chon et al. 1996, Obach et al. 2001), fish assemblages (Brosse et al. 2001) and so on. Ecological modelling approach to select the best fitting distance metric to phytoplankton community dynamics should be addressed.

In this study, a Kohonen network was prepared with M^2 artificial neurons (Fig. 5.3.2). Input for the network was data on cyanobacteria genus biovolumes i, x_i identified during the study period. All the input data were expressed in vector and input layers consisting of those species. Every node, j, of the output layer was connected to each node, i, in the input layer. A hexagonal array of neurons was selected. The weight vector, $\mathbf{w}^{(t)}$, represented the connection between input and output layers. As training preceded, each weight value, $w_{ij}^{(t)}$, was adaptively changed at each iteration t. In initial stage, $\mathbf{w}^{(t)}$ was randomly and uniformly distributed in the network architecture. As the input signal entered the network, each neuron computed the summed distance between the weight and input through the following equation:

$$\left\|\mathbf{x}-\hat{\mathbf{w}}_t\right\| = \min_i \left\{ \left\|\mathbf{x}-\hat{\mathbf{w}}_j\right\| \right\} \tag{5.3.1}$$

The neuron that exhibited a maximum response to the given input data was selected as a "winning" neuron, whose weight vector had the minimum distance to the input vector.

The winning neuron and its neighbors "learned" by changing their weights in a manner that reduced the distance between the weight and input vectors. The following equation was used for this purpose:

$$w_{ij}^{(t+1)} = w_{ij}^{(t)} + \alpha^{(t)} \left[x_j^{(t)} - w_{ij}^{(t)} \right] Z_j \tag{5.3.2}$$

where Z_j had a value of 1 for the winner and its neighbors, whereas a value of 0 was assigned to the remaining neurons. The learning rate (α) dynamically changed during the training steps. The radius value, $r^{(t)}$, was initially defined between 1 and m, where m is the integer of $(M\text{-}1)/2$. Radius gradually was reduced to zero as convergence was achieved. Further information on this modelling approach can be found in Hecht-Nielsen (1987), and Lin and Lee (1996), and the methodology of applying SOM to ecosystems is explained in Chon et al. (1996).

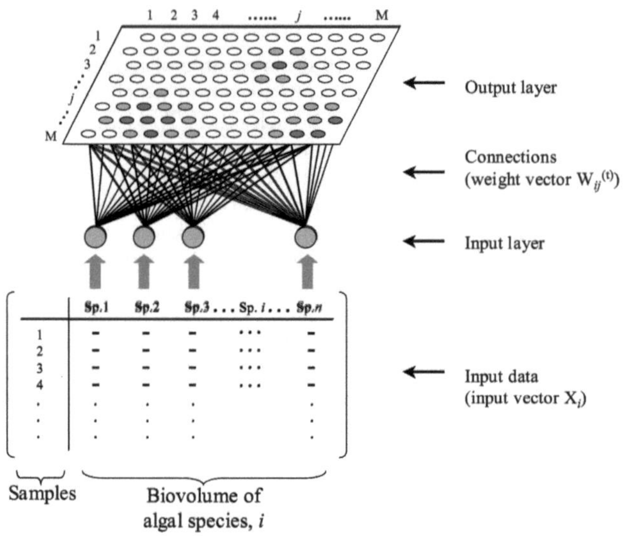

Fig. 5.3.2. Basic architecture of SOM training.

Training of the SOM was done using a series of limnological data from the study site. A total of 266 cases (sampling dates) of three algal species were gathered, and 80% (212 data) were used in the training procedure. The training data were randomly selected from the full dataset. The map quality was evaluated by two types of error values: i.e. quantization error and topographic error. The former is the average distance between each data vector and its best matching unit (BMU; the node which has the smallest of the Euclidean distances), and usually used as measuring map resolution. The latter is the proportion of all data vectors for which first and second BMUs are not adjacent units. This is for the topology preservation.

After training, data were clustered according to the calculated U-matrix which was a discrimination method for one SOM node to others according to the non-linearly calculated dissimilarity. This graphic displaying method was developed by Ultsch and Siemon (1989). It uses the average distances between neighboring nodes through shades in a gray scale. For instance, if the average distance of neighboring nodes is small, a light shade is used; and vice versa, dark shades represent large distances. Therefore a "cluster landscape" was formed over the SOM and then we could clearly visualize the classifications (Kohonen 1982).

The biovolume data were compared with each other. Limnological variables were statistically evaluated to define the important parameters. Since this neural network model of SOM is primarily intended to cluster the species data and does not include environmental variables directly into the model, sensitivity analysis was not applied on the relationship between environments and algal data. Instead of the sensitivity analysis, one-way ANOVA and Duncan's *post-hoc* test were utilized. From the result of statistical analysis, it was possible to clarify the parameters which are statistically significant to the changes of cyanobacterial species among the clusters. For developing the SOM model, Matlab 5.3 (MathWorks 1999) and SOM Toolbox for Matlab (Alhoniemi et al. 1999) were used. SPSS for Windows 11.0 (SPSS 2001) was used for statistical evaluations. For the details about SOM algorithms, readers may consult Kohonen (1982).

Results

Limnological variability and cyanobacterial dynamics

For certain limnological variables, distinct seasonal and inter-annual variations could be observed (Fig. 5.3.3). All the sampled parameters for five years from the lower Nakdong River indicated hyper-eutrophic condition. As previously observed, rainfall was concentrated during summer, and limnological seasonality seemed to be influenced by the seasonal rainfall distribution (Fig. 5.3.3A).

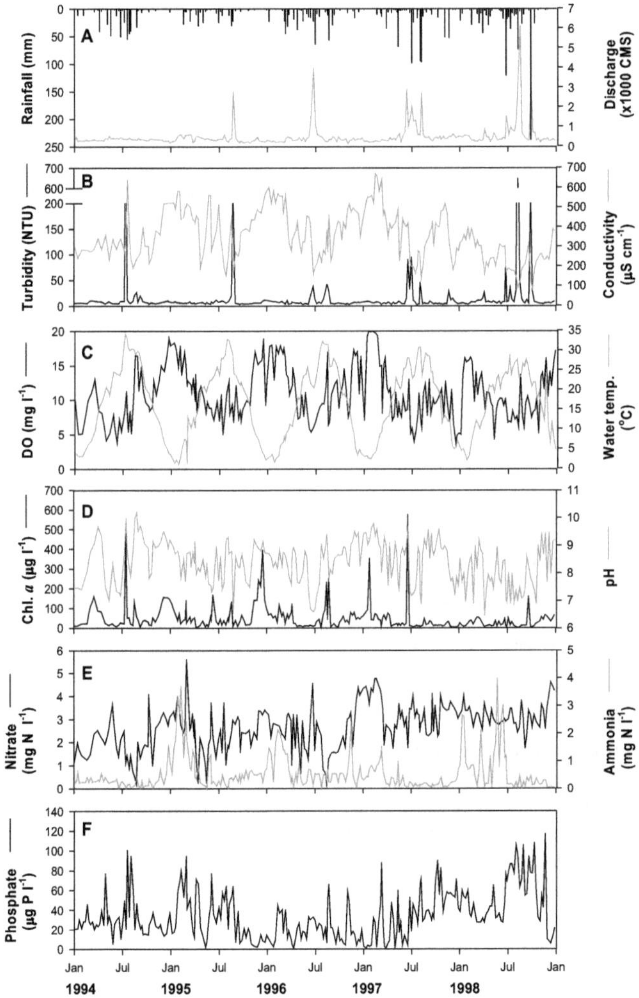

Fig. 5.3.3. Time-series limnological dynamics in the lower Nakdong River during the study period. A, rainfall and discharge; B, turbidity and conductivity; C, DO and water temperature; D, chlorophyll a concentration and pH; E, nitrate and ammonia concentrations; F, phosphate concentrations.

Peaks of discharge were observed during the rainy season. High summer discharge between 1994 and 1996 was not distinctive, but in 1997 and 1998, the summer season experienced high floods (Fig. 5.3.3A). Figure 5.3.3B displayed the changes of turbidity and conductivity. High turbidity was marked mostly during the summer rainy seasons. However, in the summer of 1994 which was a very dry year, also a sharp increase of turbidity could be measured due to the increase of phytoplankton biomass (compared with Fig. 5.3.3D). A high level of conductivity was observed in dry years, whereas lower values occurred in rainy years. High dissolved oxygen could be observed during winter and summer. This would be due to the increase of phytoplankton biomass and photosynthetic activity (Fig. 5.3.3C). In the case of winter, despite water temperature was low, DO frequently reached very high levels (over 200% of saturation) during the winter *Stephanodiscus* bloom. The river did not freeze during the study period.

Chlorophyll *a* concentration had complicated changes during the study period (Fig. 5.3.3D). A sharp increase in chl. *a* concentration was observed during summer, and high level of biomass was also detected in winter, except for 1998. Another small increase could be measured during spring when discharge was low. The pH variations did not exhibit a clear seasonality, but the changing pattern was similar to chl. *a* (Fig. 5.3.3D). Nitrate and ammonia concentrations had weak seasonality, and generally during winter those parameters increased (Fig. 5.3.3E). Phosphate varied in a complex manner, but during the rainy period of 1997 to 1998, the concentration increased considerably (Fig. 5.3.3F).

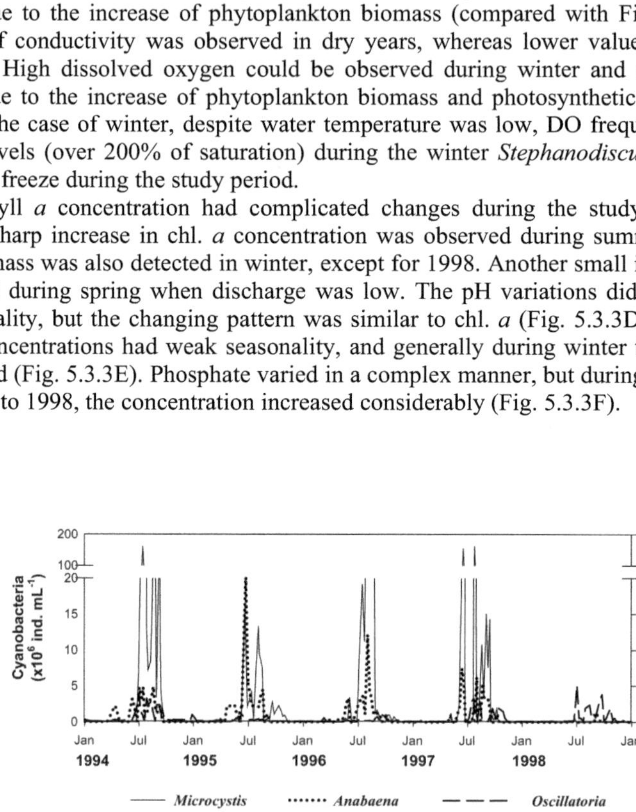

Fig. 5.3.4. Cyanobacteria proliferations in the lower Nakdong River during the study period.

Cyanobacterial communities showed inter-annual variations of dominance during the study period. In dry years, summer phytoplankton was dominated by *Microcystis* spp. (Fig. 5.3.4). However in several years, these species did not increase as much as in the extreme year (e.g. 1994). The observed species were listed in Table 5.3.1. In most cases *Anabaena* tended to increase slightly before the *Microcystis* blooms, and *Oscillatoria* had relatively lower biovolume. In 1994 and 1997, severe blooms of *Microcystis* occurred. In 1995, *Microcystis* did not proliferate, and instead, there was a large increase of *Anabaena.* In 1998, when heavy rainfall occurred, neither *Anabaena* nor *Microcystis* bloomed, and *Oscillatoria* became dominant.

Self-Organizing Map clustering

The SOM algorithm had 8 clusters on the map plane, and three species of cyanobacteria were separated into each cluster (Table 5.3.2). The map quality was reasonably high, with

0.4049 of quantization error and 0.0047 of topographic error. The clusters consisted of *Anabaena* and *Oscillatoria* (Cluster 1), *Anabaena* only (Cluster 2), *Oscillatoria* only (Cluster 3), *Oscillatoria* and *Microcystis* (Cluster 4), *Microcystis* only (Cluster 5), *Microcystis* and *Anabaena* (Cluster 6), all genera (Cluster 7), and no genus (Cluster 8).

Table 5.3.1. List of the observed cyanobacterial species during the study period. The asterisk indicates that the species was dominant within the genus.

Microcystis	***Anabaena***	***Oscillatoria***
*M. aeruginosa**	*A. flos-aquae**	*O. acutissima**
M. ichthyoblabe	*A. menderi*	*O. agardhii*
M. incerta	*A. spiroides*	*O. angustissima*
M. wesenbergii	*Anabaena sp.*	*O. limnetica*
		O. limosa

Division of clusters was based on the U-matrix, and each cluster had different abundance of phytoplankton in biovolume. The clusters contained many "0" values depending on the large seasonal variations. Even though there were differences in the number of sample units, assumptions about possible causality of presence or absence of the species could be examined through the cluster composition.

The clusters on the SOM plane were related to seasonality of cyanobacteria (Fig. 5.3.5). Seasons of moderate temperature (Cluster 1 to 3) and hot summer to autumn (Cluster 4 to 7) were well-separated. SOM Cluster 8 had had data mainly from spring and winter. The distribution of biovolume data in the clusters varied according to genus, except in Cluster 7 (Fig. 5.3.6).

When *Anabaena* distribution was compared with *Microcystis*, Clusters 1 and 2 did not have *Microcystis,* and *Anabaena* did not occur in Cluster 4. *Oscillatoria* shared most of nodes with both *Anabaena* and *Microcystis*, but it occurred alone in Cluster 3.

Table 5.3.2. Species composition in each cluster. Check marks indicate the existence of genus in the cluster.

	n	***Anabaena***	***Microcystis***	***Oscillatoria***
Cluster 1	16	☑	☐	☑
Cluster 2	8	☑	☐	☐
Cluster 3	49	☐	☐	☑
Cluster 4	17	☐	☑	☑
Cluster 5	7	☐	☑	☐
Cluster 6	7	☑	☑	☐
Cluster 7	37	☑	☑	☑
Cluster 8	71	☐	☐	☐

Comparing biovolume distributions with clustering results (Fig. 5.3.5) indicated the seasonal preferences of each algal genus. *Anabaena* mainly occurred in spring to summer, with lesser amounts in autumn. *Microcystis* had a similar seasonality to *Anabaena*, but high biovolume also were observed during summer and autumn. *Oscillatoria* occurred in all seasons, and it could be observed during winter.

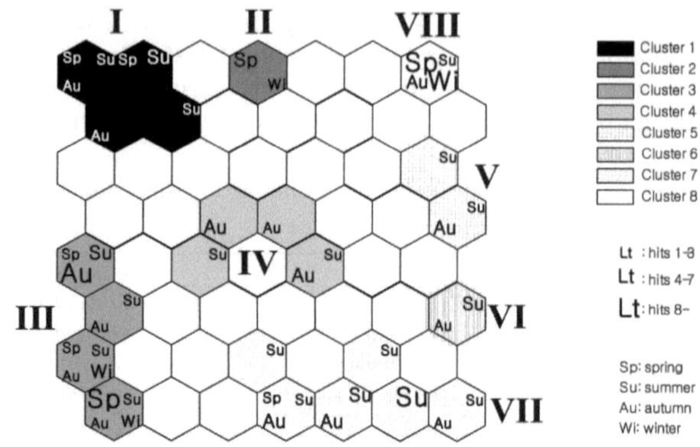

Fig. 5.3.5. SOM clustering on the cyanobacterial dataset. Each shade indicates different cluster. Letter size indicates data collection in a node.

Environmental variations for the cyanobacteria compositions

Tables 5.3.3 and 5.3.4 summarize results of a one-way ANOVA applied to the limnological variables and their means and standard deviations, based on SOM clustering. Through the ANOVA results, the statistically significant limnological variables could be found for the changes of species. When the possible sequence of species changes was given (i.e. the column "case" in Table 5.3.3), some of the limnological variables had difference of average according to clusters. Generally, irradiance, evaporation, and water temperature, grouped as "heat energy", could be seen in most of cases, and those parameters might have significant influence on the changes of phytoplankton species.

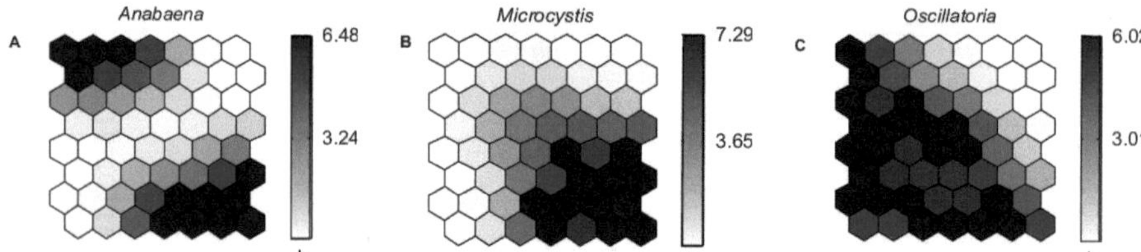

Fig. 5.3.6. SOM results for cyanobacteria genera. Dark color indicates concentrated data distribution on the SOM plane and the b-scale indicates the values of biovolume in common log (A, *Anabaena*; B, *Microcystis*; C, *Oscillatoria*).

Phosphate, nitrate, and pH were statistically different in the case of "*Anabaena* to all" and "*Oscillatoria* to all". The hydrological factors, such as rainfall and discharge, did not have significance influence statistically. Other variables could be seen as not responsible for the changes of species (i.e. statistically identical among clusters).

Table 5.3.3. Important factors for the cyanobacteria species composition according to clustering of SOM (= 0.05). ( ), cluster numbers; –,clusters gathered by Duncan test; /, clusters separated by Duncan test.

Cases		Significant parameters			
		Names	F	p	Duncan test
Microcystis to all	Mic (5) Mic & Osc (4) All genera (7)	Irradiance	4.693	0.000	5-4 / 7
		Evaporation	11.953	0.000	5-4 / 7
		Water temp.	36.431	0.000	4-5 / 7
Anabaena to all	Ana (2) Ana & Osc (1) All genera (7)	Evaproation	4.952	0.010	2-1 / 7
		Water temp.	39.139	0.000	2 / 1 / 7
		Alkalinity	6.442	0.003	1 / 7-2
	Ana (2) Ana & Mic (6) All genera (7)	Water temp.	31.514	0.000	2 / 6-7
		Phosphate	3.409	0.041	2 / 7-6
Oscillatoria to all	Osc (3) Osc & Mic (4) All genera (7)	Irradiance	19.340	0.000	3-4 / 7
		Evaporation	23.979	0.000	3-4 / 7
		pH	7.189	0.001	3-4 / 7
		Water temp.	40.464	0.000	3-4 / 7
		Nitrate	13.685	0.000	3-4 / 7
	Osc (3) Osc & Ana (1) All genera (7)	Irradiance	16.494	0.000	3 / 1 / 7
		Evaporation	20.893	0.000	3 / 1 / 7
		pH	7.008	0.001	3-1 / 7
		Water temp.	41.280	0.000	3 / 1 / 7
		Alkalinity	4.584	0.012	1-3 / 7
		Nitrate	13.312	0.000	7 / 1 / 3

Table 5.3.4. Mean and standard deviation of each limnological parameter. Values from data clustered according to SOM clustering.

	Mean±SD							
	Cluster 1	Cluster 2	Cluster 3	Cluster 4	Cluster 5	Cluster 6	Cluster 7	Cluster 8
Anabaena	1.64±0.76	0.46±	0	0	0	2.89±2.19	3.12±3.37	0
Microcystis	0	0	0	0.47±0.81	1.35±	58.88±	16.93±	0
Oscillatoria	0.55±0.85	0	0.71±1.05	0.42±0.59	0	0	0.75±1.01	0
Irradiance	45.5±11.6	44.6±	35.7±14.5	35.0±10.2	32.2±	40.4±18.3	53.8±15.6	32.6±12.9
Wind veloc-	4.1±1.0	3.9±1.2	3.9±1.0	4.1±1.3	3.0±0.6	3.9±1.4	3.6±0.9	3.9±1.1
Discharge	461±198	369±85	840±1086	461±243	433±131	538±271	515±430	471±390
Evaporation	11.1±3.3	10.3±2.2	8.8±3.3	9.3±2.4	9.2±2.9	12.7±3.7	13.9±4.1	8.2±2.9
Rainfall	10.7±18.6	4.6±7.7	18.5±40.7	4.0±7.5	4.5±6.2	9.9±20.8	9.6±13.7	7.5±15.7
Secchi depth	77±14	79±19	75±28	79±15	86±19	84±37	78±21	69±23
DO	8.3±2.3	11.7±3.9	9.8±4.4	10.0±2.5	9.4±3.0	9.1±2.4	10.0±2.8	12.5±4.5
pH	8.1±0.8	8.9±0.8	8.0±0.8	8.5±0.7	8.2±0.7	8.8±0.6	8.7±0.8	8.4±0.8
Turbidity	7.0±2.8	5.9±2.2	28.5±96.5	6.4±3.1	4.6±1.5	26.0±31.9	20.3±65.5	12.8±30.9
Water temp.	21.8±3.0	14.6±7.7	15.7±8.2	18.5±4.9	20.0±4.9	27.3±3.4	28.1±3.5	9.4±6.6
Conductivity	301±67	347±102	326±146	324±107	362±92	358±84	329±110	385±126
Alkalinity	50.4±9.9	61.2±7.6	51.7±19.2	59.3±16.6	63.7±9.9	65.7±12.4	60.9±10.2	58.0±16.0
Nitrate	2.3±0.7	2.9±1.1	2.8±0.8	2.7±0.8	2.0±0.5	2.3±1.3	1.9±0.9	3.1±0.9
Ammonia	0.3±0.2	0.3±0.3	0.6±0.7	0.3±0.3	0.4±0.3	0.2±0.2	0.3±0.3	0.6±0.6
Phosphate	34.7±24.7	13.6±9.1	45.6±27.4	37.1±25.6	21.2±	38.6±20.2	34.0±23.3	28.4±23.1
Silica	2.8±2.0	2.8±1.5	5.8±4.6	4.2±4.1	2.8±3.4	3.0±2.8	3.8±2.5	3.9±3.5
Rotifera	2816±	1143±	1796±	2809±	697±	738±731	1334±2112	1101±
Cladocera	351±872	72±141	45±93	36±52	14±16	103±165	182±290	20±52
Copepoda	103±88	122±191	48±75	17±18	17±9	75±115	94±173	47±187
Chlorophyll	50.2±52.0	47.3±	33.0±41.7	34.5±22.3	31.9±	134.2±	50.8±84.9	61.8±71.2

The results of ANOVA indicated that heat energy variables and pH/alkalinity were important for determining genus composition of the phytoplankton (Table 5.3.4). Water temperature tended to increase from 15 to near 30° C as more genera were included in the model. Irradiance and evaporation displayed similar patterns of increase (to over 50 MJ m^{-2} and 13 mm, respectively). Results were of particular interest for pH and alkalinity. In the case "*Oscillatoria* to all", the clusters with increased pH consisted of all of three genera, and also alkalinity was higher in these clusters.

Discussion

Cyanobacteria composition in the lower Nakdong River

Cyanobacteria rarely dominate in lotic ecosystems, at any point in the seasonal change of phytoplankton. Reynolds (1992) has summarized conditions which generally prevent cyanobacteria developments in rivers. Firstly, the higher flow rate relative to lakes prevents the cyanobacteria from proliferating near the water surface (i.e., high flow rate→no blooms). Lower growth rates than many eukaryotic algae, especially under conditions of frequent light fluctuations could be another reason: most cyanobacteria would be unable to accommodate the low retention times. In addition, Descy (1993) suggested that dilution process as well as continuous turbulence due to the water flow could have a role of controlling phytoplankton growth in river systems. In lentic ecosystems such as lakes and reservoirs, algal succession patterns are usually governed by solar irradiance and total nutrient conditions (Harris 1986). Reynolds (1992) suggested that diatoms dominate the algal succession in river ecosystems, whereas cyanobacteria could be observed only in pool-like regions where the river flow is slowed. Thus it has been thought that cyanobacteria are not important in river ecology; however, physical alterations which let the river systems become reservoir-like cause very different results.

A key point in this study is that the flow did not have a significant effect on the changes of cyanobacterial species in the lower Nakdong River. This would be due to the high variability of discharge in the river caused by the intensive water regulation and the unpredictable rainfall regime in summer. Many studies argued that discharge is determining the phytoplankton assemblage as well as biomass in river systems (e.g. Lack 1971, Holmes and Whitton 1981, Descy 1987). However, high regulation of river flow has become a popular practice for water resources management, and the changes of flow regime have influenced the dynamics ecological components in the regulated rivers (e.g. Gehrke et al. 1999, Growns and Growns 2001, Stanley and Doyle 2003). In the case of Nakdong River, estuarine barrage and multi-purpose dams resulted in the eutrophication and phytoplankton proliferation (Joo et al. 1997). In addition, even though concentrated rainfall and abrupt changes of discharge decreased the number of cyanobacterial cells in the river, they recovered easily and proliferated again in the lower part of the river where discharge is strongly controlled. The increased level of discharge during the monsoon and typhoon events lasted for about 3 to 5 days and the blooms are re-formed within 7 to 9 days (Ha et al. 2000, Park et al. 2002). In the early phase of bloom formation, the water column is quickly stabilized and allows high light penetration.

Although our analysis showed that the hydrological variables did not have significance, the stable river flow regime may still be a key factor for the massive proliferation of cyanobacteria. In low flow periods, elongated retention time due to the lack of concentrated rainfall became a trigger of proliferation. Whenever the conditions of stable low flow were met, cyanobacteria established large populations and other factors than hydrology governed their variations of abundance.

In the lower Nakdong River, recent studies focused on the river hydrology (Joo et al. 1997, Ha et al. 1998) and reported that abrupt changes in discharge govern the growth of cyanobacteria in summer. Ha et al. (1998, 1999) stressed that the magnitude of cyanobacterial blooms in the lower part of this river maintained a higher level than in other flow-regulated rivers in the world. However, in the Nakdong River, cyanobacteria increased in the lower part, not in the middle of the river (Ha et al. 2002). This is due to the background of elongated retention time as well as to nutrient enrichment. Neural network models of Jeong et al. (2001) and Jeong et al. (2003a) have stressed the importance of heat energy and pH dynamics for *M. aeruginosa* blooms in this river, instead of the hydrological parameters (i.e. rainfall and discharge). Therefore, in this eutrophic regulated river system, some environmental parameters related with heat energy and pH seemed to control the changes of cyanobacterial species during the summer.

Some hypotheses have been suggested for explaining the succession of cyanobacteria in lentic ecosystems. Takamura et al. (1992) documented the importance of nutrient conditions for inter-annual variability of *Microcystis* and *Anabaena* in Lake Kasumigaura, Japan. Recknagel (1997) suggested that underwater irradiance was of particular importance, and developed a neural network model based on this hypothesis. Other factors are considered to be important in affecting the succession of cyanobacteria. In the lower Nakdong River, those factors seemed to have synergic effect while the species composition of cyanobacteria changed. Generally the increase of heat energy (water temperature) led to a complex species composition. More than 2 genera occurred when temperature was over 20°C. Evaporation and irradiance are directly related to water temperature, and they can explain this phenomenon.

Nitrate concentration had slight difference among clusters (approximately 2.5 mg N L^{-1}) on average, where only one genus was contained in the clusters. Lower values of nitrate (2.0 mg N L^{-1}) occurred when two or more genera were joined. Nitrate is important for phytoplankton growth, but also is known that concentrations over 1 mg N L^{-1} can allow maximum growth. Even though statistically nitrate concentrations differed from each cluster, the overall values were not much sensitive to the changes of cyanobacterial community. The only possibility for the decline of concentration might be from the extensive uptake by those species. Ammonia and phosphate were not statistically significant, and the N:P ratio among clusters did not have distinguishable differences.

In the case of pH and alkalinity, higher values were detected as increasing the number of genera. The pH values were around 8 to 9, which is known to favor cyanobacteria (Moss 1973; Reynolds 1986). Higher alkalinity means the increased buffer capacity in the water body, and the extent of CO_2 decline is larger at the same pH variation degree in water of higher alkalinity (Talling 1985). During summer, in the lower Nakdong River, the primary factor for the increase of alkalinity could be the floods caused by concentrated rainfall with loaded nutrients (Park et al. 2002). Therefore it can be assumed that once high alkalinity conditions are maintained after the rainfall in the summer, photosynthesis by algae causes much CO_2 depletion in the river. At that time, higher water temperature as well as abundant nutrients also generates favorable conditions for cyanobacterial dominance. As pH is increasing, the algal species which are less limited to CO_2 (i.e. cyanobacteria) are able to increase easily (Shapiro 1990).

SOM applicability

In this study, a SOM algorithm exhibited good applicability to a time-series of phytoplankton abundance. Applications of SOM to river phytoplankton communities as well as cyanobacterial dynamics are still scarce, even though there were many scientific studies that dealt with phytoplankton through ANN algorithms (e.g. Recknagel 1997, Maier et al. 1998, Maier et al. 2001, Jeong et al. 2003a). Some scientific papers using SOM for addressing total phytoplankton biomass changes in lakes can be found (e.g. Chen and Mynett 2003). The SOM model in this study was successful not only for clustering complicated dataset of cyanobacterial blooms and species changes, but also finding the influencing parameters on their dynamics.

Usually the SOM algorithm has been used for clustering spatial data (e.g. Oberdorff et al. 1999), while examples with time series data are less common (e.g. Chon et al. 1996). Phytoplankton dynamics are affected by various factors, whose relative importance can change rapidly (e.g., after a rainfall flush). Park and Park (2002) suggested this in their linear models for predicting water temperature profiles in Korean peninsular systems receiving high variations of water input due to monsoon climate and summer concentrated rainfall. Thus a modelling algorithm that can rapidly adapt to a complex temporal dataset is necessary. Linear models have historically been used for algal dynamics (e.g. Dillon and Rigler 1974, Whitehead and Hornberger 1984), however, non-linear models such as neural networks can help us to investigate complex ecosystem behavior (Recknagel 1997, Jeong et al. 2003a).

The advantage of SOM application to ecological data was figured in Chon et al. (1996). They insisted that the application of SOM to stream macroinvertebrates was more fruitful than the traditional statistical approaches (e.g. cluster analysis). The latter usually uses the averaged similarity among data, and rearranges data on one-dimensional display. However, SOM uses more than two dimensions (if necessary, three dimensional arrangement could be adapted; see Kohonen 1982), and more of unseen relationships are possibly observed. In other studies, the evidence of the adequateness of machine learning techniques including SOM to ecological data are easily seen (e.g. Jeong et al. 2003b).

There have been great efforts for developing deterministic models for phytoplankton dynamics which could predict time-series changes and classifying data. These studies contributed better understanding of ecosystem dynamics (e.g. Drago et al. 2001; Lewis et al. 2002, Håkanson and Boulion 2003, James et al. 2003). Recknagel (1997) summarized the advantages of ANN models against these types of deterministic models (i.e. non-linear data processing, elucidation of unseen information), and the research of Brosse et al. (2001) encourages the use of SOM for the ecological data for clustering. Further, Whigham and Recknagel (2001b) and Jeong et al. (2003b) show the implication of other machine learning techniques such as evolutionary computations to the dynamics of phytoplankton in lotic as well as lentic ecosystems. As suggested by Pascual and Ellner (2000), the problems associated with complicated environmental and biological dataset in ecosystems through empirical computerizations (see Medsker 1996) could be solved by means of ANN models.

Compared with general ANNs that adopt commonly Multi-Layer Perceptrons (MLPs) with Backpropagation (BP) training, SOM also has advantage on data clustering. Artificial Neural Networks are able to find important patterns after training with given data, and this is similar to SOM. But the primary difference between MLP and SOM is the method of training. The former can consult data for finding errors while training (i.e. supervised training), but SOM does not need this training data (i.e. unsupervised training) (Lin and Lee 1996). Therefore, even though ANNs including MLPs are useful to dynamic systems, the developed model does not react accurately with unseen data patterns if the system has high uncertainty. From this point of view, SOM is more pliable to ecological dataset with high complexity.

Ecological models for river systems are developed and calibrated for free-flowing conditions. These models may not be useful for river-reservoir hybrids (Jeong et al. 2001). Lake-like phenomena, such as cyanobacterial blooms, often occur in flow-regulated rivers, and it is necessary to focus model ability on the way that this is facilitated by flow control. Computational algorithms, including SOM, are flexible for a variety of data conditions, and are useful for river-reservoir hybrids.

Another advantage of SOM is that it can be used to test hypotheses using field data (although ultimate causality only can be determined with controlled experiments). For the cyanobacteria blooms, some hypotheses were summarized by Shapiro (1990), and they were tested in experimental approaches as well as field survey. The test of hypotheses is one of the major objectives of ecological modelling. The cyanobacterial blooms in this point of view could be found in some examples of using non-linear models (e.g. Recknagel 1997, Jeong et al. 2003a). When pH was the major cause for cyanobacterial blooms (which was insisted by Shapiro (1990)), the result of this study could find that cyanobacterial blooms in a regulated river were closely related to the pH variations. Possibly the increased pH led the species to proliferate. Potentially photosynthesis by any abundant phytoplankton could deplete CO_2 and they used bicarbonate, which created the conditions for superior competitors to become dominant. Consequently, it seemed that pH was a driving force for the increase of cyanobacteria. Furthermore, the model results could suggest another hypothesis about the species changes within the cyanobacterial community: i.e. higher pH related with high alkalinity may cause that the community structure of cyanobacteria becomes more complicated (i.e. more species are assembled). This should be evaluated through further experimental studies.

Neural networks including SOM have the ability to synthesize information, and help us to develop a better predictive understanding of the ecosystem. This study is an example of non-linear model use by means of a field dataset. Usually the researches to find causality on a certain ecological dynamic requires controlled experiments under very restricted conditions. The control of experimental environments generally causes lack of unseen, but important, information for the changes of ecosystem. Non-linear models can fill this gap between field data and experiments.

5.4 Use of artificial intelligence (MIR-max) and chemical index to define type diatom assemblages in Rhône basin and Mediterranean region*

Rimet F†, Cauchie HM, Tudesque L, Ector L

Introduction

Benthic diatoms are unicellular brown algae that constitute a major part of the biomass in rivers. These key organisms cannot be ignored in any attempt to fully understand freshwater river ecosystems. Diatom diversity is very high and their distribution is determined by many environmental factors acting at different spatial scales (Stevenson 1997, Snyder et al. 2002). These parameters, varying at different scales in time and space, can affect diatom assemblages differently. Anthropogenic disturbances have no longer been known to be amongst the parameters controlling diatom assemblages for a long time (e.g. Butcher 1947, Fjerdingstad 1950). This observation led some authors to develop methods based on the ecology of benthic algae to assess water pollution in rivers (e.g. Patrick and Strawbridge 1963, Fjerdingstad 1964).

In France, Coste developed diatom-based stream quality assessment (Coste and Leynaud 1974, Coste 1976, Cemagref 1982), and water agencies started to use it extensively in the 1990's (Prygiel et al. 1999). A diatom index, the Biological Diatom Index (BDI) was standardised (AFNOR 2000, Prygiel and Coste 2000) in order to use it routinely in all the French territory. European standards for diatom sampling, identification and enumeration are under development or under approval (European Committee for Standardization 2002a,b).

Understanding the importance of the structuring effect of environmental parameters is a challenge in diatom ecology, their study at different spatial scales is necessary; the conclusions ensuing should show different results. This will be helpful in the choice of an appropriate strategy to develop new tools for the assessment of river quality based on diatoms. Until recently, few studies examining diatom taxa repartition and diatom assemblage structure at a large scale exist in France. The Rhône basin and Mediterranean region are grouped in a hydrographical basin managed by the same Water Agency (Agence de l'Eau Rhône-Méditerranée-Corse). Benthic diatoms have been sampled since 1995 for bioindication. This region is characterised by very different climates (from alpine to Mediterranean and continental), geologies and anthropogenic disturbances. The benthic diatom samplings carried out in this region appeared to be a good case study to explore the structure of the assemblages and to define the most important parameters affecting the structure of benthic diatom assemblages.

---

* This work is part of the PAEQANN project (EU 5th Framework Programme, contract n°: EVK1-CT1999-00026). We thank the "Rhône-Mediterranée-Corse" Water Agency, the "Directions Régionales de l'Environnement" Bourgogne, Rhône-Alpes, Provence-Alpes-Côte d'Azur, Languedoc-Roussillon, and the "Conseil Général des Alpes-Maritimes" for their important collaboration. Authors are grateful to Mr C. Bouillon for his technical support.

† Correspondence: rimet@crpgl.lu

The identification of type assemblages for benthic diatoms in unpolluted rivers is an important step in the establishment of a diatom-based stream classification. Therefore, non-impacted groups of samples were selected to propose type assemblages of benthic diatoms that could characterize different regions of the studied area. The second aim was to relate these unpolluted groups of samples to particular environmental conditions. This led to the construction of a database in this region, gathering diatom listings from several public services. On the other hand, the existing information was reinforced by new samplings in rivers of underrepresented zones. These samplings were realised in the framework of the European PAEQANN project (http://aquaeco.ups-tlse.fr/) that aims to predict biological communities in watercourses with environmental parameters using advanced modelling techniques.

An artificial intelligence technique, MIR-max (Walley and O'Connor 2001, www.cies.staffs.ac.uk) was used to cluster the samples on the basis of their diatom assemblages and to represent them according to their assemblages on a two dimensional graph; these two properties (clustering and projection) are not available with a unique analysis for multivariate analysis. The results of this new technique were compared to those obtained with a Twinspan analysis (Hill 1979b) a very often-used technique in biotypology. A forward selection was computed with Canoco 4.0 (ter Braak 1988) to find the most important environmental descriptors affecting diatom assemblages; correlation coefficients were calculated between the environmental parameters of the sampling sites. The organic pollution index IPO (Leclercq and Maquet 1987a,b) was used to select groups of samples with no organic pollution. The typology of the groups was characterised using the class boundaries of the European typological system A (European Parliament and The Council of the European Union 2000).

Materials and methods

Study area

The Rhône basin and the Mediterranean region (including Corsica) are situated in the southeastern part of France (Fig. 5.4.1). These regions encompass several hydrographical basins (area: 130,000 km^2) coordinated for water quality control by the "Rhône-Méditerranée-Corse" Water Agency. The study region contains a wide variety of landscapes, geologies and climates (alpine, continental, and Mediterranean). It is composed of high mountains (the Alps in the eastern part, the East Pyrenees in the southwestern part) and mountains of lower altitudes (e.g. the Jura in the northeastern part and the Massif Central in the western part). The two main rivers, the Rhône and its main tributary the Saône can be considered as lowland watercourses in the studied area. The Rhône flows into the Mediterranean Sea. The southern part of the region is characterised by a Mediterranean climate. Corsica, southeast of the French coast, is a large island characterised by a Mediterranean and mountainous climate and is included in the monitoring area of the "Rhône-Méditerranée-Corse" Water Agency.

Diatom sampling, preparation, identification and counting

207 benthic diatom samples were taken in the Rhône basin and the Mediterranean region by several public services (Directions Régionales de l'Environnement, Conseil Général des Alpes-Maritimes, Centre de Recherche Public - Gabriel Lippmann) from 1995 to 2001 for water quality monitoring according to the French standard of the Biological Diatom Index

BDI (Lenoir and Coste 1996, AFNOR 2000, Prygiel and Coste 2000) and also in the framework of the EU funded project PAEQANN to reinforce the existing database in underrepresented river types. These samples cover a large range of environmental conditions in these regions and allow the definition of type assemblages in relation to these conditions.

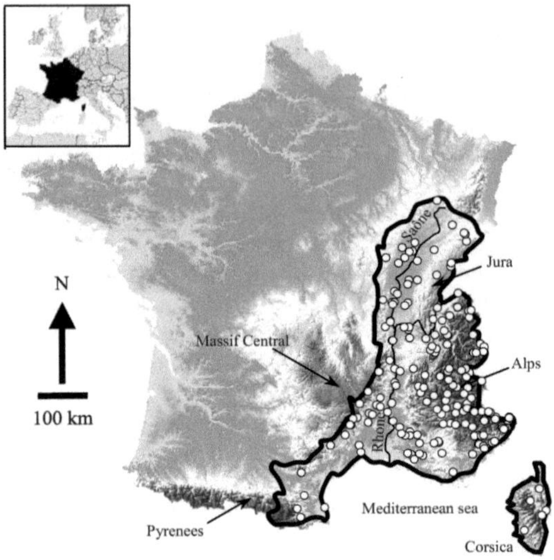

Fig. 5.4.1. Situation of the Rhône-Méditerranée-Corse basin in Europe and France. The main rivers (Saône and Rhône) and massifs are mentioned. The sampling sites are located with white spots.

In order to obtain comparable results, the French standard of the BDI was used by all the public services involved in this study. Benthic diatoms were collected from lotic parts of the sampled sites on several stones (minimum five), which cannot be moved under normal hydrological conditions. The upper surface of the stone was sampled with a clean toothbrush. The samples were fixed in 4% formaldehyde (Prygiel and Coste 2000). The diatom valves were cleaned with 40% hydrogen peroxide to eliminate organic matter and with hydrochloric acid to dissolve calcium carbonates. Clean diatom frustules were mounted in Naphrax©. Up to 400 valves were counted in each sample (Iserentant et al. 1999) with a 1000× magnification. The entire slide was investigated with a 400× magnification to check for rare taxa. The Süßwasserflora von Mitteleuropa (Krammer and Lange-Bertalot 1986, 1988, 1991a,b) was used as the basis for identification and supported by more recent books such as Krammer (2000) and Lange-Bertalot (2001) among others. Different people carried out the determinations. Therefore before gathering all the diatom counts in the same database, all the identifications were checked by the authors by looking at the permanent slides to homogenise the dataset from a taxonomical point of view.

Physical and chemical analyses

Water temperature, dissolved oxygen, conductivity and pH were measured every month by the different public services. Water samples were collected in each site and analysed in the laboratory following standard procedures (APHA 1995) for NO_3^-, NO_2^-, NH_4^+, total phosphorus, PO_4^{3-}, HCO_3^-, Na^+, Cl^-, K^+, SO_4^{2-}, Ca^{2+}, biological oxygen demand and chemical oxygen demand. In order to integrate the physical and chemical variations at each site, average values of the parameters during 3 months preceding the diatom sampling were calculated for each sample. When the data were not available, annual averages were calculated. For each site sampled, geology, altitude, slope, catchment area and distance from the source were determined using 1:20,000 topographical maps and on geographical information systems (BD-Carthage hydrographical database, BRGM geological map, IGN topographical maps).

Data analysis

The method used to explore the diatom database is an artificial intelligence technique called MIR-max (Walley and O'Connor 2001). This technique has a visualisation system similar to the self-organizing-map (Kohonen 1982), but the algorithm is considerably different. The MIR-max technique is based on two separated processes. Firstly, the samples are clustered by a pattern recognition technique based on information theory. An algorithm maximises the mutual information between the clusters and the attributes of the data (MI-max algorithm), the numbers of the groups are defined in this algorithm. Secondly, the clusters are sorted in a two-dimensional output space. In the end of these two processes, the clusters numbers of the map are not ranked in an ordered way since among other processes a random exchange procedure of the clusters intervene in the map construction algorithm. The diatom assemblages were used as input database for this technique. The 250 most abundant taxa were retained for the analysis and their absolute relative abundances were transformed to percentages in each sample.

A Twinspan analysis (Hill 1979b) was carried out using the diatom data (the same 250 most abundant taxa were used) to define clusters (pseudospecies cut level: 0, 2, 5, 10, 20, the same weights were given to all the pseudospecies). The results obtained with Twinspan and with MIR-max were compared by representing the Twinspan clusters on MIR-max maps; the percentages of each Twinspan cluster were calculated in each MIR-max cell and were represented on MIR-max maps.

A forward selection performed with a Monte-Carlo test on the environmental variables and the diatom counts was computed with Canoco v. 4.0 (ter Braak 1988). This analysis is generally carried out to select the most important environmental variables before computing a canonical correspondence analysis. The effect of each environmental variable for diatom assemblages was assessed with the conditional effects and their significance was estimated with the Monte-Carlo test. Correlation coefficients were calculated between the environmental variables measured for each sampling site; the coefficients were tested. These correlation coefficients were used to facilitate the understanding of the environmental gradients observed on the MIR-max map.

The organic pollution index IPO (Leclercq and Maquet 1987a,b) is an index giving a global value of organic pollution in rivers. It was used in this study to summarize the organic pollution information and was calculated for each MIR-max group. This index takes into account the concentrations of NH_4^+, NO_2^-, PO_4^{3-} and the biological oxygen demand. IPO has a value between 1 and 5, representing a water quality from very highly polluted to unpolluted respectively. This index was used to select the sites with a very low level of organic pollution: MIR-max groups with IPO values of 4.5 or greater were selected. The diatom assemblages of these selected groups were proposed as type assemblages for the studied rivers.

The Water Framework Directive (European Parliament and The Council of the European Union 2000) suggests choosing between two typological systems for the rivers, System A or System B. System A is simple to apply and uses 4 environmental descriptors with classes already defined: ecoregions (Alps, Pyrenees, Western highlands, Western plains for the Rhône-Méditerranée-Corse region), altitudes (lowland < 200 m, $200 \leq$ mid-altitude ≤ 800, $800 <$ high), catchment area ($10 <$ small ≤ 100 km^2, $100 <$ medium ≤ 1000, $1000 <$

large ≤ 10000, 10000 < very large), geologies (calcareous, siliceous, organic). In order to characterise the sites composing the MIR-max groups, the class boundaries of altitude and catchment area of the system A were used.

Results

Distribution of taxa in the studied area

497 taxa were identified. Some taxa such as *Fragilaria arcus*, well known to occur in cold waters with high current velocity and a low anthropogenic pollution (Krammer and Lange-Bertalot 1991a), had a regional distribution pattern.

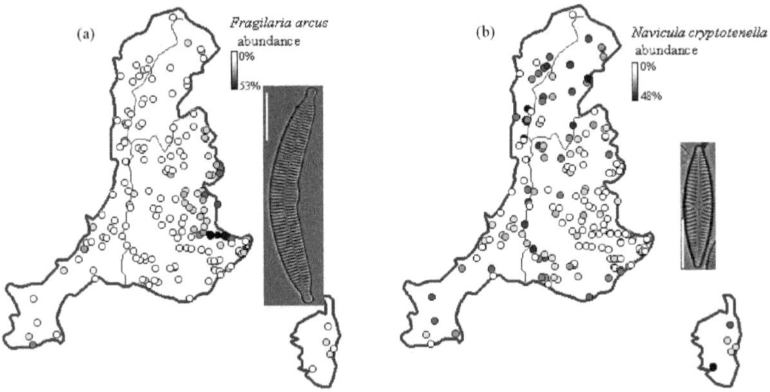

Fig. 5.4.2. Location and abundance of Fragilaria arcus and Navicula cryptotenella in the Rhône-Méditerranée-Corse basin. Scale bars on the photos correspond to 10 mm.

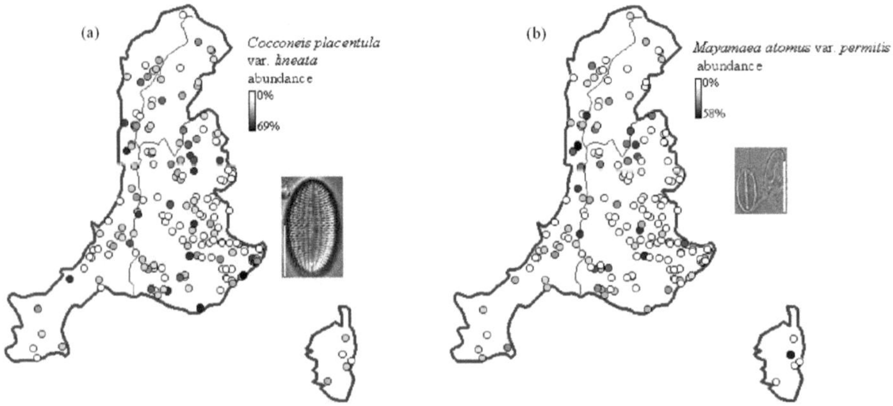

Fig. 5.4.3. Location and abundance of Cocconeis placentula var. lineata and Mayamaea atomus var. permitis in the Rhône-Méditerranée-Corse basin. Scale bars on the photos correspond to 10 mm.

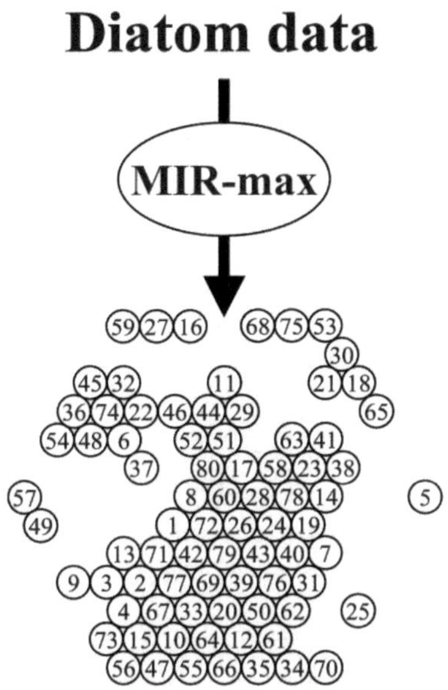

Fig. 5.4.4. MIR-max map computed on the basis of the diatom assemblages; 80 groups of diatom samplings were defined and represented on a hexagonal map.

The location of *Fragilaria arcus* (Fig. 5.4.2a) near the Italian boundary in the eastern part of the region corresponded to the high Alps and confirmed this ecology. On the contrary, the cosmopolitan taxon *Navicula cryptotenella* (Lange-Bertalot 2001) was principally located in lowland rivers, such as in the Saône and the Rhône rivers (Fig. 5.4.2b).

Other taxa such as *Cocconeis placentula* var. *lineata* were not confined to a particular region but were present throughout the study area (Fig. 5.4.3a). *Cocconeis placentula* var. *lineata* is considered as a cosmopolitan epiphytic and epilithic taxon (Krammer and Lange-Bertalot 1991b). The eutraphentic and α-meso-polysaprobous taxon *Mayamaea atomus* var. *permitis* (van Dam et al. 1994) did not show a regional distribution pattern (Fig. 5.4.3b); its distribution corresponded to sites that are always impacted by human activities.

Exploration of the diatom database structure with MIR-max software

80 groups or "clusters" were computed using the MIR-max software; these were arranged on a hexagonal output space with 127 discrete locations (Fig. 5.4.4). The MIR-max map was numbered from 1 to 80, corresponding to the 80 groups of samples. Samples with similar diatom assemblages were placed in the same group or in neighbouring groups on the map, whereas samples with very different assemblages were placed in distant groups. *Achnanthidium minutissimum* (Fig. 5.4.5b) appeared in most of the MIR-max groups, it is often the dominant diatom in upland oligo/mesotrophic rivers (see e.g. Kelly and Whitton 1995), and was also the dominant taxon in the rivers of the studied region.

The MIR-max map (Fig. 5.4.4) showed that a lot of groups were contiguous with only a few of them isolated. In the bottom part of the map were found contiguous groups dominated by taxa as *Achnanthidium biasolettianum* (Fig. 5.4.5a), *Fragilaria arcus* (Fig. 5.4.5f) or *Gomphonema pumilum* (Fig. 5.4.5g). On the upper left part were found pollution resistant taxa such as *Fistulifera saprophila* (Fig. 5.4.5e). The upper right part was dominated by β-mesosaprobous taxa such as *Amphora pediculus* (Fig. 5.4.5c) or *Navicula cryptoten-*

ella (Fig. 5.4.5h) (van Dam et al. 1994).MIR-max algorithms isolated groups with assemblages very different from each other. Group 5 is the most isolated (estuary of the Durance canal in "l'Etang de Berre") its assemblage was characterised by taxa occurring in slightly brackish environments (*Fragilaria fasciculata, Nitzschia filiformis, Gomphonemopsis obscurum*): these taxa were unusual in the database. Groups 57 and 49 were also isolated from the rest of the groups.

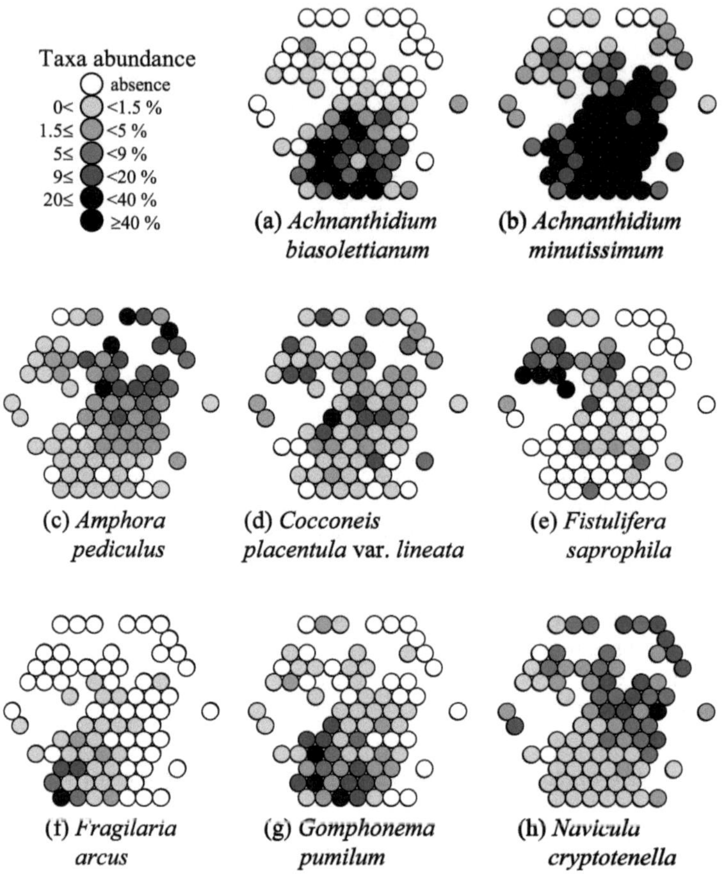

Fig. 5.4.5. Abundance of 8 taxa on the MIR-max map. Percentages are represented in grey scale.

They were both characterised by high abundances (20 and 22% respectively) of the invasive *Nitzschia* cf. *tropica* (Coste and Ector 2000) that is only abundant in these two groups, situated in the Massif Central and in the left bank of the Rhône near the Mediterranean sea. Some taxa as *Cocconeis placentula* var. *lineata* did not show any clear gradient on the MIR-max map (Fig. 5.4.5d).

Influence of environmental factors on diatom assemblages

Each MIR-max group can be characterised by physical and chemical parameters. Average values for different variables were overlaid on the MIR-max map with grey scale (Fig. 5.4.6). The range of each environmental parameter is given in Table 5.4.1 to show the extent of the environmental conditions observed in this study. The altitude showed a continuous gradient from the upper part to the bottom of the MIR-max map (Fig. 5.4.6a).

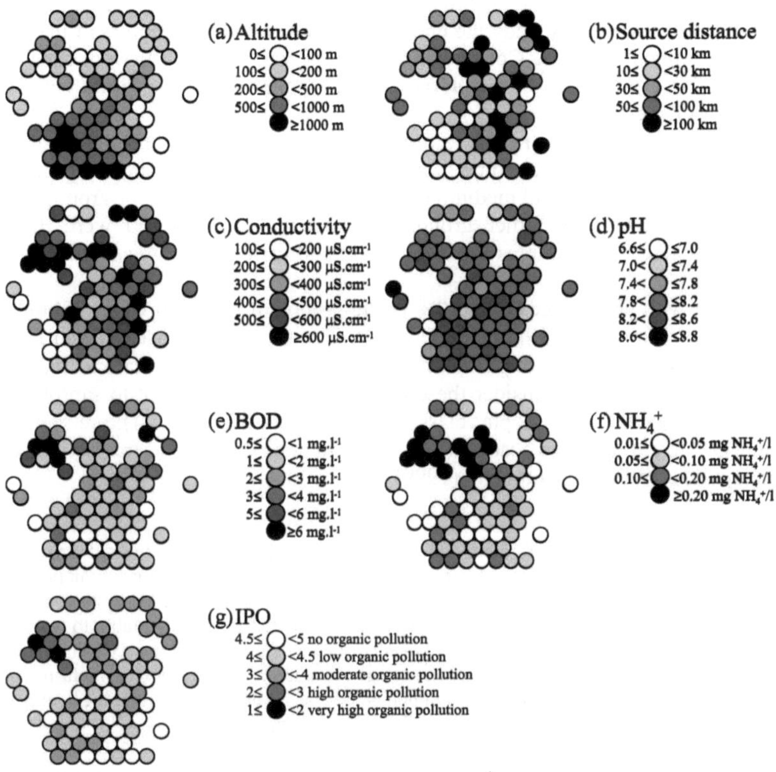

Fig. 5.4.6. Characterisation of the MIR-max groups; average values of the 6 significant physical and chemical parameters (Table 5.4.1) and of the Index of Organic Pollution IPO are overlaid on the MIR-max map.

High altitude sites (over 1000 meters) are mainly located in the bottom of the map, and low altitude sites in the upper part. The gradient of "altitude" was inversely correlated to the gradient of "distance from source" (Fig. 5.4.6b); this can be explained by the fact that headwaters were in the bottom left hand corner of the map, and sites near the estuaries were in the upper part. Tests of the correlation coefficient calculated between these two variables and slope confirmed this observation (Table 5.4.2). The forward selection showed that the altitude was the most important factor affecting the structure of diatom assemblages (Table 5.4.1). Organic pollution indicators as biological oxygen demand and ammonium as well as conductivity and pH also had important and significant effects (Table 5.4.1).

These pollution indicators were also overlaid to the MIR-max map (Fig. 5.4.6e, f). They showed a gradient from high concentrations in the upper left hand corner to low concentrations in the bottom right hand corner of the map. All the pollution indicators (NO_3^-, NO_2^-, NH_4^+, PO_4^{3-}, biological oxygen demand) were positively and significantly correlated (Table 5.4.2). Parameters expressing saprobity of rivers appeared to have a significant effect on diatom assemblages. NO_3^-, did not show any significant effect.

The results of the forward selection showed the importance of parameters related to typology (altitude, source distance). This showed the necessity to use a typological system to better understand the groups defined by the MIR-max algorithms. The characterisation of each group was also important to enable to assign some groups to precise typological levels and particular regions. In this study the class boundaries of two descriptors of the system A (European Parliament and The Council of the European Union 2000) were used (altitude and catchment area) in order to characterize the sites composing each group with IPO values of 4.5/5 or greater.

Numbers of samples in each class (altitude and catchment area) are presented in Table 5.6. This Table 5.4.3 shows that some groups are well characterized such as groups 55, 66 (class of high altitude, small and medium catchment area respectively), or group 39 (mid altitude, medium catchment). Other groups such as 1, 12, 42 or 50 were difficult to characterize because their samples belonged to several different classes of altitudes and/or catchment area.

All the taxa with abundances over 10% in the groups of table 5.6 were pollution sensitive according to Cemagref (1982) (Table 5.4.4). The saprobity classes of van Dam et al. (1994) consider these taxa as oligosaprobous or β-mesosaprobous taxa and the saprobity classes of Rott et al. (1997) consider these taxa as sensitive to tolerant to saprobity (Table 5.4.4). On the other hand the trophy classes of these taxa (Table 5.4.4), are varying from mesotraphentic to eutraphentic according to van Dam et al. (1994) and from oligotraphentic to eutraphentic according to Rott et al. (1999).

Table 5.4.1. Forward selection performed with 14 environmental variables and the 250 most abundant taxa (ter Braak 1988). The conditional effect is the variance of the species matrix explained by a new environmental variable added to the model, given the variance already explained by the environmental variables selected by the model; a Monte-Carlo test was calculated for each variable to test its significance (P value). The marginal effect is the explained variance of the species matrix using only one environmental variable at a time. Cumulative percentage variance of species-environment relation: axis 1: 28.0%, axis 2: 14.7%. The last column "Parameters values" gives the minimum, the average and the maximum values of each parameter measured in the 207 sampling sites.

	Conditional effects		**Marginal effects**	**Parameter values**		
Environmental Variables	**Additional variance explained**	**Monte-Carlo test (significant if $P \leq 0.05$)**	**Explained variance**	**Min.**	**Average**	**Max.**
Altitude (m)	0.39	**0.005**	0.39	1	547	2660
BOD* (mg l^{-1})	0.26	**0.005**	0.30	0.5	2.1	18.4
Conductivity	0.18	**0.005**	0.26	25	415	1800
pH	0.15	**0.005**	0.16	6.2	8.1	9.2
Distance to source (m)	0.14	**0.010**	0.21	0.0	53.1	437.1
NH_4^+(mg NH_4^+ l^{-1})	0.13	**0.040**	0.25	0.01	0.31	15.56
PO_4^{3-} (mg PO_4^{3-} l^{-1})	0.12	0.060	0.28	0.01	0.33	8.06
Na^+(mg l^{-1})	0.11	0.110	0.24	0.2	14.6	265.0
Cl^-(mg l^{-1})	0.07	0.380	0.22	0.1	21.48	380.0
Slope (‰)	0.07	0.370	0.15	0	10	133
NO_3^- (mg NO_3^- l^{-1})	0.07	0.210	0.31	0.02	4.39	34.00
Ca^{2+}(mg l^{-1})	0.06	0.400	0.20	0.5	69.6	269.0
Dissolved oxygen	0.05	0.835	0.09	2.2	9.5	15.6
NO_2^-(mg NO_2^- l^{-1})	0.05	0.430	0.17	0.01	0.21	3.90

*: BOD: Biological Oxygen Demand

Table 5.4.2. Correlation coefficient (Pearson product moment) calculated with the environmental parameters measured in the 207 sampling sites. Cell contents: correlation coefficient, P value: ns: not significant, significant: 0.05≤*<0.01, highly significant: 0.01≤**<0.001, very highly significant: 0.001≤***.

	Slope	**Source distance**	**pH**	**Conduc-tivity**	**Dissolved oxygen**	**NO_3^-**	**NO_2^-**	**NH_4^+**	**Na^+**	**Cl^-**	**PO_4^{3-}**	**BOD***	**Ca^{2+}**
Altitude	0.40 ***	-0.40 ***	0.15 *	-0.40 ***	0.05 ns	-0.41 ***	-0.19 **	-0.13 ns	-0.33 ***	-0.34 ***	-0.21 **	-0.19 **	-0.30 ***
Slope		-0.26 ***	0.14 *	-0.20 **	-0.17 *	-0.25 ***	0.06 ns	-0.08 ns	-0.18 **	-0.20 **	-0.10 ns	-0.22 **	-0.17 *
Source Distance			0.04 ns	0.16 *	0.08 ns	0.22 **	-0.10 ns	-0.03 ns	0.16 *	0.26 ***	-0.05 ns	0.01 ns	0.12 ns
pH				0.14 *	0.14 *	-0.12 ns	-0.08 ns	-0.12 ns	-0.09 ns	-0.09 ns	-0.06 ns	-0.10 ns	0.11 ns
Conductivity					-0.23 **	0.36 ***	0.18 *	0.17 *	0.64 ***	0.63 ***	0.26 ***	0.24 ***	0.85 ***
Dissolved Oxygen						0.05 ns	-0.28 ***	-0.22 **	-0.21 **	-0.20 **	-0.21 **	-0.01 ns	-0.10 ns
NO_3^-							0.29 ***	0.14 *	0.27 ***	0.29 ***	0.44 ***	0.36 ***	0.32 ***
NO_2^-								0.28 ***	0.20 **	0.19 **	0.47 ***	0.29 ***	0.12 ns
NH_4^+									0.25 ***	0.23 ***	0.36 ***	0.75 ***	0.12 ns
Na^+										0.92 ***	0.26 ***	0.27 ***	0.36 ***
Cl^-											0.21 **	0.27 ***	0.39 ***
PO_4^{3-}												0.49 ***	0.10 ns
BOD*													0.22 **

*: BOD: Biological oxygen demand

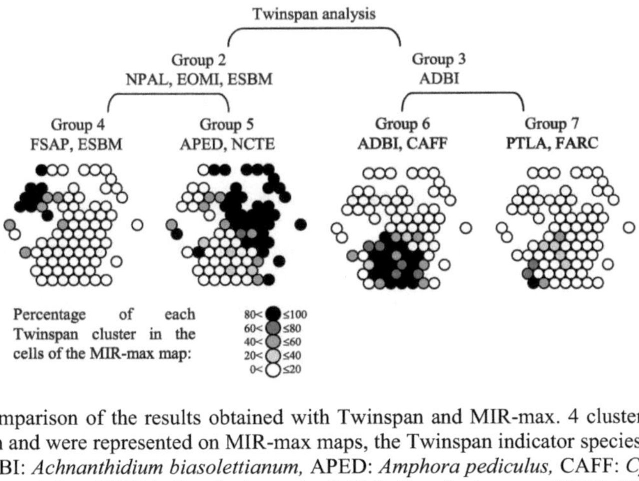

Fig. 5.4.7. Comparison of the results obtained with Twinspan and MIR-max. 4 clusters were defined with Twinspan and were represented on MIR-max maps, the Twinspan indicator species of each cluster are given (ADBI: *Achnanthidium biasolettianum,* APED: *Amphora pediculus,* CAFF: *Cymbella affinis,* EOMI: *Eolimna minima,* ESBM: *E. subminuscula,* FARC: *Fragilaria arcus,* FSAP: *Fistulifera saprophila,* NCTE: *Navicula cryptotenella,* NPAL: *Nitzcchia palea,* PTLA: *Planothidium lanceolatum*).

Table 5.4.3. Presentation of the 17 groups with an IPO of 4.5/5 or greater. Numbers of samples for each altitude and catchment area classes are given, their taxa with abundances over 5% are mentioned. Codes for geology: 1: limestone, 2: mudstone-schist-schale, 3: granitic, 4: mixed, 5: quaternary sediment, 6: other. The classes boundaries follow the system A of the Water Framework Directive (European Parliament and The Council of the European Union 2000): altitude in meters (class 1: lowland ≤ 200, class 2: 200 < medium ≤ 800, class 3: 800 < high), catchment area in km^2 (class 1: 10 < small ≤ 100, class 2: 100 < medium ≤ 1000, class 3: 1000 < large ≤ 10000, class 4: 10000 < very large). Codes signification of the taxa: ADMI : *Achnanthidium minutissimum,* ADBI : *A. biasolettianum*, ADLA : *A. latecephalum*, APED : *Amphora pediculus,* CAFF : *Cymbella affinis,* CPED : *Cocconeis pediculus*, CPLI : *C. placentula* var. *lineata*, CPPL : *C. placentula* var. *pseudolineata*, DMES : *Diatoma mesodon*, DVUL : *D. vulgaris,* ENCM : *Encyonopsis microcephala*, ENMI : *Encyonema minutum*, EOCO : *Eolimna comperei*, EOMI : *E. minima*, FCAP : *Fragilaria capucina*, FCVA : *F. capucina* var. *vaucheriae,* FDEL : *F. delicatissima,* FSAP : *Fistulifera saprophila,* GPAR : *Gomphonema parvulum*, GPUM : *G. pumilum*, GTER : *G. tergestinum,* NCTE : *Navicula cryptotenella*, NDIS : *Nitzschia dissipata,* NFON : *N. fonticola*, NPAE : *N. paleacea*, NTPT : *Navicula tripunctata,* RSIN : *Reimeria sinuata*, UULN : *Ulnaria ulna.*

Group	Geology of sampling site	Altitude classes			Catchment area classes				Number of samples	Taxa present in the group, mentioned by decreasing abundance order	
		1	2	3	1	2	3	4		100%<Abundance≤10 %	10% < Abundance ≤ 5%
1	1,2,2,5	3	1		1	1	1		4*	CPLI, ADMI, GPUM	CPPL
3	2,3		1	1	1	1			2	ADMI, FCAP	GPAR, EOMI, FCVA
7	2	1							1	ADMI	NCTE, UULN
12	1,1,1,2,3		3	2	2	2			5*	ADBI, ADMI, ENMI	GPUM
14	1,1	2			2				2	NCTE, ADMI	NTPT
25	3	1				1			1	CAFF, ADMI	DVUL, CPLI, NPAE
26	1,4,4	1		2		2			3*	ADMI, APED, CPLI	NTPT
31	2		1			1			1	ADMI	ADBI, NDIS, NCTE, NFON
33	1,1,1,1,4		3	2	1	4			5	ADMI, ADBI, GPUM	GTER
34	1,1	2				1	1		2	CAFF, ADMI, NFON	ADLA, EOCO
39	1,1,5		3			2	1		3	ADMI, GTER, ADBI	
42	1,3,4		2	1	1	2			3	ADMI, ADBI	GPUM, RSIN, DMES
50	1,1,1	1	2			1	1	1	3	ADMI, CPLI	ADBI, CAFF
55	1,3,3,3,4, 4		1	5	1	5			6	ADMI, GPUM, ADBI	RSIN, FSAP, CAFF
60	1,3	2					1		2*	ADMI, CPED, NCTE, CPLI	
61	1,1,6	1	2			1	2		3	ADMI, ADBI, FDEL	ENCM, CAFF
66	1,1,3,3,3			5	3	2			5	ADBI, ADMI, GPUM, CAFF	

*: catchment area was not defined for one sample in these groups

Representation of Twinspan clusters on MIR-max map

Twinspan clusters 4-5 were present in the upper part of the MIR-max map whereas clusters 6-7 were in the lower part of the map (Fig. 5.4.7). A difference between clusters 4 and 5 was observed since the former was in the left part of the map; similar observations were done for 6 (right part of the map) and 7 (left part of the map). Indicator species of the upper left part were pollution tolerant (Fig. 5.4.5e: *Fistulifera saprophila* and *Eolimna subminuscula*), those of the upper right part are β-mesosaprobous (Fig. 5.4.5c: *Amphora pediculus,* Fig. 5.4.5h: *Navicula cryptotenella*). Indicator taxa of the bottom right hand corner are pollution sensitive (Fig. 5.4.5a: *Achnanthidium biasolettianum,* Figure 5.4.5g: *Gomphonema pumilum*), *Fragilaria arcus* (Fig. 5.4.5f) and *Planothidium lanceolatum* are indicative taxa of the bottom left hand corner.

Table 5.4.4. Sensitivity and indicator values according to Cemagref (1982), saprobic and trophic indexes according to van Dam et al. (1994) and Rott et al. (1997, 1999) for the most abundant taxa of Table 5.4.3 (abundance over 10%). For Cemagref (1982) S is ranging from 5 (pollution sensitive taxa) to 1 (taxa tolerant to pollution) and V (indicator value of the taxa) is ranging from 3 (stenoecy) to 1 (euryecy). For van Dam et al. (1994) saprobity: 1: oligosaprobous, 2: β-mesosaprobous, to 5: polysaprobous; trophy: 1: oligotraphentic, 2: oligo-mesotraphentic, 3: mesotraphentic, 4: meso-eutraphentic, 5: eutraphentic, 6: hypereutraphentic, 7: indifferent. For Rott et al. (1997) saprobity: sensitive to saprobity < 1.3, tolerant to saprobity < 2.1, saprophilic < 2.6, saprobiontic ≥ 2.6. For trophy Rott et al. (1999): oligotraphentic < 1.4, mesotraphentic < 1.9, eutraphentic < 2.7, eutraphentic to polytraphentic ≥ 2.7, V (indicator value) is ranging from 5 (stenoecy) to 0 (euryecy). The taxa codes are given in Table 5.4.3.

	Cemagref (1982)		van Dam et al. (1994)		Rott et al. (1997)		Rott et al. (1999)	
	S	V	Saprobity	Trophy	Saprobity	V	Trophy	V
ADBI	5	2	-	3	1.4	3	1.3	1
ADMI	5	1	2	7	1.7	1	1.2	1
APED	4	1	2	5	2.1	2	2.8	2
CAFF	4	2	2	5	1.2	4	0.7	4
CPED	4	2	2	5	2	3	2.6	2
CPLI	5	1	2	5	-	-	2.3	2
ENMI	4.8	2	-	-	1.6	2	2.0	1
FCAP	4.5	1	2	3	-	-	1.8	2
FDEL	4	1	-	3	1	5	1.4	2
GPUM	5	1	-	7	1.6	3	1.1	1
GTER	4	3	1	2	1.9	4	1.4	1
NCTE	4	1	2	7	1.5	2	2.3	1
NFON	3.5	1	2	4	2.1	4	1.9	0

Discussion

Comparison of results obtained with Twinspan and MIR-max

Results obtained with these two techniques presented large similarities. Each Twinspan cluster (clusters 4, 5, 6, 7) mostly corresponds to each corner of the MIR-max map. This comparison shows that results obtained with a classical technique (Twinspan) are similar with those obtained with MIR-max. The use of the indicator species given by Twinspan can give further understanding for the MIR-max map structure. However, MIR-max remains more effective to visualize the clusters and their species composition than classic clustering techniques.

Type assemblages in unpolluted rivers of Rhône basin and Mediterranean region

In our data set, nitrate did not seem to have any significant effect on diatom assemblages (Table 5.4.1), and the classes of trophy for the taxa composing the type assemblages were rather heterogeneous (Table 5.4.4): from oligo-mesotraphentic to eutraphentic or indifferent (van Dam et al. 1994), and from oligotraphentic to eutraphentic (Rott et al. 1999). Nevertheless parameters characterising the saprobity of the rivers had a significant effect on diatom assemblages (Table 5.4.1). The taxa composing the type assemblages presented in Table 5.4.3 are pollution sensitive (Cemagref 1982) and most of them are sensitive to saprobity (van Dam et al. 1994, Rott et al. 1999, Table 5.4.4). Our work is in agreement with these studies already done in the field of classification of algal assemblages and the type assemblages proposed for the Rhône basin and Mediterranean region seem to be consistent.

In the groups proposed as unpolluted (Table 5.4.3), some as groups 1 and 60 were characterised by taxa considered as epiphyte (*Cocconeis pediculus, C. placentula* var. *euglypta,* var. *lineata,* var. *pseudolineata*). Their geographical distribution and their typology are not clear and these groups were probably characterized by the occurrence of filamentous algae (*Cladophora*) in the sampling sites.

On the contrary, group 55 was well characterized for its altitude and catchment area classes (Table 5.4.3), corresponding to high altitude rivers (770 to 1300 m) with high current velocity, relatively low conductivities (from 108 to 443 $\mu S.cm^{-1}$) and low temperature (from 8.2 to 14.4 during summer and early autumn). Most of the samples of this group were located in upstream sites of the Alps. The typical assemblage in this group was composed of small species, *Achnanthidium minutissimum, A. biasolettianum, Gomphonema pumilum* and *Reimeria sinuata.* These taxa are equipped with anchoring systems: *Achnanthidium* are attached to substrata by a mucilage stalk formed at one end of the raphe valve (Round et al. 1990), *Reimeria* has apical pore field at each end on the ventral mantle and *Gomphonema* has a basal pore field; these pore fields are involved in the secretion of mucilage for the attachment to substrata (Round et al. 1990, Hoagland et al. 1982). This attachment property and their small sizes are important characteristics, enabling resistance to high speed current and waters with a high concentration of suspended solids. Group 66 had a very similar assemblage, and also corresponded to upstream sites in the Alps. Group 12 is also rather near groups 66 and 55 but of lower altitude.

Groups 14 and 7 were composed of low altitude sites (Bourbonne river at Montbellet, altitude: 200 m, Golo at Volpajola in Corsica: 90 m). Their assemblages (Table 5.4.3) were typified by alkaliphilous taxa: *N. cryptotenella, N. tripunctata* and *Ulnaria ulna* (van Dam et al. 1994). These diatom assemblages are uncommon in the Rhône-Méditerranée-Corse database since many lowland rivers are impacted by human activities and have assemblages dominated by polysaprobous taxa. Similarly, rivers with low conductivities and weak slopes were composing group 3; taxa as *Achnanthidium minutissimum* and *Fragilaria capucina* are present in high abundance in these unpolluted watercourses (Eyrieux river and Saone river at 8 km from the sources).

Another example is provided by group 39 (Table 5.4.3). It corresponded to Alpine rivers of medium altitude (230-780 m), with a river catchment dominated by limestone (Dranse at Thonon, Leysse at Le Bourget and Durance at Embrun). The conductivity of this group was relatively high: from 394 to 438 $\mu S.cm^{-1}$. Taxa such as *Achnanthidium minutissimum, A. biasolettianum* and *Gomphonema tergestinum* were abundant in this group. *Gomphonema tergestinum* is well known to occur in unpolluted rivers with high concentrations of electrolytes in limestone Alps (Krammer and Lange-Bertalot 1986).

Groups 25 and 34 corresponded to lowland rivers (altitude: 47-100 m) of the western part of the basin (Ardèche river in Massif Central mountains). They were characterised by ubiquitous species like *Achnanthidium minutissimum, Cymbella affinis* and *Nitzschia fonticola.* Moreover group 34 had some peculiar taxa such as *Achnanthidium latecephalum* or *Eolimna comperei* (Coste and Ector 2000). *Achnanthidium latecephalum* has never been found previously in Europe; it was described only in rivers with low level of pollution in Japan (Kobayasi and Ishida 1996). *Achnanthidium latecephalum* occurred only in group 34 as a dominant species: Ardèche and Chassezac rivers in Massif Central, Tavignano river in Corsica. *Eolimna comperei* also had a local repartition in Europe and was until now known to be present only in the Adour, the Garonne, and the Rhône basins (Coste and Ector 2000, Peres et al. 2003).

This last example suggests not only that environmental parameters are structuring diatom assemblages, but also that historical processes of species dispersal can play an important role (Potapova and Charles 2002). Recent studies showed that about 15 diatom taxa (e.g. *Eolimna comperei, Gomphoneis minuta, Encyonema triangulum, Diadesmis conferva-*

cea) can be considered as invasive taxa in France (Coste and Ector 2000, Peres et al. 2003). This raises the question of whether such invasive species should rather be included in type assemblages of unpolluted rivers or should be considered as indicators of environmental modifications.

Influence of environmental factors on diatoms in the Rhône basin and Mediterranean region

According to Stevenson and Pan (1999), knowledge of the hierarchical organization of the relevant factors can help to make diatom indicators more precise. The most important parameters structuring diatom assemblages in Rhône basin and French Mediterranean region is altitude, and also, at a lower level, distance from the source. In this study, altitude, slope and source distance are correlated all together (Table 5.4.2), and are also often correlated to current velocity. Current velocity has been demonstrated to be a significant selective factor for species composition of diatom assemblages in Mediterranean springs (Roca 1990, Sabater and Roca 1990), in rivers (Biggs and Gerbeaux 1993, Lamb and Lowe 1987), and also in indoor lotic microcosms (McIntire 1966, Steinman and McIntire 1986). Temperature that has an effect on periphyton and diatom metabolism (Denicola 1996, Berges et al. 2002) is also related to altitude.

Conductivity and pH (Fig. 5.4.6c, d) were less important but significant parameters according to the Monte-Carlo test computed in the forward selection (Table 5.4.1). They can be linked to the geological substrate (Biggs 1995), which has been shown to be an important parameter affecting the structure of diatom assemblages in lotic (Rimet et al. 2004) and lentic systems (Vyverman et al. 1996). Geology is very complex in the study area, especially in the Alps; it could be reasonably expected to be an important parameter affecting the structure of diatom assemblages but was difficult to identify here. A denser sampling should be carried out to more accurately assess geology impact on diatom assemblages.

Organic pollution, measured on the basis of biological oxygen demand and NH_4^+, was also shown to have important and significant impacts on diatom assemblages, but was less influential than altitude. Numerous studies have related water quality and diatom assemblages (Stoermer and Smol 1999) and several authors developed biotic indexes to assess biological water quality of European running waters: Descy (1979), Cemagref (1982), Sládeček (1986), Leclercq and Maquet (1987a,b), Descy and Coste (1990, 1991), Schiefele and Kohmann (1993), Hofmann (1994), Kelly and Whitton (1995), Lenoir and Coste (1996), Prygiel and Coste (1998, 2000), Kelly (1998b), Coring et al. (1999), Dell'Uomo (1999), Harding and Kelly (1999), Rott et al. (1997, 1999), Rott and Pipp (1999). These biotic indices are now routinely used in many countries, for instance in France, Belgium, Luxembourg and Spain, and are standardised or are about to be (Kelly et al. 1998, AFNOR 2000, European Committee for Standardization 2002a, 2002b). Several kinds of pollution parameters and their influence on diatom taxa selection can be separated in different categories such as saprobity and trophy (Hofmann 1994, van Dam et al. 1994, Rott et al. 1997, 1999). In our study most of the parameters indicating pollution were correlated with each other (Table 5.4.2). On the MIR-max maps, these parameters had similar gradients. Parameters characterising the saprobity level of the rivers appeared to be the most important factors structuring the diatom assemblages (Table 5.4.1).

Even if most of the pollution parameters were inversely correlated with altitude (Table 5.4.2), moderate organic pollution occurred in several high altitude sites such as those in groups 56 and 4 with altitudes varying from 1245 m to 2660 m and IPO of 3.75 for both groups. *Diatoma mesodon* and *Fragilaria arcus* were present with important abundances in these two groups. *Diatoma mesodon* is considered as oligosaprobous and mesotraphentic, and, *Fragilaria arcus* as β-mesosaprobous and oligo-mesotraphentic according to van Dam et al. (1994). Both are oligotraphentic and tolerant to saprobity according to Rott et al. (1997, 1999); these species may characterise high altitude sites with moderate organic pollution. High altitude sites moderately impacted were quite rare in the databases; impacted sites were much more likely to be found in lowland rivers (groups in the upper left part of the MIR-max in Fig. 5.4.4). Lowland polluted sites had assemblages dominated by pollution resistant taxa such as *Fistulifera saprophila* or *Mayamaea atomus* var. *permitis*. On the other hand, despite their scarcity, weakly impacted lowland sites could be found in the study area, such as those in group 14, characterised by a dominance of *Navicula cryptotenella*. The effects of pollution factors are generally more complex to understand when associated to gradients of altitude, distance from source and slope (Potapova and Charles 2002).

Typology and diatoms

The results of the MIR-max map and of the forward selection (Canoco v 4.0) are comparable and both show the importance of a "downstream" gradient. This complex gradient summarizes parameters such as altitude, distance to source, slope and current speed. This gradient can be related to the typology and must be taken into account to understand diatom assemblages. The concept of typology has existed for a long time in hydrobiology. Illies and Botosaneanu (1963) defined a zonation system with three typological levels on the basis of fish fauna. More recently Vannote et al. (1980) introduced the "River Continuum Concept", which states that the change in the benthic community structure along a stream is predictable.

Concerning diatom assemblages, few studies (Descy and Coste 1991) clearly suggest the existence of a continuum or a biotypology along rivers. Studies in the United States showed the importance of this "downstream" gradient at a larger scale (Potapova and Charles 2002). These authors emphasized that this gradient is difficult to understand because many factors such as slope, elevation, concentration of nutrients, land-use and temperature intervene simultaneously. Similar observations about the importance of topography (mountain, high plateau, low plateau, valley) on diatom assemblage structure were done in the Mid-Atlantic Highlands streams in USA (Pan et al. 2000).

The limits of the River Continuum Concept were reviewed in France by Wasson (1989) who indicated that macroinvertebrate continua are different depending on the region considered. A regional or ecoregional approach must also be taken into account for the definition of the type communities for macroinvertebrates, diatoms and fishes that occur in streams. This work has already been undertaken in several countries as Spain (Munné and Prat 2000) and France (Wasson et al. 2001). The results of this study also show that a regional approach should be considred for benthic diatoms since groups corresponding to unpolluted rivers can be related to regions and to particular altitude ranges.

Conclusion and perspectives

This study is a first attempt to identify the parameters structuring diatoms and to define type assemblages of benthic diatoms in unpolluted rivers of France using artificial intelligence in combination to chemical index and multivariate analyses. It shows the importance of the downstream gradient and the regional factors for benthic diatom assemblages. Relations with a more detailed classification of geology should be also checked and comparisons of these results should be made with results from a wider spatial scale (for instance, at a European level) and with other studies realised in similar (see Tison et al. # 5.5) or in smaller areas (see Rimet et al. in # 5.7).

In order to define more precisely all the type assemblages for unpolluted rivers in Rhône basin, French Mediterranean region and high European mountains, a denser sampling especially in underrepresented river types and a more global approach should be envisaged. As the REFCOND European working group (2002) proposed, a selection of non-impacted basins or sites should be done first. New samplings should then be collected in non-impacted sites to reinforce the existing database. Comparisons of the results presented in our study with the reference conditions established following the REFCOND guidance could be carried out in the future.

5.5 Classification of stream diatom communities using a self-organizing map*

Tison J†, Giraudel JL, Park YS, Coste M, Delmas F

Introduction

One of the major axes of the new E.U. Water Framework Directive is to assess the deviation of an ecosystem with respect to the highest ecological quality awaited (non-perturbed or reference conditions), thanks to the responses of aquatic communities. By comparing diatom communities in natural and disturbed sites, indicators for different types of anthropogenic disturbance can be found. But, since diatom species composition varies among streams due to natural as well as anthropogenic factors, we should be able to increase the accuracy of assessing anthropogenic impact by first accounting for natural variability among sites. Kociolek (2000) argue that the proportion of geographically restricted diatom species is high and spatial distribution patterns of species are still poorly understood. Although many studies have focused on the effect of human pressure on diatom communities (Pan et al. 1999, Licursi and Gomez 2002, Winter and Duthie 2000, Potapova and Charles 2002, Soininen 2002), the number of studies attempting to characterise the natural patterns and the relative weights of environmental parameters influencing this natural variability is limited (Descy 1984, Leclercq and Depiereux 1987, Sabater and Roca 1992, Stevenson 1997, Pan et al. 2000).

Ordination techniques are a useful way to explore the characteristics of datasets and to find relationships between variables. Diverse linear ordination methods have been used to simplify the data including polar ordination, principal components analysis (PCA), correspondence analysis (CA) (Pearson 1901, Hill and Gauch 1980, Beals 1984, Jongman et al. 1995). The limitations are well-known, e.g. all of them present strong distortions with non-linear species abundance relations (Kenkel and Orloci 1986); horseshoe effects due to unimodal species response curves in PCA and arch effects, outliers, missing data, disjointed data matrix in CA (Giraudel and Lek 2001). Recently, as an alternative tool to deal with this problem of complexity in ecological data, artificial neural networks (ANNs) have been utilized for patterning communities in various ecosystems (i.e., aquatic, forest, agriculture, etc.) (Lek and Guégan 2000, Recknagel 2003). Among the ANN techniques, Kohonen's self-organizing map (SOM) (Kohonen 1982, 2001) is the most popular unsupervised learning algorithm, allowing the classification of data without prior knowledge and the visualisation of species assemblages in a two-dimensional space (Giraudel and Lek 2001). The SOM has been used for the classification of communities (Chon et al. 1996, Foody 1999, Park et al. 2001a, 2003a), for water quality assessment (Walley et al. 2000, Aguilera et al. 2001), and for prediction of populations and communities (Céréghino et al. 2001, Obach et al.

* Our work has been funded through the EU 5th Framework Research Programme named PAEQANN. We wish to thank the Adour-Garonne Water Agency for providing us with the physical and chemical characteristics of the sampling sites. We also thank J.G. Wasson and his colleagues from Cemagref Lyon for their helpful information concerning Adour-Garonne hydro-ecoregions.

† Correspondence: juliette.tison@bordeaux.cemagref.fr

2001). The ability of the SOM for classification and ordination in ecology has also been compared with conventional multivariate analysis (Chon et al. 1996, Foody 1999). In particular, Giraudel and Lek (2001) compared the SOM with several different multivariate analysis methods including PCA and CA, and concluded that the SOM seems fully usable in ecology and can be a perfect complement to classic techniques for exploring data and for achieving community ordination. Nijboer et al. (see # 4.5) also compared the SOM with a canonical correspondence analysis and cluster analysis for the ordination and classification of macroinvertebrate communities, showing that each method has its own strengths and weakness depending on the objectives of the study.

Our study, run on a pilot dataset (Adour-Garonne stream system, South-West of France), was the first attempt to highlight the natural spatial distribution scheme of benthic diatoms on a regional scale, and explored the performance of the SOM in diatom community studies. The main purpose is to give a practical application of unsupervised neural networks for patterning diatom community structure in reference situations, sustaining the WFD implementation. In this study, we developed an innovative methodological approach to establish a first diatom-based bio-typology of an Adour-Garonne basin stream system which makes clear headway in ecoregional zoning.

The research was carried out in the framework of the European Research Program PAEQANN (Predicting Aquatic Ecosystem Quality using Artificial Neural Networks – 5th PCRD), aiming to develop general methods, based on advanced modelling techniques, for predicting the structure and diversity of key aquatic communities under natural conditions and subjected to man-made disturbances.

Material and methods

Studied area and dataset

The Adour-Garonne hydrographic network (up to 120 000 km of streams and rivers for a total watershed of 116 000 km^2), is composed of 7 main sub-catchments (Charente river, Dordogne river, Lot river, Tarn-Aveyron rivers, Garonne river, Adour river, and coastal streams), covering a large range of altitudes (high mountains to plains and coastal areas) and geological substrates (calcareous, sedimentary, sandstone, crystalline and volcanic).

The database consisted of 49 reference sites in the basin where data was collected from 1994 to 2001 by the Cemagref (Table 5.5.1, Figure 5.5.1). "Reference sites" here represent sites with a very low level of disturbance, often located in upstream parts of the rivers above significant human activities. The IPS (Indice de Pollusensibilité Spécifique; Coste in Cemagref 1982) of the stations was always over 15 on a scale of 20, confirming their good quality status. We studied only reference sites so as to highlight natural variability, and accurately characterise the species composition expected in the least impacted streams (expected conditions for different types of habitat).

All samples were collected during summer and according to a standardised method (NFT 90-354 (AFNOR 2000)), in order to limit variability due to the season and to local factors like shadow or substrate.

Diatom species were identified at a 1000 × magnification (Leitz DMRD light microscope) according to Krammer and Lange-Bertalot (1986, 1988, 1991a,b), by examining permanent slides of cleaned diatom frustules, digested in boiling H_2O_2 (30%) and HCl (35%) and mounted in a high refractive index medium (Naphrax, Northern Biological Supplies Ltd, UK; RI = 1.74). The biological data matrix then established, expressed in presence or absence of species, is composed of 399 species × 49 sites.

Table 5.5.1. Sampling sites and their environmental conditions

Code	River	Altitude (m)	Slope (‰)	pH	Conductivity (µS/cm)	HCO3 (mg/l)
1	ADOUR	500	19	7.95	140.00	85.00
2	ARIEGE	187	5	7.15	94.50	46.50
3	CORREZE	180	24	7.00	60.00	20.65
4	DADOU	190	3	7.30	137.00	34.50
5	DORDOGNE	270	6	7.10	55.50	22.00
6	DORDOGNE	150	6	7.05	57.50	21.85
7	GARONNE	480	11	8.20	97.50	59.50
8	GARONNE	340	6	8.25	114.50	69.00
9	GAVE-DE-PAU	360	6	7.55	267.00	83.00
10	GAVE-DE-PAU	330	12	7.65	142.00	83.50
11	GAVE-DE-PAU	225	22	8.05	167.50	84.50
12	SALAT	280	6	7.90	161.00	92.00
13	SALAT	280	6	8.10	158.00	85.00
14	TARN	350	20	8.30	369.00	224.00
15	TRUYERE	230	10	7.50	68.50	39.50
16	GAVE-DE-PAU	150	8	8.10	319.00	128.00
17	PALUE	5	0	6.60	105.00	23.60
18	GAVE-D'OLORON	37	10	7.90	334.00	179.50
19	GAVE-D'OSSAU	400	40	7.95	252.00	142.00
20	NASSEY	20	1	6.20	158.00	30.00
21	LEYRE	20	1	6.50	126.00	25.00
22	NIVE	300	13	7.30	227.00	120.00
23	GAVE D'ASPE	420	12	8.30	221.00	395.00
24	NESTE	651	8	7.60	126.00	50.00
25	LE LEZ	1080	13	8.10	184.00	100.00
26	SALAT	665	45	7.90	163.00	90.00
27	ORIEGE	920	25	7.80	86.00	50.00
28	ARNETTE	480	10	7.40	533.00	20.00
29	DOURBIE	820	25	7.50	31.00	20.00
30	SERRE	668	6	8.10	474.00	305.00
31	SEYE	190	11.7	8.01	603.00	330.00
32	ANTENNE	55	3	7.50	686.00	365.00
33	AUME	80	1.5	7.80	651.00	355.00
34	DRONNE	140	2.4	7.50	89.00	40.00
35	VEZERE	690	6	7.10	95.00	25.00
36	DORDOGNE	1000	40	7.30	43.00	25.00
37	MARONNE	610	10	7.40	78.00	50.00
38	CERE	750	17	7.50	101.00	50.00
39	CELE	250	5	7.40	98.00	45.00
40	GOURGUE	20	1	6.20	161.00	30.00
41	ARREILLET	20	1	6.50	163.00	30.00
42	CIRON	110	25	6.90	123.00	25.00
43	GAVE DE PAU	420	15	6.10	117.00	65.00
44	COLAGNE	1220	22	6.11	150.00	20.00
45	VALAT DES CLOUTASSES	1250	10	6.12	18.60	10.00
46	VALAT DE LA LATTE	1250	10	6.13	18.60	10.00
47	VALAT DE LA SAPINE	1250	10	6.14	16.80	10.00
48	ALIGNON	1020	14	6.15	18.60	10.00
49	GOUDECHE	1020	10	6.16	28.60	10.00

Physical and chemical environmental variables of sampling sites measured between 1994 and 1998 were provided by the Water Agency. Mean values of each variable were calculated from 2 months' data: the sampling month and the previous one. As the new reference sites sampled in 2001 were located out of the institutional sampling station network, the physical and chemical descriptors of water quality were determined at the sampling dates in two litres of water collected in a free flowing area near the middle of the stream, preserved at 4°C then analysed by the Cemagref laboratory within 24h, according to standardised AFNOR protocols.

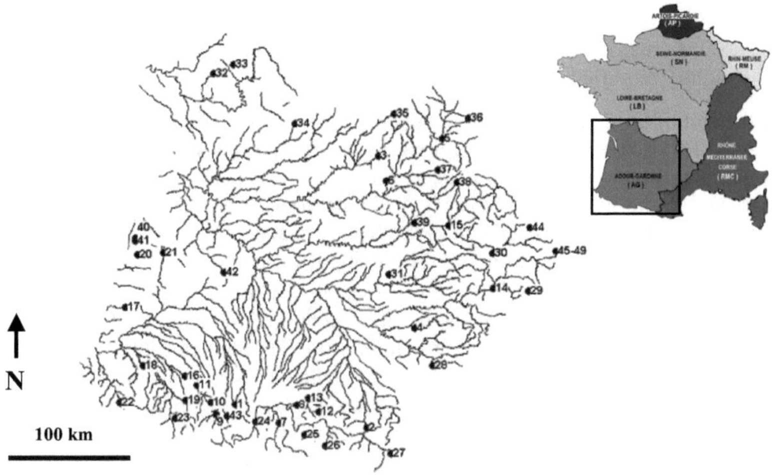

Fig. 5.5.1. Sampling site distribution over the Adour-Garonne stream system.

Modelling techniques

The SOM algorithm was used for patterning samples in the multi-dimensional database according to similarities of their species compositions and to visualise the contribution of species to the patterns. The SOM is an unsupervised neural network composed of an input layer with input neurons (computational units) (399 species in this study), and an output layer with output neurons in a two-dimensional hexagonal lattice. There are no strict rules to choose the number of output neurons (map size). To choose suitable map size, we trained the SOM with different map sizes, and following the results obtained, we chose 24 (= 4 × 6) neurons as output neurons based on advice from our ecological experts on diatom ecology and on the study areas. Output neurons act as virtual sites $(VS_k)_{1\le k\le 24}$ with species assemblages $(w_{ik})_{1\le i\le 399, 1\le k\le 24}$ to be computed (*k* for sites and *i* for species). During the learning process of the SOM, virtual sites are modified in order to approximate the probability density function of the input data.

The learning process of the SOM algorithm can be summarised as follows:

- Virtual sites are initialised with random samples drawn from the input data set.
- Virtual sites are updated iteratively:
  - A sample site is randomly chosen as an input.
  - The Euclidean distance between the new input and every virtual site is computed.
  - The virtual site closest to the input is selected as the 'best matching unit' (BMU).
  - The species assemblage of the BMU and its neighbours are changed such that the virtual sites progressively approach the input data.

In this study, training was broken down into two parts (Giraudel and Lek 2001):

- Ordering phase (the first 2,000 steps): when this first phase takes place, the composition of the virtual sites is highly modified in a wide neighbourhood of the BMU.

- Tuning phase (12,000 steps: 500 times the number of neurons in the Kohonen (SOM) map): during this phase, only the virtual sites adjacent to the BMU are slightly modified.

At the end of training, species assemblages are known for each virtual site. The BMU is determined for each site, and each real site is set in the corresponding SOM map hexagon. Neighbouring clusters of sites are expected to be represented by neighbouring hexagons on the map, and in the same way sites very different from each other (according to species assemblages) are expected to be distant in the feature space.

After the learning process of the SOM, the input components (i.e. species occurrence) of each virtual site (preserved in $(w_{ik})_{1\leq k\leq 24}$) were visualised on the SOM map in grey scales to show the contribution of each species in the structure of the map. Light indicates low frequency of occurrence, whereas dark for high values (Kohonen 2001, Park et al. 2003a). These component planes can be considered as a "sliced" version of the SOM and provide a powerful tool to analyse the community structure. However, with this method, 399 different maps (i.e., for 399 species) have to be considered. To quantify the species contribution in the SOM map, we used a Structuring Index (SI) (Park et al. 2004) in chapter 6 of this book. The SI allowed the determination of the most relevant variables for structuring the SOM map obtained. The SI was computed for each species i (SI_i) as follows:

$$SI_i = \sum_{k=1}^{24} \sum_{j=1}^{k-1} \frac{\left| w_{ij} - w_{ik} \right|}{\left\| r_j - r_k \right\|} \tag{5.5}$$

where $\left\| r_j - r_k \right\|$ is the Euclidean distance between the virtual sites VS_j and VS_k on the SOM map. The index considers the distribution gradients of each species on the SOM map. Therefore, species showing high distribution gradients display high SI values, whereas species showing low gradients have a low SI value. Thus, the higher the value of the SI, the more relevant the variable is. For details of the SI, refer to Park et al. (2004) in chapter 6 of this book. The computation of the SOM and the SI were carried out under MATLAB® on a PC with an Intel Pentium® PIII-500 and a program written by the authors.

Relationships between communities and environmental variables

According to the literature, broad scale patterns in benthic algae among streams over quite large geographic areas and between years largely reflect patterns in geology, climate and human activity (Lowe 1974, Biggs 1995, Leland 1995, Stevenson 1997, Pan et al. 2000). Human activity impacts were avoided in our dataset of reference stations, and climate was relatively homogeneous throughout the watershed. Therefore, we characterised each station with three environmental parameters (pH, bicarbonate alkalinity and conductivity) directly related to geology. To do this, sampling sites were classified using three environmental variables through a hierarchical cluster analysis with Ward's linkage method and then these clusters were presented on the SOM map obtained with diatom communities. In order to validate the geochemical-based groups, multivariate analysis of variance (MANOVA) was conducted with three environmental variables and then multiple comparison tests with unequal N HSD were also carried out for each variable using the statistical software STATISTICA (StatSoft 2001).

Ecological profiles of species

After the learning process of the SOM, virtual communities are produced in each SOM output unit (i.e. virtual unit), showing distribution gradients according to their environmental characteristics. Therefore, we can compare their distribution patterns on the SOM map with environmental patterns. We compared these distribution patterns of species according to environmental gradient in the SOM map with a classic species profile method proposed by Daget and Godron (1982). In the classic method, environmental gradients (conductivity gradients in this study) were arranged into 7 classes and the occurrence probability of each species was calculated along those 7 classes. The sum of the probabilities over the 7 classes is 1.

Results

The SOM was trained with diatom communities with presence/absence of species, and resulted in classifying sampling sites according to their diatom community similarities (Figure 5.5.2).

Stream typology

Sampling sites were classified with three environmental variables (pH, conductivity and bicarbonate alkalinity) into 4 different groups through the hierarchical cluster analysis (Fig. 5.5.3). The number of clusters was chosen at the linkage distance showing the highest distance (between 0.6 - 1.0):

- group I: quite mineralised sites mostly from the Pyrenees mountains
- group II: sites from volcanic and crystalline substrates, with low conductivities (mostly Massif Central).
- group III: stations from calcareous zones with high pH and carbonate levels.
- group IV: sandy substrates and crystalline rocks : stations with very low pH and conductivities (Landes, Cévennes).

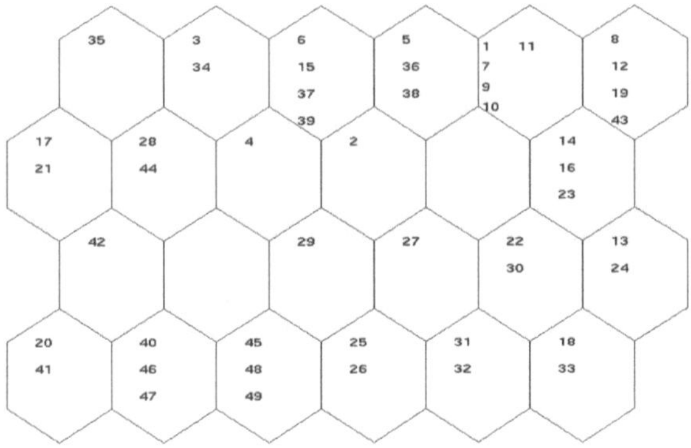

Fig. 5.5.2. Ordination of sampling sites on the SOM map.

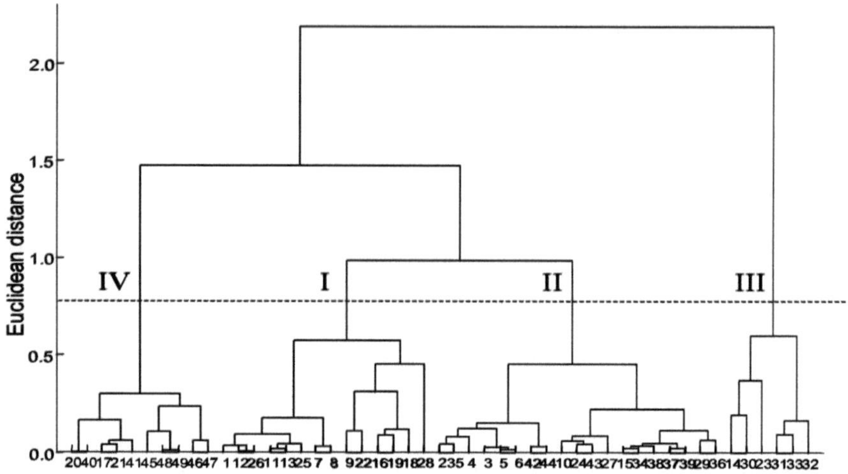

Fig. 5.5.3. Classification of sampling sites with three environmental parameters (pH, conductivity, HCO_3-). The Ward linkage method was used with Euclidean distances. The sampling site numbers are given in Fig. 5.5.1 and Table 5.5.1.

The differences of three environmental variables at different clusters are presented in Table 5.5.2. These 4 groups were significantly different from each other (MANOVA, $p<0.001$). Wilks' Lambda calculated in the MANOVA was 0.0712, showing that the groups can be easily distinguished with their environmental variables. Wilks' Lambda can range from 0 to 1, with 1 indicating no relationship of predictors (i.e., environmental variables) to responses (i.e., groups) and 0 indicating a perfect relationship of predictors to responses. Now, it is interesting to see the differences of each variable between groups. To do so, a multiple comparison test was carried out for each variable. Group III showed the highest values in all three variables and was thereby distinguished from other groups. Meanwhile, groups II and IV were not significantly different for three variables, although they were grouped differently in the cluster analysis. This is due to the differences of methods. Cluster analysis and MANOVA are multivariate analysis techniques concerning several variables in the calculation, while the multiple comparison test is a univariate analysis treating one variable in one calculation.

Table 5.5.2. Differences of three environmental variables between the 4 groups defined by the cluster analysis.

Group	pH	Conductivity (μS/cm)	HCO_3^- (mg/l)
I	7.94 (0.06)* a**	185.88 (30.25) b	117.08 (23.46) b
II	7.23 (0.06) b	123.60 (30.98) b	32.33 (3.02) c
III	7.77 (0.13) a	495.83 (75.42) a	259.92 (54.37) a
IV	6.40 (0.08) b	117.17 (50.49) b	18.86 (3.03) c
Overall	7.39 (0.09)	190.74 (25.77)	88.58 (15.25)

* Standard error

** Multiple comparison test. The same characters in each column indicate no significant difference at the 5% level of confidence according to the unequal N HSD multiple comparison test.

Classification of sampling sites with environmental variables was visualized on the SOM obtained with diatom communities (Fig. 5.5.4). The classification was also well distributed on the SOM. In the multiple comparison test, group III was strictly isolated from other clusters, and groups II and IV were quite similar compared with other groups. These characteristics were also present in the SOM map obtained with diatom communities. Group III was isolated in the lower right areas and groups II and IV were in the left areas. These facts indicate that sampling sites were classified according to the characteristics of their environment and were well presented by their diatom communities in the SOM.

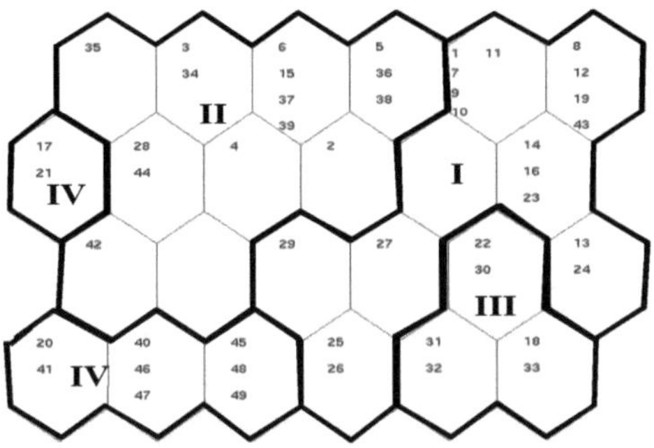

Fig. 5.5.4. The 4 geochemical-based regions (I-IV) on the SOM map. The cluster was defined based on the three environmental parameters (pH, conductivity, HCO_3-) in Fig. 5.5.3.

Evaluation of the structuring power of species

Figure 5.5.5 shows some example species displaying different distribution gradients on the SOM map according to their environmental preferences. Dark represents a high occurrence probability of each species, whereas light is a low value. According to the map, we can see that some species can be found only in limited areas based on their environment, and some species can be found in most places. In other words, the species which have a narrow ecological amplitude can be found only in map areas characterised by typical environmental gradients and so are very important for eco-regional zonation. Such species are called "structuring species" (examples: *Achnanthes oblongella* strup (AOBG), *Diatoma vulgare* Bory (DVUL), *Navicula angusta* Grunow (NAAN), *Gomphonema olivaceum* (Hornemann) Brébisson (GOLI), *Navicula rhynchocephala* Kützing (NRHY), *Navicula capitatoradiata* Germain (NCPR)).

In contrast, other types of species distribution show several clusters of dark grey hexagons indicating they can be found equally in several regions. This type of distribution does not characterise any particular environmental conditions. Such species do not provide enough explicit information about regional characteristics and so do not participate in map structuring ("non-structuring species"). These are generally ubiquitous species, like *Achnanthidium minutissimum* Kützing (ADMI). The most structuring species of the dataset studied, having the highest SI values, are listed with their corresponding cluster(s) in Table 5.5.3.

The conductivity gradient was ordered into 7 classes in the classic profiles (Fig. 5.5.7e), whereas mean values of conductivity in each SOM unit were calculated to visualize the gradient on the SOM map in the SOM profile (Fig. 5.5.7f). Both species gradient profiles showed clear gradients and agreed well with their conductivity gradients. The species *Eunotia subacicularis* Alles, Nörpel and Lange-Bertalot (ESUB) is typical from group IV (very low conductivity) defined in Figs. 5.5.4 and 5.5.5, which corresponds well to the first class of conductivity as defined in Daget and Godron's method. According to the SOM profiles, *Achnanthes oblongella* Østrup (AOBG) is found in low conductivity areas (groups II and IV). This result is also validated by the classic profiles (optimum in class 3). The distribution of *Gomphonema pumilum* (Grunow) Reichardt and Lange-Bertalot (GPUM) on the map avoids very low conductivity levels and its optimum with the classic profiles concerns the middle class of conductivity (class 4). Finally *Navicula tripunctata* (O.F.Müller) Bory (NTPT) characterises zones on the map with high conductivities and prefers the high class of conductivity. This is also observed in the classic profiles (class 5).

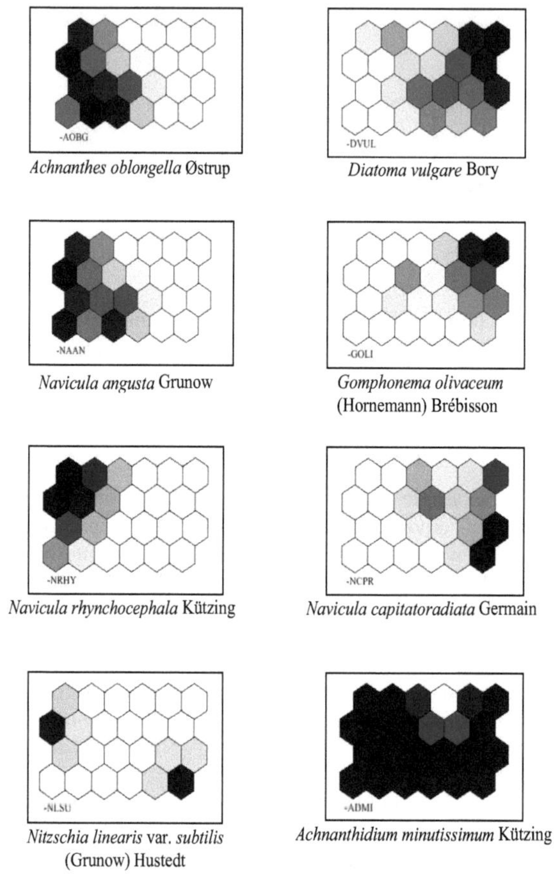

Fig. 5.5.5. Examples of SOM-based ecological profiles. Dark represents high abundance of each species, whereas light is for low values.

Table 5.5.3. List of the first 50 structuring species with their corresponding cluster(s).

Species	SI value	Cluster
Navicula gregaria Donkin	49.45	2
Gomphonema minutum (Ag.) Agardh f. minutum	47.72	1-3
Fragilaria arcus (Ehrenberg) Cleve var. arcus	46.76	1
Planothidium lanceolatum (Breb.) Round & Bukhtiyarova	45.93	2
Navicula lanceolata (Agardh) Ehrenberg	45.62	2
Navicula cryptocephala Kutzing	45.36	2-4
Tabellaria flocculosa (Roth) Kutzing	45.31	2-4
Melosira varians Agardh	44.30	2
Rhoicosphenia abbreviata (C.Agardh) Lange-Bertalot	43.69	2-3
Diatoma mesodon (Ehrenberg) Kutzing	43.62	1
Amphora pediculus (Kutzing) Grunow	43.54	3
Cymbella affinis Kutzing	43.45	1-3
Navicula tripunctata (O.F.Müller) Bory	43.39	1-3
Fragilaria capucina Desmazieres var.capucina	43.33	2
Nitzschia archibaldii Lange-Bertalot	43.14	1
Reimeria sinuata (Gregory) Kociolek & Stoermer	43.12	1
Fragilaria gracilis Østrup	42.99	2-4
Achnanthidium subatomus (Hustedt) Lange-Bertalot	42.98	1
Achnanthes oblongella Oestrup	42.95	4
Achnanthidium biasolettianum (Grunow in Cl. & Grun.) Round & Bukhtiyarov	42.68	1
Cocconeis placentula Ehrenberg var.lineata (Ehr.) Van Heurck	42.64	1-2-3
Navicula cryptotenella Lange-Bertalot	42.56	1-2-3
Diatoma vulgaris Bory 1824	42.54	1
Nitzschia fonticola Grunow in Cleve et Möller	42.41	1
Navicula cryptotenelloides Lange-Bertalot	41.99	3
Psammothidium subatomoides (Hustedt) Bukhtiyarova & Round	41.92	2
Gomphonema pumilum (Grunow) Reichardt & Lange-Bertalot	41.59	1-3
Eolimna minima (Grunow) Lange-Bertalot	41.41	2-3
Nitzschia palea (Kutzing) W.Smith	40.32	1-2-3
Planothidium frequentissimum (Lange-Bertalot) Round & Bukhtiyarova	40.23	2
Navicula angusta Grunow	40.02	2-4
Encyonema mesianum (Cholnoky) D.G. Mann	39.93	1
Nitzschia paleacea (Grunow) Grunow in van Heurck	39.64	1
Nitzschia acidoclinata Lange-Bertalot	39.45	1
Eunotia exigua (Brebisson ex Kützing) Rabenhorst	39.41	4
Gomphonema gracile Ehrenberg	38.69	2
Cocconeis pediculus Ehrenberg	38.46	1-3
Psammothidium bioretii (Germain) Bukhtiyarova & Round	37.98	2
Navicula reichardtiana Lange-Bertalot var. reichardtiana	37.92	1-3
Surirella roba Leclercq	37.89	2-4
Gomphonema exilissimum (Grun.) Lange-Bertalot & Reichardt	37.63	2
Nitzschia hantzschiana Rabenhorst	37.11	1-2
Navicula rhynchocephala Kutzing	36.91	2
Nitzschia dissipata (Kutzing) Grunow var.dissipata	36.53	1-2-3
Eunotia subarcuatoides Alles Nörpel & Lange-Bertalot	35.70	4
Nitzschia recta Hantzsch in Rabenhorst	35.56	2
Achnanthes curtissima Carter	35.27	1
Navicula exilis Kutzing	34.82	2
Cocconeis placentula Ehrenberg var.euglypta (Ehr.) Grunow	34.02	1
Eunotia minor (Kutzing) Grunow in Van Heurck	33.61	2-4

Discussion and conclusion

As diatoms are primarily autotrophic, they occupy a key position in aquatic ecosystems at the interface of the chemical-physical and biotic components of the food web. This critical link can influence the rest of the aquatic community (Lowe and Pan 1996) and is thus particularly interesting for the study of stream ecoregional zoning. In this study we used presence and absence data of diatoms instead of abundance data in order to take into account the influence of rare species, actually playing an important role in aquatic ecology (Snoeijs et al. 2002, Wunsam et al. 2002, Potapova and Charles 2002). Rare diatom species have large cells and, cell size placing heavy constraints on growth rate, they are usually much less abundant than species having smaller cells. Moreover, species quite rare at the watershed scale can be dominant in a small number of specific sites.

In this study we present a classification of streams based on the diatom communities using the SOM, an unsupervised adaptive learning algorithm. We defined 4 river types based on environmental conditions.

Our results indicate that diatom assemblages are in quite good coherence with geochemical parameters (pH, conductivity and HCO_3^-). Group I gathers sites with quite high conductivity, sites in group II are from volcanic and crystalline substrates with low conductivity, sites in group III are from calcareous zones with high pH and bicarbonate alkalinity. The sites in group IV are characterised by sandy substrates and crystalline rocks with very low pH and conductivity. The groupings also generally agreed with the geographical distribution of sampling sites. It would now be interesting to verify this relation on a larger scale: the French hydrographic network for example has already been classified into hydro-ecoregions by Wasson et al. (2002) (the hydro-ecoregion system is a physical classification of streams based on geology, altitude and climate, the basic principle being to minimize intra-regional variability and maximize inter-regional differences).

Our next study could then consist in using the SOM algorithm for patterning samples on a nationwide scale, observing the possible superimposition of this diatom-based classification with hydro-ecoregions. For each of these regions, expected conditions correspond to typical assemblages in reference sites, and the assemblages can be obtained by observing structuring species profiles on the SOM: the shape of the profile corresponds to the region(s) the species belong to. Species were classified in this way in our pilot dataset (Table 5.5.3), and on a national scale this step would be necessary to implement the Water Framework Directive.

On the other hand, diatom taxonomic diversity being very broad and complex, defining a list of species by hydro-ecoregions should be interesting for non-experts in charge of applying diatom-based indices: researchers always try to find compromises between efficiency and applicability, in order to promote correct use of diatoms for routine water quality assessment. One possible solution could be the application of genus-level identifications (Chessman et al. 1999, Hill et al. 2000), but Round (1991) cautioned that it would be "dangerous to compare streams simply on the genera recorded and using generic identifications in water quality studies is even more dubious". Another solution could consist in matching morphologically similar species (i.e. difficult to distinguish by non-experts), even if they differ from one another in ecological preferences (Lenoir and Coste 1996), but this solution introduces an important bias in stream condition assessment.

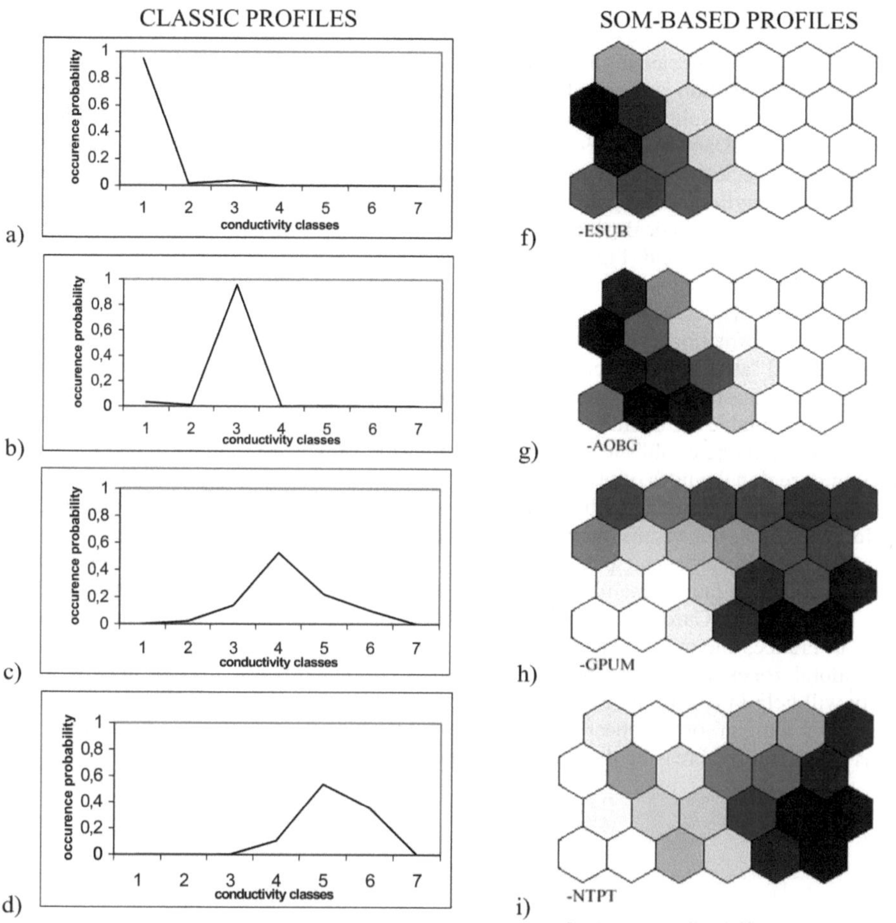

e)

Conductivity classes (μs/cm)						
1	2	3	4	5	6	7
0-100	100-200	200-400	400-600	600-800	800-1000	>1000

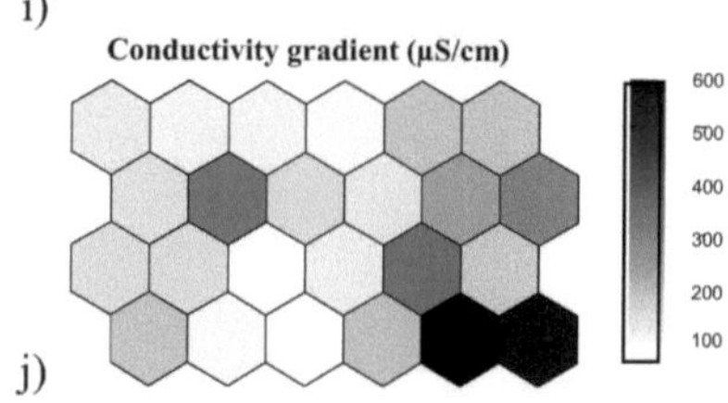

Fig. 5.5.7. Comparison of ecological profiles of 4 species between a classic method (a – e) and SOM method (f – j) based on the conductivity gradient. a) and f) *Eunotia subacicularis*, b) and g) *Achnanthes oblongella*, c) and h) *Gomphonema pumilum*, d) and i) *Navicula tripunctata*, e) and j) conductivity gradients for the classic method and the SOM, respectively. Dark in f) – i) represents high occurrence probability of each species over the range of 0-1, whereas light is low probability. Gray scale in j) indicates mean conductivity value calculated from field data of sampling sites assigned to each SOM unit.

We classified sampling sites according to their environmental gradients, each group corresponding to specific diatom communities. Distribution patterns in the SOM map give a representation of their ecological amplitude as well as their preferences. According to the cluster analysis results, species distributed in the left lower areas of the SOM map are acidobiontic, whereas species in the lower right areas are alkalophilous (Table 5.5.2 and Fig. 5.5.5). The distribution patterns of species reveal how species behave along environmental gradients. Those results are in good correspondence with generally accepted ecological preferences (van Dam et al. 1994, Hoffman 1994, Rott 1998). We compared distribution gradients of species according to the conductivity gradients. Species distribution agreed well with the environmental gradient. These species profiles in the SOM were also comparable with classic ecological profiles presenting a species occurrence gradient according to the environmental gradient (Fig. 5.5.7).

The SOM combined with the SI can be considered as an alternative method to evaluate the importance of species in communities, keeping species-level identification and avoiding the loss of information caused by matching taxa. The idea is to select relevant species which we call "structuring species", i.e. species holding the most relevant ecological information (the highest SI values) needed for ecoregional zoning and which we hope in the next step will be suitable for alteration detection. For each hydrographic basin, a list of structuring species could be established as we did here for the Adour-Garonne basin. With the production of regional guides of taxonomical identification, water managers could learn how to clearly recognise a limited number of relevant species instead of more than one thousand.

In conclusion, our results show that in the parts of the Adour-Garonne basin with little human-related disturbance, spatial patterns of diatom assemblages are well discriminated by environmental factors mainly related to geology (pH, conductivity, bicarbonate alcalinity). As the Adour-Garonne watershed is probably one of the most complex on a regional scale in France, due to its large range of environmental conditions from mountain to plain and littoral zones, and from acidic to calcareous geological substrates, we hope that this study will help in i) learning relationships between diatom communities and their environment on a national or European scale, ii) defining a biological typology of reference stations and then iii) increasing the accuracy of anthropogenic impact assessment.

5.6 Diatom typology of low-impacted conditions at a multi-regional scale: combined results of multivariate analyses and SOM*

Gosselain V†, Campeau S, Gevrey M, Coste M, Ector L, Rimet F, Tison J, Delmas F, Park YS, Lek S, Descy J-P

Introduction

Some principal characteristics qualify algae as excellent indicators of the ecological status of water bodies: (1) algae are the basis of many food webs and (2) are among the first organisms to respond to environmental changes (Lowe and Laliberte 1996, McCormick and Stevenson 1998), (3) they are very rich in species compare to other communities, and auto-ecology of species is well documented (Lowe and Laliberte 1996), and (4) most species are widely distributed among ecosystems and geographic areas (McCormick and Cairns 1994).

Intensive work has been carried out since the last century in order to recognize, describe and classify benthic diatom assemblages. In Europe, the first classification of benthic diatom assemblages, using the methods of phytosociology, has been established for the Ardennes (Belgium) and surrounding regions by Symoens (1957), who defined three water-course types characterized by specific algal "associations". Then, most studies dealing with diatom communities were conducted in the framework of water quality monitoring programs (Descy 1976a, b, c 1979, Coste 1976, 1978). Following those investigations, some regional biotypologies were defined (Descy 1980, Fabri and Leclercq 1984, 1986, Symoens et al. 1988, Ector et al. 1997), and methods based on the global sensitivity of diatoms to pollution were developed (Prygiel and Coste 1996, 1999). In the US, variation of diatom composition along various gradients, at a continental scale, has also been studied (e.g. Pan et al. 1996, 2000, Potapova and Charles 2002). Those studies, however, considered both low-impacted and disturbed conditions at the same time.

Research on diatom communities along river gradients, either at a local, regional or continental scale, has shown the prevalence of (1) downstream gradients (from fast-flowing oligotrophic highland rivers to eutrophic rivers of low-elevation plains; e.g. Symoens et al. 1988, Leland and Porter 2000, Potapova and Charles 2002), (2) chemical factors related to catchment geology, mostly alkalinity and pH (e.g. Descy 1980, Symoens et al. 1988, Fabri and Leclercq 1984, Pan et al. 1996, Potapova and Charles 2002), and (3) latitudinal and altitudinal variation of temperature (Potapova and Charles 2002).

These studies represent more than 30 years of investigations and have contributed to significant progress in the knowledge of benthic diatom auto-ecology and to the development of monitoring methods using these algae. However, there have been few studies dealing with natural or near-natural diatom assemblages and with the natural factors that determine community composition and structure, all carried out at a local or regional scale (Sabater and Roca 1992, Aboal et al. 1996, Pan et al. 2000, Cantonati et al. 2001), and

* The authors are grateful to F. Darchambeau for its help in improving the final version of the manuscript. PAEQANN project has been supported by the EU 5th Framework Programme.

† Correspondence: veronique.gosselain@fundp.ac.be

never at a large multi-regional scale. In a context of increasing environmental change, there is a need to better distinguish community changes due to natural factors from those driven by changes from human activities.

In order to set up a typology of benthic diatom assemblages in rivers from several regions of Western Europe, 467 diatom records from streams with minimal human impact were examined in relation to water chemistry, watershed characteristics, geology, and stream habitat. Ordination techniques were used to determine the major variation in species composition data and to explore relationships between diatom taxa distributions and measured environmental variables. Artificial neural networks (self-organizing maps and multilayer perceptron with back-propagation algorithm) were computed to define and predict diatom assemblages using environmental parameters. Several methods were used: results of multivariate analyses, artificial neural networks, indicator species analysis and expert knowledge were combined to define the benthic diatom typology at a multi-regional scale in Western Europe and are presented in this paper.

Material and methods

Cases and taxa selection

During the EC-funded PAEQANN project (EVK1-CT1999-00026, http://aquaeco.ups-tlse.fr/), a database was built, with the objective of structuring the records of stream benthic diatoms and corresponding environmental data (Gosselain et al. 2004). Part of these records and data were already available from previous studies carried out in several regions of Belgium, France, Luxembourg, and Austria. Another part was obtained by sampling new river sites, mostly located in regions which were not, or incompletely, investigated in the past. Diatom sampling, slide preparation, and counts under the light microscope followed standard procedures (Prygiel et al. 1996, AFNOR 2000, European standard 2001, 2002). Only diatom samples collected on rocks, usually from a lotic reach of the river sites, were considered for analyses. Harmonisation of taxonomy and identification level were carried out at the scale of the entire database; this led, in some cases, to the grouping of some taxa together prior to analysis, when they presented similar ecology (Gosselain et al. 2004). Diatom identifications were based mainly on the Süßwasserflora von Mitteleuropa (Krammer and Lange-Bertalot 1986, 1988, 1991a, 1991b) and nomenclature followed recent updates of diatom taxonomy (e.g. Round et al. 1990) compiled from recent journals like Diatom Research, Diatom Monographs, or taxonomic listings (Kusber and Jahn 2003), as provided in the OMNIDIA software (Lecointe et al. 1999).

On a database comprising nearly 3000 diatom records (Gosselain et al. 2004) from both undisturbed and disturbed environments, potential near-natural condition cases were selected. We defined a case as a single observation in a given river site, containing a diatom record and associated environmental variables. Cases were selected on the basis of the diatom index PSI (Polluo-Sensitivity Index; Coste *in* CEMAGREF 1982). PSI is a water quality index, which is calculated from relative abundances of benthic diatoms collected in a given site. In the PSI system, a large number of stream diatoms have an indicator score, according to their sensitivity to pollution and ecological amplitude. PSI has been tested several times in different countries as Luxembourg (Descy and Ector 1999), Finland (Eloranta 1999), Germany (Coring 1999), Poland (Kawecka et al. 1999), Portugal (Almeida et al. 1999), and is usually considered as a reference method for water quality assessment using diatoms (Descy and Coste 1991). It was calculated using the OMNIDIA software (Lecointe et al. 1993, 1999). Records with a PSI value of 16/20 and higher were retained, as corresponding to a good ecological status. This limit has been chosen, instead of the boundary limit between good and very good status (PSI of 17/20), in order to give more autonomy to our models to detect groups of near-natural conditions. Additionally, we considered only records with at least 380 objects counted (single diatom valves, entire frustules [2 valves], or indifferently single valves and frustules).

Table 5.6.1. Cases used as potential near-natural conditions and their distribution according to country, river basin district, and river basin system flowing to the sea ("Fluvial" basin).

Country	River basin district	"Fluvial" basin	Number of cases
Austria		Danube	33
Belgium		Scheldt	2
		Meuse	58
		Rhine	1
France	*Adour-Garonne*	Adour	9
		Charente	2
		Dordogne	9
		Garonne	17
		Gourgue	1
		Nassey	1
		Seudre	1
	Artois-Picardie	Authie	2
		Canche	1
		Escaut (Scheldt)	1
	Loire-Bretagne	Allier	4
		Loire	58
		Sèvre niortaise	2
	Rhin-Meuse	Rhine	5
	Rhône-Méditérrannée-Corse	Agly	1
		Arc	1
		Argens	1
		Artuby	5
		Cians	5
		Fium'Orbo	1
		Golo	1
		Hérault	3
		Loup	2
		Orb	1
		Rhône	76
		Roya	9
		Siagne	2
		Tech	1
		Tinée	3
		Var	6
		Vésubie	2
	Seine-Normandie	Seine	26
		Touques	1
Luxembourg		Meuse	2
		Rhine	111
TOTAL			**467**

In order to keep only significant taxa for the analysis, a selection was carried out. First, planktonic taxa were not included, as they are not part of the attached diatom assemblages, characteristic of specific spatial conditions. Furthermore, a taxon was considered when its frequency of occurrence reached a minimum of 2.5 %, which corresponded to the presence of the taxon in a minimum of 12 records in the final data matrix, and if its mean abundance was $\geq 0.1\%$. Finally, records for which selected taxa did not represent at least 75 % of the total abundance of the record were removed from the data matrix.

The data matrix finally comprised 123 taxa and 467 cases. Those cases covered four countries and 35 river basin systems ("fluvial" basins; Table 5.6.1).

Table 5.6.2. List of environmental variables collected and transformation applied prior to analyses.

Var.	Description (units)	Transformation	Remarks
Quantitative variables			
ALT	Altitude (m)	$(Alt + 1)^{1/2}$	
SLOPE	Slope (m km^{-1})	Log_{10} (Slope + 1)	
DIST	Distance from source (km)	Log_{10} (Dist +1)	
CAreaS	Catchment surface area up to the site (km^2)	Log_{10} (CAreaS + 1)	
ALK	Alkalinity (meq. l^{-1})	Log_{10} (Alk + 1)	still bi-modal!
pH	Water pH	None	
COND	Conductivity (µS cm^{-1})	$(Cond + 1)^{1/2}$	
TEMP	Water temperature (°C)	None	
DO	Dissolved oxygen (mg l^{-1})	None	
DOC	Dissolved organic carbon (mg l^{-1})	Log_{10} (DOC + 1)	
NO3	Nitrate (mg N-NO_3^- l^{-1})	Log_{10} (NO_3 + 1)	
NO2	Nitrite (mg N-NO_2^- l^{-1})	Log_{10} ((NO_2 + 0.001)·1000)	several detection limits!
NH4	Ammonium (mg N-NH_4^+ l^{-1})	Log_{10} ((NH_4 + 0.001)·1000)	several detection limits!
PO4	Phosphate (mg P-PO_4^{3-} l^{-1})	Log_{10} ((PO_4 + 0.001)·1000)	several detection limits!
Semi-qualitative or qualitative variables			
SP, SA, SW	Season	Coded as 2 dummy variables	SP = spring, SA = autumn, SW = winter
Geol	Geology	Coded as 5 dummy variables	mudstone, limestone, sandstone, granitic, quaternary, mixed and other
Morph	River morphology		1=natural 2=partly channelized 3=totally channelized
WLev	Water level		1 = lowest water levels, 2 = mid levels, 3 = flood levels
Shad	Shading at the sampling site		1=closed, 2=mid, 3=opened
Hydrpwr	Hydropower installation within 10 km upstream the sampling site		Yes or no
RedF	Reduction of flow installation within 10 km upstream the sampling site		Yes or no
Facies	Facies of the sampling point		L = lentic; R = lotic; S = semi-lotic
Vel	Water velocity		1: < 0.2 m s^{-1}; 2: 0.2 – 0.5 m s^{-1}; 3: > 0.5 m s^{-1}

NB: Ranges of the main variables can be seen on the box-plots on Fig. 5.6.17.

A total of 23 environmental variables were initially considered (Table 5.6.2). Quantitative variables were temperature, pH, dissolved oxygen [DO], slope, distance from source [Dist], catchment area [CAreaS], alkalinity [Alk], conductivity [Cond], dissolved organic carbon [DOC], nitrate [NO_3^-], nitrite [NO_2^-], ammonium [NH_4^+], ortho-phosphate [PO_4^{3-}], and altitude [Alt]. Semi-quantitative variables were coded in three categories : water velocity (Vel; < 0.2 m s^{-1}, 0.2 - 0.5 m s^{-1}, > 0.5 m s^{-1}), water level (WLev; lowest water levels, mid levels, flood levels), river morphology (Morph: natural, partly canalised, and totally canalised), shading (Shad; closed, mid-opened, opened), and facies (Facies; lentic, lotic, semi-lotic). Geology was coded as dummy variables (mudstone, limestone, sandstone, granitic, quaternary; mixed and other geology constituted the multiple zero category), as well as season, with spring [SP] defined from March till end of July, and autumn [SA] from August till end of October; double zeros were winter samples. Presence or absence of hydropower plants [Hydrpwr] or of other source of reduction of flow [RedF] within 10 km upstream the sampling site were both coded 1 (presence) or 0 (absence).

Data processing

In order to extract the structure of diatom and environmental data and to define a biotypology, two types of methods were used: multivariate analyses and artificial neural networks.

Principal components analysis (PCA) and detrended correspondence analysis (DCA) were performed to describe environmental and diatom data, respectively. Canonical correspondence analysis (CCA) was used for a first exploration of the relationship between diatoms and environmental variables.

Unsupervised neural network, the Kohonen's self-organizing map algorithm (SOM; Kohonen 1982), was used as an ordination method to define a diatom typology of near-natural conditions. Classification was performed through a go and back process using classification techniques applied on SOM results, relevance of key taxa for the groups –what we will call the diatom types in the following–, and range and contribution of environmental variables to the SOM groups.

Multivariate analyses

Normality of environmental variables was checked using Systat 10 (SPSS 2000); data transformations were applied when needed prior to analyses (Table 5.6.2).

As species abundance values often display a skewed distribution, taxon data were [Ln (10 x +1)] transformed prior to multivariate analyses.

Prior to multivariate analyses, a Pearson correlation matrix was generated on the environmental data (Statistica 5.5; StatSoft 2001), in order to detect highly correlated variables.

Multivariate analyses were carried out using Canoco version 4.0 (ter Braak and Smilauer 1998). Detrended correspondence analysis (DCA) was performed on diatom data to summarize the pattern of diatom variation among data. Principal components analysis (PCA) was performed on environmental data (centred and reduced) in order to summarize major variation patterns within environmental data, and to examine relationships among environmental data and cases. Those analyses allowed checking the quality of our reference situation data set, by showing length and distribution of gradients.

CCA is a multivariate direct gradient analytical technique as it uses taxa, cases and environmental data in a single integrated analysis (ter Braak 1994). CCA was used to (1) identify environmental variables that accounted for significant parts of the variation observed in the diatom data, and (2) quantify the variance explained by each of these significant environmental variables. As a general rule, the selection procedure of variables in the CCA was as follows. Samples were deleted if they had environmental variables with extreme (> 10 x) influence. In order to avoid redundancy, variables with high variance inflation factors (> 10) were removed successively, after running new CCA. Each CCA was tested using the Monte Carlo permutation test with 199 unrestricted permutations ($p \leq 0.05$) and non-significant variables were progressively removed through forward selection. Geology types were not included as such in the analysis but alkalinity at the sampling site was used instead. NH4+ and NO2- were removed as poorly influencing the ordination and presenting several thresholds in the normality curve, corresponding to different detection limits.

Kohonen Self-Organizing Map

The non-supervised artificial neural network Kohonen's self-organizing map (SOM) algorithm (Kohonen 1982, 2001) performs a non-linear projection of the data space on a two-dimensional space. The SOM has been computed using functions implemented in the SOM toolbox (Alhoniemi et al. 1999) for Matlab (The Mathworks 1998) developed by the Laboratory of Information and Computer Science in the Helsinki University of Technology. The functions can be implemented easily by any ecologists and the software library is available at http://www.cis.hut.fi/projects/somtoolbox.

The SOM consists of two layers: input and output layers. Each layer consists of neurons that are computing units in the algorithm. The input layer is composed of the sample units (467 cases in this study), each represented by a vector of the input data (123 taxa, each constituting a node); the output layer forms a rectangular grid. In this case, 108 neurons were organised on an array with 12 rows and 9 columns laid out on a hexagonal lattice. In the output layer, each neuron acts as a virtual unit (VU) and approximates the probability density function of the input data. During the learning process, taxon abundance assemblage is computed for each VU. The aim of the SOM is to illustrate in a two-dimensional space the case distribution by way of the VU distribution. Each neuron is connected to its nearest neighbours on the grid, and stores a set of connection intensities. VUs, which are neighbours on the grid, are expected to represent neighbouring clusters of cases. During the learning process, the algorithm allocates the samples, i.e. the cases, on the two dimensional space by minimizing the error terms between virtual units and samples. To achieve this, the algorithm finds the best matching unit by calculating distances between VU and cases. In this study, as occasional taxa were removed prior to analysis (see above), Bray and Curtis distance (Legendre and Legendre 1998) was used in order to allow a similar contribution of differences between abundant and rare species.

At the end of the training process, each case is set in the corresponding hexagon of the Kohonen map, and the taxon relative abundances are known for each VU. The taxon composition of each VU can be displayed in the component planes of the SOM (taxon distribution maps). As the SOM approximates probability density function of input data through an unsupervised learning process, the weight vectors of the SOM could be considered as approximating probabilities of taxa to be at their maximal dominances in a given site (Park et al. 2003a).

Contribution of environmental variables

In order to determine the contribution of environmental variables to the defined types, predictions were carried out using multi-layer feed-forward neural network with back-propagation algorithm (BPN; Rumelhart et al. 1986a). The BPN is based on a supervised learning, i.e. the network constructs a model based on examples of data with known outputs. It has to build the model up solely from the examples presented, which are together assumed to implicitly contain the information necessary to establish the relation. In this case, 20 environmental variables were used as input data, and output data were the number of groups defined through the SOM. As many models as defined groups were built and their performance were determined using a hold-out crossvalidation procedure. The influence of environmental variables on the different biotypes was then evaluated using a sensitivity analysis by means of a partial derivative algorithm (PaD; Dimopoulos et al. 1999, Gevrey et al. 2003).

Indicator value index

The indicator value index (IndVal; Dufrêne and Legendre 1997) was calculated for each taxon, on the relative abundances, for each group defined by the SOM (PC-ORD 4.01; McCune and Mefford 1999). The indicator value is the product of two values: one measuring the specificity of the taxon, i.e. the mean abundance across the cases pertaining to the cluster versus the sum of the mean abundance within the various clusters, whereas the other one measures the fidelity of the taxon to this cluster. The indicator value of a taxon *j* (IndValj) is commonly understood as the largest value of IndValkj observed over all clusters *k*. A Monte Carlo test was performed to test the significance of the maximum indicator value for a taxon. It is to be remembered that this indicator value index is based only on within-taxon abundance and occurrence comparison, and its value is thus not influenced by the abundance of other taxa.

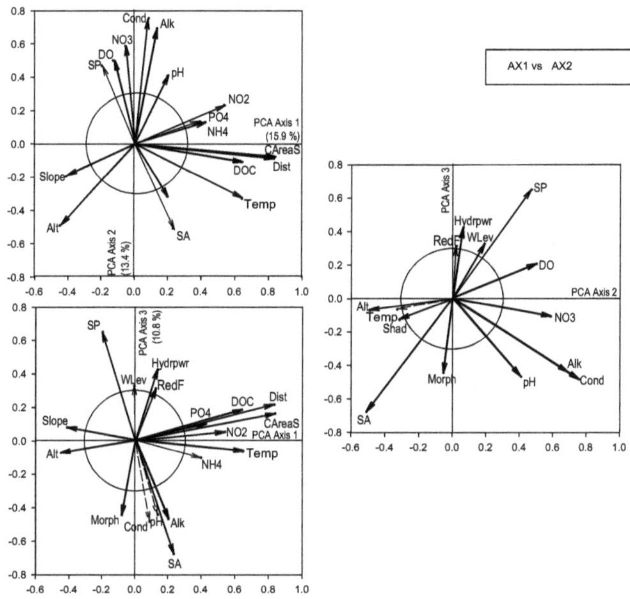

Fig. 5.6.1. Principal component analysis (PCA) on near-natural condition cases, including 467 cases and 23 environmental variables. Environmental variable ordinations: only descriptor contributions above the circle of equilibrium descriptor contribution are shown; bold arrows indicate a major contribution for the descriptor in that plane of the ordination, while dashed arrows indicate minor contribution for the descriptor. See Table 5.6.2 for the meaning of acronyms.

Classification procedure

Once the SOM set, Ward cluster (Legendre and Legendre 1998) and U-matrix (Ultsch 1993) were applied as clustering methods. The U-matrix is a map obtained after the SOM. Hexagons are inserted between original hexagons of the SOM and coloured in a grey scale according to the distance between virtual units.

Key taxa were listed for each group according to (1) the distribution maps returned by the SOM, i.e. taxa presenting their maximum probability to be at their optimum -at their maximum abundance-, exclusively in this group, (2) the indicator value of the taxa (IndVal), and (3) the expert knowledge. Finally, ranges and contribution of environmental variables to diatom typology were determined through canonical correspondence analysis (CCA), environmental variables distribution (box-plots), and sensitivity analysis using partial derivative algorithm, carried out on the defined groups.

Results

Description of the data set

The PCA (Fig. 5.6.1) showed existing gradients among environmental variables. The first axis showed a upstream-downstream gradient with distance from source/catchment area, altitude, slope, as well as temperature and DOC, being the main contributors to this axis. The second axis was related to chemistry depending on geology (alkalinity, pH, conductivity). Altitude also contributed significantly to this axis, which is to be related to numerous alkaline lowland river sites in the database. Those alkaline rivers are mainly located in agricultural areas, thus enriched in nitrate. Axis 3 corresponded to the seasonal gradient. The variables linked to water enrichment were located in the right part of the ordination (axis 1), corresponding to high distance from source. Distance from source was not positioned perfectly at the opposite of altitude, which is to be related to the origin of rivers both at high and lowland elevation.

Some variables as Morph, RedF, Hydrpwr, Facies, WLev and Vel poorly explained the ordination. The cumulative percentage of variance within the environmental data explained by the first three axes of the PCA was 40.1 %.

According to expert knowledge, axis 1 of the DCA (Fig. 5.6.2) clearly showed an acid-alkaline gradient among diatom taxa. Nevertheless, as the cases covered the whole gradient, taxa were found all along this gradient and it was difficult at this stage to identify diatom groups. Groups (biotypes) identified through the SOM (see below) are also shown on the DCA ordinations on Fig 5.6.2.

Table 5.6.3. Summary of the forward selection carried out on the canonical correspondence analysis (CCA) using CANOCO (ter Braak and Smilauer 1998). The table shows the variables in the order of their inclusion in the model, together with the additional variance each variable explains (lambda-A) and the significance of the variable (P-value) together with its test statistics (F-value). See Table 5.6.2 for the meaning of acronyms.

Conditional Effects			
Variable	**Lambda-A**	**P**	**F**
ALK	0.27	0.005	32.02
NO3	0.12	0.005	14.76
pH	0.11	0.005	13.65
DIST	0.09	0.005	12.47
DOC	0.06	0.005	7.56
TEMP	0.04	0.005	5.96
Morph	0.05	0.005	5.75
ALT	0.04	0.005	5.30
SP	0.02	0.005	3.37
PO4	0.02	0.005	3.24
SLOPE	0.03	0.005	3.17
COND	0.01	0.005	2.36
DO	0.02	0.005	2.24
SA	0.01	0.020	1.58
Hydrpwr	0.01	0.045	1.47

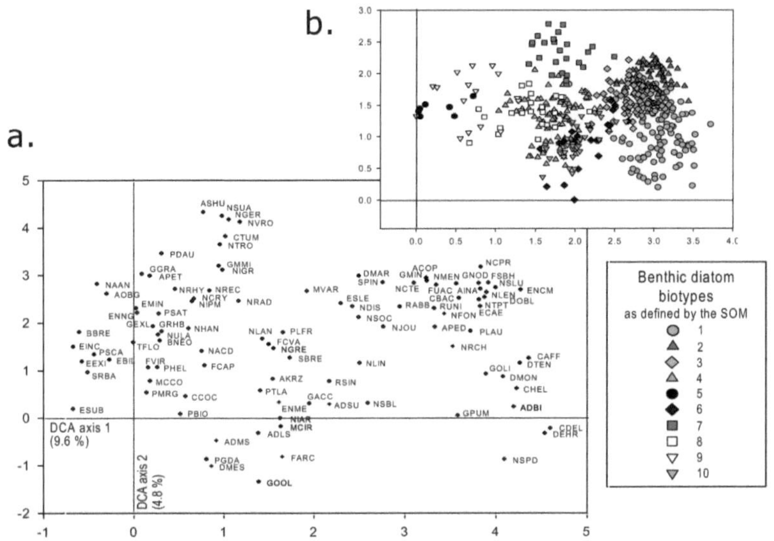

Fig. 5.6.2. Detrended correspondance analysis (DCA) on near-natural condition cases, including 467 cases and 123 diatom taxa: a. Taxon ordination; b. Case ordination.

The CCA (Figs 5.6.3, 5.6.4, Table 5.6.3) showed a major influence of water chemistry on diatom assemblages, mainly driven by alkalinity, conductivity and pH (axis 1). The second axis corresponded to nutrient and organic enrichment, with on one hand NO_3^-, and on the other hand DOC (see graph showing axes 2 and 3; Fig. 5.6.3a). The second axis also showed the gradient of temperature and the seasonal component (see graph showing axes 1 and 2; Fig. 5.6.3a). Distance from source was driven by both axes 2 and 3. Axis 3 clearly showed the gradient Altitude-Slope-Distance from source. Variables indicating nutrient enrichment and increased productivity (DOC, NO_3^- and PO_4^{3-}), as well as temperature, were associated to high distance from source (see graph showing axes 1 and 2; Fig. 5.6.3a). Morphology was in the middle of the three axes.

Cases were more numerous on the alkaline side of the chemistry axis (axis 1; Fig. 5.6.3b), indicating a poorer representation of acid river sites in our data set. Cases were well distributed on the other axes. The first three axes of the CCA accounted for 14.5 % of the cumulative variance of the diatom data. The cumulative percentage of variance of the species-environment relationship explained by the first three axes was 66.8 %.

Cases classification

Clustering of the virtual units of the Kohonen map was carried out first by considering together the Ward cluster and the U-matrix (Fig. 5.6.5b, c). Final decision on the level of clustering was based on taxon distribution maps and ranges of environmental conditions.

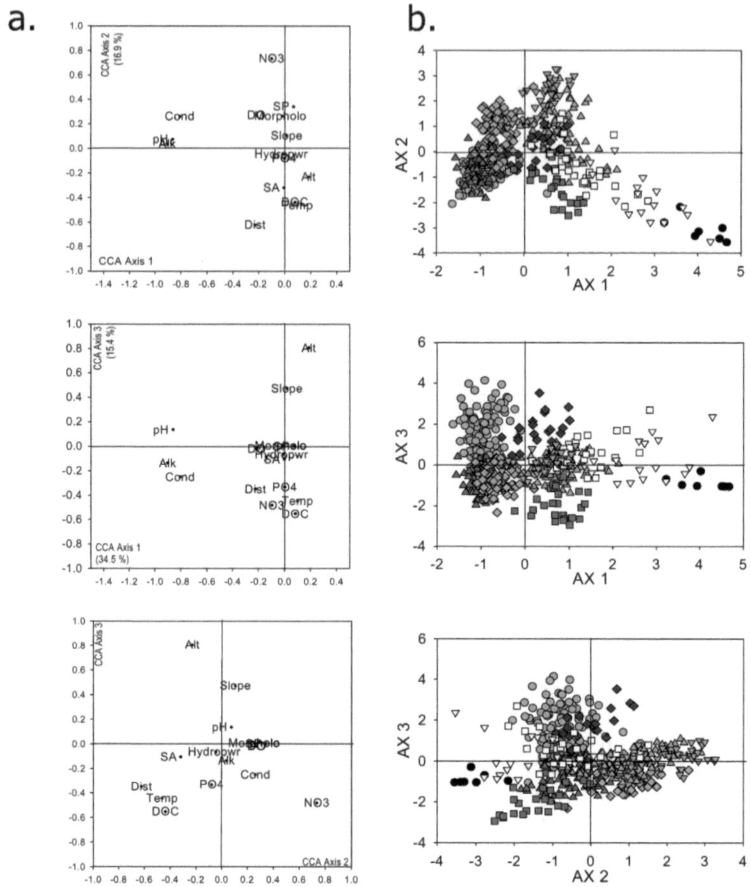

Fig. 5.6.3. Canonical correspondance analysis (CCA) on near-natural condition cases, including 467 cases, 15 environmental variables and 123 diatom taxa: a. Environmental variables ordination; percentage of variance of taxon-environment relationship shown on axes; b. Case ordination. See Table 5.6.2 for legends of variables. See Fig. 5.6.1 for legends of dots (biotypes).

Ten groups could be identified as follows (Fig. 5.6.5). The first clustering level separated cases according to a geochemistry gradient, expressed by alkalinity, conductivity and pH. This corresponded to the first axis of the CCA (Fig. 5.6.3a,b) and can be seen on the box-plots (Fig. 5.6.18): three groups (1, 2 and 3) were characterised by high pH (around 8), and high alkalinity (mean between 2,86 and 4,39 meq l^{-1}) and conductivity (mean between 330 and 560 $\mu S\ cm^{-1}$), while others presented low to medium alkalinity/conductivity (mean between 0.23 and 1.36 meq l^{-1}, and 62 and 263 $\mu S\ cm^{-1}$, respectively). The second clustering level corresponded both to altitudinal and enrichment gradients. It separated group 1 from groups 2 and 3, in the high alkalinity part of the map, and groups 5, 6, 8, 9 from groups 4, 7, and 10 in its low alkalinity part. This corresponded to axis 3 on the CCA (Fig. 5.6.3a,b) and was also shown by the box-plots (Fig. 5.6.18). Then distinction between enrichment type divided further some groups: 2 and 3; 7 from 4 and 10. Again, this is clearly revealed by the CCA (Fig. 5.6.3a,b), as well as the box-plots (Fig. 5.6.18). Distinction between groups 4 and 10 was not so clear from the CCA (Figs 5.6.3a,b) and the environmental variable ranges (Fig. 5.6.18), neither from the taxa distribution maps. It was nevertheless retained, as linkage distance of group 10 on the Ward cluster was high (Fig. 5.6.5c). Expert also recognised taxa and cases corresponding to specific conditions (see discussion). Group 5 could easily be distinguished from group 6 as the latter clearly corresponded to very acid conditions, with a median value lower than 4.5. Groups 9 and 8 then followed on a pH gradient, which is well shown on the CCA (Fig. 5.6.3a,b), better than on the SOM. From this clustering, we could already suspect the types corresponding to very good ecological status to be groups 1, 5, 6, 8 and 9.

c.

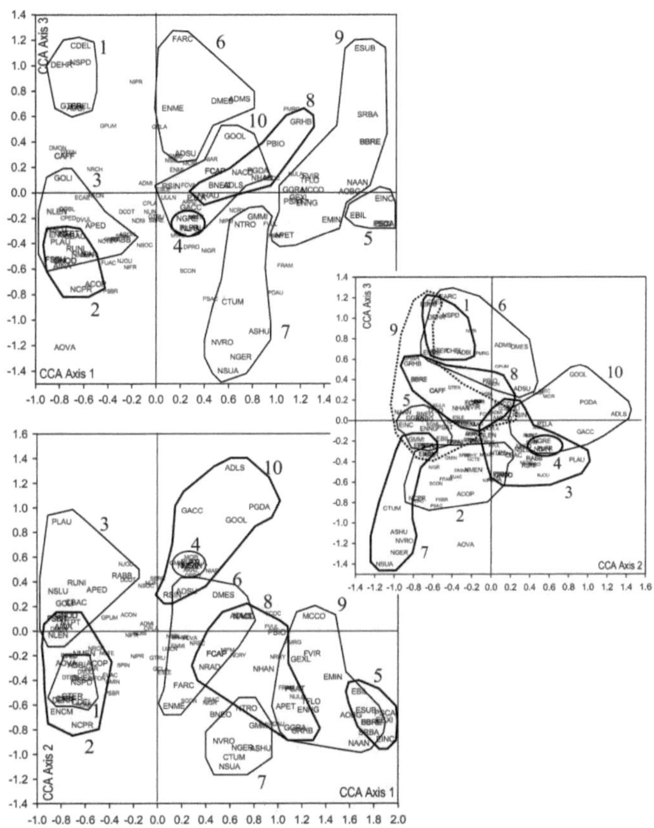

Fig. 5.6.4. Canonical correspondance analysis (CCA) on near-natural condition cases, including 467 cases, 15 environmental variables and 123 diatom taxa: c. taxa ordination, with taxa grouped according to the defined biotypes. Only taxa with acronyms in larger font were retained as key taxa of biotypes. See Figs. 5.6.7- 5.6.17 for the meaning of taxon acronyms.

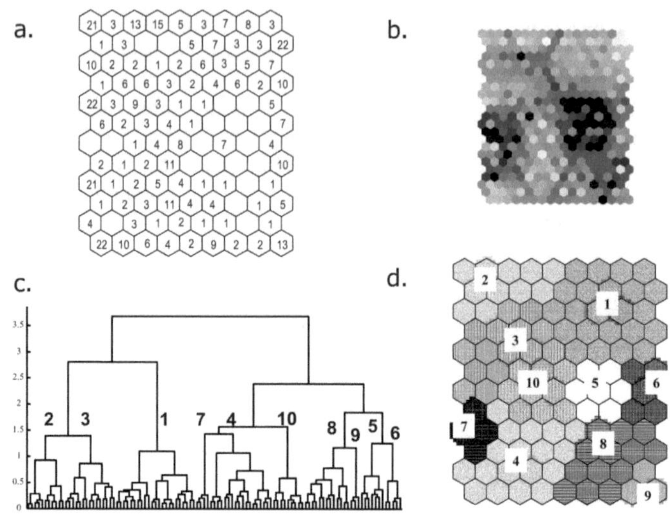

Fig. 5.6.5. Self Organizing Map carried out on 467 near-natural condition cases. a. Position and number of cases on the SOM; b. U-matrix; c. Cluster carried out on the SOM results, using Ward algorithm; d. SOM groups.

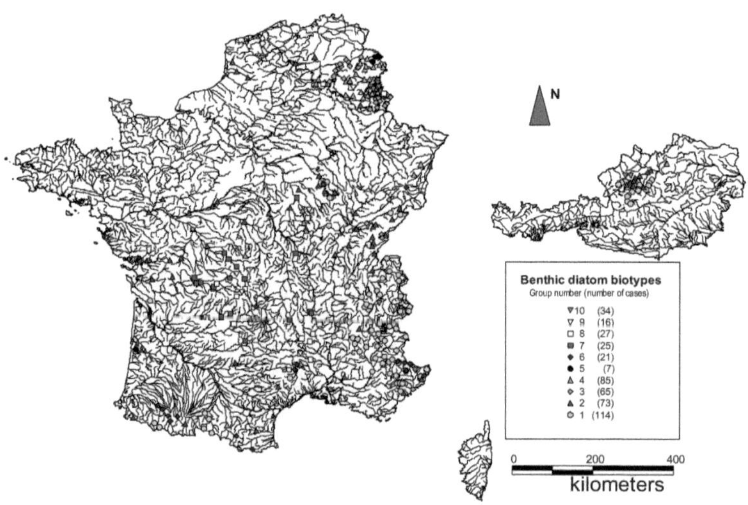

Fig. 5.6.6. Situation of near-natural condition cases and classification according to group defined from the SOM. Black bold lines show political borders.

From the individual taxon distribution maps, taxa that were present and dominant in each 10 groups were identified (Fig. 5.6.8-5.6.17). Key taxa according to distribution maps were listed for each group. Those taxa are also indicator species according to the IndVal index (numbers between brackets in Figures 5.6.7 to 5.6.16 with significant values for $p < 0.05$ in bold). The biotypes, as well as their driven environmental variables, as determined through sensitivity analyses, are discussed in the following section.

The distribution of sites per group in the studied area is shown on Fig 5.6.6.

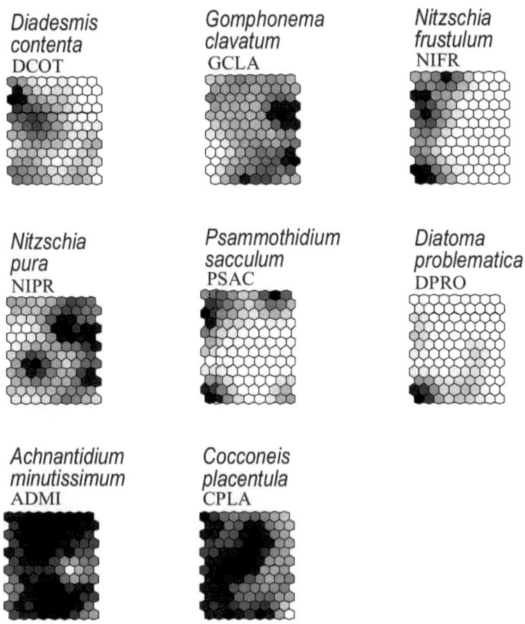

Fig. 5.6.7. Distribution maps of taxa presenting peculiar patterns

Discussion

Until recently, few studies have specifically addressed the definition of natural benthic diatom assemblages in rivers (Sabater and Roca 1992, Aboal et al. 1996, Pan et al. 2000, Cantonati et al. 2001, Rimet et al. 2003, Tison et al. 2004). In this study, the combination of multivariate analysis and ANN-based methods on a large data set of streams with good to high ecological status allowed the determination of several diatom ecological groups – or "biotypes" - in Western Europe. Although all regions of Europe were not represented, the environmental gradients covered a wide range of conditions and stream types. With the help of SOM, up to 10 groups were characterised by key taxa and environmental conditions, as shown in Figures 5.6.8 to 5.6.17.

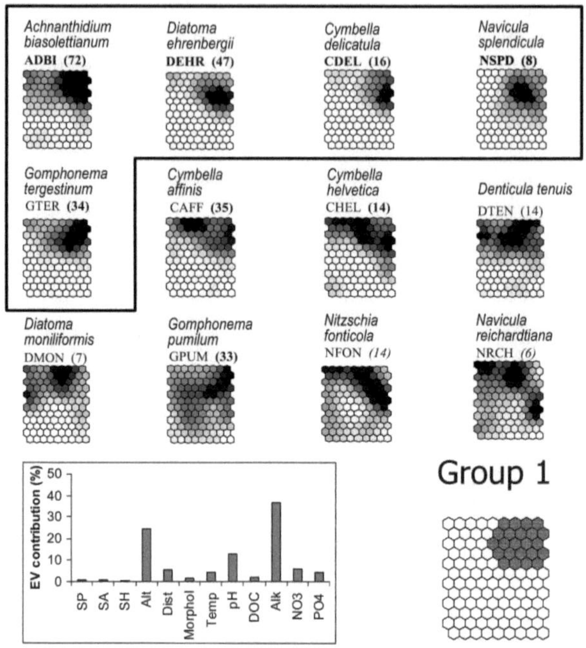

Fig. 5.6.8. Distribution maps of taxa presenting high probability of being at their maximal abundances in group 1. The grey scale used on the map indicates the probabilities of the taxon to occur in a given virtual site (VU); minimum and maximum vary according to the taxon, depending on its relative abundance and occurrence. IndVal value of taxa for this group is noted between brackets: bold indicates significant value ($p < 0.05$), italic indicates value that does not correspond to the maximum (not statistically tested). Framed taxa are retained as key taxa for the group. Sensitivity analysis on environmental variables for the group is presented on the graph (EV contribution %).

Bias and shortcomings about ecological patterns of diatom taxa

Several bias and shortcomings may hamper correct classification of taxa into biotypes and prevent obtaining clear ecological pattern, in particular when considering their distribution maps obtained from the SOM. Among the 123 diatom taxa retained for the analysis, a significant number could not be classified in the biotypes as they presented low occurrence in the database (≤ 33 cases), combined with a low relative abundance (< 1 %). This was the case for *Diadesmis contenta*, *Gomphonema clavatum*, *Nitzschia frustulum*, *Nitzschia pura*, and *Psammothidium sacculum* (Fig. 5.6.7). Others, like *Diatoma problematica*, presented a doubtful taxonomy in addition to occasional occurrence in low numbers. Another reason for taxa to occur in low numbers in most records is their large cell size, as it is the case for several species of *Cymbella, Nitzschia, Navicula, etc.* On the other hand, however, some taxa with an occurrence ≤ 20, i.e. equal or below 4 % of the cases, showed very clear distribution on taxon distribution maps and could be considered as good indicators of ecological conditions (*Brachysira brebissonii*, *B. neoexilis*, *Diploneis marginestriata*, *D. oblongella*, *Eunotia subarcuatoides*, *Navicula angusta*, *N. lenzii*, *N. sublucidula*, *N. splendicula*). In the case of *B. neoxilis* (Fig. 5.6.11), its distribution map and indicator value agreed with each other while CCA did not. Actually, this taxon was present in low relative abundances (< 1 %) and in only a few records that, all but one, belonged to other groups than group 5; this single record, from the Landes region, comprised 12 % of *B. neoxilis*. By contrast, other taxa were typically frequent and abundant in the database, which was reflected by presence in all hexagons of the taxon distribution maps.

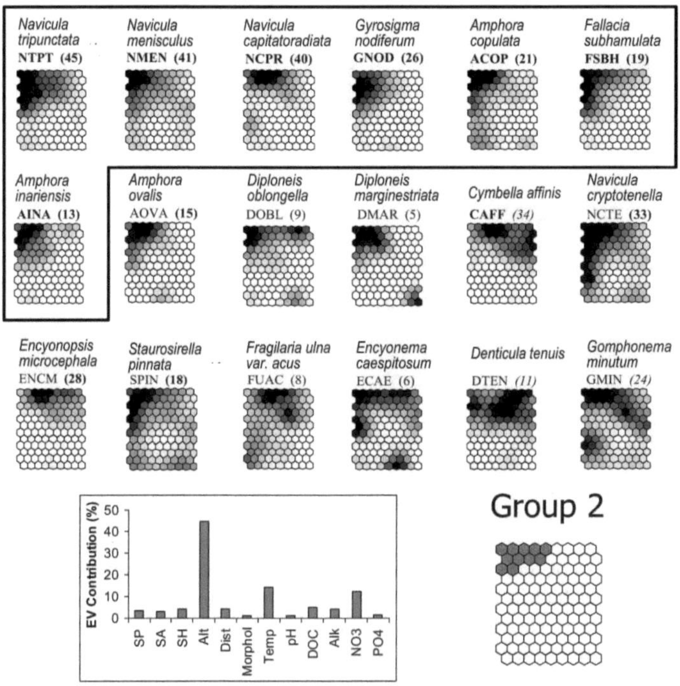

Fig. 5.6.9. As in Fig. 5.6.8, for group 2.

Examples of this ecological pattern were given by *Achnanthidium minutissimum* (Fig. 5.6.7), a small species usually abundant in fast-flowing, well-oxygenated waters, or by preferentially epiphytic taxa, like *Cocconeis placentula* (Fig. 5.6.7), which are attached on hydrophytes and filamentous algae. Abundant epiphyton in a record may also reflect sampling practice, i.e. diatom collection from stones colonised by macroscopic algae.

Unclear pattern occurred in several cases: distribution maps with two distant density spots may result from the existence of ecotypes within the same species – which could be suspected for taxa easy to identify – or, more probably, from misidentification of closely related forms. Several such cases could be detected on our data, as for *Cymbella affinis* (Fig. 5.6.8 and 5.6.9) showing high occurrence in both groups 1 and 2, both on distribution maps and through indicator values (35 and 34, respectively), and *Gomphonema truncatum* (Fig. 5.6.14) in both groups 8 and 2. Both taxa have been now divided in several new species (Krammer 2003, Reichardt 2001), while recorded as one species each in our database. In the case of *Encyonema minutum* (Fig. 5.6.14), this taxon may have been confused with *Encyonema silesiacum* (Fig. 5.6.17) by one or more observers.

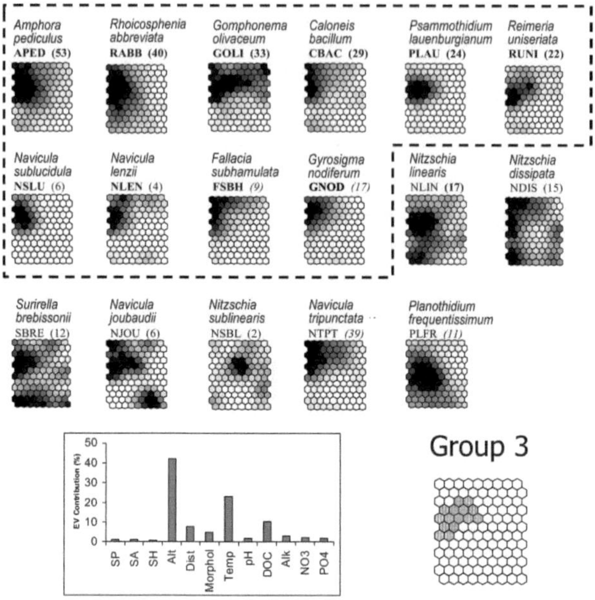

Fig. 5.6.10. As in Fig. 5.6.8, for group 3.

Another related problem occurred with the existence of several varieties within a same species, which were not distinguished in routine diatom counts for water quality monitoring. For instance, the aggregation of varieties in *Gomphonema pumilum* (Fig. 5.6.8) may explain the pattern of its distribution maps and its high indicator values for several groups (for the different varieties of *G. pumilum,* see Reichardt 1997).

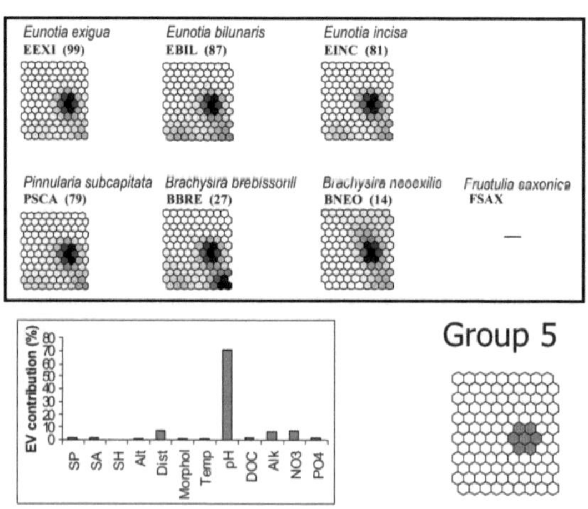

Fig. 5.6.11. As in Fig. 5.6.8, for group 5.

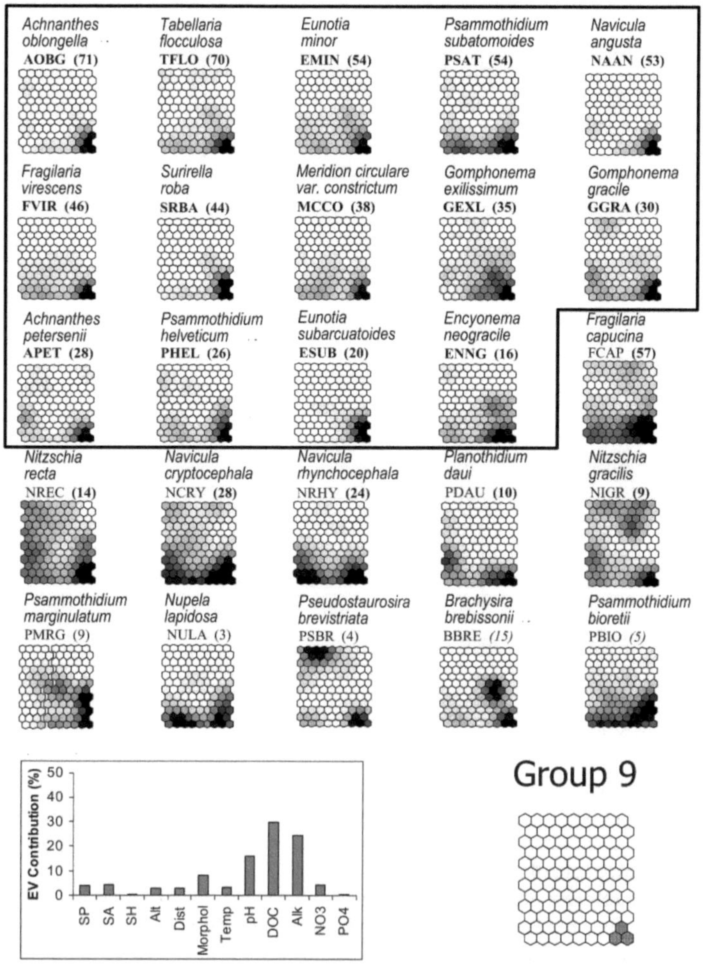

Fig. 5.6.12. As in Fig. 5.6.8, for group 9.

Similarly, *Navicula cryptotenelloides* has been grouped with *Navicula cryptotenella* (Fig. 5.6.9) in our database despite it is probably a different taxon: its distribution map showed a widespread pattern across groups 2, 3 and 7, in agreement with high indicator values in those groups (33, 17 and 26, respectively). Other examples of aggregation – often made for simplifying identification in monitoring studies, or for harmonisation of taxonomy of our database - which may have an influence on the pattern of the taxon distribution maps could be detected in our results.

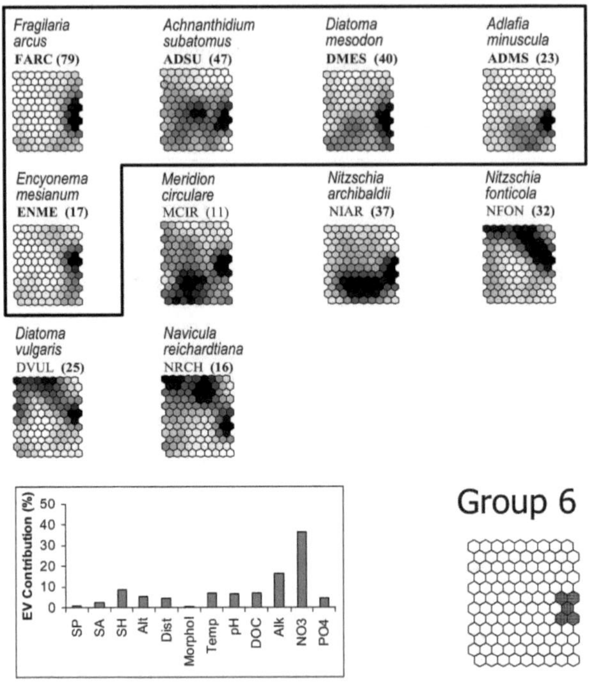

Fig. 5.6.13. As in Fig. 5.6.8, for group 6.

For example, several varieties of the *Fragilaria capucina* (Fig. 5.6.12) complex have been joined under the type variety. In Group 9 (Fig. 5.6.12), the varieties associated to pristine conditions, such as *Fragilaria capucina* var. *rumpens* and var. *gracilis* (*Synedra rumpens* sensu Hustedt) were probably dominant.

Finally, interesting complementarity of techniques can be seen for example for *Amphora ovalis* (Fig. 5.6.9). The distribution map as well as the indicator value assigned this taxon to group 2. Experts nevertheless did not agree considering it as a key taxa of group 2. The CCA actually showed how this taxon probably corresponded to more enriched conditions (Figs. 5.6.3a and 5.6.4, graphs showing axes 2 and 3).

Some of those problems would obviously be solved by further increasing the number of records in the database, but this would also require acquisition of new data with carefully applying sampling standards and with slide examination by skilled diatomists aware of all taxonomical problems. This kind of shortcomings is not exclusive to diatoms: all aquatic organisms but fish are liable to similar errors when working at the species level. Further, a common question is whether diatomists should express counts in terms of relative abundance or of relative biomass, which would give more weight to large diatom species, and less importance to the small taxa, which potentially have greater rates of population increase.

Diatom typology

Despite the biases mentioned above, we were able to define several "biotypes" from the SOM results. Hereafter we discuss the significance of these groups characterised by the key taxa, as presented in Figures 5.6.8 to 5.6.17. Our criteria for interpretation have been based on the ecological affinities of the key taxa, according to literature on diatom ecology (e.g. van Dam et al. 1994, Krammer and Lange-Bertalot 1986, 1988, 1991a,b, Lange-Bertalot 1996, 1999, Fabri and Leclerq 1984), and on sensitivity analyses on environmental variables, as well as on CCA and box-plots showing the range of these variables in the corresponding sites.

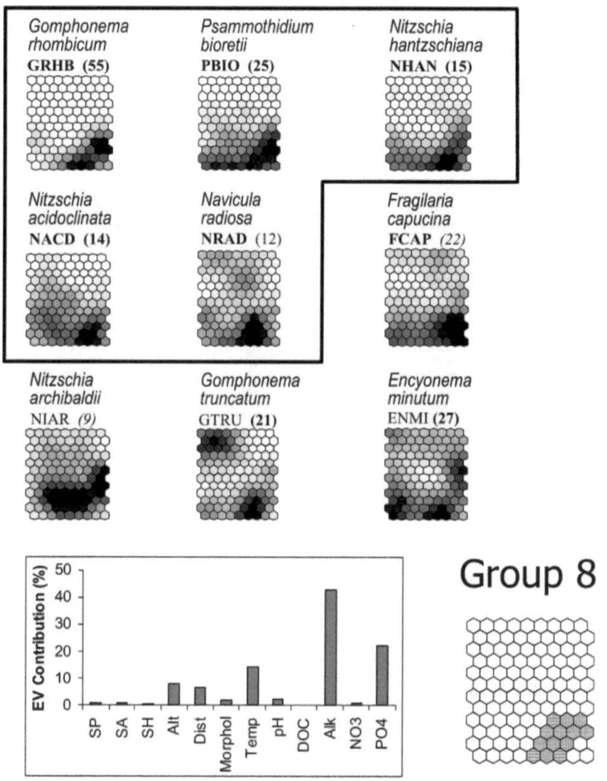

Fig. 5.6.14. As in Fig. 5.6.8, for group 8.

One should be aware, however, that some environmental variables are not unambiguous. For instance, high altitude, which was relevant to group 1 (sensitivity analysis on Fig. 5.6.8), may actually be a proxy for average low water temperature. Indeed, most key taxa defining group 1 are better known from cold water than from streams at high elevation. Moreover, high elevation streams have typically steep slopes and fast-flowing water, which are conditions that can prevail locally in rivers of lower altitude.

Therefore, group 1, despite it comprised some sites exhibiting relatively high nitrate concentration, could be considered as a very good reference group. It was mainly located in French and Austrian Alps (Fig. 5.6.8).

Three groups were characteristic of high alkalinity/conductivity and high pH, with an increasing gradient in alkalinity/conductivity and NO_3^-, and a decreasing gradient in altitude/slope, from group 1 to 3 (Fig. 5.6.18). Some taxa showed distribution map patterns corresponding to a wide ecological amplitude in alkaline conditions, as it is the case for *Denticula tenuis* (Figs. 5.6.8 and 5.6.9).

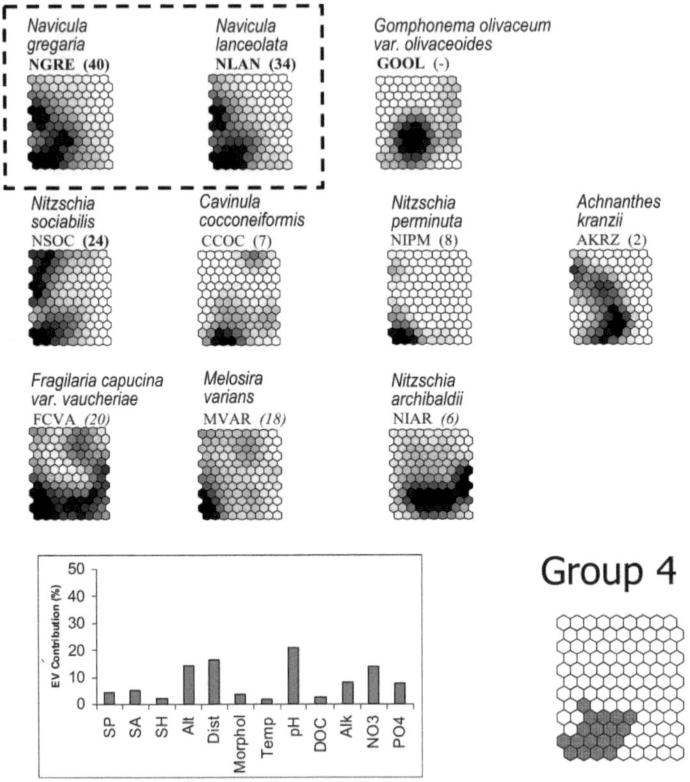

Fig. 5.6.15. As in Fig. 5.6.8, for group 4.

Group 1 presented the lower alkalinity/conductivity of the three groups, with a mean alkalinity around 3 meq l^{-1}. It was typical of high altitude and/or low temperature (Figs. 5.6.8 and 5.6.18) with phosphate and DOC in a low range, indicating good water quality according to most standards. Some taxa presented a narrow ecological amplitude and were quite restricted to this group (framed taxa distribution maps on Fig. 5.6.8).

Groups 2 and 3 were relatively close from each other, as shown on the cluster (Fig. 5.6.4c). There was no clear boundary between the distribution map patterns of several taxa, as for *Navicula tripunctata*, *Gyrosigma nodiferum* and *Fallacia subhamulata* (Figs. 5.6.9, 5.6.10). Those taxa presented a maximum, and significant, indicator value in group 2, and a high value for group 3 (not tested for significance). The intermediate position of those taxa was also shown on the CCA (axes 1 and 2; Fig. 5.6.3a). Some taxa, as *Navicula sublucidula* and *N. lenzii*, presented clear distribution pattern but low and non-significant indicator value. CCA and expert knowledge nevertheless confirmed that those taxa belong to group 3 (Fig. 5.6.10).

Low altitude was a key factor determining both groups. The range of temperature, however, was lower for group 3 (Fig. 5.6.18). The influence of temperature was pointed out by sensitivity analyses, and partial derivatives (not shown) confirmed the distinction between both groups based on their response to temperature. Group 3 was also characterised by higher nitrate and phosphate concentrations (Fig. 5.6.18), and comprised a majority of semi-canalised rivers. Nitrate concentration seemed to be less relevant for diatom typology, as this form of nitrogen was systematically above 1 mg N l^{-1} in streams located in calcareous basins where agriculture has been well developed. It did not seem, however, to affect the composition of diatom assemblages, probably because algae were never N-limited in most streams of our data set. Clearly, despite most taxa of those groups were sensitive to pollution, physical and chemical conditions indicated that they should not be classified as corresponding to pristine conditions. Nevertheless, group 2 may correspond to good ecological status for low altitude calcareous streams and be used as the reference condition of lowland alkaline rivers. By contrast, PO_4^{3-} concentration range indicated slightly degraded conditions associated with group 3.

A good example as far as the effect of altitude was concerned was the Traun river, in Austria, where a transition from group 1 to group 2 was observed as going downstream, occurring at an elevation of about 400 m. Nevertheless, group 1 could also be observed in other areas at lower elevation, with, depending on the season, shifts to group 2 in a same site.

Therefore, we propose to consider, on the basis of our data, group 1 as representing near-natural conditions in alkaline rivers at high elevation or low temperature, and group 2 at lower elevation (< 500 m) or higher temperature alkaline streams. All other groups delimited on the SOM, 4 to 10, are in the low alkalinity (mean < 1 meq l^{-1}) and low conductivity (mean < 150 µS cm^{-1}) range.

Group 5 (Fig. 5.6.11) comprised taxa typical of very acid waters such as *Eunotia exigua, E. bilunaris, E. incisa* and *Pinnularia subcapitata. Frustulia saxonica*, as well as varieties or varieties recently raised to the species level, like *Frustulia crassinervia* (associated with *Frustulia rhomboides* (Ehrenberg) De Toni in the PAEQANN database), were not in the selected species because of very low occurrence and relative abundance in the database; it could nevertheless be added to the list of typical species of group 5. These diatoms are highly sensitive to pollution. As expected, pH was the most important variable for determining the group (Fig. 5.6.11). Members of group 5 have been previously described from unpolluted stations with acid waters in Belgium by Symoens (1957) and Descy (1979). Both authors described an assemblage dominated by *Eunotia exigua, Peronia heribaudii* (*P. fibula* (Breb.ex Kutz.) Ross) and *Brachysira brebissonii* (*Anomoeoneis serians* var. *brachysira*), associated to water with a pH around 4, with very low mineral content.

A second acidophilic biotype was identified by Descy (1979), associated to rivers with a pH between 4 and 6 and mineral content slightly higher than for group 5. This community was composed of *Eunotia incisa, E. bilunaris* (*E. lunaris*), *Eunotia exigua* var. *tenella* (*E. tenella*), *Tabellaria flocculosa* and *Surirella linearis*. Some taxa of this biotype are actually included in group 5, as *Eunotia exigua* var. *tenella*, for instance, which has been aggregated with *Eunotia exigua* in the PAEQANN database. However, *Tabellaria flocculosa* have been associated to group 9 (Fig. 5.6.12), which is a transitional group to circumneutral waters. The environmental conditions associated with group 9 were low alkalinity and conductivity, short distance to source, low nitrate and phosphate, DOC variable (with probably humic compounds), and pH between all other groups and group 5. Both groups 5 and 9 were very close to source (≤ 20 km). Typically, group 5 was found in small streams running from peat bogs with an eodevonian geological substrate in the river Meuse basin in southern Belgium and in the French Landes, while group 9 was found in headwaters both in the Loire and Garonne basins (Fig. 5.6.6).

At higher pH, group 6 was found in conditions covering a rather wide range of alkalinity and conductivity values (mean: 1.25 meq l^{-1} and 160 µS cm^{-1}, respectively), and mostly in altitude (400-1300 m for 75% of the cases). The sensitivity analysis showed that nitrate was a key condition for this group (Fig. 5.6.13), in this case low concentration (mean: < 0.5 mg N l^{-1}; Fig. 5.6.18). Low phosphate and DOC were also typical, showing minor anthropogenic influence. The auto-ecology of the key taxa (Fig. 5.6.13) also indicated that group 6 corresponded to unpolluted, low alkalinity and cold-water streams, mostly found at medium to high elevation. It is interesting to note that Symoens (1957) described a benthic diatom association comprising *Diatoma mesodon* and *Meridion circulare* in rivers of the Ardennes. However, as shown by its distribution map (Fig. 5.6.13), *Meridion* has rather wide ecological amplitude and was not identified as a key taxon in our study.

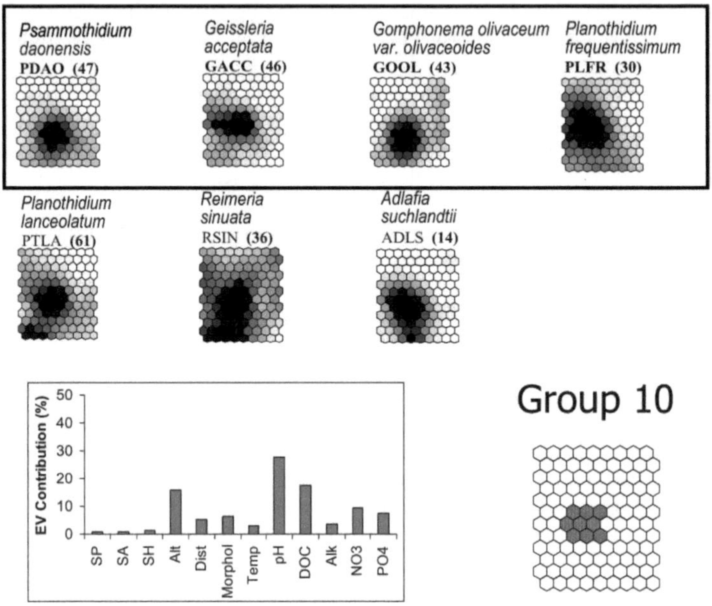

Fig. 5.6.16. As in Fig. 5.6.8, for group 10.

Close to group 6 was group 8 (Fig. 5.6.14), which occurred at lower pH and lower alkalinity (mean: 0.39 meq l^{-1}; Fig. 5.6.18), and higher temperature. Here, the key taxa seemed sensitive to phosphate (Fig. 5.6.13). These two groups appeared to correspond to near-natural conditions for circumneutral waters, which would follow group 9 in a pH / alkalinity gradient. This is well shown on the CCA. Group 9 could however also have been considered as a subgroup of group 8. Group 6 was mainly located in the Pyrenees and Alps mountains in southern France, while group 8 was found in the volcanic Central Massif and Cévennes in France and the Ardennes in Belgium and Luxembourg (Fig. 5.6.6).

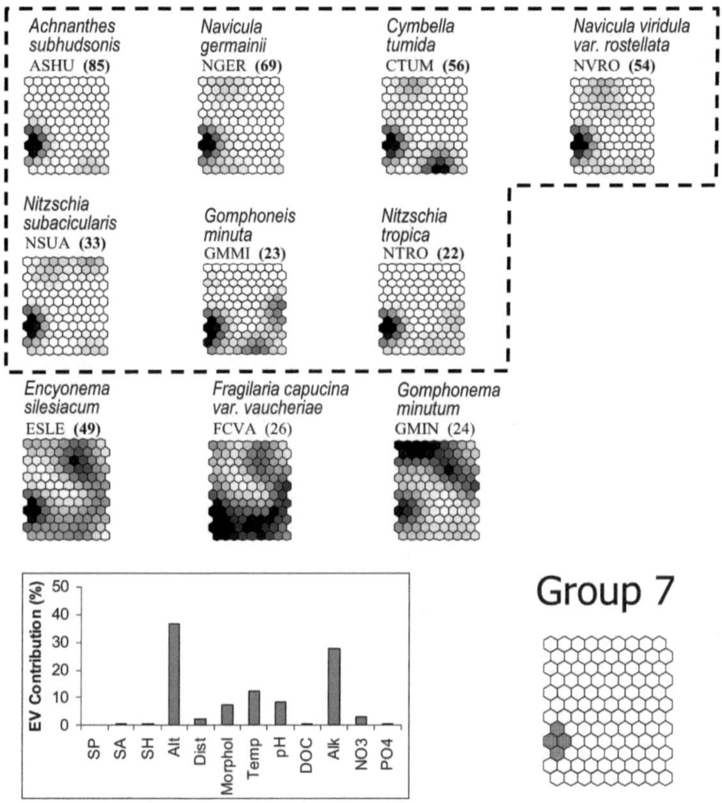

Fig. 5.6.17. As in Fig. 5.6.8, for group 7.

Group 4 (Fig. 5.6.15) was characterised by medium alkalinity (mean: 0.69 meq l^{-1}; Fig. 5.6.18) but with pH in the range 6.8 – 7.75 (mean: 7.35; Fig. 5.6.18), and low altitude (generally < 500 m; Fig. 5.6.18). Most taxa of this group show some degree of tolerance to pollution and are alkaliphilous to neutrophilous. They have rather wide ecological amplitude from springs to brackish rivers. Nevertheless, some taxa of this group are sensitive to pollution (*Cavinula cocconeiformis*, *Nitzschia perminuta*, and *Achnanthes kranzii*) and are usually found in relatively low conductivity waters (Krammer and Lange-Bertalot 1986-1991a, b).

The sensitivity analysis pointed out nitrate as a determining variable (Fig. 5.6.15). It seems clear that this group appeared in significantly degraded streams of which near-natural conditions would be group 8 or 6. As an example, a shift was observed in the Alagnon river (France) close to its source (altitude: 950 m), from group 6 in spring to group 4 in autumn; this shift was associated with an increase in nitrate (0.3 to 0.5 mg N l^{-1}) and a slight increase in alkalinity and pH. On the other hand, the most typical taxa of the group, *Navicula gregaria* and *N. lanceolata*, aretypical of spring conditions, and 66 % of the cases actually corresponded to the spring season. A seasonal effect could thus also have influenced the composition of group 4.

Group 10 (Fig. 5.6.16) was found nearly exclusively in headwater streams of northern Luxembourg (Fig. 5.6.6), thus very close to source and at low altitude (generally < 500 m;

Fig. 5.6.18). The geology of the north of this country is dominated by schists, which can account for the low conductivity of the waters (200 µS cm^{-1} as median value). DOC was lower than in group 4. NO_3^- was relatively high (above 2 mg N l^{-1} and reaching more than 5 mg N l^{-1} in some cases), and PO_4^{3-} was generally low (below 0,03 mg P l^{-1}). Most of the samples associated to this group were collected from semi-lotic facies in headwater streams (stream order from 1 to 3) while all the other samples of the data matrix were from running waters.

All indicator taxa of this group are highly polluosensitive, except *Planothidium frequentissimum*. On the other hand, *Psammothidium daonense*, *Geissleria acceptata* and *Gomphonema olivaceum* var. *minutissimum* are generally considered as representative of good ecological status. *Psammothidium daonense, Gomphonema* var. *minutissimum, Adlafia* cf. *suchlandtii* and *Geissleria acceptata* were mainly found in headwater streams of northern Luxembourg (Rimet et al. 2004). These four taxa have the maximum PSI polluosensitivity values and a medium indicator value. *Psammothidium daonense* is in the red list of Lange-Bertalot (Lange-Bertalot and Steindorf 1996, Lange-Bertalot and Genkal 1999) and is considered as an endangered species.

Finally, group 7 (Fig. 5.6.17) comprised essentially several invasive taxa such as *Achnanthes* cf. *subhudsonis*, *Gomphoneis minuta* and *Nitzschia* cf. *tropica* (Coste and Ector 2000). The ecology of these taxa is poorly known in Europe: for instance, *Gomphoneis minuta* was found in abundance by Coste (Coste et al. 1992) downstream of water treatment plants. This taxon was also found in high numbers in upstream river reaches. The occurrence of this taxon in altitude can be seen both on the CCA (Fig. 5.6.5c) and its distribution map (Fig. 5.6.17). The records associated to this group met our selection criteria probably due to the fact that taxa such as *Achnanthes* cf. *subhudsonis* and *Encyonema silesiacum* are considered as very sensitive to pollution. However, in the PAEQANN database, *Encyonema silesiacum* is certainly a complex of several taxa. *Achnanthes* cf. *subhudsonis* is invading many rivers in France and Spain, especially in the granitic regions of Massif Central and in Galice (Coste and Ector 2000). It has apparently a wide ecological amplitude and its sensitivity index may have been overestimated. Group 7 was in the same range of pH and alkalinity/conductivity as group 4 but presented higher temperature and relatively high DOC (mean above 5 mg l^{-1}) and PO_4^{3-} (90th percentile reaching 0,14 mg P l^{-1}), higher than expected for near-natural conditions or even very good ecological status, thus indicating a significant degree of water quality alteration. The cases associated with this group were mainly located in the Loire basin (Fig. 5.6.6).

Benthic diatom typologies were described previously in Europe, at a local or regional scale (Symoens 1957, Descy 1980, Symoens et al. 1988, Fabri and Leclercq 1984, 1986, Ector et al. 1997), and in the US (Pan et al. 1996, 2000, Potapova and Charles 2002). In all of these studies, the water chemistry, and more particularly pH, alkalinity and conductivity, depending on bedrock and geology, were pointed out as important variables driving diatom community composition. Our analysis covered a wide range of geology, altitude and stream orders and several eco-regions (4 *sensu* Illies 1978); we considered 23 environmental variables and 467 cases, corresponding to 375 stream sites in 255 rivers and 35 fluvial basins. Pan et al. (1999), in a study of Mid Atlantic streams, proposed that factors determining benthic algal species composition could be distinguished at the watershed or regional scale (climate geology, land use), and at the local scale (pH, DOC, nutrients, …).

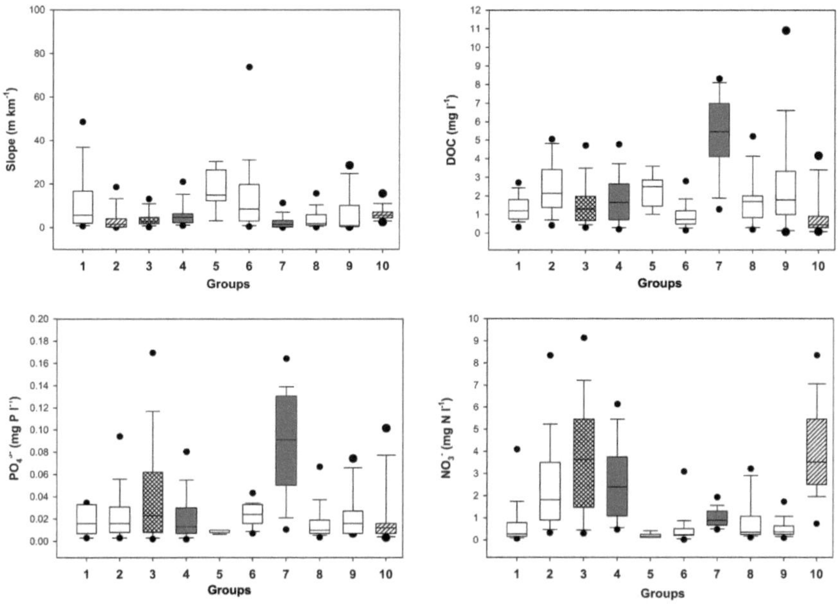

Fig. 5.6.18. Box-plots of environmental conditions, according to SOM groups (median; 10th, 25th, 75th, and 90th percentiles as vertical boxes with error bars; 5th/95th percentiles as dots). Pattern indicates the ecological status of the group as defined in conclusion of this study: ▭ very good; ▭ good; ▭ medium; ▭ lower.

From our study based on river sites of good to very good biological quality, it appears that this distinction between scales is not always straightforward: particularly, alkalinity, conductivity and pH are related to geology and rock type, so that we consider them as factors acting at the regional scale, clearly depending on natural properties of the watershed which have not been altered by human activities. Therefore, these factors (particularly alkalinity, which is the least affected by land use) could be used as predictors for assemblages of benthic diatoms in near-natural conditions. We have shown (PCA on Fig 5.6.1), however, that geology-related factors and land use are not independent: indeed, agriculture is better developed on calcareous soils at medium and low elevation, so that the nutrient loading (particularly for N) of these lowland streams is always larger than that of streams located on siliceous substrate. Again, on the North-American continent, a similar approach was used by Hill et al. (2000) who tested the effects of 47 variables on the attributes of diatom assemblages from 199 streams in the Mid-Appalachian area. In their study, a canonical correlation analysis showed that environmental determinants were basically the same as the ones we identified in our multi-regional study in Western Europe: pH, specific conductance, acid neutralizing capacity (= alkalinity). Percentage of agriculture in the catchment and total N were correlated with the proportion of eutraphentic species –terminology according to Rott et al. (1999)-, and the proportion of pollution-tolerant taxa correlated with total P. Potapova and Charles (2003), analysing a data set of more than 3000 benthic diatom samples from more than 1000 river sites throughout the USA, also pointed out the major influence of conductivity and major ions such as HCO_3^-+CO_3^{2-} (= alkalinity) in the variation of diatom assemblage composition. Those conclusions remarkably confirm the importance of factors which were pointed out systematically in our sensitivity analyses, and prove the consistency of an approach based on diatom auto-ecology across continents. As a corollary, it seems that the eco-regional approach to the determination of diatoms assemblages is less relevant, except if there is a close correspondence between eco-regions and geology and, to some extent, elevation. As pointed out by Pan et al. (2000), this allows the establishment of a precise reference system without varying metrics among regions.

Conclusion

Our analysis of benthic diatom records of the PAEQANN database, selected according to a high value of the PSI diatom index, enabled us to define several "groups" of indicator taxa, which could be organised along a gradient of alkalinity and pH. Several of these groups correspond to natural or near-natural conditions, as verified by the values of nutrients and DOC and by the auto-ecology of the taxa themselves. Moreover, in a given range of water chemistry, temperature and altitude can be discriminating factors generating distinct assemblages characterised by key taxa with a well-defined optimum. This suggests that a diatom reference system could be defined at a multi-regional or even at a continental level, rather than at the scale of eco-regions.

In this study, ANN-based techniques were combined with multivariate analyses to determine a biotypology of near-natural conditions at a multi-regional scale in Western Europe. In particular, SOM allowed giving information and an easy-to-use representation of taxa auto-ecology. Sensitivity analyses allowed identifying the variables contributing the most of to the determination of biotypes.

In order to be used for water management, this biotypology should be validated for actual diatom records and level of confidence should be defined. Then, ecological quality ratios (EQR; ratios between observed and reference biological values) could be calculated in order to estimate the distance between actual biological status and the reference, for both undisturbed and disturbed sites. This will ultimately allow providing a sound scientific basis for water restoration as commended in the European Water Framework Directive.

5.7 Prediction with artificial neural networks of diatom assemblages in headwater streams of Luxembourg[*]

Rimet F[†], Ector L, Hoffmann L, Gevrey M, Giraudel JL, Park YS, Lek S

Introduction

In rivers benthic diatom assemblages are controlled by many environmental variables, such as climate, geology or nutrients (Stevenson 1997, Snyder et al. 2002). It is therefore often difficult to understand the relationships between the different diatom taxa and the environmental parameters. Classically, multivariate analyses, such as principal component analysis, are used to project the abundance of diatom data on a two dimensional graph. The first two axes are in general used, but they represent only a part of the explained variability. Cluster analysis based on Bray-Curtis distances (Bray and Curtis 1957) or Twinspan ordinations (Hill 1979b) are also very often used techniques to sort samples of diatom assemblages into groups of homogeneous composition.

The Self Organizing Maps (SOM; Kohonen 1982) are able to cluster the sites and to summarise a database on a two dimensional graph. The projection is made in a non-linear way with an artificial neural network onto a map composed of hexagons (Fig. 5.7.1), the Kohonen map. In contrast to multivariate analysis, the Kohonen map represents all the explained variability of the dataset. The projection is made thanks to an artificial neural network composed of two layers: the first one (input layer) is connected to the samples, the second one (output layer) to the hexagons of the map. The projection is made respecting the similarities between the samples. Samples with homogeneous assemblages are placed in a same hexagon or in neighbouring hexagons and samples with very different assemblages are placed in distant hexagons. Artificial Neural Networks (ANN), are models that function somewhat like a human brain as they have a learning ability. They establish links between information and solutions. Only a few studies have already used this kind of techniques and have demonstrated their clustering properties for biological communities (Chon et al. 1996, Foody 1999, Giraudel and Lek 2001).

Artificial Neural Networks using backpropagation algorithms (ANN-BP) are models that can be used for predicting one or several variables using other, predictive, variables. This kind of model is composed of processing elements called "neurons" which are arranged in three layers (Fig. 5.7.2). The first layer, the "input layer", is corresponding to the input variables, as, for instance, environmental parameters. The last layer, the "output layer", is composed of neurons providing a prediction for the expected variables. The layer between both, the "hidden layer", is composed of neurons composed of a non-linear function. These neurons receive information from the input neurons, transform it with the non-linear function (log sigmoid for instance), and send this new information to the output neurons.

[*] This research was financed by the EU project PAEQANN (N° EVK1-CT1999-00026). We thank Dr. H.M. Cauchie for his useful comments on the manuscript.

[†] Correspondence: rimet@crpgl.lu

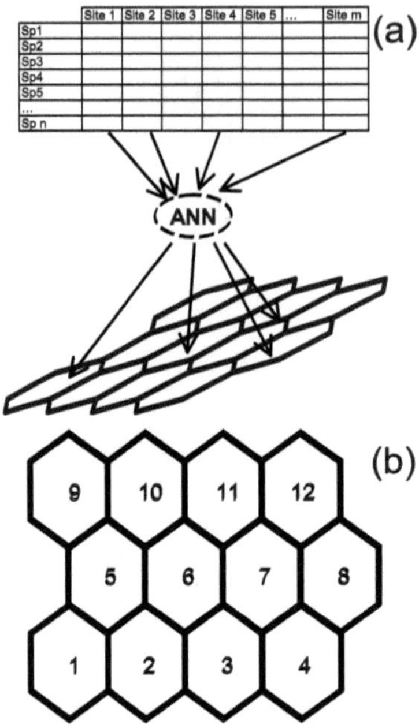

Fig. 5.7.1. The database composed of sites characterised by taxa abundances is projected in a non-linear way by means of ANN on the Kohonen map (a). A SOM with 12 hexagons was used to define 12 groups of samples with homogeneous diatom assemblages (b).

Connections between neurons are weighted in order to modulate the importance of the information fluxes. Output neurons collect the information and give a prediction. The network is trained with a part of the database. Predictions are given and compared with the reality. The aim of the training is to find the best weights for the connections in order to get a minimal error between predictions and observations. The other part of the database, which is not used for the learning phase, is used to test the network with fixed weights (validation procedure).

Several studies have already shown the ability of ANN-BP to predict the structure of macro-invertebrate or fish communities in rivers (Brosse and Lek 2000a, Chon et al. 2000c among others), density (Chon et al. 2000c) or species richness (Park et al. 2003a) of macro-invertebrate in rivers, Pacific sardine biomass (Cisneros-Mata et al. 2000) or fish species abundance in the Seine basin in France using stream order, slope, width, water quality, habitat quality and the ecoregion as input parameters (Boët and Fuhs 2000). Several studies showed that ANN-BP gives better results than other techniques such as multiple regressions (Lek et al. 1995, Brosse and Lek 2000b, Scardi 2000). This can be partly explained by the ability of artificial neural networks "to take into account the non-linear relationships between dependent variables and each independent variable" (Lek et al. 1995). Few applications exist for algal communities, besides prediction of blue-green algae blooms in rivers (Recknagel 1997) or in lakes (French and Recknagel 1994). Scardi (2000) used ANN to predict phytoplankton productivity in the Gulf of Napoli from temperature, irradiance, geographical coordinates and the day of the year as predictive variables.

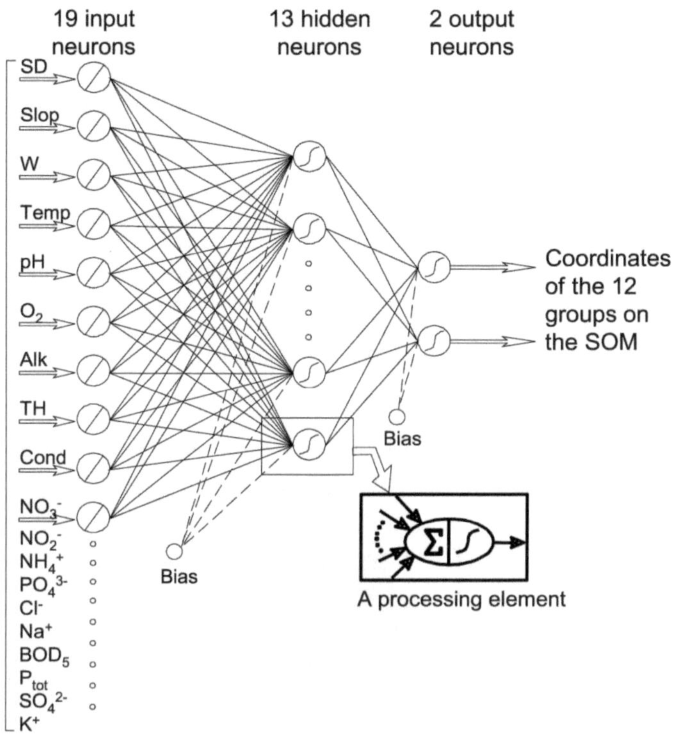

Fig. 5.7.2. Architecture of the three-layers ANN-BP used. 19 input neurons corresponding to 19 physico-chemical parameters, 13 hidden neurons and 2 output neurons to predict the coordinates of the 12 groups on the SOM, are used. Abbreviations meaning: SD: source distance, Slop: slope, W: river width, Temp: temperature, pH, O2: dissolved oxygen, Alk: alkalinity, TH: total hardness, Cond: conductivity, NO_3^-, NO_2^-, NH_4^+, PO_4^{3-}, Cl^-, Na^+, BOD_5: biological organic demand during 5 days, Ptot: total phosphorus, SO_4^{2-}, K^+.

The structure of benthic diatom assemblages is potentially determined by many environmental factors acting at different spatial scales (Stevenson 1997, Snyder et al. 2002). Several studies established correlations between diatoms and physico-chemical variables (Stoermer and Smol 1999), and, among other applications, these observations led to the development of biotic indices to assess the water quality. These biotic indices are generally calculated with species sensitivity to pollution, their indicator value and their relative abundance. They are now routinely used in Europe (Prygiel et al. 1999) and some of them are standardized (Kelly et al. 1998, AFNOR 2000).

Recently, the European Parliament (2000) required to develop new assessment tools for inland waters, which must comply to the Water Framework Directive (WFD). This directive aims to restore surface and underground waters to a good ecological status, which implies to define typologies and reference conditions for several water bodies. The WFD specifies that the ecoregional setting has to be taken into account. Ecoregions (Omernik 1987, 1995, Wasson *in* Cemagref 2001) classify landscapes on the basis of climates, geology, soil topography and potential vegetation. River typology can be defined for instance on the basis of distance to the source, river width, slope and maximal annual temperature

(Verneaux and Leynaud 1974). The reference conditions should be defined in each river type in each ecoregion, and this can be done using modelling techniques.

The first aim of this study was to explore benthic diatom assemblages in headwater streams of Luxembourg in order to propose type assemblages for unpolluted rivers of this typological level. The interest to work on a spatially reduced region and on homogeneous typological levels lies in the homogeneity of diatom and environmental data and in the precision of the classification developed. *A priori*, such an approach should provide more detail than studies carried out at a wider scale using data coming from very different regions and river types. Different assemblages were defined using SOM (Kohonen 1982) and could be related to distinct regions of Luxembourg.

The second objective was to test whether diatom assemblages of the headwater streams of Luxembourg can be accurately predicted by using environmental parameters, in order to assess the use predictive models for defining the reference conditions as required by the WFD. ANN-BP was used to predict the groups of samples formerly defined by SOM.

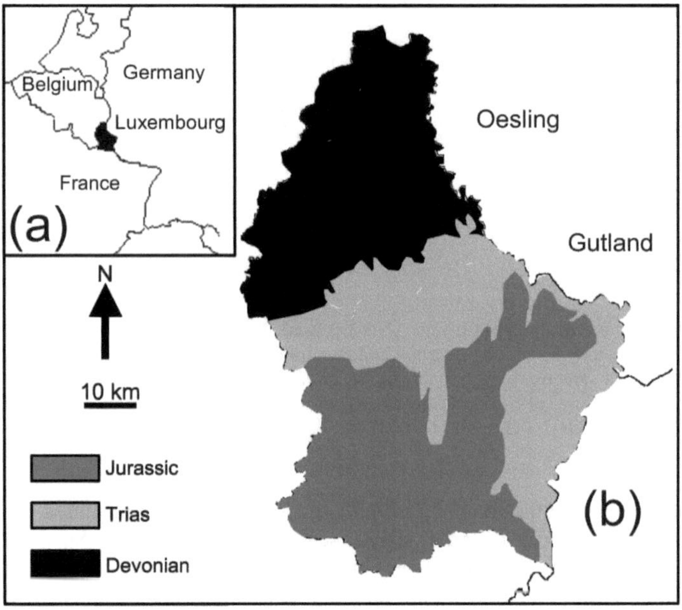

Fig. 5.7.3. Location of the Grand-Duchy of Luxembourg (a) and presentation of its geology (b).

Materials and methods

Study area

The Grand-Duchy of Luxembourg is situated between France, Germany and Belgium (Fig. 5.7.3a). The country is separated into two regions (Fig. 5.7.3b). In the northern part, the "Oesling" has a geology composed of schists and slates (Devonian), the landscape is cut by deep valleys (mean altitude around 500 m). The south part, the "Gutland", is characterised by sandstones and limestones (Triassic and Jurassic); it is flatter than the northern part and also drier (average rainfall is 782 mm for the south and up to 1050 mm for the north).

Until now, apart a study by Weckering (1953) very few data exist about the diatom flora of Luxembourg. Diatoms were also used several times to assess water quality in rivers in Luxembourg (Leclercq and Vandevenne 1987, Descy and Coste 1990, Back et al. 1994, Descy and Ector 1999).

Physical and chemical data

Water temperature, dissolved oxygen, conductivity and pH were measured in the field at the same moment as the diatom samplings. Water samples of each site were also collected and analysed in the laboratory following standard procedures (APHA 1995) for NO3-, NO2-, NH4+, total phosphorus, PO43-, Na+, Cl-, K+, SO42-, biological oxygen demand, total hardness and alkalinity. For each site sampled, altitude, slope, stream order and distance from the source were determined on 1/20000 topographical maps.

Diatom sampling, preparation, identification and counting

For this study the headwater streams (stream orders 1 to 3 according to Strahler 1963 and Leopold et al. 1964) of Luxembourg were selected. A total of 145 sites were chosen in the 17 sub-basins composing the entire territory. In each sub-basin, streams with minimal anthropic disturbances were selected. The samplings were carried out from 1994 to 1997. Each site was sampled once in late summer and autumn (from the end of August to October) and again in spring (from the end of March to June).

In order to obtain comparable results, samplings were realised according to the French standard (AFNOR 2000). Benthic diatoms were collected from fast-flowing parts of the sites on several stones (generally five). The upper surface of the stone was sampled with a toothbrush. Samples were fixed in formol 4% (Prygiel and Coste 2000). Diatom valves were cleaned with hydrogen peroxide (40%) to eliminate organic matter and with hydrochloric acid (37%) to dissolve calcium carbonates. Clean diatom frustules were mounted in a synthetic resin (Naphrax) and counted with a light microscope (up to 400 valves in each sample, Iserentant et al. 1999) with a 1000× magnification. The entire slide was scanned with a 400× magnification to check for rare taxa. Krammer and Lange-Bertalot (1986, 1988, 1991a,b) as well as Lange-Bertalot (1993) were used for diatom identifications. In addition Reichardt (1999) was used for the identification of Gomphonema C.G. Ehrenberg, Lange-Bertalot and Krammer (1989) for Achnanthes J.B.M. Bory de Saint Vincent, and Lange-Bertalot (2001) for Naviculaceae.

Data analysis

The first step of the analysis consisted in defining groups of samples with similar diatom assemblages. The physical and chemical characteristics and the geographical location in the country of these groups were identified. The second step was to predict these groups using physical chemical parameters as predictive variables; the accuracy of the prediction was assessed with the correlation coefficients of the ANN-BP.

Firstly, SOM (Kohonen 1982) were used to explore and cluster diatom assemblages, using taxa abundance as input and the proximities of the samples on the SOM as output. A total of 289 samples were collected from 1994 to 1997 and 411 taxa were recorded. Before launching the training phase of the SOM, a selection of 71 taxa was made based on the highest values of the ratio: (occurrence of the species in the samples)/(sum of the species abundance in all the samples). This selection retains the most abundant taxa and also the rare but locally abundant taxa, which can have an important ecological meaning. The other 311 taxa were removed from the dataset. Taxa percentages were calculated for each sample on the basis of these 71 taxa and the SOM was trained with the percentages of these 71 taxa.

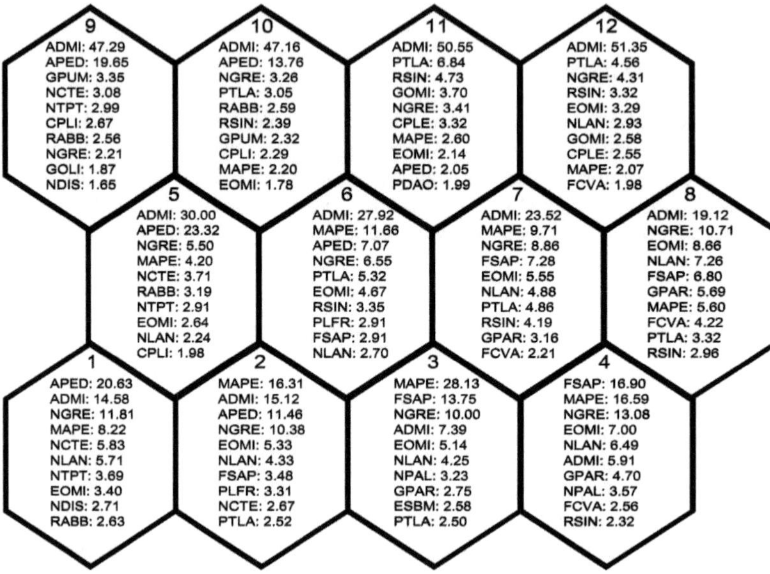

Fig. 5.7.4. Floristic composition of the 12 groups defined with the SOM. The percentages of the 10 most abundant taxa for each group are given. These assemblages represent the average of the samples assemblages composing each group of the SOM. Abbreviations meaning: ADMI: *Achnanthidium minutissimum* (Kützing) Czarnecki, APED: *Amphora pediculus* (Kützing) Grunow, CPLE: *Cocconeis placentula* Ehrenberg var. *euglypta* (Ehrenberg) Grunow, CPLI: *C. placentula* Ehrenberg var. *lineata* (Ehrenberg) Van Heurck, EOMI: *Eolimna minima* (Grunow) Lange-Bertalot, ESBM: *E. subminuscula* (Manguin) Moser, Lange-Bertalot and Metzeltin, FCVA: *Fragilaria capucina* Desmazières var. *vaucheriae* (Kützing) Lange-Bertalot, FSAP: *Fistulifera saprophila* (Lange-Bertalot and Bonik) Lange-Bertalot, GOLI: *Gomphonema olivaceum* (Hornemann) Brébisson, GOMI: *G. olivaceum* var. *minutissimum* Hustedt, GPAR: *G. parvulum* Kützing, GPUM: *G. pumilum* (Grunow) Reichardt and Lange-Bertalot, MAPE: *Mayamaea atomus* (Kützing) Grunow var. *permitis* (Hustedt) Lange-Bertalot, NCTE: *Navicula cryptotenella* Lange-Bertalot, NDIS: *Nitzschia dissipata* (Kützing) Grunow, NGRE: *Navicula gregaria* Donkin, NLAN: *N. lanceolata* (Agardh) Ehrenberg, NPAL: *Nitzschia palea* (Kützing) W. Smith, NTPT: *Navicula tripunctata* (O.F. Müller) Bory, PDAO: *Psammothidium daonense* (Lange-Bertalot) Lange-Bertalot, PLFR: *Planothidium frequentissimum* (Lange-Bertalot) Round and Bukhtiyarova, PTLA: *P. lanceolatum* (Brébisson) Round and Bukhtiyarova, RABB: *Rhoicosphenia abbreviata* (C. Agardh) Lange-Bertalot, RSIN: *Reimeria sinuata* (Gregory) Kociolek and Stoermer.

The assemblage composition and the values of the environmental variables of each group were characterised using descriptive statistics. Secondly, several ANN-BP were computed to predict diatom assemblages with environmental variables. In this study back propagation algorithms were used with environmental variables as input to predict the groups defined with the SOM. Nineteen input variables were used (temperature, slope, width, source distance, pH, dissolved oxygen, alkalinity, total hardness, conductivity, NO_3^-,

NO_2^-, NH_4^+, PO_4^{3-}, total phosphorus, Cl^-, Na^+, SO_4^{2-}, K^+, biological oxygen demand) to predict the coordinates of the 12 samples groups on the SOM (Fig. 5.7.2).

Results

Definition of diatom assemblages with SOM

After trying several sizes of SOM, a 12 hexagons map was retained because it showed the most easily explainable results (Fig. 5.7.1b). The SOM gives 12 groups of samples with homogeneous diatom assemblages. The diatom assemblages of each group of the SOM (floristic list and abundance of each taxon) were defined by calculating the average assemblage of the samples composing each hexagon and are given in Fig. 5.7.4 (the 10 most abundant taxa of each assemblage are presented in Fig. 5.7.4).

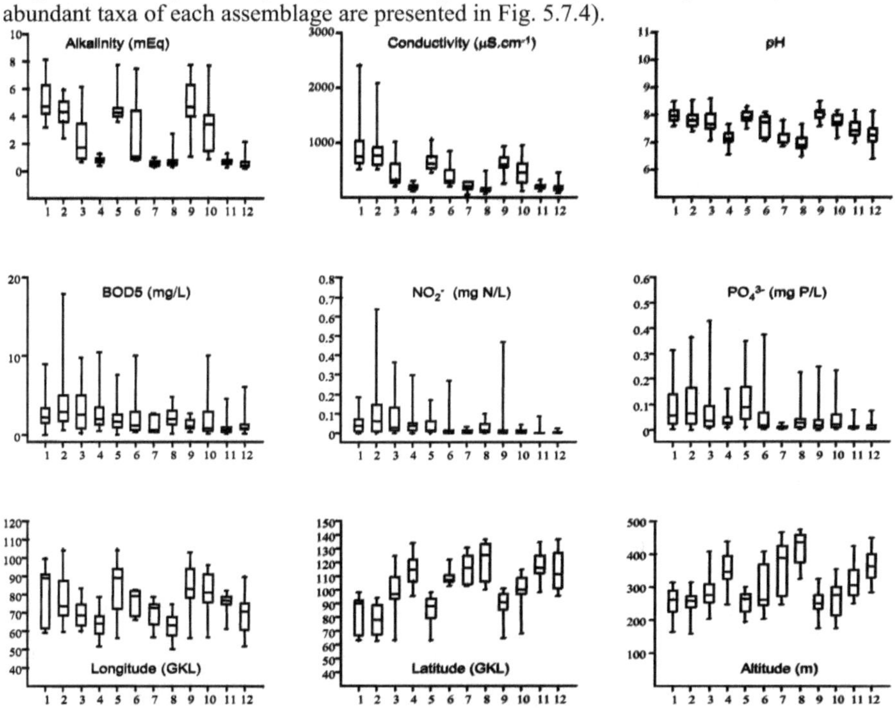

Fig. 5.7.5. Box-whisker plot of physical and chemical characteristics of the 12 groups of samples defined with the SOM. The box is corresponding to 50% of the values, the horizontal bar in the box to the median and the vertical bars to the minimum/maximum values.

The assemblages in Fig. 5.7.4 show clear differences in diatom ecology. Alkaliphilic taxa are abundant in the left part of the SOM: *Amphora pediculus, Cocconeis placentula* var. *lineata, Rhoicosphenia abbreviata, Mayamaea atomus* var. *permitis* and alkalibiontic species as *Gomphonema olivaceum*. On the right part of the SOM are found neutrophilic taxa: *Achnanthidium minutissimum, Gomphonema olivaceum* var. *minutissimum*, *Navicula gregaria, Fistulifera saprophila, Nitzschia palea, Gomphonema parvulum*.

These observations are confirmed by their physical and chemical characteristics given in Fig. 5.7.5. The box-plots (Fig. 5.7.5) show that there is a clear gradient of pH, conductivity and alkalinity between the right and the left part of the SOM: the values in groups 1, 2, 5, 6, 9 and 10 are higher than in groups 3, 4, 7, 8, 11 and 12.

Fig. 5.7.5 also shows that there is a clear geographical separation of the groups: 1, 2, 5, 6, 9 and 10 are in the north (cf. latitude) of the country, which is a hilly region. On the other hand groups 3, 4, 7, 8, 11 and 12 are in the south, which is a lowland region. There is the same tendency for the longitude in groups 1, 2, 5, 6, 9 and 10 which are eastern, and groups 3, 4, 7, 8, 11 and 12, which are western. Figure 5.7.6 illustrates on geographical maps this observation: the sites belonging to groups 1, 2, 5, 6, 9 and 10 are mainly in the Oesling. Groups 3, 4, 7, 8, 11 and 12 are mainly in the Gutland.

A pollution gradient can be observed from the top to the bottom of the map. Groups 1, 2, 3 and 4 are more polluted than groups 9, 10, 11 and 12 (e.g. biological oxygen demand, NO_2^- and PO_4^{3-} in Fig. 5.7.5). Diatom assemblages also show this gradient of pollution (Fig. 5.7.4) regarding taxa ecology. *Fistulifera saprophila, Mayamaea atomus* var. permitis, *Eolimna minima, Navicula gregaria* are α-mesosaprobic to polysaprobic eutrophic taxa and are mainly found in groups 1, 2, 3 and 4. On the other hand *Achnanthidium minutissimum*, a β-mesosaprobic taxon, is found in high abundance (over 47%) in groups 9, 10, 11 and 12.

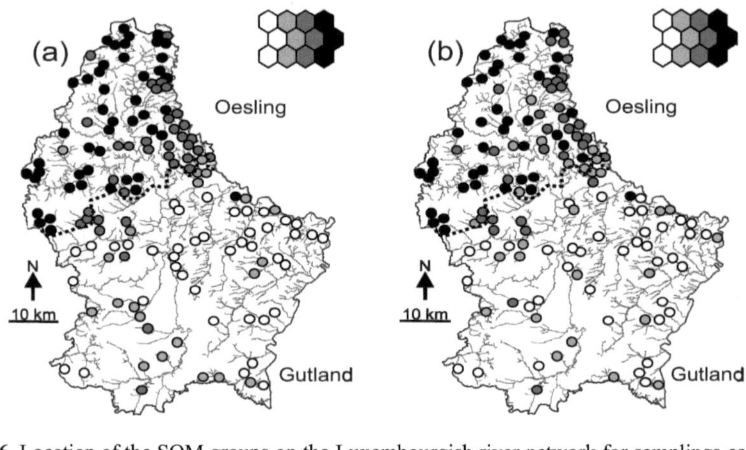

Fig. 5.7.6. Location of the SOM groups on the Luxembourgish river network for samplings carried out in spring (a) and in autumn (b). Grey scales colours are given to each group of the SOM. The samples are plotted according to these colours on the river network. Dotted line separates Gutland and Oesling regions.

Fig. 5.7.7 gives examples of 12 maps of probability of taxa presence. The maps are sorted by taxa ecology. *Achnanthidium biasolettianum, Cocconeis placentula* var. *lineata, Pinnularia subcapitata* var. *elongata, Gomphonema olivaceum* var. *minutissimum, G. pumilum* and *Neidium alpinum* are oligosaprobic to mesosaprobic taxa (Krammer and Lange-Bertalot 1991a, b, van Dam et al. 1994) and are abundant in the upper part of the maps. On the other hand, *Eolimna subminuscula, E. minima, Fistulifera saprophila, Mayamaea atomus* var. *permitis, Navicula veneta* and *Nitzschia palea* are α-mesosaprobic to polysaprobic taxa (van Dam et al. 1994) and appear at the bottom of the map. *Pinnularia subcapitata* var. *elongata* and *Neidium alpinum* are located in the top right hand corner of the map; these taxa are known to be oligosaprobic and acidophilic (Krammer and Lange-Bertalot 1986).

Achnanthidium biasolettianum and *Cocconeis placentula* var. *lineata* are oligo, β-mesosaprobic and alkaliphilic taxa (van Dam et al. 1994); they are situated in the top left hand corner of the map. *Fistulifera saprophila* and *Nitzschia palea* are polysaprobic, eutrophic species (van Dam et al. 1994) and are also located in the bottom right hand corner of the map. The abundance of *Eolimna minima* is high in many hexagons of the map (Fig. 5.7.6). This can be explained by its high abundance in headwater streams of Luxembourg. *Navicula veneta, Mayamaea atomus* var. *permitis* and *Eolimna subminuscula* are eutrophic and α-mesosaprobic taxa (van Dam et al. 1994); they are situated in the bottom left hand corner of the map.

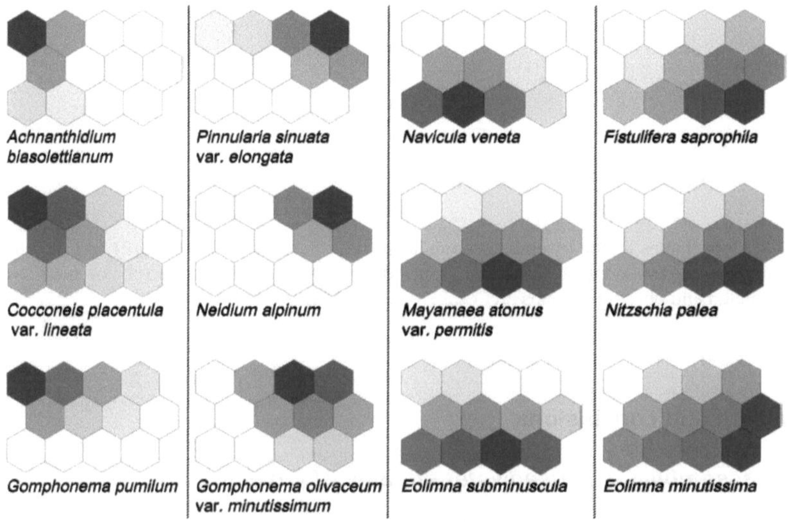

Fig. 5.7.7. Examples of taxa abundance on the SOM. Dark grey hexagons correspond to high probability of presence and white ones to low probability of presence.

Prediction of diatom assemblages with environmental descriptors using ANN-BP

The ANN-BP used is composed by three layers (19 → 13 → 2) and worked with a back propagation algorithm. Nineteen neurons in the input layer, 13 in the hidden layer, and 2 in the output layer were used to predict the coordinates (abscissa and ordinate) of the 12 groups of samples on the SOM (Fig. 5.7.2). This architecture proved to be the best after trying several others with different numbers of input and hidden neurons. A "leave-one-out" cross validation test was computed. The training phase was stopped at 500 iterations because an over-learning effect was detected (i.e. the model was becoming too specialised for the training dataset after more than 500 iterations).

The results of the predicted coordinates of the samples, presented on a Kohonen map (Fig. 5.7.8), show that the predicted samples of the group 1 fit well in hexagon 1; the same is true for group 2 and for other predicted samples. The correlation coefficients were high (0.94 for abscises and 0.96 for ordinates).

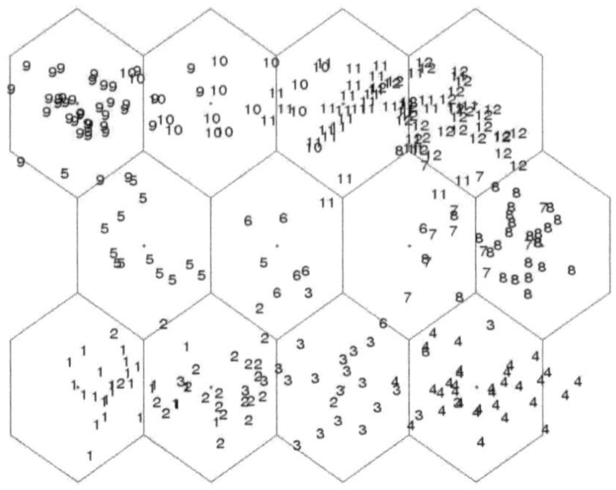

Fig. 5.7.8. Coordinates of the assemblages predicted with the ANN-BP (19-13-2) and plotted on the SOM. Numbers correspond to the observed groups defined with the SOM. Correlation coefficients between predicted and observed groups are: $r_{abscissa} = 0.94$, $r_{ordinate} = 0.96$.

Discussion and conclusions

Diatom assemblages in headwater streams of Luxembourg

With the SOM run with diatom abundances, 12 diatom assemblages corresponding to different situations found in the headwater streams of Luxembourg were characterised.

Four assemblages corresponding to streams with low pollution were defined. Group 12 is mainly located in the north of Luxembourg (Oesling) and group 9 is mainly located in the south (Gutland). The main difference between these regions is their geology (schist and slate in Oesling, sandstone and limestone in Gutland). This confers to headwater streams a different water chemistry: northern streams have lower conductivities than southern ones. This difference in water chemistry is well reflected by the difference of diatom ecology of the assemblages in groups 9 and 12, even though the most abundant species in these groups is *Achnanthidium minutissimum,* a very common taxon in fast-flowing streams that can be considered as opportunistic (Ivorra 2000). Group 9, corresponding to low polluted rivers of Gutland, is characterised by alkaliphilic and alkalibiontic taxa. Group 12, corresponding to low pollution rivers of Oesling, is characterised by neutrophilic taxa.

Ecoregions are usually defined on the basis of homogeneous ecosystems, climate, soil, geology and potential vegetation (Omernik 1987, 1995, Wasson *in* Cemagref 2001). The two regions of Luxembourg (Oesling in the north and Gutland in the south) have different geology, and the physical and chemical characteristics of their headwater streams are different. They correspond to two different climatic regions defined by Lahr (1950): the Ardennes region in the north and the Moselle/Alzette regions in the south. Similarly, Luxembourg is separated into two phytogeographical districts: 'Ardennais' in the north and 'Lorrain' in the south (Lambinon et al. 1992) with different plant associations. The same pattern is also observed for aquatic bryophytes (Werner 2001) and benthic diatoms as shown in this study. Therefore these two regions can be considered as parts of two distinct ecoregions also present in the neighbouring countries of Belgium and Germany.

Other studies also showed patterns for diatom assemblages at a regional scale (Snyder et al. 2002) or ecoregional scale (Potapova and Charles 2002). Many taxa in Luxembourg have a geographical distribution restricted to the Oesling or the Gutland regions.

The assemblages of groups 9 and 12 could be proposed as reference assemblages for unpolluted headwater streams of order 1 to 3 in the two ecoregions of Luxembourg. Groups 10 and 11 correspond to streams of intermediate situations for conductivity, pH and alkalinity (Fig. 5.7.5). The assemblages of groups 1 to 8 correspond to situations with different degrees of anthropic disturbances in the two ecoregions. This study at local scale shows that many diatom taxa can have a narrow ecological distribution (Potapova and Charles 2002), even if most of diatom taxa are considered as cosmopolitan. Similar results were already observed in over groups of algae as cyanophyceae with cosmopolitan taxa occurring in a restricted ecological niche (e.g. Hoffmann 1994, 1996).

Prediction of diatom assemblages in headwater streams of Luxembourg

ANN-BP allowed predicting, with high correlation coefficients, diatom assemblages in headwater streams of Luxembourg using physical and chemical parameters. The correlation coefficient of ordinates showed a good prediction from polluted to unpolluted situations, and the correlation coefficient of abscissas showed a good prediction between ecoregions.

The correlation coefficient of ordinates confirms the results of many studies demonstrating the importance of the correlations between benthic diatoms assemblages and water quality (Stoermer and Smol 1999). This kind of correlation led many authors to develop diatom classifications in sytems for water quality assessment: Descy (1979), Coste (*in* Cemagref 1982), Sládeček (1986), Leclercq and Maquet (1987a), Descy and Coste (1990, 1991), Schiefele and Kohmann (1993), Dell'Uomo (1999), Lenoir and Coste (1996), Prygiel and Coste (1998), Harding and Kelly (1999). These biotic indices are now routinely used in many countries (France, Belgium and Luxembourg among others) and are standardised at national and international scale (Kelly et al. 1998, AFNOR 2000).

The correlation coefficient of abscissas showed that diatom assemblages were predictable from an ecoregion to another. Nutrient experiments realised on diatom assemblages in different rivers of USA showed that large-scale factors are determinant (Snyder et al. 2002) and their integration in models is necessary to predict correctly diatom assemblages. Assemblages of sites with low level of pollution (groups 9 to 12) showed that the concept of a defined type assemblage for an ecoregion is probably a first approximation. Indeed, natural algal assemblages in streams vary along a continuum in which taxa composition is gradually changing with conductivity, pH and alkalinity and other variables. It must be kept in mind that type assemblages are groups of taxa varying from an extreme to another. This phenomenon resembles the River Continuum Concept (RCC, Vannote et al. 1980) but is nevertheless different. The RCC considers that the pattern of the changing benthic community structure in the stream size continua is predictable. In our study, this continuum is predictable in headwater streams from an ecoregion to another by means of physical and chemical parameters. Several studies at a large scale showed that diatom assemblages are also changing along rivers and showed the importance of the "downstream" gradient (Pan et al. 1999, 2000, Potapova and Charles 2002).

This study showed that these two continuums, between ecoregions and between different water qualities, control diatom assemblages. These two imbricate continuums of diatom assemblages were highly predictable by use of several environmental variables characterizing pollution as NO_3^-, NO_2^-, NH_4^+, PO_4^{2-}, BOD_5, P_{tot}, and characterizing the region as alkalinity, pH, conductivity and slope. This study showed that reference conditions for diatom assemblages can be predicted by means of SOM, ANN-BP and environmental variables in headwater streams of Luxembourg.

Perspectives and further developments

The typology of assemblages for benthic diatoms is a fundamental goal already studied since several years. For instance Descy and Coste (1991) developed a table for determining the CEC index. This table gives groups with taxa ranked following their pollution tolerance, and subgroups ranked following the natural diatom successions according to Strahler order. The European Water Framework Directive (European Parliament and The Council of the European Union 2000) formalised this concept of typology and introduced the concept of reference condition. It requires the characterisation of different surface water bodies with hydromorphological and physical and chemical variables and to define their reference conditions (which correspond to its highest ecological status, i.e. to undisturbed conditions). According to the European Parliament (2000) and the REFCOND working group (2002), the "reference conditions establishment can be either spatially based or based on modelling or a combination of these".

Our study shows that it is possible to predict with a high accuracy diatom assemblages of headwater streams in different ecoregions and different water qualities using physical and chemical variables. Artificial neural networks seem to be, at this stage, good techniques to meet the requirements of the European Water Framework Directive. Next developments could be the integration of the REFCOND working group requirements (2002) in the methodology presented here in order to correctly define reference conditions and then to develop new river quality assessment tools.

5.8 Use of neural network models to predict diatom assemblages in the Loire-Bretagne basin (France)*

Di Dato P†, Rimet F, Tudesque L, Ector L, Scardi M

Introduction

The aim of our work was to test the accuracy of the artificial neural networks (ANN) as predictive tools for benthic diatom taxa presence starting from a set of environmental variables. The river basin we studied is characterized by a huge complexity both in terms of spatial heterogeneity, as far as the environmental information is concerned, and in terms of biotic information, because of the large number of taxa that have been identified. In particular, this study focused on the application of different approaches to the reduction of the complexity of the data set and of the ANN models. In the meantime, it was also aimed at showing, as already pointed out for other kind of organisms (Scardi et. al. in # 3.8, Di Dato et al. in # 4.3), that too frequent or too rare species usually provide trivial information about the relationships of environmental variables with their presence or absence, thus affecting the accuracy of the models. The taxa that were selected according to the different approaches we tried including only those that can be actually modelled on the basis of the available environmental information. The limits and the perspectives of these species selection approaches are then thoroughly discussed.

Materials and methods

This study was carried out on a data set collected in the Loire-Bretagne (Loire-Bretain) basin, which is located in western France. In this basin a wide variety of climates and landscapes can be found, but for the sake of simplicity, it can be divided into 3 different sections. The mid and high Loire basin is rather hilly and mountainous with the Massif-Central mountains in the south-eastern part. The low Loire basin is characterised by lowland and some large rivers. The Bretagne region is a rather flat region, composed by lentic rivers and is deeply influenced by the Atlantic Ocean (Fig 5.8.1).

The diatom database has been assembled collecting 641 samples from 1996 to 2000 in the framework of national river survey, organised and funded by the Loire-Bretagne Water Agency. For each sampling, physical, chemical and geomorphological descriptors were recorded. Among the latter, some were measured in the field (width, shading, sampled substrate and current velocity), others on maps and on Geographical Information Systems (geology, altitude, distance from source, slope, catchment area, discharge). Average values computed over the last 3 months were assumed for physical and chemical descriptors (temperature, pH, dissolved oxygen, dissolved organic carbon, HCO_3^-, CO_3^{2-}, NO_3^-, NO_2^-, NH_4^+,

* This work is part of the PAEQANN project supported by the European Commission under the 5th Framework Programme (contract n°: EVK1-CT1999-00026). We thank Mr J. Durocher of the "Loire-Bretagne" Water Agency and Ms M. Leitao of the Bi-Eau society for their valuable collaboration.

† Correspondence: pdidato@mclink.it

PO_4^{2-}, Ca^{2+}, Cl^-, Na^+, biological oxygen demand, chemical oxygen demand, suspended matter, NKJ, P_{tot}). The environmental variables that were used to train the models are presented in the Table 5.8.1.

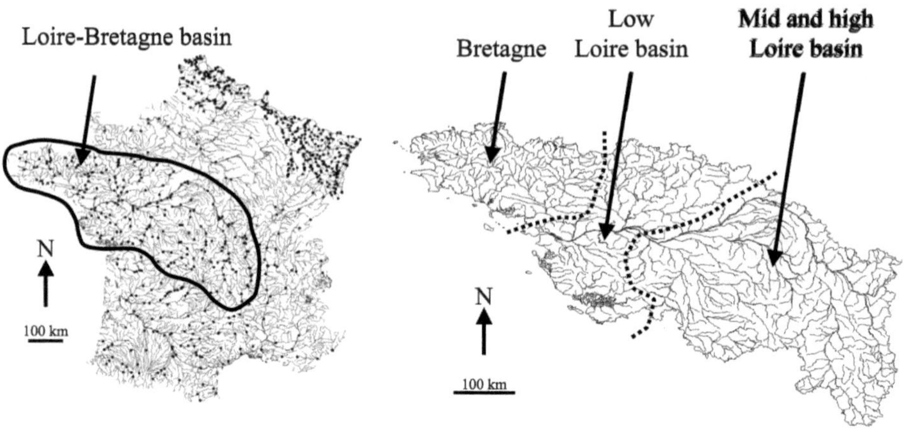

Fig. 5.8.1. The study area. Black dots in the map on the left show the sampling sites.

Table 5.8.1. List of the environmental variables measured or assessed for each sampling site.

Altitude (m)	**HCO_3^- (mg L^{-1})**
Source distance (km)	NO_3^- (mg L^{-1})
Slope (m km^{-1})	NO_2^- (mg L^{-1})
Catchment area (km^2)	NH_4^+ (mg L^{-1})
Electric conductivity 20°C (S cm^{-1})	PO_4^{2-} (mg L^{-1})
pH	Ca^{2+} (mg L^{-1})
Dissolved oxygen (DO) (mg L^{-1})	Cl^- (mg L^{-1})
Dissolved organic carbon (DOC) (mg L^{-1})	Na^+ (mg L^{-1})
Biological Oxygen Demand (BOD) (mg L^{-1})	NKJ (Kejdahl nitrogen) (mg L^{-1})
Suspended matter (mg L^{-1})	P_{TOT} (mg L^{-1})
Shading (1:closed - 3:open)	Current velocity (1:lotic - 3:semi-lotic)
Geology (limestone: 1/0)	Geology (quaternary sediment: 1/0)
Geology (sandstone: 1/0)	Geology (other:1/0)
Geology (granitic: 1/0)	Geology (mudstone-schiste-schale:1/0)
Geology (volcanic:1/0)	

The sampling of benthic diatoms was carried out by means of a brush. The epilithon was collected by brushing at least 5 stones found in lotic zones of the sampling site. In laboratory, the samples were cleaned with hydrogen peroxide and hydrochloric acid to dissolve calcium carbonates. Then the cleaned frustules were mounted in a resin (*Naphrax*).

A total of 930 diatoms taxa were determined. Therefore, the data set included 641 patterns (samples), for 32 predictive environmental variables (input variables) and 930 diatom taxa (output variables). Two subsets of taxa were then extracted from this data set. In the first subset (A), only taxa that were present in more than 5% and less than 95% of the 641 samples were included (202 taxa met this condition). In the second subset (B) all the taxa that were either present or absent in less than 100 samples were excluded, thus considering only those taxa whose frequency of occurrence ranged from 15% to 85% (91 taxa met this condition). The latter subset was also used with two different modelling strategies. The first strategy was based on the development of a single model for all the species, i.e. a single model with 91 outputs, whereas the second involved developing 91 separate models, each one with only one output. In all cases 3-layer feed-forward ANNs were used, i.e. perceptrons with a single hidden layer.

In order to evaluate how much information was retained after selecting species subsets, we compared the results of a Principal Coordinate Analysis (Gower 1966) based on the whole set of species with the ones of Principal Coordinate Analyses based on the species subsets (i.e. on 202 and 91 taxa, respectively). The Jaccard's coefficient (Jaccard 1908) was used to compute dissimilarity among samples. This asymmetrical coefficient was selected because it only takes into account the presence of species, ignoring absence data, which are not always completely reliable.

The 641 available patterns were randomly distributed among three subsets for training, validation and test. The training set included 50% of the available patterns, whereas the validation set as well as the test set included 25% of the available patterns. The validation set was used to compute the Mean Square Error (MSE) of the ANN outputs after each training epoch. The test set allowed obtaining unbiased estimates of the accuracy of the trained ANN models.

Floristic information was expressed at the lowest information level, i.e. as binary presence/absence data. The predictive environmental variables were normalized by rescaling their range of variation within the [0,1] interval.

The optimal structure of the ANN models was defined after empirical tests. In practice, the number of nodes in the hidden layer was selected by comparing the MSE for different ANNs, with up to 60 nodes in the hidden layer. The ANN structure that provided the best performance was the one with 51 hidden nodes for the subset A (202 taxa), whereas 11 hidden nodes were used for both subset B (91 taxa) modelling strategies. In the latter case the selection of the ANN structure was based on the average MSE.

During the training procedure only a random subset of training patterns (50% of the available patterns) was submitted to the ANN at each epoch in order to prevent overtraining due to the memorization of the pattern sequence. In order to better generalize the ANN learning, white noise in the [-0.01,0.01] range was also added to each input value (Györgyi 1990).

In all the nodes of the hidden and output layers of the ANN sigmoid activation functions were used. The error back propagation algorithm was selected to adjust the weights during the training procedure. In particular, we applied an early stopping procedure based on the validation set MSE. The learning rate and the momentum were respectively set to 0.90 and 0.10 and never modified during the training.

The continuous outputs values of the ANN were converted to binary using a threshold function and then compared with the observed data to obtain the percentage of Correctly Classified Instances (CCI).

For each species subset (A and B) and for each modeling strategy (1 and 2) we analyzed the accuracy of the models by testing the independence of the modeled data wit respect to observed data. Therefore, we computed the K statistic (Cohen 1960, Kraemer 1982) on the basis of contingency tables in which presence and absence data in the model output and in the target patterns (i.e. in observed data) were cross-tabulated as shown in Table 5.8.2.

Table 5.8.2. Contingency table for K statistics computation.

		Model output	
		Presence	*Absence*
Target	*Presence*	1-1 A	1-0 B
	Absence	0-1 C	0-0 D

The K statistic was then obtained as:

$$K = \frac{\mathrm{O_a - E_a}}{\mathrm{N - E_a}} \tag{5.8.1}$$

where O_a is the observed count of CCI (O_a = A+D), E_a is the count of CCI that are expected if the model is independent of the observed data ($E_a=[A+B][A+C]/N+[C+D][B+D]/N$) and N is total number of cases ($N=A+B+C+D$).

Results

Reducing the number of species in a floristic data set has a cost in terms of information about the overall structure of the species assemblages. In other words, it is not possible to preserve the whole amount of information about the relationships among samples when only a subset of species is considered. However, a smart selection of the most relevant species might reduce the information loss to a minimum. In order to provide a very rough estimate of the degree of approximation about the diatom assemblage structure that we accepted when we decided to reduce the number of species to be modelled, the results of a Principal Coordinates Analysis performed on the whole data set and on the two subset of species were compared (Fig 5.8.2).

The two scatter plots show that the distortion of the first Principal Coordinate for each sample due to the reduction of the number of species is quite limited both in the case of subset A (202 taxa, left) and subset B (91 taxa, right). In fact, in both cases the correlation between the Principal Coordinates is very high (Spearman's r= 0.969 for Subset A and r=0.992 for subset B). Of course, the Mantel statistics between the dissimilarity matrices is highly significant in both cases, so we rejected the null hypothesis of independence of the subsets from the whole data set. These evidences suggest that reducing the number of species does not affect significantly the ability to represent the main features of the diatom assemblage and to reproduce the relationships among samples.

The results of the ANN models were evaluated by taking into account the test data set only (i.e. 25% of the available patterns). In other words, the accuracy of the predictions was estimated on the basis of information that is completely independent of that on which the ANN models were developed. This strategy, that is often overlooked when conventional statistical models are considered, allows both to obtain unbiased estimates of the modelling errors and to effectively compare different modelling approaches.

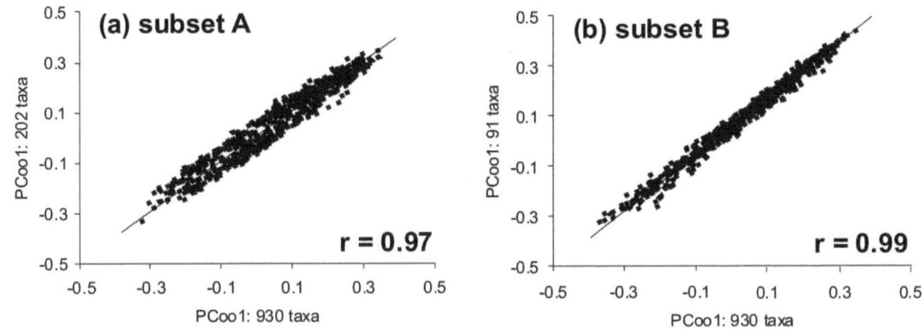

Fig. 5.8.2. First Principal Coordinate of the diatom assemblages: all the species vs. reduced species set. Subset A, 202 taxa, on the left (a), and subset B, 91 taxa, on the right (b).

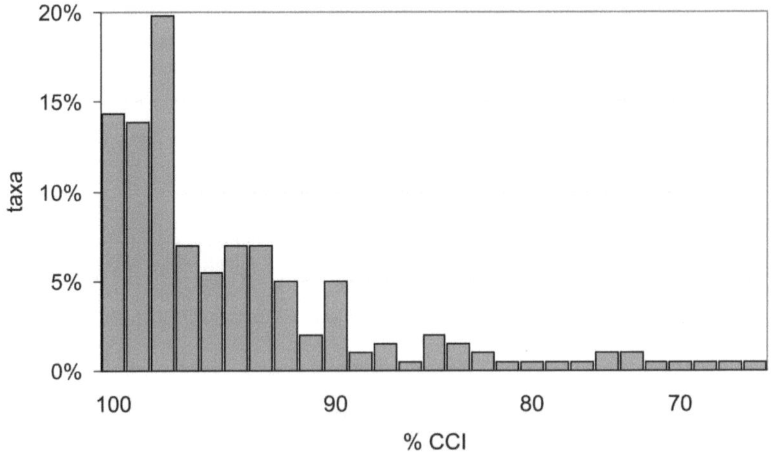

Fig. 5.8.3. Percentage of correctly classified instances (CCI) for the subset A (202 taxa).

The percentage of correctly classified instances (CCI) for the larger subset of species (subset A, 202 taxa) was quite high, as the average value was 94.5% (Fig 5.8.3). More than 10% of the taxa had a percentage of CCI of 100%, whereas almost 20% had a percentage higher than 97%. In Figure 5.8.3 the frequency distribution of the CCI percentages is shown, and it is very clear that almost 90% of the taxa had CCI values larger than 90%.

Although the model performed very well according to this criterion, the values of the K statistic did not confirm that result. In fact, only 5 taxa out of 202 were effectively predicted by the model, i.e. in only 5 cases the prediction about species presence or absence was significantly different from random according to the K statistics (Fig. 5.8.4a).

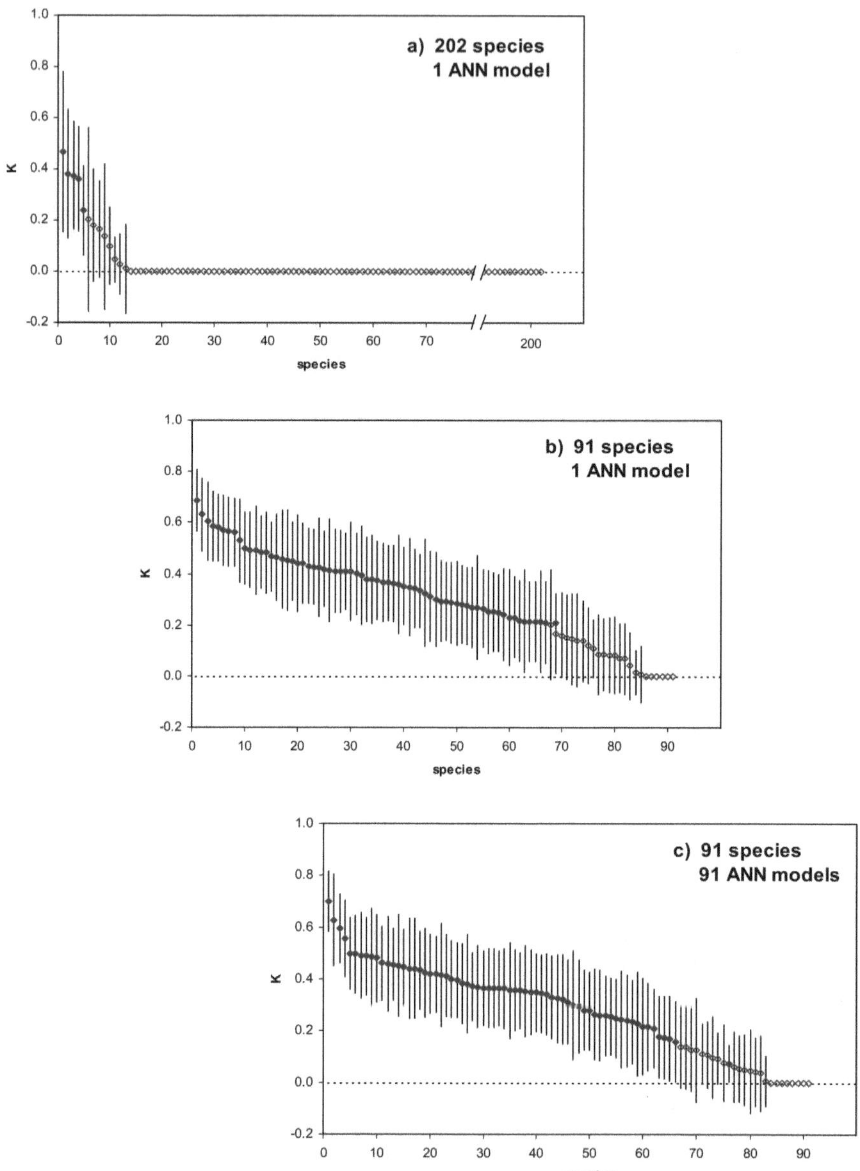

Fig. 5.8.4. K statistics values with confidence intervals for the different modelling strategies: a) species subset A (202 species, 1 ANN model); b) species subset B, strategy 1 (91 species, 1 ANN model); c) species subset B, strategy 2 (91 species, 91 ANN models). Solid diamonds indicate significant K values (i.e. cases in which the lower P=0.95 confidence bar does not intersect the K=0 line). White diamonds at K=0 with no bars stand for species that were never predicted by the model. In this case the K value was actually undetermined.

The disagreement between CCI percentages and K statistics results might seem surprising at first glance, but it can be easily explained if the errors in model prediction are carefully analysed. In fact, most species are present only in a very limited number of samples (on the average 6.6 samples out of 125, i.e. slightly more than 5%, in the test set) and it is very difficult for the ANN model to correctly predict their presence on the basis of a handful of known cases. Therefore, when dealing with rare species, ANN models learn to predict only species absence independently of their inputs, easily attaining low mean square errors and very high CCI percentages. Needless to say, the predictions of those models are useless from a practical point of view, and the high CCI percentages they attain are meaningless.

On the contrary, the K statistics is able to provide a reliable estimate of the predictive ability of a model, as it takes into account the relative frequency of presence and absence records. Of course, predictions about rare species are inherently unreliable because of the lack of "examples" about the relationships between species presence and environmental variables, but a "false" model cannot obtain an high (and significant) K statistics value if it is not able to predict enough instances of presence with respect to the real frequency of the species. It is not surprising that the five species that were associated to significant K statistics were more frequently found than the average (21.4 cases out of 125, i.e. 17.1%).

Reducing the number of species to be modelled by excluding the most rare ones seems therefore a sound choice, especially considering that the relationships among samples are well described even when rare species are excluded. Therefore using a smaller subset of species, selected on the basis of their frequency and excluding the rarest ones, makes definitely sense and that is the reason why the subset B (91 taxa) was selected. In this case, however, two modelling strategies were adopted, respectively involving a single model with 91 outputs (strategy 1) and 91 models with a single output (strategy 2). The results for the two strategies are shown in Figure 5.8.4b,c as far as the K statistics is concerned.

The CCI percentages, although not relevant in the light of the evaluation of the predictive capabilities of the models, are lower than in the model for species subset A (on the average 75.2% and 74.2%, whereas the first model attained 79.9%). In particular, 68 out of 91 species had significant K statistics values with strategy 1 and 67 out of 91 with strategy 2. Thus, the two strategies for the species subset B returned similar results, although the average K value for strategy 1 was slightly larger than the one for strategy 2 (0.30 and 0.28, respectively).

A more detailed comparison of the K statistics values obtained for strategies 1 and 2 is shown in Figure 5.8.5. The unit slope line corresponds to a perfect agreement between strategy 1 and strategy 2 K statistics values, and it is evident that the overall agreement between the two series is rather good (r=0.77). This implies that species that some species are intrinsically more predictable than others, as they tend to have high K statistics values independently of the modelling strategy.

However, there are also differences in the accuracy of the predictions that depend on the modelling strategy. In fact, points below the unit slope line correspond to species that were more accurately predicted using strategy 1 (i.e. $K_1>K_2$), whereas strategy 2 was a better option for species represented by points above that line (i.e. $K_2>K_1$). Since 50 points are located below the unit slope line and only 36 above it, strategy 1 seems more effective than strategy 2, but the difference between the two is not dramatic. The K values are exactly the same for both strategies in 5 cases.

Finally, the case of *Fragilaria capucina* var. *vaucheriae*, the outlier in the lower right corner of the plot, is worth commenting, as it was the only case in which there was a very large difference between the K statistics values for the two modelling strategies. In particular, this taxon was accurately predicted in the strategy 1 model, whereas the strategy 2 model completely failed.

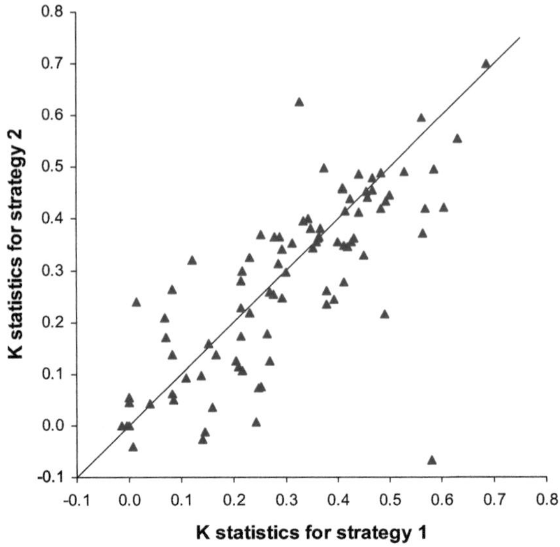

Fig. 5.8.5. K statistics values for the modelling strategy 1 (one model for all the species) are compared to those for modelling strategy 2 (a different model for each species). The overall agreement between the two strategies (r=0.77) indicates that some species are intrinsically more predictable than others. The unit slope line represents the perfect agreement between the two strategies that was observed in 5 cases. Strategy 1 provided larger K statistics values in 50 cases (points below the unit slope line), while strategy 2 was more effective in 36 cases (points above the unit slope line). The outlier in the lower-right corner represents Fragilaria capucina var. vaucheriae, which was accurately predicted only by strategy 1.

The only likely reason for this difference is that information about species interactions (or association) that is implicitly embedded in the strategy 1 model (that predicts all the species simultaneously and has as a much more complex structure) may play a role in cases in which the relationships between environmental variables and species distribution are so weak that the latter cannot be reliably modelled by a single species model (strategy 2). The alternate hypothesis, of course, is that strategy 2 modelling failed by chance, and a more effective model could have been obtained by further iterating the ANN training procedure.

Another comparison between the two modelling strategies for species subset B was carried out on the basis of the results of Principal Coordinate Analyses performed on Jaccard's dissimilarity matrices. In particular, the first Principal Coordinates for the predicted species composition of the samples in the test set were plotted against the first Principal Coordinates for the observed species composition of the same samples for both strategies (Fig. 5.8.6). The agreement between Principal Coordinates obtained from analyses involving observed and predicted diatom assemblage compositions is a proxy for the agreement between predicted and observed dissimilarity matrices, which, in turn, is a proxy for the resemblance of the predicted and observed assemblage composition.

The best strategy, according to this comparison criterion, was the one based on single species model, i.e. strategy 2 (see Fig. 5.8.6b). The rank correlation between Principal Coordinates based on predicted and observed data (Spearman's r=0.726) was higher than in that case of strategy 1 (Spearman's r=0.812), but the difference between the two values of the correlation coefficient was not significant (n=148, p=0.070).

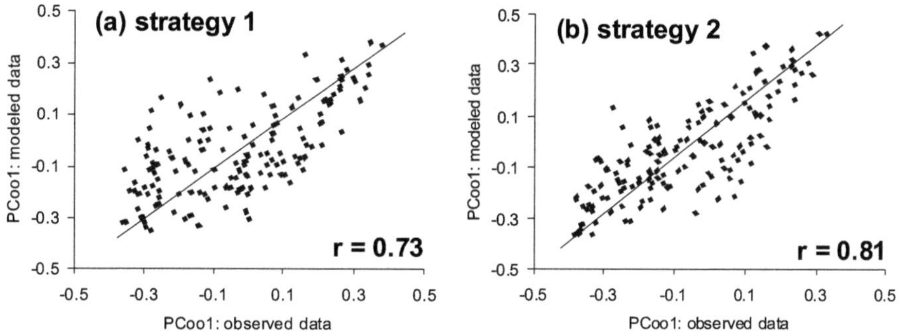

Fig. 5.8.6. Principal Coordinates obtained from Jaccard's dissimilarity matrices computed from predicted data were compared to those obtained from observed data according to the same procedure. Results based on modelling strategy 1, i.e. one model for all the species is shown on the left (a), whereas those based on strategy 2 (a separate model for each species) are shown on the right (b).

These results are not in agreement with those based on the K statistics, which but in both cases the differences between the two modeling strategies were minor. In fact, the Mantel test allowed rejecting the null hypothesis of independence between the dissimilarity matrices based on predicted and observed data in the case of both modeling strategies.

Discussion and conclusions

This study provided further evidence about the problems in modelling rare species, as well as ubiquitous, if any, that has been pointed out elsewhere (e.g. Scardi et. al, this book). In fact, even in the case of diatom assemblages such species significantly affect the performance of ANN models.

In particular, very high percentages of correctly classified instances (CCI) are often a false indication of model accuracy. As CCI only take into account the number of cases in which the model predictions match observed data, they are strongly affected by the relative frequency of presence and absence records. Therefore they are biased indicators of model accuracy when dealing with rare or very common species and other approaches, like the one based on K statistics, should be selected.

Excluding rare species from a data set may cause a significant reduction in the number of species, and the ability to correctly represent the complex relationships among samples might be seriously impaired. However, if properly selected, even a subset of taxa that represent a significant part of the assemblage structure may preserve enough information. Our results showed that 91 taxa out of 930, selected in a way that excluded the less frequent ones, were able to provide an adequate representation of the differences in the diatom assemblage structure among many different sites.

Most of the taxa that were well predicted by the model are found in eutrophic waters: *Cocconeis placentula* var. *placentula, Fistulifera saprophila, Gomphonema minutum, Gyrosigma attenuatum, Mayamaea atomus* var. *permitis, Navicula tripunctata* according to van Dam et al. (1994), *Diatoma vulgaris* according to Hoffman (1994), *Navicula antonii, N. capitatoradiata* according to Lange-Bertalot (2001). No taxa characterizing oligotrophic waters were well predicted by the model.

The most accurately predicted species, independently of the model, was *Navicula tripuncata,* which is considered as good indicator specie for eutrophic waters with average to high electrolyte content (Lange-Bertalot 2001). *Navicula antonii* is also a species considered as good indicator for waters that are often affected by anthropic sources of perturbation (Lange-Bertalot 2001). The high efficiency of the ANN models in predicting the above-mentioned species on the basis of physical and chemical variables, including pollution indicators, supports the hypotheses about an environmental control of their distribution.

Taxa as *Fistulifera saprophila* and *Mayamaea atomus* var. *permitis* have particular ecological characteristics. In fact, they are found in heavily polluted water and are α meso-polysaprobic taxa according to van Dam et al. (1994). As environmental variables that are linked to pollution parameters play an important role among the models inputs, it is not surprising that the models are able to accurately predict the distribution of these species.

Nitzschia cf. *tropica* is an invasive species (Coste and Ector 2000) and its presence is associated to a well-defined area within the studied region, i.e. the upper part of the mid and high Loire basin. The accuracy of the ANN models in predicting its presence is probably related to the use of geographical predictive variables (e.g. elevation, distance form source) that allow recognizing sites within the boundaries of the area where this species can be found.

In practice, however, in many cases the model had not enough information to learn the taxa response to environmental variables. This is probably the case especially for pollution sensitive taxa, since the sampling network of this study is adapted to the assessment of pollution in the Loire-Bretagne basin. The database is composed only by 7% of very good biological quality samples according to the diatom index SPI (Specific Pollution sensitivity Index; Coste 1982).

A similar case is the one involving *Bacillaria paradoxa, Nitzschia clausii, N. filiformis.* These taxa, which are uncommon in the Mid and High Loire basin, indicate brackish waters (van Dam et al. 1994) even if their abundance in the assemblage is low. Despite this remarkable stenoecy, ANN models are not able to accurately predict their presence in such particular environments because of the relative rarity of the occurrence of brackish waters (less than 1% of the samples were collected in waters with conductivity above 1000 µS/cm).

On the other hand, several very common taxa were not well predicted. For instance, this was the case of *Achnanthidium minutissimum*, which is absent in only 9% of the samples and is considered as pollution sensitive specie in the diatom indices (Coste 1982, Descy 1979, Leclercq and Maquet 1987a, b). However, this taxon has also an opportunistic behaviour (Ivorra 2000) and develops rapidly in the biofilm when clear space is available (Sabater 2000) and when inter-specific competition has decreased (Rodríguez 1994). As biotic factors play a relevant role in controlling the abundance of this taxon, its behaviour is rather difficult to predict. Moreover, the lack of an adequate number of records in which this species is absent made it even more difficult for the ANN models to predict its distribution.

In fact, the results improved by reducing the number of taxa and by defining subsets of species that accounted for a very large part of the variation within the ecological data set. Starting from this reduced set of suitable data, we obtained very similar results with two different modeling strategies. However, strategy 1 (a single model predicting all the species) can be considered theoretically more effective, because it can take profit of the information that is provided by the inter-specific relationships, even though they are not known.

An alternate approach to the reduction of the number of taxa to be modelled might involve the reduction of the taxonomic resolution to the genus level. Some diatom indices based on information at the genus level were developed for bioindication (Coste and Ayphassorho 1991). However, this approach is probably not the best one, and it was not included in our tests. In fact, pollution tolerant and pollution sensitive taxa often occured in the same genus in the dataset of the Loire basin.

An interesting perspective for reducing the complexity of diatom databases could be in the use of life forms that have been described in several papers (e.g. Hoagland et al. 1982), but the applicability of this approach in biomonitoring studies is still to be evaluated.

In conclusion, our results showed that accurate models for the prediction of diatom assemblages on the basis of environmental variables can be developed using ANNs. Unfortunately, these models are not able to predict all the species, because in many cases the information about the relationships that link species distribution to environmental variables or to other species distribution (e.g. via competitive interactions) is not sufficient. Rare species and almost ubiquitous ones belong to this category, and the only way to improve the accuracy of the predictions about their distribution is to develop models *ad hoc*, using very focused data sets. To reduce the complexity of the data, dividing the dataset of the Loire basin in sub-datasets by use of an adequate typology should be tested. Then developing models for each sub-dataset corresponding to precise river types should improve the efficiency of the prediction of diatom species abundances.

However, even a subset of species can be very effective in reproducing the ecological relationships among sampling sites and in pointing out sites in which the diatom assemblage structure is impaired because of various types of disturbance. Models that are able to predict the main features of the expected diatom assemblages will effectively support the detection of ecological perturbations and the evaluation of the environmental quality in freshwater ecosystems, and ANNs will certainly play a major role in this scenario.

6 Development of community assessment techniques

Editor: Park YS*

6.1 Introduction

Data mining is currently an important topic in ecosystem management. It is necessary to characterize the sequence of community dynamics in spatial and/or temporal terms if the ecosystem needs to be assessed after the impacts of natural or anthropogenic stresses. The structures of their assemblages are potentially determined by many environmental factors acting at different spatial and time scales (Stevenson 1997, Snyder et al. 2002). However, it is not an easy task to find patterns embedded in community datasets because ecological communities consist of a large number of species and many sampling sites at different times and/or locations. Each species shows spatial-temporal dynamics with different occurrences and abundances.

Several methods exist for quickly producing and visualizing simple summaries of data sets (Tukey 1977, Kaski 1997). Several graphical means have been proposed for visualizing high-dimensional data items directly by letting each dimension govern some aspect of the visualization and then integrating the results into one figure (du Toit et al. 1986, Jain and Dubes 1988). These methods can be used to visualize any kind of high dimensional data vector: either the data items themselves or vectors formed from descriptors of the data set (Tukey 1977). Techniques for dimension reduction including classification methods, projection methods, and visualization methods were well documented by Kaski (1997) and Carreira-Perpinan (2001).

In this chapter, we present 5 papers concerning techniques for exploratory data analysis of aquatic communities:

1. Evaluation of relevant species in communities: development of structuring indices for the classification of communities using a self-organizing map by Park et al. They propose a computational method to determine the most relevant variables for structuring the organization of the self-organising map. It provides a quantitative evaluation of the relative importance of input variables in the map patterns.
2. Projection pursuit with robust indices for the analysis of ecological data by Werner et al. The projection pursuit was implemented in the software "Autonomous Projection Mapping" (APM). This software enables the user to analyse data interactively and also facilitates largely autonomous data analysis without user interaction. The software is available from the authors.
3. A framework for computer-based data analysis and visualisation by pattern recognition by O'Connor and Walley. They describe a framework for decision support systems based on data analysis and visualisation by pattern recognition. A software system, River Pollution Diagnostic System (RPDS) and its generic form (MIR-max) are used

* Correspondence: park@cict.fr

as concrete examples to demonstrate how the framework can be used in practice. The software is available from the authors.

4. A rule-based vs. a set-covering implementation of the knowledge system LIMPACT and its significance for maintenance and discovery of ecological knowledge by Neumann and Baumeister. They present a new model-based implementation of the existing knowledge system LIMPACT, and estimate the pesticide contamination of small lowland streams within agricultural catchment areas using LIMPACT. The system is available over the internet via http://www.limpact.de.
5. Predicting macro-fauna community types from environmental variables by means of support vector machines by Akkermans et al. They applied SVM to the prediction of macroinvertebrate community types and compared its performance with that of multinomial logistic regression. Additionally they provide details of SVM algorithms and bibliography for further reading.

6.2 Evaluation of relevant species in communities: development of structuring indices for the classification of communities using a self-organizing map*

Park YS, Gevrey M†, Lek S, Giraudel JL

Introduction

Ecological data are mostly multivariate, characterizing complexity and nonlinearity and some information in the data is only interpretable indirectly (Jongman et al. 1995). Ecological interpretation and especially the explanation of the structure of several descriptors (i.e., multivariate data) can be carried out following two approaches: direct or indirect comparison schemes (Legendre and Legendre 1998) which refer to direct gradient analysis and indirect gradient analysis respectively (ter Braak 1987). The former includes principal component analysis (PCA) and correspondence analysis, whereas the latter includes canonical correspondence analysis, redundancy analysis, and canonical correlation analysis. However, these conventional multivariate methods are mainly based on the linear data matrix limiting the usages due to strong distortions with nonlinear relations in the dataset (Kenkel and Orloci 1986, Bunn et al. 1986, Ludwig and Reynolds 1988, Legendre and Legendre 1998).

Due to the nonlinearity and complexity of ecological data, nonlinear analyzing methods are preferred (Blayo and Demartines 1991, Lek and Guegan 2000). One of these methods is artificial neural networks (ANNs), which are versatile tools to extract information out of complex data, and which could be effectively applicable to classification and association (i.e., presentation with fewer space dimensions). Among ANNs, recently a self-organizing map (SOM) has become more and more popular in ecological studies. The SOM approximates the probability density function of the input data, and it is a method for clustering, visualization, and abstraction, the idea of which is to show the data set in another, more usable, representation (Kohonen 2001). These characteristics have been used efficiently in various ecological areas (Lek and Guégan 1999, 2000, Recknagel 2003): classification of communities (Chon et al. 1996, Park et al. 2001a, 2003a); identification of community patterns (Brosse et al. 2001), water quality assessments (Walley et al. 2000, Aguilera et al. 2001), and prediction of population and community structure (Céréghino et al. 2001, Obach et al. 2001). Recently Park et al. (2003a) proposed a method to integrate the relationships between sampling sites (or clusters), biological attributes, and environmental variables in the SOM map. They showed the distribution gradient of each variable on the SOM map, presenting the importance of variables concerning communities. However, there are difficulties to quantify the importance of each variable in the patterns defined by the SOM. Therefore, in this study we aim to develop a method to quantify the importance of each variable in the patterns defined in the SOM map. These quantified values can be used as an

* Funding for this research was provided by the EU project PAEQANN (N° EVK1-CT1999-00026).

† Correspondence: gevrey@cict.fr

index of the relative importance of the variables, and will be helpful for the interpretation of ecological data.

Material and Methods

Ecological data

We used the vegetation dataset reported in Jongman et al. (1995). The dataset collected in the Dutch island of Terschelling using the Braun-Blanquet method (Batterink and Wijffels 1983) was recorded according to the ordinal scale of van der Maarel (1979). From 80 sites, 20 were selected with 30 species (Table 6.2.1), representative of the variation in the complete set of data. Jongman et al (1995) used this extracted dataset to demonstrate the performance of different kinds of ordination and cluster analysis techniques. The dataset is very useful to present the characteristics of multivariate analysis methods. Therefore, in this study we used this data matrix consisting of 20 sites reporting the abundance of 30 species. Species acronyms are given in Table 6.2.1 and hereinafter we will use the acronyms to simplify expression. In the modelling process, abundance data were scaled between 0 and 1 in the range of the minimum and maximum values of each species. Furthermore, species presence and absence data were used to verify the responses of the model developed.

Modelling procedure

SOM algorithm The community data (presence/absence or abundance data) were used to classify samples through training the SOM. Formally the SOM consists of input and output layers connected with weight vectors (connection intensities). The array of neurons (i.e., computational units) in the input layer receives the input vectors, whereas the output layer consists of a two-dimensional network of neurons (N output neurons; 20=5×4 in this study) arranged on a hexagonal lattice. The best arrangement for the output layer is a hexagonal lattice, because it does not favour horizontal or vertical directions as much as rectangular or triangular arrays (Kohonen 2001). In the learning process of the SOM, when an input vector $\boldsymbol{x}$ (i.e., a species) is given through the network, each output neuron k of the network computes the distance between the weight vector $\boldsymbol{w}$ and the input vector $\boldsymbol{x}$. Among all N output neurons, the best matching unit (BMU), which has the minimum distance between weight and input vectors, is the winner. For the BMU and its neighbourhood neurons, the weight vectors are updated by the SOM learning rules. This results in training the network to classify the input vectors by the weight vectors they are closest to. After the learning process of the SOM, we used a hierarchical cluster analysis with the Ward linkage method to define the cluster boundaries in the units of the SOM.

The detailed algorithm of the SOM can be found in Kohonen (2001) for theoretical considerations and Park et al. (2003a) for ecological applications.

Species distribution planes During the learning process of the SOM, neurons that are topographically close in the array of output neurons activate each other to learn something from the same input vector. This results in a smoothing effect on the weight vectors of neurons (Kohonen 2001). Thus, these weight vectors tend to approximate the probability density function of the input vector. The visualization of elements of input vectors is convenient to understand the contribution of each input variable with respect to the clusters in the SOM (Park et al. 2003a). Therefore, to present the contribution of input variables (i.e., species) in cluster structures defined in the SOM, calculated values (weights) of each input variable (species) during the training process were visualized in each neuron of the SOM on a grey scale. Based on these species distribution plans, quantified indices were computed as explained in the following sections.

Table 6.2.1. Dataset used in the modelling process. The data were collected in the Dutch island of Terschelling using Braun-Blanquet method and recorded according to the ordinal scale of van der Maarel (1979) (data from Jongman et al. (1995)

Species name	Acronym	Sampling sites																			
		1	2	3	4	5	6	7	8	9	10	11	12	13	14	15	16	17	18	19	20
Achillea millefolium	Achmil	1	3	0	0	2	2	2	0	0	4	0	0	0	0	0	0	2	0	0	0
Agrostis stolonifera	Agrsto	0	0	4	8	0	0	0	4	3	0	0	4	5	4	4	7	0	0	0	5
Aira praecox	Airpra	0	0	0	0	0	0	0	0	0	0	0	0	0	0	0	0	2	0	3	0
Alopecurus geniculatus	Alogen	0	2	7	2	0	0	0	5	3	0	0	8	5	0	0	4	0	0	0	0
Anthoxanthum odoratum	Antodo	0	0	0	0	4	3	2	0	0	4	0	0	0	0	0	0	4	0	4	0
Bellis perennis	Belper	0	3	2	2	2	0	0	0	0	2	0	0	0	0	0	0	0	2	0	0
Brachythecium rutabulum	Brarut	0	0	2	2	2	6	2	2	2	2	4	4	0	0	4	4	0	6	3	4
Bromus hordaceus	Brohor	0	4	0	3	2	0	2	0	0	4	0	0	0	0	0	0	0	0	0	0
Calliergonella cuspidata	Calcus	0	0	0	0	0	0	0	0	0	0	0	0	0	4	0	3	0	0	0	3
Chenopodium album	Chealb	0	0	0	0	0	0	0	0	0	0	0	0	1	0	0	0	0	0	0	0
Circium arvense	Cirarv	0	0	0	2	0	0	0	0	0	0	0	0	0	0	0	0	0	0	0	0
Eleocharis palustris	Elepal	0	0	0	0	0	0	0	4	0	0	0	0	0	4	5	8	0	0	0	4
Elymus repens	Elyrep	4	4	4	4	4	0	0	0	6	0	0	0	0	0	0	0	0	0	0	0
Empetrum nigrum	Empnig	0	0	0	0	0	0	0	0	0	0	0	0	0	0	0	0	0	0	2	0
Hypochaeris radicata	Hyprad	0	0	0	0	0	0	0	0	0	0	2	0	0	0	0	0	2	0	5	0
Juncus articulatus	Junart	0	0	0	0	0	0	0	4	4	0	0	0	0	0	3	3	0	0	0	4
Juncus bufonius	Junbuf	0	0	0	0	0	0	2	0	4	0	0	4	3	0	0	0	0	0	0	0
Leontodon autumnalis	Leoaut	0	5	2	2	3	3	3	3	2	3	5	2	2	2	2	0	2	5	6	2
Lolium perenne	Lolper	7	5	6	5	2	6	6	4	2	6	7	0	0	0	0	0	0	2	0	0
Plantago lanceolata	Plalan	0	0	0	0	5	5	5	0	0	3	3	0	0	0	0	0	2	3	0	0
Poa pratensis	Poapra	4	4	5	4	2	3	4	4	4	4	4	0	2	0	0	0	1	3	0	0
Poa trivialis	Poatri	2	7	6	5	6	4	5	4	5	4	0	4	9	0	0	2	0	0	0	0
Potentilla palustris	Potpal	0	0	0	0	0	0	0	0	0	0	0	0	0	2	2	0	0	0	0	0
Ranunculus flammula	Ranfla	0	0	0	0	0	0	0	2	0	0	0	0	2	2	2	2	0	0	0	4
Rumex acetosa	Rumace	0	0	0	0	5	6	3	0	2	0	0	2	0	0	0	0	0	0	0	0
Sagina procumbens	Sagpro	0	0	0	5	0	0	0	2	2	0	2	4	2	0	0	0	0	0	3	0
Salix repens	Salrep	0	0	0	0	0	0	0	0	0	0	0	0	0	0	0	0	0	3	3	5
Trifolium pratense	Tripra	0	0	0	0	2	5	2	0	0	0	0	0	0	0	0	0	0	0	0	0
Trifolium repens	Trirep	0	5	2	1	2	5	2	2	3	6	3	3	1	6	1	0	0	2	2	0
Vicia lathyroides	Viclat	0	0	0	0	0	0	0	0	0	1	2	0	0	0	0	0	0	1	0	0

Global structuring index (GSI) The GSI was developed to define species showing the strongest influences on the structure of the SOM map organization. In other words, the GSI is the value indicating the relative importance of each species in determining the distribution patterns of the samples in the SOM. Therefore, the index values of each species indicate the relative importance of that species, and in turn the set of species showing high GSI can be considered as the indicator species. The GSI is calculated from the sum of the ratios of the distances between the weights (i.e., connection intensities) of all species in the SOM and the topological distance of two SOM units (Fig. 6.2.1). This results in representing distribution gradients for each species in the trained SOM. It is expressed in the equation as follows:

$$GSI_i = \sum_{k=1}^{S}\sum_{j=1}^{k-1}\frac{\left|w_{ij} - w_{ik}\right|}{\left\|r_j - r_k\right\|} \tag{6.2.1}$$

where GSI_i is a GSI value of species i, w_{ij} and w_{ik} are the connection weights of species i in SOM units j and k respectively, r_j and r_k are the coordinates of units j and k, and $\|r_j - r_k\|$ is the topological distance between units j and k. S is the total number of SOM output units.

a)

		Virtual Units					
		VU$_1$	**VU$_2$**	**...**	**VU$_k$**	**...**	**VU$_S$**
Species	sp$_1$	w_{11}	w_{12}	...	w_{1k}	...	w_{1S}
	sp$_2$	w_{21}	w_{22}	...	w_{2k}	...	w_{2S}
	:	:	:		:		:
	sp$_i$	w_{i1}	w_{i2}	...	w_{ik}	...	w_{iS}
	:	:	:		:		:
	sp$_m$	w_{m1}	w_{m2}	...	w_{mk}	...	w_{mS}

b)

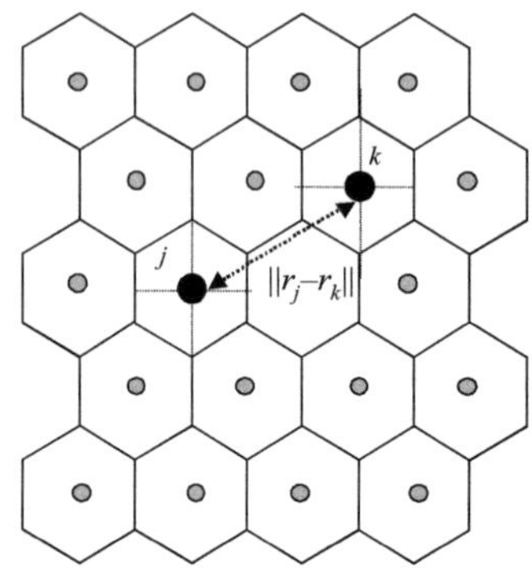

Fig. 6.2.1. The data matrix with virtual units (i.e., output units of the SOM) (**a**) and topological distance between virtual units (**b**).

Cluster structuring index (CSI) Each cluster defined in the SOM represents typical community types with a specific community composition. Therefore, the question is how to define key species in clusters considering distribution patterns of species in the SOM map. The values indicating the distribution patterns of species in each cluster can be considered to represent the importance of each species in each cluster. The idea was that a cluster defined in the SOM is an assembly of neighbouring units of the SOM, and that when a species is specific to a certain cluster, the difference of weights between SOM units is smaller. We named the indicating value as the cluster structuring index. It is implemented in the equation as follows:

$$CSI_i = \sum_{k=1}^{S}\sum_{j=1}^{k-1}\varepsilon_{jk}\frac{\left|w_{ij} - w_{ik}\right|}{\left\|r_j - r_k\right\|} \tag{6.2.2}$$

where CSI_i is a CSI value of species i, and $\varepsilon_{jk} = -1$ if units j and k are in the same cluster and $\varepsilon_{jk} = 1$ if units j and k are not in the same cluster. Other variables are defined in Eq. (6.2.1). The CSI of each species was calculated in different clusters.

Results

Presence and absence data

First, the SOM patterned 20 presence and absence data samples according to the occurrence similarities of species (Fig. 6.2.2). For instance, samples 6, 7 and 10 were grouped in the same cell (k). As shown in Table 6.2.1, these three samples have similar species assemblages. In contrast, samples 2 and 20 were assigned respectively to *a* and *t* cells far from each other, representing large differences in their species assemblages (Table 6.2.1). A hierarchical cluster analysis with the Ward linkage method showed 5 clusters (I-V) in the units of the SOM (Fig. 6.2.2). Units with similar species assemblages were grouped together in the same cluster.

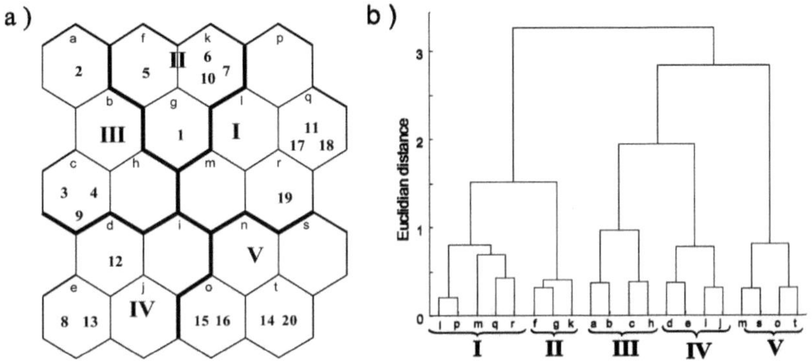

Fig. 6.2.2. Classification of samples in the SOM using presence-absence data (**a**) and hierarchical clustering of the SOM units (**b**). The numbers in the SOM units represent the samples listed in Table 6.2.1.

Fig. 6.2.3 shows the distribution patterns of each species in the units of the SOM and the corresponding GSI. Dark represents a high occurrence probability of each species in samples assigned to the units of the SOM, whereas light is low. Therefore, it represents the relative importance of each species in each unit (and in samples) of the map. Based on this distribution map, GSI values were calculated for each species mentioned as acronyms. The GSI values ranged from 4.82 to 33.91. Based on the distribution maps and GSI values, we defined 4 typical distribution patterns.

- **Type A** Species showing high GSI. These species have preferences in their distribution areas with high occurrence probabilities. For example the species *Agrsto* showing the highest GSI (33.91) displayed a strong gradient of distribution in the map. The highest occurrence probability of the species was >0.99 in the lower left areas of the map, whereas the lowest value was <0.01 in the upper right areas. Therefore, the species in this group strongly contribute to determining the patterns in the SOM.
- **Type B** Species showing low GSI. These species are limited in their distribution with low occurrence probability. For instance, species *Chealb* has the lowest GSI of 4.82. The highest occurrence probability was <0.279. This species was rare and particular to few samples. Therefore, the contribution of this species to the patterning in the SOM was relatively low.
- **Type C** Species showing an intermediate value. For example species *Rumace* displayed a mid-range GSI of 15.17. The highest occurrence probability was 0.517 in the upper left areas. Normally these species are observed in samples scattered over wide areas of the SOM map.
- **Type D** Species showing low GSI. Type D is distinguished from type B which also have a low GSI. Species in type D occur in most samples, therefore they do not play important roles in community classifications in the SOM. For example species *Brarut* showed

a GSI of 5.86, although it displayed relatively high occurrence probabilities in most units of the map in the range of 0.58 - 0.86. Actually, this was observed in most samples as shown in Table 6.2.1.

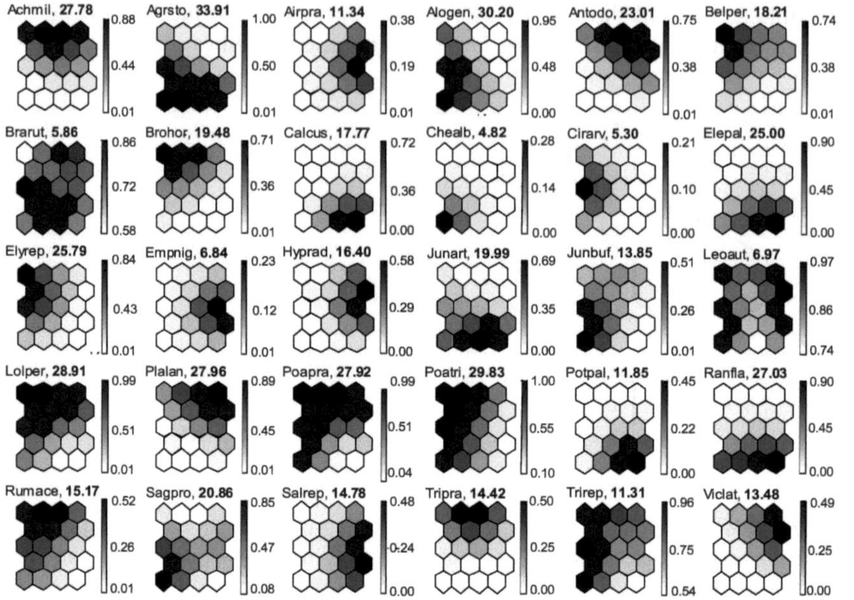

Fig. 6.2.3. Distribution patterns of each species in the SOM units based on the presence-absence data. Species acronyms are given in Table 6.2.1. The values following species acronyms are the global structuring index (GSI) of the corresponding species.

The CSI were also calculated for each species in different groups in the presence-absence data (Table 6.2.3). In Table 6.2.3, the CSI values for each species are given in each cluster. The cluster numbers showing the maximum CSI for each species are given in Roman numerals (I-V) defined in Fig. 6.2.2, and are marked in corresponding clusters in grey. The mean occurrence probabilities (i.e., mean weights) of each species in each cluster (i.e., mean values of weights of each species in each cluster) are also given in Table 6.2.2. The clusters displaying the highest occurrence probability were marked in grey, and if the cluster showing the highest CSI had the lowest (or the second lowest) probability, the cluster was marked in dark grey to differentiate it from the others. Overall the CSI and the mean occurrence probability showed similar results to each other, indicating that the characteristics of occurrence probabilities are well reflected in the CSI. For example, species *Airpra* showed the highest CSI and highest mean occurrence probability in cluster I, *Achmil* in cluster II, *Belper* in cluster III, *Alogen* in cluster IV, and *Elepal* in cluster V. However, there were also differences in 6 species (*Agrsto*, *Lolper*, *Poapra*, *Poatri*, *Rumace*, *Trirep*, and *Brarut*) marked in dark grey scale on the mean values. This was due to the method used for calculating the CSI.

Table 6.2.2. Global structuring index (GSI) and cluster structuring index (CSI) of each species based on the presence-absence data. The CSI was calculated in different clusters (I-V) defined in the SOM. The clusters showing the highest CSI are indicated in bold underlined. Mean connection intensities were calculated from the SOM weights in each cluster, and clusters defined by the SCI were indicated in bold underlined. If the mean weight is not maximum in the cluster showing the highest CSI, the clusters are indicated in grey. Based on this indication, a + or - sign was attributed to each species. Species acronyms are given in Table 6.2.1.

Species	GSI	CSI					Cluster*	Sign	Connection intensity				
		I	II	III	IV	V			I	II	III	IV	V
Achmil	27.78	25.24	**28.48**	19.75	24.88	24.17	II	+	0.54	**0.94**	0.49	0.04	0.06
Agrsto	33.91	**38.38**	25.06	23.20	35.23	27.55	I	-	**0.12**	0.09	0.55	0.94	0.78
Airpra	11.34	**15.45**	5.40	7.97	7.64	7.47	I	+	**0.72**	0.13	0.03	0.05	0.38
Alogen	30.20	31.18	16.66	27.56	**32.99**	20.35	IV	+	0.06	0.16	0.71	**0.83**	0.23
Antodo	23.01	**24.65**	19.90	15.58	22.05	16.70	I	+	**0.70**	0.85	0.28	0.04	0.20
Belper	18.21	14.54	12.44	**20.76**	13.36	18.63	III	+	0.34	0.59	**0.73**	0.19	0.04
Brarut	5.86	3.94	2.74	**4.88**	3.91	3.50	III	-	0.64	0.64	**0.58**	0.75	0.60
Brohor	19.48	15.01	**18.78**	18.51	14.42	17.81	II	+	0.27	**0.75**	0.62	0.11	0.01
Calcus	17.77	13.50	9.29	11.92	11.97	**30.71**	V	+	0.08	0.01	0.02	0.21	**0.76**
Chealb	4.82	3.68	2.25	2.93	**8.67**	2.84	IV	+	0.01	0.00	0.07	**0.56**	0.04
Cirarv	5.30	4.14	2.38	**6.75**	3.70	3.58	III	+	0.04	0.12	**0.65**	0.33	0.02
Elepal	25.00	21.01	14.92	18.30	21.07	**36.38**	V	+	0.09	0.01	0.05	0.45	**0.81**
Elyrep	25.79	20.75	15.40	**34.60**	16.15	21.78	III	+	0.15	0.52	**0.88**	0.28	0.03
Empnig	6.84	**7.32**	3.17	4.61	4.34	5.22	I	+	**0.58**	0.10	0.04	0.07	0.48
Hyprad	16.40	**23.70**	7.80	11.36	11.05	10.33	I	+	**0.70**	0.13	0.02	0.04	0.30
Junart	19.99	19.50	13.57	13.67	18.55	**22.18**	V	+	0.09	0.04	0.25	0.65	**0.77**
Junbuf	13.85	11.37	6.76	9.93	**15.88**	12.06	IV	+	0.19	0.35	0.52	**0.78**	0.07
Leoaut	6.97	**4.92**	3.23	3.97	4.18	4.90	I	+	**0.71**	0.48	0.68	0.60	0.37
Lolper	28.91	22.84	21.64	24.56	22.34	**36.89**	V	-	0.64	0.94	0.87	0.40	**0.08**
Plalan	27.96	**33.38**	22.28	19.02	24.98	21.12	I	+	**0.75**	0.75	0.23	0.02	0.11
Poapra	27.92	21.68	18.17	20.56	20.30	**39.99**	V	-	0.77	0.97	0.89	0.55	**0.12**
Poatri	29.83	29.43	17.49	26.11	20.71	**32.03**	V	-	0.34	0.89	0.97	0.81	**0.16**
Potpal	11.85	9.16	6.35	8.07	8.35	**19.27**	V	+	0.09	0.01	0.03	0.28	**0.80**
Ranfla	27.03	24.35	17.15	20.49	26.41	**34.90**	V	+	0.09	0.01	0.07	0.62	**0.82**
Rumace	15.17	11.77	13.22	12.31	9.83	**17.38**	V	-	0.38	0.93	0.73	0.47	**0.05**
Sagpro	20.86	14.66	13.66	13.71	**23.17**	13.85	IV	+	0.29	0.05	0.42	**0.73**	0.15
Salrep	14.78	**15.23**	7.96	11.72	10.41	14.83	I	+	**0.61**	0.10	0.03	0.09	0.71
Tripra	14.42	11.78	**17.37**	9.78	11.24	11.34	II	+	0.36	**0.89**	0.34	0.02	0.01
Trirep	11.31	8.88	5.72	9.42	8.05	**14.02**	V	-	0.50	0.68	0.84	0.75	**0.14**
Viclat	13.48	**19.68**	8.05	9.33	10.47	8.59	I	+	**0.71**	0.40	0.08	0.01	0.13

* Cluster showing the highest CSI value for each species.

The stronger the gradient in the distribution patterns of each species in the SOM units, the higher the CSI obtained. Therefore, a high CSI represents a high association with a corresponding cluster without considering positive or negative effects on the cluster. A positive effect means high occurrence frequency (or high density) of species in a cluster, whereas a negative one represents low frequency (or low density) of species in a cluster. To determine the direction of the effects, we used the differences between CSI and mean probability marked in dark grey, representing negative effects. Based on the differences, the CSI was assigned positive or negative signs as shown in Table 6.2.2. Therefore, a high occurrence frequency for most species played a positive role in determining the cluster in which they showed the highest CSI, whereas the low occurrence frequency of 6 species showing negative signs were important in determining the characteristics of their clusters. Finally, based on the CSI, the relevant species were summarized in each cluster as shown in Table 6.2.3. The species showing negative effects in clusters are indicated on a grey scale.

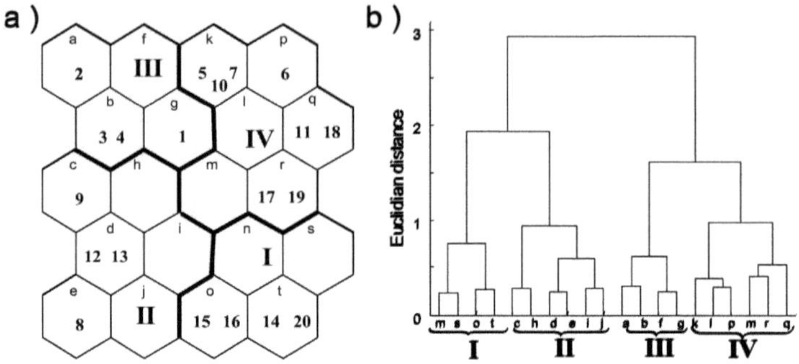

Fig. 6.2.4. Classification of samples in the SOM using abundance data (**a**) and hierarchical clustering of the units of the SOM (**b**). The numbers in the SOM units represent the samples listed in Table 6.2.1.

Abundance data

The SOM patterned the samples of abundance data according to the density differences of species (Fig. 6.2.4). Based on hierarchical cluster analysis, we found 4 clusters (I-IV) in the units of the SOM. The distribution patterns of samples on the map were similar to those of the presence-absence data (Fig. 6.2.2). For example, samples 2, 3, and 4 were classified in the same cluster (clusters III from both abundance and presence-absence data), samples 8, 12, 13 in cluster IV from presence-absence data and in cluster II from abundance data, and samples 14, 15, 16, and 20 were in cluster V from presence-absence data and in cluster I from abundance data. Samples "5, 6, 7, and 10" and samples "11, 17, 18, and 19" were grouped in the same cluster IV from abundance data, and they could be made into subgroups based on the cluster analysis dendrogram (Fig. 6.2.4b). Similarly they were grouped in cluster II and cluster I from presence-absence data.

Fig. 6.2.5 shows the distribution patterns of species on the map using weight vectors of the SOM. Dark represents a high occurrence probability of each species in samples assigned to the units of the SOM, whereas light is low. The weight vectors were calculated during the learning process of the SOM. The weight vector is considered as the approximation of the probability density function. Therefore, the higher the weight value, the higher the probability of a species being observed in a cluster (or sample) (with high density for abundance data).

The GSI was calculated for each species based on the SOM weight ranging from 5.80 to 25.23 (Fig. 6.2.5, Table 6.2.4). The results were very similar to those of the presence-absence data (Fig. 6.2.2, Table 6.2.2) and their typical patterns are summarized as follows:

- **Type A** Species showing a high GSI. They are abundant species and have preferences in their distribution areas like those of presence-absence data. For example, species *Lolper* showed the highest GSI with 25.23 in abundance data. This species was abundant in samples 1-11 and 18 which were allocated in the upper areas of the SOM map, while absent in other samples. Therefore, the species played an important role in determining distribution patterns of samples in the map during the learning process of the SOM and their importance was effectively reflected in the GSI.

Table 6.2.3. Relevant species in each cluster based on the cluster structuring index (CSI) in the presence-absence data. Species showing negative effects in clusters are indicated in grey. Species acronyms are given in Table 6.2.1.

Cluster	Species
I	Agrsto, Airpra, Antodo, Empnig, Hyprad, Leoaut, Plalan, Salrep, Viclat
II	Achmil, Brohor, Tripra
III	Belper, Cirarv, Elyrep, Brarut
IV	Alogen, Chealb, Junbuf, Sagpro
V	Elepal, Junart, Lolper, Poapra, Poatri, Potpal, Ranfla, Rumace, Trirep, Calcus

Table 6.2.4. Global structuring index (GSI) and cluster structuring index (CSI) of each species based on the abundance data (see Table 6.2.2 for detailed legend).

Species	GSI	CSI				Cluster*	Sign	Connection intensity			
		I	II	III	IV			I	II	III	IV
Achmil	16.83	13.43	**20.23**	15.23	19.61	II	-	0.08	**0.06**	0.39	0.39
Agrsto	24.56	17.78	31.07	17.12	**36.66**	IV	-	0.58	0.68	0.35	**0.10**
Airpra	10.70	8.18	9.66	7.04	**10.65**	IV	+	0.18	0.03	0.02	**0.19**
Alogen	24.19	16.96	**35.32**	16.31	29.67	II	+	0.17	**0.63**	0.38	0.07
Antodo	20.24	13.11	25.36	12.88	**30.21**	IV	+	0.21	0.05	0.26	**0.53**
Belper	17.41	16.03	16.22	**23.28**	16.43	III	+	0.03	0.12	**0.53**	0.26
Brarut	8.30	5.20	6.44	**8.55**	8.30	III	-	0.55	0.51	**0.41**	0.59
Brohor	16.78	14.62	16.03	**22.25**	15.11	III	+	0.01	0.08	**0.47**	0.23
Calcus	14.90	**25.42**	12.54	10.43	13.34	I	+	**0.46**	0.10	0.00	0.04
Chealb	6.57	4.15	**11.93**	4.19	6.44	II	+	0.02	**0.17**	0.02	0.01
Cirarv	5.80	4.26	4.59	**7.77**	5.15	III	+	0.00	0.06	**0.16**	0.02
Elepal	19.38	**26.46**	17.94	15.53	19.89	I	+	**0.54**	0.25	0.01	0.05
Elyrep	21.66	19.11	17.14	**27.52**	18.77	III	+	0.02	0.27	**0.61**	0.18
Empnig	5.97	4.57	5.39	3.93	**5.94**	IV	+	0.10	0.02	0.01	**0.10**
Hyprad	11.99	8.42	11.37	8.02	**13.55**	IV	+	0.18	0.03	0.03	**0.24**
Junart	18.19	18.22	20.94	14.73	**22.38**	IV	-	0.45	0.38	0.07	**0.05**
Junbuf	15.83	11.86	**26.10**	9.71	14.70	II	+	0.05	**0.43**	0.16	0.09
Leoaut	8.76	6.18	8.26	5.16	**10.70**	IV	+	0.54	0.54	0.60	**0.69**
Lolper	25.23	**29.98**	24.31	26.58	24.11	I	-	**0.08**	0.35	0.79	0.60
Plalan	22.94	15.57	26.55	15.19	**35.50**	IV	+	0.11	0.04	0.26	**0.58**
Poapra	23.98	**31.98**	20.65	23.53	21.27	I	-	**0.14**	0.52	0.84	0.64
Poatri	23.83	**27.12**	23.17	20.38	21.53	I	-	**0.14**	0.65	0.73	0.38
Potpal	10.99	**17.46**	9.19	7.94	10.10	I	+	**0.33**	0.09	0.00	0.03
Ranfla	19.50	**23.94**	20.22	16.72	21.77	I	+	**0.53**	0.33	0.02	0.05
Rumace	10.74	**11.78**	8.68	6.70	10.50	I	-	**0.02**	0.17	0.20	0.26
Sagpro	15.07	11.26	**22.17**	8.82	13.07	II	+	0.14	**0.48**	0.24	0.19
Salrep	11.13	**12.31**	11.37	8.60	10.52	I	+	**0.28**	0.03	0.02	0.18
Tripra	9.46	6.86	9.82	6.85	**12.94**	IV	+	0.01	0.01	0.12	**0.21**
Trirep	9.04	**9.54**	7.31	6.85	8.05	I	-	**0.37**	0.47	0.56	0.54
Viclat	9.42	5.87	10.15	5.82	**14.89**	IV	+	0.04	0.01	0.06	**0.23**

* Cluster showing the highest CSI value for each species

- **Type B** Species showing low GSI. The rare species belong to type B. For instance, species *Chealb* showed a GSI of 6.57. This species was observed only in sample 13 assigned to lower left areas of the map with a very low density.
- **Type C** Species showing intermediate values of GSI. For example, the mid-range GSI of 10.74 for species *Rumace*. The maximum occurrence probability of this species was about 0.81. Normally these species are observed in samples scattered over wide areas of the SOM map.
- **Type D** Species showing low GSI. These species are distinguished from the species in type B which also have low GSI. Species in type D occur in most samples. Therefore, the effects of these species are not important in patterning communities. For example, species *Brarut* showed a very low GSI of 8.30, being observed in most samples.

Table 6.2.5. Relevant species in each cluster based on the CSI in the abundance data. Species showing negative effects in clusters are indicated in grey. Species acronyms are given in Table 6.2.1.

Cluster	Species
I	Elepal, Lolper, Poapra, Poatri, Potpal, Ranfla, Rumace, Salrep, Trirep, Calcus
II	Achmil, Alogen, Chealb, Junbuf, Sagpro
III	Belper, Brohor, Cirarv, Elyrep, Brarut
IV	Agrsto, Airpra, Antodo, Empnig, Hyprad, Junart, Leoaut, Plalan, Tripra, Viclat

The CSI values were calculated in different clusters for each species (Table 6.2.4). The mean occurrence probabilities of each species calculated in the SOM learning process are also presented in Table 6.2.4. Overall, the results are similar to those obtained with presence-absence data (Table 6.2.2). For instance, species *Elyrep* characterized cluster III, showing a high CSI of 27.52. Species *Argsto* showed the highest CSI in cluster IV followed by cluster II. It indicated an importance of this species in cluster IV, but this species was present in low density in cluster IV, but abundant in cluster II. Therefore, the low density of this species was important in determining cluster IV, whereas the high density in cluster II had little importance. Some species such as *Chealb* and *Leoaut* showed low CSI values, representing relatively less importance in determining classifications in the SOM. Relevant species in each cluster are summarized in Table 6.2.5 and species showing negative effects in clusters are indicated in grey.

Discussion and conclusion

In this study we developed two indices (GSI and CSI) to indicate the relative importance of species in determining classifications in the SOM. The GSI can be used as an index for the selection of the most relevant species in classifications in the SOM, and the CSI can also be a useful indicator to determine the key species assemblages in each cluster defined in the SOM. To evaluate the feasibility of species selection based on the GSI, we selected 18 relevant species from the dataset and calculated the correlation coefficient between species richness in samples in the original dataset and in the reduced dataset with only the 18 selected species. The correlation coefficient was 0.81, representing a fairly good selection of species. Therefore, it showed the possibility of the GSI to be used as an indicator to reduce the data matrix size without losing too much information. The structuring index was applied to benthic diatoms, characterizing typical species assemblages in different ecoregions (Tison et al. 2004).

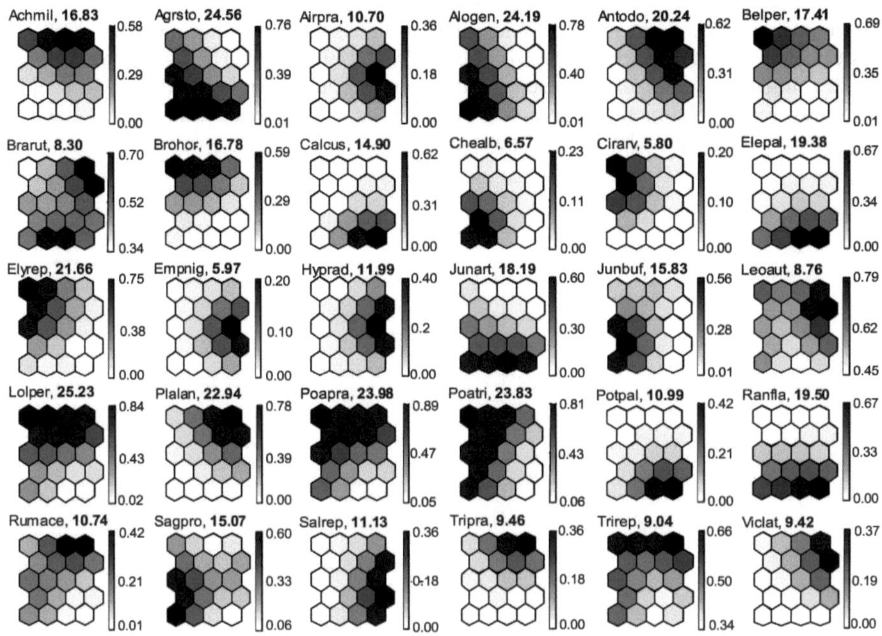

Fig. 6.2.5. Distribution patterns of each species in the SOM units based on the abundance data. Species acronyms are given in Table 6.2.1. The values following species acronyms are the global structuring index (GSI) of the corresponding species.

The calculation procedures of GSI and CSI are based on the weight values of the SOM. Due to the characteristics of the mathematical formula of the GSI, the GSI values are dependent on two properties: distribution patterns of species (limited areas or wide areas) and degree of occurrence probabilities of each species (high occurrence (or abundance) species or rare species). Therefore, the GSI and CSI show high values when species are observed in limited samples assigned to the same areas in the SOM with high occurrence (or abundance). They have very low values when observed in many samples in different clusters or in low densities (or occurrence frequencies).

The indices are highly dependent on the training resolution of the SOM. In this study we assumed that the SOM was smoothly trained in topology. Another weakness of these indices comes from the use of the distance between units of the SOM in the computation, because the SOM does not give strict gradient distribution. Most species show strong gradient distributions from one area to other areas in the SOM map. In the simple dataset, the gradient can be found easily. However, when the complexity of the dataset increases, the distribution gradients of species decrease in the map. Therefore, it is also important to choose optimum SOM map size. Another disadvantage of the CSI is its sensitivity to the cluster choice as in general analysis. Actually, the number of clusters is chosen by the user, although several different statistical techniques (i.e., U-matrix, hierarchical cluster analysis, k-means, etc.) can be used. Different defining boundaries in the SOM units can lead to different results of the CSI. Therefore, prior to calculation of the CSI, clusters should be defined properly based on ecological knowledge. In further studies, the weaknesses of the indices should be investigated before practical application to ecological issues.

In conclusion, the global structuring index (GSI) is useful in determining the importance of each species for structuring the SOM map, and the cluster structuring index (CSI) defines the most relevant species in each cluster defined in the SOM. The indices worked well in both abundance and presence-absence data; therefore, they can be used in most ecological studies. Finally the methods proposed here will be useful to quantitatively evaluate the importance of input variables in SOM patterns.

6.3 Projection pursuit with robust indices for the analysis of ecological data*

Werner H†, Rohatsch T, Pöppel G, Obach M, Wagner R

Introduction

The aim of projection pursuit (PP) (Friedman and Tukey 1974, Huber 1985, Jones and Sibson 1987) is to find low (1-3)-dimensional projections showing the most interesting views of high-dimensional data. To reach this goal, many projections are calculated and rated by an objective function, a so called projection index *I*. Dependent on *I*, this method is able to characterise the high-dimensional data from different points of view (Fig. 6.3.1).

To find good projections it is necessary to maximize *I*. There are several ways to do this: Gradient Algorithms (Friedman 1987), Simulated Annealing (Montanari and Guglielmi 1996) or Genetic Algorithms (Crawford 1991, Guo et al. 2000).

The application of PP has some advantages over other methods. Although the projections are linear, the index may be non-linear, facilitating the detection of non-linear connections in the data. By choosing different and suitable indices, PP is able to provide various views of the data. One important feature of the index functions is robustness; this enables PP to be insensitive to outliers. Projection vectors from previous analysis can be used to view and analyse another data set from the same viewpoint.

The multivariate analysis method PP can be used to classify data, comparable to supervised and unsupervised learning in artificial neural networks, but with reproducible results.

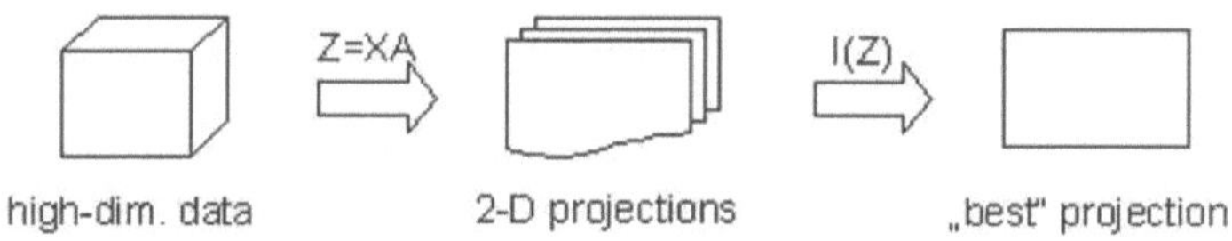

Fig. 6.3.1. How projection pursuit works: two-dimensional projections are calculated from high-dimensional data, and rated by an index function. Dependent on this resulting index-value, a specific projection can be determined, which shows an "interesting" view of the data set.

In ecological applications we are faced with the problem of small data sets of high dimensionality, where statistical confidence is not achievable. Even in this situation PP is a suitable tool for data analysis, combining computational power (fast production and preselection of two-dimensional views of the data) with human intuition (detection of regularities and dependencies in two-dimensional diagrams). PP can generate hypotheses on the underlying data and provide insight regarding experimental design for hypotheses testing.

* We thank Dr. Hans-Heinrich Schmidt, Limnologische Fluss-Station Schlitz, for providing environmental data.

† Correspondence: heinrich.werner@uni-kassel.de.

Methods

Generally PP indices can be divided into subgroups. Two essential subgroups are

1. Exploratory Projection Pursuit (EPP) (Friedman, 1987) Indices: This kind of PP-indices are characterised by their exploratory characteristics. The aim of these indices is to find structured projections without former knowledge of any characteristics of the data.
2. Projection Pursuit Discriminant Analysis (PPDA) (Posse, 1992) Indices: This group of PP indices aims to find projections in which data points belonging to different categories are most separated. This task requires knowledge of information for at least one category for the data points to be analysed, so this kind of index is also known as a supervised index.

   The indices presented here can be applied to both EPP and PPDA.

Definitions and notation

Let $\boldsymbol{X}$ be a NxK data matrix, with has N cases and K variables (dimensions). Let $\boldsymbol{A}$ be a KxP projection matrix. For a linear projection onto P dimension with $P << K$ the matrix of the projected data is calculated.

$$\boldsymbol{Z}=\boldsymbol{XA}$$

For all the following indices let $P\equiv 2$. The length of all column vectors $\vec{a}_i$ of $\boldsymbol{A}$ is normalized to 1 and they are chosen to be orthogonal.

Some definitions in statistics which are used in robust data analysis.

Trim Cases in which values for a previously sorted variable are greater or less then k extreme values are deleted from the dataset. k is often based on a percentage of all cases in the dataset. With many variables (dimensions), this kind of data manipulation may often result in the rejection of a large number of cases, leaving few cases for analysis.

Winsorise To be robust against outliers, k extreme values of a previously sorted variable are changed: the i smallest values are changed to the $(i+1)$ smallest value and the j greatest values are changed to the $(j+1)$ greatest value. k is often based on a percentage of all cases in the data set.

Rank Each variable is sorted in ascending order and the position in the sorted list is used rather than the actual value of the variable. This softens the effect of outliers, but preserves the basically structure of the data.

Rank Nearest-Neighbour (NN)

To detect "near-order-constellations" within the data, an index is used which concentrates on Nearest-Neighbour distances. The nearer the data points (with the wanted category) and the corresponding Nearest Neighbours lie together the higher is the resulting index value. To be robust against outliers and to make the calculation of the Nearest-Neighbour distance computationally efficient, the data are ranked (Fig. 6.3.2). It should be noted that a category could be specified which would make this index either a member of EPP or of PPDA. Here as an example, we use the category information to consider only cases with the specified category. Without the category information, considering all cases, this index would be a member of EPP. This is equally true of all the indices presented here.

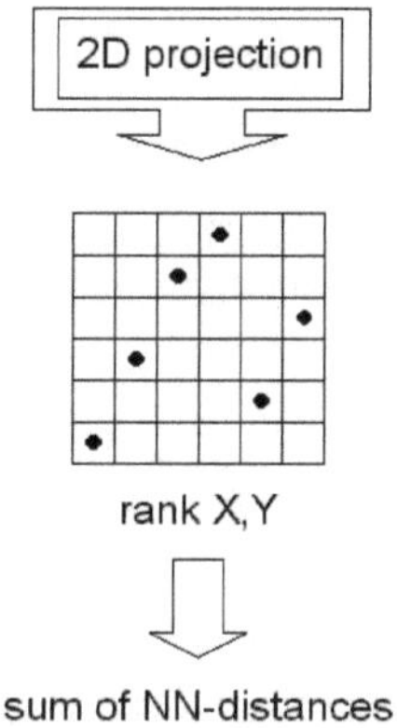

Fig. 6.3.2. Rank NN-index. Starting from a two-dimensional projection the data is previously ranked in the *x*- and *y*-dimension (rank X,Y), then the sum of the Nearest-Neighbour distances is calculated as a measure for this index.

Let $\boldsymbol{X}^c$ be the matrix of data which contains only N_c cases with the category c. This category c is chosen by the user. Let $\boldsymbol{S}^c$ be the matrix containing the ranked values of the projected data matrix $\boldsymbol{Z}^c$. d^{NN} is the distance between a data point and its Nearest-Neighbour, e.g. measured by the Manhattan distance. To be robust against outliers, the d^{NN} are sorted in ascending order and the extreme distance values (e.g. the greatest 5%) are trimmed. As a result, the number of cases N_c also decreases by 5%. The sum of all d^{NN} over all N_c cases gives the Nearest-Neighbour distance sum.

$$S_{NN} = \sum_{i=1}^{N_c} d_i^{NN} \tag{6.3.1}$$

The closer all different Nearest-Neighbours, the smaller is this sum. To get a maximizing index in the interval [0,1] the following transformation is used:

$$I = \frac{2N_c}{S_{NN}} \tag{6.3.2}$$

It is straightforward to generalize this index to k-Nearest-Neighbour.

Different Distance Distribution (DDD)

This index uses the difference between the cumulative frequencies of point distances (Läuter and Pincus 1989) from previously trimmed projected data compared with a reference distribution (point distances from a trimmed projected normal distribution). This difference is measured by the maximum distances of the two distributions (e.g in analogy with the Kolmogorov-Smirnov test (Sachs, 1999)) (Fig. 6.3.3).

Let $\boldsymbol{T}$ be a matrix with N rows and 2 columns, consisting of random data from a normal distribution. Let $\boldsymbol{Z}$ be the matrix of the projected original data. Both matrices $\boldsymbol{T}$ and $\boldsymbol{Z}$ are trimmed in the *x*- and *y*-dimension to be robust against outliers. The point distances can be calculated using any metric, here for example the Manhattan distance d_M:

$$d_M(\vec{r}_i, \vec{r}_j) = |x_i - x_j| + |y_i - y_j| \tag{6.3.3}$$

The set of all point distances d_P of $\boldsymbol{T}$ (respectively, $\boldsymbol{Z}$) is given by

$$d_p(\mathbf{T}) = \{ d_M(\vec{r}_i, \vec{r}_j) \mid \vec{r}_i, \vec{r}_j \in \mathbf{T}, i < j \}$$
$$d_p(\mathbf{Z}) = \{ d_M(\vec{r}_i, \vec{r}_j) \mid \vec{r}_i, \vec{r}_j \in \mathbf{Z}, i < j \} \tag{6.3.4}$$

Let $F(d_P(\boldsymbol{T}))$ (respectively, $F(d_P(\boldsymbol{Z}))$) be the empirical distribution of the point distances. The index value is the maximum distance D_{max} between the two distributions of the point distances.

$$D_{\max} = \max(|F(d_p(\mathbf{T})) - F(d_p(\mathbf{Z}))|)$$
$$I = D_{\max}. \tag{6.3.5}$$

This index could also be easily extended to a member of PPDA. Therefore different reference distributions could be used e.g. the distances of data points with category A. Calculating the index value considering only data points with category B for the distance distribution of the actual projection, projections could be found, in which data points with category B are as much different in means of a different distances distribution as possible from data points with category A.

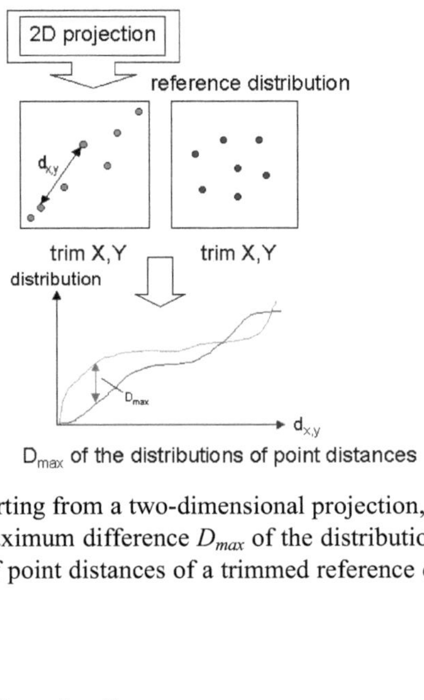

Fig. 6.3.3. DDD-index: starting from a two-dimensional projection, which is trimmed in the x- and y-dimension, the maximum difference D_{max} of the distribution of this point distances $d_{x,y}$ from the distribution of point distances of a trimmed reference distribution is calculated as an index value.

Separate handling of projection axes

To be able to analyse data from different data groups and correlations between them at once, two different data matrixes $\boldsymbol{X}_1$, $\boldsymbol{X}_2$ could be used at the same time. $\boldsymbol{X}_1$ is then projected onto the x-dimension and $\boldsymbol{X}_2$ onto the y-dimension, of the two-dimensional projection.

This has some advantages compared to the projection of the same data matrix $\boldsymbol{X}$ for both projection dimensions.

- simultaneous projections of different ($\boldsymbol{X}_1$, $\boldsymbol{X}_2$) preprocessed data
- $\boldsymbol{X}_1$ and $\boldsymbol{X}_2$ could consist of different kinds of data and could also have different numbers of dimensions

- inter- and intra-data dependencies (e.g. structure, clusters etc. within the data of one axis or between the different data of the two axes) could be analysed at the same time (dependent on the projection index)

Implementation

PP with the indices presented here (NN and DDD), and some more indices and expansions, was implemented in the software-package "Autonomous Projection Mapping" (APM). This software enables the user to analyse data interactively and also facilitates a largely autonomous data analysis without user interaction (Fig. 6.3.4). APM was realised as a OO-classpackage using the programming language Delphi. This provides rapid prototyping and enables a fast translation to a program.

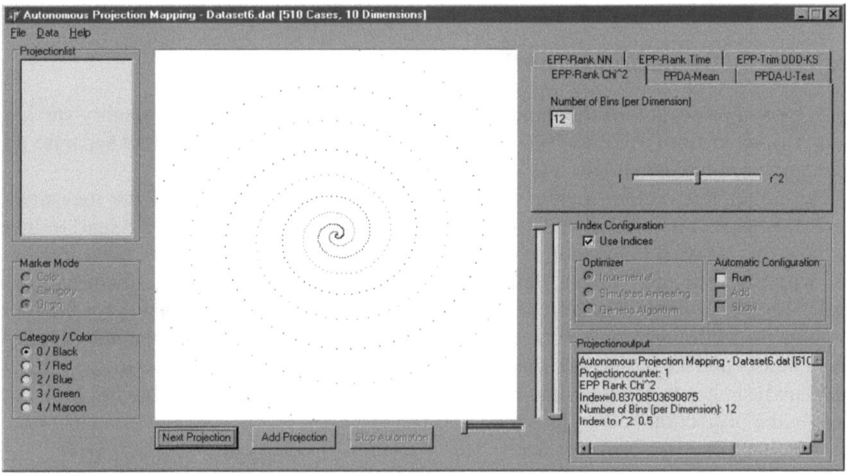

Fig. 6.3.4. Main window of the software APM (Autonomous Projection Mapping).

Data

Data are from the period 1982 to 1993 and are part of a long-term study of the Limnologische Fluss-Station Schlitz (Central Germany) on the small mountain stream Breitenbach, which was extensively described by Ringe (1974) and Cox (1990).

Variables are the monthly abundance of adults of the caddis fly *Agapetus fuscipes* (Benedetto 1975, Becker 2001) in four emergence traps along a two-km study stretch. The environmental parameters are monthly maxima of water temperature and discharge measured on the Breitenbach and the monthly sum of precipitation (measured at a gauging station, 2 km distance) (Wagner et al. 2000a).

Data were mainly standardised to zero mean and one standard deviation. With standardization it was possible to compare the standardized values on a comparable numerical scale independent of the original data.

Results and discussion

The APM program detects and quantifies regularities, irregularities, and interconnections in large high-dimensional data sets to a high degree of certainty; it was originally designed for technical applications (Rohatsch et al. 2002).

In environmental studies we are faced with similar problems, but the amount of data available tends to be small. Our goal is to show that APM can also be useful in ecological research. With a small number of high dimensional vectors we are always in danger that regularities we might detect are accidental and cannot be generalized. We may still use this tool to find regularities to generate hypotheses, although at a much lower level of certainty. We can also test hypotheses with APM projections and new test-data. The real power of this tool lies in the interaction of a fast calculation of two-dimensional projections with the human observer who can see and select the most promising results for further investigation. We explain this procedure with four examples.

Example 1. Detection of outliers

Fig. 6.3.5 is a projection of the monthly discharge maxima (x-axis) and monthly sums of *A. fuscipes* (y-axis) from 1982 to 1993. A dense cloud of data-points and two separate points is recognized, a typical situation to get a hint to outliers in the data set.

The projection vectors in Fig. 6.3.5 show with respect to which of the measured parameters (months) why 1989 and 1991 behave as outliers and so we have a lead to investigate how this unusual behaviour of data points arises. Typical reasons are measurement errors, wrongly estimated values, failures or unusual climatic conditions, disturbances etc. Even though December, January, and February seem to have high relevance for the outliers (compare projection vectors), it is not true, since all examples (species abundance) are 0 in these months and hence do not contribute anything to the high distance to the cluster. A closer analysis has to focus on the remaining months, particulary those having high coefficients in the projection vectors.

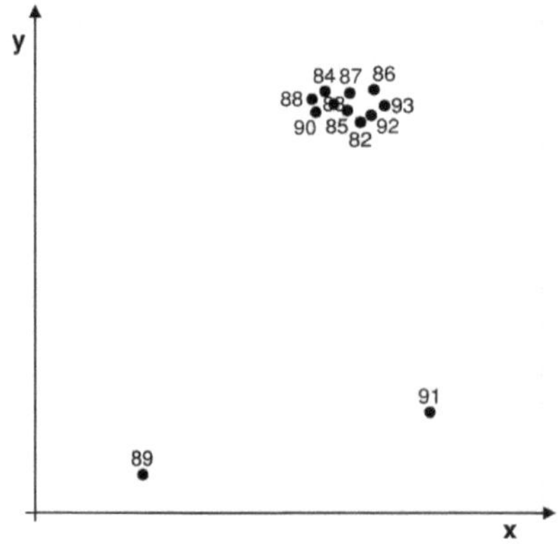

Projection vectors ($\cdot 10^{-2}$)

month	X	Y
Jan	5.74	-2.11
Feb	-3.96	-0.68
Mar	0.25	0.38
Apr	-1.38	-0.62
May	-3.98	0.15
Jun	3.19	0.22
Jul	3.52	-2.38
Aug	-2.71	-2.01
Sep	-1.32	-4.71
Oct	-0.15	-4.19
Nov	-1.25	-6.55
Dec	0.59	-1.45

Fig. 6.3.5. Projection of the monthly discharge maxima (x-axis) and monthly sums of *A. fuscipes* (y-axis) from 1982 to 1993

In this case the outliers promote an interesting interpretation: in 1989 an extreme flood occurred in February and very low discharge in April (shift in x-direction). Due to low flow the emergence period of *A. fuscipes* in 1989 and 1991 was extraordinarily long with large specimen numbers from May to September resulting in a *y*-shift downwards because of the negative coefficients.

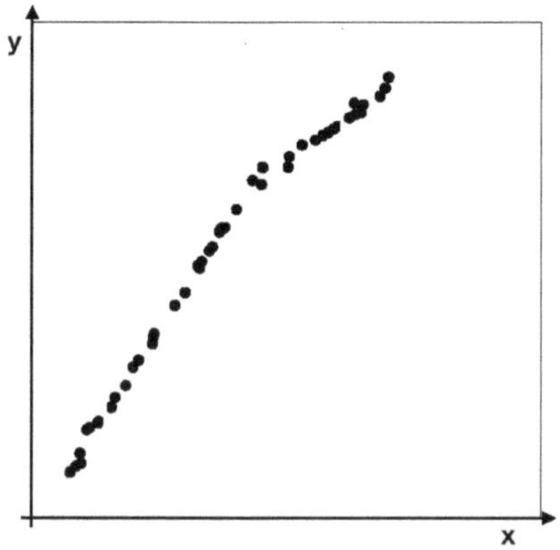

Projection vectors ($\cdot 10^{-2}$)

month	x	y
Apr	-2.90	0.65
Mai	-1.02	1.12
Jun	5.70	4.90
Jul	-4.19	-1.44
Aug	-0.41	-0.90
Sep	0.39	2.14
Oct	2.52	3.89
Nov	5.82	-7.20

Fig. 6.3.6. Projection of the *A. fuscipes* abundance data from the four traps just from April to November

Example 2. Dimension reduction

Fig. 6.3.6 shows a projection of the *A. fuscipes* abundance data from the four traps just from April to November. We observe a simple linear relation between the different data vectors. This implies a strong intercorrelation between the measured parameters. The data points appear approximately linear, slope 1.9. Hence the combination y-1.9x of the two projection vectors is orthogonal to the data set:

month	Apr	May	Jun	Jul	Aug	Sep	Oct	Nov
y-1.9·x	6.16	3.058	-5.93	6.521	-0.121	1.399	-0.898	-18.258

This is very useful in respect to a dimension reduction on the data set. The correlation was established mainly without August, September, and October. Since November had hardly any values greater than 0, it was almost without effect on the position of any data point and thus was ignored. The remaining months, April, May, June, and July, almost exclusively determined the position of the points - a substantial dimension reduction.

This result is biologically sound but not surprising, since these are the months of maximum emergence in which similarities and dissimilarities between different environmental conditions become greatest. Nevertheless, the method can lead to obvious or concealed interconnections between the data and to substantial dimension reduction.

Example 3. Detection of functional dependencies

The projection in Fig. 6.3.7 shows a clear sigmoid functional (S-shape, saturation curve) dependency between the temperature (x-axis) and the population (y-axis). The two projection vectors show which linear combination of temperature presents the input to this sigmoid function and which linear combination of abundance values is produced by this function.

Maximum summer water temperature increased from 1989 to 1991, discharge decreased. This combination of environmental factors provided optimal conditions for the development of large numbers of *A. fuscipes* in any trap. [remplacer virgules par des points dans la fig 367

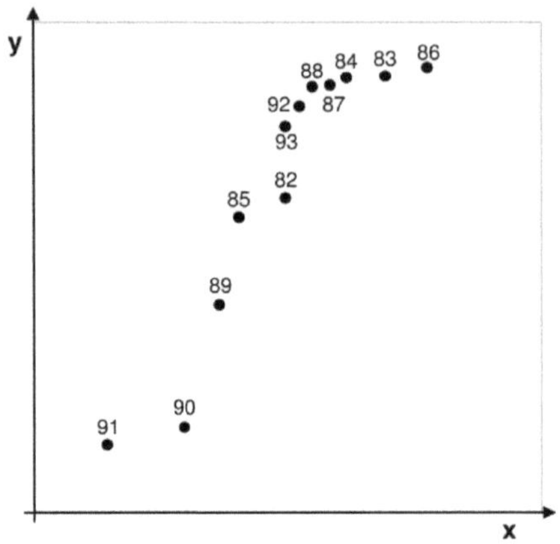

Projection vectors ($\cdot 10^{-2}$)		
month	X	Y
Jan	-0,37	1,17
Feb	0,19	1,14
Mar	-4,48	-1,92
Apr	-0,54	-0,59
May	1,52	-2,88
Jun	0,94	-2,41
Jul	1,32	-2,82
Aug	4,68	-0,56
Sep	1,44	-0,07
Oct	0,43	4,47
Nov	2,67	-7,09
Dec	6,57	0,77

Fig. 6.3.7. Temperature (x-axis) and the population (y-axis) of *A. fuscipes* from 1982 to 1993

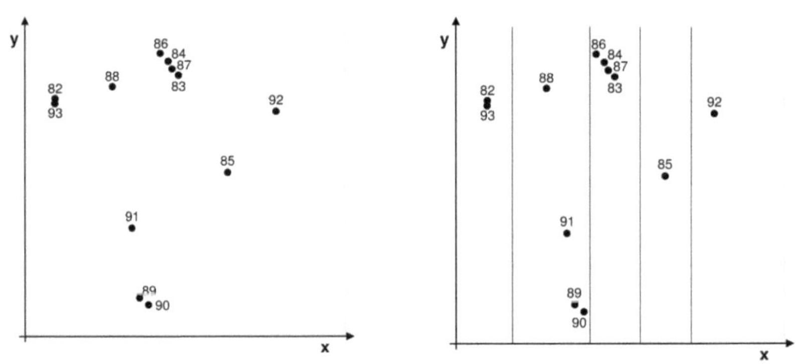

Fig. 6.3.8. Precipitation (x-axis) and the population (y-axis) of *A. fuscipes* from 1982 to 1993

Example 4. Clusters and functional dependencies

Fig. 6.3.8 shows a less clear functional dependency compared with Fig. 6.3.7. However, the x-axis (precipitation) shows a clear subdivision into separate intervals each with a particular characteristic. The first and the third interval contain dense clusters, hence in this precipitation range the abundance is almost constant. Intervals four and five are singletons and as such are difficult to interpret. The second interval shows no clear pattern and therefore needs further investigation. This projection suggests that it might be sensible not to expend too much effort on very precise precipitation measurements, but rather only to consider rough ranges.

As precipitation affects indirectly -via discharge- populations of aquatic insects, and as particularly in summer precipitation is not necessarily followed by increased discharge, this functional dependency appeared less clear than the direct effects of water temperature or discharge. However, the detection of the dependency precipitation – population even though small, was a good example for the capability of the method.

Summary

It is common to all four examples that the number of points in the projections were far too small to permit any generalisation. However, we want to stress the fact that PP still produces meaningful figures, allowing users to generate sound hypotheses about the data. The projection vectors give hints how these hypotheses can be verified in further experiments. Using the same projections, every new measurement can be tested whether it fits the figure produced by the given data points or the underlying hypothesis. The same method combined with new measurements can also serve to verify hypotheses.

Conclusion

The APM-program based on PP is designed for the visual inspection of data in high-dimensional spaces by constructing interesting two-dimensional projections. APM rapidly computes large numbers of projections. With a given projection index, it automatically selects the most interesting projections, i.e. those of the highest index values.

The figures and the corresponding projection vectors can be inspected for the detection of exceptional data, clusters, and functional dependencies. Even though it was designed for, and has been successfully tested with, very large data sets, APM also produces valuable views on small data sets typical for environmental research. These views can help to generate sound hypotheses and to design experiments for their verification.

6.4 A framework for computer-based data analysis and visualisation by pattern recognition

O'Connor MA*, Walley WJ

Introduction

Overview

In a wide variety of situations, masses of data must be analysed and condensed in order to extract meaningful information. Computer-based methods of data analysis and visualisation offer a means by which such data can be used to its full potential. Research in the domain of river water quality monitoring led the authors, together with colleagues, to develop a software system for classification and diagnosis of biological samples (Walley et al. 2002). The software comprises a 'training' element – MIR-max (Walley and O'Connor 2001) - and a user-friendly interface for presentation and analysis of results – River Pollution Diagnosis System (RPDS) (O'Connor and Walley 2001). Although these systems were developed for the specific domain of biological river water quality monitoring, they are almost entirely data-driven, and thus it is clear that the same methods could be readily adapted to many other applications. It would be beneficial for researchers to define a unified framework within which to approach broadly similar problems and to present and compare results.

Definitions

Most of the terminology used when describing the proposed framework derives either from the original application domain (biological river water quality monitoring) or from standard pattern recognition terminology. The following definitions are used throughout:

An *indicator* is a variable associated with the application domain. For RPDS, the indicators include abundance levels of various river dwelling macroinvertebrates (worms, leeches, snails, insects, etc), environmental variables such as river width, depth and the nature of the substrate, severity of perceived stresses at the site, and various chemical measures (alkalinity, biochemical oxygen demand, ammonia, etc). Indicators may be either *training* or *non-training*. Training indicators are those that will be used in the initial data analysis, whilst non-training indicators are not used in the initial analysis.

A *sample* is an instantiation of a set of indicators. For example, in RPDS each sample is a vector whose elements are the values of the indicators recorded for a particular river site on a particular date.

An *archive* sample is one that was used in the initial training process; an *input* sample is one that was not used in training the system and thus has not been previously 'seen' by the system. Generally, the set of archive samples should have full data coverage for all the

* Correspondence: m.a.oconnor@staffs.ac.uk

training indicators and good coverage of the non-training indicators, whilst input samples are only required to have full coverage of the training indicators. Non-training indicators may thus be defined as those for which data are not expected to be routinely available for any new samples.

A *cluster* is a set of samples that have been grouped together because they exhibit 'similar' characteristics.

A *map* is an output space in which clusters are arranged to indicate their relative similarity (i.e. clusters that are considered similar should be positioned close together on the map and dissimilar ones far apart). As well as showing the relative positions of the clusters, the map may also be colour-coded to show the cluster values for particular indicators: such a map is referred to as a *feature map*, a visualisation technique used in SOM (Kohonen 2001).

A *template* is a graphical representation of a set of indicator values for a particular sample or cluster – a bar chart that displays each sample or cluster as a *pattern*. The bars must be suitably scaled so that each template is directly comparable. A template may also be used to compare samples and clusters, by using two bars for each indicator, one for the sample and one for the cluster.

Pattern recognition and clustering

Modelling an expert

A computer-based system for data analysis and interpretation should act as an expert assistant to provide decision support. Thus, a good starting point when developing such a system is to attempt to model the mental processes of a human expert in the domain. When human experts process data, they use two complementary techniques:

- *Plausible reasoning*, based on scientific knowledge of facts and causal relationships, together (possibly) with heuristics built up from experience.
- *Pattern recognition*, based on the expert's knowledge of past cases, treating the data holistically and using experience of meaningful patterns previously encountered in similar data sets.

The framework described in this paper is based on a pattern recognition approach. However, researchers at Staffordshire University's Centre for Intelligent Environmental Systems (CIES) are also using Bayesian belief networks (BBN) (Jensen 1996) to model the plausible reasoning approach to data interpretation by experts (Trigg et al. 2000, Walley et al. 2002). It is hoped that in future these two approaches, pattern recognition and plausible reasoning, may be integrated into a single framework, to provide a model for a 'true' expert system.

Automated pattern recognition

A computer-based pattern recognition process takes a new data sample (e.g. for RPDS, a set of abundance data for creatures found at a river site) and attempts to find the 'best match' between it and one member of a set of clusters derived from analysis of previously known data. The sample is then assigned to this best matching cluster. The set of clusters is derived from a large set of previously known data (the training data), which is partitioned in such a way that samples representing very similar conditions are clustered together. The average values of samples associated with each cluster can be regarded as exemplar patterns; thus, each cluster represents a particular type or class of sample. The process is summarized in Fig. 6.4.1, where each sample is represented as a template; no information is provided on the meaning of each sample (i.e. no labels are given for the axes) so as to focus on the patterns produced by the data, and to emphasise the generic nature of the process. When a new sample is assigned to a cluster, previous knowledge of the samples used to construct that cluster can be used to draw inferences about the new sample. Thus, the system uses its experience of past cases to assess new cases, a process analogous to the way human experts often work.

This clustering of the data samples can be achieved by a wide variety of methods, perhaps the best-known being the partitional methods k-means clustering (MacQueen 1965) and ISODATA (Hall and Ball 1965); for environmental applications, hierarchical methods such as TWINSPAN (Hill 1979b) are also popular. However, the authors do not recommend the use of hierarchical techniques because they do not classify the data holistically, but place greater emphasis on specific features, thus making them more vulnerable to incorrect classification.

Most clustering techniques require that distance or dissimilarity between samples can be measured using a well-defined metric (for example, Euclidean distance in multidimensional data space or, commonly in ecological applications, Bray-Curtis distance). For RPDS, a novel information-theoretic technique developed at CIES, called Mutual Information Maximisation (MI-max), is used. MI-max is particularly suited to discrete categorical data such as the abundance level data available from Environment Agency biological surveys; it does not assume normality on the distribution of the data, and does not require a distance metric (which would be difficult to define rigorously since each data sample consists of values for a variety of environmental, chemical and biological indicators). However, the most appropriate clustering technique for any dataset is dependent upon the nature of the data and the application domain.

In most cases (when using a partitional method), the user will specify the number of clusters, based on the number of archive samples available and the application domain. The number of clusters can then be revised by 'trial and error'. It may be possible to determine a theoretical optimum number of clusters from an initial statistical analysis of the data set, although a user-defined number of clusters (adapted and refined by experience) should suffice. For RPDS, approximately 6000 river site archive samples are grouped into 250 clusters, giving an average of 24 samples per cluster: this is felt to be sufficient to enable 'meaningful' clusters that can be regarded as exemplar patterns for differing types of sample, whilst not 'over-generalising'.

Data visualisation

Effective dissemination of information

An essential feature of any data analysis and decision support system is its ability to convey complex multivariate data in a way that is meaningful and informative to a wide variety of possible end-users. Software should be as user-friendly as possible, so that all potential interested parties can use it effectively: not just computer specialists, data analysts or statisticians, but also managers, politicians, or those involved in public relations, for example. The use of templates (as in Fig. 6.4.1) is a simple way in which individual samples or clusters can be compared and contrasted. Clustering the large initial set of data makes it more manageable whilst retaining, or even enhancing, its information value. However, with a large number of clusters (as will be the case in many practical applications) it is still difficult to gain an informative overview of the entire data set: it would be impractical, for example, to use a large number of templates simultaneously.

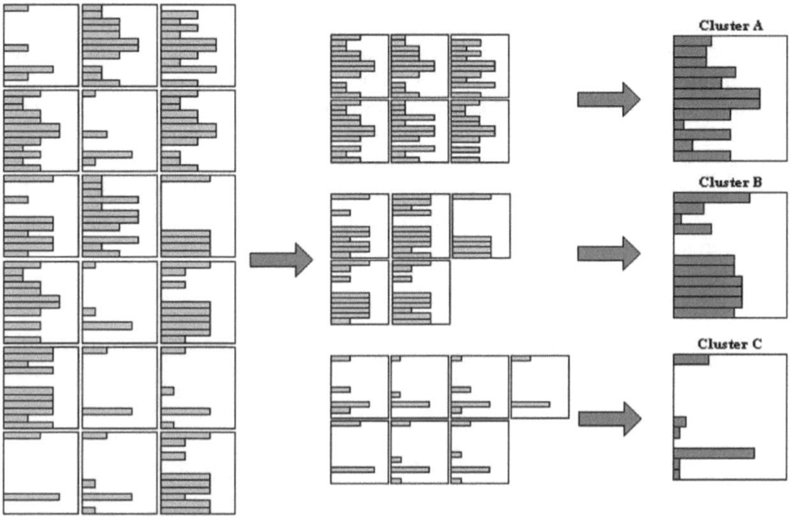

Fig. 6.4.1. The clustering process. A set of samples, here represented as templates (bar charts representing the value of each indicator, rescaled so as to be comparable) is arranged into groups – clusters – according to some measure of similarity. Labels on the *x*- and *y*-axes have been omitted in order to focus on the *patterns* represented by the data. The resultant clusters can then be characterized by the average values of each indicator, to provide a cluster template or exemplar pattern.

Feature maps

The clusters can be arranged in an output space in such a way that those clusters that are most similar are positioned close together, whilst those that are very different are far apart. For RPDS, this ordering in output space was achieved by a novel algorithm developed by the authors, Regression Maximisation (R-max), which together with MI-max forms the integrated pattern recognition and visualisation package MIR-max (Walley and O'Connor 2001). R-Max requires that some measure of distance or dissimilarity can be defined between samples or clusters: however, this is not a restriction in practice (indeed, the concept of a map is itself dependent on distance between clusters carrying some meaning) because the ordering process is regarded as secondary to the clustering and so only an approximation is required. Alternative techniques for producing a visualisation of the data in a 2-d output space include ordination methods such as PCA (Hotelling 1933), Sammon mapping (Sammon Jr. 1969) or a variety of multidimensional scaling (MDS) methods (Kruskal and Wish 1978). However, these techniques plot points in a continuous output space, rather than feature maps based on the lattice structures used by MIR-max and SOM, making them less intuitive to interpret and possibly not so useful in practice. The discrete nature of the MIR-max output space produces a powerful and practical visual overview and user-friendly 'point and click' capabilities for data exploration using a computer interface. There is, however, no theoretical restriction to rigid lattice structures within the framework. The idea of maps based on discrete lattice structures derives from the output of the SOM (Kohonen 2001), an unsupervised neural network; indeed, SOM was originally used in the early stages of development of RPDS (Walley et al. 2000). SOM or related techniques, such as generative topographic mapping (GTM) (Bishop et al. 1996), can also be used for both clustering and visualisation. However, the fact that these techniques combine the clustering and ordering processes into one process is likely to lead to sub-optimal clustering. MIR-max separates these processes, as it is the authors' belief that the clustering is of prime importance (O'Connor and Walley 2000). The map produced by the chosen ordering technique provides an overview of the relationships between the clusters. By colour-coding the map, it can be used to represent the average values of any particular indicator. Such feature maps can then be compared and contrasted to gain information about the relationships between different indicators.

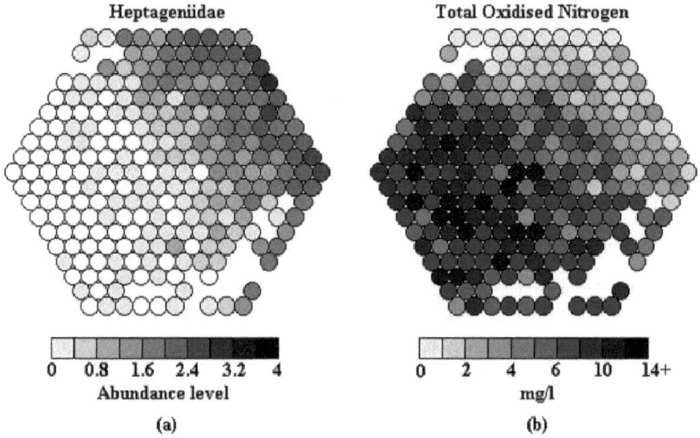

Fig. 6.4.2. RPDS feature maps for (**a**) Heptageniidae and (**b**) Total Oxidized Nitrogen

Fig. 6.4.2 shows two feature maps from RPDS, one for the abundance level of the mayfly family Heptageniidae (a training indicator) and one for the Total Oxidized Nitrogen (TON) level (a non-training indicator). Each cluster is represented by a circle, colour-coded (converted to grey-scale for this paper) to show the average value of the indicator for samples within the cluster. The inverse relationship between these two indicators is clearly shown by the feature maps. The map shown in Fig. 6.4.2 is based on a discrete hexagonal lattice structure, with a limited number of possible locations for the clusters to occupy. The shape and size of the output lattice can be varied according to the user's needs or the particular application area, provided there are at least as many possible output locations as there are clusters. For example, using a map with more output locations allows for greater disaggregation, so that the differences between clusters can be more accurately represented.

The maps are also useful for analysing individual input samples. When an input sample is presented to the system, the cluster to which that sample should be allocated can be highlighted. However, the designated cluster may not represent a perfect match. The 'next best' clusters are implied by the relative locations of clusters on the map (i.e. neighbouring or nearby clusters represent similar conditions), but can also be made explicit by colour-coding the map to show a gradation of possible clusters to which the input sample may belong.

Other visualisation techniques

Templates, as discussed in section 4.1, are incorporated in the framework. Whilst feature maps facilitate visualisation of indicator values in all clusters, templates are an ideal means by which to view the underlying pattern within individual clusters, and to compare the patterns with particular samples. The user can select a subset of indicators (in any order) to use as a template, and can also produce text reports and a variety of charts and graphs to present the data in more detail, although the exact format of such reports is dependent on the application domain. The generalised user interface – based on that developed for RPDS – incorporates a number of overlapping tabbed panels, such as a panel for displaying templates and a panel for displaying text reports, that can be 'user-defined' to show more specifically application-dependent information, e.g. in RPDS one panel is used to display the location of river sample sites on a map of England and Wales. Thus, the system is flexible enough to be adapted to a wide variety of applications.

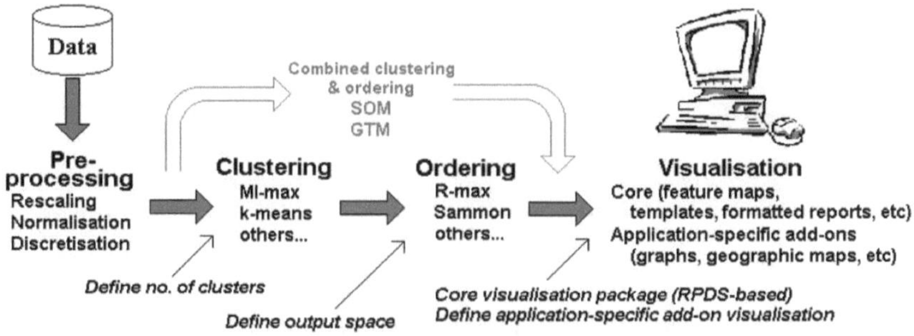

Fig. 6.4.3. Summary of data analysis and visualisation framework

The framework

The framework is summarised in Fig. 6.4.3. Each of the main components – pre-processing, clustering, ordering, and visualisation – is described below. Each step is illustrated by reference to the corresponding component in a prototype generic pattern recognition system, MIR-max, developed by the authors, based on the discrete information-theoretic clustering (MI-max) and ordering (R-max) algorithms discussed previously. The MIR-max interface and visualisation features are based on those used in the RPDS software.

Data pre-processing

It is unlikely that 'raw' data will be in a form suitable for analysis; for example, the units of each indicator may not be comparable, data may consist of textual category information (e.g. 'small', 'medium', 'large'), or data may be continuous when discrete data is required. Some pre-processing of the raw data is almost inevitable. Often the data will need to be rescaled and/or normalised so that each indicator is directly comparable, or so that a similarity metric can be defined in the sample space. Continuous data may need to be discretised, for example if the discrete information-theoretic clustering technique MI-max is to be used.

For MIR-max, there are two stages of pre-processing. First the data (having been validated) must be stored in a standard format. MIR-max accepts delimited text files (for example, comma-separated values format) with all indicator values stored numerically that can be read and edited using either a standard text editor or a spreadsheet. The user specifies the precise form of the data (Fig. 6.4.4a). Secondly, because MIR-max uses a clustering process based on discrete information theory, the data is discretised by definition of category bounds for each indicator (Fig. 6.4.4b). This second stage also allows the user to specify whether each indicator is to be 'training' or 'non-training', and to define the units of measurement for each indicator.

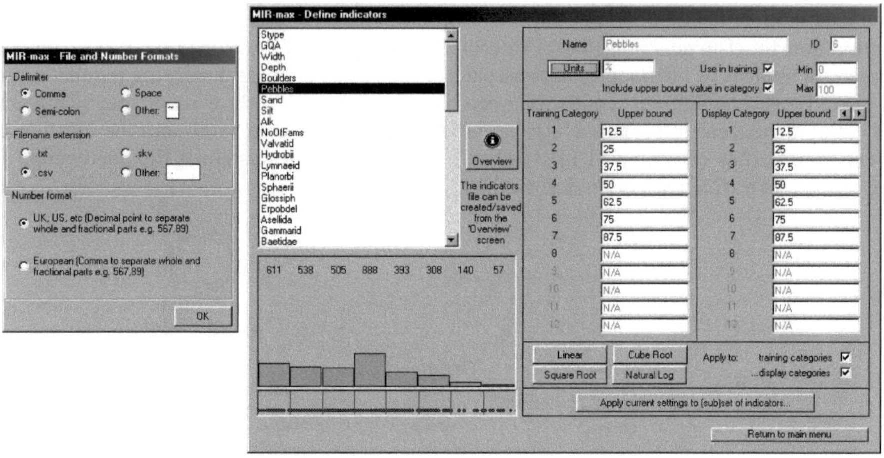

(a) (b)

Fig. 6.4.4. MIR-max pre-processing: (**a**) selection of standard file format and (**b**) definition of category bounds

Clustering and ordering

As discussed above, a number of possible clustering methods can be used; the choice will largely be dependent on the nature of the data under consideration. In most cases, it will be necessary to first define the number of clusters to be used.

Similarly, the specific ordering procedure to be used is not defined. An output space will first need to be defined. In the case of, for example, PCA, this is a continuous space. However, a discrete output space based on a regular lattice structure (such as would be used in SOM or MIR-max, for example) provides advantages in that it is easy to visualise and interpret, and enables easy functional user interaction via a 'point and click' interface.

The clustering and ordering could also be combined in a single process such as SOM or GTM, but this is not recommended by the authors because it is likely to result in sub-optimal clustering. In the case of RPDS, the clustering (classification) process was considered more important, whilst ordering provided a useful and important form of visualisation but was considered to be of secondary importance. By separating the two processes priority can be given accordingly. The separation of the two processes also allows for easier evaluation of results: for example, there is no clear measure of 'quality' for SOM output, whilst most dedicated clustering techniques are designed to optimise some objective function (in the case of MI-max, the mutual information between indicator values and cluster membership is optimised). The generic MIR-max software thus does not support a combined clustering and ordering procedure; however this is included in the theoretical framework for completeness.

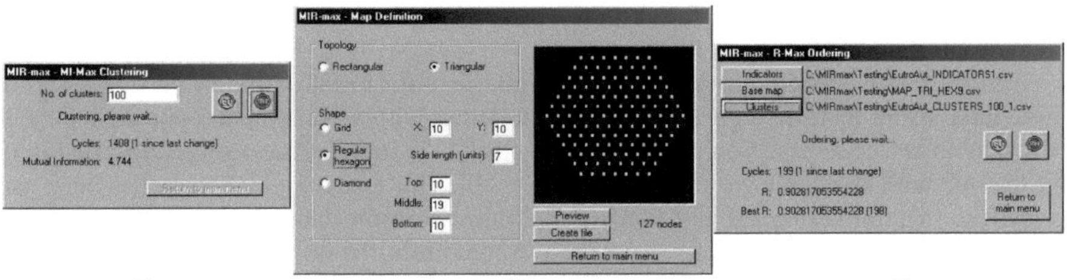

(a) (b) (c)

Fig. 6.4.5. MIR-max analysis: (**a**) clustering; (**b**) definition of output space; (**c**) ordering

Both clustering and ordering are simple processes using the MIR-max software, with much of the technical detail hidden from the end-user. Clustering (using the MI-max algorithm) requires the user first to define the number of output clusters (Fig. 6.4.5a); ordering (using the R-max algorithm) first requires definition of an output space (Fig. 6.4.5b,c). All the processes in MIR-max produce output files in standard format ready to be used as input to the visualisation component; these files can also be viewed and edited easily using a spreadsheet if further detailed analysis is required.

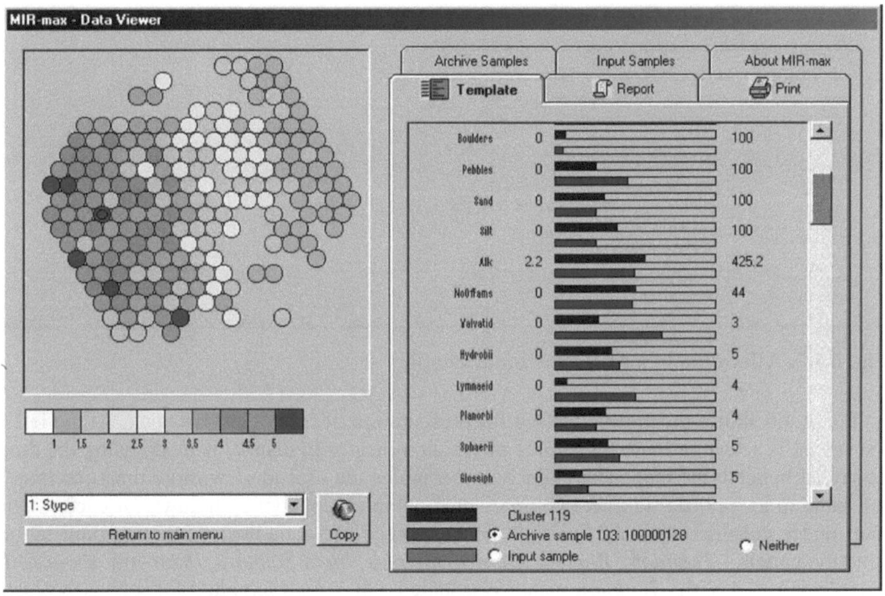

Fig. 6.4.6. MIR-max data visualisation interface

Visualisation

An essential part of any computer-based decision support system is its ability to present the complex data in a meaningful way, so that it can be readily and effectively used by the variety of personnel involved in the decision-making process – e.g. scientists, managers, and politicians. Effective dissemination of complex information requires suitable data visualisation techniques – such as the feature maps and templates described above – and a user-friendly interface.

The MIR-max system provides a generic interface for visualisation of clustered and ordered data, using feature maps and templates together with textual reports and easy data access. The interface is based on that developed for RPDS, but is necessarily simplified in order to cope with as wide a variety of data as possible (for example, MIR-max has been used to analyse macroinvertebrate and fish communities, forestry data, chemical data, and even for optical character recognition). However, to maximise the utility of the software, it will normally be useful to define a more application-specific interface.

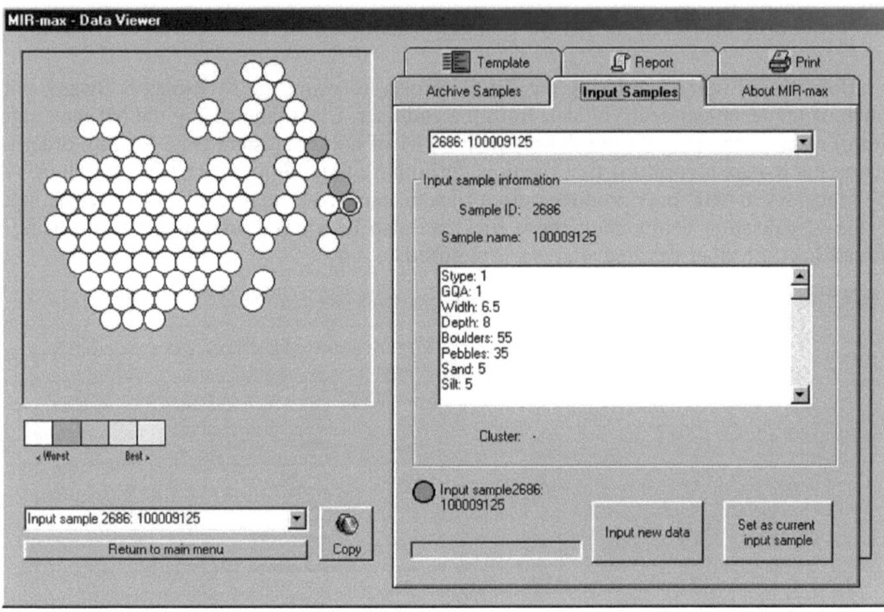

Fig. 6.4.7. Allocation of a MIR-max input sample

Fig. 6.4.6 shows the main interface for presentation of results in MIR-max. At the left of the screen is a feature map; the choice of which indicator to display is made using the drop-down list beneath the map. The *Copy* button enables the user to view more than one map at the same time (in order to make comparisons) using independent windows for each feature map under consideration. At the right of the screen are a number of overlapping tabbed function panels – *Template*, *Report*, *Archive Samples*, *Input Samples*, *Print* and *About MIR-max*. In Fig. 6.4.6, the *Template* panel is selected, showing a template for a chosen cluster and archive sample. Clicking the colour-coded circle on the feature map chooses the cluster, whilst archive samples (from the original data file) and input samples (loaded from a file in standard format) are selected using the corresponding function panels. In the case of input samples, a graded colour scheme represents the degree to which a sample fits within each cluster (based on mutual information, but currently displayed on a simple 'best' to 'worst' scale; in future, an objective scale, e.g. a percentage for 'goodness of fit', will be utilised) (Fig. 6.4.7). The *Report* panel provides a text summary of information regarding any cluster.

The user-friendly RPDS interface (Fig. 6.4.8) is considered a key part of the success of the system. The main visualisation features are the same as those provided by MIR-max (although with some increased functionality), but additional features such as graphs of specific indicator values and a map of site locations (Fig. 6.4.9) are also provided. A detailed description of the functions and features of RPDS is given by Walley et al. (2002).

Conclusions

A generic framework for decision support systems, based on data analysis and visualisation by pattern recognition, is beneficial for a wide range of possible ecological applications, where a large amount of multi-dimensional data needs to be analysed and understood. The framework described is currently supported by general-purpose data analysis software (MIR-max) and has been demonstrated in a full application (RPDS). A generic framework introduces increased potential for data sharing, and – if accepted as a 'standard' methodology - provides a basis for enhanced dissemination of information by using accepted and well-understood visualisation and analysis techniques. There is sufficient flexibility within the framework to allow for a wide variety of specific pattern recognition (clustering) and ordination techniques, but also unifies these by providing a common software interface for visualisation and interpretation, thus allowing direct comparison of outputs produced by different techniques.

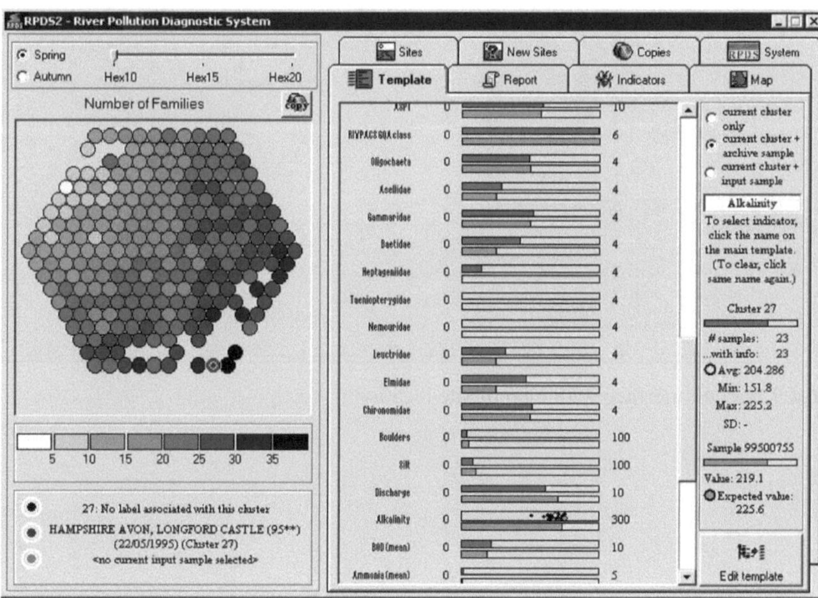

Fig. 6.4.8. Main RPDS interface, with feature map and template

The need for advanced visualisation methods to aid data interpretation is clear, but currently most analysis is performed on an 'ad hoc' basis requiring specialised knowledge and/or software for each individual scenario. It is hoped that the adoption of a common framework will further the acceptance of computer-based techniques for analysis and visualisation of ecological data, and allow for quicker development and a common basis for presentation of results. RPDS and its generic form (MIR-max) have successfully demonstrated the framework in practice, although it is acknowledged that a truly generic system (i.e. one that incorporates as wide a variety of clustering/classification and ordination/visualisation methods as possible) is yet to be developed. Additionally, the current framework regards each sample as independent, in that there is no spatial or temporal reasoning involved. It is also intended to extend the framework to incorporate not only pattern recognition processes, but also those based on plausible reasoning, to produce a 'true' expert system modelling both aspects of human expert intelligence. It is clear that such a system would have considerable potential for use in the interpretation and visualisation of multivariate ecological data.

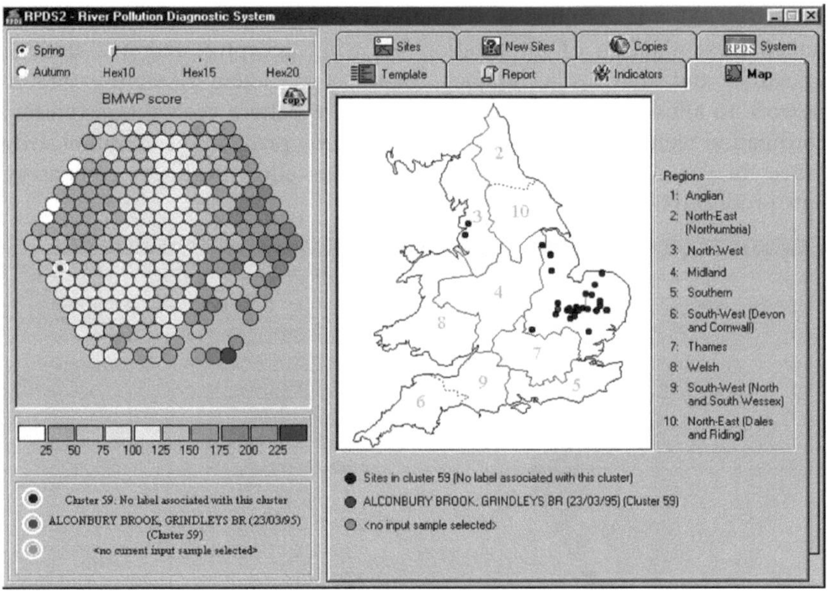

Fig. 6.4.9. RPDS interface with map of site locations

6.5 A rule-based vs. a set-covering implementation of the knowledge system LIMPACT and its significance for maintenance and discovery of ecological knowledge*

Neumann M†, Baumeister J

Introduction

Small streams collectively add up to an enormous length on the landscape level, so that the conservation and protection of their aquatic community should be a major concern. In catchment areas with agricultural activities, these streams are subject to various stressors. During heavy rainfall, runoff from agricultural fields may introduce soil, nutrients, and pesticides and increases discharge (Cooper 1993, Neumann and Dudgeon 2002). It has been shown that the impact of pesticides is an important parameter of influence for the aquatic fauna (Liess and Schulz 1999, Schulz and Liess 1999). No regular monitoring systems have been established for these agricultural non-point sources of pesticides. Because of its short-term character (Kreuger 1995), only rainfall event-controlled sampling methods can reflect such transient pesticide contamination (Liess et al. 1999), which makes its detection via chemical analysis costly.

In this field the use of a biological indicator system brings a number of benefits. The main advantage is its easy, cost-efficient application. When used to monitor toxic contamination, it additionally indicates the ecotoxicological effect of the contaminant. A biological indicator system also provides information on the long-term effects of contamination, whereas information from each chemical measurement applies only to the time the measurement was taken.

There is a wide range of biological indicator systems to evaluate water-quality parameters. In Great Britain RIVPACS (Wright et al. 1998) predicts the macroinvertebrate fauna to be expected at a site in the absence of environmental stress and can be used to evaluate the present fauna. In the Netherlands, a similar approach is used for STOWA (Peeters et al. 1994). In Scotland, the integrated evaluation system SERCON (Boon et al. 1998) and in USA the Rapid Bioassessment Protocols (Resh et al. 1995) were developed. In Germany, the saprobic index is well established to evaluate the biodegradable organic pollution in running waters (Friedrich 1990). Systems to monitor heavy metals (Wachs 1998) and acidification (Brakke et al. 1994) have been developed. However, no biological indicator system

* The former rule-based implementation of LIMPACT was financed by the Deutsche Bundesstiftung Umwelt (German Federal Environmental Foundation), An der Bornau 2 in 49090 Osnabrück, Germany and was supported by Matthias Liess and Ralf Schulz of the Zoological Institute at the Technical University of Braunschweig, Germany. The authors would like to thank Frank Puppe and Dietmar Seipel of the Department of Computer Science at the University of Würzburg, Germany for their helpful discussions and comments

† Correspondence: m.neumann@uni-jena.de

has yet estimated the pesticide contamination of small streams via benthic macroinvertebrate indicators.

To fill this gap we developed a biological indicator system that estimates the pesticide contamination of small streams. In order to consider the ecological complexity and the uncertain knowledge in this domain, we implemented a diagnostic knowledge system. The advantages are that knowledge systems utilize uncertain expert knowledge and ideally come to the same solution as would the expert. The user has full control over the knowledge system, can scrutinize the solution, and interactively change the question trail. The database and the development of the rule-based knowledge system LIMPACT (from limnology and impact) was presented in (Neumann et al. 2002a, b).

Here, we present a new implementation of the knowledge base using a set-covering approach. The new implementation was motivated by the complex maintainability of the rule base of the former implementation. Thus, it was difficult to extend the knowledge base by rules for new taxa in conjunction with an appropriate scoring of these rules. In contrast to the rule-based approach, set-covering models are intended to minimize the knowledge acquisition costs, since models can be built and extended incrementally in a simple manner. In this paper a performance comparison of the former rule-based and a new set-covering implementation, based on the classification accuracy, the complexity of the knowledge base, and knowledge acquisition costs, is presented. Furthermore, with the new set-covering approach we have been able to extract ecological knowledge about the common appearance of the macroinvertebrate biocoenosis in small, pesticide contaminated streams.

The knowledge system LIMPACT

We developed the knowledge system LIMPACT using the shell-kit D3 (http://www.d3web.de), which is applicable for diagnostic tasks, provides a web-based user interface (d3web) and offers a visual knowledge acquisition environment for a wide range of knowledge types (Puppe 1998, Puppe et al. 2001). The diagnoses of LIMPACT estimate the pesticide contamination of small streams. They represent a calculated annual toxic sum (for details on types of pesticides see Neumann et al. 2002a) without any specification of the chemical agents. Therefore, the vital diagnoses of LIMPACT are four classes of pesticide contamination named: Not Detected (ND), Low (L), Moderate (M) and High (H) pesticide contamination. The required input parameters (observations) of LIMPACT are abundance data for aquatic macroinvertebrate taxa in a stream. We established four time frames for which information about abundance is requested. The time frames are T1 (March/April), T2 (May/June), T3 (July/August), and T4 (September/October). LIMPACT allows abundance values to be entered for these four periods of the year for the 39 taxa named in Table 6.5.1. Additionally, LIMPACT interprets the increasing or decreasing abundance dynamics of each taxon.

We differentiate between *positive indicator* (PI) taxa, which indicate contamination by high abundance values and positive abundance dynamics, and *negative indicator* (NI) taxa, which exclude contamination and indicate none or low contamination by high abundance values and positive abundance dynamics.

Besides abundance data, LIMPACT evaluates 9 basic water-quality and morphological parameters, such as stream size or conductivity of the water, to characterise a given stream. For simplification, these parameters are abstracted to qualitative values. These abstractions are used for determining the type of stream (for details see Neumann et al., 2002a), because LIMPACT only contains knowledge applicable to small lowland streams within agricultural catchments and cannot make a distinction between pesticides and other types of impact. Hence streams affected by any of the latter factors are excluded to ensure that the impact of pesticide is the main stressor to the aquatic macroinvertebrate fauna. At this stage, such interfering factors include industrial waste impact, severe organic contamination, and extreme chloride or pH values. Additionally, no highland or large streams are considered.

Table 6.5.1. The 39 indicator taxa of the knowledge system LIMPACT and the type of indicator (N = negative; P = positive indicator). For the covering relations, taxa with more than 30% appearance in the considered class are marked and for the exclusions, taxa which activated more than 10% of the sum of the rest of the classes are marked.

Order	Taxon	Type of indicator	Covering relations				Exclusions			
			ND	L	M	H	Not ND	Not L	Not M	Not H
Turbellaria	*Dugesia gonocephala*	N	X	X						X
Oligochaeta	*Erpobdella octoculata*	P	X	X	X	X				
	Glossiphonia complanata	N	X	X	X	X				
	Glossiphonia heteroclita	P	X		X	X				
	Tubificidae	P	X	X	X	X	X			
	other Oligochaeta	N	X							X
Gastropoda	*Pisidium* sp.	N	X	X	X	X				X
	Potamopyrgus antipodarum	P				X	X			
	Radix ovata	P	X	X	X	X				
Amphipoda	*Gammarus pulex*	P	X	X	X	X				
Isopoda	*Asellus aquaticus*	N	X	X	X				X	
Plecoptera	*Nemoura cinerea*	N								X
Coleoptera	Dytiscidae	N	X							X
	Agabus sp.	N							X	
	Platambus maculatus	N								
	Elmis sp.	N		X	X					
	Haliplus sp.	N								
	Helodes sp.	N	X	X	X				X	X
Diptera	Ceratopogonidae	P			X	X				
	Chironomidae "white"	N	X		X	X				X
	Chironomidae "red"	N	X	X	X	X			X	X
	Limoniidae	N		X	X					X
	Ptychopteridae	N								X
	Simuliidae	N	X	X	X					X
	Tipulidae	N						X		
	Other Diptera	N	X							X
Ephemeroptera	*Baetis vernus*	N			X					X
	Baetis sp.	N			X					
	Ephemera danica	N								X
Megaloptera	*Sialis lutaria*	N		X	X					X
Trichoptera	*Hydropsyche angustipennis*	N								
	Anabolia nervosa	N			X					
	Chaetopteryx villosa	N	X	X						X
	Halesus radiatus/digitatus	P								
	Ironoquia dubia	P	X							
	Limnephilus lunatus	N						X		
	Limnephilus extricatus	N						X		
	Limnephilus rhombicus	N								
	Plectrocnemia conspersa	N								

The potential application of LIMPACT is for annual monitoring of streams and would reduce costly chemical analysis to the mandatory cases. Furthermore, it could be used to evaluate the success of risk mitigation strategies in the catchment designed to reduce the impact of pesticides. The system is available over the internet via http://www.limpact.de.

Methods of knowledge engineering

Ontological knowledge about diagnoses, parameters, and abstractions was used when we implemented the two versions (rule-based and set-covering) of the diagnostic knowledge. In general, diagnostic knowledge states relations between diagnoses and observations and describes how to obtain a diagnosis for a given set of observed parameters. For the acquisition of diagnostic knowledge we have to consider the following aspects and requirements.

As mentioned in Puppe (1998), developing diagnostic knowledge systems is still a time- and cost-intensive task. A variety of knowledge representations have been designed and evaluated to build diagnostic systems effectively, but practical maintenance of such systems is still a difficult issue. In general, we can emphasize the following requirements for a successful knowledge engineering project:

- *Understandability of the knowledge representation*: The representation is easily and quickly understood by the domain specialist (expert). This property enables a quick initiation of the development project.
- *Incremental development characteristics*: For a rapid development cycle it is helpful to start with extremely simple knowledge, which can be extended incrementally to increase the diagnostic quality.
- *Maintainability of the implemented knowledge*: The implemented knowledge base needs to be manageable even if the size of the system increases.
- *Explanation facilities*: Furthermore, the representation should allow for the generation of comprehensive explanations to scrutinise the resulting diagnoses.

In the following we present the methods of the two knowledge representations we used for developing two versions of the LIMPACT knowledge system. We also compare their characteristics with respect to their maintainability and reasoning accuracy.

The former rule-based approach

For the first development of the LIMPACT system we applied a heuristic rule-based formalism called *diagnostic scores* (Puppe 2000) and implemented the rules with the shell-kit D3. Here heuristic classification is based on rules of the following kind:

IF observation OBS_i then give diagnosis D the score Z

The observations OBS_i were clearly defined as the abundances of taxa, whereas the diagnoses are the graded amount of pesticide contamination in the stream, i.e. *Not Detected* (ND), *Low* (L), *Moderate* (M) and *High* (H). The domain expert estimated certain scores (negative or positive) to characterise types of stream contamination on the basis of given abundance data or combinations of them. D3 provides a fixed range of seven positive (P1=+5% to P7 =+100%) and seven negative (N1=-5% to N7 -100%) scores, which has been proven to be useful in various previous applications of D3. Reasoning with scores is easy and understandable for the expert: Given a true condition, the corresponding rule fires and adds the stated score to the specified diagnosis. The sum of two equal categories is aggregated to the next higher category (e.g. P3+P3=P4). A diagnosis about the pesticide contamination is established (confirmed), if the aggregation of the given scores exceeds the category P5.

For a detailed description of the development and evaluation of the rule-based version of LIMPACT we refer to Neumann et al. (2002a, b). The system has been operational since February 2001 and can be used via the web (http://www.limpact.de).

The new set-covering approach

It has been shown that model-based representations are more appropriate for developing maintainable and explanatory knowledge systems (David et al. 1993). For the development of a model-based approach of LIMPACT we applied set-covering models, which allow for an incremental development of diagnostic systems (Baumeister and Seipel 2002, Baumeister et al. 2003). Set-covering models describe relations like

Diagnosis D typically covers observation OBS_i.

These relations are called covering relations and we say that OBS_i is covered by diagnosis D. As in the former rule-based implementation of LIMPACT, the diagnoses D were defined by the four different contamination classes, whereas the observations OBS_i are the abundances of taxa.

After implementing simple covering relations for the most typical diagnosis-parameter relations we added *weights* for parameters to the model. With weights we can emphasize that some parameters have a more significant diagnostic importance than other parameters, e.g. parameters stating clear positive indicators. During a second improvement phase we extended the set-covering model by exclusion conditions, which contain knowledge about a categorical exclusion of specific contamination classes (e.g. if we did not find an increasing abundance of a negative indicator taxon in a highly contaminated stream).

Reasoning with set-covering models is very simple: Given a set of observed parameters *OBS*, it uses a simple hypothesize–and–test strategy, which picks a hypothesis *H* (set of diagnoses) in the first step and tests it against the given observations in a second step. The test is defined by calculating a quality measure, which expresses the covering degree of the hypothesis *H* with respect to the observed findings *OBS*. The quality measure q of a hypothesis *H* is defined as follows

$$q(H, OBS) = \frac{\varpi(OBS_{\text{cov}}^{+}, H)}{\varpi(OBS_{\text{cov}}^{all}, H) + \varpi(OBS^{un\exp l}, H)}, \qquad (6.5.1)$$

where $\varpi(OBS_{\text{cov}}^{all}, H)$ is the weighted sum of all covered and observed parameters of hypothesis H and $\varpi(OBS_{\text{cov}}^{+}, H) \subseteq \varpi(OBS_{\text{cov}}^{all}, H)$ is the weighted sum of all covered and correctly observed parameters of hypothesis *H*. A parameter is correctly observed if the observed value of the parameter corresponds to the value specified in the covering relation. $\varpi(OBS^{un\exp l}, H)$ sums all parameters that are observed but not covered by the hypothesis *H*. Clearly, for a given hypothesis *H* it holds that $OBS = OBS_{\text{cov}}^{all} \cup OBS^{un\exp l}$. A hypothesis is not considered for a given observation if one of its exclusion conditions evaluates true.

Besides weights and exclusion conditions, set-covering models can be extended by similarity measures, complex covering relations and constrained covering relations (Baumeister and Seipel 2002).

Results

Size and complexity

For the implementation of LIMPACT we defined 9 variables (see Neumann et al. 2002a) describing the stream (i.e. structural parameters) and 39 variables representing abundances of different taxa. Each abundance variable can record abundances for the four defined time frames. Furthermore, we specified four diagnoses for the contamination classes of a stream as well as a diagnosis for detecting unsuitable streams. This ontological knowledge was augmented by diagnostic knowledge represented either by heuristic rules (former approach) or by set-covering relations (new approach).

The former rule-based version of LIMPACT contains 921 diagnostic rules (see Table 6.5.2) with scores to establish or to de–establish a diagnosis. Diagnostic rules are of the following kind:

IF (Rule Condition C) THEN give diagnosis D score S.

The complexity of these rules is moderate, which means that the rule condition mostly contains between two and four combined single conditions connected by Boolean operators (e.g. and, or, not). A single condition evaluates whether a taxon's abundance is above a given threshold, i.e. a single observation. Additionally, for each rule an appropriate diagnosis score was defined by the expert.

For the set-covering knowledge base, we implemented 816 simple covering relations (see Table 6.5.2) of the following kind:

Diagnosis D covers the observation of taxon T with abundance A.

We can see that these relations are simpler than the implemented rules described above. In contrast to the rule-base, we only consider one taxon's abundance information, disregarding other taxa also covered by the same diagnosis. We also do not consider scores for diagnoses.

Table 6.5.2. Size of the two implemented rule-based and set-covering knowledge bases for each diagnosis. The left side of the table shows the complexity of the rule conditions for the rule-based approach in more detail. The last two columns give an overview of the size of the rule-based and the set-covering knowledge base.

Contamination	Rule-based knowledge base								Set-covering knowledge base
	Number of evaluatable symptoms in rule condition							Total	Total
	1	2	3	4	5	6	7		
Not detected	0	113	82	39	13	3	1	251	212
Low	0	85	75	38	7	4	1	210	202
Moderate	1	105	75	44	5	2	2	234	206
High	1	112	76	28	8	1	0	226	195
Sum	3	417	311	153	38	16	11	921	815

Table 6.5.2 gives the complexities of the implemented knowledge in more detail. Whereas the last column shows the number of set-covering relations for each contamination class, we extended the presentation for the rule-based version. Thus, we depict the overall

number of diagnostic rules besides the number of covering relations, and display more precisely the number of rules with 1 to 7 single conditions in the first columns of the table. Rules with 2-4 conditions dominate the rule-based version.

Knowledge acquisition costs

Comparing the size of the two knowledge bases, Table 6.5.2 shows that the number of implemented knowledge elements is comparable. The size of the set-covering knowledge base is even a little bit smaller. These characteristics are illustrated by the fact that the expert required about six weeks to implement the rule-based version of LIMPACT versus two weeks for implementing the set-covering counterpart.

Table 6.5.3. Result of the classification of 146 investigations per stream and year using the rule-based (RB) and set-covering (SC) implementation of LIMPACT. The measured real contamination is given according to the four classes and compared with the percentage of cases classified by LIMPACT into the four classes plus not classified. Correct classifications are indicated by bold values. The number of cases per contamination class is given in parentheses.

Real contamination	Classification result (%)									
	Not detected		Low		Moderate		High		Not classified	
	RB	SC	RB	SC	RB	SC	RB	SC	RB	SC
Not detected	**90.4**	**96.2**	0	0	1.9	0	0	0	7.7	3.8
(52)	(47)	(50)	(-)	(-)	(1)	(-)	(-)	(-)	(4)	(2)
Low	16.7	0	**80.0**	**93.3**	0	6.7	0	0	3.3	0
(30)	(5)	(-)	(24)	(28)	(-)	(2)	(-)	(-)	(1)	(-)
Moderate	2.5	0	0	2.5	**72.5**	**87.5**	7.5	0	17.5	10
(40)	(1)	(-)	(-)	(1)	(29)	(35)	(3)	(-)	(7)	(-)
High	0	0	0	4.2	0	12.5	**87.5**	**79.1**	12.5	4.2
(24)	(-)	(-)	(-)	(1)	(-)	(3)	(21)	(19)	(3)	(1)

Classification results

The classification result of both rule-based and set-covering implementation was calculated with the same cases that were used to develop the system. This was necessary because no independently obtained stream investigations, including macroinvertebrate abundance data and chemical pesticide measurements, were available.

A detailed evaluation is presented by Neumann et al. (2002b) for the rule-based (RB in Table 6.5.3) implementation. For RB Table 6.5.3 shows that the correct diagnosis of the 146 cases is established by LIMPACT in 72.5 to 90.4% of the cases, with better results for uncontaminated sites. The evaluation showed a very good classification result. Most errors occur between ND and Low and on the other hand between Moderate and High contamination. A high percentage of cases were not classified. Because of our conservative approach, LIMPACT established no diagnosis instead of a wrong one for cases with insufficient data availability. Possible reasons for classification errors and not classified cases can be related to uncertainty in the sampling and identification methods and the number of sampling dates within a year. The more data the user provides, the more rules can be activated. Consequently, the chance of a correct classification increases.

For the set-covering (SC in Table 6.5.3) implementation Table 6.5.3 gives the classification result for the same 146 cases as for the rule-based approach. The correct diagnosis is found in 79.1 to 96.2 cases, which is a better classification result than for the rule-based implementation. Only the highly contaminated cases show a decline in classification result and at the same time an increase in wrong classifications. Additional errors occur between the Low and Moderate contamination classes.

Explanatory characteristics

The two implementations differ not only in the way the knowledge is represented but also in the way new knowledge can be extracted and discovered from the knowledge bases. For the rule-based implementation the domain expert found it difficult to gain any new insights. The explanatory characteristics are complex because the knowledge is represented in small pieces (rules) and is weighted with different scores. In the following, we give only four rules as example:

- IF Agabus at T2 in [2; 9] THEN Contamination High P3
- IF Agabus at T2 > 9 THEN Contamination High N4
- IF Anabolia at T1 in [0; 80] THEN Contamination High P2
- IF Anabolia at T1 > 80 THEN Contamination High N3

The different scores to establish (here: P2 and P3) or to de-establish (here: N3 and N4) a diagnosis (here: High) make it difficult to obtain a general overview. To extract from the rule-based knowledge base how the aquatic community of an average stream with e.g. High pesticide contamination appears, the domain expert has to interpret the rule for and against the High diagnosis and has to interpret the different scores.

The set-covering implementation has a better explanatory characteristic, because of its more straightforward design. In the following, we give only two covering relations as example:

- Contamination High: Agabus at T2 in [2; 9]
  Anabolia at T1 in [0; 80]

Each covering relation represents a characteristic of the considered contamination class (here High). The domain expert simply looks at all covering relations of one specific diagnosis and gains an overview. The same is true for exclusion conditions. They represent those characteristics that are not the case for the considered contamination class.

Discovery of ecological knowledge

Using the set-covering implementation we were able to discover the common macroinvertebrate community of an average stream. For each of the four diagnosis classes, we analysed which covering relation and which exclusion was activated most frequently. For the covering relation we considered those activated in more than 30% of the cases within the contamination class and for the exclusion we considered those activated in more than 10% of the sum of the rest of the classes. Table 6.5.1 indicates which taxa activated the most covering relations and exclusions. For the sake of simplicity we do not indicate abundance values and do not itemise each single covering relation. Generally speaking, the type of the indicator specifies whether the taxon is found in higher abundance in more highly contaminated streams (positive indicator) or in uncontaminated streams (negative indicator). Bearing in mind all this information, Table 6.5.1 illustrates a theoretical average community in the four contamination classes.

As Table 6.5.2 shows, we implemented only 8% (212 to 195) fewer covering relations for the High contamination class vs. the ND class. Table 6.5.1 shows that these covering relations are activated by 35% (17 to 11) fewer common taxa in the High contamination class than in the ND class. This indicates that in highly contaminated streams fewer taxa are common. At the same time the large number of exclusions indicates that in this contamination class 16 taxa cannot be found with high abundances. Most common taxa are found in streams classified as Moderate, which may indicate highly variable conditions in this type of stream.

The analysis shows that considering the appearance of the common taxa, the stream classes look very similar. Eleven taxa appear at least in three diagnoses classes, separated by the abundance only. Four taxa clearly indicate the ND class (e.g. Oligochaeta, Dytiscidae), but none the H class. Only a few taxa appear in the ND and/or L classes and exclude the H class (e.g. *Dugesia gonocephala*, Oligochaeta) and only *Potamopyrgus antipodarum* indicates the H class and excludes the ND class. Some taxa indicate a specific class by their low abundance and exclude the same class by high abundances (e.g. Tubificidae, *Pisidium sp.*). Overall, we found a wide range of common taxa with a tendency towards more taxa in less severely contaminated streams. For the exclusion conditions a clear trend, with more taxa excluding the more highly contaminated streams, was likewise found.

Discussion

Size and complexity The reduction of size and complexity of knowledge bases is the main focus of knowledge engineering research. Both aspects are crucial for developing and maintaining successful knowledge systems. It has been shown that knowledge bases tend to be confusing and unmanageable if their size increases and the complexity of the embedded knowledge develops excessively.

Comparison of the knowledge bases presented here shows that the number of covering relations in the set-covering approach is only slightly smaller than the number of implemented rules in our rule-based system. Nevertheless, the complexity of the modeled set-covering knowledge is significantly simpler than the implemented rules-based knowledge. When adding rules for taxa to the rule base, we also have to consider the associated diagnosis scores. These scores interact with other rules deriving the same diagnosis and therefore have to be obtained by thorough analysis. Thus, adding a new rule to the knowledge base can demand reconsideration of all rules (and of the associated scores) deriving the same diagnosis. In contrast to these interwoven rules, set-covering relations can be viewed as isolated knowledge elements without mutual interdependencies. For a new taxon we only have to define relevant covering relations for the four diagnoses, i.e. contamination classes, and the new taxon. In general, this means that we have to define the abundance of the new taxon for each diagnosis, if we expect the taxon to occur with the given diagnosis. If available, we can additionally define abundance trends (positive or negative) between the time frames T1, T2, T3, T4 for the new taxon and each diagnosis.

Knowledge acquisition costs The costs of knowledge acquisition often can be measured only by the time the domain specialist (expert) or engineer had spent in developing the knowledge system. For maintenance purposes we also need to consider the time the developer needs to change or extend the knowledge base. In our experiences with LIMPACT, the modular characteristics of the set-covering relations had a direct impact on knowledge acquisition costs. The expert found the set-covering representation easy to understand and to apply to the diagnosis problem. In contrast to the rule-based version of LIMPACT, he did not need to consider the interconnections between rules deriving the same diagnosis. This experience is emphasized by the time the expert spent to develop the two knowledge bases: implementing the rule-based knowledge took about 6 weeks vs. 2 weeks for defining the set-covering model.

Classification results It is obvious that the classification accuracy of a diagnostic system is the key factor for its user acceptance. A user is more likely to accept that the system cannot supply a diagnosis for a particular case, but will lose confidence if the system derives wrong diagnoses in some cases. For this reason, a system should not only provide a solution for a given case, but furthermore should deliver a "confidence level" for the diagnosis that is obtained. This confidence level can depend on the score of the diagnosis or an overall "believability" function defined by the developer of the knowledge base.

As described in the previous section, the classification accuracies of the rule-based and the set-covering system are comparable. Nevertheless, the diagnostic system applying set-covering knowledge outperforms the rule-based version for contamination classes Not Detected, Low, and Moderate. The rule-based implementation only outperforms the set-covering implementation for highly contaminated streams. One can say the rule-based version of LIMPACT has no (high) confidence level for streams with contamination classes Low and Moderate, while the set-covering implementation has a lower confidence level for the diagnosis of highly contaminated streams. Reasons include the fact that we have not implemented any covering relations interpreting the absence or the decreasing abundance dynamic, which could indicate highly contaminated streams. For the domain expert the absence of a taxon or its decreasing abundance is difficult to interpret, because the causal connection explicitly to pesticide contamination is uncertain.

Explanatory characteristics The set-covering knowledge base is much more suitable for discovering ecological knowledge than the rule-based implementation. The covering relation and the exclusions can be easily interpreted as characteristics of the group of streams considered. By analysing the frequently used relations we found the common taxa for each contamination class. This procedure was simple and fast. For the rule-based knowledge base this would have been a time-consuming process, because of the interpretation of the rules and the scores.

Other knowledge representations, such as case-based reasoning, also cause problems in finding common and average characteristics of the considered diagnoses classes. For implementation they represent a set of characteristics at the same time and therefore cannot activate each characteristic separately. In summary, we can say that the model-based knowledge representation using a set-covering interpretation is easy to implement. It outperforms the rule-based implementation in size, complexity, and maintainability and helped the domain expert to discover new ecological knowledge at a higher level.

6.6 Predicting macro-fauna community types from environmental variables by means of support vector machines*

Akkermans W†, Verdonschot P, Nijboer R, Goedhart P, ter Braak C

Introduction

This paper compares the performance of two classification methods: Support vector machines (SVMs) and multinomial logistic regression (MLR). Why classification? Already more than a century ago ecologists realized that each distinction into classes is artificial. It divides parts in nature which in reality are connected through a number of transitions. Several disadvantages of using strict classes have been recognised. Classes have been called arbitrary (Maitland 1966) and subjective (Armitage 1961). Macan (1961) argues that a rigid framework creates a cage, whereas we need freedom of thought. Since then, many ecologists must have been aware of the continuum in the surroundings they still tried to classify.

The main argument for using classes is a need for more explicit and sharp-cut terms. Identification and arrangement of ecological classes has furthermore been defended because it is an intellectual challenge (Hawkes 1975); because it is necessary for understanding, describing and explaining the enormous diversity of the mixed species populations (Rietz 1965); because it is helpful in comparing different waters worldwide (Pennak 1971); and because it is of practical value, especially with respect to water management (Deusen 1954, Hawkes 1975). Examples of the practical value are utilitarian applications (the first classification schemes were introduced by fish biologists), the prediction of effects of projected water management policies, and the assessment of water quality and pollution. So although it should be borne in mind that ecological classes are not natural entities and that their definition is always influenced by the ecologist's choices, the plea for using and distinguishing classes, from a practical, water management point of view, is very clear.

Classification is the process of generating a rule for recognising whether a particular object belongs to one of a certain number of predefined classes; and, in a narrower sense, to apply this rule to objects (sites) with unknown class membership (see Section 6.3).

Classical methods for classification rely on distributional assumptions. These methods include linear discriminant analysis and (multinomial) logistic regression. The latter is the modern standard approach. The term machine learning is sometimes used to refer to more recently developed, distribution-free methods such as neural networks. In the tradition of machine learning, Support Vector Machines are a relatively new and modern tool. In this paper, the performance of SVMs will be compared to Multinomial Logistic Regression.

These two techniques will be applied to empirical data on Streams in the Netherlands. The data are described in Section 6.2. Section 6.3 gives a brief description of Support Vector Machines and Multinomial Logistic Regression. The emphasis here will be on an intui-

* This work has been supported by the EU through the PAEQANN project (5th Framework Programme, contract EVK1-CT-00026).

† Correspondence: wies.akkermans@wur.nl

tive explanation of the methods. A more detailed description of SVMs is given in the Appendix. Section 6.4 contains the results.

The data

Two data sets will be analysed: Streams and Canals. As the main purpose of this paper is to demonstrate the usefulness of SVMs, we will only describe the Streams data in detail, and mention results on the Canals data in passing.

Streams data The streams data have been collected by water district managers. In general, the samples were taken in different types of streams, but there is a bias towards moderately polluted sites. Natural sites and sites in more extreme conditions, like intermittent streams, are underrepresented. The small size of several classes therefore does not mean that these classes are unimportant. The classification model has to be able to correctly predict the smaller classes too.

The data set consists of 563 sample sites, belonging to the following 6 classes:

1. 156 *hill streams*: fast flowing upper to middle course of natural to semi-natural hill streams
2. 270 *middle to lower courses*: middle to lower courses of lowland streams either semi-natural to channelised with a developed aquatic vegetation
3. 32 *small natural upper courses*: small, neutral to weakly acid, natural to semi-natural small upper courses of lowland streams
4. 29 *acid upper courses*: acid, sometimes intermittent, small, natural to semi-natural upper courses of lowland streams
5. 60 *polluted streams*: heavily to moderately polluted upper to middle courses of slow to fast running lowland streams
6. 13 *small, natural hill streams*; small, almost natural upper courses of hill streams.

Table 6.6.1. Variables used with streams

Nr.	Name	Unit	Type	Transf.	Mean	Sd
1	Meandering	-	Factor 0/1	-	0.40	0.45
2	**Natural transv. profile**	-	Factor 0/1	-	0.48	0.47
3	Land use: natural area	-	Factor 0/1	-	0.44	0.46
4	**Permanency**	-	Factor 0/1	-	0.91	0.27
5	Regulation	-	Factor 0/1	-	0.26	0.40
6	Spring	-	Factor 0/1	-	0.43	0.48
7	Winter	-	Factor 0/1	-	0.03	0.16
8	**Shading**	%	Percentage	logit	1.67	2.30
9	Substrate: silt	%	Percentage	logit	-3.23	2.37
10	**Depth**	m	Continuous	log	-1.24	1.10
11	**Current velocity**	m/s	Continuous	log	-2.23	1.25
12	**Width**	m	Continuous	log	0.92	1.03
13	Substrate: sand	%	Percentage	logit	-0.99	2.10
14	**Acidity**	-	Continuous	-	7.19	0.62
15	**Chloride**	mg/l	Continuous	log	3.65	0.51
16	Conductivity	μS	Continuous	log	6.01	0.55
17	**Ammonium**	mgN/l	Continuous	log	-0.56	1.39
18	Kjehldal-N	mgN/l	Continuous	log	0.87	0.83
19	Oxygen content	mg/l	Continuous	log	7.45	2.48
20	**Total phosphorus**	mgP/l	Continuous	log	-1.19	1.10
21	**Nitrate**	mgN/l	Continuous	log	0.68	1.50

Transf: transformation applied. logit(p)=log[p/(100- p)]

The assignment of the sites to these classes is not part of the present research; the classes have been formed (Verdonschot 1996) by means of a cluster analysis using the macrofauna of the sites and measurements on a large number of environmental variables. Hence the species composition of the sites was very important in deriving these clusters, even though their names do not directly reflect this fact.

In the present analysis the class membership of a site is treated as given. The present goal is to predict this class membership from only a relatively small subset of the available environmental variables.

This small subset consists of 21 abiotic and biotic variables (Table 6.6.1). Of these, 7 variables were factors with 2 levels, that is, they are 0/1 variables. The other variables are continuous. To decrease the influence of possible outliers and to, hopefully, better meet the linearity assumptions in the MLR model, most of the continuous variables were subjected to logit (for percentages; see the line below the table) or logarithmic transformations. The last two columns of the table contain the mean and standard deviation of the variables. For the 2-level factors the mean is the proportion of 1's. Some variables are printed in boldface; the reason for this will become clear in Section 6.4.

Hence, 21 environmental predictor variables are available, to predict the (given) membership in one of the classes A - F. No species data will be used in the prediction (and no distinct species occurrences will be predicted).

Canals data The Canals data set consists of 408 samples divided over 7 classes, with 25 predictor variables.

Methods

The present section contains a paragraph on the notation used, and a brief explanation of classification. Then the general idea of SVMs is presented, with a graphical example. Finally a description of multinomial logistic regression is given.

Notation

A sample will be considered consisting of N sampling sites. Boldface type will be used for vector-valued variables, ordinary typeface for single-valued variables. Hence, x is a vector consisting of the variables $x_1 \ldots x_D$, where D is the dimension of the input space X. With interactions or with categorical predictors, D may be larger than the number of variables actually measured. Subscripts i and j will be used to refer to the value of x for observations i and j respectively, so x_i and x_j contain the measurements for sites i and j. The symbol X will denote the design matrix. Class membership will be denoted by y; if there are only two classes, y will have values -1 and 1. Estimated values will be denoted by a hat, so $\hat{y}$ will be the predicted (estimated) class membership. The sample on which the model is fitted will be referred to as the training sample.

Classification

To derive a classification rule, a training sample of size N is obtained. For this training sample both class membership y and a number of explanatory or predictor variables $x = (x_1, x_2, \ldots, x_D)$ are known. The variables x and y are used to generate a classification rule, which relates class membership y to x. The classification rule thus obtained can be used to predict class membership for new cases, whose y is unknown. For example, the training sample might consist of patients with a tumour, the x-es might be tumour characteristics, and y might be the result: patient did or did not die of cancer. In our application the sample will consist of water sites, the explanatory variables will consist of physical and chemical characteristics such as width, depth and acidity; and y will denote the 'type' of the site (A - F), which represents both the macrofauna composition and the environmental characteristics. As often in this kind of research (see e.g. the paper on RIVPACS by Clarke et al. 2003), in this paper too the type is the result of a previous cluster analysis of macrofauna data (Section 6.2).

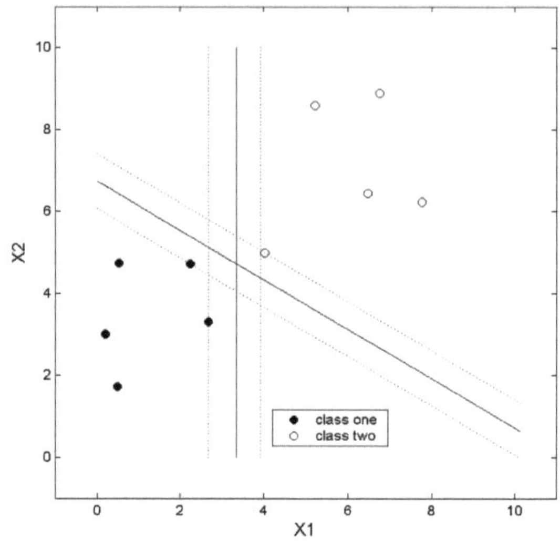

Fig. 6.6.1. Example of two separating lines with their margins. The data points exactly on the margin border are called support vectors.

Support vector machines: an intuitive description

SVMs can be used for classification and regression. Here the emphasis will be on their use as a tool for classification. The basic SVM distinguishes two classes, i.e. y has either the value 1 or -1. Classification into $K > 2$ classes is achieved using a combination of several 2-class SVMs. As an example, consider the fictitious data in Fig. 6.6.1. The black dots represent sardines, the open circles are herrings, and the two x-variables are size and weight, respectively. The data in this figure are linearly separable, i.e. it is possible to draw a straight line, such that all open circles are on one side of the line, and all black dots are on the other side. Because of the gap between the two classes it is possible to construct many lines that all correctly separate the data. Two of these have been drawn in the figure. With SVMs, the criterion to decide on the best separating line is follows. Any separating line can be moved, parallel to itself, until it collides with a data point. Doing this in both directions, a 'street' appears. This street is known as the *margin*, and its width as the *margin width*. The margins for the two separating lines in Fig. 6.6.1 have also been drawn. In the SVM context, the best separating line is defined as that line that has the widest margin. The best separating line and its margin, for the data in Fig. 6.6.1, are depicted in Fig. 6.6.2. The meaning of the extra point in this figure is explained below.

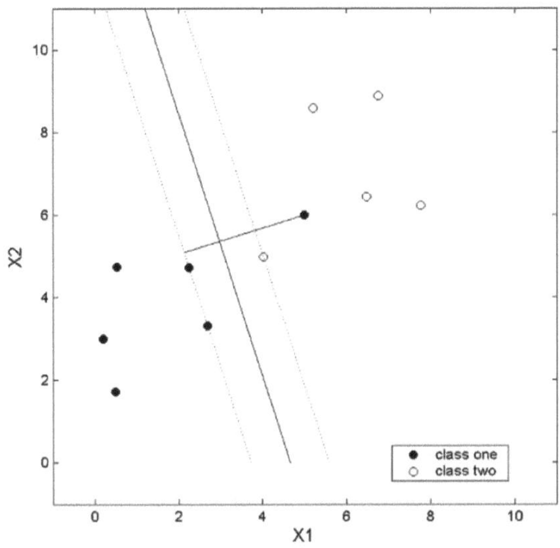

Fig. 6.6.2. Maximal margin for the points in Fig. 6.6.1. Also: the penalty of a misclassified point is proportional to its distance from the margin.

In real life the data usually are not linearly separable. In the SVM context, two strategies are used to cope with this problem: penalties and kernels.

Penalties When the data are not linearly separable, every point landing on the wrong side of the margin can be given a penalty which is proportional to how far it is from the margin, see Fig. 6.6.2; it is denoted as ξ_i. The penalty of an object is referred to as its *slack*. The objective becomes to simultaneously keep the margin width maximal and the sum of (squared) slack minimal. In this optimization process, a parameter R governs the importance of the margin width relative to the total squared slack; its value has to be set by the user. $1/R$ should be small compared to 1 (see the discussion below Eq. B9 in Appendix B).

Kernels It would also help if a (nonlinear) transformation could be found, such that the transformed data were linearly separable, or at least 'more' linearly separable. In the upper part of Fig. 6.6.3, for example, the data cannot be separated by a straight line, so we cannot apply the procedure described above. But we can let, for example, $z_1 = x_1$ and $z_2 = 2x_1^2 - 20x_1 + 58$, and, in the lower part of Fig. 6.6.3, plot z_1 and z_2 instead of x_1 and x_2. In the figure it can be seen that the transformation has made the data linearly separable.

In this particular example the dimensionality of $z = (z_1, z_2)$ is equal to the dimensionality of x: both have dimension dim = 2. In practice, however, with SVMs there is no need for dim(z) to be equal to dim(x). In fact, it might be useful to have dim(z) > dim(x). To illustrate this, consider the data in upper left part of Fig. 6.6.4, with dim(x) = 1. These data are not linearly separable in X, the space of the data x. Now consider the function $f(x) = x^3 - 20x^2 + 124x - 240$ (lower left part of Fig. 6.6.4). Then let $z_1 = x$ and $z_2 = f(x)$, i.e. z_2 is some cubic function of x, so dim(z) = 2. As can be seen in the lower right part of Fig. 6.6.4, this results in linear separability in the 2-dimensional space of z. The space of z is known as the *feature space F*.

Hence, the basic idea of SVMs when the data are not linearly separable is to:

- transform the data to a (possibly high-dimensional) feature space F, where, hopefully, they do become 'more', or even completely linearly separable; and then

- perform an optimal linear separation in the feature space *F*.

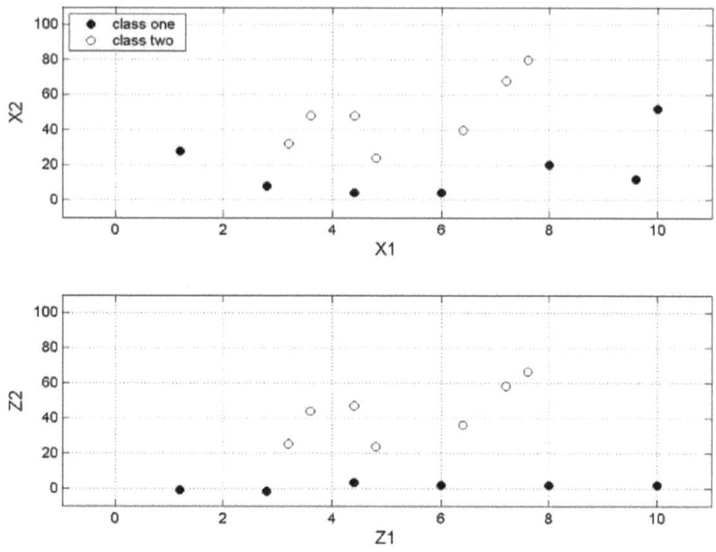

Fig. 6.6.3. Above: Linear separation is not possible in the data space. Below: After suitable transformation of x, linear separation becomes possible.

Upon returning to the lower dimensional data space X, a nonlinear separation will have been performed. Fortunately, there is no need to explicitly find a suitable transformation that makes the data linearly separable. Suitable transformations can be written as *kernels*. Kernel functions are well known in mathematics, they have been studied among others by Mercer (1909); we only say that kernels can be interpreted as similarity measures (see Appendix). Therefore, analyzing a matrix of similarities - instead of analyzing the raw data - will implicitly perform the desired transformation. Note that, although every kernel is a similarity measure (Legendre and Legendre 1998), not all similarity measures are kernels, so the similarity measure should be chosen with care. It can be shown that the Gaussian function of the Euclidean distance between data points (Eq. C4 in Appendix C) is a kernel; it will be used in this paper. The Gaussian kernel requires a user-specified parameter σ, governing the smoothness of the solution: small values for σ will give a solution that closely follows the data and hence is highly curved; larger values for σ will result in a smoother solution. With σ small enough, it is usually possible to obtain perfect separation of the training data. But when a model thus derived will be applied to new data, the performance will probably be less than perfect. Hence, in choosing a value for σ (and so implicitly increasing the number of dimensions), care must be taken to avoid overfitting: usually it is better to allow some violation of the separability also in *F*. Values for σ may be chosen in the range of the distance of the closest points with different classifications (Cristianini and Shawe-Taylor 2000).

The objective function of the SVM can be written as a convex quadratic programme (see the Appendix). This means the optimization function has no isolated local minima, which is a great advantage over other modern techniques such as for example artificial neural networks. More information on constrained optimization can be found in Lasdon (1970), Beale (1988), and Chong and Zak (2001).

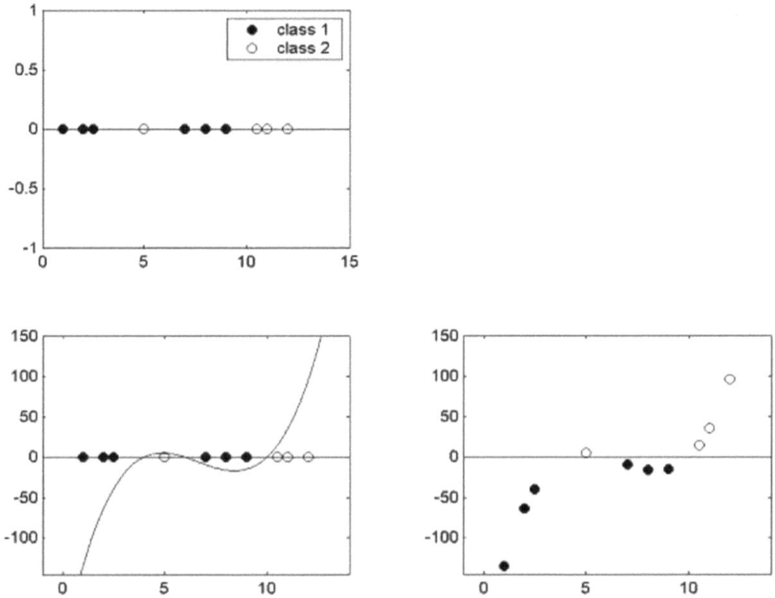

Fig. 6.6.4. Transforming linearly inseparable data to a space of higher dimensionality than dim(x) may result in linear separability. For more explanation see the text.

Multiple classes An SVM for more than 2 classes, say K, consists of a combination of 2-class SVMs. One could either construct K SVMs, each of these considering one class against all the rest; or $\binom{K}{2}$ = K(K - 1)/2 SVMs, each of these contrasting one class versus one other class. In both cases, the assignment of a site to a particular class is made after comparing its outputs for all the basic SVMs. Platt et al. (2000) constructed the so-called DAGSVM algorithm which makes navigating through the $\binom{K}{2}$ classes very fast.

Illustration We illustrate the nonlinear separation with SVMs by re-analyzing Fisher's Iris data. These are measurements of petal and sepal length and width of 150 Iris flowers, 50 of which are of the Setosa variety, 50 Versicolor, and 50 Virginica (Fisher 1936). The data were collected by Anderson (1935); they are also available in Splus and Minitab. Only the petal measurements will be used here, so that a 2-dimensional graphical representation of the data is possible (with more than 2 predictor variables a graphical representation of the solution in only 2-d would not be very instructive). The 50 Versicolor flowers will be taken as one class, and the Setosa and Virginica flowers will together constitute the second class. In Fig. 6.6.5 the Versicolor flowers are indicated by a square, and the other flowers by a circle. The squares are all in the centre, and a single straight line would be incapable of separating the two classes. Fig. 6.6.5 displays the result, in the data space, of applying an SVM with Gaussian kernel to these data. The middle line represents the decision boundary, i.e. all future flowers with measurements within the central circle will be classified as Versicolor. As the Setosa and Virginica varieties have been lumped into one single class for this particular analysis, all future flowers with measurements outside the central circle will be classified as *not Versicolor*. The other two lines are the margins: they indicate the region where the classifier is relatively unsure of the decision.

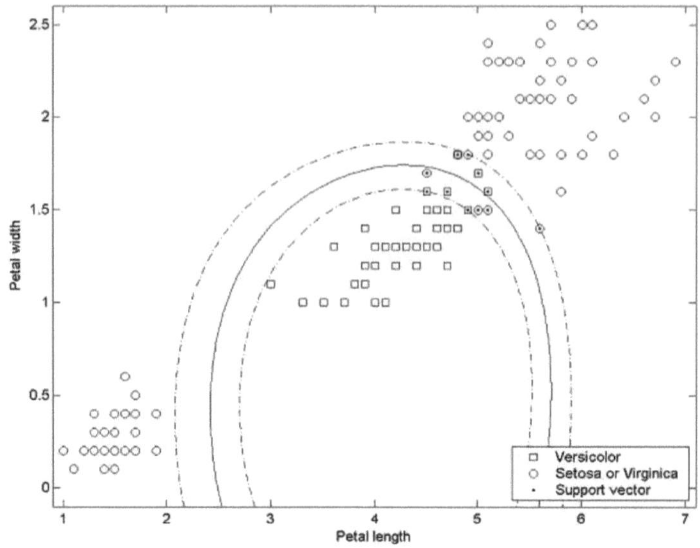

Fig. 6.6.5. Petal length and width of Fisher's Iris Data. A 2-class SVM is used to distinguish the Versicolor from the Virginica and Setosa varieties. The lines are the decision boundary and its two margins. Support vectors are those irises lying on the wrong side of their margin; they are indicated by a black dot.

Fig. 6.6.6 gives the decision lines and margins resulting from separating the 3 varieties of Iris in Fisher's Iris data, again using only petal length and width, but now the separation was done using 3 1-versus-rest SVMs. The lines in the figure are the decision boundary and margins for each of the three separate SVMs. Note that the lines separating Versicolor from Setosa and Virginica are the same lines as in Fig. 6.6.5.

Further reading The founder of SVMs is Vapnik (Vapnik and Chervonenkis 1971, Vapnik 1982, Boser et al. 1992). An introductory book has been written by Cristianini and Shawe-Taylor (2000). The book by Hastie et al. (2001) gives an overview of many relatively modern methods for predicting output from inputs, relates SVMs to more classical methods, and identifies SVMs as a penalized regression method. Introductory papers on SVMs have been written by Bennett and Campbell (2000) and Burges (1998). Finally, websites with useful information are being maintained at http://www.support-vector.net and http://www.kernel-machines.org.

Multinomial logistic regression (MLR)

Let $z_i = (z_1, z_2, \ldots, z_K)$ be an indicator vector for the class membership of site i, and let the conditional probability for site i belonging to class k, given the data, follow a multinomial distribution with parameter $\pi_i = (\pi_{i1} \ldots \pi_{iK})$. With MLR a linear regression model is assumed for each $\log(\pi_i/\pi_K)$:

$$E\left[\log\left(\frac{\pi_{ik}}{\pi_{iK}}\right)\right] = f(x_i)\cdot \tag{6.6.1}$$

In this expression, $f(x_i)$ is a linear regression function; and $\pi_{ik} = 1 - \sum_{k=1}^{K-1} \pi_{ik}$. With only 2 classes, no subscript k is needed, and π_{iK} reduces to 1 - π.

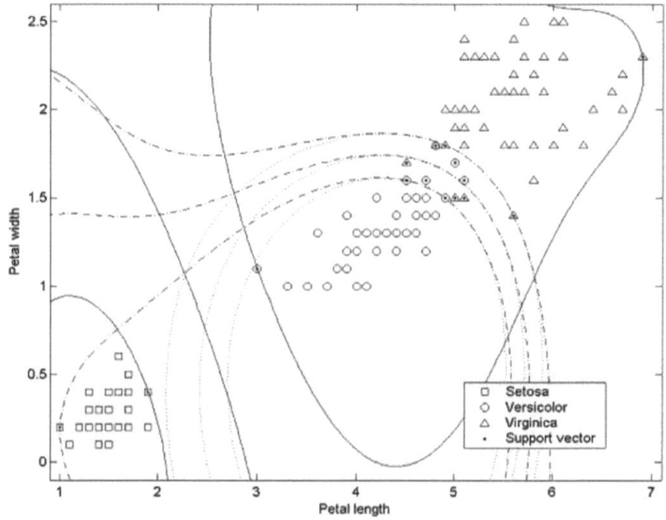

Fig. 6.6.6. Application of a 1-versus-rest 3-class SVM to Fisher's Iris data (only petal length and width). Let the Setosa variety be Class 1, Versicolor Class 2, and Virginica Class 3. The continuous lines (--) are the decision boundary and margins for the separation of class 1 from classes 2 and 3; the dotted lines (...) separate class 2 from classes 1 and 3; and the dash-dotted lines (-.-.) separate class 3 from classes 1 and 2. The support vectors are indicated by a black dot.

Contrary to SVMs and neural networks, which give a hard classification, MLR gives a so-called soft classification: for each site *i* the conditional probability of class membership given x_i is obtained for all classes. If a hard classification is required, sites may be assigned to the class with the highest conditional probability. We have not done so: in the presentation of our results, the entire distribution is retained. We will return to this point in Section 6.6.3. As an aside, note that the dependent variable in Eq. (6.6.1) is continuous, which is the reason for this technique to be known as regression rather than classification.
Further reading More information on MLR can be found in McCullagh and Nelder (1990), Hosmer and Lemeshow (1989), and Hastie et al. (2001).

Model selection

One important difference between MLR and SVMs is the way in which the model is selected. When the predictors are highly correlated, MLR requires selecting the subset of 'best' predictor variables $X_1 \ldots X_M$. For SVMs multicollinearity does not seem to be so problematic, so in theory all variables could be used for the prediction, even though they might be (highly) correlated. But with SVMs values for the penalty parameter R and the kernel width σ have to be selected.
MLR We will start our investigation by selecting a 'best' model for the MLR case. This problem will be tackled by first looking for the best subset of 4 explanatory variables, then

for the best subset of 5 variables and so on; and then finally to choose, from all these best subsets, the very best. The decisions on what is 'best' will be based on the results of cross-validations (Section 6.3). For comparison purposes with the SVM case, it was also decided to fit an MLR model on all 21 variables.

SVM Upon selection of the best MLR model, two Gaussian kernel SVMs of $\binom{K}{2}$ 1-vs-1 nets will be fitted to the data: one using only the subset of best variables selected with MLR, and a second one using all 21 variables available.

Here the parameters σ and R have to be chosen by the user. As R must be small relative to 1, we will consider R=10, 15 and 25. A possible range for σ will be derived from the distance between the 2 closest points in different classes. The final values for R and σ will be decided on by comparing results for different combinations of these 2 parameters. The results for the 'best' combination of R and σ will then be compared to an SVM that uses just the midpoints of the a priori ranges.

Finally, the effect, in the SVM case, of standardizing the x-variables will be investigated. For logistic regression this is not necessary as it has no influence of the conditional distributions.

Presentation of results As mentioned above, the MLR output for each site is a set of K probabilities, summing to 1, indicating the probability that the site belongs to class K. These probabilities can be averaged over the sites in a class, yielding a table with average classification probabilities. As an example, assume there are 5 sites and 3 classes. Let 2 of the sites belong to class A, and let these two sites have conditional distributions (0.80, 0.20, 0.10) and (0.40, 0.40, 0.20) for membership in classes A, B and C respectively. The numbers reported for class A then will be (0.60, 0.30, 0.15).

Crossvalidation and computation

After fitting a model on a data set, new data are needed to evaluate the performance of the model. Often however, data is scarce. As a second best, one might apply cross validation: in a data set consisting of N objects, the model is fitted using only N-1 of these, and used to predict the outcome of the N'th. This procedure is repeated N times, with each of the N objects left out in turn. In the end one has a set of N predictions which may be compared to their observed values, to obtain an indication of the goodness of fit. The final model is fitted on all data.

The calculations for the MLR analysis will be performed using Genstat (Payne 1997); for the SVM a collection of MATLAB and C++ routines written by Cawley (2000) is used, with some modifications and extensions. Cawley's program uses Platt's Sequential Minimal Optimization algorithm (SMO) for performing separation in the feature space (Platt 1999).

Results

Variable selection with MLR

The best model found with MLR uses 11 of the 21 variables; these are the variables numbered 2 4 8 10 11 12 14 15 17 20 21, which are indicated by boldface in Table 6.6.1. So we will investigate SVMs with only these 11 variables, and also with all 21 variables available, in either case both with and without standardization of the x-variables.

Selection of ranges for R *and* σ *with SVM*

With SVMs, values for the parameters R and σ have to be decided on. As mentioned before, values of R = 10, 15 and 25 will be considered. Appropriate values for σ lie in the range of the minimum distance between points from different classes. For each of the $\binom{6}{2}$ = 15 possible combinations of two classes, the minimum distance between a point from one class to a point from the other class was calculated. For the 21 raw variable case these 15 minima range from 0.79 to 3.14, with a mean of 2.01. Hence the value 2 could be used for σ. As the classes overlap to a considerable extent, however, we will also investigate some larger values for σ. It was decided to consider values in the range of 2 to 5. Our software wants the input to be given as $1/(2\sigma^2)$; we will denote this quantity as τ. A range of 2 to 5 for σ roughly corresponds to 0.10 to 0.02 for τ.

The mean of the minimum distances between two points from different classes for the 11 raw variables is 1.80, so here too 2 seems a good value for σ. The mean of the minimum distance for the two standardized sets of variables equals approximately 1, so here some smaller σ's (larger τ's) will also be considered.

A preliminary investigation showed that larger values for R did not perform better than the smaller values 10, 15 and 25; nor did larger τ's seem to perform better than τ = 0.10. Hence, in deciding on τ and R, all combinations of τ = 0.02, 0.03, 0.05, 0.075, 0.10 and R = 10, 15, 25 were investigated.

Model selection for SVMs

A possible criterion for the evaluation of the results would be the overall percentage of correct classifications. The variation in class size is rather large, however, and therefore the percentage correctly classified in each class was also taken into account.

For the 11-variable model with raw X-variables, τ = 0.02 and R = 25 was one of the best for the smaller class sizes, and it also happened to have the best overall performance. In Table 6.6.2 it can be seen that the overall performance of this model is 83.5 % correct (i.e. 93 out of the 563 sites were incorrectly classified in the crossvalidation). Classes E and F were most difficult to learn: only 47 and 46 % correct, respectively. The best combination for standardized X-variables was τ= 0.05 and R = 25, whose performance is more or less comparable to that of the raw X-variables (Table 6.6.2, second line). The difference between 54 and 46 % of the 13 cases in class F constitutes exactly one extra misclassification. Of the two models, the raw data model, with τ = 0.02 and R = 25, is preferred over the standard score model as it is slightly better overall, and it performs comparably in the smaller classes. However, if one would have a preference for using standardized scores instead of raw scores, not much is lost.

Simply taking the midpoints of the investigated ranges would give τ = 0.05 and R = 15. The percentages for these midpoint parameter values are also given in the table, and it can be seen that although the performance is somewhat less, the differences are not large.

With all 21 variables in the model, the pattern was more or less similar: for raw X the best overall percentage was 79.2; also taking into account the performance in the smaller classes gave a model with τ = 0.02 and R = 25, having 78.7 % correct overall, which is only marginally less than 79.2 For standardized X the difference between the two criteria was also small. With the 21 variable model the difference between standardized and raw X is larger than for the model with 11 variables: it amounts to 81.5 - 78.8 $\approx$ 3 percent overall. But seeing that the unstandardized model is better with the 3 small classes C, D and F, our preference is for the model with raw X, τ = 0.02 and R = 25. Here too, simply choosing the midpoints $\tau = 0.05$ and $R = 15$ would not do very much worse, as can be seen in the lowest part of the table.

Table 6.6.2. Selection of parameters τ and R for SVMs: Percentage correct in crossvalidation for Streams.

Model #Vars	**Param.**	**X**	**τ**	**R**	**A N=159**	**B 270**	**C 32**	**D 29**	**E 60**	**F 13**	**Tot 563**	**#Err.**
11	Best	Raw	0.02	25	90	93	50	86	47	46	83.5	93
		Stand.	0.05	25	91	90	50	90	40	54	82.2	100
	Midpoints	Raw	0.05	15	91	90	50	79	47	46	82.1	101
		Stand.	0.05	15	91	91	47	90	37	46	81.7	103
21	Best	Raw	0.02	25	86	87	56	76	43	54	78.7	120
		Stand.	0.075	15	91	91	50	69	45	46	81.5	104
	Midpoints	Raw	0.05	15	87	90	56	52	42	46	79.2	117
		Stand.	0.05	15	91	90	50	66	45	38	80.6	109

#Vars: number of variables in model. Param.: parameters used.
Raw/Stand: model using raw or standardized X.
A-F: classes of dependent variable. N=159 ...: number of sites per class.
Tot: first line: total number of sites; next lines: total percentage correct.
#Err.: Number of incorrectly classified sites.

Table 6.6.3. Crossvalidation results for Streams: Percent correct per class

	Model with 11 variables							**Model with 21 variables**					
Class	**A N=159**	**B 270**	**C 32**	**D 29**	**E 60**	**F 13**	**Ntot**	**A N=159**	**B 270**	**C 32**	**D 29**	**E 60**	**F 13**
MLR	Average 75 %							Average - %					
A	**80**	6	6	1	3	4	159	-	-	-	-	-	-
B	2	**85**	2	1	10	0	270	-	-	-	-	-	-
C	30	8	**49**	2	10	0	32	-	-	-	-	-	-
D	1	9	5	**82**	4	0	29	-	-	-	-	-	-
E	12	41	5	3	**38**	2	60	-	-	-	-	-	-
F	36	0	3	0	0	**61**	13	-	-	-	-	-	-
Npred	-	-	-	-	-	-	563	-	-	-	-	-	-
SVM	Overall 84 %							Overall 79 %					
	Raw X, τ= 0.02, 1/R = 25							Raw X, τ= 0.02, 1/R = 25					
A	**90**	4	1	0	2	3	159	**86**	6	4	1	3	2
B	1	**93**	1	1	3	0	270	2	**87**	2	1	8	0
C	31	16	**50**	0	3	0	32	19	13	**56**	6	6	0
D	0	7	7	**86**	0	0	29	0	10	10	**76**	3	0
E	12	40	2	0	**47**	0	60	8	45	2	2	**43**	0
F	54	0	0	0	0	**46**	13	31	0	15	0	0	**54**
Npred	170	290	25	27	41	10	563	157	277	35	30	54	10

A-F: classes of dependent variable. N=159 ... 13: Class size.
Ntot: number of sites per class / in entire sample.
Npred: number of sites predicted per class (middle column: entire sample).

Full results for the selected models

Full results for the models decided on, both with 11 and 21 variables, are presented in Table 6.6.3. The numbers reported for SVMs are percentages correct, i.e. of all 159 sites in class A the 11 variable SVM predicted 90 % correctly as A, 4 % incorrectly as B and so on. Multinomial logistic regression yields, for every site, a set of numbers denoting the probability of belonging to each class. These probabilities have been averaged over all sites in a class (Section 6.3). Consequently it is not meaningful to report the 'size' of the predicted classes, as is done with SVMs. Although, strictly speaking, the numbers reported for MLR and SVM do not have exactly the same meaning, both will be referred to in the text as percentage correct.

Comparing results for MLR and SVM for the 11-variable model, first of all it is clear that the overall percentage correct is some 10 percent higher for the SVM (84 versus 75 percent). The SVM seems to be better in all classes, except class F, which is the smallest class. Furthermore, the pattern of misclassification is much the same for both methods: with both methods most of the incorrectly classified sites belonging to class C go into class A (some 30 %), and the incorrectly classified class E sites go into class B (some 40 %). Class F is subsumed for a large part under class A.

Unfortunately, comparing the 21-variable model for MLR and SVM proved impossible as an MLR model with all 21 variables could not be estimated. With MLR, for each class more than 20 parameters have to be estimated; but the smallest class only has 13 sites. This causes overparametrisation, which prohibited finding a realistic solution. (This situation is analogous to fitting a 3rd order polynomial through only 2 data points: no unique solution then exists).

The final comparison now is between the two SVM models. Again the classification patterns are much the same, but the 21-variable model does not perform as well as the 11-variable model. The 21-variable model seems to perform exactly in between the MLR and SVM models for 11 variables.

Interpretation of the results

The cross validation results for the streams are different in classes A to F. Classes A and B are both large. Class A represents 156 sites of fast flowing upper and middle courses of streams. This class is quite distinct within the dataset. Class B represents 270 sites, all disturbed either channelised or polluted with organic waste. Beside the number of sites also their environmental conditions characterise both classes A and B as quite discrete units. Class C, with only 32 sites, is small and the streams in this class are more often intermittent. Intermittent streams can have a rather heterogeneous taxon composition. The heterogeneity of a class makes accurate prediction or re-assignment of sites less reliable. Class D is composed of only 29 sites, the majority of which is acid. This class is quite explicit and homogeneous in taxon composition. The cross-validation results were high. Class E, with 60 sites, is larger and the sites are more often organically polluted upper courses. The cross validation results are low due to the overlap with the large class B, which also contains organically polluted sites. Class F, with 13 sites, is very small and the sites are all fast flowing, near natural small upper courses of streams, a distinct and reasonably homogeneous group of sites. Still there is some overlap with class A which lowers the cross validation results.

Canals data

The Canals data are more difficult to classify, because environmental differences among streams are much larger than among canals. The best MLR model uses 14 of the 25 predictors. The overall percentage correct obtained with this model is 56. The 14 variable SVM has 65 percent correct, and an SVM with all 25 predictors shows 69 percent correct in the crossvalidation. Hence, with the Canals data, an SVM model using all available variables produces better results than all other models fitted.

Computation times

Estimating one multiclass support vector machine with Gaussian kernels, for one combination of R and σ, using the DAGSVM algorithm for predicting class membership, took about 15 minutes on a 1000 MHz PC, including the crossvalidation, for one combination of *R* and σ. Carrying out one multinomial logistic regression analysis, including the crossvalidation, took about 30 minutes.

Discussion and conclusion

SVM seems a promising tool for classification. Especially for ecological data, where the 'curse of dimensionality' is sometimes heavily felt, a technique that can deal with many predictors is very useful. Another advantage of SVMs for ecological data is that they do not assume relationships to be linear. With the data sets examined here they give good results, the overall correct classification rate is some 10 % higher than with multinomial logistic regression. The performance is also reasonably good in small classes. Multicollinearity does not seem to be a large problem: it was possible to find good models without having to go through extensive variable selection procedures. With the Canals data set, using all available 25 variables resulted in better performance than using only the 14 variables that had been selected for the best MLR model. This was not so with the Streams data set; here the full 21 variable model performed halfway between an MLR and the 'best' SVM. For both these data sets the penalized regression approach (SVM with all available variables) showed a better performance than the best subset selection approach (multinomial logistic regression). Results with standardized X-variables were very similar to results with unstandardized input. This might be explained by the fact that most X-variables already had been subject to log- or logit transformations, which can be considered as standardizing transformations too. Finally it must be mentioned that, on the data sets examined here, simply choosing values of σ and R in the middle of the range that seemed acceptable a priori, would have resulted in models performing only marginally less than the ones selected as best models.

Generalization The entire SVM methodology is the result of Vapnik's search for classifiers with good generalization properties (Vapnik 1998, 2000). Even if a classifier were perfect on a training data set, it would be of little practical use if it performs poorly on new data. The central concept here is the notion of the so-called VC-dimension (Vapnik and Chervonenkis 1971), which is equal to the number of points the classifier can be *guaranteed* to learn, during training, without error. Vapnik (1982, 1998) has shown that there exists an upper bound on the generalization error if the VC-dimension is finite. For infinite VC-dimension, no such bound has yet been established. It can be shown that the VC-dimension of an SVM with Gaussian kernel is infinite, so during training it can correctly learn any number of points. In practice, however, the generalization performance of Gaussian SVMs is quite often very good.

Relation to ANNs Just like the SVM, the ANN is a relatively modern tool for classification and regression (Bishop 1995, Haykin 1999). The main difference between the two lies in the behaviour of the minimizing function. ANNs have many parameters, and more often than not, many distinct local minima. SVMs on the other hand, being convex quadratic programmes, are guaranteed to have no isolated local minima, which is a very desirable property. As the SVM objective function is convex, it will be clear when the minimum has been reached. In other words, contrary to many ANNs, the stop criterion is evident and unambiguous.

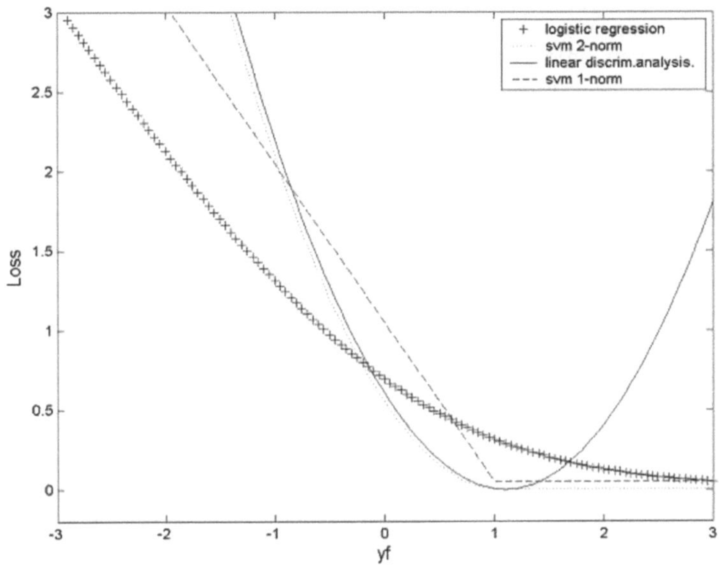

Fig. 6.6.7. Loss functions for 4 different models: (1) logistic regression (+++); (2) 2-norm SVM (…); (3) linear discriminant analysis (--); and (4) 1-norm SVM (- - -). For *yf* < 1 the loss for LDA is equal to the loss for the 2-norm SVM, and for *yf* > 1 the loss for the 1-norm SVM is equal to that for the 2-norm SVM. The functions have been drawn slightly apart to show them better.

In the case of ANNs the user can specify the functional form of the neuron, e.g. sigmoid, radial basis (RBF or Gaussian), or hyperbolic tangent (tanh) functions. Analogously, with SVMs the user may specify the functional form of the kernel: for example polynomial, Gaussian (RBF), or linear. In addition to the above, with ANNs all decisions regarding the architecture of the network also have to be taken by the researcher. The mathematical formulation for the prediction of the output of a 2 layer RBF neural network for a new site p is given by (see Bishop 1995)

$$\hat{y}(p) = \sum_{j=1}^{M} \hat{v}_j \phi(p, \mu_j) + \hat{v}_0 \tag{6.6.2}$$

with $\phi(p,\mu_j)$ the Gaussian function $\phi(p) = \exp\left[-\tfrac{1}{2}(p-\mu_j)^T \sum_j^{-1} (p-\mu_j)\right]$ and M the number of centres. Comparing this to the classification function for the Gaussian kernel SVM in (C5), which is repeated here:

$$\hat{f}(p) = \sum_{i=1}^{N} y_i \hat{\alpha}_i K(p, x_i) + \hat{b} = \sum_{i=1}^{N} y_i \hat{\alpha}_i \phi(p, x_i) + \hat{b}. \tag{6.6.3}$$

We see that, as $\hat{\alpha}_i \neq 0$ only for support vectors (Equation A8 in the appendix), these two equations are equivalent, with $y_i\alpha_i$ playing the role of v_j, and x_i that of μ_j, that is, the support vectors have the role of the centres in the ANN. So Gaussian SVMs are radial basis ANNs with the following property (Burges 1998):

"For the RBF case, the number of centres, the centres themselves, the weights, and the thresholds are all produced automatically by the SVM training and give excellent results compared to classical RBF neural networks (Schölkopf et al. 1997)".

An SVM analog to a sigmoidal (instead of radial basis) transfer function could be obtained by taking as kernel the function $K(p,x) = \tanh(\kappa p^T x - \delta)$; however, the kernel matrix associated with this function does not appear to be positive definite, so it is not guaranteed to have all the desirable SVM properties (Burges 1999).

Logistic regression and discriminant analysis Hastie et al. (2001) compared SVMs to logistic regression and linear discriminant analysis by means of their respective loss functions for classification into two classes. These loss functions are given by:

$$LR: l_i = \log\left\{\frac{1+\exp(y_i f_i)}{\exp(y_i f_i)}\right\} = \log[1+\exp(-y_i f_i)]$$

$$LDA: l_i = \frac{1}{2}(1-y_i f_i)^2$$

$$SVM: l_i = \frac{1}{2}\xi_i^2 = \frac{1}{2}[\max(0,1-y_i f_i)]^2. \tag{6.6.4}$$

Fig. 6.6.7 contains the three different loss functions, and also the loss for an SVM with 1-norm on ξ. It can be seen that the two SVM loss functions have zero loss for $y_i f_i > 1$, i.e. for correctly classified sites. This is in contrast both to the LDA and logistic regression loss, where correct classification may still increase the loss.

Apart from having different loss functions, the MLR and SVM approach also differ in their treatment of the 'curse of dimensionality': SVMs inherently are a penalization technique, and with MLR we adopted a 'best subset selection' approach. For a fairer comparison we might have used a ridge (penalized) form of MLR rather than the 'best subset selection' approach.

One definite drawback of SVMs is that decisions regarding the value of σ and R are not (yet) automatically made. Trying to find good values for these parameters is an elaborate process. Also, it would be more elegant if classification into $K > 2$ classes could be achieved in a more direct way than by having to compare the outputs of several 2-class SVMs. However, our **general conclusion** is that SVMs have many desirable properties both when compared to ANNs and also when compared to more traditional methods for classification. They can deal with many predictors, do not suffer much from multicollinearity, are not restricted to linear relationships, and do not display isolated local minima.

Appendix

This appendix contains a somewhat more mathematical description of SVMs.

A Linear separation of linearly separable data

In 2-d, a linear separator or decision line S is a straight line, and it might be given, for example, by $x_2 = -2x_1 + 3$, or equivalently, by $2x_1 + x_2 - 3 = 0$. For the development of the SVM it is convenient to consider not only the *line* $2x_1 + x_2 - 3 = 0$ but the entire *function* $f(x_1, x_2) = 2x_1 + x_2 - 3$, or, in general, the function

$$f(x,w,b) = \sum_{d=1}^{D} w_d x_d + b = w^T x + b,$$

where the coefficients $w = (w_1, w_2, ..., w_D)$ are referred to as weights, and the index d runs from 1 to D, the dimensionality of x. The superscript T denotes the transpose. The decision line or separator S is then given by those x for which f(x) = 0, i.e. S is the solution to f(x) = 0 for the case of only 2 predictors). Furthermore, f(x) > 0 for all x on one side of S, and f(x) < 0 for all x the other side of S. Therefore, once suitable values for w and b have been found, the function f can be used as a classification function for future observations: letting y denote class membership and p be a vector of future observations, then

$$\text{if } f(\hat{p}) \geq 0 : \text{predict } \hat{y} = 1;$$

$$\text{if } f(\hat{p}) < 0 : \text{predict } \hat{y} = -1.$$

So the problem to solve is to find the coefficients w and b of the maximal margin separator S.

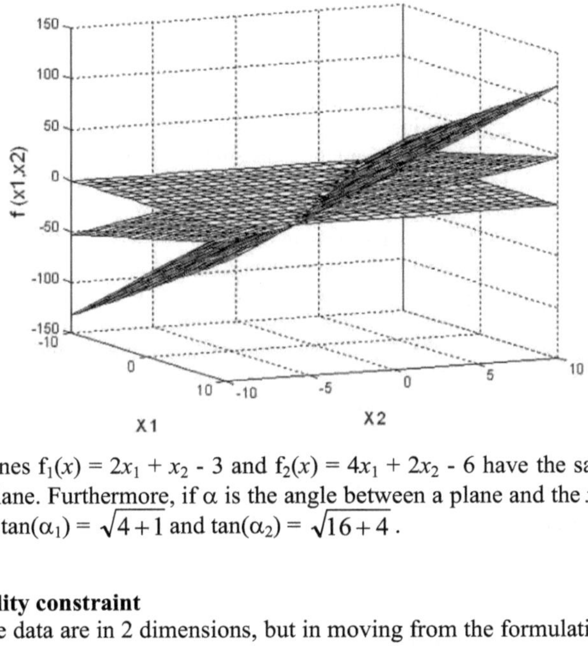

Fig. A. The planes $f_1(x) = 2x_1 + x_2 - 3$ and $f_2(x) = 4x_1 + 2x_2 - 6$ have the same intersection with the x_1/x_2-plane. Furthermore, if α is the angle between a plane and the x_1/x_2 plane, then $\tan(\alpha) = ||w||$, so $\tan(\alpha_1) = \sqrt{4+1}$ and $\tan(\alpha_2) = \sqrt{16+4}$.

A.1 Identifiability constraint

In Fig. 6.6.1, the data are in 2 dimensions, but in moving from the formulation $x_2 = -2x_1 + 3$ to $f(x) = 2x_1 + x_2 - 3$ an extra dimension has been added: $f(x) = 2x_1 + x_2 - 3$ can also be written as $x_3 = 2x_1 + x_2 - 3$. In fact, we have been moving from a 1-d line in the 2-d data space to a 2-d plane in a 3-d space. This leads to indeterminacy: there exist, in 3-d, many planes having the same intersecting line with the x_1/x_2 plane. In Fig. A, for example, two of these are illustrated:

1. $f(x) = 2x_1 + x_2 - 3$. The intersection with the x_1/x_2 plane is given by f(x) = 0, which gives the separating line defined by $x_2 = -2x_1 + 3$; and
2. $f'(x) = 4x_1 + 2x_2 - 6$, whose intersection again follows from f'(x) = 0, which gives the same separating line $x_2 = -2x_1 + 3$.

In general: if $f(x; w, b) = w^T x + b$ has S as separator, then all functions f(x; w', b'), with $w' = kw$, $b' = kb$, and $k \in \Re \backslash 0$, have the same solution S. In what follows, the notation f' will be used as a shorthand for f(x; w', b'), so f' = kf, with $k \in \Re \backslash 0$. For the classification of future sites it is of no consequence which of the many planes f' is used; but one has to decide on one, because otherwise the problem of finding a maximal margin separator would not be identifiable. An example of an identifiability constraint would be $||w||=1$, but with SVMs another constraint is used.

Note that, S being in the middle of the margin, there is exactly one of the functions f', say f^0, which has function values of 1 and -1 for the support vectors: $|f^0(SV_S)|=1$. The plane formed by this function is known as the *canonical hyperplane* for the line S (Cristianini and Shawe-Taylor 2000). The weight vector corresponding to the canonical hyperplane will be denoted as w^0, and be called the *canonical weight vector*. Given the data, every separator S has a unique canonical weight vector w_S^0. In the SVM context,

the identifiability constraint is to choose $w=w^0$,

so that with any particular separator S we will associate only its unique canonical hyperplane f^0.

A.2 Usefulness of the identifiability constraint w=||w^0||

The reason for choosing the identifiability constraint $w = w^0$ may become clear upon noting the following. Again consider the two functions $f(x) = 2x_1 + x_2 - 3$ and $f'(x) = 4x_1 + 2x_2 - 6$ introduced above, and note that $f(0,0) = -3$, and $f'(0,0) = -6$. Hence the plane described by f' is steeper than the plane described by f. This is caused by the coefficients of f' being larger than those of f. So, loosely speaking:

a plane with gentler slope has smaller coefficients. (A1)

Now again consider Fig. 6.6.1. This figure contains 'just some' separating line S_1, which is not the line having maximal margin. Let SV_1 be the support vectors associated with S_1, and $f_1^0(x) = f(x; w_1^0, b_1^0)$ be the canonical hyperplane for S_1, i.e. $|f_1^0(SV_1)|=1$. Fig. 6.6.2 contains another separating line for the same data points, S_2, having support vectors SV_2. The function $f_2^0(x) = f(x; w_2^0, b_2^0)$ describes the canonical hyperplane for S_2. Note that the separating line S_2 in Fig. 6.6.1 has a wider margin than the line S_1 in Fig. 6.6.2. Hence, because both $|f_1^0(SV_1)|=1$, and $|f_2^0(SV_2)|=1$, the plane f_2^0 associated with S_2 is less steep than f_1^0; or, more general:

The wider the wider margin,
The gentler the slope of the canonical hyperplane.(A2)

Recalling (A1), a wider margin therefore corresponds to smaller canonical weights w^0. The exact relation is given in the following statement: if S is a separator, w_S^0 are the weights of its canonical hyperplane, and $\|w\| = \sqrt{\sum w^2}$, then the size of the margin associated with S equals the inverse of $||w_S^0||$, i.e.

$$\gamma_S = 1/\|w_S^0\|.$$

This will now be demonstrated for the case of 2-dimensional x.

Derivation of $\gamma_S = 1/ ||w_S^0||$ Consider the separating line S in Fig. B, given by $x_2 = -2x_1$, and imagine a plane through S, not necessarily the canonical hyperplane, but for example the plane given by $f(x) = 2x_1 + x_2$. The line S has $b = 0$ and hence goes through the origin. For the result we are to derive, this transformation can be made without loss of generality and it simplifies the presentation. In this section we consider only one line S, so there is no need for the cumbersome subscripts on f, on w and on γ. The weight vector $\binom{2}{1}$ is the normal of the plane formed by f. This vector has been drawn in the figure as well. The weight vector of any other plane f' through S is a multiple of $\binom{2}{1}$.

The tangent of the angle α between the plane f and the x_1/x_2 plane is given by

$$\tan(\alpha) = \frac{f(w)}{\|w\|} = \frac{w_1 w_1 + w_2 w_2}{\sqrt{w_1^2 + w_2^2}} = \sqrt{w_1^2 + w_2^2} = \|w\|. \quad \text{(A3)}$$

As a steeper plane means a larger angle and hence a larger tangent, we have: the steeper the plane, the larger $||w||$. This holds for all possible f'.

On the other hand, letting d(*SV*, *S*) be the distance of the line S to its associated support vectors, the tangent is also given by tan(α) = f(*SV*)/d(*SV*, *S*). But the distance of the support vectors to S is the halfwidth of the margin, which is denoted as γ, so

$$\tan(\alpha) = \frac{f(SV)}{d(SV,S)} = \frac{f(SV)}{\gamma}. \tag{A4}$$

This too holds for all possible f.

Now let f be not just some plane, but the canonical hyperplane $f^0 = f(x; w^0)$. Then in (A3) the value w^0 has to be inserted for *w*; and $| f^0(SV) | = 1$, so the numerator of (A4) becomes 1 or -1. Combining this yields that $||w^0|| = 1/\gamma$, so for every line S its canonical weight vector is inversely proportional to its margin:

$$\left\| w_S^0 \right\| = 1/\gamma_s$$

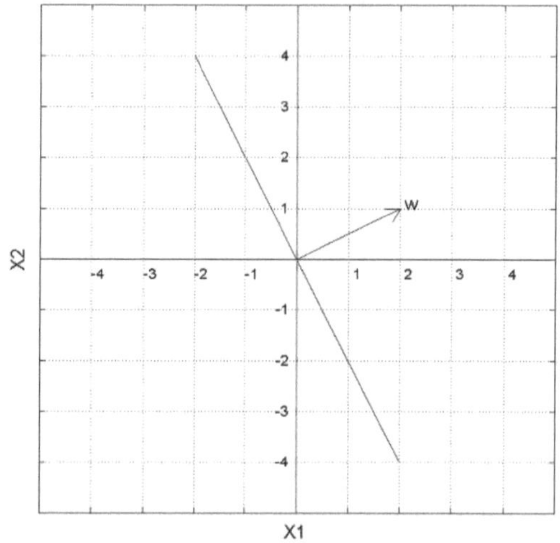

Fig. B. The plane $f(x) = 2x_1 + x_2$ has the line S given by $x_2 = -2x_1$ as intersection with the x_1/x_2 plane. This line and its normal vector $w = \binom{2}{1}$ have been drawn in the figure. The tangent of the angle between the plane f and the x_1/x_2 plane is given by $f(w)/ ||w||$.

A.3 Objective function

Summarizing the discussion so far we have the following. Let *S* be the set of all possible candidate separators S, and let each S ∈*S* have an associated margin with half-width γ_S, a set of associated support vectors SV_S and, given the data, a unique associated canonical weight vector w_S^0, defining a canonical hyperplane $f_S^0 = f(x; w_S^0, b_S^0)$ with the property that $|f_S^0(SV_S)| = 1$. Then for every candidate separator S we have $\gamma_S = 1/||w_S^0||$. Therefore the separator S with maximal margin, say this is S^*, is the separator that has the smallest canonical weight vector: $w_{s*}^0 \le w_s^0$ for all S∈*S*. Hence finding the maximal margin separator is equivalent to finding the smallest canonical weight vector. So we have to minimize $||w||$, subject to w being a canonical weight vector.

Now when is w a canonical weight vector? The canonical hyperplane associated with S has $|f^0(x)| = 1$ for those x that are the support vectors associated with S. It also has $|f^0(x)| = 1$ for all x on the margins. From now on, we will only consider the canonical hyperplane, and the superscript 0 will be dropped from f^0. As the support vectors are, by definition, those data points closest to S, the identifiability condition

$$|f(SV_s)| = 1 \quad \text{implies that} \; |f(x_i)| \geq 1$$

for all x_i in the training set. Furthermore, because the training data are separable (linearly inseparable data are covered in the next section), all x_i with $y_i = 1$ have $f(x_i) > 1$, and all x_i with $y_i = -1$ have $f(x_i) < -1$. So

$$|f(x_i)| > 1 \quad \text{is equivalent to} \; y_i f(x_i; w, b) \geq 1, \quad i = 1 \ldots N.$$

We are therefore looking for the smallest w having y_i $f(x_i; w, b) > 1$, for all i in the training set. The mathematical formulation of this optimization problem is the following objective function:

$$\text{minimize} \; \|w\|^2, \quad \text{subject to} \; y_i f(x_i; w, b) \geq 1, \quad i = 1 \ldots N. \tag{A5}$$

The problem in (A5) is a constrained optimisation problem, known in mathematics as a convex quadratic programme. Convex quadratic programmes have been extensively studied and their properties are well known, see for example Dantzig (1963), Beale (1988), Lasdon (1970) or Chong and Zak (2001). One of the most convenient properties of convex quadratic programmes is that they have no isolated local minima, which is a very convenient property indeed.

A.4 Estimation equation

In (A5), $f(x_i) = w^T x_i + b$, and $\|w\|^2 = \sum_{d=1}^{D} w_d^2$. The Lagrangean for this problem, with Lagrange multipliers $\alpha_1 \ldots \alpha_N$, gives

$$L_p(w, b, \alpha) = \frac{1}{2}\|w\|^2 - \sum_{i=1}^{N} \alpha_i \left[y_i (w^T x_i + b) - 1 \right]. \tag{A6}$$

As L_P is a convex function subject to only convex constraints, stationarity of L_P with respect to w and b, together with three conditions on the constraints, solves (A5):

$$\begin{aligned}
&(a) \quad \frac{\partial L_p}{\partial w} = 0 \Leftrightarrow w = \sum_{i=1}^{N} \alpha_i y_i x_i \\
&(b) \quad \frac{\partial L_p}{\partial b} = 0 \Leftrightarrow \sum_{i=1}^{N} \alpha_i y_i = 0 \\
&(c) \quad \alpha_i \geq 0, \; i = 1 \ldots N \\
&(d) \quad y_i (w^T x_i + b) - 1 \; \geq 0, \; i = 1 \ldots N \\
&(e) \quad \alpha_i \left[y_i (w^T x_i + b) - 1 \right] = 0, \; i = 1 \ldots N.
\end{aligned} \tag{A7}$$

The conditions *a* - *e* in (A7) are known as the Karush-Kuhn-Tucker (KKT) conditions. Condition *d* is the original inequality constraint. Condition c arises because the condition in d is an inequality constraint; if it had been an equality constraint, such as e.g. $y_i f(x_i) = 0$, the unconstrained optimization of the Lagrangian with respect to w, b and α would have sufficed to find the solution to (A5). Condition e is the KKT complementarity condition; it ensures that $\alpha_i = 0$ when the constraint is inactive, and that the constraint is active when $\alpha_i > 0$. That is, if $\alpha_i > 0$ then $y_i f(x_i) = 1$, so

$$\alpha_i > 0 \; \text{only for support vectors.} \tag{A8}$$

It now follows from (A7a) that the weights of the maximal margin separating hyperplane depend only on the support vectors and not on any other data. Therefore the solution to the classification problem only depends on those observations in the 2 classes that are closest to each other, i.e. that are close to the 'border'. Rerunning the problem with only part of the data will result in exactly the same solution, as long as the data points removed are no support vectors. In this sense SVMs are parsimonious models.

A.5 Dual formulation

With linearly separable data, maximising the margin can also be achieved in another way. This alternative method will prove very useful in the construction of a feature space F.

Substituting (A7a) and (A7b) back into L_P, the resulting function is known as L_D, the dual Lagrangean. The dual depends only on α, and no longer on w and b. As L_P is a convex function with convex constraints, there is no duality gap and maximizing the dual with respect to α, subject to some simple constraints, will also give the solution to the problem in (A5), see the literature on constrained optimization mentioned above.

The dual is most easily obtained by writing both problem (A5) and the primal (A6) in matrix notation. Let Y = diag(y) be an N × N diagonal matrix containing the values y_i on the diagonal (so $Y^T = Y$), let 1 be an N-vector of ones, and let 0 be an N-vector or an N × N matrix of zeros. The problem then reads:

$$\text{minimize } \|w\|^2, \quad \text{subject to } Y(Xw + b1) \geq 1, \tag{A9}$$

so that the primal Lagrangian can be written as

$$L_p(w,b,\alpha) = \frac{1}{2}\omega^{\mathrm{T}}\omega - \omega^{\mathrm{T}}[Y(X\omega + b1) - 1]. \tag{A10}$$

Equating the derivatives of this function with respect to w and b to 0 gives

$$\begin{cases} \text{(a)} \quad \dfrac{\partial L_p}{\partial w} = 0 \Leftrightarrow w = X^T Y^T \alpha \\ \text{(b)} \quad \dfrac{\partial L_p}{\partial b} = 0 \Leftrightarrow \alpha^T Y1 = 0 \end{cases} \tag{A11}$$

which is the matrix formulation of (A7). Substituting (A11a) into (A10) gives

$$\frac{1}{2}\alpha^T YXX^T Y^T \alpha - \alpha^T YXX^T Y^T \alpha - \alpha^T Yb1 + \alpha^T 1,$$

which using (A11b) reduces to the dual:

$$\mathrm{L_D}(\alpha) = -\frac{1}{2}\alpha^T YXX^T Y^T \alpha + \alpha^T 1.$$

So maximizing, with respect to $\alpha_1 \ldots \alpha_N$,

$$\mathrm{L_D}(\alpha) = \sum_{i=1}^{N} \alpha_i - \frac{1}{2}\sum_{i=1}^{N}\sum_{j=1}^{N} \alpha_i \alpha_j y_i y_j x_i^T x_j$$

$$\text{subject to } \Sigma\, y_i\alpha_i = 0 \text{ and } \alpha_i > 0,\ i = 1 \ldots N \tag{A12}$$

yields the same function value that is obtained by the KKT conditions in (A7). On the whole, maximizing L_D is an easier task than minimizing L_P. This is mainly so because the constraints in (A12) are constraints on the Lagrange multipliers, and these are much easier to handle than the constraints c - e in (A7), which are constraints on function values.

Having found the $\hat{\alpha}$ that maximizes L_D, the estimated hyperplane weights $\hat{w}$ are obtained upon applying (A7a) or (A11a), and $\hat{b}$ can be found from any i for which $\hat{\alpha}_i \neq 0$, upon noting in (A7e) that if $\alpha_i \neq 0$, then $y_i(w^T x_i + b) - 1 = 0$. In this last equation b is the only unknown.

B Linear separation when the data are not linearly separable: soft margin approach

When the data are not linearly separable, a perfectly separating linear hyperplane does not exist. In this case a so-called soft margin hyperplane can be found upon relaxing the constraints y_if(x_i) > 1: these constraints are now replaced by

$$y_i\text{f}(x_i) + \xi_i \geq 1,\ \xi_i \geq 0,\ i = 1 \ldots N. \quad \text{(B1)}$$

The variables ξ_i (one per observation) are known as slack variables; they indicate the degree of violation of linear separability. The slack of an observation is the extent to which it fails to have a function value > 1 or < -1 on the canonical hyperplane. So

$$\begin{cases} \text{if } y_i f(x_i) < 1, \text{i.e. if } 1 - y_i f(x_i) > 0 & \text{then } \xi_i = 1 - y_i f(x_i) \\ \text{if } y_i f(x_i) \geq 1, \text{i.e. if } 1 - y_i f(x_i) \leq 0 & \text{then } \xi_i = 0. \end{cases} \quad \text{(B2)}$$

Hence,

$$\xi_i = \max[0, 1 - y_i\text{f}(x_i)). \quad \text{(B3)}$$

As the slack is the extent to which a point fails to have function value >1 or <-1, it is proportional to the distance of the point from the margin, i.e. proportional to the length of the line segment in Fig. 6.6.2. Fig. A may help to see this for the case of 2-dimensional x; the constant of proportionality equals the tangent of the angle between f and the x_1/x_2 plane.

The amount of slack can be kept minimal by incorporating it in the equation with a penalty term R, so the problem to solve becomes:

$$\text{minimize } \|w\|^2 + R\sum_{i=1}^{N}\xi_i^2$$

$$\text{subject to } y_i\text{f}(x_i) + \xi_i > 1 \text{ and } \xi_i > 0,\ i = 1 \ldots N. \quad \text{(B4)}$$

This equation must be minimized w.r.t. to both w and ξ. The constraint $\xi_i > 0$ is superfluous: if $\xi_i < 0$ then increasing it to 0 will always decrease the value of the objective function; and if the first constraint were met by some $\xi_i < 0$, it will still be met by $\xi_i = 0$. Hence it is not necessary to explicitly include the constraint $\xi_i > 0$. The primal Lagrangian for linearly inseparable data therefore becomes

$$L_p(w,b,\alpha,\xi) = \frac{1}{2}\|w\|^2 + \frac{1}{2}R\sum_{i=1}^{N}\xi_i^2 - \sum_{i=1}^{N}\alpha_i\left[y_i(w^T x_i + b) + \xi_i - 1\right]$$

$$= \frac{1}{2}w^T w + \frac{1}{2}R\xi^T\xi - \alpha^T(YXw + Yb1 + \xi - 1)\cdot \quad \text{(B5)}$$

Again the Lagrangean problem is quadratic with only convex constraints so stationarity of (B5) with respect to w, b and ξ, together with several conditions arising from the inequality constraints, solves the soft margin problem:

$$(a)\ \frac{\partial L_p}{\partial w} = 0 \Leftrightarrow w = \sum_{i=1}^{N}\alpha_i y_i x_i \qquad \text{or } w = X^T Y^T \alpha$$

$$(b)\ \frac{\partial L_p}{\partial w} = 0 \Leftrightarrow \sum_{i=1}^{N}\alpha_i y_i = 0 \qquad \text{or } \alpha^T Y1 = 0$$

$$(c)\ \frac{\partial L_p}{\partial w} = 0 \Leftrightarrow R\xi_i = \alpha_i \qquad \text{or } R\xi = \alpha$$

$$(d)\ \qquad \alpha_i \geq 0$$

$$(e)\ \qquad y_i(w^T x + b - 1 + \xi_i \geq 0$$

$$(f)\ \qquad \alpha_i[y_i(w^T x + b - 1 + \xi_i] = 0\cdot \quad \text{(B6)}$$

Again a dual formulation in α can be developed. Substituting a and c of (B6) into the second line of (B5) gives

$$-\frac{1}{2}\alpha^T YXX^T Y^T\alpha+\frac{1}{2}\alpha^T\alpha\frac{1}{R}-\alpha^T YXX^T Y^T\alpha-\alpha^T Yb1-\alpha^T\xi+\alpha^T 1$$

which, because of (B6b), and again using (B6c), is equal to

$$-\frac{1}{2}\alpha^T YXX^T Y^T\alpha-\frac{1}{2R}\alpha^T\alpha+\alpha 1,$$

so that the problem in (B2) is solved by maximizing, with respect to $\alpha_1 \ldots \alpha_N$, and subject to $\Sigma y_i\alpha_i = 0$, the following dual:

$$L_D(\alpha)=\sum_{i=1}^{N}\alpha_i-\frac{1}{2}\sum_{i=1}^{N}\sum_{j\neq i}\alpha_i\alpha_j y_i y_j{x_i}^T x_j-\frac{1}{2R}\sum_{i=1}^{N}\alpha_i^2$$

$$=\sum_{i=1}^{N}\alpha_i-\frac{1}{2}\sum_{i=1}^{N}\sum_{j\neq i}\alpha_i\alpha_j y_i y_j\left\{{x_i}^T x_j+\frac{1}{R}\delta_{ij}\right\},\qquad\text{(B7)}$$

where the symbol δ_{ij} is the Kronecker δ which is 1 if $i = j$ and 0 otherwise. R is the same user specified parameter as in (B5). The difference between (B7) and (A12) is in the term δ_{ij}/R: in the soft margin case the diagonal of the N × N matrix XX^T is increased by an amount $1/R$, so that the soft margin SVM with 2-norm on ξ can be interpreted as a ridge technique. The parameter R has to be chosen by the user; Appendix C will give a heuristic to help in this choice.

With non-separable data too, maximizing the dual is easier than minimizing the primal. But the dual formulation has another advantage. The data appear in it only in the form of inner products ${x_i}^T x_j$. This property makes the dual formulation very attractive for use in the feature space, as will become clear in the nexet section.

C Nonlinear separation: Kernels and feature space

Recall that nonlinear separation of the data may be achieved by means of linear separation in a suitably defined feature space. We now first give a definition:

Definition *If $K(a, b) = [\varphi(a)]^T\varphi(b)$, with $\varphi(a) = (\varphi_1(a), \varphi_2(a), \varphi_2(a),\ldots, \varphi_M(a))$ some function of* a *and M usually larger than dim*(a), *then K*(a, b) *is a kernel function.*

If a kernel function K and associated transformation φ have been identified, then of course also $[\varphi(a)]^T\varphi(b) = K(a, b)$. This may seem trivial and of little practical use, but it is not. Recall that in the data space X, the dual is given by

$$L_D(\alpha)=\sum_{i=1}^{N}\alpha_i-\frac{1}{2}\sum_{i=1}^{N}\sum_{j=1}^{N}\alpha_i\alpha_j y_i y_j({x_i}^T x_j+\frac{1}{R}\delta_{ij}).$$

with δ_{ij} the Kronecker delta. In the feature space F however we would have

$$L_D^*(\alpha)=\sum_{i=1}^{N}\alpha_i-\frac{1}{2}\sum_{i=1}^{N}\sum_{j=1}^{N}\alpha_i\alpha_j y_i y_j\left\{[\varphi(x_i)]^T\varphi(x_j)+\frac{1}{R}\delta_{ij}\right\}.\qquad\text{(C1)}$$

If it were possible to find a simple kernel function $K(x_i, x_j)$, this kernel function could be substituted for the inner product $[\varphi(x_i)]^T\varphi(x_j)$ in (C1). Instead of having to work in the high dimensional feature space F we could then just minimize in ordinary N-dimensional data space, with respect to $\alpha_1, \ldots, \alpha_N$:

$$L_D^*=\sum_{i=1}^{N}\alpha_i-\frac{1}{2}\sum_{i=1}^{N}\sum_{j=1}^{N}\alpha_i\alpha_j y_i y_j\left\{K(x_i,x_j)+\frac{1}{R}\delta_{ij}\right\}$$

subject to $\Sigma y_i \alpha_i = 0$ and $\alpha_i > 0$, $i = 1 \ldots N$. (C2)

For the interpretation of a kernel function, note that the inner product $x_i^T x_j$ can be seen as a similarity measure between the two data points x_i and x_j. Generalizing this, any kernel function $K(x_i, x_j) = [\varphi(x_i)]^T \varphi(x_j)$ could be regarded as a similarity measure.

For suitable kernels the optimization problem in (C2) is still convex, so it still only has a single optimum in the feature space F, i.e. there are no local optima. To find the global optimum, Platt's Sequential Minimal Optimization or SMO-algorithm (Platt 1999) can be used. This algorithm is also described in Section 7.5 of Cristianini and Shawe-Taylor (2000).

The question now is how to choose a suitable measure $K(x_i, x_j)$ for the similarity between two objects or sampling sites. A good help here is Mercer's theorem, which says (Cristianini and Shawe-Taylor 2000) that a symmetric function $K(x_i, x_j)$ on X is a kernel if and only if the matrix K, given by $K_{ij}=K(x_i, x_j)$ is positive semidefinite for all possible i and j. Mercer's theorem implies that similarity measures which may yield negative eigenvalues with PCA are not suitable kernels, see e.g. Pielou (1984), Legendre and Gallagher (2001) or Legendre and Anderson (1999).

It can be shown that the Gaussian function

$$K(x_i, x_j) = \exp[-\|x_i - x_j\|^2/\sigma^2] \tag{C3}$$

is a kernel, see, e.g. Cristianini and Shawe-Taylor (2000), who give examples of kernels and associated transformation functions φ. The Gaussian kernel is very often used. It is also used in the analyses in this paper.

From (C2), one sees that 1/R has to be small relative to the diagonal values of the kernel matrix. With a Gaussian kernel, whose formula is given in (C3), these diagonal values can be calculated as $K(x_i, x_i) = \exp[-\|x_i - x_i\|^2/\sigma^2]=1$. So when using a Gaussian kernel the value $1/R$ should be small relative to 1.

Unfortunately, there is no once-and-for all best value for $1/R$. This can best be seen when the objective function in (B4) is divided by R, and hence becomes to

$$\text{minimize} \sum_{i=1}^{N} \xi_i^2 + 1/R\|w\|^2 \cdot$$

The equation is now cast in the formulation of a ridge technique, and the best choice for $1/R$ can be seen to depend on the total amount of slack, i.e. on the separability of the data.

Once the parameters $\hat{\alpha}$ and $\hat{b}$ have been estimated, the classification function for a future observation p becomes

$$\hat{f} = \sum_{i=1}^{N} y_i \hat{\alpha}_i K(p, x_i) + \hat{b}, \tag{C4}$$

and p will be classified as class 1 whenever $\hat{f}(p) > 0$, and as class -1 whenever $\hat{f}(p) < 0$.

7 User interface tool*

Park YS†, Lek S

7.1 Introduction

The goal of the PAEQANN project was to develop general methodologies, based on advanced modelling techniques, for predicting structure and diversity of key aquatic communities under natural and man-made disturbances. This allowed the detection of the significance of various environmental variables that structure these aquatic communities. These have been shown to reveal predictable changes due to natural variability and human disturbances. Natural conditions are described as undisturbed by human activities and man-made disturbances are defined as various pollutants, discharge regulation etc.

Such an approach to the analysis of aquatic communities made it possible to:

1. set up robust and sensitive ecosystem evaluation procedures that will work across a large range of running water ecosystems throughout Europe;
2. point out the cause and effect relationships between environmental conditions (physical, chemical, due to management actions) and certain relevant aquatic communities (diatoms, macroinvertebrates, and fish) and subsequently,
   - to predict biocenosis structures in disturbed ecosystems, taking into account all the relevant ecological variables,
3. test ecosystem sensitivity to disturbances, and
4. explore specific actions to be taken for the restoration of ecosystem integrity.

The long-term aim of these investigations was therefore to help to define strategies for conservation and restoration of ecosystems, compatible with local and regional development, and supported by a strong scientific backup. The development of these general methodologies allowed the:

1. provision of predictive tools that can be easily applied to define the most effective policies and institutional arrangements for resource management;
2. application of the most effective and innovative techniques (mainly Artificial Neural Networks) to identify problems in ecosystem functioning, resulting from ecosystem degradation from human impact, and to model relevant biological resources;
3. full exploitation of existing information, reducing the amount of field work (that is both expensive and time consuming) needed in order to assess the health of freshwater ecosystems;
4. exploration of specific actions to be taken for restoration of ecosystem integrity,
5. promotion of collaboration among scientists of different interested countries and research fields, by encouraging collaboration and dissemination of results and techniques.

This is why the structure of the software developed in the PAEQANN project is presented here, as well as indications as to how the software is working.

---

* Funding for this research was provided by the EU project PAEQANN (N° EVK1-CT1999-00026).

† Correspondence: park@cict.fr

7.2 Software aims

The main objective of the software is to suggest a set of tools for water management and water policies in order to facilitate the assessment of ecological quality and perturbations of stream ecosystems. These tools will provide information about running water quality as well as community structure, and allow identifying measures which should be taken to restore biological integrity in running waters. Hopefully, the study can be considered as a first step toward linking the improvement of water quality through specific management measures (*e.g.* waste water treatment, habitat restoration, etc.) with the expected improvement in ecological and biological value of running water systems. It will also allow scientists and ecosystem managers to consult the occurrence patterns of organisms in streams based on the database used in the tool, visualise the results of patterning and predicting models with existing data, and provide the possibility of testing the new data based on models developed with existing data.

7.3 System requirements

The program has been implemented for the Microsoft Windows operating system and is available as a single compressed archive containing all the files required for installation. It is recommended to be operated under Windows XP with 64-plus megabyte memory, high resolution graphic video, and at least a Pentium III 500 MHz.

7.4 Installing/Uninstalling

The installation procedure is based on standard Windows procedure, familiar to most users. To install, run setup.exe and follow all steps required to install the software correctly. The installation process is automatic and the default path is recommended (c:\Program files\paeqann). There is no need to restart the computer, once installation is complete. The icon of the program is created in the desktop window.

To uninstall the software, there are two possibilities:

1. "Start" in the taskbar → "PAEQANN folder" → "uninstall"
2. "Start" in the taskbar → "Control panel" → "Add or Remove Programs" → "PAEQANN"

7.5 Models implemented in the tool

Two different artificial neural networks (ANNs) - self-organizing map (SOM) and multilayer perceptron (MLP) with a backpropagation algorithm - were used to develop the models in the tool. The SOM is the most well known ANN with unsupervised learning rules and performs a topology-preserving projection of the data space onto a regular two-dimensional space. In the project, the SOM as an ordination method was applied to summarize the variability of the data. Thus, sampling sites could be arranged on the reduced dimensions, so that these arrangements optically summarize the spatial variability of their biological and environmental features. The MLP is based on a supervised learning rule, .*i.e.*, the network is built with a dataset where the outputs are known. In the project, the MLP was used to predict the biological attributes with environmental variables and to evaluate the influence of the environmental variables on communities using a sensitivity analysis with a partial derivative algorithm (PaD). Details of the model structures and their results are given in chapters 3, 4 and 5.

Data used in the tool

Three main aquatic taxonomic groups (diatoms, benthic macroinvertebrates, and fish) were used to implement the models. Diatom data are available in France, Luxemburg, Belgium, and Austria, benthic macroinvertebrates in France, the Netherlands, Luxemburg, and Austria, and fish in France (nationwide and Garonne basin), Belgium, Luxemburg, and Italy (Table 7.1). Diatom data were analysed on a European scale after building one database, while macroinvertebrates and fish were analysed nationally for each country. Details on datasets of organisms are given in section 7.7.

Table 7.1. Data implemented in the tool

Organisms	Country available where available	Data type
Diatoms	France, Luxemburg, Belgium, Austria	Species
Macroinvertebrates	France, Netherlands, Luxemburg, Austria	Species, functional feeding group
Fish	France, Luxemburg, Belgium, Austria	Species, trophic guild

Structure and function

The tool functions basically consist of three parts: *community data visualization, community data ordination, and prediction of community structure*. The latter two functions can handle user inputs, whereas the first one only performs a database query based on the graphical selection of a sampling site on a scalable map, there are two ways to use the tool: by selecting an organism picture. Secondly, by selecting a country, the organisms or the country's flags with available data are activated. These two ways are associated. To aid users, a HELP document, which gives the users direct context-related supports, is written in HTML format and available in the tool. Details of the usages are explained in section 7.6.

Community data visualization can be activated by selecting a sampling site on water network systems in the scalable map. The scalable map window will change into an alphanumerical window where all the data about environmental data as well as species or broader groups of organisms are displayed. By using the function 'zoom in' or 'zoom out', users can find more precise locations of sampling sites on the geographical map. With the latter, users can choose a sampling site by clicking on it/by pointing the cursor of the mouse. On selecting a site, users can recognize the characteristics of sampling sites with environmental variables measured and community composition in the corresponding site. In case more than a single replicate of the sampling is available at a given site (especially in diatoms), the user will be allowed to select the sample to be visualized using a drop-down list box.

The prediction of community structure function is based on the MLP that has been trained in order to predict the presence/absence or the abundance of each species or group of organisms on the basis of environmental data. Two options are available after selecting the prediction function. In fact, the user can either visualize information about the underlying model or obtain predictions based on his/her data.

Firstly, users can analyse the sensitivity of the model with respect to predictive variables based on the PaD and the overall agreement between predicted and observed data for each species or group of organisms However, it is also possible to compare the predicted and observed data at each sampling site by selecting them on a scalable map. Users can chose a species (or group) using a drop-down list box. By changing the species concerned the relative contribution of environmental variables and predictability of species are also modified correspondingly. Furthermore, it is possible to analyse the response of the species according to the changes of a variables in a scatter plot. It is useful to observe the sensitive range of the species on the environmental changes.

Secondly users are allowed to predict the community composition (at species or at broader group level, depending on the available models) by entering environmental data into the fields of a dialog box. In the column to input new environmental variables, mean values of each environmental variable are provided as default. The new environmental values and the results of the prediction can be saved in different files. If users wish to predict the communities of several sites, they have access to the environmental data saved in the data file in the tool.

The results of the model implemented in the tool do not give perfect predictions of real data, although the best results are implemented in the tool. Generally, the results of the learning data showed high predictability with greater than 0.9 of correlation coefficients between observed and estimated values. However, the prediction results with the new datasets showed an overall correlation coefficient of 0.6 between observed and predicted values and high variations in predictability depending on the datasets with correlation coefficients in the range 0.2~0.9. When the species are predicted, normally abundant species show high predictabilities, whereas rare ones display low values.

The community data ordination function relies on the hexagonal lattice of the SOM, which is displayed at two different levels of detail as far as the recognition of clusters of units is concerned. The clusters of communities are also presented on the geographical map. By selecting a group in the SOM map, only corresponding sampling sites are displayed on the geographical map. It is always possible to display the structure of each SOM unit as well as the correspondence between SOM units and real observations. The structure of the SOM units is displayed as a bar chart representing the abundance or the probability of presence for each species or group of organisms, whereas the correspondence between SOM units and sampling sites is represented on a scalable map by colour and hatching coding the site symbols. Each environmental variable can be also mapped onto the SOM, where it is represented in grey scale. To visualize the environmental variables on the SOM map, the mean values of the each variable were calculated from the raw data of each sample assigned in each map unit. The users are also allowed to input their community composition data to be projected onto the SOM, where the best matching unit will be displayed. The sizes of the SOM maps were chosen for convenience of visualization and interpretation. The ordinations of communities with the SOM were in good agreement with those of classical multivariate analyses, published in the scientific papers in the PAEQANN project.

7.6 How to use the software

Main window

The main window of the program consists of three parts: selection of organism, selection of country, and command (Fig. 7.1). Each part is presented in the following sections. Dataset and model consists of different organisms in different countries or regions. Users can visualise the datasets and results of the models from the visualization window. There are two ways to access the visualisation window:

1. Select organism → Select country → visualisation window
2. Select country → Select organism → visualisation window

Selection of an organism

Three biological organisms are considered: fish, diatoms and macroinvertebrates. Users can choose one by selecting a picture (Fig. 7.1A).

Selection of a country

Eight European partners collaborated in the PAEQANN project. The flags of the countries are activated if the dataset is available. Users can choose any activated national flag (Fig. 7.1B).

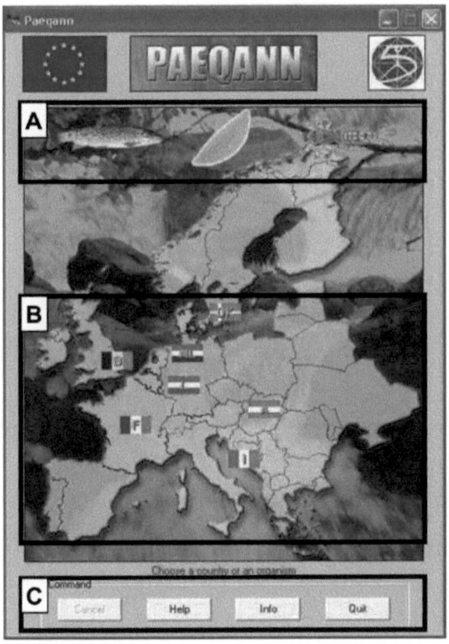

Fig. 7.1. Main window of the PAEQANN tool. (A) Selection of an organism, (B) selection of a country, and (C) common commands

Common commands

Users can cancel the last action or return to the previous screen by pressing the *Cancel* or *Back* button, or access the help file giving them direct context-related support from the *Help* button (Fig. 7.1C). The *Info* button displays information about PAEQANN's partners, programmers and project organizers. *Quit* is used to exit the program. These command instructions are also used in the visualization window.

Visualisation window

This window consists of three parts: action, geographical visualization, and common command (Fig. 7.2). The action and visualization functions are presented here, and the command function can be seen in the previous section.

Action

Users can choose different models. The *Prediction* button visualizes results of predicting models developed through MLP and provides a function to predict new values based on the existing dataset, whereas the *Ordination* button displays classification of the datasets defined by the SOM on a geographical map (Fig. 7.2A). They are explained in detail in the proceeding sections.

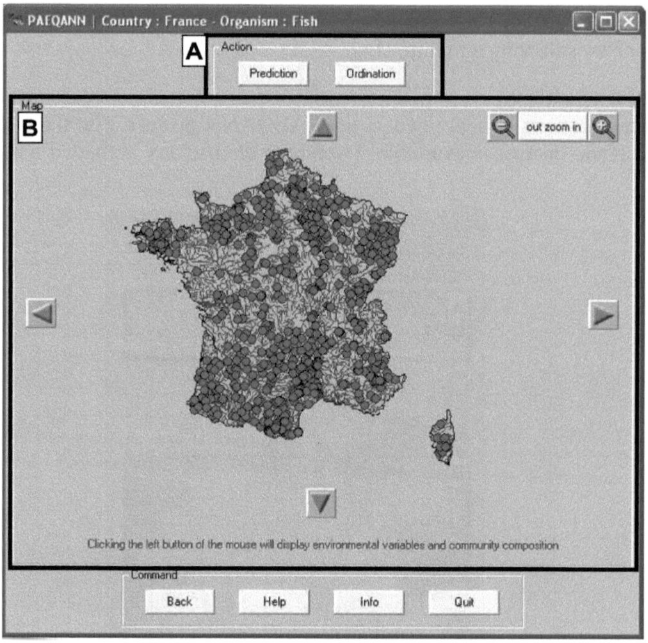

Fig. 7.2. Visualisation window. (A) Selection of models and (B) visualisation of sampling sites

Visualisation

Sampling sites are plotted on the geographical map with the water system network (Fig. 7.2B). To observe a more precise position of the sampling site, the *zoom in* and *zoom out* buttons can be used. The map can be also moved to four directions (left, right, up and down) by using the triangle buttons. When users chose a sampling site on the map, the sampling site displays its environmental variables and community composition as shown in Figure 7.3.

Prediction window

Through *Prediction*, users can visualize the results of predictive models. There are two categories: (i) community and (ii) species (Sp), functional feeding group (FFG), species richness (SR) or guild according to the availability in the PAEQANN project database. Users also can predict (or test) their new data (Fig. 7.4).

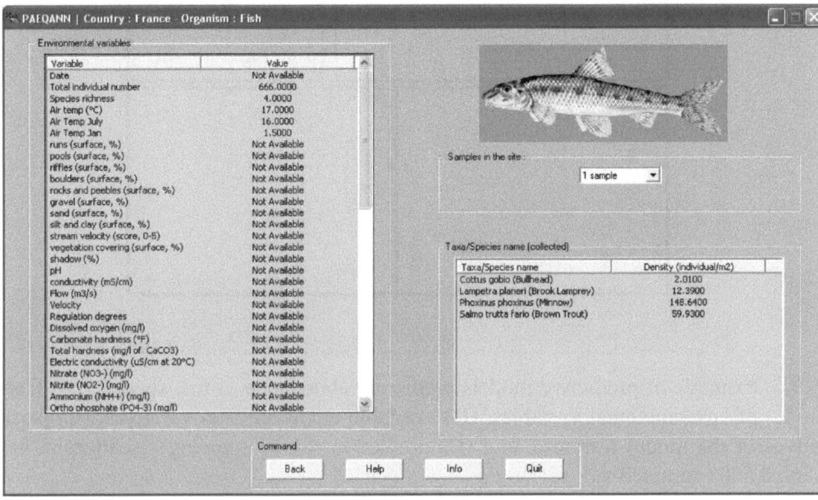

Fig. 7.3. Example of visualization of environmental parameters and community composition in a sampling site chosen from the geographical map

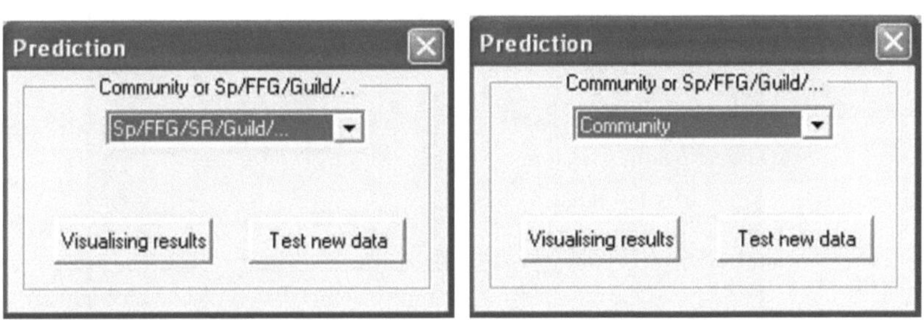

Fig. 7.4. Selection of prediction parameters

Sp/FFG/SR/Guild/ --> visualising results

When this category is chosen, predicted results of Sp, FFG, SR, or Guild one of these parameters according to the availability of the datasets. Figure 7.5 shows an example of visualization of predictive models predicting abundance of fish species in France. Different target variables (*i.e.*, species in this case) can be chosen in Figure 7.5A. When the target variables are changed, corresponding values are also changed. Users can evaluate the contribution (%) of the environmental variables through a sensitivity analysis using a PaD algorithm (Fig. 7.5A); its partial derivatives of the ANN model response with respect to each environmental variable are provided (Fig. 7.5C). Prediction results are given as scatterplots between observed and estimated values (Fig. 7.5D).

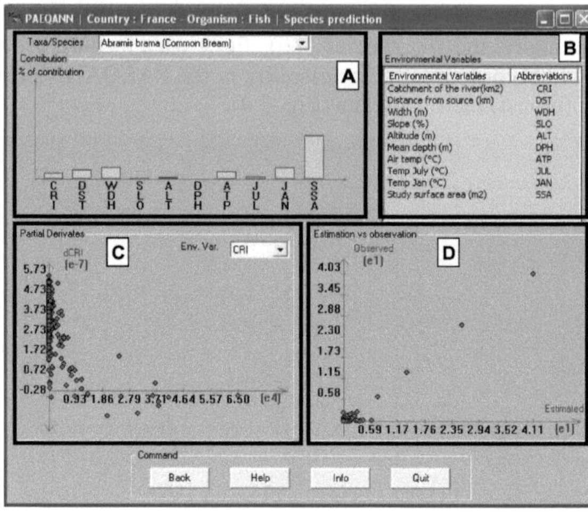

Fig. 7.5. Example of predictive models predicting abundance of fish species. (A) Contribution (%) of environmental variables, (B) available environmental variables, (C) partial derivatives of the model response, and (D) prediction results showing a scatterplot between observed and estimated values

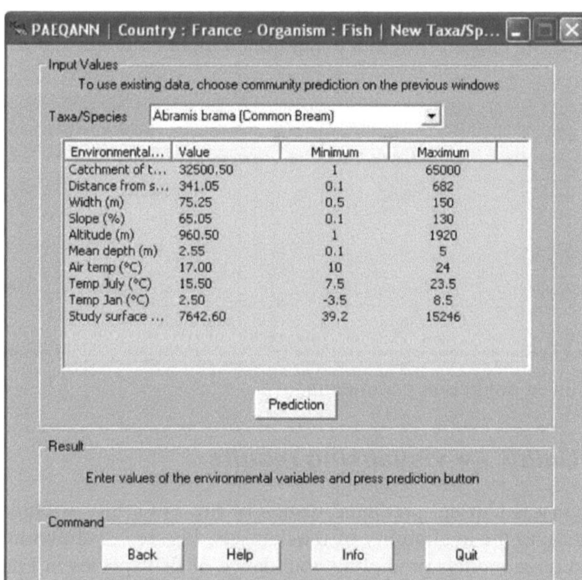

Fig. 7.6. Example window to predict abundance of fish species with given environmental variables. The predicted value is given in the Result area of the window.

Sp/FFG/SR/Guild/... --> Test new data

New data can be tested with MLP models trained with available PAEQANN datasets. After choosing the Sp/FFG/SR/Guild/... to be predicted, the user can enter new values of environmental variables in the column "value" by simply clicking on the default value as shown in Figure 7.6.

Community --> Visualising results

Through these selections, users can compare predicted values with observed values of community composition in a given sampling site. Firstly, a sampling site has to be selected from the geographical map (Fig. 7.7), and the results are given in a bar chart (Fig. 7.8).

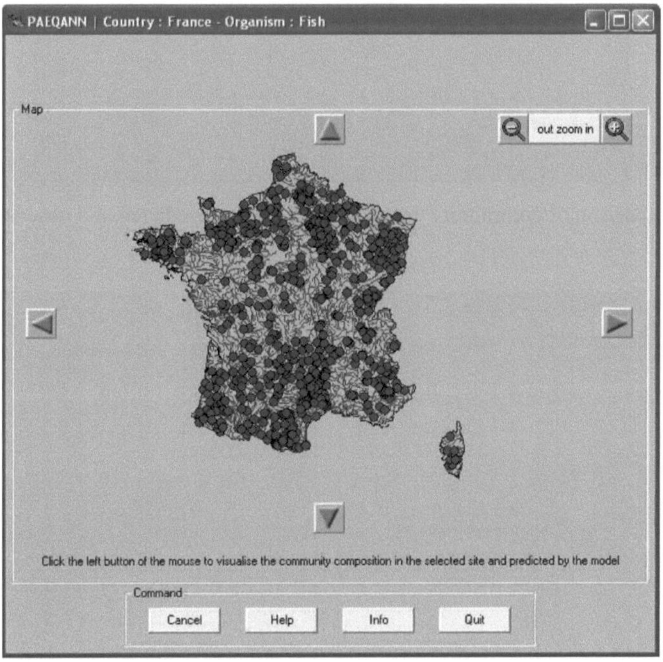

Fig. 7.7. Visualization of sampling sites to present results of prediction models for community composition

Community --> Test new data

Community composition is predicted with new values of environmental variables. The values of the variables should be given in the column *value* in the given range of minimum and maximum values by a simple click on the default value. The prediction results appear in the corresponding frame *Result* (Fig. 7.9). Through the *Create data structure* button, users can create the structure of the input file needed by the program to predict the communities at several sites. Through the *Open input data* button, the environmental input saved in a file can be also used to predict community composition. Figure 7.10 displays the community prediction results of a selected data file. In the Figure, the program displays the results of the first sample (*i.e.*, the sample in the first line of the data file). The results of the of the first sample (*i.e.*, the sample in the first line of the data file). The results of the predicted models can be saved in a file through the Save Result button (Fig. 7.11).

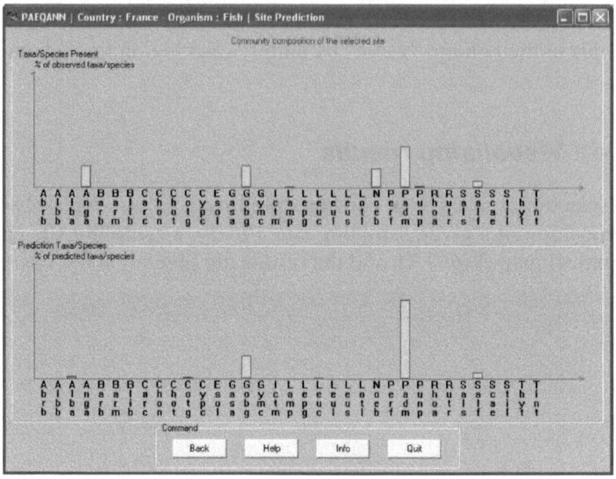

Fig. 7.8. Comparison of community composition between predicted and observed values

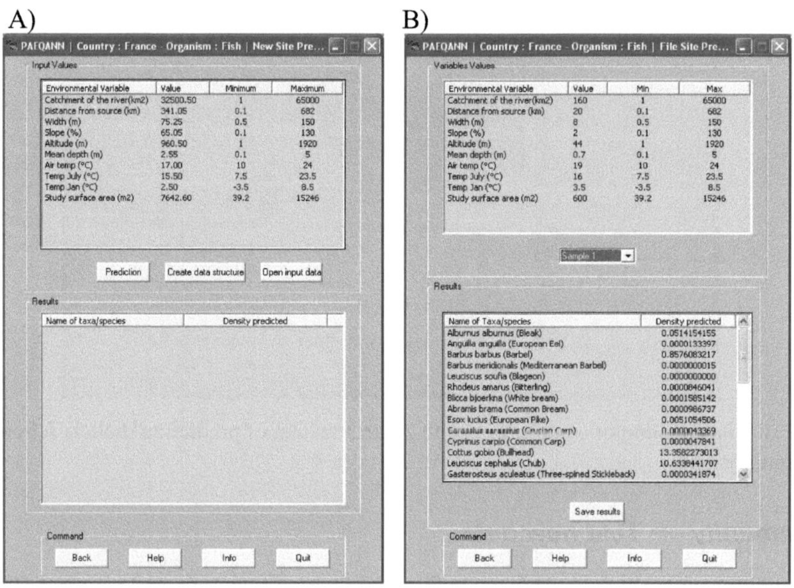

Fig. 7.9. Example of prediction of community composition with environmental variables (A) and results of community predictions at several sampling sites (B)

Ordination window

Visualization of ordination

Classification of sampling sites is presented on the geographical map (Fig. 7.10) through the choice of the *Ordination* button in the Visualization window. Samples were classified in the SOM map through the SOM learning process with community data, and then the SOM units were classified into several corresponding groups using hierarchical cluster analysis by means of the Ward linkage method. Different colours of SOM units represent different clusters and samples assigned to the cluster are presented on the geographical map in the same colour. Thus, the sampling sites can be visualised at two different cluster levels. At the first level, all units of the SOM map are visualised on the geographical map in different colours. At the second level, units of the map are clustered in several groups according to their similarities (community similarity) and the groups are visualised on the geographical map in different colours. Different SOM units represent different community types.

To see the position of the sampling site more precisely, the *zoom in* and *zoom out* buttons can be used, and the position of the map can be adjusted using four moving buttons (left, right, up and down).

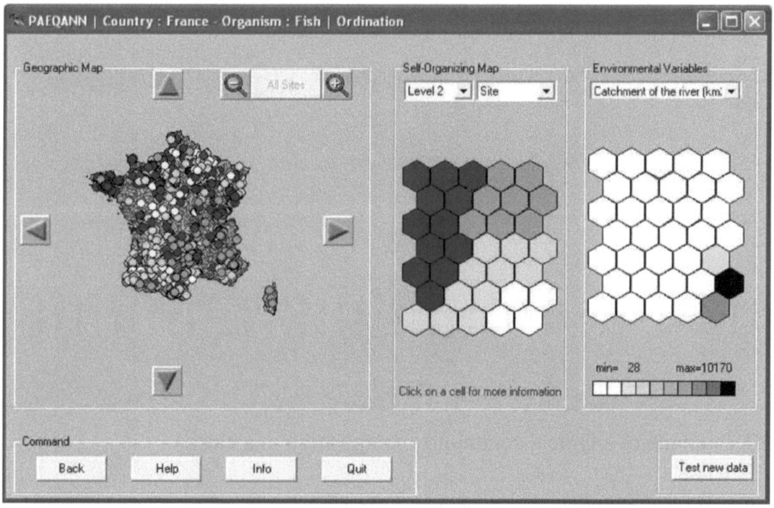

Fig. 7.10. Example of an ordination window indicating fish communities in France

Additionally, sampling sites in each SOM unit or in each cluster can be visualized on the geographical map separately, depending on the selection of the visualization level (Fig. 7.11). To do so, users should select *Site* which is located in the upper right areas of the SOM map part, and choose one concerning an SOM cell or cluster. By choosing the *All Sites* button, all sampling sites are displayed on the geographical map.

Furthermore, typical community types can be visualized in each SOM unit or in each cluster (Fig. 7.12). To do so, *Community* should be chosen instead of *Site*, and one SOM unit or cluster can be chosen. The values in a SOM unit were estimated through the learning process of the SOM with community data, and the values in a cluster were calculated as a mean from values in SOM units in the cluster.

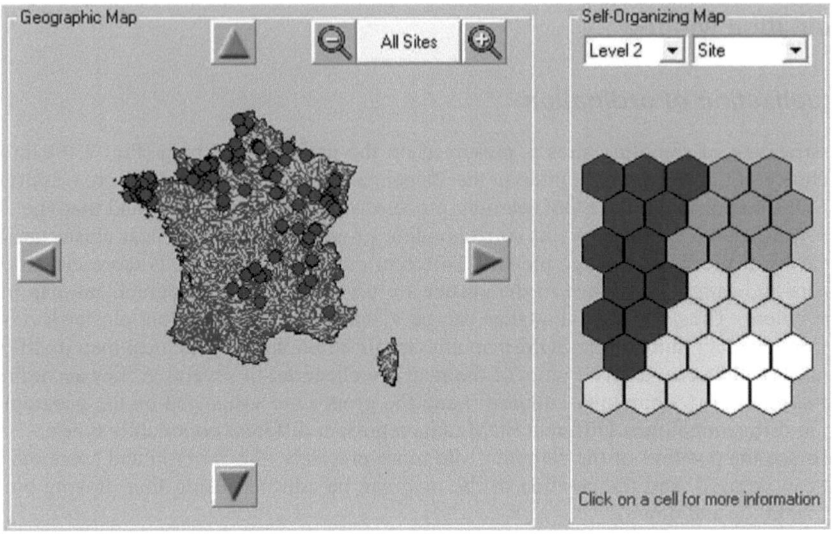

Fig. 7.11. Example of visualization of sampling sites assigned in a cluster

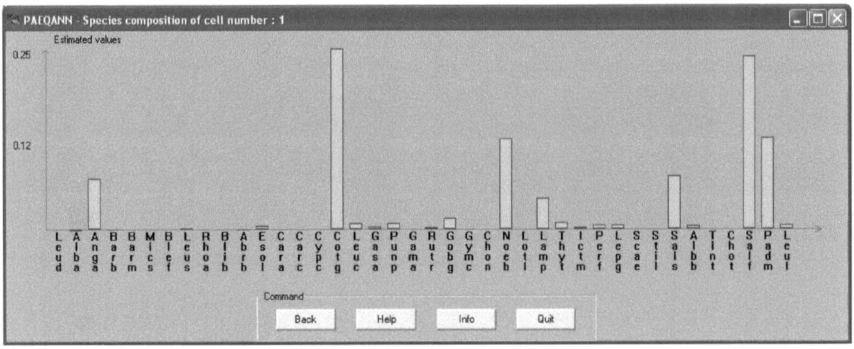

Fig. 7.12. Example of a typical community composition in a cluster

Visualization of environmental variables

Differences of environmental variables are presented on the SOM map trained with communities in the grey scale in the range of minimum and maximum values (Fig. 7.10). Dark represents high values, and light low values. Visualization of environmental variables can be selected variable by variable. The values in each SOM unit were calculated as mean values from sampling sites assigned in the SOM unit. In the scroll list displayed on the upper areas, users can choose between environmental variables. This visualization is efficient to find distribution gradients of environmental variables on the SOM map.

Test new data

Users can test new community data with trained SOM by presenting corresponding values in the column *Density* in the given ranges (Fig. 7.13A). The tested results are indicated on the corresponding unit of the SOM map with a black circle (Fig. 7.13B).

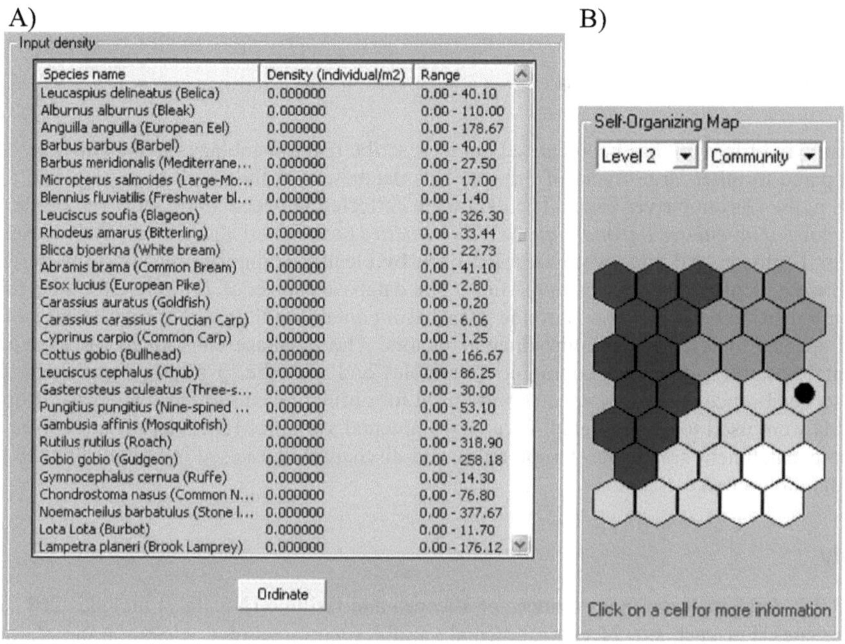

Fig. 7.13. Example of testing new community data in ordination. (A) Input values and (B) predicted result

7.7 Organisms used in the PAEQANN software

Fish

Belgium

The fish database of Belgium is composed of 804 sampling stations in streams. Four different institutes sampled these stations: CSP (398 stations), FUNDP (153 stations), RIVO (36 stations), IBW (217 stations). In the modelling, 47 species and 9 environmental variables were used.

France

The database is a set coming from the database held by the *Conseil Supérieur de la Pêche* (*Banque Hydrobiologique et Piscicole*), covering a 13 year survey period. The database consists of 688 reference sites, which are fairly evenly distributed across French rivers and which contain the occurrence of 40 fish species and regional and local environmental factors. The data were sampled between 1985and 1998 by the *Conseil Supérieur de la Pêche* by means of electrofishing during low-flow periods (from August to October). Fish were identified to species in the field, and released in the water. In the modelling, 40 species and 8 environmental variables were used.

Garonne basin in France

Relatively little work has been carried out to describe fish assemblages in the Garonne river compared to other large rivers of France. This database was thus built to pattern fish richness in the Garonne river basin. The data were collected between 1986 and 1996 by the *laboratoire d'ingénierie agronomique, ENSAT (mettre en extenso)* and University Paul Sabatier in Toulouse. All sites were sampled once by electrofishing, during low-flow periods. Fish were identified to species in the field. The database is a set of 239 reference sites fairly evenly distributed across the Garonne river basin containing the presence or absence of 44 fish species in relation to environmental factors. The database constituted of 6 environmental variables, 3 being geographical variables and the other 3 physical variables. The presence/absence data of 44 species were used for ordination of samples, whereas 7 trophic guilds were used to be predicted. Five environmental variables (altitude, distance from the source, catchment area, water temperature, and discharge) were used in both ordination and prediction models.

Italy

The fish database from the provinces of Vicenza and Belluno (NE Italy) includes 264 sampling sites. For each site 21 environmental variables were recorded: elevation in m?, mean depth, % of surface for runs, pools and riffles (3 variables), mean width, % of surface with boulders, rocks and pebbles, gravel, sand, silt and clay (5 variables), stream velocity (score 0-5), % of surface with vegetation covering, % of surface with shadow, anthropic disturbance (score 0-4), water temperature, pH, conductivity, gradient, distance from headwater source, surface of drainage basin. The fish community is represented as the presence/absence of 36 species. These data were obtained from two small consulting cooperative enterprises (*Aquaprogram* scrl and *Bioprogram scrl*), which collected them on behalf of the provincial Administration from 1987 to 1994.

Luxemburg

The Luxemburg fish database is composed of 34 stream-sampling stations. These stations were sampled from 1994 to 1996. Diatom and macroinvertebrates were also sampled in these sites for the same period. The Water and Forest Administration helped the *CRP-GL* to sample the fishes. They were sampled in the framework of the biocenotic study of the rhithral part of Luxemburg streams.

Physical variables (geographical position, altitude, geology, source distance, slope, river width, water level, speed current, shading, temperature) were measured by the *Centre de Recherche Public-Gabriel Lippmann (CRP-GL)* in the field and on maps (1/25000). Some chemical variables were also measured in the field by the *CRP-GL* (pH, oxygen, conductivity), and in laboratories (PO_4^{2-}). The other variables were measured by the *Ministère de l'Environnement* (BOD_5, NH_4^+, NO_2^-, NO_3^-, carbonate hardness, total hardness, Cl^-, Ptot, SO_4^{2-}, K^+).

Macroinvertebrates

Austria

The Upper Austrian macroinvertebrate database includes 225 sampling sites. All sampling sites are within the Danube basin. The samples were taken between 1996 and 1999, once per site, synchronously with diatom and ciliata samples. All samples and measurements were taken by the Upper Austrian Water Authority (UAWA). Sampling of organisms is part of a long-term survey program of the water authority. Due to some geographical differences, only data from 102 sites can be used for combination with biological data. As a result of the agreement on data combination within the PAEQANN project only a reduced number of sites could be used for further calculation (77). The database includes 536 macroinvertebrate taxa and 25 chemical and physical variables.

France

The matrix contains 425 taxa with different levels of identification from 157 running-water sites. A total of 5 environmental variables (stream order, elevation, slope, annual maximal water temperature, and distance to source) are associated with each sampling site.

Luxemburg

The Luxemburg macroinvertebrate database includes 147 different sampling stations. Macroinvertebrate samplings were carried out in the same sampling sites as the diatoms samplings, and on the same date. Each station was sampled twice: the first time in autumn, and the second in spring. The samplings were carried out from 1994 to 1997. The Luxemburg macroinvertebrate database is composed of 292 records. The determinations are for all the taxonomic groups made at a specific level. These records came from the small Luxemburg streams and were sampled in the framework of the biocenotic study of the rhithral part of these streams. This study has two aims: to establish the most complete faunistic inventory (to a specific level), and to provide a quantitative and qualitative analysis with chemical and biological indices.

Physical variables (geographical position, altitude, geology, source distance, slope, river width, water level, speed current, shading, and temperature) were measured by the *Centre de Recherche Public-Gabriel Lippmann (CRP-GL)* in the field and on maps (1/25000). Some chemical variables were also measured in the field by the *CRP-GL* (pH, oxygen, conductivity), and in the laboratories (PO_4^{2-}). The other variables were measured by the *Ministère de l'Environnement* (BOD_5, NH_4^+, NO_2^-, NO_3^-, carbonate hardness, total hardness, Cl^-, Ptot, SO_4^{2-}, K^+).

The Netherlands

The Dutch database consists of samples which were collected from 664 sites situated in the province of Overijssel (the Netherlands); only 609 sites were visited in one season and 55 sites in two seasons. The objective was to capture the majority of the species and their relative abundances present at a given site. At each site, major habitats were selected over a 10 to 30 m long stretch of the waterbody and were sampled by means of the same sampling effort. The sampling effort was thus standardised for each site.

At shallow sites, vegetation habitats were sampled by sweeping a pond-net (200 mm × 300 mm, mesh size 0.5 mm) several times through each vegetation type over a length of 0.5-1 m. Bottom habitats were sampled by vigorously pushing the pond net through the upper few centimetres of each bottom type over a length of 0.5 to 1m. The habitat samples were then combined for the site to give one sample with a standard area of 1.5 m^2 1.2m^2 of vegetation & 0.3 m^2 of bottom). At sites lacking vegetation, the standard sampling was confined to the bottom habitats. At deeper sites, five samples were taken from the bottom habitats with an Ekman-Birge sampler. These five grabs were equivalent to one 0.5 m pond net bottom sample. Vegetation habitats were sampled with a pond net as described above. Again the total sampling area was standardised as 1.5 m^2. Macroinvertebrate samples were taken to the laboratory, sorted without any external aid, counted and identified to species level.

The sampling dates were spread over the four seasons as well as over several years (1981 up to and including 1985). Season was taken into account by defining sampling periods as nominal "environmental" variables within the analysis.

A data sheet was used to note a number of abiotic and some biotic variables in the field. Some were measured directly (width, depth, surface area, temperature, transparency, percentage of vegetation cover, percentage of sampled habitat), others (such as regulation, substratum, bank shape) were classified. Field instruments were used to measure oxygen, electrical conductivity, stream velocity and pH. Surface water samples were taken to determine chemical variables. Other variables, such as land-use, bottom composition, and distance from source, were gathered from additional sources (data from water boards, maps). In total, 70 abiotic variables were measured at each site.

Diatoms

Several teams of the PAEQANN network were involved in providing diatom records and the corresponding environmental data: *CEMAGREF* (France), *CRPGL* (Luxembourg), *ARCS* (Austria) and *LFE-URBO-FUNDP* (Belgium), coordinator of the "diatom group".

The PAEQANN Diatom Database comprises 2847 records in total. To make a list of sufficiently representative and/or significant taxa and to guarantee sufficient homogeneity among samples, only records comprising at least 380 counted objects, and originating from sampling carried out on stony substrates were selected for further analyses. Consequently, 2147 records were finally available for further analysis, among which 467 were identified as reference, according to their IPS‡ value (equal or higher than 16). 1719 different taxa names were recorded for the whole database, among which 1255 different taxa could be identified. After grouping, 1051 taxa were potentially available for further analysis. A selection of taxa was nevertheless made prior to analysis, in order to remove occasional taxa; this led to a list of 123 taxa for the reference data matrix, and 283 for analysis of the 2147 records constituting the whole database data matrix.

Finally, it must be mentioned that in order to implement the PAEQANN tool, a simplified MS Access Diatom database was produced.

‡ IPS = Index of Pollution Sensitivity

8 General conclusions and perspectives

Scardi M*

Understanding the way community structure is affected by environmental conditions is a key issue in modern ecology, especially since the advent of ordination techniques and, in particular, of multivariate statistical methods for direct gradient analysis. New statistical tools have been developed by ecologists to address this problem more effectively, and they have been widely accepted and used in many studies. Canonical Correspondence Analysis (ter Braak 1987), which constrains coenocline analysis to physical, chemical and other environmental variables, is an example of such an ecologically-inspired data analysis procedure.

Although data analysis techniques made it possible to infer relationships between community structure and environmental variables, they are mainly useful for descriptive purposes or, to a limited extent, to test ecological hypotheses. Of course, some results of data analysis procedures can be used as a basis for assessing species' responses to environmental variables, but this is only possible to a limited extent and from a strictly qualitative point of view.

Therefore, if predictions about community structure are needed, suitable modelling techniques have to be used. In particular, statistical methods that directly relate environmental variables to species presence and/or abundance can play a role, ranging from very simple regression models to more complex ones (e.g. Partial Least Squares). However, only a few applications have proven to be useful, such as those based on logistic regression for predicting species presence or absence given adequate environmental data.

Modelling single species distributions as a function of environmental (mostly abiotic) information is certainly an interesting task, but predicting community structure is a more complex problem that usually cannot be solved just by assembling single species models into a more complex composite model. In fact, in many cases the available information about the environmental relationships (both biotic and abiotic) that determine species distributions is too limited, and it cannot support the development of reliable models.

Therefore, it is obvious that the efficiency of the modelling approach plays a fundamental role in predicting community structure in such data-limited situations. This is the reason why the number of ecological applications involving Artificial Intelligence (A.I.) techniques and Machine Learning methods has grown significantly during the last ten years.

These new modelling methods rely on computing power that is now easily available to extract as much useful information as possible from the existing - and usually insufficient - data. Sometimes these approaches do not provide significant advantages over conventional methods, but they are often much more effective than the latter and some applications among those that are presented in this book provide clear evidence for their superiority.

About ten years since the first attempts, ecological applications of A.I. and Machine Learning modelling methods are now mature and, in many cases, they are presented without comparisons to conventional counterparts, as their improved performance is accepted. Readers who are interested in understanding to what an extent these methods may be beneficial can find complete, yet easy introductions and examples in Fielding (1999) as well as in Lek and Guegan (2000).

A common factor in many applications aimed at modelling freshwater community structure is that the number of field records is usually limited with respect to the ecological

* Correspondence: mscardi@mclink.it

complexity of the problem, thus making it very difficult to reconstruct the causal relationships that link species distribution to environmental variables. However, in such data-poor situations other sources of information are often available, although deeply embedded in the data sets, and they can be exploited by appropriate modelling methods.

For instance, many species tend to occur in association with others, while other species seldom coexist at the same site. Obviously, this can be due to the biotic interactions that make a set of species a community, but in some cases species assemblages are modelled, rather than real communities, and therefore the relevance of biotic interactions cannot be taken for granted. Nevertheless, relationships between species are still useful from the modeller's point of view as they narrow the number of independent combinations of species to be predicted. Basically, if two species respond in the same (or in opposite) way to a given set of environmental conditions, then the actual dimensionality of the modelling problem to be solved is lower than expected and less data are needed to build a good model.

Moreover, morphodynamic features play a fundamental role in defining the ecological characteristics of freshwater ecosystems, specifically in streams and rivers. In other words, given a set of physical constraints (e.g. elevation, slope, etc.) only a very limited range of ecological conditions is likely to be observed, thus reducing the "degrees of freedom" of models aimed at predicting the structure of communities or other biotic assemblages, at least within a given ecoregion.

Such ecological relationships, however, are usually complex, and linear or unimodal responses that are the basis for most statistical approaches are seldom observed. On the contrary, complex non-linear relationships are often involved, and this is probably the more important reason for the success of new modelling strategies. In particular, Artificial Neural Networks (ANNs) have been successfully applied in many cases, both for predictive modelling (usually via supervised methods) and for descriptive modelling, i.e. for revealing underlying patterns in data sets (usually by means of unsupervised methods). The majority of the case studies that are presented in this book are based on ANN applications and this is clear - although not unbiased - evidence for the role they play in modelling community structure.

Independent of the modelling technique, however, there are limits to the predictability of species distributions that strictly depend on the intrinsic nature of ecological data sets. In fact, different species may occur with very different frequencies in data sets, according to their actual density or as a consequence of the sampling strategy or spatial scale. Since species presence (as well as abundance) depends on environmental conditions, rare species, as well as nearly ubiquitous ones, are usually unpredictable, because in both cases it is virtually impossible to detect significant correlations between environmental variables and species presence (or abundance).

This limitation cannot be overcome by improving modelling algorithms or by introducing new techniques. The only viable solution is to modify sampling strategies, making them more suited to the modelling needs. As a matter of fact, most attempts at modelling community structure are usually carried out on the basis of data sets collected for other purposes, e.g. for mapping species distributions via GIS tools or for applying multivariate statistical methods for indirect gradient analysis. A typical feature in such data sets is a regular or random sampling design that is certainly adequate when no prior information is available, but that often fails to reveal essential information if small scale coenoclines also play a role. In these cases, sampling strategies that address variable spatial scales would provide much more relevant information, especially when previous data or pilot surveys are available and the sampling design can be effectively stratified. In this framework, modelling community structure, independently of the accuracy of the results obtained in the first attempts, may also induce both a significant optimisation in sampling strategies and a better understanding of the factors that control species distributions.

New modelling techniques that are able to exploit existing information as efficiently as possible (such as ANNs) will certainly catalyze a better ecological understanding of factors controlling community structure because of their ability to reproduce complex non-linear responses. In particular, sensitivity analysis of such models may provide useful insights into the ecological relationships that control species distributions and biotic interactions, thus offering clues to improve sampling designs and sampling scales to resolve relevant biotic signals. This task is not trivial, of course, and, while several methods have been developed (as presented in this book), no well established standards are available at present. However, this topic is certainly among the most stimulating and new approaches to sensitivity analysis are emerging. The successful application of methods that elucidate second and possibly higher order interactions between abiotic factors in determining species distributions, as well as complex interactions among species, will probably be the main goal for the next decade in modelling community structure.

Sensitivity analysis is an example of indirect use of models as tools for stimulating advances not only in computational methods, but also in the ecological background of the modelling applications, thus involving the way data are collected, coded, analyzed, etc. On the other hand, ecological issues may play a significant role in improving modelling techniques and especially in adapting them to properly handle the peculiar characteristics of ecological data. For instance, using Mean Square Error (MSE) as a measure for goodness of fit is a common practice in modelling, and it is certainly adequate for many quantitative variables. Therefore, it is usually adopted as a default choice and, in the case of many software packages, is the only available option. It is obvious, however, that this way of measuring the distance between observed and modelled data is seldom appropriate when species abundance data (not to mention species presence data, i.e. binary data) are taken into account. The vast amount of similarity and distance coefficient that have been developed for measuring differences in community structure are a clear evidence for the inadequacy of Euclidean distances and related coefficients from an ecological viewpoint. In particular, when community structure is concerned, it is obvious that the role of each species must be interpreted in the light of its ecological context.

A very simple example of this need is in the different weight that should be given to the same error in predicting the presence or the abundance of a given species in the case of a very simple community (low species richness) and in the case of a more complex one (high species richness): MSE obviously fails in this task, because it does not scale the errors with respect to the complexity of the community, whereas other coefficients, such as Jaccard or Bray-Curtis similarity, do. Thus, significant improvements in community structure modelling could be achieved by adapting existing algorithms to their ecological framework, e.g. by adopting procedures for measuring modelling errors that are based on appropriate metrics. Adaptations of modelling algorithms are probably beyond the capabilities of most ecologists, but some are actively working in this field, developing new strategies and methods for ecological modelling that more closely match their specific needs.

The need for ecologically sound metrics is not only a problem in predictive modelling, of course. As many community ecologists already know, the outcomes of multivariate analyses are very deeply influenced by the selected metrics. Not only the results of quantitative analyses often are quite different from those of qualitative ones, but even among the results of analyses based on the same type of data there might be significant differences. The most obvious example is that the meaning of absence data (zeroes) is not the same in all the cases: it can express real absence of a species in a given site, but it can also depend on the frequency of its occurrence (i.e. by its density) with respect to the characteristics of the sampling design and devices. Thus, the selection of appropriate metrics is a key issue even in descriptive models, such as those based on Self-Organizing Maps (see the applications in this book) as well as those based on conventional ordination techniques.

Given the available methods and data, community structure modelling is really feasible only in case of "toy" problems or when complexity is somehow reduced. Good results can be achieved by focusing on subsets of the whole community (e.g. on assemblages of species that are not too complex and that can be sampled in an effective and straightforward way, like fish), or by simplifying the way the community structure to be modelled is represented (e.g. by using a few trophic guilds instead of a complex list of species). In any case, the number of successful applications is rapidly increasing, and it is clear that the ability of these models to address complex problems is only limited by the availability of adequate information, i.e. by the lack of field data, either in general or with respect to the spatial scale that is relevant to the distribution of the species to be modelled. Obviously, expectations for species distribution and community structure modelling are also growing, but potential users have to bear in mind that the reliability of mathematical models depends on the adequacy of the data bases that support their development, e.g. no reliable weather forecasts would be available if the underlying models were not supported by large meteorological observation networks and data bases.

Despite the difficulties in developing community structure models, the demand for such tools in applied ecology is certainly growing, as comparing observed community structure with some reference conditions may form the basis for assessing environmental quality. This will be the case, for instance, in those European countries that will follow guidelines indicated by the EU Water Framework Directive (also known as Directive 2000/60/EC), which sets restoration targets and clearly points out that changes in community structure with respect to reference conditions are related to changes in the ecological status of a water body. However, in many cases natural reference conditions are not available because of the lack of genuinely unperturbed water bodies, and only models can provide estimates about the expected community structure in such pristine conditions.

In this rapidly evolving scenario, mathematical models aimed at predicting community structure will certainly play a key role in many applied and basic research tasks. Conventional models (e.g. statistical models) will continue to be widely used, but ecological modellers who want to be on the leading edge will explore new approaches. This book is both a showcase of successful applications and a useful reference for those who want to get started in this field.

In this rapidly evolving scenario, mathematical models aimed at predicting community structure will certainly play a key role in many applied and basic research tasks. Conventional models (e.g. statistical models) will continue to be widely used, but ecological modellers who want to be really on the leading edge might want to explore new approaches. This book is both a showcase of successful applications and a useful reference for those who want to get started in this field.

References

Aboal M, Puig, MA. Soler G (1996) Diatoms assemblages in some Mediterranean temporary streams in Southeastern Spain. Arch Hydrobiol 136: 509-527

Adkison MD, Peterman RM (1996) Results of Bayesian methods depend on details of implementation: an example of estimating salmon escapement goals. Fish Res 25: 155-170

AFNOR (1992) Qualité de l'eau - Détermination de l'Indice Biologique Global Normalisé (IBGN). Norme NF T 90-350. AFNOR, Paris

AFNOR (2000) Détermination de l'Indice Biologique Diatomées (IBD). Norme NFT 90-354. AFNOR, Paris

AFNOR (2004) Qualité de l'eau - Détermination de l'indice poisons rivière (IPR). NF T90-344. AFNOR, Paris

Agences de l'eau (1998) SEQ milieu physique: système d'évaluation de la qualité du milieu physique des cours d'eau. - Les études des Agences de l'Eau (testing version 0). Agences de l'eau

Aguilar Ibarra A (2004) Les peuplements de poissons comme outil pour la gestion de la qualité environnementale du réseau hydrographique de la Garonne. PhD Thesis, Ecole National Supérieur Agronomique de Toulouse, Institut National Polytechnique, Toulouse, France

Aguilar Ibarra A, Gevrey M, Park YS, Lim P, Lek S (2003) Modelling the factors that influence fish guilds composition using a back-propagation network: assessment of metrics for indices of biotic integrity. Ecol Model 160: 281-290

Aguilera PA, Frenich AG, Torres JA, Castro H, Vidal JLM, Canton M (2001) Application of the Kohonen neural network in coastal water management: methodological development for the assessment and prediction of water quality. Water Research 35: 4053–4062

Alba-Tercedor J, Sanchez-Ortega A (1988) Un metodo rapido y simple para evaluar la calidad biologica de las aguas corrientes basado en el de Hellawell (1978). Limnetica 4: 51-56

Alhoniemi E, Himberg J, Parviainen J, Vesanto J (1999) SOM Toolbox 2.0, a software library for Matlab 5 implementing the Self-Organizing Map algorithm. [online] http://www.cis.hut.fi/projects/somtoolbox/

Allan JD (1995) Stream ecology: structure and function of running waters. Chapman & Hall, London

Allen TFH, Hoekstra TW (1992) Toward a unified ecology. Columbia University Press, New York

Allen TFH, Starr TB (1982) Hierarchy: perspectives for ecological complexity. The University of Chicago Press, Chicago

Almeida SFP, Pereira MJ, Gil MC, Rino JM (1999) Freshwater algae in Portugal and their use for environmental monitoring. In: Prygiel J, Whitton BA, Bukowska J (eds) Use of algae for monitoring rivers III, Agence de l'Eau Artois-Picardie, Douai pp 10-16

Andersen MM, Riget FF, Sparholt H (1994) A modification of the Trent index for use in Denmark. Water Research 18: 145-151

Anderson A (1935) The irises of the Gaspe Peninsula. Bull Am Iris Soc 69: 2-5

Angelier E (2001) Ecologie des Eaux Courantes. Tec & Doc, Paris

Angermeier PL (1994) Does diversity include artificial diversity? Conserv Biol 8: 600-602
Angermeier PL, Schlosser IJ (1989) Species area relationships for stream fishes. Ecology 70: 1450-1462
Angermeier PL, Schlosser IJ (1995) Conserving aquatic biodiversity: beyond species and populations. Am Fish Soc Symp 17: 402-414
Angermeier PL, Winston MR (1998) Local vs regional influences on local diversity in stream fish communities of Virginia. Ecology 79: 911-927
Aoki I, Komatsu T (1997) Analysis and prediction of the fluctuation of the sardine abundance using a neural network. Oceanol Acta 20: 81-88
Aoki I, Komatsu T, Hwang K (1999) Prediction of response of zooplankton biomass to climatic and oceanic changes. Ecol Model 120: 261-270
APHA (1995) Standard methods for examination of water and wastewater. In: Eaton AD, Clesceri LS, Greenberg AE (eds) American Public Health Association, Washington
Armand C, Bonnieux F, Changeux T (2002) Evaluation économique des plans de gestion piscicole. Bull Fr Pêche Piscic 365/366: 565-578
Armitage K (1961) Distribution of riffle insects of the Firehole River, Wyoming. Hydrobio 17: 152-174
Armitage PD, Moss D, Wright JF, Furse MT (1983) The performance of a new biological water quality score system based on macroinvertebrates over a wide range of unpolluted running water sites. Water Research 17: 333-347
Arzet K, Krause-Dellin D, Steinberg C (1986) Acidification of four lakes in the Federal Republic of Germany as reflected by diatom assemblages, cladoceran remains and sediment chemistry. In: Smol JP, Battarbee RW, Davis RB, Merilainen J (eds) Diatoms and lake acidity. Dr. W. Junk Publishers, Dordrecht pp 227-250
Aussem A, Hill D (1999) Wedding connectionist and algorithmic modelling towards forecasting *Caulerpa taxifolia* development in the north-western Mediterranean sea. Ecol Model 120: 225-236
Back M, Dohet A, Molitor M (1994) La qualité de l'eau de la Haute-Sûre luxembourgeoise. In: Groupe Loutre Luxembourg (eds) La loutre au Luxembourg et dans les pays limitrophes, Graphic press, Mamer, Luxembourg pp 41-51
Backiel T, Penczak T (1989) The fish and fisheries in the Vistula River and its tributary, the Pilica River. In: Dodge DP (ed) Proceedings of the International Large River Symposium, Honey Harbour, Ontario, Canada. Can Spec Publ Fish Aquat Sci pp 488-503
Baeck T (1996) Evolutionary algorithms in theory and practice. Oxford University Press, New York
Bahls LR, Burkantis R, Tralles S (1992) Benchmark biology of Montana reference streams. Department of Health and Environmental Science, Water Quality Bureau, Helena, Montana
Balloch D, Davies CE, Jones FH (1976) Biological assessment of water quality in three British rivers: The North Esk (Scotland), The Ivel (England) and The Taf (Wales). Water Pollution Control 75: 92-110
Balon EK (1975) Reproductive guilds of fishes: a proposal and definition. J Fish Res Board Can 32: 821-864
Balon EK (1990) Epigenesis of an epigeneticist: the development of some alternative concepts on the early ontogeny and evolution of fishes. Guelph Ichthyol Rev 1: 1-48
Balon EK, Coche AC (1975) Lake Kariba, a man made lake ecosystem in Central Africa. Monographiae biologicae, 24 Junk, The Hague, The Niederlands

Balon EK, Crawford SS, Lelek A (1986) Fish communities of the upper Danube River (Germany, Austria) prior to the new Rhein-Main-Donau connection. Environ Bio Fish 15: 243-271

Baraldi A, Alpaydin E (1998) Simplified ART: A new class of ART algorithms. International Computer Science Institute, Berkeley, California

Baran P, Dauba F, Delacoste M, Lascaux JM (1993a) Essais d'évaluation quantitative du potentiel halieutique d'une rivière à salmonidés à partir des données de l'habitat physique. In: Gascuel D, Durand JL, Fonteneau A (eds) Les recherches françaises en évaluation quantitative et modélisation des ressources et des systèmes halieutiques. Premier forum halieumétrique, IRD Editions, Paris pp 15-38

Baran P, Delacoste M, Dauba F, Lascaux JM, Belaud A, Lek S (1995a) Effects of reduced flow on brown trout (*Salmo trutta* L.) populations downstream dams in French Pyrenees. Reg Riv Res Manage 10: 347-361

Baran P, Delacoste M, Lascaux JM (1997) Variability of mesohabitat used by brown trout populations in the French Central Pyrenees. Trans Am Fish Soc 126: 747-757

Baran P, Delacoste M, Lascaux JM, Belaud A (1993b) Relations entre les caractéristiques de l'habitat et les populations de truites communes (*Salmo trutta* L.) de la vallée de la Neste d'Aure. Bull Fr Pêche Piscicol 331: 321-340

Baran P, Delacoste M, Lascaux JM, Dauba F, Segura G (1995b) La compétition interspécifique entre la truite commune (*Salmo trutta* L.) et la truite arc-en-ciel (*Oncorhynchus mykiss* Walbaum): influence sur les modèles d'habitat. Bull Fr Pêche Piscic 337-339: 283-290

Baran P, Lek S, Delacoste M, Belaud A (1996) Stochastic models that predict trout population density or biomass on a mesohabitat scale. Hydrobiologia 337: 1-9

Barbour MT, Gerritsen J, Griffith GE, Frydenborg R, McCarron E, White JS, Bastian ML (1996) A framework for biological criteria for Florida streams using benthic macroinvertebrates. J N Am Benthol Soc 15: 158-211

Barbour MT, Gerritsen J, Snyder BD, Stribling JB (1999) Rapid bioassessment protocols for use in streams and Wadeable rivers: Periphyton, benthic macroinvertebrates and fish, 2nd edn. EPA 841-B-99-002. US Environmental Protection Agency, Office of Water, Washington, DC

Barbour MT, Plafkin JL, Bradley BP, Graves CG, Wisseman RW (1992) Evaluation of EPA's rapid bioassessment benthic metrics: metric redundancy and variability among reference stream sites. Environ Tox Chem 11: 437-449

Barbour MT, Swietlik WF, Jackson SK, Courtemanch DL, Davies SP, Yoder CO (2000) Measuring the attainment of biological integrity in the USA: a critical element of ecological integrity. Hydrobiologia 422/423: 453-464

Barciela RM, Garcia E, Fernandez E. (1999) Modelling primary production in a coastal embayment affected upwelling using dynamic ecosystem models and artificial neural networks. Ecol Model 120: 199-211

Barinaga M (1996) A recipe for river recovery? Science 273: 1648-1650

Bath NV, McAvoy TJ (1992) Determining model structure for neural models by network stripping. Comput Chem Eng 115: 271-281

Battarbee RW, Flower RJ, Juggins S, Patrick ST, Stevenson AC (1997) The relationship between diatoms and surface water quality in the Hoylandet area of Nord-Trondelag, Norway. Hydrobiologia 348: 69-80

Batterink M, Wijffels G (1983) Een vergelijkend vegetatiekundig onderzoek naar de typologie en inveloeden van het beheer van 1973 tot 1972 in de Duinweilanden op

Terschekking. Report Agricultural University, Department of Vegetation Science, Plant Ecology and Weed Science, Weignengen
Baumeister J, Seipel D (2002) Diagnostic reasoning with multilevel set-covering models. Proceedings of 13th International Workshop on Principles of Diagnosis (DX-02), Semmering (Austria) pp 58-64
Baumeister J; Seipel D; Puppe F (2003) Incremental development of diagnostic set-covering models with therapy effects. Int J Uncert Fuzz Knowl-Based Sys (IJFUKS) 11: 25-49
Beale E (1988) Introduction to optimization. Wiley, New York
Beals EW (1984) Bray-Curtis ordination: an effective strategy for analysis of multivariate ecological data. Adv Ecol Res 14: 1-55
Becker G (2001) Larval size, case construction and crawling velocity at different substratum roughness in three scraping caddis larvae. Arch Hydrobiol 151: 317-334
Belaud A, Baran P (1997) Influence et détermination des débits réservés en rivières à salmonidés. C R Acad Agric Fr 83: 65-74
Belaud A, Bengen D, Lim P (1989a) Observations sur la faune de poissons de la moyenne Garonne. Rev Géog Pyrén Sud-Ouest 60: 625-634
Belaud A, Bengen D, Lim P (1990) Approche de la structure du peuplement ichthyologique de six bras morts de la Garonne. Ann Limnol 26: 81-90
Belaud A, Chaveroche P, Lim P, Sabaton C (1989b) Probability-of-use curves applied to brown trout (*Salmo trutta fario* L.) in rivers of southern France. Reg Riv Res Manage 3: 321-336
Belaud A, Dautrey R, Labat R, Lartigue JP, Lim P (1985) Observations sur le comportement migratoire des aloses (*Alosa alosa* L.) dans le canal artificiel de l'usine de Golfech. Ann Limnol 21: 161-172
Belaud A, Labat R (1992) Etudes ichtyologiques préalables à la conception d'un ascenseur à poissons à Golfech (Garonne, France). Hydroécol Appl 4: 65-89
Belkessam D, Oberdorff T, Hugueny B (1997) Unsatured fish assemblages in rivers of the North-Western France: potential consequences for species introductions. Bull Fr Pêche Piscic 344/345: 193-204
Bellariva JL, Belaud A (1998) Environmental factors influencing the passage of allice shad *Alosa alosa* at the Golfech fish lift on the Garonne River, France. In: Jungwirth M, Schmutz S, Weiss S (eds) Fish migration and fish bypasses. Fishing News Books, Oxford pp 171-179
Belliard J, Boët P, Tales E (1997) Regional and longitudinal patterns of fish community structure in the Seine River basin, France. Envir Biol Fishes 50: 133-147
Belpaire C, Smolders R, Vanden Auweele I, Ercken D, Breine J, Van Thyne G, Ollevier F (2000) The index of biotic integrity characterizing fish populations and ecological quality of Flandrian water bodies. Hydrobiologia 434: 17-33
Benchaken M, Loftus K, Wattanadilokgul C, Nuttasarin J (1989) Development of improved fisheries management program at Ubolratana reservoir, Thailand. Canada North-East Project. D.O.F./C.I.D.A. 906/11415, Mimeo
Bendell BE, McNicol DK (1987) Cyprinid assemblages, and the physical and chemical characteristics of small northern Ontario lakes. Environ Biol Fish 19: 229-234
Benedetto L (1975) Ökologie und Produktionsbiologie von *Agapetus fuscipes* CURT. im Breitenbach 1971-1972: Schlitzer produktionsbiologische Studien (11). Arch Hydrobiol Suppl 45: 305-375

Bengen D, Belaud A, Lim P (1992) Structure et typologie ichtyenne de trois bras morts de la Garonne. Ann Limnol 28: 35-56

Bennett K, Campbell C (2000) Support vector machines: Hype or hallelujah? SIGKDD Explorations 2: 1-13

Berges JA, Varela DE, Harrison PJ (2002) Effects of temperature on growth rate, cell composition and nitrogen metabolism in the marine diatom *Thalassiosira pseudonana* (Bacillariophyceae). Mar Ecol Pro Ser 225: 139-146

Berkman HE, Rabeni CF (1987) Effect of siltation on fish communities. Environ Biol Fish 50: 133-147

Bernascek G (1997) Management of reservoir fisheries in the Mekong basin project. Pub Mekong River Commission, Bangkok, Thailand

Berrebi-dit-Thomas R, Belliard J, Boët P (1998) Caractéristiques des peuplements piscicoles sensibles aux altérations du milieu dans les cours d'eau du bassin de la Seine. Bull Fr Pêche Piscic 348: 47-64

Biggs BJF (1995) The contribution of flood disturbance, catchment geology and land use to the habitat template of periphyton in stream ecosystems. Fresh Biol 33: 419-438

Biggs BJF, Gerbeaux P (1993) Periphyton development in relation to macro-scale (geology) and micro-scale (velocity) limiters in two gravel-bed rivers, New Zealand. N Z J Mar Fresh Res 26: 39-53

Billen G, Garnier J, Hanset P (1994) Modelling phytoplankton development in whole drainage networks: the RIVERSTRAHLER Model applied to the Seine river system, Hydrobiologia 298: 119-137

Billheimer D, Cardoso T, Freeman E, Guttorp P, Ko HW, Silkey M (1997) Natural viability of benthic species composition in the Delaware Bay. Environ Ecol Stat 4: 95-115

Birks HJB, Juggins S, Line JM (1990a) Lake surface-water reconstructions from paleolimnological data. In: Mason BJ (ed) The surface waters acidification program pp 301-313

Birks HJB, Line JM, Juggins S, Stevenson AC, ter Braak C (1990b) Diatoms and pH reconstruction. Phil Trans Roy Soc London B 327: 263-278

Bishop CM (1995) Neural networks for patterns recognition. Oxford University Press, Oxford

Bishop CM, Svensén M, Williams CKI (1996) GTM: a principled alternative to the self-organising map. In: von der Malsburg C, von Seelen W, Vorbrüggen JC, Sendhoff B (eds) Proc ICANN96, Int Conference on Artificial Neural Networks, Lecture Notes in Computer Science vol 1112, Springer, Berlin pp 165-170

Blayo F, Demartines P (1991) Data analysis: how to compare Kohonen neural networks to other techniques? In: Prieto A (ed) Proceedings of International Workshop on Artificial Neural Networks, Springer, Berlin pp 469-476

BMWP (1979) Biological monitoring working party. The 1978 national testing exercise. Technical Memorandum 19. Water Data Unit, Reading, UK

Bobbin J, Recknagel F (1999) Mining Water Quality Time Series for Predictive Rules of Algal blooms by Genetic Algorithm. In: Oxley L, Scrimgeour F (eds) MODSIM'99 International Congress on Modelling and Simulation. Modelling and Simulation Society of Australia and New Zealand Inc, Hamilton, New Zealand pp 691-696

Boddy L, Morris CW, Wilkins MF, Luan Al-Haddad GA, Tarran RR, Jonker P, Burkill H (2000) Identification of 72 phytoplankton species by radial basis function neural network analysis of flow cytometric data. Mar Ecol Prog Ser 195: 47-59

Boesch DF (1977) Application of numerical classification in ecological investigations of water pollution. Special scientific report no 77. Virginia Institute of Marine Science, Gloucester Point, VA

Boët P, Fuhs T (2000) Predicting presence of fish species in the Seine river basin using artificial neuronal networks. In: Lek S, Guégan JF (eds) Artificial neuronal networks, application to ecology and evolution. Springer-Verlag, Berlin, Heidelberg pp 131-142

Bogaczewicz-Adamczak B, Klosinska D, Zgrundo A (2004) Diatoms as indicators of water pollution in the coastal zone of the gulf of Gdansk (Southern Baltic sea). Oceanol Stud 30: 59-75

Bollabas B, Das G, Gunopulos D (1997) Time-Series Similarity Problems and Well-separated Geometric Sets. In: Proc 13th Annual Symposium on Computational Geometry, Nice, France

Boon PJ, Holmes, NTH, Maitland PS, Rowell TA, Davies J (1998) A system for evaluating rivers for conservation (SERCON): development, structure and function. In: Boon PJ, Howell DL (eds) Freshwater quality: defining the undefinable? The Stationary Office, Edingburgh pp. 299-326

Boone RB, Krohn WB (2000) Partitioning sources of variation in vertebrate species richness. J Biogeo 27: 457-470

Bormans M, Webster IT (1999) Modelling the spatial and temporal variability of diatoms in the River Murray. J Plankton Res 21: 581-598

Bos F, Wasscher M (1997) Veldgids libellen [Field guide for Odonata]. Stichting Uitgeverij KNNV, Utrecht, The Netherlands (in Dutch)

Boser BE, Guyon IM, Vapnik VN (1992) A training algorhthm for optimal margin classifiers. In: Haussler D (ed) Proc 5th annual ACM Workshop on Computational Learning Theory, ACM Press, New York pp 144-152

Bossel H (1992) Real-structure process description as the basis of understanding ecosystems and their development. Ecol Model 63: 261-276

Boulton AJ (1999). An overview of river health assessment: philosophies, practice, problems and prognosis. Fresh Biol 41: 469-479

Bovee KD (1982) A guide to stream habitat analysis using the instream flow incremental methodologies. US Wildlife Service Instream Flow Information Paper no. 12. Report no. FWS/OBS-82/26. USFWS Instream Flow Group, Fort Collins, USA

Box GEP, Jenkins GM (1970) Time series analysis: forecasting and control. Holden-Day, San Francisco, USA

Boyle TP, Smillie GM, Anderson JC, Beeson DR (1990) A sensitivity analysis of nine diversity and seven similarity indices. J Wat Poll Cont Fed 62: 749-762

Brakke DF, Baker JP, Böhmer J, Hartmann A, Havas M, Jenkins A, Kelly C, Ormerod SJ, Paces T, Putz R, Rosseland BO, Schindler DW, Segner H (1994) Group report: physiological and ecological effects of acidification on aquatic biota. In: Steinberg CEW, Wright RF (eds) Acidification of freshwater ecosystems: implications for the future. John Wiley and Sons, Vol 1: 275-312

Brann DM,. Thurman DA, Mitchell CM (1995) Case based reasoning as a methodology for accumulating human expertise for discrete system control. In: Proc IEEE International Conference on Systems, Man and Cybernetics. Vancouver, Canada pp 4219-4223

Brasquet C, Bourges B, Le Cloirec P (1999) Quantitative structure-property relationship (QSPR) for the adsorption of organic compounds onto activated carbon cloth: comparison between multiple linear regression and neural network. Environ Sci Tech 33: 4226-4231

Bray JR, Curtis JT (1957) An ordination of upland forest communities of Southern Wisconsin. Ecol Monog 27: 325-349

Breiman L, Freidman J, Olshen R, Stone C (1984) Classification and regression trees. Wadsworth, Belmont, California, USA

Brey T, Gerdes D (1998) High Antarctic macrobenthic community production. J Exp Mar Biol Ecol 231: 191–200

Brey T, Jarre-Teichmann A, Borlich O (1996) Artificial neural network versus multiple linear regression: predicting P/B ratios from empirical data. Mar Ecol Prog Ser 140: 251-256

Bridge D (1998) Defining and Combining Symmetric and Asymmetric Similarity Measures. In Smyth B, Cunningham P (eds) EWCBR-98, Lecture Notes in Artificial Intelligence. Spinger-Verlag, Berlin pp 52-63

Brittain JE, Saltveit SJ (1989) A review of the effect of river regulation on Mayflies (Ephemeroptera). Reg Riv Res Manage 3: 191-204

Brosse S, Giraudel JL, Lek S (2001) Utilisation of non-supervised neural networks and principal component analysis to study fish assemblages. Ecol Model 146: 159-166

Brosse S, Guegan J, Tourenq J, Lek S (1999a) The use of artificial neural networks to assess fish abundance and spatial occupancy in the littoral zone of a mesotrophic lake. Ecol Model 120: 299-311

Brosse S, Lek S (2000b) Linear and non-linear methods to predict the microhabitat of 0+roach (*Rutilus rutilus* L.) in the littoral zone of a large reservoir. Ver Int Ver Theor Ange Limnol 27: 811-814

Brosse S, Lek S (2000a) Modelling roach (*Rutilus rutilus*) microhabitat using linear and nonlinear techniques. Fresh Biol 44: 441-452

Brosse S, Lek S, Dauba F (1999b) Predicting fish distribution in a mesotrophic lake by haydroacoustic survey and artificial neural networks. Limnol Oceanog 44: 1293-1303

Buslé J, Quignard JP (2001) Biologie des poissons d'eau douce européens. Editions Tec & Doc, Paris

Bryce SA, Omernik JM, Larsen DP (1999) Ecoregions: a geographic framework to guide risk characterization and ecosystem management. Environ Pract 1: 141-155

Buckland ST, Elston DA (1993) Empirical models for the spatial distribution of wildlife. J Appl Ecol 30: 478-495

Bunn SE, Edward DH, Loneragan NR (1986) Spatial and temporal variation in the macroinvertebrate fauna of streams of the northern jarrah forest, Western Australia: community structure. Fresh Bio 16: 67-91

Burges C (1998) A tutorial on support vector machines for pattern recognition. Data Min Knowe Disc 2: 121-167

Burges CJ (1999) Geometry and invariance in kernel based methods. In: Schölkopf B, Burges C, Smola A (eds) Advances in kernel methods: support vector learning. MIT Press, Cambridge. pp 89-116

Burrough PA, McDonnell RA (1998) Principles of geographical information systems. Oxford University Press, New York

Butcher RW (1947) Studies in the ecology of rivers. IV. The algae of organically enriched water. J Ecol 35: 186-191

Cairns DM (2001) A comparison of methods for predicting vegetation type. Plant Ecol 156: 3-18

Cairns JJr, Pratt JR (1993) A history of biological monitoring using benthic macro-invertebrates. In: Rosenberg DM, Resh VH (eds) Freshwater biomonitoring and benthic macro-invertebrates. Chapman & Hall, New York pp 10-27

Campeau S, Pienitz R, Hequette A (1999) Diatoms from the Beaufort Sea, southern Arctic Ocean (Canada). Modern analogues for reconstructing environments and relative sea levels. Biblio Diatom 42: 1-244

Cantonati M, Corradini G, Jüttner I, Cox EJ (2001) Diatom assemblages in high mountain streams of the Alps and the Himalaya. Nova Hedw Beih 123: 37-61

Carchon P, de Pauw N (1997) Development of a methodology for the assessment of surface waters. Study by order of the Flemish Environment Agency (VMM), Ghent University, Laboratory of Environmental Toxicology and Aquatic Ecology, Ghent, Belgium (in Dutch)

Carpenter GA, Gopal S, Macomber S, Martens S, Woodcock CE, Franklin J (1999) A neural network method for efficient vegetation mapping. Remote Sens Environ 70: 326–338

Carpenter GA, Grossberg S (1987) ART2: self-organization of stable category recognition codes for analog input patterns. Applied Optics 26: 4919-4930

Carpenter SR, Kitchell JF, Cottingham KL, Schindler DE, Christensen DL, Post DM, Voichick N (1996) Chlorophyll variability, nutrient input, and grazing: Evidence from whole-lake experiments. Ecology 77: 725-735

Carreira-Perpinan MA (2001) Continuous latent variable models for dimensionality reduction and sequential data reconstruction. PhD thesis, Department of Computer Science, University of Sheffield, Sheffield, UK

Cassie V (1989) A contribution to the study of New Zealand diatoms. J Cramer, Berlin

Cattaneo F, Lim P, Belaud A (1999) Approche de la structuration spatiale du peuplement piscicole de la zone de transition de la Garonne. Ichtyophys Acta 22: 61-74

Cawley GC (2000) MATLAB support vector machine toolbox (v0.55 β). [http://theoval.sys.uea.ac.uk/~gcc/svm/toolbox]. University of East Anglia, School of Information Systems, NR4 7TJ, Norwich, Norfolk, U.K

Cayrou J, Compin A, Giani N, Céréghino R (2000) Associations spécifiques chez les macroinvertébrés benthiques et leur utilisation pour la classification des cours d'eau. Cas du réseau hydrographique Adour – Garonne (France). Ann Limnol 36: 189-202

CBAG (Comité de Bassin Adour Garonne) (1996) Cahier géographique: Garonne. Comité de Bassin Adour Garonne, Toulouse

Cemagref (1982) Etude des méthodes biologiques quantitatives d'appréciation de la qualité des eaux. Rapport Q.E. Lyon-A.F. Bassin Rhône-Méditerranée-Corse, Agence de l'Eau Rhône-Méditerranée-Corse, Lyon, France

Cemagref (2001) Définition des hydroécorégions françaises: Méthodologie de détermination des conditions de référence au sens de la Directive cadre pour la gestion des eaux. Rapport de phase 1, Cemagref, Lyon, France

Céréghino R, Giraudel JL, Compin A (2001) Spatial analysis of stream invertebrates distribution in the Adour-Garonne drainage basin (France), using Kohonen self organising maps. Ecol Model 146: 167-180

Céréghino R, Giraudel JL, Compin A (2001) Spatial analysis of stream invertebrates distribution in the Adour-Garonne drainage basin (France), using Kohonen self organizing maps. Ecol Model 146: 167-180

Chambers J, Hastie T (1991) Statistical models in S. CRC Press, Boca Raton, FL, USA

Chandler RJ (1970) A biological approach to water quality management. Wat Poll Cont 69: 415-422
Changeux T, Bonnieux F, Armand C (2001) Cost benefit analysis of fisheries management plans. Fish Manage Ecol 8: 425-434
Chen Q, Mynett AE (2003) Integration of data mining techniques and heuristic konledge in fuzzy logic modeling of eutrophication in Taihu Lake. Ecol Model 162: 55-67
Chen Y, Breen PA, Andrew NL (2000) Impacts of outliers and mis-specification of priors on Bayesian fisheries-stock assessment. Can J Fish Aquat Sci 57: 2293-2305
Chen Y, Fournier D (1999) Impacts of atypical data on Bayesian inference and robust Bayesian approach in fisheries. Can J Fish Aquat Sci 59: 1525-1533
Chessman BC (1999) Predicting the macroinvertebrate faunas of rivers by multiple regression of biological and environmental differences. Fresh Biol 41: 747-757
Chessman BC, Growns IO, Plunket-Cole N (1999) Predicting diatom communities at genus level for the biological management of rivers. Fresh Biol 41: 317-331
Chesters RK (1980) Biological monitoring working party: the 1978 national testing exercise (Technical memorandum). Water Data Unit, Department of the Environment, London
Chhetri DB, Fowler G.W (1996) Prediction models for estimating total heights of trees from diameter at breast height measurements in Nepal's lower temperate broad-leaved forests. Forest Ecol Manage 84: 177-186
Cho HJ, Han KY, Kim SH (1996) Water quality impact assessment due to dredging in the downstream of the Nakdong River. Kor J Wat Res 29: 177-186 (in Korean)
Cho SB (1997) Self-organizing map with dynamical node-splitting: application to handwritten digit recognition. Neural Comput 9: 1345-1355
Chon TS, Kwak IS, Park YS (2000b) Pattern recognition of long-term ecological data in community changes by using artificial neural networks: Benthic macroinvertebrates and chironomids in a polluted stream. Kor J Ecol 23: 89-100
Chon TS, Kwak IS, Park YS, Kim TH, Kim Y (2001) Patterning and short-term predictions of benthic macroinvertebrate community dynamics by using a recurrent artificial neural network. Ecol Model 146: 181–193
Chon TS, Kwak IS, Song MY, Park YS, Cho HD, Kim MJ, Cha EY, Lek S (2002) Characterizing the effects of water quality on benthic stream macroinvertebrates in South Korea using a self-organizing mapping model. In: Lee D (ed) Ecology of Korea. Bumwoo Publishing, Seoul, Korea pp 356-384
Chon T-S, Park Y-S, Cha EY (2000c) Patterning of community changes in benthic macroinvertebrates collected from urbanized streams for the short time prediction by temporal artificial neural networks. In: Lek S, Guegan J-F (eds) Artificial neural networks: application to ecology and evolution. Springer-Verlag, Berlin pp 99-114
Chon TS, Park YS, Kwak IS, Cha EY (2003) Non-linear approach to grouping, dynamics and organizational informatics of benthic macroinvertebrate communities in streams by artificial neural networks, In: Recknagel F (ed) Ecological informatics: understanding ecology by biologically-inspired computation. Springer-Verlag, New Yourk pp127-178
Chon TS, Park YS, Moon KH, Cha EY (1996) Patternizing communities by using an artificial neural network. Ecol Model 90: 69-78
Chon TS, Park YS, Park JH (2000a) Determining temporal pattern of community dynamics by using unsupervised learning algorithms. Ecol Model 132: 151-166
Chong EKP, Zak SH (2001) An introduction to optimization. Wiley, New York

Cisneros-Mata MA, Brey T, Jarre-Teichmann A (2000) Performance comparison between regression and neuronal network models for forecasting Pacific sardine (Sardinops caeruleus) biomass. In: Lek S, Guegan J-F (eds) Artificial neural networks: application to ecology and evolution. Springer-Verlag, Berlin pp 157-164

Ciutti F, Cappelletti C, Monauni C, Siligardi M, Dell'Uomo A (2000) Qualità biologica e funzionalità del Torrente Ferina (Trentino). Dendronatura 2: 12-22

Ciutti F, Cappelletti C, Torrisi M (2001) Le diatomee come indicatori della qualità biologica dei corsi d'acqua. Il Pescatore Trentino 3: 32-35

Civco DL (1993) Artificial neural networks for land cover classification and mapping. Int J Geog Infor Sys 7: 173-186

Clarke RT, Wright JF, Furse MT (2003) RIVPACS models for predicting the expected macroinvertebrate fauna and asessing the ecological quality of rivers. Ecol Model 160: 219-233

Cleveland WS (1979) Robust locally- weighted regression and scattered plot smoothing. J Am Statist Soc 74: 829-836

Cohen J (1960) A coefficient of agreement for nominal scales. Educ Psychol Meas 20: 37-46

Colasanti RL (1991) Discussions of the possible use of neural network algorithms in ecological modelling. Binary 3: 13–15

Compin A, Céréghino R (2003) Sensitivity of aquatic insect species richness to disturbance in the Adour-Garonne stream system (France). Ecol Indic 3: 135-142

Connell JH (1978) Diversity in tropical rain forests and coral reefs. Science 199: 1302-1309

Cooper CM (1993) Biological effects of agriculturally derived surface-water pollutants on aquatic systems - a review. J Env Qual 22: 402-408

Copp GH (1992) An empirical model for predicting micro habitat of 0+ juvenile fishes in a lowland river catchment. Oecologia 91: 338-345

Coring E (1993) Zum indikationswert bentischer diatomeengesellschaften in basenarmen Fliessgewassern. Verlag Shaker, Aachen

Coring E (1999) Situation and developments of algal (diatom)-based techniques for monitoring rivers in Germany. In: Prygiel J, Whitton BA, BukowskavJ (eds) Use of algae for monitoring rivers III, Agence de l'Eau Artois-Picardie, Douai pp 122-127

Coring E, Hamm A, Schneider S (1999) Durchgehendes trophiesystem auf der Grundlage der trophieindikation mit Kieselalgen. - DVWK Mitteilungen Nr. 6/1999. Deutscher Verband für Wasserwirtschaft und Kulturbau e.V, Bonn

Coste M (1976) Contribution à l'écologie des diatomées benthiques et périphytiques de la Seine: Distribution longitudinale et influence des pollutions. Soc Hydrot France 14: 1-6

Coste M (1978) Sur l'utilisation des diatomées benthiques pour l'appréciation de la qualité biologique des eaux courantes. Thèse Biol Végét. Université de Besançon, Besançon, France

Coste M (1982) Etude des méthodes biologiques quantitative d'appréciation de la qualité des eaux. Cemagref, Lyon

Coste M, Ayphassorho H (1991) Etude de la qualité des eaux du bassin Artois-Picardie à l'aide des communautés de diatomées benthiques (Application des indice diatomiques). Cemagref, Bordeaux

Coste M, Ector L (2000) Diatomées invasives exotiques ou rares en France: principales observations effectuées au cours des dernières décennies. Syst Geogr Plant 70: 373-400

Coste M, Le Cohu R, Bertrand J (1992) Sur l'apparition d'espèces du genre Gomphoneis en France. Distribution, caractéristiques morphologiques et écologiques. In: Agence de l'Eau Artois Picardie (ed) Actes du XIème Colloque ADLAF, Agence de l'Eau Artois Picardie, Douai pp 71-77

Coste M, Leynaud G (1974) Etudes sur la mise au point d'une méthodologie biologique de détermination de la qualité des eaux en milieu fluvial. CTGREF, Paris

Cottingham KL, Schindler DE (2000) Effects of grazer community structure on phytoplancton response to nutrient pulses. Ecology 81: 183-200

Cow-Rogers S (1997) Inseason forecasting of skeena River sockeye run size using Bayesian probability theory. Canadian Report of Fisheries and Aquatic Sciences, 2412: 1-43

Cowx IG, Welcomme RL (1998) Rehabilitation of rivers for fish. Food and Agricultural Organization of the United Nations and Fishing News Books, Oxford, UK

Cox EJ (1990) Studies on the algae of a small softwater stream I. Occurrence and distribution with particular reference to the diatoms. Arch Hydrobiol 83: 525-552

Crawford SL (1991) Genetic optimisation for exploratory projection pursuit. In: Keramida EM (ed) Computer science and statistics. Proc 23rd Symp Interface, Fairefax Station. pp. 318-321

Crespin BV, Usseglio-Polatera P (2002) Traits of brown trout prey in relation to habitat characteristics and benthic invertebrate communities. J Fish Biol 60: 687-714

Crisp DT (2000) Trout and salmon: ecology, conservation and rehabilitation. Blackwell Publishing, Oxford

Cristianini N, Shawe-Taylor J (2000) An introduction to support vector machines and other kernel-based learning methods. Cambridge University Press, Cambridge

Crül RCM (1992) Models for estimating potential fish yields of African inland waters. FAO/ IFA Occasional Paper, Roma

Cuinat R (1971) Principaux caractéres démographiques observés dans 50 rivières à truites françaises: Influence de la pente et du calcium. Ann Hydrobiol 2: 187-167

Culp JM, Davies RW (1982) Analysis of longitudinal zonation and the river continuum concept of the Oldman – South Saskatchewan River system. Can J Fish Aquatic Sc 39: 1258-1266

Cumming BF, Smol JP (1993) Development of diatom-based salinity models for paleoclimatic research from lakes in British Columbia (Canada). Hydrobiologia 269/270: 179-196

Cummins KW (1974) Structure and function of stream ecosystems. Bioscience 24: 631-641

Cummins KW (1979) The natural stream ecosystem. In: Ward JV, Stanford JA (eds) The ecology of regulated streams. Plenum Press, New York pp 7-24

Cummins KW, Minshall G.W, Sedell JR, Cushing CE, Petersen RC (1984) Stream ecosystem theory. Ver Int Ver Theor Ange Limnol 22: 1818-1827

Cummins KW, Petersen RC, Howard FO, Wuycheck J, Cand Holt VI (1973) The utilization of leaf litter by stream detritivores. Ecology 54: 336-345

Cummins KW, Wilzbach MA (1985) Field procedures for analysis of functional feeding groups of stream macroinvertebrates. Appalachian Environmental Laboratory, University of Maryland, Frostburg

Daget J, Le Guen JC (1975) Dynamique des populations exploitées de poissons. In: Lamotte M, Bourlière F (eds) Dynamique des populations de Vertébrés Publ. Masson. Paris, France

Daget P, Godron M (1982) Analyse fréquentielle de l'écologie des espèces dans les communautés. Masson, Paris

Dane AD, Rea GJ, Walmsley AD, Haswell SJ (2001) The determination of moisture in tobacco by guided microwav spectroscopy and multivariate calibration. Analy Chim Acta 429: 185-194

d'Angelo DJ, Howard LM, Meyer JL, Gregory SV, Ashkenas LR (1995) Ecological uses for genetic algorithms: predicting fish distributions in complex physical habitats. Can J Fish Aquat Sci 52: 1893-1908

Dantzig GB (1963) Linear programming and extensions. Princeton University Press, Princeton, NJ

Dauba F, Lek S, Mastrorillo S, Copp GH (1997) Long-term recovery of macrobenthos and fish assemblages after water pollution abatement measures in the river petite Baïse (France). Arch Environ Contam Toxicol 33: 277-285

David J-M, Krivine J-P, Simmons R (1993) Second-generation expert systems Springer

Davies NM, Norris RH, Thoms MC (2000) Prediction and assessment of local stream habitat features using large scale catchment characteristics. Fresh Biol 45: 343-369

Davis WS, Simon TP (1995) Biological Assessment and criteria: Tools for water resource planning and decision-making. CRC Press, Boca Raton

de Iong HH, Van Zon JCJ (1993) Assessment of impact of introduction of exotic fish species in north-east Thailand. Aquacul Fish Manag 24: 279-289

de Pauw N, Ghetti PF, Manzini DP, Spaggiari DR (1992) Biological assessment methods for running water. In: Newman PJ, Piavaux MA, Sweeting RA (eds) River water quality: ecological assessment and control. European Communities / Union (EUR-OP/OOPEC/OPOCE) pp. 217-248

de Pauw N, Lambert V, van Kenhove A, Bij de Vaate A (1994) Performance of two artificial substrate samplers for macroinvertebrates in biological monitoring of large and deep rivers and canals in Belgium and the Netherlands. Environ Monit Assess 30: 25-47

de Pauw N, Vanhooren G (1983) Method for biological assessment of watercourses in Belgium. Hydrobiologia 100: 153-168

de Silva SS, Moreau J, Amarasinghe US, Chookajorn T, Guerrero R (1991) A comparative assessment of the fisheries in lacustrine inland waters in three Asian countries based on catch and effort data. Fish Res 11: 177-189

De'ath G (2002) Multivariate regression trees: a new technique for modeling species-environment relationships. Ecology 85: 1105-1107

De'ath G, Fabricius KE (2002) Classification of regression trees: a powerful tool yet simple technique for ecological data analysis. Ecology 81: 3178-3192

Deacon JR, Mize SV (1997) Effects of water quality and habitat on the composition of fish communities in the upper Colorado river basin. US Geological Survey Fact Sheet FS-122-97. U.S. Geological Survey, Denver, Colorado

Dean T, Richardson J (1997) Native fish survival during exposure to low levels of dissolved oxygen. Water Atmosph 5: 12-14

Décamps H (1968) Vicariances écologiques chez les Trichoptères des Pyrénées. Ann Limnol 4: 1-50

Décamps H, Naiman RJ (1989) L'écologie des fleuves. La Recherche 208: 310-319

Dedecker AP, Goethals PLM, de Pauw N (2002) Comparison of artificial neural network (ANN) model development methods for prediction of macroinvertebrate communities in the Zwalm river basin in Flanders, Belgium. TheScientificWorldJOURNAL 2: 96-104

Dedecker AP, Goethals PLM, Gabriels W, de Pauw N (2004) Optimisation of artificial neural network (ANN) model design for prediction of macroinvertebrates in the Zwalm river basin (Flanders, Belgium). Ecol Model 174: 161-173

Delacoste M, Baran P, Dauba F, Belaud A (1993) Etude du macrohabitat de reproduction de la truite commune (*Salmo trutta* L.) dans une rivière pyrénéenne, la Neste du Louron. Evaluation d'un potentiel de l'habitat physique de reproduction. Bull Fr Pêche Piscic 331: 341-356

Delalieux F, Cardell-Fernandez C, Torfs K, Vleugels G, van Grieken R (2002) Damage functions and mechanism equations derived from limestone weathering in field exposure. Wat Air Soil Poll 139: 75-94

Dell'Uomo A (1996) Assessment of water quality of an Apennine river as a pilot study for diatom-based monitoring of Italian watercourses. In: Whitton BA, Rott E (eds) Use of Algae for monitoring rivers II, Innsbruck Austria 17-19 Sept. 95. Studia Student. G.m.b.H., Innsbruck pp 65-72

Dell'Uomo A (1999) Use of algae for monitoring rivers in Italy: current situation and perspectives. In: Prygiel J, Whitton BA, Bukowska J (eds) Use of algae for monitoring rivers III, Agence de l'Eau Artois-Picardie, Douai pp 17-25

Dell'Uomo A (2004) L'indice diatomico di eutrofizzazione/polluzione (EPI-D) nel monitoraggio delle acque correnti. Linee guida. APAT Agenzia per la protezione dell'ambiente e per I servizi tecnici, Roma

Denicola DM (1996) Periphyton responses to temperature at different ecological levels. In: Stevenson RJ, Bothwell ML, Lowe RL (eds) Algal ecology. Academic Press Inc, San Diego pp 149-181

Denys L (1991a) A check-list of the diatoms in the holocene deposits of the Western Belgian coastal plain with a survey of their apparent ecological requirements.I. Introduction, ecological code and complete list. Ministère des Affaires Economiques - Service Géologique de Belgique

Denys L (1991b) A check-list of the diatoms in the holocene deposits of the Western Belgian coastal plain with a survey of their apparent ecological requirements. II. Centrales. Ministère des Affaires Economiques - Service Géologique de Belgique

Descy JP (1976a) Utilisation des algues benthiques comme indicateurs biologiques de la qualité des eaux courantes. In: P. Pesson (ed) La pollution des eaux continentales, 1e edition, Gauthier-Villars, Paris pp 149-172

Descy JP (1976b) Etude quantitative du peuplement algal benthique en vue de l'établissement d'une méthodologie d'estimation biologique de la qualité des eaux courantes. Application au cours d'eau belge de la Meuse et de la Sambre. Rech Tech Serv l'Environ: 159-206

Descy JP (1976c) Value of aquatic plants in the characterization of water quality and principles of the method used. In: Amavis R, Smeets J (eds) Principles and methods for determining ecological criteria on Hydrobiocenoses (Commission of the European communities), Luxembourg, 1975, Pergamon Press, Oxford pp 125-183

Descy JP (1979) A new approach to water quality estimation using diatoms. Nova Hedwigia 64: 305-323

Descy JP (1980) Utilisation des algues benthiques comme indicateurs biologiques de la qualité des eaux courantes. In : Pesson P (ed), La pollution des eaux continentales, Gauthier-Villars, Paris pp 169-194

Descy JP (1984) Ecologie et distribution des diatomées benthiques dans le bassin Belge de la Meuse. Documents de travail de l'IRScNB, Institut Royal des Sciences Naturelles de Belgique, Bruxelles

Descy JP (1987) Phytoplankton composition and dynamics in the river Meuse (Belgium). Arch Hydrobiol 78: 225-245

Descy JP (1993) Ecology of the phytoplankton of the River Moselle: effects of disturbances on community structure and diversity. Hydrobiologia 249: 111-116

Descy JP, Coste M (1990) Utilisation des diatomées benthiques pour l'évaluation de la qualité des eaux courantes. Rapport final, UNECED, Namur, Cemagref, Bordeaux

Descy JP, Coste M (1991) A test of methods for assessing water quality based on diatoms. Verhandlungen der Internationalen Vereinigung Für Theoretische und Angewandte Limnologie 24: 2112-2116

Descy JP, Ector L (1999) Use of diatoms for monitoring rivers in Belgium and Luxembourg. In: Prygiel J, Whitton BA, Bukowska J (eds) Use of algae for monitoring rivers III, Agence de l'Eau Artois-Picardie, Douai pp 128-137

Descy JP, Everbecq E, Gosselain V, Viroux L, Smitz JS (2003) Modelling the impact of benthic filter-feeders on composition and biomass of river plankton, Fresh Biol 48: 404-417

Dethier M, Castella E (2002) A ten year survey of longitudinal zonation and temporal changes of macrobenthic communities in the Rhône River, downstream from lake Geneva (Switzerland). Ann Limnol 38: 151-162

DEV (Deutsches Institut für Normung e.V.) (1992) Biologisch-ökologische Gewässergüteuntersuchung: Bestimmung des saprobienindex (M2). In: Deutsche Einheitsverfahren zur Wasser-, Abwasser- und Schlammuntersuchung. - VCH Verlagsgesellschaft mbH, Weinheim pp 1-3

d'Heygere T, Goethals PLM, de Pauw N (2002) Use of genetic algorithms to select input variables in artificial neural network models for the prediction of benthic macroinvertebrates. In: Rizzoli AE, Jakeman AJ (eds) Integrated Assessment and Decision Support proceedings of the 1st biennial meeting of the International Environmental Modelling and Software Society, Vol 2, SEA, Como pp 136-141

d'heygere T, Goethals PLM, De Pauw N (2003) Use of genetic algorithms to select input variables in decision tree models for the prediction of benthic macroinvertebrates. Ecol Model 160: 291-300

Dillon PJ, Rigler FH (1974) The phosphorus-chlorophyll relationship in lakes. Limnol Oceanogr 19: 767-773

Dimopoulos I, Chronopoulos J, Chronopoulos-Sereli A, Lek S (1999) Neural network models to study relationships between lead concentration in grasses and permanent urban descriptors in Athens city (Greece). Ecol Model 120: 157-165

Dimopoulos Y, Bourret P, Lek S (1995) Use of some sensitivity criteria for choosing networks with good generalisation ability. Neural Process Lett 2: 1-4

Dixit SS, Dixit AS, Smol JP (1990) Paleolimnological investigation of three manipulated lakes from Sudbury, Canada. Hydrobiologia 214: 245-252

Dobson AJ (1983) An introduction to statistical modelling. Chapman & Hall, London

Drago M, Cescon B, Iovenitti L (2001) A three-dimensional numerical model for eutrophication and pollutant transport. Ecol Model 145: 17-34

Drost MBP, Cuppen HPJJ, van Nieukerken EJ, Schreijer M (eds) (1992) De waterkevers van Nederland [The water beetles of the Netherlands]. Uitgeverij KNNV, Utrecht, The Netherlands (in Dutch)

du Toit SHC, Steyn AGW, Stumpf RH (1986) Graphical exploratory data analysis. Springer-Verlag. New York

Dufrêne M, Legendre P (1997) Species assemblages and indicator species: the need for a flexible asymmetrical approach. Ecol Monogr 67: 345-366

Dunkerley D (1999) Cellular automata. In: Fielding A (ed) Machine learning methods for ecological application. Kluwer Academic, Bostom, MA pp 145-183

Dunkerley DL (1997) Banded vegetation: development under uniform rainfall from a simple cellular automaton model. Plant Ecol 129: 103-111

Dupias G, Rey P (1985) Document pour un zonage des regions phyto-écologiques. Centre National de la Recherche Scientifique, Toulouse, France

Dynesius M, Nilsson C (1994) Fragmentation and flow regulation of river systems in the northern third of the world. Science 266: 753-762

Ector L, Dohet A, Dolisy D, Hoffmann L (1997) Les diatomées épilithiques meilleures bioindicateurs que les macroinvertébrés pour l'évaluation de la qualité de la partie rhithrale des cours d'eau: cas des ruisseaux de l'Our (Luxembourg). Cryptogamie Algol 18: 74-75

Ector L, Kingston JC, Charles DF, Denys L, Douglas MSV, Manoylov K, Michelutti N, Rimet F, Smol JP, Stevenson RJ, Winter JG (2004) Workshop report. Freshwater diatoms and their role as ecological indicators. Pages 469-480 in M. Poulin. 7th International Diatom Symposium 2002, Ottawa, Canada, Biopress Limited, Bristol

Efron B (1983) Estimating the error rate of a prediction rule: improvement on cross-validation. J Am Stat Asso 78: 316-330

Efron B, Tibshirani RJ (1995) Cross-validation and the bootstrap: estimating the error rate of the prediction rule. Rep Tech Univ Toronto, Tronro, Canada

Einsle U (1993) Crustacea, copepoda, calanoidia and cyclopoida. Susswasserfauna von Mitteleuropa, Vol. 8, Part 4-1. J Fisher, Stuttgart

Eliott JA, Irish AE, Reynolds CS, Tett P (1999) Sensitivity analysis of PROTECH, a new approach in phytoplankton modelling, Hydrobiologia 414: 45-51

Elizondo DA, McClendon RW, Hoongenboom G (1994) Neural network models for predicting flowering and physiological maturity of soybean. Trans ASAE 37: 981-988

Ellingsen BK (1994) A comparative analysis of backpropagation and counterpropagation neural networks. Neural Network World 4: 719-733

Eloranta P (1990) Periphytic diatoms in the Acidification Project Lakes. In: Kauppi (ed) Acidification in Finland. Springer-Verlag, Berlin pp 985-994

Eloranta P (1999) Application of diatom indices in Finnish rivers. In: Prygiel J, Whitton BA, Bukowska J (eds) Use of algae for monitoring rivers III, Agence de l'Eau Artois-Picardie, Douai pp 138-144

Eloranta P (1999) Applications of diatom indices in Finnish rivers. In: Prygiel J, Whitton BA, Bukowska J (eds) Use of algae for monitoring rivers III, Agence de l'Eau Artois-Picardie, Douai pp 138-144

Emmons E, Jennings MJ, Edwards C (1999) An alternative classification method for northern Wisconsin lakes. Can J Fish Aqua Sc 56: 661-669

EPA (Environmental Protection Agency) (1988) WQS: draft framework for the water quality standards program. Office of Water, United States Environmental Protection Agency, Washington DC

Eriksson L, Hermens JLM, Johansson E, Verhaar HJM, Wold S (1995) Multivatiate-analysis of aquatic toxicity data with PLS. Aqua Sci 57: 217-241

Etchanchu D, Probst JL (1988) Evolution of the chemical composition of the Garonne river during the period 1971-1984. Hydrol Sci J 33: 243-256
Ette EI, Ludden TM (1996) Population pharmacokinetic modeling: the importance of informatics graphics. Pharm Res (NY) 12: 1845-1855
European Committee for Standardization (2002a) Water quality - Guidance standard for the identification and enumeration of benthic diatom samples from rivers, and their interpretation, prEN 14407
European Committee for Standardization (2002b) Water quality - Guidance on quality assurance aspects of the sampling and analysis of benthic diatoms
European Parliament (2000) Directive 2000/60/EC of the European Parliament and of the Council establishing a framework for Community action in the field of water policy. O.J.L327
European Standard (2001) Water quality - Guidance standard for the identification and numeration of benthic diatom samples from rivers, and their interpretation, prEN14407 CEN TC230
European Standard (2002) Water quality - Guidance standard for the routine sampling and pretreatment of benthic diatoms from rivers, prEN13946, Final Draft
Everbecq E, Gosselain V, Viroux L, Descy JP (2001) POTAMON: a dynamic model for predicting phytoplankton composition and biomass in lowland rivers, Wat Res 35: 901-912
Fabri R, Leclercq L (1984) Étude écologique des rivières du nord du massif Ardennais (Belgique): Flore et végétation de Diatomées et physico-chimie des eaux. Thèse de doctorat. Université de Liège, Liège
Fabri R, Leclercq L (1986) Végétation de diatomées des rivières du nord de l'Ardenne (Belgique): types naturels et impact des pollutions. In: Ricard M (ed) Proceedings of the 8th. International Diatom Symposium, Koeltz Publisher, Koenigstein pp 337-346
Fausch K.D., Lyons J, Karr JR, Angermeier PL (1990) Fish communities as indicators of environmental degradation. Am Fish Soc Sym 8: 123-144
Faush KD, Hawkes CL, Parsons MG (1988) Models that predict the standing crop of stream fish from habitat variables: 1950–85. Gen. Tech. Rep. PNW-GTR-213. U.S. Department of agriculture, Forest service, Pacific north reaserch station, Portland, OR
Fayyad U, Piatetsky-Shapiro G, Smyth P (1996) The KDD process for extracting useful knowledge from volumes of data. Comm ACM 39: 27–34
Fielding AH (1999) An introduction to machine learning methods. In: A. H. Fielding (editor). Machine learning methods for ecological applications. Kluwer Academic, Boston, MA pp 1-35
Fielding AH, Bell JF (1997) A review of methods for the assessment of prediction errors in conservation presence/absence models. Environ Conser 24: 38-49
Fink MH, Moog O, Wimmer R (2000). Fließgewässer-aturräume Österreichs. Monographien des Umweltbundesamtes, Wien
Firth D. (1991) Generalized linear models. In: Hinkley DV, Reid N, Snell EJ (eds) Statistical theory and modelling. Chapman & Hall, London pp 55-82
Fisher RA (1936) The use of multiple measurements in taxonomic problems. Ann Eugeneics 7: 179-188
Fjerdingstad E (1950) The microflora of the river Molleaa with special reference to the relation of benthic algae to pollution. Folia Limnol Scand 5: 1-123

Fjerdingstad E (1964) Pollution of stream estimated by benthal phytomicro-organisms. I. A saproby system based on communities of organisms and ecological factors. Int Revue Gesam Hydrobiol 49: 63-131

Fodor IK, Kamath C (2002) Dimension reduction techniques and the classification of bent double galaxies. Comp Stat Data Analy 41: 91-122

Foged E (1978) Diatoms in Eastern Australia, J Cramer, Berlin

Foody GM (1999) Applications of the self-organizing feature map neural-network in community data-analysis. Ecol Model 120: 97-107

Fore SA, Guttman SI, Bailer AJ, Altfater DJ, Counts BV (1995) Exploratory analysis of population genetic assessment as water quality indicator. Ecotox Environ Safe 30: 36-46

Fox PJA, Naura M, Raven P (1996) Preceding habitat components for semi-natural rivers in the United Kingdom. Proc 2nd Int Symp on Habitat Hydraulics pp 227-238

French M, Recknagel F (1994) Modelling algal blooms in freshwater using artificial neuronal networks. In: Zanetti P (ed) Computer techniques in environmental studies V, vol II. Environmental systems. Computational Mechanics Publications, Southampton, Boston

Friberg N, Johnson RK (eds) (1995) Biological monitoring of streams: methods used in the Nordic countries based on macroinvertebrates. TemaNord 1995

Friedman JH (1987) Exploratory projection pursuit. J Am Stat Ass 82: 249-266

Friedman JH (1997) On bias, variance, o/1-loss and the curse-of-dimensionality. Data Min Know Disco 1: 55-77

Friedman JH, Tukey JW (1974) A projection pursuit algorithm for exploratory data analysis. IEEE Trans Comp C 23(9): 881-890

Friedrich G (1990) Eine revision des saprobiensystems. Zeits Wass Abwass 23: 141-152

Frissell CA, Liss WJ, Warren CE, Hurley MC (1986) A hierarchical framework for stream habitat classification: viewing streams in a watershed context. Environ Manage 10: 199-214

Froese R, Pauly D (eds) (2003) FishBase. World Wide Web electronic publication. http: //www.fishbase.org

Gardarsson A (1979) Vistfrædileg flokkun íslenskra vatna (a classification of Icelandic freshwaters). Týli 9: 1-10

Gardeniers JJP, Tolkamp HH (1976) Hydrobiologische kartering, waadering en schade aan de beekfauna in Achterhoekse beken. In: Nes TVD (ed) Modelonderzoek 71-74, Comm Best Waterhuish Gld pp 26-29

Gardner MW, Dorling SR (2000) Statistical surface ozone models: an improved methodology to account for non-linear behaviour. Atmos Environ 34: 21-34

Garrison GD (1991) Interpreting neural network connection weights. Art Intell Exper 6: 47-51

Gaston KJ (1996) Spatial covariance in the species richness of higher taxa. In: Hochberg ME, Clobert J, Barbault R (eds) Aspects of the genesis and maintenance of biological diversity. Oxford University Press, Oxford pp 221-242

Gauch HG Jr (1982) Multivariate analysis in community ecology, Cambridge University Press, Cambridge

Gehrke PC, Astles KL, Harris JH (1999) Within catchment effects of flow alteration on fish assemblages in the Hawkesbury-Nepean River system, Australia. Reg Riv Res Manage 15: 181-198

Geijskes DC, van Tol J (1983) De libellen van Nederland [The Odonata of the Netherlands]. Uitgeverij KNNV, Utrecht, The Netherlands (in Dutch)

Geladi P, Kowalski BR (1986) Partial least squares regression: a tutorial. Anal Chim Acta 185: 1-17

Gell PA (1997) The development of a diatom database for inferring lake salinity, western Victoria, Australia: towards a quantitative approach for reconstructing past climates. Austral J Botany 45: 389-423

Gelman A, Carlin JB, Stern HS, Rubin DB (1995) Bayesian data analysis. Chapman & Hall, London

Geman S, Bienenstock E, Doursat R (1992) Neural networks and the bias/variance dilema. Neural Comput 4: 1-58

Gevrey M, Dimopoulos I, Lek S (2003) Review and comparison of methods to study the contribution of variables in artificial neural network models. Ecol Model 160: 249-264

Ghetti PF (1997) Indice biotico esteso (IBE). I macroinvertebrati nel controllo della qualità degli ambienti di acque correnti. Manuale di applicazione. Provincia Autonoma di Trento, Agenzia Provinciale per la Protezione dell'Ambiente, 222

Gillman M, Hails R (1997) An introduction to ecological modelling: putting practice into theory. Blackwell Science, Oxford

Giraudel JL, Aurelle D, Berrebi P, Lek S (2000) Application of the self-organizing mapping and fuzzy clustering microsatellite data: how to detect genetic structure in brown trout (*Salmo trutta*) populations. In: Lek S, Guegan JF (eds) Artificial neural networks: application to ecology and evolution. Springer, Berlin pp 187-202

Giraudel JL, Lek S (2001) A comparison of self-organizing map algorithm and some conventional statistical methods for ecological community ordination. Ecol Model 146: 329-339

Giske J, Huse G, Fiksen O (1998) Modelling spatial dynamics of fish. Rev Fish Biol Fish 8: 57–91

Gittenberger E, Janssen AW, Kuijper WJ, Kuiper JGJ, Meijer T, van der Velde G, de Vries JN (1998) De Nederlandse zoetwatermollusken. Recente en fossiele weekdieren uit zoet en brak water [The Dutch freshwater molluscs. Recent and fossil molluscs from fresh and brackish water]. Nederlandse Fauna 2. KNNV Uitgeverij, EIS-Nederland and Nationaal Natuurhistorisch Museum, Leiden, The Netherlands

Giudicelli J, Bouzidi A, Ait Abdelaali N (2000) Contribution à l'étude faunistique et écologique des simulies (Diptera: Simuliidae) du Maroc. IV Les simulies du Haut Atlas. Description d'une nouvelle espèce. Ann Limnol 36: 57-80

Glansdorff P, Prigogine I (1971) Thermodynamic theory of structure, stability and fluctuations. Wiley, New York

Gledhill T, Sutcliffe DW, Williams WD (1993) British freshwater crustacea Malacostraca: a key with ecological notes. Freshwater Biological Association vol 52

Goethals P, Dedecker A, Gabriels W, de Pauw N (2003) Development and application of predictive river ecosystem models based on classification trees and artificial neural networks. In: Recknagel F (ed) Ecological informatics: understanding ecology by biologically-inspired computation. Springer-Verlag, New York pp 91-107

Goethals P, Gasparyan K, de Pauw N (2001) River restoration simulations by ecosystem models predicting aquatic macroinvertebrate communities based on J48 classification trees. Meded Facul Land Toeg Biol Weten 66: 213-217

Goethals PLM, de Pauw N (2001) Development of a concept for integrated ecological river assessment in Flanders, Belgium. J Limnol 60: 7-16

Goldberg DE, Holland JH (1988) Genetic algorithms and machine learning. Machine Learn 3: 95-99

Gomà J, Ortiz R, Cambra J, Ector L (2004) Water quality evaluation in Catalonian Mediterranean rivers using epilithic diatoms as bioindicators. Vie Milieu 54 (in press)

Gorman OT, Karr JR (1978) Habitat structure and stream fish community. Ecology 59: 507-515

Gosselain V, Coste M, Campeau S, Ector L, Fauville C, Delmas F, Knoflacher M, Licursi M, Pfister P, Rimet F, Tison J, Tudesque L, Gevrey M, Park YS, Lek S, Descy JP (2004) A large-scale multi-regional diatom database: Structure, applications and feedback on methodologies. Hydrobiologia (in press)

Gouraud V, Baglignière JL, Baran P, Sabaton C, Lim P, Ombredane D (2001) Factors regulating brown trout populations in two french rivers: application of a dynamic population model. Reg Riv Res Manage 17: 557-569

Gouraud V, Sabaton C, Baran P, Lim P (1999) Dynamics of a population of brown trout (*Salmo trutta*) and fluctuations in physical habitat conditions -experiments in a stream in the Pyrenees; first results. In: Cowx IG (ed) Management and ecology of river fisheries. Blackwell Publishing, London pp 126-142

Gower JC (1966) Some distance properties of latent root and vector methods used in multivariate analysis. Biometrika 53: 325-338

Gozlan RE, Mastrorillo S, Dauba F, Tourenq JN, Copp GH (1998) Multi-scale analysis of habitat use during late summer for 0+ fishes in the River Garonne (France). Aqua Sci 60: 99–117

Green RE (1996) Factors affecting the population density of the corncrake *Crex crex* in Britain and Ireland. J Appl Ecol 33: 237-248

Grenouillet G, Pont D, Hérissé C (2004) Within-basin fish assemblage structure: the relative influence of habitat versus stream spatial position on local species richness. Can J Fish Aqua Sci 61: 93-102

Gross L, Thiria S, Frouin R (1999) Applying artificial neural network methodology to ocean color remote sensing. Ecol Model 120: 237-246

Grossberg S (1969) On the production and release of chemical transmitters and related topics in the cellular control. J Theor Biol 22: 325-364

Grossberg S (1982) Studies of mind and brain: neural principals of learning, perception, development, cognition, and motor control. Reidel Press, Boston

Grossman G, Ratajczac RE, Crawford M, Freeman MC (1998) Assemblage organization in stream fishes: effects of environmental variation and interspecific interactions. Ecol Monog 68: 395-420

Grossman GD, Freeman MC, Moyle PB, Whittaker JO (1985) Stochasticity and assemblage organization in an Indiana stream fish assemblage. Am Natur 126: 275-285

Growns IO, Growns JE (2001) Ecological effects of flow regulation on macroinvertebrate and periphytic diatom assemblages in the Hawkesbury-Nepean River, Australia. Reg Riv Res Manage 17: 275-293

Guegan JF, Lek S, Oberdorff T (1998) Energy availability and habitat heterogeneity predict global riverine fish diversity. Nature 391: 382–384

Gunn S. (1998) Support vector machines for classification and regression. ISIS Technical Report, Image-Speech & Intelligent Systems Group, University of Southampton

Guo Q, Wu W, Massart DL, Boucon C, de Jong S (2000) Sequential projection pursuit using genetic algorithms for data mining of analytical data. Anal Chem 72(13): 2846-2855

Györgyi G (1990) Inference of a rule by a neural network with thermal noise. Phys Rev Lett 64: 2957-2960

Ha K, Cho EA, Kim HW, Joo GJ (1999) *Microcystis* bloom formation in the lower Nakdong River, South Korea: importance of hydrodynamics and nutrient loading. Mar Fresh Res 50: 89-94

Ha K, Jang MH, Joo GJ (2002) Spatial and temporal dynamics of phytoplankton communities along a regulated river system, the Nakdong River, Korea. Hydrobiologia 470: 235-245

Ha K, Kim HW, Jeong KS, Joo GJ (2000) Vertical distribution of *Microcystis* population in the regulated Nakdong River (S. Korea). Limnol 1: 225-230

Ha K, Kim HW, Joo GJ (1998) The phytoplankton succession in the lower part of hypertrophic Nakdong River (Mulgum), South Korea. Hydrobiologia 369/370: 217-227

Hagan MT, Demuth HB, Beale M (1996) Neural network design. PWS Publishing Company, Boston

Håkanson L, Boulion VV (2003) A general dynamic model to predict biomass and production of phytoplankton in lakes. Ecol Model 165: 285-301

Håkansson S (1993) Numerical methods for the inference of pH variations in mesotrophic and eutrophic lakes in Southern Sweden: A progress report. Diatom Res 8 (2): 349-370

Hall DJ, Ball GB (1965) ISODATA: a novel method of data analysis and pattern classification. Technical report, Stanford Research Institute, Menlo Park, CA

Hammond T (1997) A Bayesian interpretation of target strength data from the Grand Banks. Can J Fish Aquat Sci 54: 2323-2333

Harding JPC, Kelly MG (1999) Recent developments in algal-based monitoring in the United Kingdom. In: Prygiel J, Whitton BA, Bukowska J (eds) Use of algae for monitoring rivers III, Agence de l'Eau Artois-Picardie, Douai pp 26-34

Harding JS, Benfield EF, Bolstad PV, Helfman GS, Jones EBD (1998) Stream biodiversity: the ghost of land use past. Proc Nat Acad Sci US 95: 14843-14847

Harris GP (1986) Phytoplankton ecology: structure, function and fluctuation. Chapman & Hall, New York

Harris JN (1995) The use of fish in ecological assessments. Austral J Ecol 20: 65-80

Hart BT, Maher B, Lawrence I (1999) New generation water quality guidelines for ecosystem protection. Fresh Biol 41: 347-359

Hart DD, Robinson CT (1990) Resource limitation in a stream community: Phosphorous enrichment effects on periphyton and grazers. Ecology 71: 1494-1502

Hastie JT, Tibshirani RJ (1986) Generalized additive models. Stat Sci 1: 297-318

Hastie T, Tibshirani R (1990) Generalized additive models. Chapman & Hall, New York

Hastie T, Tibshirani R, Friedman J (2001) The elements of statistical learning. Springer, New York

Hawkes HA (1975) River zonation and classification. In: Whitton BA (ed) River ecology. University of California Press, Berkeley, California pp 312-374

Hawkes HA (1979) Invertebrates as indicators of river water quality. In: James A, Evison L (eds) Biological indicators of water quality. John Wiley and Sons, Chichester, UK pp 2.1-2.45

Hawkins CP, Norris RH, Gerritsen J, Hughes RM, Jackson SK, Johnson RK, Stevenson RJ (2000a) Evaluation of the use of landscape classifications for the prediction of freshwater biota: synthesis and recommendations. J N Am Benthol Soc 19: 541-556

Hawkins CP, Norris RH, Hogue JN, Feminella JW (2000b) Development and evaluation of predictive models for measuring the biological integrity of streams. Ecol Appl 10: 1456-1477

Haykin S (1994) Neural networks. Macmillian College Publishing Company, New York

Haykin S (1999) Neural networks: a comprehensiv foundation. Prentice Hall, New Jersey

Hecht-Nielsen R (1987) Counterpropagation networks. In: Proc Int Conf on Neural Networks II. IEEE Press, New York pp 19-32

Hecht-Nielsen R (1990) Neurocomputing. Addison-Wesley, Reading, UK

Heiland I (1990) Partial least squares regression and statistical models, Scand J Statis 17: 97-114

Heino J (2002) Concordance of species richness patterns among multiple freshwater taxa: a regional perspective. Biodiv Conser 11: 137-147

Hellawell JM (1978) Biological surveillance of rivers. Water Research Center. Stevenage Laboratory, England

Hellawell JM (1986) Biological indicators of freshwater pollution and environmental management. Elsevier, London

Helu SL, Sampson DB, Yin Y (2000) Application of statistical model selection criterion to the stock synthesis assessment program. Can J Fish Aqua Sci 57: 1784-1793

Henderson HF, Welcomme RL (1974) The relationship of yield to morpho-edaphic index and number of fishermen in African inland fisheries. FAO/CIFA Occasional Paper no 1, FOA, Rome

Hendry K, Cragg-Hine D, O'Grady M, Sambrook H, Stephen A (2003) Management of habitat for rehabilitation and enhancement of salmonid stocks. Fish Res 62: 171-192

Henrikson L, Medin M (1986) Biologisk bedömning av försurningspåverkan på Lelångens tillflöden och grundområden 1986. Aquaekologerna, Rapport till länsstyrelsen i Älvsborgs län

Hering D, Buffagni A, Moog O, Sandin L, Sommerhäuser M, Stubauer I, Feld C, Johnson RK, Pinto P, Skoulikidis N, Verdonschot PFM, Zahrádková S (2003) The development of a system to assess the ecological quality of streams based on macroinvertebrates – design of the sampling programme within the AQEM project. Internat. Rev. Hydrobiol. 88: 345-361

Hershey AE, Hitner AL, Hullar MAJ, Miller MC, Vestal JR, Lock MA, Rundle S, Peterson BJ (1988) Nutrient influence on a stream grazer: orthocladius microcommunities respond to nutrient input. Ecology 69: 1383-1392

Hickley P, Tompkins H (1998) Recreational fisheries: social, economical and management aspects. FAO, Fishing News Books, Cornwall

Hilborn R, Liermann M (1998) Standing on the shoulders of giants: learning from experience in fisheries. Rev Fish Boi Fish 8: 273-283

Hill BH, Herlihy AT, Kaufmann PR, Stevenson RJ, McCormick FH, Burch Johnson C (2000) Use of periphyton assemblage data as an index of biotic integrity. J N Am Benthol Soc 19: 50-67

Hill BH, Stevenson RJ, Pan Y (2001) Comparison of correlations between environmental characteristics and stream diatom assemblages characterized at genus and species levels. J N Am Benthol Soc 20: 299-310

Hill MO (1979a) DECORANA: a FORTRAN program for detrended correspondence analysis and reciprocal averaging. Ecology and Systematics, Cornell University. Ithaca, New York

Hill MO (1979b) TWINSPAN: a FORTRAN program for arranging multivariate data in an ordered two-way table by classification of the individuals and attributes. Cornell University, Dept. of Ecology and Systematics, Ithaca, New York

Hill MO, Gauch HG (1980) Detrended correspondence analysis, an improved ordination technique. Vegetation 42: 47-58

Hoagland KD, Roemer SC, Rosowski JR (1982) Colonization and community structure of two periphyton assemblages, with emphasis on the diatoms (Bacillariophyceae). Am J Botany 69: 188-213

Hoang H, Recknagel F, Marshall J, Choy S (2001) Predictive modelling of macroinvertebrate assemblages for stream habitat assessments in Queensland (Australia). Ecol Model 195: 195-206

Hoffmann L (1994) Biogeography of marine blue-green algae. Algol Stud 75: 137-148

Hoffmann L (1996) Geographical distribution of freshwater blue-green algae. Hydrobiologia 336: 33-40

Hofmann G (1994) Aufwuchs-diatomeen in seen und ihre eignung als indikatoren der trophie. Biblio Diatom 30: 1-241

Hoijtink H (1998) Constrained latent class analysis using the Gibbs sampler and posterior predictive p-values: Applications to educational testing. Statistica Sinica 8: 691-712

Hoijtink H (2001) Confirmatory latent class analysis: model selection using Bayes factors and (pseudo) likelihood ratio statistics. Multiv Behav Res 36: 563-588

Hoijtink H, Molenaar IW (1997) A multidimensional item response model: Constrained latent class analysis using the Gibbs sampler and posterior predictive checks. Psychometrika 62: 171-190

Holland JH (1975) Adaptation in natural and artificial systems. Addison-Wesley, New York

Holler C, Borgemeister C, Haardt H, Powell W (1993) The relationship between primary paraitoids and hyperparasitoids of cereal aphids: An analysis of field data. J Anim Ecol 62: 12-21

Holmes NTH, Whitton BA (1981) Phytoplankton of four rivers, the Tyne, Wear, Tees and Swale. Hydrobiologia 80: 111-127

Hornik K, Stinchcombe M, White H (1989) Multilayer feed forward neural networks are universal approximators. Neural Networks 2: 359-366

Horwitz RJ (1978) Temporal variability patterns and the distributional patterns of stream fishes. Ecol Monog 48: 307-321

Höskuldsson A (1988) PLS regression methods. J Chemom 2: 211-228

Hosmer DW, Lemeshow S (1989) Applied logistic regression. Wiley, New York

Hotelling H (1933) Analysis of a complex of statistical variables into principal components. J Edu Psych 24: 417-441

Huber PJ (1985) Projection pursuit (with discussion). Ann Stat 13(2): 435-525

Huet M (1949) Aperçu des relations entre la pente et lespopulations piscicoles des eaux courantes. Schweiz Z Hydrol 11: 331-351

Huet M (1954) Bioloie, profiles en long et en travers des eaux courantes. Bull Fr Piscic 175: 41-53

Huet M (1959) Profiles and biology of Western European streams as related to fisheries management. Trans Am Fish Soc 88: 155-163

Hughes RM (1995) Defining acceptable biological status by comparing with reference conditions. In: Davis WS, Simon TP (eds) Biological assessment and criteria: tools for water resource planning and decision making. Lewis Press, Boca Raton pp 31-47

Hughes RM, Gammon JR (1987) Longitudinal changes in fish assemblages and water quality in the Willamette River, Oregon. Trans Am Fish Soc 116: 196-209

Hughes RM, Larsen DP, Omernik JM (1986) Regional reference sites: a method for assessing stream potentials. Environ Manage 10: 629-635

Hughes RM, Oberdorff T (1999) Applications of IBI concepts and metrics to waters outside the United States and Canada. In: Simon TP (ed) Assessment approaches for estimating biological integrity using fish assemblages. Lewis Press, Boca Raton, FL, USA pp 79-83

Hughes RM, Rexstad E, Bond CE (1987) The relationship of aquatic ecoregions, river basins, and physiogeographic provinces to the ichthyogeographic regions of Oregon. Copeia 2: 423-432

Hugueny B, Paugy D (1995) Unsaturated fish communities in African rivers. Am Natur 146: 162-169

Huntley B (1999) Species distribution and environmental change. Ecosystem management: questions for science and society. In: Maltby E, Holdgate M, Acreman M, Weir A (eds) Royal Holloway Institute for Environmental Research, University of London, London. pp 115-129

Hürlimann J, Elber F, Niederberger K (1999) Use of algae for monitoring rivers: an overview of the current situation and recent developments in Switzerland. In: Prygiel J, Whitton BA, Bukowska J (eds) Use of algae for monitoring rivers III, Agence de l'Eau Artois-Picardie, Douai pp 39-56

Hürlimann, J, Niederhauser, P (2002) Méthode d'étude et d'appréciation de l'état de santé des cours d'eau: Diatomées - niveau R (région). OFEFP, Berne

Huston M (1979) A general hypothesis on species diversity. Am Natur 113: 81-101

Hutagalung RA (1998) Evolution du peuplement piscicole de la Garonne à Toulouse dans un environnement enthropisé: analyses biologique et écologique. Thèse Doctorale, Institut National Polythecnique, Ecole National Supérieure Agronomique, Toulouse

Hutagalung RA, Lim P, Belaud A, Lagarrigue T (1997) Effets globaux d'une agglomération sur la typologie ichtyenne d'un fleuve: cas de la Garonne à Toulouse (France). Ann Limnol 33: 263-279

Hutchinson GE (1957) Concluding remarks. Cold Spring Harbor Symposia on Quantitative Biology 22: 415-427

Hynes HBN (1960) The biology of polluted waters. Liverpool Univ Press, London

IGN (Institut Géographique National) (2003) Site internet de l'Institut Géographique National, http: //www.ign.fr. Version du 22 août 2003

Illies J (1961) Versuch einer allgemeinen biozönotischen Glienderung der Fliessgewässer. Int Revue Gesam Hydrobiol 46: 205-213

Illies J (1978) Limnofauna Europaea. Eine zusammenstellung aller die europäischen Binnengewässer bewohennden mehrzellingen Tierart mit Angaben irhe Verbreitung und ökologi. G Fisher Verlag, Stuttgart

Illies J, Botosaneanu L (1963) Problèmes et méthodes de classification et de la zonation écologique des eaux courantes, considérées surtout du point de vue faunistique. Int Ver Theor Ange Limnol 12: 1-57

Iserentant R, Ector L, Straub F, Hernández-Becerril DU (1999) Méthodes et techniques de préparation des échantillons de diatomées. Cryptog Algol 20: 143-148

Ivorra N (2000) Metal indiced succession in benthic diatom consortia. Univ Amsterdam. Amsterdam

Jaccard P (1908) Nouvelles recherches sur la distribution florale. Bull Soc Vaudoise Sc Nat 44: 223-270

Jackson DA, Harvey HH (1989) Biogeographic associations in fish assemblages: local vs regional processes. Ecology 70: 1472-1484

Jackson DA, Peres-Neto PR, Olden JD (2001) What controls who is where in freshwater fish communities: the roles of biotic, abiotic, and spatial factors. Can J Fish Aqua Sci 58: 157-170

Jain AK, Dube RC, Chen C (1987) Bootstrap techniques for error estimation, TEEE Trans Patt Anal Mach Intell PAMI 9: 628-633

Jain AK, Dubes RC (1988) Algorithms for clustering data. Prentice Hall, Englewood Cliffs, New Jersey

James A, Evison L (1979) Biological indicators of water quality. John Wiley, Chichester

James A, Pitchford JW, Brindley J (2003) The relationship between plankton blooms, the hatching of fish larvae, and recruitment. Ecol Model 160: 77-90

James FC, MacCulloch CE (1990) Multivariate analysis in ecology and systematics. panacea or Pandora's box? Ann Rev Ecol Syst 21: 129-166

Jang MH (2002) Ecological study of freshwater fish in Korea: fish fauna, Prey-predator interaction and the response of cyanobacteria to fish grazing. PhD dissertation, Pusan National Univ, Busan

Jarmulak J, Craw S, Rowe R (2000) Genetic algorithms to optimise CBR retrieval. In: Blanzieri E, Portinale L (eds) EWCBR 2000, Lecture notes in artificial intelligence. Springer-Verlag, Berlin pp 136-147

Jarre-Teschmann A, Brey T, Halto H (1995) Exploring the use of neural networks for biomass forecasts in the Peruvian upwelling ecosystem. Naga. The ICLARM Quartely Review 18: 38-40

Jenerette GD, Lee J, Waller DW, Carlson RE (2002) Multivariate analysis of the ecoregion delineation for aquatic systems. Environ Manage 29: 67-75

Jensen FV (1996) An introduction to Bayesian networks. Springer-Verlag, New York

Jeong KS, Joo GJ, Kim HW, Ha K, Recknagel F (2001) Prediction and elucidation of algal dynamics in the Nakdong River (Korea) by means of a recurrent artificial neural network. Ecol Model 146: 115-129

Jeong KS, Kim DK, Whigham P, Joo GJ (2003b) Modelling *Microcystis aeruginosa* bloom dynamics in the Nakdong River by means of evolutionary computation and statistical approach. Ecol Model 161: 63-75

Jeong KS, Recknagel F, Joo GJ (2003a) Prediction and elucidation of population dynamics of a blue-green Alga (*Microcystis aeruginosa*) and diatom (*Stephanodiscus hantzschii*) in the Nakdong River-Reservoir System (South Korea) by a recurrent artificial neural network. In: Recknagel F (ed) Ecological informatics. Springer, Berlin pp 196-213

Johnson RA, Wichern DW (1992) Applied multivariate statistical analysis. Prentice-Hall Inc, Englewood Cliffs

Johnson RK (1998) Classification of Swedish lakes and rivers using benthic macroinvertebrates. In: Wiederholm T (ed) Bakgrundsrapport 2 till bedömningsgrunder för sjöar och vattendrag - biologiska parametrar. Swedish Environmental Protection Agency Report 4921

Jon T, Cass A, Richards LJ (2000) A Bayesian decision analysis to set escapement goals for Fraser River sockeye salmon (*Oncorhynchus nerka*). Can J Fish Aquat Sci 57: 962-979

Jones MC, Sibson R (1987) What is projection pursuit? (with discussion). J Royal Stat Soc Ser A 150: 1-36

Jongman RHG, ter Braak CJF, van Tongeren OFR (eds) (1995) Data analysis in community and landscape ecology. Cambridge University Press, Cambridge

Joo GJ, Kim HW, Ha K (1997) The development of stream ecology and current status in Korea. Kor J Ecol 20: 69-78

Jørgensen SE (1982) A holistic approach to ecological modelling by application of thermodynamics. In: Mitsch WJ, Ragade RK, Bosserman RW, Dillon Jr JA (eds) Energetics and systems. Ann Arbor Press, Ann Arbor, MI

Jørgensen SE (1986) Structural dynamic model. Ecol Model 31: 1-9

Jørgensen SE (1988) Use of models as experimental tools to show that structural changes are accompanied by increased exergy. Ecol Model 41: 117-126

Jørgensen SE (1990) Ecosystem theory, ecological buffer capacity, uncertainty and complexity. Ecol Model 52: 125-133

Jørgensen SE (1997) Integration of ecosystem theories: a pattern. Kluwer Academic, Dordrecht

Jørgensen SE, de Bernardi D (1998) The use of structural dynamic models to explain successes and failures of biomanipulation. Hydrobiologia 359: 1-12

Jørgensen SE, Mejer HF (1977) Ecological buffer capacity. Ecol Model 3: 39-61

Jørgensen SE, Padisak J (1996) Does the intermediate disturbance hypothesis comply with thermodynamics? Hydrobiologia 323: 9-21

Joy MK, Death RG (2002) Predictive modelling of freshwater fish as a biomonitoring tool in New Zealand. Fresh Biol 47: 2261-2275

Joy MK, Death RG (submitted) Assessing biological quality: predicting freshwater fish and macro-crustacean assemblages using habitat selection functions

Joy MK, Henderson IM, Death RG (2000) Diadromy and longitudinal patterns of upstream penetration of freshwater fish in Taranaki, New Zealand. New Zealand J Mar Fres Res 34: 531-543

Kallis G, Butler D (2001) The EU water framework directive: measures and implications. Water Poll 3: 125-142

Kamath C, Musick R (2000) Scalable data mining through fine-grained parallelism: the present and the future. In: Kargupta H, Chan P (eds), Advances in distributed and parallel knowledge discovery. AAAI Press/MIT Press, Cambridge, MA pp 29–77

Kamp-Nielsen L (1978) Modelling the vertical gradients in sedimentary phosphorus fractions. Verh Int Verein Limnol 20: 720-727

Kang DH, Chon TS, Park YS (1995) Monthly changes in benthic macroinvertebrate communities in different saprobities in the Suyong and Soktae streams of the Suyong River. Kor J of Ecol 18: 157-177

Karr JR (1981) Assessment of biological integrity using fish communities. Fisheries 6: 21-27

Karr JR (1991a) Biological integrity: a long-neglected aspect of water resource management. Ecol Appl 1: 66-84

Karr JR (1991b) Ecological integrity: protecting earth's life support systems. In: Costanza R, Norton BG, Haskell BD (eds) Ecosystem health: goals for environmental management. Island Press, California pp 223-238

Karr JR (1995) Protecting aquatic ecosystems: clean water is not enough. In: Davis WS, Thomas TP (eds) Biological assessment and criteria: tools for water resource planning and decision making. CRC Press, Boca Raton pp 7-13

Karr JR (1999) Defining and measuring river health. Fresh Biol 41: 221-234

Karr JR, Chu EW (1999) Restoring life in running waters: better biological monitoring. Island Press, Washington, DC, USA

Karr JR, Fausch KD, Angermeier PL, Yant PR, Schlosser IJ (1986) Assessing biological integrity in running waters: a method and its rationale. Special Publ 5. Illinois Natural History Survey, Urbana

Karul C, Soyupak S, Germen E (1998) A new approach to mathematical water quality modeling in reservoirs: neural networks. Int Rev Hydrobiol 83: 689-696

Kaski S (1997) Data exploration using self-organizing maps. Acta Polytechnica Scandinavica, Mathematics, Computing and Management in Engineering Series No. 82. Finish Academy of Technology, Espoo, Filand

Kass RE, Raftery AE (1995) Bayes factors. J Am Stat Asso 90: 773-795

Kastens TL, Featherstone AM (1996) Feedforward back-propagation neural networks in prediction of farmer risk preference. Am J Agri Econ 78: 400–415

Kawecka B, Kwandrans J, Szyjkowski A (1999) Use of algae for monitoring rivers in Poland - Situation and development. In: Prygiel J, Whitton BA, Bukowska J (eds) Use of algae for monitoring rivers III, Agence de l'Eau Artois-Picardie, Douai pp 57-65

Keiner LE, Yan X-H (1998) A neural network model for estimating sea surface chlorophyll and sediments from Thematic Mapper imagery. Remote Sens Environ 66: 153–165

Keith P (1998) Evolution des peuplements ichtyologiques de France et strategies de conservation. Doctoral Thesis, Biological Sciences, Université de Rennes, France

Keith P (2000) The part played by protected areas in the conservation of threatened French freshwater fish. Biol Conserv 92: 265-273

Keith P, Allardi J (2001) Atlas des poissons d'eau douce de France. Patrimon Nat 47: 1-387

Kelly MG (1998b) Use of community-based indices to monitor eutrophication in rivers. Environ Conserv 25: 22-29

Kelly MG (1998a) Use of the trophic diatom index to monitor eutrophication in rivers. Wat Res 36: 236-242

Kelly MG, Cazaubon A, Coring E, Dell'Uomo A, Ector L, Goldsmith B, Guasch H, Hürlimann J, Jarlman A, Kawecka B, Kwandrans J, Laugaste R, Lindstrom EA, Leitao M, Marvan P, Padisak J, Pipp E, Prygiel J, Rott E, Sabater S, van Dam H, Vizinet J (1998) Recommendations for the routine sampling of diatoms for water quality assessments in Europe. J Appl Phycol 10: 215-224

Kelly MG, Whitton BA (1995) The Trophic Diatom Index: a new index for monitoring eutrophication in rivers. J Appl Phycol 7: 433-444

Kemper T, Sommer S (2002) Estimate of heavy metal contamination in soils after a mining accident using reflectance spectroscopy. Environ Sci Tech 36: 2742-2747

Kenkel NC, Orloci L (1986) Applying metric and nonmetric multidimensional scaling to ecological studies: some new results. Ecology 67: 919-928

Kerans BL, Karr JR (1994) A benthic index of biotic integrity (B-IBI) for rivers of the Tennessee Valley. Ecol Appl 4: 768-785

Kesminas V, Virbickas T (2000) Application of an adapted index of biotic integrity to rivers of Lithuania. Hydrobiologia 422/423: 257-270

Kestemont P, Didier J, Depiereux E, Micha JC (2000) Selecting ichtyological metrics to assess river basin ecological quality. Arch Hydrobiol 121: 321-348

Keukelaar JHD (2002) Topics in soft computing. Royal Institute of Technology, Stockholm

Kim HW, Ha K, Joo GJ (1998) Eutrophication of the lower Nakdong River after the construction of an estuarine dam in 1987. Int Rev Hydrobiol 83: 65-72

Kim HW, Joo GJ, Walz N (2001) Zooplankton dynamics in the hyper-eutrophic Nakdong River system (Korea) regulated by an estuary dam and side channels. Int Rev Hydrobiol 86: 127-143

Kimes DS, Holben BN, Nickeson JE, McKee WA (1996) Extracting forest age in a Pacific Northwest forest from Thematic Mapped and topographic data. Remote Sens Environ 56: 133-140

Kinas PG (1996) Bayesian fishery stock assessment and decision making using adaptive importance sampling. Can J Fish Aquat Sci 53: 414-423

Kiviluoto K (1996) Topology preservation in self-organizing maps. In: Pro. ICNN'96, IEE Int Conf on Neural Networks. IEEE Service Center, Piscataway pp 294-299

Klemm DJ, Lewis PA, Fulk F, Lazorchak JM (1990) Macroinvertebrate field and laboratory methods for evaluating the biological integrity of surface waters. US Environmental Protection Agency, Office of Water, Washington, DC

Klolodner J (1993) Case-based reasoning. Morgan Kaufmann Publishers, San Francisco, CA

Knoben RAE, Roos C, van Oirschot MCM (1995) Biological assessment methods for watercourses. UN/ECE task force on monitoring and assessment, Lelystad 3: 1-86

Kobayasi H, Ishida N (1996) Three diatom species newly found in Japan and their occurence in the Inabe-gawa (Inabe river), and Tama-gawa (Tama river), central Japan. Diatom 12: 27-33

Kociolek JP (2000) Freshwater diatom biogeography. Nova Hedwigia 71: 223-241

Koel TM (1997) Distribution of Fishes in the Red River of the North Basin on Multivariate Environmental Gradients. Ph.D. thesis, North Dakota State University, Fargo, North Dakota. USA

Kohavi R (1995) A study of cross-validation and bootstrap for estimation and model selection. Proc 14th Int Joint Conf on AI, Vol 2, Canada

Kohavi R, John GH (1998) The wrapper approach. In: Liu H, Motoda H (eds) Feature selection for knowledge discovery and data mining. Kluwer Academic, New York pp33-50

Köhler J (1993) Growth, production and losses of phytoplankton in the lowland River Spree. I. population dynamics. J Plankton Res 15: 335-349

Kohonen T (1982) Self-organized formation of topologically correct feature maps. Biol Cybern 43: 59-69

Kohonen T (1989) Self-organization and associative memory. Springer-Verlag. Berlin

Kohonen T (1995) Self-organizing maps. Springer, Berlin

Kohonen T (2001) Self-organizing maps. Springer, Berlin

Kolding J (1994) On the ecology and exploitation of fish in fluctuating tropical freshwater systems. Dsc thesis, Dept of Fisheries and Marine Biology, Univ Bergen, Norway

Kolozsvary MB, Swihart RK (1999) Habitat fragmentation and the distribution of amphibians: patch and landscape correlates in farmland. Can J Zool 77: 1288-1299

Koste W (1978) Rotatoria. Die Radertiere Mitteleuropes. Ein Bestimmungswerk begrunder von Max Voigt. 2nd ed. Borntrager, Stuttgart, Vol. 1, Textband 673 pp, Vol. 2. Tafelband

Kraemer HC (1982) Kappa coefficient. In: Kotz S, Johnson NL (eds) Encyclopedia of statistical sciences. John Wiley & Sons, New York

Krammer K (2000) The genus *Pinnularia*. In: Lange-Bertalot H (ed) Diatoms of Europe, volume 1: Diatoms of the European inland waters and comparable habitats. Gantner Verlag, Ruggell

Krammer K (2003) *Cymbella.* In: Lange-Bertalot H (ed) Diatoms of Europe, volume 3: Diatoms of the European inland waters and comparable habitats. Gantner Verlag, Ruggell

Krammer K, Lange-Bertalot H (1986) Bacillariophyceae 1. Teil: Naviculaceae. In: Ettl H, Gerloff J, Heyning H, Mollenhauer D (eds) Sübwasserflora von Mitteleuropa. Gustav Fisher Verlag, Stuttgart

Krammer K, Lange-Bertalot H (1988) Bacillariophyceae 2. Bacillariaceae, Epithemiaceae, Surirellaceae. Sübwasserflora von Mitteleuropa. In: Ettl H, Gerloff J, Heyning H, Mollenhauer D (eds) Sübwasserflora von Mitteleuropa. Gustav Fisher Verlag, Stuttgart

Krammer K, Lange-Bertalot H (1991a) Bacillariophyceae 3. Centrales, Fragilariaceae, Eunotiaceae. In: Ettl H, Gerloff J, Heyning H, Mollenhauer D (eds) Sübwasserflora von Mitteleuropa. Gustav Fisher Verlag, Stuttgart

Krammer K, Lange-Bertalot H (1991b) Bacillariophyceae 4. Achnanthaceae. Kritische Ergänzungen zu Navicula (Linolatae) und Gomphonema. In: Ettl H, Gerloff J, Heyning H, Mollenhauer D (eds) Sübwasserflora von Mitteleuropa. Gustav Fisher Verlag, Stuttgart

Kreuger J (1995) Monitoring of pesticides in subsurface and surface water within an agricultural catchment in southern Sweden. In: Walker A, Allen R, Bailey SW, Blair AM, Brown CD, Günther P, Leake CR, Nicholls PH (eds) Pesticide movement to water. British Crop Protection Council Monograph No. 62, Farnham, Surrey, UK pp 81-86

Kromkamp J, Walsby AE (1990) A computer model of buoyancy and vertical migration in cyanobacteria. J Plankton Res 12: 161-183

Krstic S, Levkov Z, Stojanovski P (1999) Saprobiological characteristics of diatom microflora in river ecosystems in Macedonia as a parameter for determination of the intensity of anthropogenic influence. In Prygiel J, Whitton BA, Bukowska J (eds) Use of algae for monitoring rivers III, Agence de l'Eau Artois-Picardie, Douai pp 145-153

Kruskal JB, Wish M (1978) Multidimensional scaling. Sage University paper series on Quantitative applications in the social sciences, 07-011. Sage Publications, Newbury Park, CA

Kuikka S, Hilden M, Gislason H, Hansson S, Sparholt H, Varis O (1999) Modeling environmentally driven uncertainties in Baltic cod (*Gadus morhua*) management by Bayesian influence diagrams. Can J Fish Aquat Sci 59: 629-641

Kung SY (1993) Digital neural networks. Prentice Hall, Englewood Cliffs, New Jersey

Kusber WH, Jahn R (2003) Annotated list of diatom names by Horst Lange-Bertalot and co-workers. – Version 3.0. [http: //www.algaterra.org/Names_Version3_0.pdf]

Kwak IS, Liu GC, Chon TS, Park YS (2000) Community patterning of benthic macroinvertebrates in streams of South Korea by utilizing an artificial neural network. Kor J Limnol 33: 230-243

Kwon TS, Chon TS (1991) Ecological studies on benthic macroinvertebrates in the Suyong River II: Investigations on distribution and abundance in its main stream and four tributaries. Kor J Limnol 24: 179-198

Kwon TS, Chon TS (1993) Ecological studies on benthic macroinvertebrates in the Suyong River III: Water quality estimations using chemical and biological indices. Kor J Limnol 26: 105-128

Lack TJ (1971) Quantitative studies on the phytoplankton of the Rivers Thames and Kennet at Reading. Fresh Biol 1: 213-224

Lae R, Lek S, Moreau J (1999) Predicting fish yield of African lakes using neural networks. Ecol Model 120: 325-335

Lagarrigue T, Baran P, Lascaux JM, Delacoste M, Abad N, Lim P (2001) Taille à 3 ans de la truite commune (*Salmo trutta* L.) dans les rivières des Pyrénées françaises : rélations avec les caractéristiques mésologiques et influence des aménagements hydroéléctriques. Bull Fr Pêche Piscicol 357/360: 549-571

Lahr E (1950) Temps et climat au Grand-Duché de Luxembourg. Service de météorologie et hydrographique national, Luxembourg

Lamb MA, Lowe RL (1987) Effects of current velocity on the physical structuring of diatom (Bacillariophyceae) communities. Ohio J Sc 87: 72-78

Lambinon J, de Langhe JE, Delvosalle L, Duvigneaud J (1992) Nouvelle flore de la Belgique, du Grand-Duché de Luxembourg, du Nord de la France et des Régions voisines. Editions du Patrimoine du Jardin botanique national de Belgique, Meise

Lammert M, Allan JD (1999) Assessing biotic integrity of streams: effects of scale in measuring the influence of land use/cover and habitat structure on fish and macroinvertebrates. Environ Manage 23: 257–270

Lamon EC, Clyde MA (2000) Accounting for model uncertainty in prediction of chlorophyll a in Lake Okeechobee. J Agr Bio Env Stat 5: 297-322

Lamouroux N, Capra H (2002) Simple predictions of instream habitat model outputs for target fish populations. Fresh Biol 47: 1543-1556

Lamouroux N, Souchon Y (2002) Simple predictions of instream habitat model outputs for fish habitat guilds in large streams. Fresh Biol 47: 1531-1542

Lange-Bertalot H (1979) Toleranzgrenzen und populationsdynamik benthischer diatomeen bei unterschiedlich starker Abwasserbelastung. Arch Hydrobiol 56: 184-219

Lange-Bertalot H (1993) 85 neue Taxa, und über 100 weitere neu definierte Taxa ergänzend zur Sübwasserflora von Mitteleuropa. Biblio Diatom 27: 1-454

Lange-Bertalot H (2001) *Navicula sensu stricto*, 10 genera separated from *Navicula sensu lato*, *Frustulia*. In: Lange-Bertalot H (ed) Diatoms of Europe Vol 2: Diatoms of the European inland waters and comparable habitats. Gantner Verlag, Frankfurt

Lange-Bertalot H, Genkal SI (1999) Diatomeen aus Sibirien I. Inseln im Arktischen Ozean (Yugorsky-Shar Strait). Iconog Diatom 6: 1-271

Lange-Bertalot H, Krammer K (1989) Achnanthes eine monographie der Gattung. Biblio Diatom 18: 1-393

Lange-Bertalot H, Steindorf A (1996) Rote liste der limnischen Kieselalgen (Bacillariophyceae) Deutschlands. Schr.-R.f. Vegetationskde, 28: 633-677

Larinier M (2001) Environmental issues, dams and fish migration. In: Marmulla G (ed) Dams, fish and fisheries: opportunities, challenges and conflict resolution. FAO Fish Tech Paper pp 45-89

Lasdon L (1970) Optimization theory for large systems. Macmillan, New York

Läuter H, Pincus R (1989) Mathematisch-statistische datenanalyse, volume 73 of Mathematische Monographien. Akademie Verlag, Berlin

Lauters F, Lavandier P, Lim P, Sabaton C, Belaud A (1996) Influence of hydropeaking on invertebrates and their relationship with fish feeding habits in a Pyrenean river. Reg Riv Res Manage 12: 563-573

Le Roch C, Mollard A (1996) Les intruments économiques de réduction de la pollution diffuse en agriculture. Cah Econ Sociol Rur 39/40: 64-92

Leclercq L, Depiereux E (1987) Typologie des rivières oligotrophes du massif Ardennais (Belgique) par l'analyse multivariée de relevés de diatomées benthiques. Hydrobiologia 153: 175-192

Leclercq L, Maquet B (1987a) Deux nouveaux indices chimiques et diatomiques de qualité d'eau courante. Application au Samson et ses affluents (Bassin de la Meuse Belge). Comparaison avec d'autres indices chimiques biocénotiques et diatomiques. Institut Royal des Sciences Naturelles de Belgique. Document de Travail 38: 1-113

Leclercq L, Maquet B (1987b) Deux nouveaux indices diatomique et de qualité chimique des eaux courantes. Comparaison avec différents indices existants. Cahiers de Biologie Marine 28: 303-310

Leclercq L, Vandevenne L (1987) Impact d'un rejet d'eau chargée en sel et d'une pollution organique sur les peuplements de diatomées de la Gander (Grand-Duché du Luxembourg). Cahiers de Biologie Marine 28: 311-317

Lecointe C, Coste M, Prygiel J (1993) "OMNIDIA" software for taxonomy, calculation of diatom indices and inventories management. Hydrobiologia 269/270: 509-513

Lecointe C, Coste M, Prygiel J, Ector L (1999) Le logiciel OMNIDIA version 2, une puissante base de données pour les inventaires de diatomées et pour le calcul des indices diatomiques européens. Cryptogamie Algol 20: 132-134

Lee D, Shepard B, Sanborn B, Ulmer L (1996) Assessing extinction risks for westslope cutthroat trout in the upper Missouri River Basin using BayVAM model. Bul Ecol Soc Am 77: 258

Lee DA, Rieman BE (1997) Population viability assessment of salmonids by using probabilistic networks. N Am J Fish Manage 17: 1144-1157

Legendre P (1987) Constrained clustering. In: Legendre P, Legendre L (eds) Developments in numerical ecology. Springer-Verlag, Berlin. pp 289-307

Legendre P, Anderson JJ (1999) Distance-based redundancy analysis: testing multi-species responses in multi-factorial ecological experiments. Ecol Monog 69: 1-24

Legendre P, Dallot S, Legendre L (1985) Sucession of species within a community: chronological clustering, with applications to marine and freshwater zooplankton. Am Natur 125: 257-288

Legendre P, Gallagher E (2001) Ecologically meaningful transformations for ordination of species data. Oecologia 129: 271-280

Legendre P, Legendre L (1987) Developments in numerical ecology. Springer-Verlag, Berlin

Legendre P, Legendre L (1998) Numerical ecology. Elsevier, Amsterdam

Lek S, Belaud A, Baran P, Dimopoulos I, Delacoste M (1996a) Role of some environmental variables in trout abundance models using neural networks. Aquat Liv Res 9: 23-29

Lek S, Belaud A, Dimopoulos I, Lauga J, Moreau J (1995) Improved estimation, using neural networks, of the food consumption of fish populations. Mar Fresh Res 46: 1229-1236

Lek S, Delacoste M, Baran P, Dimopoulos I, Lauga J, Aulagnier S (1996b) Application of neural networks to modelling nonlinear relationships in ecology. Ecol Model 90: 39-52

Lek S, Giraudel JL, Guegan JF (2000) Neuronal networks: algorithms and architectures for ecologists and evolutionary ecologists. In: Lek S, JF Guegan (eds) Artificial neuronal networks: application to ecology and evolution. Springer-Verlag, Berlin pp. 3-27

Lek S, Guégan JF (1999) Artificial neural networks as a tool in ecological modelling, an introduction. Ecol Model 120: 65-73

Lek S, Guegan JF (eds) (2000) Artificial neuronal networks: application to ecology and evolution. Springer-Verlag, Berlin

Leland HV (1995) Distribution of phytobenthos in the Yakima River basin, Washington, in relation to geology, land use and other environmental factors. Can J Fish Aqua Sci 52: 1108-29

Leland HV, Porter SD (2000) Distribution of benthic algae in the upper Illinois River basin in relation to geology and land use. Fresh Biol 44: 279-301

Lenat DR (1988) Water quality assessment of streams using a qualitative collection method for benthic macroinvertebrates. J N Am Benthol Soc 7: 222-233

Lenoir A, Coste M (1996) Development of a practical diatom index of overall water quality applicable to the French national water Board network. Whitton BA, Rott E (des) Use of Algae for monitoring rivers II. Innsbruck Austria 17-19 Sept. 95: Studia Student. G.m.b.H, Innsbruck pp 29-43

Leopold LB, Wolman MG, Miller JP (1964) Fluvial processes in geomorphology. Freeman, San Francisco

Lévêque C (1995) L'habitat : être au bon endroit au bon moment? Bull Fr Pêche Pisci 337/338/339: 9-20

Lévêque C (1999) Etat de santé des écosystèmès aquatiques: l'intérêt des variables biologiques. In: GIP Hydrosystèmes (ed) Indicateurs de l'état de santé écologique des hydrosystèmes: Résultats du programme "Variables biologiques", 17 mai 1999, Paris. GIP Hydrosystèmes & Ministère de l'aménagement du territoire et de l'environnement pp 13-26

Levin SA (1992) The problem of pattern and scale in ecology. Ecology 73: 1943-1967

Lewis DM, Elliott JA, Lambert MF, Reynolds CS (2002) The simulation of an Australian reservoir using a phytoplankton community model: PROTECH. Ecol Model 150: 107-116

Leynaud G, Trocherie F (1980) Effets toxiques des pollutions sur la faune piscicole. In: Pesson P (ed) La Pollution des eaux continentales: incidence sur les biocénoses aquatiques. Gauthier-Villars, Paris pp 147-169

LfU BW (Landesanstalt für Umweltschutz Baden-Württemberg, ed) (1998) Regionale Bachtypen in Baden-Württemberg. Arbeitsweisen und exemplarische Ergebnisse an Keuper- und Gneisbächen. (= Handbuch Wasser 41). Karlsruhe: 1-276

Licursi M, Gomez N (2002) Benthic diatoms and some environmental conditions in three lowland streams. Ann Limnol 38: 109-118

Liebmann H (1962) Handbuch der Frischwasser und Abwasserbiologie. Band I, R. Oldenburg, Munich

Liermann M, Hilborn R (1997) Depensation in fish stocks: a hierarchic Bayesian meta-analysis. Can J Fish Aqua Sci 54: 1976-1984

Liess M, Schulz R (1999) Linking insecticide contamination and population response in an agricultural stream. Environ Tox Chem 18: 1948-1955

Liess M, Schulz R, Liess MH-D, Rother B, Kreuzig R (1999) Determination of insecticide contamination in agricultural headwater streams. Wat Res 33: 239-247

Lilliefors HW (1967) The Komogorov-Smirnov test for normality with mean and variance unknown. J Am Statist Assoc 62: 399-402

Lim P, Belaud A, Labat R (1985) Peuplement piscicole de la Garonne entre St Gaudens et Agen. Ichthyophys Acta 9: 187-201

Lin C-T, Lee CSG (1996). Neural fuzzy systems. Prentice Hall PTR, Upper Saddle River, New Jersey

Lindgren F, Geladi P, Ränner S, Wold S (1994) Interactive variable selection (Ivs) for PLS. I: theory and algorithms. J Chenometr 8: 349-363

Lippmann RP (1987) An Introduction to computing with neural nets. IEEE ASSP Magazine, April. pp 4-22

Lotka AJ (1956) Elements of mathematical biology. Dover Books, New York

Lowe RL (1974) Environmental requirements and pollution tolerance of freshwater diatoms. EPA-670/7-74-005. US Environmental Protection Agency, Cincinnati, Ohio

Lowe RL, LaLiberte GD (1996) Benthic stream algae: distribution and structure. In: Lamberti G, Hauer FR (eds) Stream ecology: field and laboratory exercises. Academic Press, Oxford pp 269-293

Lowe RL, Pan Y (1996) Benthic algal communities as biological monitors. In: Stevenson RJ, Bothwell ML, Lowe RL (eds) Algal ecology of freshwater benthic ecosystems. Academis Press, Boston pp 705-739

LUA NW (Landesamweltamt Nordrhein-Westfalen, ed) (1999a) Referenzgewässer der Fließgewässertypen Nordrhein-Westfalens. Teil I: Kleine bis mittelgroße Fließgewässer. - (= LUA-Merkblätter 16). Düsseldorf

LUA NW (Landesumweltamt Nordrhein-Westfalen, ed) (1999b) Leitbilder für kleine bis mittelgroße Fließgewässer in Nordrhein-Westfalen. Gewässerlandschaften und Fließgewässertypen. - (= LUA-Merkblätter 17). Düsseldorf

Ludwig JA, Reynolds JF (1988) Statistical ecology: a primer of methods and computing. John Wiley and Sons, New York

Macan TT (1961) A review of running water studies. Ver Intt Verein Limnol 14: 587-602

MacCune B, Mefford MS (1992) PcOrd multivariate analysis of ecological data, version 2.0. MjM Software Design, Gleneden Beach, Oregon

MacGillivray K (2000) The mires of Southeast South Island: a comparison of predictive modelling methods. University of Otago, Dunedin

MacNeil C, Dick JTA, Bigsby E, Elwood RW, Montgomery WI, Gibbins CN, Kelly DW (2002) The validity of the Gammarus: Asellus ratio as an index of organic pollution: abiotic and biotic influences. Wat Res 36: 75-84

MacQueen J (1965) On convergence of k-means and partitions with minimum average variance. Ann Math Stat 36: 1084

Magnuson JJ, Tonn WM, Banerjee A, Toivonen J, Sanchez O, Rask M (1998) Isolation vs extinction in the assembly of fishes in small northern lakes. Ecology 79: 2941-2956

Mahon R (1984) Divergent structure in fish taxocenes of north temperate streams. Can J Fish Aqua Sci 41: 330-350

Maier HR, Dandy GC (2000) Neural networks for the prediction and forecasting of water resources variables: a review of modelling issues and applications. Environ Model Soft 15: 101-124

Maier HR, Dandy GC, Burch MD (1998) Use of artificial neural networks for modelling cyanobacterial *Anabaena* spp. in the River Murray, South Australia. Ecol Model 105: 257-272

Maier HR, Sayed T, Lenc BJ (2001) Forecasting cyanobacterium Anabaena spp in the River Murray, South Australia, using B-spline neurofuzzy models. Ecol Modelling 146: 85-96

Maitland PS (1966) Studies on Loch Lomond. 2. The fauna of the River Endrick. Blackie and Son, Glasgow

Maitland PS (1995) The conservation of freshwater fish: past and present experience. Biol Conserv 72: 259-270

Malmqvist B, Hoffsten P-O (1999) Influence of drainage from old mine deposits on benthic macroinvertebrate communities in central Swedish streams. Wat Res 33: 2415-2423

Malmqvist B, Otto C (1987) The influence of substrate stability on the composition of stream benthos: an experimental study. Oikos 48: 33-38

Manel S, Dias J, Ormerod SJ (1999) Comparing discriminant analysis, neural networks and logistic regression for predicting species distributions: a case study with a Himalayan river bird. Ecol Model 120: 337-347

Manel S, Wililiams HC, Ormerod SJ (2001) Evaluating presence-absence models in ecology: the need to account for prevalence. J Appl Ecol 38: 921-931

Mangiameli P, Chen SK, West D (1996) A comparison of SOM neural network and hierarchical clustering mehods. Eur J Oper Res 93: 402-417

Mann S, Benwell GL (1996) The integration of ecological, neural and spatial modelling for monitoring and prediction for semi-arid landscapes. Comput Geosci 22: 1003-1012

Manté C, Dauvin JC, Durbec JP (1995) Statistical method for selecting representative species in multivariate analysis of long-term changes of marine communities. Applications to a macrobenthic community from the Bay of Morlaix. Mar Ecol Prog Ser 120: 243-250

Mantel N (1967) The detection of desease clustering and a generalized regression approach. Cancer Res 27: 209-220

Marchant R (2002) Do rare species have any place in multivariate analysis for bioassessment? J N Am Benthol Soc 21: 311-313

Marchant R, Hirst A, Norris R, Metzeling L (1999) Classification of macroinvertebrate communities across drainage basins in Victoria, Australia: consequences of sampling on broad spatial scale for predictive modelling. Fresh Biol 41: 253-268

Margalef R (1958) Information theory in ecology. General Systems 3: 36-71

Margalef R (1968) Perspectives in ecological theory. University of Chicago Press, Chicago

Marshall BE (1984) Towards predicting ecology and fish yields in African reservoirs from pre-impoundment physico-chemical data. FAO/CIFA Technical Paper 12

Mastrorillo S, Dauba F, Oberdorff T, Guegan JF, Lek S (1998) Predicting local fish species richness in the Garonne river basin. C R Acad Sci Paris, Life Science, 321: 423-428

Mastrorillo S, Lek S, Dauba F (1997a) Predicting the abundance of minnow *Phoxinus phoxinus* (Cyprinidae) in the River Ariege (France) using artificial neural networks. Aqua Liv Resour 10: 169-176

Mastrorillo S, Lek S, Dauba F, Belaud A (1997b) The use of artificial neural networks to predict the presence of small-bodied fish in a river. Fresh Biol 38: 237-246

MathSoft (1997) S-Plus programmer's guide. Mathsoft, Inc. Seattle, Washington

MathWorks Inc. (1998) MATLAB Version 5.3, Natik, Massachusetts

Matthews EM, Matthews WJ (2000) Geographic, terrestrial and aquatic factors: which most influence the structure of stream fish assemblages in the midwestern United States? Ecol Fresh Fish 9: 9-21

Matthews WJ (1985) Distribution of midwestern fishes on multivariate environmental gradients, with emphasis on Notropis lutrensis. Am Natur 113: 225-237

Matthews WJ (1998) Patterns in freshwater fish ecology. Chapman & Hall, New York

Matthews WJ, Robinson HW (1988) The distribution of the fish of Arkansas: a multivariate analysis. Copeia 1988: 358-374

Mau B, Newton MA, Larget B (1999) Bayesian phylogenetic inference via Markov chain Monte Carlo methods. Biometrics 55: 1-12

Mc Intire CD (1966) Some effects of current velocity on periphyton communities in laboratory streams. Hydrobiologia 27: 559-570

McAllister MK, Ianelli JN (1997) Bayesian stock assessment using catch-age data and the sampling importance resample algorithm. Can J Fish Aqua Sci 54: 284-300

McAllister MK, Kirkwood GP (1998a) Bayesian stock assessment: a review and example application using the logistic model. ICES J Mar Sci 55: 1031-1060

McAllister MK, Kirkwood GP (1998b) Using Bayesian decision analysis to help achieve a precautionary approach for developing fisheries. Can J Fish Aqua Sci 55: 2642-2661

McAllister MK, Pikitch EK (1997) A Bayesian approach to choosing a design for surveying fishery resources: application to the eastern Bering Sea trawl survey. Can J Fish Aqua Sci 54: 301-311

McArthur RH, Wilson EO (1963) An equilibrium theory of insular zoogeography. Evolution 17: 373-387

McArthur RH, Wilson EO (1967) The theory of island biogeography. Princeton University Press, Princeton

McCormick FH, Peck DV, Larsen DP (2000) Comparison of geographic classification schemes for mid-Atlantic stream fish assemblages. J N Am Benthol Soc 19: 385–404

McCormick PV, Cairns J (1994) Algal as Indicators of Environmental Change. J Appl Phycol 6: 509-526

McCormick PV, Stevenson RJ (1998) Periphyton as a tool for ecological assessment and management in the Florida everglades. J Phycol 34: 726-733

McCullagh P, Nelder JA (1990) Generalized linear models. Chapman & Hall, London

McCullagh P, Nelder JA (1994) Generalized linear models. Chapman & Hall, London

McCune B, Mefford MJ (1999) PC-ORD: Multivariate analysis of ecological data, version 4. MjM Software Design, Gleneden Beach, Oregon

McDowall RM (1990) New Zealand freshwater fishes: a natural history and guide. Heinemann Reed, Auckland

McDowall RM, Taylor MJ (2000) Environmental indicators of habitat quality in a migratory freshwater fish fauna. Environ Manage 25: 357-374

Meador MR, Cuffney TF, Gurtz ME (1993) Methods for sampling fish communities as part of the national water-quality assessment program. U.S. Geological Survey Open-File Report 93–104

Medsker LR (1996) Microcomputer applications of hybrid intelligent systems. J Netw Comput Appl 19: 213-234

Mejer HF, Jørgensen SE (1979) Energy and ecological buffer capacity. In: Jørgensen SE (ed) State-of-the-art of ecological modelling. Int Soc Ecol Model, Copenhagen pp 829-846

Mercer, J. 1909. Functions of positive and negative type and their connection with the theory of integral equations. Philos Trans Roy Soc London A 209: 415-446

Merino V, Garcia J, Hernandez-Marine M (1995) Use of diatoms for pollution monitoring in the Valira basin (Andorra). In: Marino D, Montresor M (eds) Proc 13th Int Diatom Symp 1994. Biopress limited, Bristol pp 107-119

Metcalfe JL (1989) Biological water quality assessment of running water based on macroinvertebrate communities: history and present status in Europe. Environ Poll 60: 101-139

Metzeling L, Wells F, Newall P, Tiller D, Reed J (2001) Biological objectives for rivers and stream – ecosystem protection. Policy background paper. Environmental Protection Authority, Victoria

Meyer R, Millar R (1999) Bayesian stock assessment using a state-space implementation of the delay difference model. Can J Fish Aqua Sci 59: 37-52

Michel P, Oberdorff T (1995) Feeding habits of fourteen European freshwater fish species. Cybium 19: 5-46

Minshall GW (1988) Stream ecosystem theory: a global perspective. J N Am Benthol Soc 7: 263-288

Minshall GW (1993) Stream-riparian ecosystems: rationale and methods for basin-level assessments of management effects. In: Jensen ME, Bourgeron PS (eds) Eastside forest ecosystem health assessment. Volume II. Ecosystem management: principles and applications. US Forest Service Pacific Northwest Research Station, Portland, Oregon. General Technical Report PNW-GTR-318 pp 153-177

Minshall GW, Petersen RC, Nimz CF (1985) Species richness in streams of different size from the same drainage basin. Am Natur 125: 16-38

Moatar F, Fessant F, Poirel A (1999) pH modelling by neural networks: application of control and validation data series in the Middle Loire River. Ecol Model 120: 141-156

Moisen GG, Frescino TS (2002) Comparing five modelling techniques for predicting forest characteristics. Ecol Model 157: 209-225

Mol AWM (1984) Limnofauna Neerlandica. Een lijst van meercellige ongewervelde dieren aangetroffen in binnenwateren van Nederland [A list of invertebrates collected in the inland waters of the Netherlands]. Nieuwsbrief European Invertebrate Survey-Nederland 15: 1-124 (in Dutch)

Montanari A, Guglielmi N (1996) The role of projection indices in projection pursuit. Statistica -Bologna- 56(1): 63-86

Montesanto B, Ziller S., Coste M (1999) Communautés diatomiques épilithiques et qualité des ruisseaux du Mont Stratoniko, Chalkidiki (Grèce): premiers resultants. Crypt Algol 20: 137-138

Moog O (ed) (1995) Fauna aquatica austriaca: a comprehensive species inventory of Austrian aquatic organisms with ecological notes. 1. Edition. Wasserwirtschaftskataster, Bundesministerium für Land- und Forstwirtschaft, Wien

Moog O (ed) (2000) Erstellung typspezifischer benthoszönotischer Leitbilder österreichischer Fließgewässer. - Bundesministerium für Land- und Forstwirtschaft, Umwelt und Wasserwirtschaft, Wasserwirtschaftskataster. Studie i.A. BM: LFUW und UBA, Wien

Moog O, Chovanec A, Hinteregger J, Römer A (1999) Richtlinie zur Bestimmung der saprobiologischen Gewässergütebeurteilung von Fließgewässer (Guidelines for the saprobiological water quality assessment in Austria; in German). - Bundesministerium für Land - und Forstwirtschaft, Wasserwirtschaftskataster, Wien

Moreau J (ed.) (1997) Advances in the ecology of Lake Kariba. Publ Univ Zimbabwe Harare, Zimbabwe

Moreau J, de Silva SS (1991) Predictive fish yield models for lakes and reservoirs of the Philippines, Sri Lanka and Thailand. FAO Fisheries Technical Paper, 319

Morris CW, Autreta A, Boddy L (2001) The SVM has been applied for identifying organisms: a comparison with strongly partitioned radial basis function networks. Ecol Model 146: 57-67

Moss B (1973) The influence of environmental factors on the distribution of freshwater algae: an experimental study. II: the role of pH and the carbon dioxide-bicarbonate system. J Ecol 61: 157-177

Moss D, Furse MT, Wright JF, Armitage PD (1987) The prediction of the macroinvertebrate fauna of unpolluted running-water sites in Great Britain using environmental data. Fresh Biol 17: 41-52

Moyle PB, Light T (1996) Biological invasions of fresh water: empirical rules and assembly theory. Biol Conserv 78: 149-161

Munné A, Prat N (2000) Delimitación de regiones ecológicas para el establecimiento de tipos de referencia y umbrales de calidad biológica: Propuesta de aplicación de la nueva Directiva Marco del Agua en la cuenca del Ebro. Actas del II Congreso Ibérico sobre Gestión y Planificación del Agua. Oporto

Myers RA., Bowen KG, Barrowman NJ (1999) Maximum reproductive rate of fish at low population sizes. Can J Fish Aqua Sci 56: 2404-2419

Naiman RJ, Décamps H, Pastor J, Johnston CA (1988) The potential importance of boundaries to fluvial ecosystems. J N Am Benthol Soc 7: 289-306

Nakano H, et al. (1991) Identification of plankton using a neural network with a function of unknown species detection. Rep Measure Res Group, IM-91-30: 47-56

Nakhaeizadeh G (1993) Learning prediction of time series: a theoretical and empirical comparions of CBR with some other approaches. In: Wess S, Althoff K, Richter M (eds) Lecture notes in artificial intelligence. Springer-Verlag, Berlin

NBN (1984) Biological water quality: determination of the biotic index based on aquatic macroinvertebrates. Norme Belge T 92-402, Institut Belge de Normalisation (IBN) (in Dutch and French)

Nelder JA, Wedderburn RWM (1972) Generalized additive models. J Roy Stat Soc A 135: 370-384

Nelson RL, Platts WS, Larsen DP, Jensen SE (1992) Trout distribution and habitat in relation to geology and geomorphology in the North Fork Humboldt river drainage, northeastern Nevada. Trans Am Fish Soc 121: 405-426

Nelson WG (1990) Prospects for development of an index of biotic integrity for evaluating habitat degradation in coastal systems. Chem Ecol 4: 197-210

Nestler JM, Milhous RT, Layzer JB (1989) Instream habitat modelling techniques. In Gore JA, Petts GE (eds) Alternatives in regulated river management. CRC Press, Boca Raton pp 295-315

Neumann M, Baumeister J, Liess M, Schulz R (2002b) An expert system to estimate the pesticide contamination of small streams using benthic macroinvertebrate as bioindicators, Part 2: The knowledge base of LIMPACT. Ecol Ind 2: 391-401

Neumann M, Dudgeon D (2002) The impact of agricultural runoff on stream benthos in Hong Kong, China. Wat Res 36: 3103-3109

Neumann M, Liess M, Schulz R (2002a) An expert system to estimate the pesticide contamination of small streams using benthic macroinvertebrate as bioindicators, Part 1: The database of LIMPACT. Ecol Ind 2: 379-389

Newman K (1997) Bayesian averaging of generalized linear models for passive integrated transponder tag recoveries from salmonids in the Snake River. N Am J Fish Manage 17: 362-377

Newman PJ (1988) Classification of surface water quality. Heinnema, Oxford

Nijboer RC, Schmidt-Kloiber A (2004). The effect of excluding taxa with low abundances or taxa with small distribution ranges on ecological assessment. Hydrobiologia 516: 347-363

Nixon SC, Mainstone CP, Iversen TM, Kristensen P, Jeppensen E, Friberg N, Papathanassiou E, Jensen A, Pedersen F (1996) The harmonised monitoring and classification of ecological quality of surface waters in the European Union. WRc Report No. CO 4150, Medmenton

Norris RH (1995) Biological monitoring: The dilemma of data analysis. J N Am Benthol Soc 14: 440-450
Norusis MJ (1986) SPSS/PC+ advanced statistics. SPSS Inc., Chicago, IL
NRA (1996) River habitats in England and Wales, a national overview. Report Number 1, Environment Agency, Bristol
O'Connor MA, Walley WJ (2000) An information-theoretic self-organising map with disaggregation of output classes. In: Proc 2nd Int Conf on Enterprise Information Systems, Stafford, UK pp 108-115
O'Connor MA, Walley WJ (2001) River pollution diagnostic system (RPDS) – computer-based analysis and visualisation for bio-monitoring data. In: Proc 2nd World Water Congress of the International Water Association, Berlin
O'Neill RV (1976) Ecosyste; persistence and heterotrophic regulation. Ecology 57: 1244-1253
O'Neill RV (1989) Perspectives in hierarchy and scale. In: May RM, Levin SA (eds) Perspectives in Ecological Theory. Princeton University Press, Princeton pp 140-156
O'Neill RV, d'Angelis DL, Allen TFH (1986) A hierarchical concept of ecosystems. Princeton University Press, Princeton
Obach M, Wagner R, Werner H, Schmidt HH (2001) Modelling population dynamics of aquatic insects with artificial neural networks. Ecol Model 146: 207-217
Oberdorff T, Gilbert E, Lucchetta JC (1993) Patterns of fish species richness in the Seine River basin, France. Hydrobiologia 259: 157-167
Oberdorff T, Guégan JF, Hugueny B (1995) Global scale patterns of fish species richness in rivers. Ecography 18: 345-352
Oberdorff T, Hughes RM (1992) Modification of an index of biotic integrity based on fish assemblages to characterize rivers of the Seine basin, France. Hydrobiologia 228: 117-130
Oberdorff T, Hugueny B, Compin A, Belkessam D (1998) Non-interactive fish communities in the coastal streams of North-Western France. J Anim Ecol 67: 472-484
Oberdorff T, Hugueny B, Guégan JF (1997) Is there an influence of historical events on contemporary fish species richness in rivers? Comparaisons between Western Europe and North America. J Biogeog 24: 461-467
Oberdorff T, Lek S, Guégan JF (1999) Patterns of endemism in riverine fish of the Northern Hemisphere. Ecol Lett 2: 75-81
Oberdorff T, Pont D, Hugueny B, Belliard J, Berrebi dit Thomas R., Porcher JP (2002b) Adaptation et validation d'un indice poisson (FBI) pour l'évaluation de la qualité biologique des cours d'eau français. Bull Fr Pêche Pisci 365/366: 405-433
Oberdorff T, Pont D, Hugueny B, Chessel D (2001) A probabilistic model characterizing fish assemblages of French rivers : a framework for environmental assessment. Fresh Biol 46: 399-415
Oberdorff T, Pont D, Hugueny B, Porcher JP (2002a) Development and validation of a fish-based index for the assessment of 'river health' in France. Fresh Biol 47: 1720-1734
Odum EP (1971) Fundamentals of ecology. Saunders Company, Philadelphia
Odum EP (1980) Ecology. Holt-Saunders, London
Odum EP (1983) Basic ecology. Saunders College Publishing, Florida
Odum HT, Pinkerton RC (1955) Time's speed regulator: the optimum efficiency for maximum power output in physical and biological systems. Am Sci 43: 331-343

Ogle DH, Pruitt RC, Spangler GR, Cyterski MJ (1996) A Bayesian approach to assigning probabilities to fish ages determined from temporal signatures in growth increments. Can J Fish Aquat Sci 53: 1788-1794

Ohio EPA (1987/1989) Biological criteria for the protection of aquatic life. Vol. I, II, III. Ohio Environmental Protection Agency, Columbus, OH

Olden JD, Jackson DA (2001) Fish-habitat relationships in lakes: gaining predictive and explanatory insight by using artificial neural networks. Trans Am Fish Soc 130: 878-897

Olden JD, Jackson DA (2002) A comparison of statistical approaches for modeling fish species distributions. Fresh Biol 47: 1976-1995

Omernik JM (1987) Ecoregions of the conterminous United States. Ann Assoc Am Geog 77: 118-125

Omernik, J.M. (1995) Ecoregions: a spatial framework for environmental management. In: Davis WS, Simon TP (eds) Biological assessment and criteria. Tools for water resource planning and decision making. Lewis Publishers, Boca Raton, Florida pp 49-62

Österreichisches Normungsinstitut (1997) ÖNORM M 6232 - Guidelines for the ecological study and assessment of rivers (bilingual edition), Wien

Oswood MW, Reynolds JB, Irons JG, Milner AM, Rabeni CF, Doisy KE (2000) Distributions of freshwater fishes in ecoregions and hydroregions of Alaska. J N Am Benthol Soc 19: 405-418

Özesmi S, Özesmi U (1999) An artificial neural network approach to spatial habitat modelling with interspecific interaction. Ecol Model 116: 15-31

Paasavirta L (1990) The macrozoobenthos studies in the upper part of the Vanajavesi catchment area in the years of 1985 and 1988, with a comparison to earlier data. Ass. Wat. Poll. Control (the Kokemaenjoki river), Publ., 225: 1-24

Paerl HW, Bowles NN (1987) Dilution bioassays: their application to assessments of nutrient limitation in hypereutrophic waters. Hydrobiologia 156: 265-273

Paller MH (1994) Relationships between fish assemblage structure and stream order in South Carolina coastal plain stream. Trans Am Fish Soc 123: 150-161

Paller MH, Reichert MJM, Dean JM, Seigle JC (2000) Use of fish community data to evaluate restoration success of a riparian stream. Ecol Eng 15: S171-S187

Palmer M (2000) Ordination methods for ecologists. http://www.okstate.edu/artsci/botany/ordinate

Palomares ML, Yulianto B, Lim P, Bengen D, Belaud A (1993) A preliminary model of the Garonne river (Toulouse, France) ecosystem in Spring. In: Christensen V, Pauly D (eds) Trophic models of aquatic ecosystems. ICLARM Conf Proc pp 172-179

Pan Y, Stevenson RJ, Hill BH, Herlihy AT (2000) Ecoregions and benthic diatom assemblages in Mid-Atlantic streams, USA. J N Am Benthol Soc 19: 518-540

Pan Y, Stevenson RJ, Hill BH, Herlihy AT, Collins G (1996) Using diatoms as indicators of ecological conditions in lotic systems: a regional assessment. J N Am Benthol Soc 15: 481-495

Pan Y, Stevenson RJ, Hill BH, Kaufmann PR, Herlihy AT (1999) Spatial patterns and ecological determinants of benthic algal assemblages in mid-atlantic streams, USA. J Phycol 35: 460-468

Pantle E, Buck H (1955) Die biologische Überwachung de Gewässer und die Darstellung der Ergebnisse. Gas und Wasserfach 96: 1-604

Pao YH (1989) Adaptive pattern recognition and neural networks. Addison-Wesley Publishing Company, Inc., Reading, Massachusetts

Park OR, Park SS (2002) A time variable modelling study of vertical temperature profiles in the Okjung Lake. Kor J Limnol 35: 79-91 (in Korean)
Park SB (1998) Basic water quality of the mid to lower part of Nakdong River and the influences of the early rainfall during monsoon on the water quality. MS thesis, Pusan National Univ, Busan (in Korean)
Park SB, Lee SK, Chang KH, Jeong KS, Joo GJ (2002) The impact of Jangma (monsoon rainfall) on the changes of water quality in the lower Nakdong River (Mulgeum). Kor J Limnol 35: 161-170
Park Y-S, Céréghino R, Compin A, Lek S (2003a) Applications of artificial neural networks for patterning and predicting aquatic insect species richness in running waters. Ecol Model 160: 165-280
Park Y-S, Chang J, Lek S, Cao W, Brosse S (2003b) Conservation strategies for endemic fish species threatened by the Three Gorges Dam. Cons Biol 17(6): 748-1758
Park Y-S, Chon TS, Kwak IS, Kim JK, Jørgensen SE (2001a) Implementation of artificial neural networks in patterning and prediction of exergy in response to temporal dynamics of benthic macroinvertebrate communities in streams. Ecol Model 146: 143-157
Park Y-S, Kwak I-S, Cha EY, Lek S, Chon T-S (2001b) Relational patterning on different hierarchical levels in communities of benthic macroinvertebrates in an urbanized steam using an artificial neural network. J Asia-Pacific Entomol 4: 131-141
Park Y-S, Verdonschot PFM, Chon TS, Lek S (2003c) Patterning and predicting aquatic macroinvertebrate diversities using artificial neural network. Water Research 37: 1749–1758
Parsons M, Norris R (1996) The effect of habitat-specific sampling on biological assessment of water quality using a predictive model. Fresh Biol 41: 43-49
Paruelo JM, Tomasel F (1997) Prediction of functional characteristics of ecosystems: a comparison of artificial neural networks and regression models. Ecol Model 98: 173-186
Pascual M, Ellner SP (2000) Linking ecological patterns to environmental forcing via nonlinear time series models. Ecology 81: 2767-2780
Paszkowski CA, Tonn WM (2000) Community concordance between the fish and aquatic birds of lakes in northern Alberta: the relative importance of environmental and biotic factors. Fresh Biol 43: 421-437
Patrick R (1949) A proposed biological measure of stream condition based on a survey of the Cenestoga basin, Lancaster country, Pennsylvania. Proc Acad Natur Sci Philad 101: 277-341
Patrick R, Matthew HH, Wallace JH (1954) A new method for determining the pattern of diatom flora. Notulae Naturae 252: 1-12
Patrick R, Strawbridge D (1963) Variation in the structure of natural diatom communities. Am Natur 97: 51-57
Patterson KR (1999) Evaluating uncertainty in harvest control law catches using Bayesian Markov chain Monte Carlo virtual population analysis with adaptive rejection sampling and including structural uncertainty. Can J Fish Aqua Sci 56: 208-221
Pawaputanon O (1987) Management of fish populations in Ubolratana reservoir. Arch Hydrobiol 28: 309-317
Payne AI, Crombie J, AS Halls, Temple SA (1993) Synthesis of simple predicting models for tropical river fisheries. MRAG, London
Payne AJ (1986) The ecology of tropical lakes and rivers. Wiley and Sons, New-York

Payne RW (1997) Genstat 5 Release 4.1: Reference manual supplement. Lawes Agricultural Trust, Rothamsted Experimental Station. Rothansted

Pearson K (1901) On lines and planes of closest fit to systems of points in space. Philos Magazine 2: 559-572

Peeters ETMH, Gardeniers JJP, Tolkamp HH (1994) New methods to assess the ecological status of surface waters in the Netherlands. Ver Int Ver Limnol 25: 1914-1916

Pella JJ, Masusa MM, Chen DG (1998) Forecast methods for inseason management of the southeast Alaska chinook salmon troll fishery. In: Funk f, Quinn TJ, Heifetz J, Ianellí JN, Powers J, Schweigert JF, Sullivan P, Zhang CI (eds) Lowell Wakefield fisheries Symposium Series; Fishery stock assessment models. Alaska Sea Grant College Program, Fairbanks pp 287-314

Penczak T (1967) The biological and technical principles of the fishing by use of direct-current field. Prz Zool 11: 114-131

Penczak T (1988) The ichthyofauna of the Pilica drainage basin. Part I. Preimpoundment study. Sci Ann Pol Angl Assoc 1: 23-59

Penczak T (1989) The ichthyofauna of the Pilica drainage basin. Part II. Postimpoundment study. Sci Ann Pol Angl Assoc 2: 71-99

Penczak T, Godinho F, Agostinho AA (2002) Verification of the dualism ordering method by the canonical correspondence analysis: fish community samples. Limnologica 32: 14-20

Penczak T, Kruk A (1999) Applicability of the abundance/biomass comparison method for detecting human impacts on fish populations in the Pilica River, Poland. Fish Res 39: 229-240

Penczak T, Kruk A (2000) Threatened obligatory riverine fishes in human-modified Polish rivers. Ecol Fresh Fish 9: 109-117

Penczak T, Marszał L, Kruk A, Koszaliński H, Kostrzewa J, Zaczyński A (1996) Monitoring of fish fauna in the Pilica drainage basin. Part II. Pilica. Sci Ann Pol Angl Assoc 9: 91-104

Penczak T, Zaczyński A, Marszał L, Koszaliński H (1995) Monitoring of the fish fauna in the Pilica drainage basin. Part I. Tributaries. Sci Ann Pol Angl Assoc 8: 5-12

Pennak RW (1971) Towards a classification of lotic habitats. Hydrobiologia 38: 321-324

Peres F, Eulin-Garrigue A, Coste M (2003) Evolution spatio-temporelle des diatomées invasives et exotiques ou rares en France de 1996 à 2002 dans les stations du RNB du bassin Adour-Garonne. Livre des résumés du 22ème Colloque de l'Association des Diatomistes de Langue Française, Espot pp 33

Persat H, Keith P (1997) La répartition géographique des poissons d'eau douce en France: qui est autochtone et qui ne l'est pas? Bull Fr Pêche Pisci 344/345: 15-32

Peterman RM, Peters CN, Robb CA, Frederick SW (1999) Bayesian decision analysis and uncertainty in fisheries management. In: Pitcher TJ, Hart PJ, Pauly D (eds) Reinverting fisheries management. Kluwer Academic Publishers, Dordrecht pp 387-398

Petersen RC, Gíslason GM, Vought LBM (1995) Rivers of the Nordic countries. In: Cushing CE, Cummins KW, Minshall GW (eds) River and stream ecosystems. Ecosystems of the world Vol. 22. Elsevier, Amsterdam pp 295-341

Petit P (1996) Les pêcheries du secteur nord du Lac Tanganyika, situation actuelle et évolution récente. phD Thesis, Institut National Polytechnique, Toulouse, France

Piegay H, Dupont P, Faby A (2002) Questions of water resources management. Feedback on the implementation of the French SAGE and SDAGE plans (1992–2001). Wat Poll 4: 239–262

Pielou EC (1984) The interpretation of ecological data: a primer on classification and ordination. Wiley, New York

Pineda F (1987) Generalization of backpropagation to recurrent neural networks. Phys Rev Lett 19: 2229-2232

Pinelalloul B, Niyonsenga T, Legendre P (1995) Spatial and environmental components of fresh-water zooplankton structure. Ecoscience 2: 1-19

Pitcher T, Hart PJB (eds) (1995) Impact of species changes in African Great Lakes. Chapman and Hall. London

Plafkin JL, Barbour MT, Porter KD, Gros SK, Hughes RM (1989) Rapid bioassessment protocols for use in streams and rivers: benthic macroinvertebrates and fish. EPA 444/4-89-001. U.S. Environmental Protection Agency. Washington

Platt J, Cristianini N, Shawe-Taylor J (2000). Large margin DAGs for multiclass classification. In: Solla S, Leen TK, Müller KR (eds) Advances in neural information processing systems 12. MIT Press, Cambridge, Massachussets pp 547-553

Platt JC (1999) Fast Training of support vector machines using sequential minimal optimization. In: Schölkopf B, Burges C, Smola A (eds) Advances in kernel methods: Support Vector Learning. MIT Press, Cambridge, Massachussetts pp 185-208

Poff NL, Allan D, Bain MB, Karr JR, Prestegaard KL, Richter BD, Sparks RE, Stromberg JC (1997) The natural flow regime: a paradigm for river conservation and restoration. BioSci 47: 769-784

Poff NL, Allan JD (1995) Functional organization of stream fish assemblages in relation to hydrological variability. Ecology 76: 606-627

Poff NL, Ward JV (1989) Implications of streamflow variability and predictability for lotic community structure: a regional analysis of streamflow patterns. Can J Fish Aqua Sci 46: 1805-1818

Posse C (1992) Projection pursuit discriminant analysis for two groups. Commun Stat-Theor M 21(1): 1-19

Potapova MG, Charles DF (2002) Benthic diatoms in USA rivers: distribution along spatial and environmental gradients. J Biogeog 29: 167-187

Potapova MG, Charles DF (2003) Distribution of benthic diatoms in U.S. rivers in relation to conductivity and ionic composition. Fresh Biol 48: 1311-1328

Pouilly M, Souchon Y, LeCoarer Y, Jouve D (1996) Methodology for fish assemblages habitat assessment in large rivers: application in the Garonne river (France). Proc 2nd Int Symposium on Hydraulic Habitats Ecohydraulique 2000 pp 323-229

Preston FW (1962) The canonical distribution of commonness and rarity: I and II. Ecology 43: 185-215, 410-432

Prygiel J, Coste M (1996) Recent trends in monitoring French rivers using algae, especially diatoms. In: Whitton BA, Rott E (eds) Use of algae for monitoring rivers II. Institut für Botanik, Universität Innsbruck pp 87-97

Prygiel J, Coste M (1998) Mise au point de l'Indice Biologique Diatomée, un indice diatomique pratique applicable au réseau hydrographique français. L'Eau, l' Industrie, les Nuisances 211: 40-45

Prygiel J, Coste M (1999) Progress in the use of diatoms for monitoring rivers in France. In: Prygiel J, Whitton BA, Bukowska J (eds) Use of algae for monitoring rivers III, Agence de l'Eau Artois-Picardie, Douai pp165-179

Prygiel J, Coste M (2000) Guide méthodologique pour la mise en œuvre de l'Indice Biologique Diatomées. NF T 90-354. Agences de l'eau – Cemagref, Douai

Prygiel J, Coste M, Bukowska J (1999) Review of the major diatom-based techniques for the water quality assessment of rivers: state of art in Europe. In: Prygiel J, Whitton BA, Bukowska J (eds) Use of algae for monitoring rivers III, Agence de l'Eau Artois-Picardie, Douai pp 224-238

Prygiel J, Leveque L, Iserentant R (1996) L'IDP: Un nouvel Indice Diatomique Pratique pour l'évaluation de la qualité des eaux en réseau de surveillance. Revue des Sciences de l'Eau 9: 97-113

Pudmenzky A, Marshall J, Choy S (1998) Preliminary application of artificial neural networks model for predicting macroinvertebrates in rivers. Freshwater Biological Monitoring Report No. 9, The State of Queensland, Department of Natural Resources, Queensland

Punt AE, Hilborn R (1997) Fisheries stock assessment and decision analysis: the Bayesian approach. Rev Fish Biol Fisher 7: 35-63

Punt AE, Walker TI (1998) Stock assessment and risk analysis for the school shark (*Galeorhinus galeus*) of southern Australia. Mar Fresh Res 49: 719-731

Puppe F (1998) Knowledge reuse among diagnostic problem solving methods in the shell-kit D3. Int J Human-Comp Stud 49: 627-649

Puppe F (2000) Knowledge formalization patterns. Proc PKAW, Sydney, Australia

Puppe F, Ziegler S, Martin U, Hupp J (2001) Wissensbasierte diagnosesysteme im service-support. Konzepte und Erfahrungen, Springer Verlag, Berlin

Pusey BJ, Arthington AH, Read MG (1995) Species richness and spatial variation in fish assemblage structure in two rivers of the wet tropics of northern Queensland, Australia. Env Biol Fish 42: 181-199

Qian SS, Lavine M, Stow CA (2000) Univariate Bayesian nonparametric binary regression with application in environmental management. Environ Ecol Stat 7: 77-91

Quensen JFI, Woodruff DS (1997) Associations between shell morphology and land crab predation in the land snail Cerion. Funct Ecol 11: 464-471

Quinn MA, Halbert SE, Williams III L (1991) Spatial and temporal changes in aphid (Homoptera: Aphididae) species assemblages collected with suction traps in Idaho. J Econ Entomol 84: 1710-1716

Racca JMJ, Philibert A, Racca R, Prairie YT (2001) A comparison between diatom-based pH inference models using artificial neural networks (ANN), weighted averaging (WA) and weighted averaging partial least squares (Wa-PLS) regressions. J Paleolimnol 26: 411-422

Rahel FJ (2000) Homogenization of fish faunas across the United States. Science 288: 854-856

Rahel FJ, Hubert WA (1991). Fish assemblages and habitat gradients in a rocky mountain-great plain stream: biotic zonation and additive patterns of community change. Trans Am Fish Soc 120: 319-332

Ram A, Santamaria JC (1997) Continuous case-based reasoning. Artif Intell 90: 25-77

Ramos-Nino ME, Ramirez-Rodriguez CA, Clifford MN, Adams MR (1997) A comparison of quantitative structure-activity relationships for the effect of benzoic and cinnamic acids on *Listeria monocytogenes* using multiple linear regression, artificial neural network and fuzzy systems. J Appl Microbiol 82: 168-176

Recknagel F (1997) ANNA - artificial neural network model predicting blooms and succession of blue-green Algae. Hydrobiologia 349: 47-57

Recknagel F (2001) Applications of machine learning to ecological modelling. Ecol Model 146: 303-310

Recknagel F (ed) 2003. Ecological informatics: understanding ecology by biologically-inspired computation. Springer-Verlag, Berlin

Recknagel F, Wilson H. (2000) Elucidation and prediction of aquatic ecosystems by artificial neural networks. In: Lek S, Guegan JF (eds) Artificial neuronal networks: application to ecology and evolution. Springer-Verlag, Berlin

Recknagel F, Bobbin J, Whigham P, Wilson H (2002) Comparative application of artificial neural networks and genetic algorithms for multivariate time-series modelling of algal blooms in freshwater lakes. J Hydroinf 4: 125-134

Recknagel F, French M, Harkonen P, Yabunaka K (1997) Artificial neural network approach for modelling and prediction of algal blooms. Ecol Model 96: 11-28

Recknagel F, Fukushima T, Hanazato T, Takamura N, Wilson H (1998) Modelling and prediction of phyto- and zooplankton dynamics in Lake Kasumigaura by artificial neural networks. Lake Reser Res Manage 3: 123-133

Recknagel F, Wilson H (2000) Elucidation and prediction of aquatic ecosystems by artificial neuronal networks. In: Lek S, Guegan JF (eds) Artificial neuronal networks: application to ecology and evolution. Springer-Verlag, Berlin pp 143-155

REFCOND (European working group) (2002) Guidance on establishing reference conditions and ecological status class boundaries for inland surface waters. 6th draft. Water Framework Directive. Common Implementation Strategy

Reichardt E (1997) Taxonomische Revision des Artenkomplexes um Gomphonema pumilum (Bacillariophyceae). Nova Hedwigia 65: 99-129

Reichardt E (1999) Zur Revision der Gattung Gomphonema. Iconog Diatom 8: 1-203

Reichardt E (2001) Revision of the species around *Gomphonema truncatum* and *G. capitatum*. In: Jahn R, Kociolek JP, Compère P (eds) Lange-Bertalot-Festschrift, Gantner, Ruggell pp 187-224

Rejwan C, Collins NC, Brunner LJ, Shuter BJ, Ridway MS (1999) Tree regression analysis on the nesting habitat of small mouth bass. Ecology 80: 341-348

Renberg I, Hellberg T (1982) The pH history of lakes in southwestern Sweden, as calculated from the subfossil diatomflora of the sediments. Ambio 11: 30-33

Resh VH, Brown AV, Covich AP, Gurtz ME, Li HW, Minshall GW, Reice SR, Sheldon AL, Wallace JB, Wissmar R. (1988) The role of disturbance in stream ecology. J N Am Benthol Soc 7: 433-455

Resh VH, Jackson JK (1993) Rapid assessment approaches to biomonitoring using benthic macroinvertebrates. In: Rosenberg DM, Resh VH (eds) Freshwater biomonitoring and benthic macroinvertebrates. Chapman & Hall, New York pp 195-233

Resh VH, Meyers MJ, Hannaford MJ (1996) Macroinvertebrates as indicators of environmental quality. In: Hauer FR, Lamberti GA (eds) Methods in stream ecology. Academic Press, San Diego, California pp 647–698

Resh VH, Norris RN, Barbour MT (1995) Design and implementation of rapid assessment approaches for water resource monitoring using benthic macroinvertebrates. Aust J Ecol 20: 108-121

Revenga C, Murray S, Abramovits J, Hammond A (1998) Watersheds of the world: ecological value and vulnerability. World Resources Institute, Washington DC

Reyjol Y, Compin A, Aguilar Ibarra A, Lim P (2003) Longitudinal diversity patterns in streams: comparing invertebrates and fish communities. Arch Hydrobiol 157: 525-533

Reyjol Y, Lim P, Belaud A, Lek S (2001b) Modelling of microhabitat used by fish in natural and regulated flows in the river Garonne (France). Ecol Model 146: 131-142

Reyjol Y, Lim P, Dauba F, Baran P, Belaud A (2001a) Role of temperature and flow regulation on the Salmoniform-Cypriniform transition. Arch Hydrobiol 152: 567-582

Reynolds CS (1984) The Ecology of Freshwater Phytoplankton. Cambridge University Press, New York

Reynolds CS (1986) Experimental manipulations of phytoplankton periodicity in large limnetic enclosures in Blelham Tern, English Lake District. In: Munawar M, Talling JF (eds) Seasonality of freshwater phytoplankton. Junk, Dordrecht pp 43-64

Reynolds CS (1992) Algae. In: P. Calow and G. E. Petts (Editors). The river handbook: hydrological and ecological principles. Vol. 1. Blackwell Scientific Publication, Oxford pp 195-215

Reynoldson TB, Bailey RC, Day KE, Norris RH (1995) Biological guidelines for freshwater sediment based on benthic assessment of sediment (the BEAST) using a multivariate approach for predicting biological state. Aust J Ecol 20: 198-219

Reynoldson TB, Norris RH, Resh VH, Day KE, Rosenberg DM (1997) The reference condition: a comparison of multimetric and multivariate approaches to assess water-quality impairment using benthic macroinvertebrates. J N Am Benthol Soc 16: 833-852

Reynoldson TB, Wright JF (2000) The reference condition: problems and solutions. In: Wright JF, Sutcliffe DW, Furse MT (eds) Assessing the biological quality of freshwaters. RIVPACS and other techniques. Freshwater biological association, Ambleside, UK pp 293-303

Richardson J (1997) Acute ammonia toxicity for eight New Zealand indigenous freshwater species. New Zealand J Mar Fres Res 31: 185-190

Richardson J, Boubee J, Dean T, Rowe D, West D (1998) Effects of suspended solids on migratory native fish. Wat Atmosph 6: 22-23

Richardson J, Boubee JAT, West DW (1994) Thermal tolerance and preference of some native New Zealand freshwater fish. New Zealand J Mar Fresh Res 28: 399-407

Richardson J, Rowe DK, Smith JP (2001) Effects of turbidity on the migration of juvenile banded kokopu (*Galaxias fasciatus*) in a natural stream. New Zealand J Mar Fresh Res 35: 191-196

Rietz GE (1965) Biozonosen und synusien in der planzensoziologie. Biosoziologie. Junk, Den Haag: 23-39

Rimet F, Ector L, Cauchie HM, Hoffmann L (2004) Regional distribution of diatom assemblages in the headwater streams of Luxembourg. Hydrobiologia 520: 105-117

Rimet F, Tudesque L, Peeters V, Vidal H, Ector L (2003) Assemblages-types de diatomées benthiques des rivières non-polluées du bassin Rhône-Méditerranée-Corse (France). Actes de 21ème Colloque de l'ADLaF, Nantes, 10-13 septembre 2002. Bulletin de la Société des Sciences Naturelles de l'Ouest de la France, 2ème supplément hors série pp 272-287

Ringe F (1974) Chironomiden-Emergenz 1970 in Breitenbach und Rohrwiesenbach. Schlitzer produktionsbiologische Studien (10). Arch Hydrobiol Suppl 45: 212-304

Roadknight CM, Balls GR, Mills GE, Palmer-Brown D (1997) Modeling complex environmental data. IEEE Trans Neural Network 8: 852-862

Robb CA, Peterman R (1998) Application of Bayesian decision analysis to management of a sockeye salmon (*Oncorhynchus nerca*) fishery. Can J Fish Aqua Sci 55: 86-98

Roberts D, McMinn A (1998) A weighted-averaging regression and calibration model for inferring lake water salinity from fossil diatom assemblages in saline lakes of the Vestfold Hills: a new tool for interpreting Holocene lake histories in Antarctica. J Paleolimnol 19: 99-113

Roca JR (1990) Tipología fisico-química de las fuentes de los Pirineos centrales: síntesis regional. Limnetica 6: 57-78

Rodríguez MA (1994) Succession, environmental fluctuations, and stability in experimentally manipulated microalgae communities. Oikos 70: 107-120

Rogers DJ, Tanimoto TT (1960) A computer program for classifying plants. Science 132: 1115-1118

Rohatsch T, Pöppel G, Werner H (2002) Robuste indizes für projection pursuit. Informatik Forschung und Entwicklung 17: 53-59

Rosenberg DM, Resh VH (eds) (1993) Freshwater biomonitoring and benthic invertebrates. Chapman and Hall, New York

Rott E, Duthie HC, Pipp E (1998) Monitoring organic pollution and eutrophication in the Grand River, Ontario, by means of diatoms. Can J Fish Aqua Sci 55: 1443-1453

Rott E, Hofmann G, Pall K, Pfister P, Pipp E (1997) Indikationslisten für Aufwuchsalgen in österreichischen Fliessgewässern. Teil 1: Saprobielle Indikation (Indicator species lists for periphyton in Austrian rivers. Part 1: Saprobic indication), 73 p. - Wasserwirtschaftskataster, Bundeministerium f. Land- u. Forstwirtschaft, Wien

Rott E, Pfister P, van Dam H, Pipp E, Pall K, Binder N, Ortler K (1999) Indikationslisten für Aufwuchsalgen in Österreichischen Fliessgewässern. Teil 2: Trophieindikation und autökologische Anmerkungen. Bundesministerium für Land- und Forstwirtschaft, Wasserwirtschaftskataster, Vienna

Rott E, Pipp E (1999) Progress in the use of benthic algae for monitoring rivers in Austria. In: Prygiel J, Whitton BA, Bukowska J (eds) Use of algae for monitoring rivers III, Agence de l'Eau Artois-Picardie, Douai pp 110-112

Rott E, Pipp E, Pfister P (2003) Diatom methods developed for river quality assessment in Austria and a cross-check against numerical trophic indication methods used in Europe. Algolog Stud 110: 91-115

Round FE (1990) Diatom communities - their response to changes in acidity. In: Battarbee RW, Mason SJ, Renberg I, Talling JF (eds) Palaeolimnology and lake acidification. Philos Trans Roy Soc London B, London

Round FE (1991) Diatoms in river monitoring studies. J Appl Phycol 3: 129-145

Round FE, Crawford RM, Mann DG (1990) The diatoms: biology and morphology of the genera. Cambridge University press, Cambridge

Rowe DK, Boubee JAT, Richardson J (1999a) Effects of suspended solids on native fish. Draft NIWA Technical Report, NIWA, Aucklandm New Zealand

Rowe DK, Chisnall BL, Dean TL, Richardson J (1999b) Effects of land use on native fish communities in east coast streams of the North Island of New Zealand. New Zealand J Mar Fresh Res 33: 141-151

Rowe DK, Hicks M, Richardson J (2000) Reduced abundance of banded kokopu (Galaxias fasciatus) and other native fish in turbid rivers of the North Island New Zealand. New Zealand J Mar Fresh Res 34: 547-558

Roydhouse A, Jone EK (1995) Case retriever for historical meteorological data. New Zealand J Comp 6: 1B 261-267

Ruiz ME, Srinivasan P (1997) Automatic text categorization using neural networks. In: Efthimiadis E (ed) Proc 8th ASIS/SIGCR Workshop on Classification Research. American Society for Information Science, Washington pp 59-72

Rumeau A, Coste M (1988) Initiation à la systématique des diatomées d'eau douce. Bull Fr Pêche Pisci 309: 1-69
Rumelhart DE, Hinton G.E, Williams RJ (1986b) Learning internal representations by error propagation. In: Rumelhart DE, McCelland JL (eds) Parallel distributed processing: explorations in the microstructure of cognition, Vol. I. Foundations, MIT Press, Cambridge pp 318-362
Rumelhart DE, Hinton GE, Williams GE (1986a) Learning representations by back-propagating errors. Nature 323: 533-536
Rumelhart DE, Mc Clelland JL (1986) Parallel distributed processing: explorations in the microstructure of cognition. MIT Press, Cambridge
Ryder RA (1982) The morpho-edaphic index: use, abuse and fundamental concepts. Trans Am Fish Soc 111: 154-164
Sabater S (2000) Diatom communities as indicators of environmental stress in the Guadiamar River, S-W. Spain, following a major mine tailings spill. J Appl Phycol 12: 113-124
Sabater S, Guasch H, Picon A, Romani AM, Muñoz I (1996) Using diatom communities to monitor water quality in a river after the implementation of a sanitation plan (River Ter, Spain). In: Whitton BA, Rott E (eds) Use of algae for monitoring rivers II. Innsbruck Austria 17-19 Sept. 95: Studia Student. G.m.b.H., Innsbruck pp 97-103
Sabater S, Roca JR (1990) Some factors affecting distribution of diatom assemblages in Pyrenean springs. Fresh Biol 24: 493-507
Sabater S, Roca JR (1992) Ecological and biogeographical aspects of diatom distribution in Pyrenean springs. Brit Phycol J 27: 203-213
Sachs L.(1999) Angewandte Statistik. 9.Auflage. Springer-Verlag, Berlin
Sammon Jr JW (1969) A non-linear mapping for data structure analysis. IEEE Trans on Comp 18: 401-409
Sandin L, Johnson RK (2000) Ecoregions and benthic macroinvertebrate assemblages of Swedish streams. J N Am Benthol Soc 19: 462-474
Sand-Jansen K (2001) Freshwater ecosystems, human impact on. Encycl Biodiv 3: 89-108
Sanz J, Margarita P, Martinez MT, Plaza M (1999) Experimental design methodologies to optimize monobutyltin chloride determination by hybride generation gas phase molecular absorption spectrometry. Talanta 50: 149-164
SAS (1999) SAS enterprise miner. SAS Institute Inc. SAS Campus Drive, Cary, North Carolina, USA
Sauvage S, Teissier S, Vervier P, Améziane T, Garabétian F, Delmas F, Caussade B (2003) A numerical tool to integrate biophysical diversity of a large regulated river: hydrobiogeochemical bases. The case of the Garonne River (France). River Res Appl 19: 181-198
Scardi M (1996) Artificial neural networks as empirical models of phytoplankton production. Mar Ecol Prog Ser 139: 289-299
Scardi M (2000) Neuronal network models of phytoplankton primary production. Pages 115-129 In: Lek S, Guegan JF (eds) Artificial neuronal networks: application to ecology and evolution, Springer-Verlag, Berlin
Scardi M (2001) Advances in neural network modeling of phytoplankton primary production. Ecol Model 146: 33-45
Scardi M, Harding Jr LW (1999) Developing an empirical model of phytoplankton primary production: a neural network case study. Ecol Model 120: 213-223

Schank R, Abelson R (1977) Scripts, plans, goals and understanding. Erlbaum, Hillsdale, New Jersey

Schiefele S, Kohmann F (1993) Bioindikation der trophie in Fliessgewässern. Umweltforschungsplan des Bundesministers für Umwelt, Naturschutz und Reaktorsicherheit. Forschungsbericht Nr. 102 01 504

Schiefele S, Schreiner C (1991) Use of diatoms for monitoring nutrient enrichment, acidification and impact of salt in rivers in Germany and Austria. In: Whitton BA, Rott E, Friedrich G (eds) Use of algae for monitoring rivers. Düsseldorf, Germany. Institut für Botanik Universität Innsbruck, Studia Student. G.m.b.H., Innsbruck pp 103-110

Schleiter IM, Borchardt D, Wagner R, Dapper T, Schmidt K, Schmidt H, Werner H (1999) Modelling water quality, bioindication and population dynamics in lotic ecosystems using neural networks. Ecol Model 120: 271-286

Schleiter IM, Obach M, Borchardt D, Werner H (2001) Bioindication of chemical and hydromorphological habitat characteristics with benthic macro-invertebrates based on artificial neural networks. Aquat Ecol 35: 147-158

Schlosser IJ (1982) Fish community structure and function along two habitat gradients in a headwater stream. Ecol Monog 52: 395-414

Schlosser IJ (1990) Environmental variation, life history attributes, and community structure in stream fishes: implications for environmental management and assessment. Environ Manage 14: 621-628

Schlosser IJ, Ebel KK (1989) Effects of flow regime and cyprinid predation on a headwater stream. Ecol Monog 59: 41-57

Schmedtje U, Sommerhäuser M, Braukmann U, Briem E, Haase P, Hering D (2001) Top down - bottom up -Konzept einer biozönotisch begründeten Fließgewässetypologie. DGL-Tagungsbericht 2000 pp 147-151

Schmidt R, Gierl L (2000) Experiences with prototype designs and retrieval methods in medical case-based reasoning systems. In: Smyth B, Cunningham P (eds) EWCBR-98, Lecture notes in artificial intelligence. Springer-Verlag, Berlin pp 370-381

Schmilovitch Z, Mizrach A, Hoffman A, Egozi H, Fuchs Y (2000) Determination of mango physiological indices by near-infrared spectrometry. Posth Biol Technol 19: 245-252

Schöl A, Kirchesch V, Bergfeld T, Schöll F, Borcherding J, Müller D (2002) Modelling the chlorophyll a content in the River Rhine: Interrelation between riverine algal production and population biomass of grazers, rotifers and the zebra mussel, Dreissena polymorpha. Int Rev Hydrobiol 87: 295-317

Schölkopf B, Sung K, Burges C, Girosi F, Niyogi P, Poggio T, Vapnik V (1997) Comparing support vector machines with gaussian kernels to radial basis function classifiers. IEEE Trans Sign Proc 45: 2758-2765

Schulz R, Liess M (1999) A field study of the effects of agriculturally derived insecticide input on stream macroinvertebrate dynamics. Aqua Tox 46: 155-176

Schwefel HP (1995) Evolution and optimum seeking. John Wiley and Sons, New York

Semhi K, Suchet PA, Clauer N, Probst JL (2000) Impact of nitrogen fertilizers on the natural weathering-erosion processes and fluvial transport in the Garonne basin. Appl Geochem 15: 865-878

Shannon CE, Weaver W (1949) The mathematical theory of communication. University of Illinois Press, Urbana, IL

Shapiro J (1984) Blue-green dominance in lakes: the role and management significance of pH and CO2. Int Revue Ges Hydrobiol 69: 765-780

Shapiro J (1990) Current beliefs regarding dominance by blue-greens: the case for the importance of CO2 and pH. Ver Int Verein Limnol 24: 38-54
Sheldon AL (1968) Species diversity and longitudinal succession in stream fishes. Ecology 49: 193-198
Sherman BS, Webster IT, Jones GJ, Oliver RL (1998) Transition between Aulacoseira and Anabaena dominance in a turbid river weir pool. Limnol Oceanogr 43: 1902-1915
Shin SK, Park CK, Song KO (1998) Evaluation of water quality using principal component analysis in the Nakdong River estuary. J Kor Env Sci Soc 7: 171-176 (in Korean)
Silveira VF, Khator SK, Barcia RM (1996) An information management system for forecasting environmental change. Comp Ind Eng 31: 289-292
Silverton J, Holtier S, Johnson J, Dale P (1992) Cellular automation models of interspecific competition for space: the effect of pattern on process. J Ecology 80: 527-534
Simpson J, Norris RH (2000) Biological assessment of water quality: development of AUSRIVAS models and outputs. In: Wright JF, Sutcliffe DW, Furse MT (eds) Assessing the biological quality of freshwaters. RIVPACS and other techniques. Freshwater Biological Association, Ambleside, UK pp 125-142
Simpson R, Culverhouse PF, Williams R, Ellis R (1993) Classification of Dinophyceae by artificial neural networks. Dev Mar Biol 3: 183-190
Simpson R, Williams R, Ellis R, Culverhouse PF (1992) Biological pattern recognition by neural networks. Mar Ecol Prog ser 79: 303-308
Skoulikidis NT (1993) Significance evaluation of factors controlling river water composition. Environ Geol 22: 178-185
Skriver J, Friberg N, Kirkegaard J (2000) Biological assessment of watercourse quality in Denmark: introduction of the Danish stream fauna index (DSFI) as the official biomonitoring method. Ver Int Ver Theor Ange Limnol 27: 1822-1830
Sládeček V (1973) System of water quality from the biological point of view. Ergeb Limnol 7: 1-128.
Sládeček V (1979) Continental systems for the assessment of river water quality. In: James A, Evison L (eds) Biological indicators of water quality. Ch 3. Wiley, Chichester pp 3.1-3.22
Sládecek V (1986) Diatoms as indicators of organic pollution. Acta Hydroch Hydrobiol 14: 555-566
SMEAG (Syndicat Mixte d'Etudes et d'Aménagement de la Garonne) 2003. L'Agenda Garonne: un développement durable pour un fleuve européen. EPTB Garonne, Toulouse
Smirnov NN, Timms BV (1983) A revision of the Australian Cladocera (Crustacea). Rec Austral Museum Suppl 1: 1-132
Smit H, van der Hammen H (2000) Atlas van de Nederlandse watermijten (Acari: Hydrachnidia) [Atlas of the Dutch Hydrachnidia]. Nederlandse Faunistische Mededelingen 13. EIS-Nederland and Nationaal Natuurhistorisch Museum Naturalis, Leiden, The Netherlands (in Dutch)
Smith ADM, Punt AE (1998) Stock assessment of gemfish (*Rexea solandri*) in eastern Australia using maximum likelihood and Bayesian methods. In: Quinn II TJ, Funk F, Heifetz J, Ianelli JN, Powers JE, Schweigert JF, Sullivan PJ, Zhang C-I (eds) Fisheries stock assessment models. Alaska Sea Grant College Program, AK-SG-98-01, Fairbanks pp 245-286
Smith M (1994) Neural networks for statistical modelling. Van Nostrand Reinhold, New York

Smith MA (1990) The ecophysiology of epilithic diatom communities of acid lakes in Galloway, southwest Scotland. In: Battarbee RW, Mason SJ, Renberg I, Talling JF (eds) Palaeolimnology and lake acidification. Philos Trans Roy Soc London B, London pp 103-110

Smith MJ, Kay WR, Edward DHD, Papas PJ, Richardson KSJ, Simpson JS (1999) AusRivAs: using macroinvertevrates to assess ecological conditions of rivers in Western Australia. Fresh Biol 41: 269-282

Smogor RA, Angermeier PL (1999) Effects of drainage basin and anthropogenic disturbance on relations between stream size and IBI metrics in Virginia. In: Simon TP (ed) Assessment approaches for estimating biological integrity using fish assemblages. Lewis Press, Boca Raton, FL pp 249-272

Smogor RA, Angermeier PL (2001) Determining a regional framework for assessing biotic integrity of Virginia streams. Trans Am Fish Soc 130: 18-35

Snoeijs P (1994) Distribution of epiphytic diatom species composition, diversity and biomass on different macroalgal hosts along seasonal and salinity gradients in the Baltic sea. Diatom Res 9: 189-211

Snoeijs P, Busse S, et al (2002) The importance of diatom cell size in community analysis. J Phycol 38: 265-272

Snyder EB, Robinson CT, Minshall GW, Rushforth SR (2002) Regional patterns in periphyton accrual and diatom assemblages structure in a heterogeneous nutrient landscape. Can J Fish Aqua Sci 59: 564-577

Soininen J (2002) Responses of epilithic diatom communities to environmental gradients in some Finnish rivers. Int Rev Hydrobiol 87: 11-24

Soininen J, Paavola R, Muotka T (2004) Benthic diatom communities in boreal streams: community structure in relation to environmental and spatial gradients. Ecography 27: 330-342

Sommer U, Gliwicz ZM, Lampert W, Duncan A (1986) The PEG-model of seasonal succession of plankton events in fresh waters. Arch Hydrobiol 106: 433-471

Song KO, Park HY, Park CG (1993) Water quality modeling in the Nakdnog river (I): a study on the characteristics of nutrients distribution. J Kor Soc Wat Qual 9: 41-53

Sørensen T (1948) A method of establishing groups of equal amplitude in plant sociology based on similarity of species content and its application to analyses of the vegetation on Danish commons. Biologiske Skrifter 5: 1-34

Soresma (2000) Environmental impact assessment report on the development of fish migration channels and natural overflow systems in the Zwalm River basin, Soresma advies-en ingenieursbureau, Antwerp (in Dutch)

Southwood TRE (1977) Habitat, the templet for ecological strategies? J Anim Ecol 46: 337-365

Southwood TRE (1988) Tactics, strategies and templets. Oikos 52: 3-18

Spellerberg IF (1991) Monitoring ecological change. Cambridge University Press, Cambridge

Spitz F, Lek S (1999) Environmental impact prediction using neural network modelling. An example in wildlife damage. J Appl Ecol 36: 317-326

SPSS Inc (2001) SPSS for Windows 11.0. Chicago, IL

SPSS Inc (2000) Systat 10. Statistical software for Windows. Chicago, IL

Stankovski V, Debeljak M, Bratko I, Adamic M (1998) Modelling the population dynamics of red deer (*Cervus elaphus* L) with regard to forest development. Ecol Model 108: 143-153

Stanley EH, Doyle MW (2003) Trading off: the ecological effects of dam removal. Front Ecol Environ 1: 15-22

StatSoft Inc (2001) STATISTICA for Windows [Computer program manual]. Tulsa, OK

Statzner B, Resh VH, Roux AL (1994) The synthesis of long-term ecological research in the context of concurrently developed ecological theory: design of a research strategy for Upper Rhône River and its floodplain. Fresh Biol 31: 253-263

Steiger J, James M, Gazelle F (1998) Channelization and consequences on floodplain system functioning on the Garonne river, SW France. Reg Riv Res Manage 14: 13-23

Steinberg C, Schiefele S (1988) Biological indication of trophy and pollution of running waters. Zeit Wasser Abwas Fors 21: 227-234

Steinberg LJ, Reckhow KH, Wolpert RL (1996) Bayesian model for fate and transport of polychlorinated biphenyl in upper Hudson River. J Env Eng 122: 341-349

Steinman AD, Mc Intire CD (1986) Effects of current velocity and light energy on the structure of periphyton assemblages in laboratory streams. J Phycol 22: 352-361

Stevenson MM, Schnell GD, Black R (1974) Factor analysis of fish distribution patterns in western and central Oklahoma. Syst Zool 23: 202-218

Stevenson RJ (1997) Scale-dependent determinants and consequences of benthic algal heterogeneity. J N Am Bent Soc 16: 248-262

Stevenson RJ, Pan Y (1999) Assessing environmental conditions in river and streams with diatoms. In: Stoermer EF, Smol JP (eds) The diatoms: application for the environmental and earth sciences. Cambridge University Press, Cambridge pp 11-41

Stoermer EF, Smol JP (1999) The diatoms. Applications for environmental and earth sciences. Cambridge University press, Cambridge

STOWA (1992) Ecologische Beoordeling en beheer van oppervlaktewater: Beoordelingssystem voor stromende wateren op basis van macrofauna. STOWA Report 92-8

Straskraba M (1979) Natural control mechanisms in models of aquatic ecosystems. Ecol Model 6: 305-322

Strayer DL (1993) Macrohabitats of freshwater mussels (Bivalvia, Unionacea) in streams of the northern Atlantic slope. J N Am Benthol Soc 12: 236-246

Suarez-Seoane S, Osborne PE, Alonso JC (2002) Large-scale habitat selection by agricultural steppe birds in Spain: identifying species-habitat responses using generalized additive models. J Appl Ecol 39: 755-771

Symoens JJ (1957) Les eaux douces de l'Ardenne et des régions voisines: les milieux et leur végétation algale. Bull Soc Roy Bot Belg 89: 111-314

Symoens JJ, Kusel-Fetzmann E, Descy JP (1988) Algal communities of continental waters. In: Symoens JJ (ed) Vegetation of inland waters, 15/1, Kluwer Academic Publishers, Dordrecht pp 183-221

Szabados M, Bethem T, Evans M (2003) Data quality monitoring using embedded intelligence. Sea Technol 44: 5

Szabo K, Kiss KT, Ector L, Kecskés M, Acs E (2004) Benthic diatom flora in a small Hungarian tributary of river Danube (Rákos-stream). Algolog Stud 111: 79-94

Takamura N, Otsuki A, Aizaki M, Nojiri Y (1992) Phytoplankton species shift accompanied by transition from nitrogen dependence to phosphorus dependence of primary production in Lake Kasumigaura, Japan. Arch Hydrobiol 124: 129-148

Talling JF (1985) Inorganic carbon reserves of natural waters and eco-physiological consequences of their photosynthetic depletion: Microalgae. In: Lucas WJ, Berry JA (eds) Inorganic carbon uptake by aquatic photosynthetic organisms. Proc Int Workshop on

Bicarbonate use in photosynthesis, University of California, 18–22 August 1984. American Society of Plant Physiologists, Rockville pp 176
Tan SS, Smeins FE (1996) Predicting grassland community changes with an artificial neural network model. Ecol Model 84: 91-97
Tang X, Stewart WK, Vincent L, Huang H, Marra M, Gallager SM, Davis CS (1998) Automatic plankton image recognition. Artif Intell Review 12: 177-199
Taylor CM, Winston MR, Matthews WJ (1993) Fish species-environment and abundance relationships in a Great Plain river system. Ecography 16: 16-23
Tegelmark DO (1998) Site factors as multivariate predictors of the success of natural regeneration in Scots pine forests. Forest Ecol Manage 109: 231-239
Ten Brink BJE, Hosper SH, Colin F (1991) A quantitative method for description and assessment of ecosystems: the AMOEBA-approach. Mar Poll Bull 23: 265-270
ter Braak C (1987) The analysis of vegetation-environment relationships by canonical correspodence analysis. Vegetatio 69: 69-77
ter Braak C (1988) CANOCO - A Fortran program for canonical ordination by [partial] [detrended] [canonical] correspondence analysis, principal components analysis and redundancy analysis. Agricultural Mathematics Group, Wageningen
ter Braak CJF (1994) Canonical community ordination. Part I: Basic theory and linear methods. Ecoscience 1: 127-140
ter Braak CJF, Smilauer P (1998) CANOCO reference manual and user's guide to Canoco for Windows: Software for canonical community ordination (version 4). Microcomputer Power, Ithaca, New York
ter Braak CJF, Šmilauer P (2002) CANOCO Reference manual and CanoDraw for Windows user's guide: software for canonical community ordination (version 4.5). Microcomputer Power, Ithaca, New York
ter Braak CJF, van Dam H (1989) Inferring pH from Diatoms: a comparison of old and new calibration methods. Hydrobiologia 178: 209-223
ter Braak CJF, Verdonschot PFM (1995) Canonical correspondence analysis and related multivariate methods in aquatic ecology. Aqua Sci 57: 255-289
Thébault JM, Qotbi A (1999) A model of phytoplankton development in the Lot River (France): simulations of scenarios. Wat Res 33: 1065-1079
Thienemann A (1925) Die Binnengewasser Mitteleuropas. Die Binnengewasser 1: 54-83
Thienemann A (1954) Ein drittes biozonotisches Grundprinzip. Arch Hydrobiol 49: 421-422
Thorne RSTJ, Williams WP (1997) The response of benthic macroinvertebrates to pollution in developing countries: a multimetric system of bioassessment. Fresh Biol 37: 671-686
Tison J, Park YS, Coste M, Delmas F, Giraudel JL (2004b) Use of unsupervised neural networks for eco-regional zonation of hydrosystems through diatom communities: case study of Adour-Garonne watershed (France). Arch Hydrobiol 159: 409-422
Titterington DM, Smith AFM, Makov UE (1985) Statistical analysis of finite mixture distributions. Wiley, New York
Tockner K, Malard F, Ward JV (2000) An extension of the flood pulse concept. Hydrol Proc 14: 2861-2883
Tomassone R, Lesquoy E, Miller C (1983) La régression, nouveaux regards sur une ancienne méthode statistique. INRA, Paris
Tonn WM (1990) Climate change and fish communities: a conceptual framework. Trans Am Fish Soc 119: 337-352

Torrisi M (2003) Monitoraggio biologico dei corsi d'acqua Appenninici mediante l'Indice Algale di Eutrofizzazione/Polluzione (EPI-D) comparato ad altri indici diatomici europei. Università di Camerino

Tourenq JN, Dauba F (1978) Transformation de la faune des poissons dans la rivière Lot. Ann Limnol 14: 133-138

Trigg DJ, Walley WJ, Ormerod SJ (2000) A prototype Bayesian belief network for the diagnosis of acidification in Welsh rivers. In: Brebbia CA, Ibarra-Berastegui G, Zanetti P (eds) Development and application of computer techniques to environmental studies VIII, WIT Press, Southampton, UK

Tuffery G, Verneaux J (1968) Méthode de détermination de la qualité biologique des eaux courantes. CERAFER, Paris

Tukey J (1977) Exploratory data analysis. Addison-Wesley, Reading, MA

Ulanowicz RE (1980) An hypothesis of the development natural communities. J Theor boil 85: 223-245

Ultsch A (1993) Self-organizing neural networks for visualization and classification. In: Opitz O, Lausen B, Klar R (eds) Information and classification. Springer-Verlag, Berlin pp 307-313

Ultsch A (1999) Data mining and knowledge discovery with emergent self-organizing feature maps for multivariate time series. In: Oja E, Kaski S (eds) Kohonen maps, Elsevier, Amsterdam pp 33-45

Ultsch A, Siemon H (1989) Technical Report 329. Univ. of Dortmund, Dortmund

Ultsch A, Siemon HP (1990) Kohonen's Self-Organizing feature maps for exploratory data analysis. In: Proc. INNC'90, Int. Neural Network Conf., Dordrecht, Netherlands. Kluwer pp 305-308

Umetrics (2001) Simca-P 9, user guide and tutorial. Umetrics AB, Umea, Sweden

Underwood GJ, Phillips J, Saunders K (1998) Distribution of estuarine benthic diatom species along salinity and nutrient gradients. Eur J Phycol 33: 173-183

UNPF (2000) Site internet de l'Union Nationale pour la Pêche en France et la Protection du Milieu. http: //www.unpf.fr

Urban DL, O'Neill RV, Shugart HH (1987) Landscape ecology: a hierarchical perspective can help scientists understand spatial patterns. BioScience 37: 119-127

Utans J, Moody JE (1991) Selecting neural network architectures via the prediction risk: application to corporate bond rating prediction. In: Proc First Int Conf on Artificial intelligence applications on Wall Street. IEEE Computer Society Press, Los Alamitos

Utermöhl H (1958) Zur Vervollkommnung der Quantitativen Phytoplankton. Method Mitt Int Ver Limnol 9: 1-38

van Dam H (1997) Partial recovery of moorland pools from acidification: indications by chemistry and diatoms. Netherlands J Aqua Ecol 30: 203-218

van Dam H, Mertens A, Janmaat LM (1993) De invloed van atmosferische deposite op diatomeeén en chemische samenetelling van het water in sprengen, beken en bronnen. IBN-rapport 052, ibn-dlo, Wageningen

van Dam H, Mertens A, Sinkeldam J (1994) A coded checklist and ecological indicator values of freshwater diatoms from the Netherlands. Netherlands J Aqua Ecol 28: 117-133

van Den Brink P, Roelsma J, van Nes E, Scheffer M, Brock T (2002) Perpest Model, A case-based reasoning approach to predict ecological risks of pesticides. Environ Toxicol Chem 21: 2500-2506

van Densen W, Burgis M (1999) Lakes and reservoir fisheries management in South east Asia and Africa. Wetsbury Academy and Scientific Publishing, Otley, West Yorkshire
van der Maarel E (1979) Transformation of cover-abundance values in phytosociology and its effects on community similarity. Vegetatio 39: 97-114
van Deusen RD (1954) Maryland freshwater stream classification by watersheds. Chesap Biol Lab 106: 1-30
van Dobben H.F, ter Braak CJF (1998) Effects of atmospheric NH3 on epiphytic lichens in the Netherlands: the pitfalls of biological monitoring. Atmosp Environ 32: 551-557
van Sickle J, Hughes RM (2000) Classification strengths of ecoregions, catchments, and geographic clusters for aquatic vertebrates in Oregon. J N Am Benthol Soc 19: 370 – 384
van Tongeren O (1986) Flexclus, an interactive flexible cluster program. Acta Botan Neerlan 35: 137-142
Vannote RL, Minshall G.W, Cummins KW, Sedell JR, Cushing CE (1980) The river continuum concept. Can J Fish Aqua Sci 37: 130-137
Vapnik V (1995) The nature of statistical learning theory. Springer, New York
Vapnik VN (1982) Estimation of dependences based on empirical data. Springer, New York
Vapnik VN (1998) Statistical learning theory. Wiley, New York
Vapnik VN, Chervonenkis AY (1971) On the uniform convergence of relative frequencies of events to their probabilities. Theor Prob Appl 25: 103-109
Varis O (1997) Bayesian decision analysis for environmental and resource management. Environ Model Soft Envi Data News 12: 177-185
Varis O, Kuikka S (1997) Joint use of multiple environmental assessment models by a Bayesian meta-model: the Baltic salmon case. Ecol Model 102: 341-351
Vasko K, Toivonen HTT, Korhola A (2000) A Bayesian multinomal Gaussian response model for organism-based environmental reconstruction. J Paleolimnol 24: 43-250
Verdonschot PFM (1984) The distribution of aquatic oligochaetes in the fenland area of NW Overijssel (The Netherlands). Hydrobiologia 115: 215-222
Verdonschot PFM (1990) Ecological characterization of surface waters in the province of Overijssel (The Nether-lands), Province of Over-ijs-sel, PhD thesis, Alterra, Wageningen, The Netherlands
Verdonschot PFM (1991) The web-approach: a tool in water management. In: Ecological water managemenet in practice (Proc Technical Meeting, Ede, The Netherlands, 3 October 1990). Proc and Inf CHO-TNO, 45: 59-76
Verdonschot PFM (1994) Water typology: A tool for water management and nature conservation. Ver Int Ver Limnol 25: 1911-1913
Verdonschot PFM (1996) Ecological characterization of surface waters in the province of Overijssel (The Neteherlands). PhD thesis, Wageningen University, The Netherlands
Verdonschot PFM (2000) Integrated ecological assessment methods as a basis for sustainable catchment management. Hydrobiologia 422/423: 389-412
Verdonschot PFM, Goedhart PW (2000) RISTORI. Effecten van ingrepen in het waterbeheer op aquatische levensgemeenschappen. Cenotypenbenadering, fase 2: Verfijning van het prototype. Alterra, Wageningen, 172: 1-109
Verdonschot PFM, Nijboer RC (2000) Typology of macrofaunal assemblages applied to water and nature management: a Dutch approach. In: Wright JF, Sutcliffe DW, Furse MT (eds) Assessing the biological quality of fresh waters. RIVPACS and other techniques. Freshwater Biological Association, Ambleside, Cumbria, UK pp 241-262

Verneaux J (1973) Cours d'eau de Franche-Comté (massif du Jura). Recherches écologiques sur le réseau hydrographique du Doubs - essai de biotypologie. Ph-D Thesis, University of Besançon, Besançon

Verneaux J (1976a) Biotypologie de l'écosystème " eau courante ". la structure typologique. Compte Rendu de l'Académie des Sciences de Paris 282D: 1663-1666

Verneaux J (1976b) Biotypologie de l'écosystème " eau courante ". la structure typologique. Compte Rendu de l'Académie des Sciences de Paris 283D: 1791-1793

Verneaux J (1977) Biotypologie de l'écosytème 'eau courante'. Détermination approchée de l'appartenance typologique d'un peuplement ichtyologique. C R Acad Sc Paris 284: 675-678

Verneaux J, Galmiche P, Janier F, Monnot A (1982) Une nouvelle methode pratique d'evaluation de la qualité des eaux courantes: un indice biologique de qualité générale (I.B.G.). Ann Sci Université Besancon Biol Anim 3: 11-21

Verneaux J, Leynaud G (1974) Note sommaire sur la définition d'objectifs et de critères de la qualité des eaux courantes. CTGREF-DQEP, Paris

Vesanto J, Alhoniemi E (2000) Clustering of the self-organizing map. IEEE Trans Neural Networks 11: 586-600

Vesanto J, Himberg J, Alhoniemi E, Parhankangas J (2000) SOM toolbox for Matlab 5. Technical report. Helsinki University of Technology, Neural Networks Research Centre. http: //www.cis.hut.fi/projects/somtoolbox/documentation/

Vignaux M, Vignaux GA, Lizamore S, Gresham D (1998) Fine-scale mapping of fish distribution from commercial catch and effort data using maximum entropy tomography. Can J Fish Aqua Sci 55: 1220-1227

Vila-Gispert A, García-Berthou E, Moreno-Amich R (2002) Fish zonation in a Mediterranean stream: effects of human disturbances. Aqua Sci 64: 163–170

Villmann T, Bauer HU (1998) Applications of the growing self-organizing map. Neurocomputing 21: 91-100

Vinçon G, Clergue M (1988) Etude hydrobiologique de la vallée d'Ossau (Pyrénées Atlantiques, France). III Simuliidae (Diptera, Nematocera): leur originalité biogéographique et écologique. Ann Limnol 24: 67-81

Vinçon G, Thomas, AGB (1987) Etude hydrobiologique de la vallée d'Ossau (Pyrénées Atlantiques). I Répatition et écologie des Ephéméroptères. Ann Limnol 23: 95-113

VMM (2000) Water quality – water discharges 1999. Flemish Environmental Agency, VMM, Erembodegem (in Dutch)

Vyverman W, Vyverman R, Rajendran VS, Tyler P (1996) Distribution of benthic diatom assemblages in Tasmanian highland lakes and their possible use as indicators of environmental changes. Can J Fish Aqua Sci 53: 493-508

Wachs B (1998) A qualitative classification for the evaluation of the heavy metal contamination in river ecosystems. Ver Int Ver Limnol 26: 1289-1294

Wagner R, Dapper T, Schmidt H.H (2000a) The influence of environmental variables on the abundance of aquatic insects: a comparison of ordination and artificial neural networks. Hydrobiologia 422-423: 143-152

Wagner R, Dapper T, Schmidt H.H (2000b) The influence of environmental variables on the abundance of aquatic insects: a comparison of ordination and artificial neural networks. Hydrobiologia 422-423: 511-520

Walker, H.M and Lev, J, 1969. Elementary Statistical Methods, New York

Wallace RA, Sanders GP, Ferl RJ (1991) Biology. 3rd edition. HarperCollins Publishers, New York

Walley WJ, Martin RW, O'Connor MA (2000) Self-organizing maps for classification of river quality from biological and environmental data. In: Denzer R, Swayne DA, Purvis M, Schimak G (eds) Environmental Software Systems: Environmental Information and Decision Support, IFIP Conference Series, Kluwer Academic Publishers pp 27-41

Walley WJ, Dzeroski S (1996) Biological monitoring: a comparison between Bayesian, neural and machine learning methods of water quality classification. In: Environmental software systems. Chapman & Hall, London pp 229-240

Walley WJ, Fontama VN (1997) Bio-monitoring of rivers: an AI approach to data interpretation. Toxic Ecotox News 4: 182-184

Walley WJ, Fontama VN (1998) Neural network predictors of average score per taxon and number of families at unpolluted river sites in Great Britain. Water Research 32: 613-622

Walley WJ, Fontama VN (2000) New approaches to river quality classification based upon Artificial Intelligence. In: Wright JF, Sutcliffe DW, Furse MT (eds) Assessing the biological quality of fresh waters. RIVPACS and other techniques. Freshwater Biological Association, Ambleside pp 263-280

Walley WJ, Martin RW, O'Connor MA 2000. Self-organising maps for classification of river quality from biological and environmental data. In: Denzer R, Swayne DA, Purvis M, Schimak G (eds) Environmental software systems: environmental information and decision support, IFIP Conference Series, Kluwer Academic Publishers, Amsterdam pp 27-41

Walley WJ, O'Connor MA (2001) Unsupervised pattern recognition for the interpretation of ecological data. Ecol Model 146: 219-230

Walley WJ, O'Connor MA, Trigg DJ, Martin RW (2002). Diagnosing and predicting river health from biological survey data using pattern recognition and Plausible reasoning. Environment Agency Technical Report E1-056/TR2, Water Research Centre

Wang L, Lyons J, Kanehl P, Bannerman R, Emmons E (2000) Watershed urbanization and changes in fish communities in southeastern Wisconsin streams. J Am Wat Res 36: 1173-1175

Ward JV, Stanford JA (1979) Ecological factors controlling stream zoobenthos with emphasis on thermal modification of regulated streams. In: Ward JV, Stanford JA (eds) The ecology of regulated streams. Plenum Press, New York pp 35-55

Ward JV, Stanford JA (1983a) The serial discontinuity concept of lotic ecosystems. In: Fontaine TD, Bartell SM (eds) Dynamics of lotic ecosystems. Ann Arbor Sci, Michigan pp 29-42

Ward JV, Stanford JA (1983b) The intermediate disturbance hypothesis: an explaination for biotic diversity patterns in lotic ecosystems. In: Fountain TD, Bartell SM (eds) Dynamics of lotic ecosystems. Ann Arbor Sci, Michigan pp 347-356

Ward JV, Tockner K (2001) Biodiversity: towards a unifying theme for river ecology. Fresh Biol 46: 807-819

Ward JV, Tockner K, Schiemer F (1999) Biodiversity of floodplain river ecosystems: ecotones and connectivity. Reg Riv Res Manage 15: 125-139

Wasson JG (1989) Eléments pour une typologie nouvelle des eaux courantes : une revue critique de quelques approches existantes. Bulletin d'Ecologie 20: 109-127

Wasson JG, Chandesris A, Pella H (2002) Définition des hydro-écorégions de France métropolitaine. Approche régionale de typologie des eaux courantes et éléments pour la définition des peuplements de référence d'invertébrés. Technical report, Cemagref Lyon BEA/LHQ. Cemagref, Lyon

Wasson JG, Chandesris A, Pella H, Souchon Y (2001) Définition des hydroécorégions françaises. Méthodologie de détermination des conditions de référence au sens de la Directive cadre pour la gestion des eaux. Rapport de phase 1. Cemagref, Lyon

Watanabe T, Asai K, Houki A (1988) Numerical water quality monitoring of organic pollution using diatom assemblage. In: Round FE (ed) Proc Ninth International Diatom Symposium 1986, Koeltz Scientific Books pp 123-141

Watson I (1997) Applying case-based reasoning: techniques for enterprise systems. Morgan Kaufmann Publishers, San Francisco

Watters G, Deriso R (2000) Catch per unit of effort of bigeye tuna: A new analysis with regression trees and simulated annealing. Int Am Trop Tuna Comiss Bull 21: 531-571

Weckering S (1953) Les diatomées de la région Lorraine (Gutland) du Grand-Duché du Luxembourg. Bull Soc Natur Luxem 53: 151-211

Welcomme RL (1985) River Fisheries. Fao Fisheries Technical Paper 262. FAO, Rome

Welcomme RL (1986) The effect of the sahelian drought on the fisheries of the Central Delta of the Niger River. Aqua Fish Manage 17: 147-164

Welcomme RL, Hagborg H (1977) Towards a general model for floodplain ecology and fisheries. Environ Biol Fish 2: 7-22

Wen CG, Lee CS (1998) A neural network approach to multiobjective optimization for water quality management in a river basin. Wat Resour Res 34: 427-436

Werner J (2001) Aperçu sur les bryophytes (sub-)aquatiques des rivières luxembourgeoises. Bull Soc Natur Luxem 101: 3-18

Wetzel RG, Likens GE (1991) Limnological analyses. Springer-Verlag, New York

Whigham PA (2000) Induction of a marsupial density model using genetic programming and spatial relationships. Ecol Model 131: 299-317

Whigham PA, Recknagel F (1999) Predictive modelling of plankton dynamics in freshwater lakes using genetic programming. In: Oxley L, Scrimgeour F (eds) MODSIM '99 Int Congress on Modelling and Simulation. The Modelling and Simulation Society of Australia and New Zealand Inc, Hamilton, New Zealand pp 679-685

Whigham PA, Recknagel F (2001a) An inductive approach to ecological time series modelling by evolutionary computation. Ecol Model 146: 275-287

Whigham PA, Recknagel F (2001b) Predicting chlorophyll-a in freshwater lakes by hybridising process-based models and genetic algorithms. Ecol Model 146: 243-251

Whitehead P, Hornberge G (1984) Modelling algal behaviour in the River Thames. Wat Res 18: 945-953

Whittaker RH (1977) Evolution of species diversity in land communities. Evol Biol 10: 1-67

Whittaker RH; Woodwell GM (1971) Evolution of natural communities. In: Weins JA (ed) Ecosystem structure and function. Oregon State University Press, Corvallis pp 137-159

Whittier TR, Hughes RM, Larsen D (1988) Correspondence between ecoregions and spatial patterns in stream ecosystems in Oregon. Can J Fish Aqua Sci 45: 1264-1278

Wichert GA, Rapport DJ (1998) Fish community structure as a measure of degradation and rehabilitation of riparian systems in an agricultural drainage basin. Environ Manage 22: 425-443

Wiederholm T (1984) Responses of aquatic insects to environmental pollution. In: . Resh VH Rosenberg DM (eds) The ecology of aquatic insects. Praeger Publishers, New York pp 508-557

Wilppu E (1997) The visualisation capability of self-organizing maps to detect deviations in distribution control. TUCS Technical Report No 153, Turku Centre for Computer Science, Turku

Wilson H, Recknagel F (2001) Towards a generic artificial neural network model for dynamic predictions of algal abundance in freshwater lakes. Ecol Model 146: 69-84

Wilson SE, Cumming BF, Smol JP (1994) Diatom-salinity relationships in 111 lakes from the Interior Plateau of British Columbia, Canada: the development of diatom-based models for paleosalinity reconstructions. J Paleolim 12: 197-221

Wilson SE, Smol JP, Sauchyn DJ (1997) A Holocene paleosalinity diatom record from southwestern Saskatchewan, Canada: Harris lake revisited. J Paleolimnol 17: 23-31

Wimmer R, Chovanec A, Moog O, Fink MH, Gruber D (2000) Abiotic stream classification as a basis for a surveillance monitoring network in Austria in accordance with the EU Water Framework Directive. Acta Hydroch Hydrobiol 28: 177-184

Winter JG, Duthie HC (2000) Epilithic diatoms as indicators of stream total N and total P concentration. J N Am Benthol Soc 19: 32-49

Witten IH, Frank E (2000) Data mining: practical machine learning tools and techniques with Java implementations. Morgan Kaufmann Publishers, San Francisco

Wold S, Albano C, Dunn III WJ, Esbensen K, Hellberg,S, Johansson E, Sjöström H (1983) Pattern recognition: finding and using regularities in multivariate data. In: Martens J (ed) Proc IUFOST Conf on Food Research and Data Analysis. Applied Science Publications, London

Woodiwiss FS (1964) The biological system of stream classification used by the Trent River Board. Chem Indus 11: 443-447

Woodiwiss FS (1981) Biological water assessment methods. Nottingham-Abriged report of working group expert, Commission of the European Communities, ENV/416/80

Wright DH (1983) Species-energy theory: an extension of species-area theory. Oikos 41: 495-506

Wright DH, Currie DJ, Maurer BA (1993a) Energy supplies and patterns of species richness on local and regional scales. In: Ricklefs RE, Schulter D (eds) Species diversity in ecological communities. University of Chicago Press, Chicago pp 66-74

Wright JF (1995) Development and use of a system for predicting macroinvertebrates in flowing water. Australian Journal Ecology 20:181-197

Wright JF (2000) An introduction to RIVPACS In: Wright JF, Sutcliffe DW, Furse MT (eds) Assessing the biological quality of fresh waters. RIVPACS and other techniques. Freshwater Biological Association, Ambleside, Cumbria, UK pp 1-24

Wright JF, Armitage PD, Furse MT (1989) Prediction of invertebrate communities using stream measurements. Reg Riv Res Manage 4: 147-155

Wright JF, Furse MT, Armitage PD (1993b) RIVPACS: a technique for evaluating the biological quality of rivers in the U.K. Wat Res 3: 15-25

Wright JF, Furse MT, Moss D (1998) River classification using invertebrates: RIVPACS applications. Aqua Conserv Mar Fresh 8: 617-631

Wright JF, Moss D, Armitage PD, Furse MT (1984) A preliminary classification of running water-sites in Great Britain based on macro-invertebrate species and the prediction of community type using environmental data. Fresh Biol 14: 221-256

Wright JF, Sutcliffe DW, Furse MT (2000) Assessing the biological quality of fresh waters: RIVPACS and other techniques. Freshwater Biological Association, Ambleside, UK

Wunsam S, Cattaneo A, et al (2002) Comparing diatom species, genera and size in biomonitoring : a case study from streams in the Laurentians (Québec, Canada). Fresh Biol 47: 325-340

Zegers SL, Karim MR (1991) Generalized linear models with random effects: a Gibbs sampling approach. J Am Stat Assoc 86: 79-86

Zelinka M, Marvan P (1961) Zur Präzisierung der biologischen Klassification der Reinheit fliessender Gewässer. Arch Hydrobiol 57: 389-407

Zhu ML, Fujita M, Hashimoto N (1994) Application of neural networks to runoff prediction. In: Hipel KW, McLeod AI, Panu US, Singh VP, Fang L (eds) Stochastic and statistical methods in hydrology and environmental engineering. Kluwer Academic Publishers, Norwell

Ziemann H (1971). Die wirkung des salzgehaltes auf die diatomeenflora als grundlage für eine biologische analyse und klassifikation der binnengewässer. Limnologica 8: 505-525

Ziemann H (1991) Veranderungen der diatomeenflora der werra unter dem einfluss des salzgehaltes. Acta Hydroch Hydrobiol 19: 159-174

Ziller S, Montesanto B (2004) Phytobenthos (diatoms) and water frame directive implementation: the case of two Mediterranean rivers in Greece. Fresen Environ Bull 13: 128-138

Zurada JM (1992) Introduction to artificial neural systems. West Publishing, St Paul

Subject index

adaptive learning algorithms, 44, 189
adaptive resonance theory (ART), 132, 208, 210, 211, 213, 214, 215, 218, 219, 240, 251
Adour-Garonne, 221, 222, 224, 225, 226, 235, 237, 304, 305, 307, 316, 319
advanced modelling techniques, 1, 289, 305, 435
ANN, *see* artificial neural network
APM, *see* Autonomous Projection Mapping
ART, *see* adaptive resonance theory
artificial neural network (s), ANNs, 2, 4, 21, 22, 26, 27, 31, 33, 35, 37, 39, 44, 54-59, 61, 63, 68, 79-82, 84, 87, 88, 93-96, 98, 100, 103, 110, 114, 115, 117-125, 127, 128, 132-138, 140-146, 148-151, 153-158, 160-164, 166, 168, 190, 192, 206-208, 210, 218, 219, 221-223, 225-237, 239, 251-254, 256, 257, 259-263, 272, 274, 285, 286, 304, 305, 318, 321, 329, 342-348, 351-355, 358, 360-364, 369, 381, 416, 424 - 436, 441, 452
Artois Picardie Diatom Index, 10
assemblage, 12, 14, 16, 41, 43, 45-50, 52, 54, 63, 69, 72, 77, 100, 101, 111, 115, 119-123, 127, 157, 158, 189, 206, 222, 254-256, 258-260, 284, 288, 294, 300, 302, 307, 322, 337, 341, 348, 349, 353, 358, 362-365
assessment techniques, 13
assessment tools, 2, 7, 19, 345
AUSRIVAS, *see* Australian River Assessment Scheme
Australian River Assessment Scheme (AUSRIVAS), 15, 148, 234
Australian Water Resources Council (AWRC), 253
Autonomous Projection Mapping (APM), 367, 385, 386, 389
AWRC, *see* Australian Water Resources Council
backfitting, 24
backpropagation, 27, 29, 54, 56, 79, 93, 118, 120, 127, 132, 135, 136, 140, 145, 150, 160, 162, 180, 190, 195, 203, 224, 286, 318, 322, 343, 436
back-propagation algorithm, *see* backpropagation
Bayesian models, 35, 36
BEAST, *see* benthic assessment of sediment
benthic assessment of sediment (BEAST), 15, 148, 234
benthic diatom typologies, 340
best matching unit, 27, 47, 169, 192, 209, 278, 307, 322, 370, 438
bicarbonate alkalinity, 308, 309, 314
biocenotic analysis, 10
bioindicators, 5, 7
Biological Diatom Index, 10, 12, 288, 289
biological index, 168
biosurveillance, 221, 222, 237
biotic indices, 1, 23, 14, 19, 147, 157, 209, 301, 345, 353
biotic zonation, 16, 43
Boolean operators, 406
BPN, *see* backpropagation
Bray-Curtis distance, 343, 392
canals, 38, 191, 199, 200, 412, 413, 423, 424
CANOCO, 172
canonical correspondence analysis (CCA), 44, 106, 132, 167, 172, 184, 186, 188, 291, 305, 321, 324-327, 331, 334-336, 338, 340, 369
CART, 22, 25, 93, 94, 98
case-based reasoning (CBR), 262-266, 268, 270-272, 410
catchment, 1, 4, 7, 17, 26, 45, 49-53, 65, 69, 77, 89, 117, 207, 225, 253, 275, 291, 296, 298, 300, 317, 320, 324, 341, 355, 356, 368, 401, 403, 448
CBR, *see* case-based reasoning
CCA, *see* canonical correspondence analysis
CCI, *see* correctly classified instances
CEE index, 10, 12

cenotypes, 14, 22
Chi-Squared Automated Interaction Detection, 22
classification, 8, 13, 15, 16, 22, 24-27, 29, 32, 34-37, 44, 45, 47-50, 56, 65, 67, 76, 93, 105, 108-110, 114, 132, 144, 147, 161, 168, 171-173, 176, 178-184, 186-190, 195-197, 199, 200, 204, 206-208, 212-214, 216, 218, 222, 227, 231, 234, 235, 238, 251-254, 261-263, 289, 299, 302, 304, 310, 314, 317, 321, 323, 325, 328, 330, 346, 367, 369, 373, 376, 390, 392, 396, 400, 402, 404, 407, 409-414, 419, 420, 423-427, 430, 434, 440, 445
classification methods, 29, 367, 411
classification system, 8, 236
classification tree, 144, 252
cluster analysis, 21, 36, 44, 47, 167, 170, 234, 286, 305, 308-310, 316, 343, 370, 373, 376, 379, 413, 414, 445
cluster structuring index, 372, 375, 377, 380
clustering, 22, 27, 29, 32, 39, 47, 131, 167-170, 173, 175, 178, 180, 182, 207, 208, 211, 214, 215, 219, 220, 225, 228, 229, 237, 239, 253, 258, 274, 277, 280-283, 285, 286, 289, 299, 323, 325, 326, 343, 369, 373, 376, 391-393, 395-397, 399, 400
Coleoptera, 168, 172, 190, 223, 225, 230, 232, 235, 403
community assessment, 14, 367
community dynamics, 205, 206, 239, 244, 250, 251, 274, 277, 367
community indices, 104, 107, 190, 196, 197, 201
community structure, 1, 2, 3, 14, 21, 34, 52, 73, 114, 131, 147, 155, 168, 205, 262, 287, 302, 305, 308, 353, 369, 436-438, 451-454
community types, 15, 147, 368, 372, 411, 445
comparison, 3, 7, 11, 14, 22, 30, 33, 50, 66, 80, 90, 93-95, 107, 126, 152-156, 162-167, 169, 170, 172, 178, 179, 182, 184, 203, 215, 216, 219, 229, 232, 252, 254, 268, 297, 299, 308, 310, 315, 323, 361, 362, 369, 399, 402, 409, 420, 423, 426, 444
competitive artificial neural network, 27, 32
complexity, 21, 31, 34, 36, 39, 60, 70, 93, 119, 131, 147-149, 156, 158, 159, 168, 205, 206, 210, 239, 286, 304, 355, 365, 369, 379, 402, 406, 409, 410, 452-454
computer-based data analysis, 367, 390
conventional statistical models, 21, 358
correctly classified instances (CCI), 118-120, 123, 128, 135-140, 143-145, 150-155, 357-359, 361, 363
correspondence analysis, 44, 132, 167, 304, 369
counterpropagation network (CPN), 31, 32, 34, 190, 192, 193, 194, 195, 197, 199, 203, 204, 205
CPN, *see* counterpropagation network
crossvalidation, 23, 26, 56, 93, 322, 361, 420-424
cyanobacteria, 261, 262, 273, 274, 277, 280, 281-286
DAIpo, 10, 12
data mining, 45, 367
DCA, *see* detrended correspondence analysis
detrended correspondence analysis (DCA), 100, 101, 104, 109, 110, 111, 112, 321, 324, 325
Diatom Assemblage Index, 10
diatom assemblages, 19, 39, 262, 288, 289, 291, 293, 295, 296, 299, 300, 301, 302, 314, 316, 317, 318, 319, 325, 329, 337, 341, 343, 344, 345, 346, 347, 348, 349, 350-355, 359, 363, 365
diatom community, 305, 309, 340
diatom ecology, 288, 307, 335, 349, 352
diatom ecology, 9
diatom index, 288, 318, 342, 364
Differentiating Species System, 10
dimension reduction, 240, 251, 367, 387
discriminant analysis, 21, 23, 426
dissolved organic carbon, 320, 355, 356
diversity index, 104, 132, 189, 190, 195, 198, 201, 203, 205, 209
diversity indices, 10, 13, 14, 111, 189, 196, 203, 204, 205
ecological modelling, 34, 39, 115, 190, 205, 238, 274, 277, 286, 453

ecological quality, 2, 4, 75, 147, 304, 342, 436
ecological status, 7, 18, 114, 145, 147, 205, 262, 317, 319, 327, 329, 337, 340, 341, 345, 354, 454
ecological system, 21, 40, 131, 263, 273
ecoregions, 7, 64, 65, 148, 157, 291, 304, 314, 345, 352-354, 378
ecosystem evaluation, 1, 435
EKOO, 167, 190, 195
Enumeration indices, 14
Ephemeroptera, 149, 168, 172, 222, 223, 225, 230, 232, 235
EPI-D, 10, 12
EPT, 14, 222, 223
EPTC, 168, 222-225, 227-229, 231-238
European Community, 4, 7, 54
Eutrophication Pollution Index Diatoms, 10, 12
Evaluation of relevant species, 367, 369
exergy, 34, 39, 40
extreme values, 382
factor analysis, 44
FFGs, *see* functional feeding groups
fish assemblage, 17, 42, 43-45, 47-55, 63, 64, 68, 69, 72-74, 87, 100, 112, 114, 115, 118, 120, 121, 123, 125, 127, 252, 277, 448
fish community, 5, 16, 18, 41, 54, 63, 66, 68, 111, 448
fish population, 70, 71, 75, 98, 100, 110, 111, 115
freshwater bodies, 189
functional feeding groups (FFGs), 14, 132, 158, 159, 160, 164, 166
functional indices, 14
GAM, *see* generalised additive model
Garonne, 42, 62, 64, 65, 66, 68, 69, 70, 71, 72, 73, 74, 75, 225, 226, 227, 300, 304, 305, 316, 319, 337, 437, 448
GDI, *see* Generic Diatom Index
generalisation, 4, 33, 92, 184, 389
generalised additive model (GAM), 22, 24, 92-94, 98, 99, 158, 161-166
generalised linear models, 24, 92
Generic Diatom Index (GDI), 10, 12
genetic algorithms, 21, 38, 132, 252, 263
global structuring index, 371, 374, 375, 377, 379, 380
Grossberg layer, 31, 192, 193, 194, 197, 201
halobiont index, 9
headwater streams, 68, 339, 340, 343, 346, 347, 351-354
hierarchical patterning, 132, 206, 210, 218, 219
hydropower, 74, 235, 320, 321
IBD, 10, 11, 12
IBI, *see* Index of Biotic Integrity
IDAP, 10, 12
ILM, 10, 11
Index of Biotic Integrity (IBI), 3, 14, 17, 18, 77
indicator value, 13, 299, 323, 324, 330, 331, 332, 333, 334, 336, 340, 345, 391, 395, 396, 399
Indice de Pollusensibilité Spécifique (IPS), 10, 11, 305, 450
Indices assessment, 13
Intermediate Disturbance Hypothesis, 10
Invertebrate Community Index, 14
IPO, 289, 291, 295, 296, 298, 301
IPS, *see* Indice de Pollusensibilité Spécifique
ISODATA, 392
K statistic, 118, 120-122, 124-126, 128, 129, 150-153, 357-363
Kernel-induced nonlinear model, 27, 33
k-means, 28, 219, 224, 228, 229, 254, 379, 392
K-means, 32, 225, 256, 258
knowledge acquisition costs, 402, 407, 409
knowledge system, 368, 401, 402, 403, 404, 405, 409
learning, 27, 28, 30, 31, 33, 37, 44, 45, 46, 47, 49, 51, 53, 57, 61, 63, 93, 103, 104, 106, 123, 127, 149, 160, 162, 163, 164, 169, 170, 193-195, 197, 198, 200-203, 205-211, 217, 219, 224, 232, 234, 258, 261, 263, 264, 273, 277, 286, 307, 308, 309, 314, 316, 322, 343, 344, 351, 357, 370, 376-378, 411, 438, 445, 451
leave-one-out, 23, 31, 93, 96, 351
Leclercq and Maquet Index, 10, 11
LIMPACT, 368, 401, 402, 403, 404, 405, 406, 407, 409, 410
linear discriminant analysis, 411, 425, 426
linear vector quantisation, 32

logistic regression, 23, 24, 119, 411, 420, 424-426, 451
Loire-Bretagne basin, 355, 364
macroinvertebrate community, 5, 131, 190, 368, 408
MANOVA, 308, 310
MIR-max, 39, 288, 289, 291-299, 301, 302, 367, 390, 393, 395-400
Mixture models, 21, 35, 36
MLP, *see* multilayer perceptron
MLR, 24, 26, 33, 94, 95, 98, 158, 161-166, 411, 413, 418-420, 422-424, 426
model evaluation, 80
model validation, 56, 133, 135, 143
modelling techniques, 5, 21, 53, 115, 131, 147, 158, 160, 166, 204, 221, 234, 253, 307, 346, 451, 453
Monte-Carlo test, 291, 296, 301
multi-layer feed-forward neural network, 322
multilayer perceptron (MLP), 27, 29-31, 34, 55-57, 59, 60, 62, 63, 118, 120, 132, 149, 180, 190, 224, 225, 232, 236, 253, 254, 257, 259, 286, 436, 438, 440, 443
multimetrics, 13, 14
multinomial logistic regression, 21-23, 368, 411, 413, 418, 422, 424
multinomial regression, 26
multiple classes, 417
multiple linear regression, 24, 54, 92, 132, 158, 161
mutual information, 21, 39, 291, 396, 398
negative indicator, 402, 405, 408
non-hierarchical clustering, 132, 167, 169, 170, 187
optimisation, 39, 114, 136, 140, 145, 146, 267, 271, 430, 452
ordination, 22, 34, 90, 100, 104, 131, 167-169, 172, 173, 184-187, 203, 233, 234, 236, 239, 252, 304, 309, 318, 321-327, 370, 393, 399, 400, 436-438, 440, 445, 447, 448, 451, 453
organic pollution, 10, 12, 13, 14, 17, 70, 167, 174-177, 207, 218, 289, 291, 296, 301, 401
organic pollution index, 289, 291
overfitting, 37, 57, 96, 224, 416
overtrained, 31
PaD, *see* partial derivatives
PAEQANN, 19, 22, 27, 43, 54, 114, 128, 147, 158, 167, 189, 206, 221, 262, 288-290, 304, 305, 317, 318, 337, 340, 342, 343, 355, 369, 411, 435, 436, 438, 439, 441, 443, 447, 449, 450
parameter, 7, 24, 29, 30, 36, 98, 159, 170, 211, 231, 232, 255, 283, 295, 296, 301, 401, 405, 415, 416, 418, 419, 421, 433
partial derivative algorithm (PaD), 56, 58, 59, 60, 61, 96, 162, 322, 324, 437, 438, 441
partial least square (PLS), 22, 25, 26, 132, 161-166
pattern recognition, 26, 31, 34, 291, 367, 390, 391, 393, 395, 399, 400
PCA, *see* principal component analysis
pesticides, 401, 402, 403
phytoplankton, 37, 40, 261, 263, 273-275, 277, 280-287, 344
phytoplankton, 275, 285
plausible reasoning, 391, 400
Plecoptera, 149, 152, 153, 168, 222, 223, 225, 229, 230, 232, 235, 403
PLS, *see* partial least square
Polluo-Sensitivity Index (PSI), 318, 340, 342
pollution, 10, 13, 14, 69, 70, 73, 74, 134, 140, 153, 202, 206, 207, 212-216, 218, 219, 239, 251, 288, 289, 291-293, 295, 296, 298-301, 317, 318, 337, 339, 340, 341, 345, 350, 352, 353, 354, 364, 390, 411, 450
polysaprobic, 350, 351, 364
positive indicator, 402, 403, 405, 408
predicting structure and diversity, 435
prediction, 1, 3-5, 7, 12, 14-16, 22, 23, 25, 32, 34, 35, 39, 54, 56-62, 76, 79-82, 87, 88, 90, 93, 96-98, 122, 123, 127, 131, 132, 135, 138, 139, 141, 143, 145-147, 157-159, 164, 165, 180, 189, 190, 195, 204, 206, 232, 234, 237-239, 251-256, 259-263, 265-272, 304, 343, 344, 347, 351, 353, 359, 361, 365, 368, 369, 411, 413, 419, 423, 425, 437, 438, 440-444, 448

principal components analysis (PCA), 28, 204, 237, 304, 321, 323, 324, 341, 369, 393, 396, 434
projection methods, 367
projection pursuit, 367, 381
PSI, *see* Polluo-Sensitivity Index
qualitative environmental data, 199
Rank Nearest-Neighbour, 382
rapid assessment, 2, 13, 221, 237
reduced taxa data, 172, 173, 181, 182
REFCOND, 303, 354
reference conditions, 3, 7, 11, 18, 19, 76, 78, 114, 127, 303, 304, 345, 346, 353, 354, 454
regression maximisation, 21, 39
regression trees, 25, 161
reservoir, 42, 90, 92, 98, 101, 207, 215, 274, 284, 286
restoration, 2, 3, 4, 14, 15, 64, 71, 72, 74, 133, 138, 143, 145, 167, 217, 242, 342, 435, 436, 454
Rhône river, 293
richness indices, 14
River Continuum Concept, 52, 53, 54, 62, 63, 204, 236, 302, 353
river morphology, 320
River Pollution Diagnostic System, 367
river quality assessment, 5, 7, 9, 136, 139, 354
river restoration scenarios, 133, 138, 143, 146
RIVPACS, 1, 13, 15, 16, 19, 76, 87, 147, 148, 167, 189, 234, 401, 414
R-max, 393, 395, 397
robust indices, 367, 381
robustness, 133, 136, 381
RPDS, 367, 390-396, 398-400
saprobic indices, 13
saprobity, 9, 215
self-organizing map (SOM), 27-29, 31, 32, 34, 39, 42, 44-53, 56, 59, 63, 100, 101, 103-112, 132, 167-188, 190, 192-204, 207-219, 223, 225-235, 238, 240, 242-259, 262, 273, 274, 277, 278, 280-287, 304, 305, 307-318, 321-324, 327-330, 335, 337, 341-352, 354, 367, 369-391, 393, 396, 436, 438, 440, 445, 446, 447, 453
sensitivity analysis, 42, 54, 56, 58, 59, 62, 96, 97, 114, 138, 139, 140, 144, 145, 160, 225, 233, 235, 236, 261, 278, 322, 324, 329, 330, 335, 337-339, 341, 342, 437, 441, 453
Shannon (-Weaver) index, 10, 104
SHE index, 10
similarity/loss indices, 14
SLA, 10, 11
software, 11, 93, 128, 136, 253, 293, 308, 318, 322, 367, 385, 390, 392, 395, 396-399, 421, 435, 436, 439, 447, 453
SOM, *see* self-organizing map
spatial variation, 42, 100, 244, 246
species assemblages, 52, 127, 222, 234, 236, 304, 307, 308, 358, 373, 378, 452
species richness, 1, 16-18, 26, 43, 48, 50, 52, 54-57, 59, 60, 62, 63, 65, 66, 69, 77, 87, 115, 132, 158, 160, 163, 164, 189, 190, 195, 198, 201, 203-205, 218, 221-223, 225, 227, 229, 230, 231, 234, 235, 236, 237, 344, 378, 441, 453
Specific Pollution Index, *see* Indice de Pollusensibilité Spécifique
STOWA, 8, 15, 401
stream, 7, 8, 15, 55, 187, 225, 228, 231, 234, 235, 242, 309
stream classifications, 222, 234, 235
stream typology (ies), 7, 8, 309
structural dynamic models, 21, 39, 40
structure and function, 52, 79, 437
supervised learning, 27, 190, 205, 322, 436
support vector machines (SVM), 21, 37, 368, 411, 414
taxonomic richness, 26, 232
TBI, *see* Trent biotic index
TDI, *see* Trophic Diatom Index
temporal variation, 212, 218
testing, 25, 30, 56, 57, 117, 135, 136, 148, 156, 160, 162-165, 201, 204, 225, 232, 255, 357, 381, 436, 447
tolerance/intolerance indices, 14
tool (s), 2-5, 7, 9, 11, 14, 15, 18, 19, 25-27, 31, 32, 34, 37, 39, 53, 63, 76, 87, 88, 93, 98, 99, 112, 116, 126, 128, 132, 148, 162, 166, 168, 170, 189, 190, 205, 206, 221, 222, 234, 235, 237-239, 251, 252, 264, 288, 304, 308, 354, 355, 369, 381, 386, 411, 414, 424, 435-454

training, 25, 27, 29-33, 41, 47, 56-59, 62, 71, 79-82, 87, 88, 98, 104, 117-121, 123, 124, 127, 128, 133, 135-143, 145, 146, 149, 150, 155-157, 160-163, 165, 169, 170, 172, 173, 182, 192-195, 203, 204, 207, 209-211, 224, 228, 232, 238, 240-244, 251, 255, 268, 277, 278, 286, 307, 308, 322, 344, 347, 351, 357, 362, 370, 371, 379, 390, 391, 394, 413, 416, 424, 426, 430
transformation, 37, 45, 54, 94, 191, 320, 383, 412, 415, 416, 428, 433, 434
tree-based method, 22, 24, 25, 161
tree-based regression models, 24
Trent biotic index, 209, 216
Trichoptera, 140, 143, 149, 152, 153, 168, 222, 223, 225, 227, 230, 232, 235, 403
Trophic Diatom Index (TDI), 10, 11, 12
TWINSPAN, 392
typifying taxa, 171, 181
U-matrix, *see* unified-matrix algorithm
unified-matrix algorithm (U-matrix), 28, 47, 104, 105, 194, 195, 196, 199, 207, 214, 215, 219, 224, 254, 255, 256, 278, 281, 323, 325, 328, 379
unsupervised learning, 27, 45, 100, 103, 168, 169, 207, 208, 219, 304, 322, 381, 436
user interface, 5, 395, 402, 435
validation, 4, 31, 37, 56, 57, 58, 59, 62, 80-82, 87, 88, 93, 96, 117, 118, 121, 123, 128, 131, 135-138, 144, 146, 149, 150, 165, 243, 244, 246, 255, 344, 357, 423
visualization, 27, 28, 29, 39, 44, 46, 47, 49, 52, 103, 169, 194, 204, 208, 219, 228, 229, 231, 234, 238, 291, 304, 367, 369, 370, 390-393, 395-399, 437-441, 443, 445, 446
Ward linkage, 310, 370, 373, 445
Water Framework Directive (WFD), 3, 7, 11, 18, 64, 65, 70, 114, 127, 145, 190, 291, 298, 304, 305, 314, 342, 345, 346, 354, 454
water management, 1-4, 13, 21, 34, 70-72, 75, 167, 168, 188, 221, 342, 411, 436
water policies, 2, 436
weight, 27, 28, 30, 32, 37, 39, 46-48, 56, 103, 122, 162, 169, 171, 172, 192, 194, 195, 208-211, 234, 240, 266, 268, 277, 322, 334, 370, 375, 376, 379, 414, 428, 429, 453
WFD, *see* Water Framework Directive

MIX
Papier aus verantwortungsvollen Quellen
Paper from responsible sources
FSC® C105338

If you have any concerns about our products,
you can contact us on
ProductSafety@springernature.com

In case Publisher is established outside the EU,
the EU authorized representative is:
Springer Nature Customer Service Center GmbH
Europaplatz 3, 69115 Heidelberg, Germany

Printed by Libri Plureos GmbH
in Hamburg, Germany